SCIENTIFIC AND TECHNICAL INFORMATION SOURCES

Second Edition

CHING-CHIH CHEN

SCIENTIFIC AND TECHNICAL INFORMATION SOURCES

Second Edition

The MIT Press
Cambridge, Massachusetts
London, England

Copyright © 1977, 1987 by Massachusetts Institute of Technology

All rights reserved. No part of this book may be reproduced in any form or by any means, electronic or mechanical, including photocopying, recording, or by any information storage and retrieval system, without permission in writing from the publishers.

This book was set in Baskerville by The MIT Press Computergraphics Department and printed and bound by Halliday Lithograph in the United States of America.

Library of Congress Cataloging-in-Publication Data

Chen, Ching-chih, 1937–
  Scientific and technical information sources.

  Bibliography: p.
  Includes indexes.
  1. Science—Bibliography. 2. Engineering—Bibliography. 3. Technology—Bibliography. I. Title.
  Z7401.C48 1986 [Q158.5]   016.5   86-7310   ISBN 0-262-03120-5

To Sow-Hsin,
whose devotion to science inspires me

CONTENTS

PREFACE	xxiii
ACKNOWLEDGMENTS	xxvii
REVIEW SOURCES	xxix
Science Journals	xxix
Nonscience Journals	xxxiv
Book Sources	xxxvii
1 SELECTION TOOLS	1
Basic Tools	1
Booklists and Serial Union Lists	3
Subject Library Catalogs for Retrospective Selection	5
General Science	5
Astronomy	6
Mathematics	6
Chemistry	6
Physics	7
Biological Sciences	7
Agriculture	7
Botany	8
Zoology	9
Earth Sciences	9
Oceanography	10
General Engineering	10
Civil Engineering-Transportation	11
Computer Technology	12
Environmental Sciences	12
2 GUIDES TO THE LITERATURE	14
General Science	14
Astronomy	17
Mathematics	17
Physics	18
Chemistry	18
Biological Sciences	18
Earth Sciences	21
General Engineering	21
Chemical Engineering	22
Civil Engineering	23
Electrical and Electronics Engineering	23
Computer Technology	23
Industrial Engineering	24

Nuclear Engineering | 24
Energy | 25
Environmental Sciences | 26
Transportation | 28

3 BIBLIOGRAPHIES | 29
General Science | 29
Astronomy | 31
Mathematics | 31
Physics | 32
Chemistry | 35
Biological Sciences | 36
  Agriculture | 38
  Botany | 39
  Nutrition | 41
  Zoology | 41
Earth Sciences | 42
  Oceanography | 46
General Engineering | 46
Aeronautical and Astronautical Engineering | 47
Chemical Engineering | 48
Electrical and Electronics Engineering | 49
  Computer Technology | 50
Industrial Engineering | 51
Materials Science | 51
Mechanical Engineering | 52
Metallurgy | 52
Military Science | 53
Nuclear Engineering | 53
Energy | 54
Environmental Sciences | 57
  General | 57
  Pollution | 60
Transportation | 61

4 ENCYCLOPEDIAS | 62
General Science | 62
Astronomy | 63
Mathematics | 65
Physics | 65
Chemistry | 66
  Specific | 67
Biological Sciences | 68
  Botany | 69
  Zoology | 72

Earth Sciences	76
Oceanography	79
General Engineering	80
Aeronautical and Astronautical Engineering	80
Chemical Engineering	81
Food Technology	83
Polymer Technology	85
Textile Technology	86
Architectural Engineering	86
Electrical and Electronics Engineering	87
Computer Technology	88
Industrial Engineering	89
Materials Science	89
Mechanical Engineering	90
Energy	90
Environmental Sciences	91
Transportation	92
Military Science	93
5 DICTIONARIES	**95**
Bibliographical Tools	95
Abbreviations and Acronyms	95
Subject	96
Nomenclatures and Thesauri	97
Multilingual	100
Bibliographies	100
English-Chinese	100
English-French	101
English-German	102
English-Russian	104
English-Spanish	106
More than Two Languages	107
Subject Dictionaries	114
General—Units and Measurements	114
General Science	114
Astronomy	117
Mathematics	118
Physics	119
Chemistry	121
Chemical Names	123
Biological Sciences	124
Botany	125
Zoology	127
Microbiology	129
Agriculture	129
Earth Sciences	130

General Engineering	133
Aeronautical and Astronautical Engineering	133
Chemical Engineering	133
Nutrition and Food Technology	135
Civil Engineering	136
Electrical and Electronics Engineering	137
Computer Technology	138
Communication	144
Materials Science	144
Mechanical Engineering	145
Automotive	146
Military Science	146
Energy	147
Environmental Sciences	148
Transportation	151
6 HANDBOOKS	**152**
General Science	152
Astronomy	153
Mathematics	155
Physics	157
Chemistry	163
Biological Sciences	170
Agriculture	173
Botany	174
Zoology	177
Earth Sciences	180
Oceanography	181
General Engineering	181
Aeronautical and Astronautical Engineering	182
Chemical Engineering	183
Food Technology	185
Polymer Technology	187
Textile Technology	188
Civil Engineering	189
General	189
Concrete and Construction Engineering	190
Other	193
Electrical and Electronics Engineering	194
Electronics	196
Radio, TV, Etc.	201
Computer Technology	203
Industrial Engineering	208
Materials Science	211

Mechanical Engineering	214
General	214
Specific	215
Metallurgy	220
Nuclear Engineering	223
Energy	223
Environmental Sciences	227
Pollution	232

7 TABLES, ALMANACS, DATABOOKS, AND STATISTICAL SOURCES

	234
Tables	234
General Science	234
SI Units	234
Astronomy	235
Mathematics	236
Statistics	238
Physics	239
Chemistry	241
Biological Sciences	243
Earth Sciences	243
General Engineering	244
Subject Engineering	244
Almanacs, Databooks	246
General Science	246
Astronomy	246
Mathematics	248
Physics	248
Chemistry	250
Biological Sciences	253
Earth Sciences	254
General Engineering	256
Chemical Engineering	257
Civil Engineering	257
Electronics Engineering	258
Materials Science	259
Mechanical Engineering	259
Metallurgy	260
Nuclear Engineering	260
Energy	261
Environmental Sciences	262
Transportation	263
Statistical Sources	264
Guide to Sources of Statistical Information	264
Governmental Statistical Sources	264
Nongovernmental Statistical Sources	267

8 MANUALS, SOURCE BOOKS, LABORATORY MANUALS AND WORKBOOKS, AND HOW-TO-DO-IT MANUALS   272

Manuals   272
  Science   272
  Biological Sciences   273
    Botany   274
    Zoology   275
  Earth Sciences   275
  General Engineering   276
  Aviation   276
  Chemical Engineering   277
  Civil Engineering   278
  Electrical and Electronics Engineering   280
  Industrial Engineering   282
  Mechanical Engineering   282
  Metallurgy   283
  Environmental Sciences   284
  Navigation   285
Source Books   286
  Science   286
  Engineering   288
  Energy and Environmental Sciences   291
Laboratory Manuals and Workbooks   293
  Science   293
  Mathematics   293
  Physics   293
  Chemistry   294
  Biological Sciences   295
  Earth Sciences   296
  Engineering   297
How-to-Do-It and Popular Manuals   299
  General   299
  Science   300
  Biological Sciences   300
  Engineering   301
    Construction Engineering   301
    Electrical and Electronics Engineering   302
      Computer   302
      Electronics   304
      Radio and Stereo   305
      Television   305
    Mechanics   305
      Automobiles   306

9 GUIDES AND FIELD GUIDES	307
Guides	307
General Science	307
Mathematics	307
Physics	307
Chemistry	308
Biological Sciences	309
Agriculture	311
Botany	312
Nutrition	314
Zoology	316
Earth Sciences	317
Engineering	319
Aeronautical and Astronautical Engineering	319
Chemical Engineering	320
Textile Technology	321
Civil Engineering	322
Electrical and Electronics Engineering	324
Computer Technology	326
Microcomputer Hardware	326
Microcomputer Software	327
Programming	329
Other Related Topics	329
Industrial Engineering	331
Marine Engineering	332
Materials Science	332
Metallurgy	333
Mechanical Engineering	333
Transportation	335
Nuclear Engineering	336
Energy	337
Environmental Sciences	341
Field Guides	342
Astronomy	342
Botany	344
Flowers	351
House Plants	351
Weeds and Wildflowers	352
Zoology	355
Birds	357
Fishes	363
Insects	364
Mammals	366
Reptiles	369
Earth Sciences	370
Minerals and Rocks	370

10 ATLASES AND MAPS	373
Bibliographical Tools	373
Astronomy	375
Star Atlases and Catalogues	377
Physics	378
Chemistry	379
Biological Sciences	380
Zoology	382
Earth Sciences	384
Atmospheric Science	388
Oceanography	389
Engineering	391
Energy and Environmental Sciences	392
11 DIRECTORIES, YEARBOOKS, AND BIOGRAPHICAL SOURCES	393
Directories	393
Bibliographical Sources	393
General	393
General Science	397
Science and Engineering Career and Education	399
Astronomy	403
Mathematics	403
Physics	404
Chemistry	404
Biological Sciences	405
Botany	406
Zoology	407
Agriculture	408
Earth Sciences	409
General Engineering	410
Aeronautical and Astronautical Engineering	411
Chemical Engineering	412
Food Technology	414
Petroleum Engineering	415
Civil Engineering	415
Electrical and Electronics Engineering	416
Computer Technology	417
Computer Hardware	420
Computer Software	420
Marine Engineering	425
Mechanical Engineering	427
Metallurgy	428
Nuclear Engineering	428
Energy	430

Environmental Sciences	436
Pollution	439
Transportation	439
Yearbooks	440
Subject Science	441
Engineering	442
Aeronautical and Astronautical Engineering	443
Aeronautical and Aerospace Engineering	443
Civil Engineering	443
Electrical and Electronics Engineering	444
Mechanical Engineering	444
Mining and Metallurgy	445
Energy and Environmental Sciences	445
Transportation	446
Biographical Sources	448
Reference Tools	448
General Science	448
Astronomy	454
Mathematics	454
Physics	455
Chemistry	457
Biological Sciences	457
Earth Sciences	459
General Engineering	459
Aeronautical Engineering	461
Chemical Engineering	461
Civil Engineering	462
Electrical and Electronics Engineering	462
Mechanical Engineering	463
Nuclear Engineering	463
Energy	463
Environmental Sciences	464
Membership Directories	464
12 HISTORY	**465**
General Science	465
Astronomy	469
Mathematics	470
Physics	471
Chemistry	476
Biological Sciences	477
Agriculture	480
Earth Sciences	480
General Engineering	482
Aeronautical and Astronautical Engineering	483
Chemical Engineering	484
Civil Engineering	484

Electrical and Electronics Engineering	485
Computer Engineering	486
Mechanical Engineering	487
Nuclear Engineering	487
Environmental Science	488

13 IMPORTANT SERIES AND OTHER REVIEWS OF PROGRESS — 490

Guide to Series Publications	490
Important Series	490
Astronomy	490
Mathematics	490
Physics	491
Biophysics	494
Nuclear Physics	495
Chemistry	495
Biochemistry	499
Inorganic Chemistry	501
Organic Chemistry	502
Polymer Chemistry	504
Biological Sciences	505
General	505
Agriculture and Food Science	508
Botany	508
Cell and Molecular Biology	509
Enzymology	510
Genetics	510
Microbiology and Immunology	511
Physiology	512
Zoology	513
Earth Sciences	513
Aeronautical and Astronautical Engineering	514
Bioengineering	514
Chemical Engineering	515
Civil Engineering	515
Electrical and Electronics Engineering	516
Materials Science	517
Mechanical Engineering	517
Nuclear Engineering	518
Energy	518
Environmental Sciences	519
Other Reviews of Progress	520

14 TREATISES — 522

Astronomy	522
Mathematics	522
Physics	523

Chemistry	524
Biochemistry	527
Biological Sciences	528
Earth Sciences	531
Chemical Engineering	532
Materials Science	533
Mechanical Engineering	534
Nuclear Science & Engineering	534
Energy	534
Environmental Science	534

15 ABSTRACTS AND INDEXES, AND CURRENT-AWARENESS SERVICES

	536
Abstracts and Indexes	536
Guides to Abstracts and Indexes	536
General Science	537
Astronomy	540
Mathematics	540
Physics	542
Chemistry	543
Biological Sciences	545
Agriculture	547
Botany	548
Zoology	550
Earth Sciences	552
General	552
Oceanography	554
General Engineering	555
Aeronautical and Astronautical Engineering	556
Chemical Engineering	557
Civil Engineering	558
Electrical and Electronics Engineering	559
General	559
Computer Technology	560
Marine Engineering	562
Materials Science	562
Mechanical Engineering	562
Metallurgy	563
Military Science	564
Nuclear Engineering	564
Energy	565
Environmental Sciences	566
General	566
Pollution	568
Current-Awareness Services	569
General	569
Subject	569

16 PERIODICALS	572
Reference Sources	572
Abbreviations	574
Cumulative Indexes	575
Selective Titles	576
General Science	576
Astronomy	577
Mathematics	577
Physics	578
Chemistry	579
Biological Sciences	580
General	580
Agriculture	580
Biochemistry and Biophysics	581
Botany	581
Genetics	581
Microbiology	581
Nutrition	582
Physiology	582
Zoology	582
Earth Sciences	582
Atmospheric Sciences	583
Oceanography	583
General Engineering	584
Aeronautical and Astronautical Engineering	584
Chemical Engineering	585
Civil Engineering	585
Electrical and Electronics Engineering	585
Computer Technology	586
Industrial Engineering	587
Mechanical Engineering	587
Nuclear Engineering	588
Energy	588
Environmental Sciences	589
17 TECHNICAL REPORTS AND GOVERNMENT DOCUMENTS	591
Technical Reports	591
General Reference Tools	591
Subject Reference Tools	594
Governmental Documents	595
General Bibliographic Sources	596
Subject Bibliographic Sources	599
Sample Government Documents	600
Bureau of Mines	600
Department of Agriculture	600
Environmental Protection Agency	601

Federal Aviation Administration	601
Geological Survey	601
National Highway Safety Bureau	601

18 CONFERENCE PROCEEDINGS, TRANSLATIONS, DISSERTATIONS AND RESEARCH IN PROGRESS, PREPRINTS, AND REPRINTS — 602

Conference Proceedings	602
Calendars and Forthcoming Meetings	602
General	602
Subject	603
Published Proceedings	603
Major Bibliographical Tools	603
Sample Subject Bibliographical Tools	605
Conference Publications	607
Astronomy	607
Mathematics	607
Physics	607
Chemistry	608
Biological Sciences	609
Earth Sciences	610
Engineering	610
Aeronautical Engineering	610
Chemical Engineering	610
Civil Engineering	611
Electrical Engineering	611
Marine Engineering	611
Materials Engineering	612
Mechanical Engineering	612
Naval Engineering	612
Nuclear Engineering	612
Energy	612
Translations	612
General Sources	613
Abstracting and Indexing Sources	614
Subject Sources	614
Cover-to-Cover Translations	615
Dissertations and Research in Progress	616
Dissertations	616
General Tools	616
Subject Guides	618
Research in Progress	619
General Science	619
Mathematics	622
Physics	622
Chemistry	623

Biological Sciences	623
Agricultural Sciences	624
Earth Sciences	624
General Engineering	624
Aeronautical and Astronautical Engineering	625
Civil Engineering	625
Electrical and Electronics Engineering	625
Nuclear Engineering	625
Energy	626
Environmental Sciences	627
Pollution	627
Preprints	627
Reprints	628
Reference Tools	628
Subject Tools	629

19 PATENTS AND STANDARDS — 630

Patents	630
Indexes to US and International Patents	631
Sources about Patents	631
US Patents	633
International Patents	634
Subject Guide to Patents (Selective)	634
Science	634
Engineering	636
Standards	637
Guides to Standards	638
Major Bibliographic Tools—General	639
Subject Guides to Standards	641
Units of Measurement	641
Terminology and Symbolic Representation	643
Safety of Persons and Goods	644
Products and Processes (Sample Listing)	645
Biology	645
Chemistry	646
Aeronautical Engineering	646
Chemical Engineering	646
Civil Engineering	646
Electrical Engineering	647
Computer Technology	648
Marine Engineering	649
Mechanical Engineering	650
Nuclear Engineering	651

20 TRADE LITERATURE	653
Guide to Trade Literature	653
Trade Reference Tools	654
Selected Trade Journals	655
Commercial Product Information	656
21 NONPRINT MATERIALS	657
General Reference Sources	657
Guides and Indexes	657
Audiovisual Equipment	662
Subject Sources	663
General	663
Subject Specific	664
22 PROFESSIONAL SOCIETIES AND THEIR PUBLICATIONS	667
Directories	667
Selective List of Professional Organizations	669
Society Publications	672
Guide	672
Sample Society Catalogs and Bibliographic Tools	673
Catalogs	673
Indexes	673
23 DATABASES	675
Reference Source	675
Database Search Guides and Reference Manuals	678
Selective Databases	679
NLM Databases (Selected)	712
Electronic Publishing and Videodisc and CD-ROM Products	712
NAME INDEX	715
TITLE INDEX	753

PREFACE

This book is intended primarily as a basic one-volume reference guide for science and engineering information professionals and their assistants and as a textbook for library and information science school students engaged in the study of the structure, properties, and output of scientific and technical literature. The work should also be useful as a handy reference manual for scientists, engineers, science students, and other users of science and technology library and information resources.

Given the vast amount of published material in the fields of science and technology the titles included in this second edition, as in the first edition, are necessarily selective, although in most categories no obvious attempt has been made to present only the best of the sources available. While the user may find a surprising variety of information (how-to-do-it manuals, field guides) suitable for a general audience, the majority of the information sources are probably more suited to a research collection.

The first edition was published in 1977. Given the dynamic nature of the scientific and technical information sources, with relatively short "half-life," the first edition needed a complete update and overhaul. This second edition is completely revised and greatly expanded. In fact, aside from the classical reference tools and those information sources with new editions, supplements, or continuing volumes, the second edition includes few titles published prior to 1976. Since scientific and technical information must be up-to-date, the primary criterion for inclusion is that the reference must be a current source of information. Thus the majority of titles have an imprint date after 1980 and a closing date of February 1986.

Most books of this type tend to include secondary sources only, yet active scientists and engineers in the field rely more heavily on informal and primary sources of information. This pattern has been substantiated by a significant number of use studies of the information needs of scientists. For this reason, this book provides substantial coverage of the primary sources.

Each type of subject literature is used for a distinctive purpose. Thus the sources are grouped first by types of material and then by subject within each type, in contrast with the more commonly used groupings by subject first for many library reference books. Altogether, 5,300 sources are grouped under twenty-three categories. It is hoped that by this arrangement users of scientific libraries and information centers may have easier access to available information sources, and the science and technical information professionals and library and information science school students may be made more keenly aware of the varied uses of each type of information as well as the characteristics and properties of scientific and technical literature. The necesary discussion of a few types of information has been kept extremely brief since lengthy essays on the structure of scientific and technical literature and the use of these information sources can be located in several books listed in chapter 2. These books also provide rather extensive bibliographic sources for each type of information source, while both current and retrospective bibliographic sources can also be easily searched, manually or via online retrieval, in standard library and information science indexing and abstracting tools, such as *Library Literature* and *Library and Information Science Abstracts*. Thus, no separate reference list is provided in this edition.

In terms of subject coverage, it is fair to point out that this edition has reflected well the fast evolution and development of scientific and technical fields. While conventional subject disciplines have remained, new headings with extensive new titles have been added to new subsubject fields since 1977: most noticeably, computer technology, microcomputer technology, communications, etc. In chapter 23, the number of online databases in science and technology listed has been greatly increased to reflect the proper state of the art. It is also important to point out that generally no clinical material is included here except those few of particular interest to bioscientists and bioengineers. Interested readers for clinical information sources are referred to my companion volume, *Health Sciences Information Sources* (Cambridge, MA: MIT Press, 1981). Although some foreign titles are included, this edition, like the first, has a heavy American emphasis.

All entries within each subject have been arranged by title rather than by the more conventional main-entry or author approach. There are several justifications for such a treatment. First, most users, including informational professionals who are not very familiar with scientific and technical information sources, cannot usually remember authors' precise names. Second, many sources are more commonly known by their titles (such as *Handbook of Physics and Chemistry*), and many secondary sources change editorship frequently. Third, scientists, many of whose key sources (such as *Science* and *Physics Today*) list reviewed books by title, are probably more comfortable with such an approach. Finally, the arrangement makes it possible for readers to locate a needed title easily without the use of a title index.

Most entries are annotated. These annotations, generally brief, have been designed whenever possible to include critical as well as descriptive information, with emphasis on the former. While I am familiar with most of the sources, some, particularly more current ones, have been included without personal inspection. These annotations are generally brief abstracts from various review sources.

A major feature of this book is that review sources, when available, are provided at the end of the annotations, introduced by the letter "R". The order of the review information is science journal sources by alphabetical order first, then nonscience journal sources, and finally book sources. A list of these sources, greatly expanded from that of the first edition, is included in the beginning of this book. For purposes of comparison, the review citations also include, whenever possible, those for the earlier editions of the titles listed.

To facilitate the use of the second edition, a detailed table of contents is provided for easy and quick subject approach, and extensive indexes have been added: a complete author index that includes personal as well as corporate names, and a title index. This is in direct response to many gracious letters received from readers of the first edition requesting the provision of more indexes.

It is a monumental attempt to search, select, annotate, and group in one handy volume the major useful tools from an abundant pool of available scientific and technical information sources; therefore a book of this size and scope cannot possibly be perfect. I wish to thank those readers who have communicated with me on the first edition, and it is obvious that some of their gracious suggestions have influenced the making of this new book. I would be most grateful to receive any suggestions, corrections, and additions that readers may be inclined to offer on this second edition.

ACKNOWLEDGMENTS

I wish to express my deep appreciation to all those who have made this second edition possible. I am grateful to the Simmons College Emily Hollowell Fund for Research for its partial support. Several research assistants have participated in this time-consuming project, and all their help is deeply appreciated. In particular, Cindy Hadad was very helpful in the typing of the entries during her tenure as undergraduate research assistant. My sincere thanks go especially to Xia Wen Zhen, of the Academy of Science Library of China, who devoted much of his one-year research period at the Graduate School of Library and Information Science, Simmons College, to learning about the new reference information sources in science and technology and to performing detailed data collection work, which made this completely revised and greatly expanded edition possible. Thanks also go to Judith Sacknoff and Barbara Gergely for proofreading. Finally, I am forever indebted to every member of my family for their constant and unselfish support, love, understanding, and encouragement. My two daughters, Anne and Cathy, who are both science majors, deserve my continuing thanks for their active participation in the preparation of the first edition and for their continuing critical work and review on this second edition.

REVIEW SOURCES

Abbreviations will be given only when necessary (in parentheses).

SCIENCE JOURNALS

AAPG Bulletin

Abstracts on Hygiene

Acta Crystallographica

Advances in Colloid and Interface Science

Aeronautical Journal

AIA Journal

Aircraft Engineering

American Association of Petroleum Geologists Bulletin

American Forests

American Horticulturist

American Institute of Chemical Engineers Journal (AIChE J)

American Ink Maker

American Journal of Physics

American Journal of Science

American Machinist

American Scientist

American Society of Heating, Refrigerating and Air Conditioning Engineers Journal (ASHRAE J)

Analytica Chimica Acta

Analytical Biochemistry

Analytical Chemistry

Analyst

Angewandte Chemie; International Edition in English

Animal Kingdom

Annals of Allergy

Anti-Corrosion Methods and Materials

Applied Optics

Applied Spectroscopy

Association of Official Analytical Chemists Journal

ASTM Standardization News

Astronomy

Atmospheric Environment

Audiology

Australian Mining

Automotive Engineering

Biometrics

Bioscience

Blair and Ketchum's Country Journal

British Chemical Engineering

British Journal of Non-Destructive Testing

British Plastics

Building Services Engineer

Bulletin of the American Meteorological Society

Byte

California Geology

Canadian Aeronautics & Space Journal

Canadian Journal of Spectroscopy

Ceramic Abstracts

Chartered Mechanical Engineer
Chemical and Engineering News
Chemical Engineering
Chemical Engineering Science
Chemical Geology
Chemistry
Chemistry and Industry
Chemistry in Australia
Chemistry in Britain
Chemistry in Canada
Chemtech
Chronicle of the Early American Industries Association
Civil Engineering
Civil Engineering and Public Works Review
Clean Air
Clinical Chemistry
Composites
Computer
Computer Bulletin
Condor
Conservation
Consulting Engineer
Cosmetics and Toiletries
Data Processing Digest
Datamation
Deep Sea Research
Drug and Cosmetic Industry
Earth Science
Earch Science Reviews
Ecology

Economic Geology
Education in Chemistry
Elastomerics
Electrical Apparatus
Electrical Engineer
Electrical Review
Electrical World
Electronics
Electronics and Power
Endeavour
Energy World
Engineer
Engineering
Engineering Materials and Design
Engineers' Digest
Environmental Research
Farm Chemicals
Food Engineering
Food Processing Industry
Food Technology
Forestry Abstracts
Foundry Trade Journal
Fuel
Garden
Gems and Minerals
General Engineer
Geological Magazine
Geology
Geophysics
Geotechnique
Geotimes
Ground Engineering

Library World (LW)
New Library Science Annual
New Library World
New Technical Books (NTB)
Online
Online Review
Quarterly Bulletin of the International Association of Agricultural Librarians and Documentalists (Quarterly Bulletin of the IAALD)
Recorder
RQ
Reference Services Review (RSR)
Saturday Review
School Library Journal (SLJ)
Science & Technology Libraries
Science Reference Notes
Scientific Information Notes
Sci-Tech News
Serials Librarians
Special Libraries (SL)
Stechert-Hafner Book News
Subscription Books Bulletin
Technical Book Review Index (TBRI)
Ulrich's Quarterly
Unesco Bulletin for Libraries (UBL)
Unesco Journal of Information Science, Librarianship and Archives Administration
Wilson Library Bulletin (WLB)

Heat/Piping/Air Conditioning
Highways and Bridges and Engineering Works
Horticulture
IEEE Proceedings
IEEE Spectrum
Industrial Chemist
Industrial Engineering
Industrial Lubrication and Tribology
Instruments and Control Systems
Interavia
International Journal of Heat and Mass Transfer
International Journal of Rock Mechanics and Mining Sciences
International Sugar Journal
Iron and Steel International
Ironmaking and Steelmaking Journal
ISIS
JASIS
Journal of Aerosol Science
Journal of American Leather Chemists Association
Journal of American Veterinary Medical Association
Journal of Applied Bacteriology
Journal of Applied Photographic Engineering
Journal of Chemical Documentation
Journal of Chemical Education
Journal of Chromatographic Sciences
Journal of Chromatography
Journal of Coatings Technology
Journal of Dairy Science
Journal of Environmental Health
Journal of Fluid Mechanics
Journal of Food Technology
Journal of Forestry
Journal of Gemmology
Journal of Geological Education
Journal of Geology
Journal of Indiana State Medical Association
Journal of Invertebrate Pathology
Journal of Materials Science
Journal of Metals
Journal of Molecular Structure
Journal of New England Water Works Association
Journal of Nuclear Energy
Journal of Nutrition Education
Journal of Paleontology
Journal of Parasitology
Journal of Petroleum Technology
Journal of Pharmaceutical Sciences
Journal of Physics-Section E
Journal of Polymer Science-Part B
Journal of Sedimentary Petrology
Journal of Sound and Vibration
Journal of South Africa Institute of Mining and Metallurgy
Journal of the Acoustical Society of America
Journal of the American Chemical Society

Journal of the American Dietetic Association

Journal of the American Oil Chemists' Society

Journal of the American Statistical Assocation

Journal of the American Veterinary Medical Association

Journal of the Association of Official Analytical Chemists

Journal of the Electrochemical Society

Journal of the Franklin Institute

Journal of the Institution of Water Engineers and Scientists

Journal of the Medical Society of New Jersey

Journal of the Oil and Color Chemists' Association

Journal of the Royal Astronomical Society of Canada

Journal of the Royal Horticultural Society

Journal of the Society of Dyers and Colourists

Journal of the Society of Environmental Engineers

Journal of the Society of Motion Picture and Television Engineers (JSMPTE)

Journal of the Washington Academy of Sciences

Journal of Veterinary Research

Journal of Water Pollution Control Federation

Laboratory Management

Laboratory Practice

Lancet

Lapidary Journal

Laser Focus

Lighting Design and Application

Limnology and Oceanography

Listener

Machine Design

Marine Engineering/Log

Marine Engineers Review

Marine Geology

Materials and Methods

Materials Engineering

Mathematical Gazette

Mathematics of Computation

Measurements and Control

Mechanical Engineering

Metal Finishing

Metal Science

Metals Technology

Microchemical Journal

Microwaves

Mine and Quarry

Minerological Magazine

Mining Magazine

National Safety News

Nature

Nature and Resources

Naval Engineers Journal

New Scientist

New Scientist and Science Journal

NSS News

Nuclear Applications and Technology
Nuclear Science and Engineering
Observatory
Ocean Engineering
Oceans
Optical Engineering
Optical Society of America Journal
Optical Spectra
Optics and Laser Technology
Petroleum Engineer International
Pharmaceutical Journal
Physical Electronics
Physics Bulletin
Physics Education
Physics in Technology
Physics Teacher
Physics Today
Phytochemistry
Plant Engineering
Plastics Engineering
Plating and Surface Finishing
Polymer
Polymer News
Post Office Electrical Engineers Journal
Power
Proceedings of the Royal Australian Chemical Institute
Product Engineering
Professional Safety
Public Works
Quality Progress

Quarterly of Applied Mathematics
Quarterly Journal of Experimental Physiology
Quarterly Review of Biology
Radio Electronics
Reactor Technology
Reclamation Era
Refrigeration and Air Conditioning
RIBA Journal
Royal Microscopical Society Proceedings
Royal Statistical Society Journal
Rubber Chemistry and Technology
Rubber Plastic Age
Rubber World
School Science and Mathematics
Science
Science Digest
Science Journal (SJ)
Science News
Science Progress
Scientific American
Sea Frontiers
Sky and Telescope
Soil Science
Soil Science Society of America Proceedings
Solid State Technology
Sound and Vibration
Sources of Invention
South African Journal of Science
Space World
Structural Engineer

Tappi
Technology and Culture
Telecommunications Journal
Textile Research Journal
Theoretica Chimica Acta
Transactions and Journal of the British Ceramic Society
Transactions of the American Microscopical Society
United States Naval Institute Proceedings
Urban Design Newsletters
Water, Air and Soil Pollution
Water Power
Water and Waste Treatment Journal
Welding and Metal Fabrication
World Corps

British Standard Institution Yearbook
Bulletin of the Medical Library Association (BMLA)
Catholic Library World
Choice
College and Research Libraries (CRL)
Current Geographical Publications
Geographical Journal
Geographical Magazine
Geographical Review
Government Information Quarterly
Government Publications Review
Incorporated Linguist
Information
Information Technology and Libraries
International Bibliography Information Documentation (IBID)
Journal of Academic Librarianship
Journal of American Society for Information Science (JASIS)
Journal of Documentation
Librarians' Newsletter
Library Association Record
Library Hi-Tech News
Library Journal (LJ)
Library of Congress Information Bulletin (LCIB)
Library Professional Publications
Library Resources and Technical Services (LRTS)
Library Trends (LT)

NONSCIENCE JOURNALS

American Documentation
American Libraries (AL)
Aslib Booklist (ABL)
Aslib Proceedings
Assistant Librarian
Australian Library Journal
Babel
Bibliographical Society of America Papers
Bibiography, Documentation, Terminology
Booklist (BL)
British Book News

BOOK SOURCES

Abbreviation used	Title
ARBA	*American Reference Book Annual.* Vol. 1–. Littleton, CT: Libraries Unlimited, 1970–.
Chen	Chen, Ching-chih, *Health Sciences Information Sources.* Cambridge, MA: MIT Press, 1981.
Jenkins	Jenkins, Frances Briggs. *Science Reference Sources.* 5th ed. Cambridge, MA: MIT Press, 1969.
Katz	Katz, William. *Magazines for Libraries.* 2d ed. New York: Bowker, 1972. Also, suppl. 1, 1974.
Owen	Owen, Dolores, and Hanchey, Marguerite M. *Indexes and Abstracts in Science and Technology.* Metuchen, NJ: Scarecrow Press, 1974.
Sheehy	Sheehy, Eugene P. *Guide to Reference Books.* 9th ed. Chicago, IL: American Library Association, 1977. (Eighth edition under Winchell)
Wal	Walford, A. J., ed. *Guide to Reference Material.* 3d ed. Vol. I: *Science and Technology.* London: Library Association, 1973.
Win	Winchell, Constance M. *Guide to Reference Books.* 8th ed. Chicago, IL: American Library Association, 1967. Also, supps. 1, 1965–1966 (1968); 2, 1967–1968 (1970); 3, 1969–1970 (1972).

SCIENTIFIC AND TECHNICAL INFORMATION SOURCES

Second Edition

CHAPTER 1 SELECTION TOOLS

Publishers' blurbs and catalogs are usually the most up-to-date information sources of newly available scientific and technical books. But journals are probably the most important sources for critical evaluative information on new scientific books. For better understanding and easy identification of the major reviewing journals in the fields of science and technology, the following book should be consulted:

Biomedical, Scientific, and Technical Book Reviewing. **Ching-chih Chen.** Metuchen, NJ: Scarecrow Press, 1976.

Presents results of a comprehensive study of about 500 biomedical, scientific, and technical reviewing journals. The book identifies the major reviewing journals in terms of their quantitative coverage of book reviews; explores the effectiveness of the review media in terms of speed of reviewing, comprehensiveness of review treatment, and authority; etc.

BASIC TOOLS

Aslib Book List: A Monthly List of Recommended Scientific and Technical Books, with Annotations. London: Aslib, 1935–. Monthly.

All entries have evaluative annotations by specialists and are graded into categories for users of general, intermediate, highly technical, and reference books. Of special interest to librarians for acquisition purposes. Arranged by universal decimal system categories.
R: Wal (p. 5).

Books in Series. 4th ed. New York: Bowker, 1984.
Second edition, 1979.
A complete up-to-date listing of over 57,500 new titles and more than 3,500 series that have more recently been published. All are arranged by Library of Congress series headings in the main series index and by author and title. Includes a publishers and distributors index and more. An outstanding reference book for any library.

Books Out-of-Print, 1980–1984. 2 vols. New York: Bowker, 1984.
A useful tool that answers the question, "Is this book out of print?" It provides bibliographic information for about 200,000 titles.
R: *WLB* 58: 593 (Apr. 1984); *ARBA* (1985, p. 5).

British Books in Print 1985. London: Whitaker, 1985, and distr. by New York: Bowker, 1985. Annual.

A basic ordering tool that provides access to about 400,000 titles from over 10,000 British publishers. Entries are published in one alphabetical sequence of authors, titles, and subjects. Separate listing of publishers' ISBN prefix and a separate directory of publishers and Book Trade Bibliography.

Canadian Books in Print: Author and Title Index, 1979. **Martha Pluscauskas and Marian Butler**, eds. Toronto: University of Toronto Press, 1979.
R: *CRL* 41: 396 (July 1980).

International Books in Print 1985: English-Language Titles Published outside the United States and the United Kingdom. **Archie Rugh, ed.** Munich, New York: Saur, 1985.
Part 1, *Author-Title Index*, 2 volumes; part 2, *Subject Guide*, 2 volumes.
A useful tool that lists about 140,000 titles from 5,000 publishers in 95 countries. The subject guide, organized by Dewey classification scheme, with alphabetical country arrangement, is a welcome addition in this edition. Useful to large research libraries with substantial holdings.
R: *ARBA* (1985, p. 8).

New Technical Books: A Selective List of Descriptive Annotations. Vols. 1–. New York: New York Public Library, 1951–.
Useful current bibliography of scientific and technical books. Entries arranged by subject with annual author and subject indexes. Dewey decimal classification used.
R: Jenkins (p. 28).

NLL Announcement Bulletin. Vols. 1–. Boston Spa, England: National Lending Library for Science and Technology, 1971–.

Paperbound Books in Print: Spring 1985. 3 vols. New York: Bowker, 1985. Annual.
Over 260,000 titles from 12,000 publishers are included, with bibliographic information including author or editor, title, pages and/or volumes, publishers, year of publication, Library of Congress number, and ISBN number given to each entry.
R: *ARBA* (1985, p. 7).

Pure and Applied Science Books: 1876–1982. 6 vols. New York: Bowker, 1982.
A bibliography of more than 220,000 titles in all the physical and biological sciences, and all the technologies, published in the last 107 years. Entries are listed under 56,000 LC subject headings, with author and title indexes.

Reference Sources for Small and Medium-sized Libraries. 4th ed. **Ad Hoc Committee for the Fourth Edition of "Reference Sources for Small and Medium-sized Libraries," comp.** Edited by Jovian P. Lang and Deborah C. Masters. Chicago: American Library Association, 1984.
A basic reference tool that provides information on fundamental works in each subject area. Entries are listed by 22 major subjects with concise annotation given to each entry.
R: *LJ* 109: 1430 (Aug. 1984); *WLB* 59: 226–227 (Nov. 1984); *ARBA* (1985, p. 5).

Science Books and Films. Washington, DC: American Association for the Advancement of Science, 1965–. Quarterly.

A classified listing. Critical reviews.
R: *BL* 77: 907 (Mar. 1, 1981).

Science Books: A Quarterly Review. Vols. 1–. Washington, DC: American Association for the Advancement of Science, 1965–. Quarterly.

Annotated classified list of new books in the pure and applied sciences.
R: *L J* 92: 1577 (Apr. 1967); Jenkins (A228); Wal (p. 9); Win (1EA6).

Scientific and Technical Books and Serials in Print. 3 vols. New York: Bowker, 1972–. Annual.

Formerly entitled *Scientific and Technical Books in Print.* Latest edition, 1986.
Provides full bibliographic information for titles in the biological sciences, engineering, technology, etc. Contains a directory of over 2,000 publishers. Author/title index.
R: *BL* 75: 501 (Nov. 1978); *Catholic Library World* 49: 406 (Apr. 1978); 50: 408 (Apr. 1979); *Choice* 19: 1016 (April 1982); *ARBA* (1973, p. 533; 1974, p. 542).

Scientific, Engineering, and Medical Societies Publications in Print, 1980–1981. 5th ed. **James M. Kyed and James M. Matarazzo**. New York: Bowker, 1983.

Updated every 2 years.
A standard regerence to publications of 365 professional societies of the United States, Canada, Great Britain, and international societies. Societies are arranged alphabetically with name, address, telephone number, and payment policy. Publications are divided into 3 broad categories; books, periodicals, nonprint material. Books provide information on author, edition, publication date and price; serials include title, frequency and price. Nonprint materials cover information such as films, slides, cassettes, filmstrips, etc. Author, subject, and periodical indexes are given. An essential addition to all library collections.
R: *BL* 77: 349 (Oct. 15, 1980); *CRL* 41: 189 (Mar. 1980); *ARBA* (1980, p. 599).

Technical Book Review Index. New York: Special Libraries Association, 1935–. Monthly.

Printed excerpts from reviews appearing in scientific and trade publications. Time lag of often up to 9 months. Author indexes.
R: Katz (p. 27); Wal (p. 10); (Win EA68; EA69).

Technical Books in Print: A Reference Catalogue of Books in Print and on Sale in Great Britain. London: Whitaker, 1964–.
R: Win (1EA7).

BOOKLISTS AND SERIAL UNION LISTS

The AAAS Science Book List: A Selected and Annotated List of Science and Mathematics Books for Secondary School Students, College Undergraduates, and Nonspecialists. 3d ed. **Hilary J. Deason, comp.** Washington, DC: American Association for the Advancement of Science. 1970–.

Titles numbering over 2,500 are arranged under Dewey decimal classification categories. Descriptive annotations and evaluative critiques.

AAAS Science Book List Supplement. **Kathryn Wolff and Jill Storey, comps.** Washington, DC: American Association for the Advancement of Science, 1978 and later supps.

This supplement of over 2,800 entries updates previous editions. Arranged under Dewey decimal system; each entry contains annotation and full bibliographic information. Covers a broad spectrum of the scientific literature.
R: *BL* 75: 1234 (Apr. 1, 1979); *ARBA* (1979, p. 639); Sheehy (EA4).

AGLINET Union List of Serials. Rome: Food and Agriculture Organization of the United Nations, 1979 and later editions.

A listing of 6,837 serial titles dealing with the agricultural sciences. Entries are alphabetically organized by title and also contain frequency, place of publication, ISSN. Designed to assist AGLINET library users.
R: *Quarterly Bulletin of the IAALD* 24: 112 (Fall/Winter 1979).

The Best Science Books for Children: A Selected and Annotated List of Science Books for Children Ages Five through Twelve. **Kathryn Wolff et al., comps. and eds.** Washington, DC: American Association for the Advancement of Science, 1983.

A selected and annotated list of science, mathematics, and social and behavioral science books for children in grades kindergarden through 12. Author, and subject/title indexes are available. A good supplement to *AAAS Science Book List for Children.*
R: *ARBA* (1985, p. 488).

International Serials Catalogue. 2 vols. **International Council of Scientific Unions, Abstracting Board.** Paris: International Council of Scientific Unions, Abstracting Board, 1978.

An alphabetical listing of serial applications abstracted by ICSUAB. Includes full bibliographic information.
R: Sheehy (EA5a).

Monographic Series. Washington, DC: Library of Congress, 1976–. Quarterly.

A collection of catalog cards listing all monographs cataloged by the Library of Congress as parts of series.

Research Catalog of the Library of the American Museum of Natural History: Authors. 13 vols. **American Museum of Natural History Library, New York.** Boston: Hall, 1977.

Complete access to the catalog is facilitated by personal, corporate, joint author, compiler, editor, and illustrator index. Includes bibliographic and journal citations.
R: Sheehy (EC16).

Research Catalog of the Library of the American Museum of Natural History: Classed Catalog. 12 vols. Boston: Hall, 1978.

Lists over 325,000 volumes that deal with a variety of natural science fields including geology, zoology, anthropology, entomology, mineralogy, paleontology, etc. Also includes a listing of 17,000 serial titles.

Science Fiction Book Review Index, (1974–1979). **H. W. Hall, ed.** Detroit: Gale Research, 1981.

Earlier edition—1923–1973, 1975.

A complete record of all books (including non-science-fiction books) reviewed in the science-fiction magazines from 1974–1979 and a record of all science-fiction and fantasy books reviewed in that period in selected general magazines. Indexed.

R: *LJ* 100:1535 (Sept. 1, 1975).

Scientific and Technical Books and Serials in Print. Books: Subject Index, Author Index, Title Index. Serials: Subject Index, Title Index. New York: Bowker, 1981–. Annual.

R: *CRL* 42: 183 (Mar. 1981); Wal (p. 133).

Scientific and Technical Information Resources. **Krishna Subramanyam.** New York: Dekker, 1981.

A combined reference of several publications along with a narrative description of each. Most helpful to those in library schools as a textbook and to those information professionals working with the science and technology collections.

R: *ARBA* (1983, p. 596).

Scientific Books, Libraries and Collectors: Supplement, 1969 –75. **John L. Thornton and R. I. J. Tully.** London: Library Association, 1978.

First edition, *Scientific Books, Libraries and Collectors*, 1954; second edition, 1962; third edition, 1971.

A standard British reference, updated by this supplement. Covers material on the history of science between 1969–1975. Meticulously arranged. Considered an essential reference.

R: *Library Association Record* 80: 575 (Nov. 1978).

Sources of Serials: An International Publisher and Corporate Author Directory. New York: Bowker, 1977.

A standard reference to 63,000 serial publishers and corporate authors. Organized into 181 countries with 90,000 entries providing address, published serial titles, and ISSN. An essential aid for all libraries.

Titles in Series: A Handbook for Librarians and Students. 3d ed. 4 vols. **Eleanora A. Baer.** Metuchen, NJ: Scarecrow Press, 1978.

Updated edition includes 69,700 book titles listed by series. Contains complete bibliographic citations. For college and university libraries.

R: *LJ* 104: 816 (Apr. 1, 1979).

SUBJECT LIBRARY CATALOGS FOR RETROSPECTIVE SELECTION

GENERAL SCIENCE

Catalog of Books in the American Philosophical Society. 28 vols. **American Philosophical Society.** Westport, CT: Greenwood Press, 1970.

Emphasizes books on the history of science and publications of scientific academies.
R: Win (3EA27).

John Crear Library [Chicago] Catalog. 78 vols. Boston: Hall, 1967.

Photoreproduction of the library catalog of perhaps the most comprehensive scientific and technical library in the world. *Author/Title Catalog, Classified Subject Catalog,* and *Subject Index to the Classified Subject Catalogs.*
R: Jenkins (A24); Wal (p. 7); Win (2EA5).

John Crear Library Classified Subject Catalog. 42 vols. Boston: Hall, 1967.

The published catalog of an outstanding library with rich scientific holdings.

International Catalogue of Scientific Literature. London: Royal Society of London. Annual.

Astronomy

Astronomical Catalogues 1951–75. **Mike Collins**. Old Working, England: Institute of Electrical Engineers, 1977.

Includes 2,500 catalogs printed between 1951 and 1975. Contains a bibliography that lists celestial objects and phenomena. All information is complete and thorough.
R: Sheehy (EB2).

Catalog of the Naval Observatory Library, Washington, DC. 6 vols. **US Naval Observatory Library.** Boston: Hall, 1976.

A comprehensive catalog of 75,000 volumes of the Naval Observatory Library, including many rare books on the history of astronomy. Many of the cards included are in Library of Congress format.
R: *RSR* 5: 18 (Apr./June 1977).

Mathematics

A Basic Library List for Four-Year Colleges. 2d ed. **Mathematical Association of America. Committee on the Undergraduate Program in Mathematics**. Washington, DC: Mathematical Association of America, 1976.

First edition, 1966.
Contains a list of 700 books and journals arranged under subject area.
R: Sheehy (EF2).

Chemistry

Selected Titles in Chemistry: An Annotated Bibliography of Moderately Priced Books for the Student, the Teacher, and the General Reader. 4th ed. **American Chemical Society**. Washington, DC: American Chemical Society, 1977. Also, later edition.

Selective listing of moderately priced, paperback books on various topics of the chemical sciences. Brief annotations and reading classification level accompany each entry. Useful guide to elementary, junior high and senior high school teachers.
R: *ARBA* (1979, p. 652).

Physics

Dictionary Catalog of the Princeton University Plasma Physics Laboratory Library. 4 vols. Boston: Hall, 1970. Also, supps. 1–, 1973–.

Comprehensive collection of technical journals and reports.

Biological Sciences

Agriculture

AGRIS Forestry World Catalogue of Information and Documentation Services. **Susan Lederer Bewer, ed. and comp.** Rome: Food and Agriculture Organization of the United Nations. Distr. New York: Unipub, 1979.

A trilingual catalog (English, Spanish, French) covering information sources to use in locating forestry literature. Lists entries by country and includes such information as address, type of service, parent organization, area covered, collection size, subject coverage, etc. Entries in one language only. An invaluable reference aid to all science and faculty libraries.

R: *ARBA* (1981, p. 741).

Catalogue of the Imperial College of Tropical Agriculture, University of the West Indies, Trinidad. 8 vols. Boston: Hall, 1975.

International collection of information about tropical and subtropical agriculture; includes many early imprints.

R: *ARBA* (1976, p. 741).

Dictionary Catalog of the National Agricultural Library, 1862–1965. 73 vols. **US National Agricultural Library**. New York: Rowman & Littlefield, 1967–1970.

Monthly supplements since 1966.
Over 1.5 million entries.

R: *Bibliography, Documentation, Terminology* 10: 210 (Sept. 1970); Jenkins (J10); Wal (p. 397); Win (1EK2; 2EK3; 3EK4).

Guide to Manuscripts in the National Agricultural Library. (US Department of Agriculture Miscellaneous Publication no. 1374). **Alan E. Fusonie, comp.** Beltsville, MD: US Department of Agriculture, Science, and Education Administration, Technical Information Systems, 1979.

Outlines the manuscript holdings and their contents to the researcher. The collection features account books, diaries, letters, personal papers, memoirs, photographs, and other historical items.

R: *Quarterly Bulletin of the IAALD* 24: 54 (Spring–Summer 1979).

List of Available Publications of the United States Department of Agriculture. **US Department of Agriculture.** A21.9/8; item 91. 1841–. Annual.

A valuable selection tool, listing titles that are available free from the Department of Agriculture. Arranged by broad subject categories. Contains full bibliographical data. A useful reference tool.

R: *ARBA* (1978, p. 736).

World List of Aquatic Sciences and Fisheries Serial Titles. **Food and Agriculture Organization of the United Nations.** New York: Unipub, 1975.

Presents bibliographic details on 646 serial publications dealing with the aquatic sciences and fisheries. International in scope.
R: *IBID* 6: 406 (Dec. 1978).

Botany

Catalog of the Farlow Reference Library of Cryptogamic Botany, Harvard University. 6 vols. Boston: Hall, 1979.

Includes over 60,000 entries of books, reprints, and periodicals covering nonflowering plants.

Catalog of the Manuscript and Archival Collections and Index to the Correspondence of John Torrey. **New York Botanical Garden Library**. Boston: Hall, 1973.

Some 180,000 entries related to botany and horticulture. Items include papers, diaries, unpublished manuscripts, and scientific correspondence.

Catalog of the Royal Botanic Gardens, Kew, England. 9 vols. Boston: Hall, 1973.

A listing of the collection of the Kew Herbarium, which includes material on plant taxonomy, botany, plant cytology, etc. For specific subjects, catalog is classified by the Dewey decimal system.

Dictionary Catalog of the Library of the Massachusetts Horticultural Society. 3 vols. Boston: Hall, 1963.

Supplement. Nos. 1–. Boston: Hall, 1972–.

A catalog of the 31,000 volumes contained in the Massachusetts Horticultural Society Library. Covers all topics in gardening including herbals, pomology, home landscaping, garden design.

Flower and Fruit Prints of the 18th and Early 19th Centuries, Their Histories, Makers, and Uses, with a Catalogue Raisonné of the Works in Which They Are Found. Reprint of 1938 ed. **Gordon Dunthorne**. New York: Da Capo, 1970.

A definitive work, bound to be important for art, botanical, and horticultural collections. Textual material and bibliographic data as well as 80 full-page plates.
R: *Choice* 7: 1014 (Oct. 1970); *ARBA* (1971, p. 497).

Plant Science Catalog: Botany Subject Index. 15 vols. **US National Agricultural Library**. Boston: Hall, 1958.

Over 315,000 cards from the subject catalog section of the *Plant Science Catalog*.
R: Wal (p. 206); Win (EC41).

Printed Books, 1481–1900, in the Horticultural Society of New York. **Elizabeth Cornelia Hall**. New York: Horticultural Society of New York, 1970.

Unannotated alphabetical listing of some 3,000 titles, many of which are classics in the field.
R: *ARBA* (1972, p. 568).

Zoology

Blacker-Wood Library of Zoology and Ornithology Dictionary Catalog. 9 vols. **McGill University (Montreal)**. Boston: Hall, 1966.
Reproduction of the card catalog of over 60,000 volumes and 20,000 notes and correspondences.
R: *LJ* 92: 726 (1967); Jenkins (G192); Wal (p. 222); Win (2EC14).

Catalogue of the Library of the Royal Entomological Society of London. 5 vols. Boston: Hall, 1979.
A listing of approximately 9,000 monographs, 600 journals, and 50,000 pamphlets covering the subject of entomology.

Catalogues of the Library of the Marine Biological Association of the United Kingdom, Plymouth, England. 16 vols. Boston: Hall, 1977.
An extensive listing of over 13,000 books, 40,000 bound periodicals, 1,450 current periodicals, and 50,000 reprints and pamphlets covering such topics as marine biology, fisheries, and oceanography. Main catalog is alphabetically organized under author/name; 14,500 subject index entries are arranged under broad subject headings.

Harvard University Museum of Comparative Zoology. Library Catalogue. 8 vols. Boston: Hall, 1968.
Some 250,000 volumes on zoology and paleontology from the nineteenth century. Useful historical guide to older periodicals.
R: Jenkins (G190); Win (2EC14a).

Earth Sciences

Arctic Institute of North America. Montreal Library Catalog. 4 vols. Boston: Hall, 1968.
Some 30,000 books and pamphlets dealing with all aspects of the polar regions.
R: Wal (p. 154).

Catalogs of the Glaciology Collection. 3 vols. **American Geographical Society, New York, Department of Exploration and Field Research**. Boston: Hall, 1971.
A library catalog containing 20,000 authors, 18,000 subject entries, and 15,000 regional entries in the fields of geography, geophysics, geology, and earth science.

Earthquake Engineering Research Center Library Catalog. Berkeley: University of California, College of Engineering, 1975.
Includes slides, movies, and maps, as well as the usual print material.

Geography and Earth Sciences Publications: An Author, Title, and Subject Guide to Books Reviewed, and an Index to the Reviews. **John Van Balen, comp.** Ann Arbor: Pierian Press, 1978.

Volume 1, 1968–1972; volume 2, 1973–1975. A bibliographic listing of books published in the earth sciences. The author section is arranged alphabetically by author's last name and is followed by subject, geographical, and title indexes.

US Department of the Interior Library. *Dictionary Catalog of the Department Library.* Boston: Hall, 1968–.

Guide to specialized collections on minerals, petroleum, fisheries, land management, etc.

US Geological Survey Library Catalog. 25 vols. Boston: Hall, 1964. Also, supps. 1–. Boston: Hall, 1972–.

Indexes by author, title, and subject. A variety of volumes, maps, and pamphlets in the fields of geology, paleontology, petrology, mineralogy, etc. Over 400,000 catalog cards.
R: Wal (p. 145); Win (1EE4).

Oceanography

The Defense Mapping Agency Hydrographic Center Catalog of Publications. Washington, DC: Defense Mapping Agency Hydrographic Center, 1974.

A listing of publications of interest to the scientific, oceanographic, and maritime communities, arranged by subject matter with a brief description of each publication.

Dictionary Catalog of the Water Resources Center Archives, University of California, Berkeley. 5 vols. **University of California, Water Resources Center Archives**. Boston: Hall, 1970. Also, supps. 1–. Boston: Hall, 1971–.

Emphasizes a century of report literature relating to water, its utilization, and industrial uses.
R: Win (3E114).

ESSA Libraries Holdings in Oceanography and Marine Meteorology, 1710–1967. 4 vols. **US Environmental Science Services Administration**. Rockville, MD: US Environmental Science Services Administration, Scientific Information and Documentation Division, 1969.

A computer-generated catalog of approximately 3,000 references. Three volumes are composed of indexes.
R: Win (3EE9).

Marine Biological Laboratory and Woods Hole Oceanographic Institution [Woods Hole, MA] Library Catalog. 12 vols. Boston: Hall, 1971.

Volume 12 is the *Journal Catalog.*
Author catalog covering 12,000 books, 138 expeditions, and some 300,000 journal articles.
R: *ARBA* (1973, p. 550); Wal (p. 162).

Ocean Engineering and Oceanography Technical Literature Collection, Water Resources Center Archives, University of California, Berkeley. **Michael Poniatowski**. Washington, DC: US National Oceanic and Atmospheric Administration, 1974.

Some 1,500 items.
R: *ARBA* (1975, p. 806).

Scripps Institution of Oceanography Library. *University Library Catalog.* 12 vols. Boston: Hall, 1970.

Author/Title Catalog, 7 volumes; *Subject Catalog,* 2 volumes; *Shelf List,* 2 volumes; *Shelf List of Documents, Reports and Translations,* 1 volume.

Comprehensive library catalog of some 80,000 volumes in the fields of oceanography and marine biology.
R: *ARBA* (1971, p. 522); Wal (p. 163).

GENERAL ENGINEERING

Bibliographic Guide to Technology, 1978. 2 vols. Boston: Hall, 1979.

A thorough 2-volume annual subject bibliography that deals with engineering and technology. Designed for use in conjunction with other bibliographic guides, the listing consists of current publications cataloged by the Library of Congress and the Research Libraries of the New York Public Library. All entries are arranged alphabetically and include title, subject heading, and main and added entries. Of interest to acquisition, reference, and cataloging departments of science libraries.
R: *Choice* 17: 48 (Mar. 1980).

Classed Subject Catalog of the Engineering Societies Library. **Engineering Societies Library [New York].** 13 vols. Boston: Hall, 1963. Also, supps. 1964–. Annual.

Together, these volumes represent a retrospective and current bibliographic effort. Volumes consist of library cards in the universal decimal system by subject.

Volumes 1–13 include the library's 185,000-volume collection. Supplements include information on subjects of peripheral interest such as geology and geophysics.
R: Jenkins (K14); Wal (p. 284).

CIVIL ENGINEERING—TRANSPORTATION

Building Technology Publications Supp. 3, 1978. **JoAnne R. Debelius, ed.** Washington, DC: National Bureau of Standards Distr. Washington, DC: US Government Printing Office, 1979.

The third supplement to *Building Technology Publications,* which catalogs all Center for Building Technology publications. Includes titles and abstracts of all reports. Entries provide title, author, date of publication, keywords, and abstracts. Contains author and keyword indexes. Also lists non-National Bureau of Standards papers and reports.
R: *ARBA* (1981, p. 757).

Northwestern University Transportation Center Catalog. 12 vols. Boston: Hall, 1972.

Some 79,000 books and reports and 1,000 serials that emphasize transportation management and engineering.

COMPUTER TECHNOLOGY

Book Bytes: The User's Guide to 1200 Microcomputer Books. **Cris Popenoe, ed.** New York: Pantheon Books/Random House, 1984.

A review publication that categorizes books on microcomputers and software in 5 major sections—general, system-oriented, applications, programming, and theoretical topics. A title and author index.
R: *LJ* 109: 1233 (June 15, 1984); *ARBA* (1985, p. 588).

Computer Books and Serials in Print 1985. New York: Bowker, 1985. Annual.

A single-volume reference tool that includes about 15,000 United States and international books and pamphlets and almost 1,800 serial titles on the subject of computers and/or computing. Useful for all types of libraries.
R: *LJ* 109: 1840 (Oct. 1, 1984); *WLB* 59:145 (Oct. 1984); *ARBA* (1985, p. 581).

Computer Publishers and Publications 1985–86: An International Directory and Yearbook. 2d ed. **Efrem Siegel and Frederica Evan, eds. Communications Trends, Distr.** Detroit: Gale Research, 1985. Annual.

Provides updated information on over 275 publishers of books on computers and more than 600 periodicals concerned with computers. With several indexes.
R: *Choice* 21:1438 (June 1984); *RQ* (Summer 1984, p. 468); *Science Technology Libraries* 5: 2 (Winter 1984); *WLB* 59: 290–291 (Dec. 1984); *ARBA* (1985, p. 585).

Science Software Quarterly. Vols. 1–. Tempe, AZ: Arizona State University, Center for Environmental Studies, 1984–. Quarterly.

Provides critical and descriptive reviews of scientific software.
R: *LJ* 111: 66 (Feb. 1, 1986).

Software Reviews on File. Nos. 1–. New York: Facts on File, 1985–. Monthly.

A type of *Book Review Digest* for about 500 new software programs a year. Reviews are cited from over 60 periodicals.
R: *LJ* 111: 66 (Feb. 1, 1986).

ENVIRONMENTAL SCIENCES

Asbestos: An Information Resource. **Richard J. Levine, ed.** Bethesda, MD: US National Institutes of Health, 1978.

Good introduction to asbestos and its biological effects. Lists some 400 references to published reports. Information sources deal with biological effects of asbestos fibers, descriptions of fibers, history of illness among asbestos workers, and asbestos substitutes.
R: *National Safety News* 119: 194 (June 1979); *TBRI* 45: 257 (Sept. 1979).

Catalog of the Conservation Library, Denver Public Library. 6 vols. **Denver Public Library.** Boston: Hall, 1974.

Emphasizes pollution control, economics of ecology, and the history of the environmental movement.

ECOL: Book Catalog of the Environmental Conservation Library. **Minneapolis Public Library.** Chicago: American Library Association, 1974.

Book catalog for this special collection as of 1973. Main entry, title, and subject catalogs. Emphasizes books bearing on environmental matters in the midwestern United States.
R: *AL* 5:299 (June 1974); *ARBA* (1975, p. 693).

CHAPTER 2 GUIDES TO THE LITERATURE

The following are examples of sources that analyze the structure and availability of scientific and technical literature:

On Documentation of Scientific Literature. 2d ed. **T. P. Loosjes**. London: Butterworth. Distr. Hamden, CT: Archon Books, 1973.

Concentrates on theoretical problems of bibliographic control and information retrieval.

Pilot Study on the Use of Scientific Literature by Scientists. **Ralph R. Shaw**. Metuchen, NJ: Scarecrow Press, 1971.

R: *LJ* 96: 2747 (Sept. 15, 1971); *ARBA* (1972, p. 543).

GENERAL SCIENCE

A Brief Guide to Sources of Scientific and Technical Information. 2d ed. **Saul Herner**. Arlington, VA: Information Resources Press, 1980.

First Edition, 1970.

A revised and expanded edition of a practical guide. Contains major sources of information with an emphasis on directories, research in progress, and important American research collections. Includes coverage of computer-generated information and machine-readable data bases. Arrangement is by time covered. General information as well as specific reference tools are discussed. A useful tool for scientists, engineers, reference libraries.

R: *American Scientist* 59: 639 (Sept./Oct. 1971); *Journal of Applied Photographic Engineering* 7: 158A (Oct. 1981); *AL*: 620 (June 1970); *CRL* 41: 567 (Nov. 1980); *LCIB* 29: 165 (1970); *LJ* 95: 2448 (July 1970); 105: 1723 (Sept. 1, 1980); *UBL* 24: 282 (Sept./Oct. 1970); *WLB* 55: 221 (Nov. 1980); *ARBA* (1971, p. 470; 1981, p. 617); Wal (p. 1); Win (3EA3).

European Sources of Scientific and Technical Information. 5th ed. **Anthony P. Harvey and Ann Pernet, eds.** Essex, England: Longman Group. Distr. Detroit: Gale Research, 1981.

First edition, 1957; first to fourth editions, entitled *Guide to European Sources of Technical Information.*

A geographical arrangement of sources concerning pure sciences and applied technology information. A fine attempt to identify key national centers of information.

R: *ARBA* (1983, p. 595).

Finding Answers in Science and Technology. **Alice Lefler Primack**. New York: Van Nostrand Reinhold, 1984.

A general overview of and introduction to how to find information and how to formulate a search strategy. Also includes chapters on specific information sources. For

the undergraduate science student and the nonscientist librarian or teacher as well as for all science libraries.
R: *Science Technology Libraries* 5: 2 (Winter 1984).

Guide to Reference Material. Science and Technology. 4th ed., vol. 1. **A. J. Walford, ed.** London: Library Association. Distr. Chicago: American Library Association, 1980.

Second edition, 1966; third edition, 1973–1977.
A standard reference of some 5,000 main entries with more than 1,000 subsumed entries. Expanded coverage includes additional information on biochemistry, environmental pollution, microcomputers, and alternative technology. Annotations are generally descriptive rather than critical. Recommended for academic and special libraries.
R: *BL* 70: 393 (Dec. 15, 1973); *Choice* 10: 1536 (Dec. 1973); *Library Association Record* 80: 123 (Feb. 1978); *L J* 91: 3687 (1966); *New Library World* 81: 207 (Oct. 1980); *ARBA* (1974, p. 542; 1981, p. 619); Jenkins (A9).

A Guide to US Scientific and Technical Resources. **Rao Aluri and Judith Robinson.** Littleton, CO: Libraries Unlimited, 1983.

A helpful guide in accessing federal, scientific and technical information used in the United States and abroad. Each category of information evaluates its form in the flow of scientific and technical communication and describes major sources of information for each category.

Information Resources for Engineers and Scientists: Workshop Notes. 4th ed. **Charlie Maiorana.** White Plains, NY: Knowledge Industry, 1985.

A guide to technical information written for engineers and the librarians who work with them. About 50 categories of reference sources are covered, including technical reports, conference proceedings, abstracts and indexes, etc. Indexed.

Information Sources: Physical Sciences and Engineering. Washington, DC: Library of Congress, 1974 and updates.

Japanese Scientific and Technical Literature: A Subject Guide. **Robert W. Gibson, Jr. and Barbara K. Kunkel.** Westport, CT: Greenwood Press, 1981.

A reference guide, written in 2 parts, to Japanese scientific and technical literature. Part 1 covers bibliographic control, monographs, technical reports, current periodicals. Part 2 lists over 9,100 periodicals under 64 subjects. Alphabetized title index is included. Recommended to all academic libraries.
R. *ARBA* (1983, p. 600).

LC Science Tracer Bullet. **US Library of Congress, Science and Technology Division, Reference Section.** Washington, DC: US Library of Congress, 1974–. Irregular.

An irregular series of reference aids which provides guides to systematically searching the literature of specific topic areas. Intended to aid those who seek a clearer understanding of a specific topic. Some of the sample reference aids include *Infrared Applications* (10 pages); *Ocean Thermal Energy Conversion (OTEC): A Brief Guide to Materials in the Library of Congress* (6 pages); *Acid Rain: A Brief Guide to Materials in the Library of Congress* (6 pages); *Agent Orange Dioxin: TCDD* (9 pages); *Solar Energy* (12 pages); *Elec-*

tric and Hybrid Vehicles (8 pages); *Industrial Robots: A Brief Guide to Materials in the Library of Congress* (5 pages); *Automotive Electronics* (6 pages); *Lasers and Their Applications* (10 pages); *Low-Level Ionizing Radiation—Health Effects* (13 pages); *Synthetic Fuels* (10 pages); *Science and Technology: Toward the 21st Century* (10 pages).
R: *LCIB* 39: 157 (May 9, 1980); 39: 341, 342 (Sept. 5, 1980); 39: 440 (Nov. 7, 1980); 39: 482 (Dec. 12, 1980); 39: 487 (Dec. 19, 1980); 40: 142 (Apr. 24, 1981); 40: 172 (May 22, 1981).

Range Science: A Guide to Information Sources. **John F. Vallentine and Phillip L. Sims.** Detroit: Gale Research, 1980.

A praiseworthy guide to range science information sources such as periodicals, organizations, government publications, handbooks, etc. Discusses literature searching, both manually and online. Indexes include author; organizations, agencies, and services; periodicals, serials, abstracts, and bibliographies; and subject. Invaluable reference tool for academic, research, and public libraries.
R: *Choice* 18: 513 (Dec. 1980); *ARBA* (1981, p. 742).

Science and Engineering Literature: A Guide to Reference Sources. 3d ed. **H. Robert Malinowsky and Jeanne M. Richardson.** Littleton, CO: Libraries Unlimited, 1980.

First edition, 1967, entitled *Science and Engineering Sources: A Guide for Students and Librarians*; second edition, 1976.
A rewritten and updated version of an invaluable reference tool for information sources in science and engineering. Includes over 1,270 sources and new editions that have appeared since 1976. Contains new material covering databases, abstracting services, and bibliographies. A must for all libraries.
R: *Choice* 14: 658 (July/Aug. 1977); *CRL* 38: 69 (Jan. 1977); *LJ* 93: 55 (Jan. 1, 1968); *SL* 59: 208 (1968); *WLB* 51: 267 (Nov. 1976); 55: 542 (Mar. 1981); *ARBA* (1977, p. 626; 1981, p. 618); Jenkins (A7); Sheehy (EA2); Wal (p. 8); Win (2EA2).

Science and Technology: An Introduction to the Literature. 4th ed. **Denis Grogan.** Hamden, CT: Linnet Books, 1982.

Second edition, revised 1973. First edition, 1970.
Primarily a student's guide: reference materials arranged by type rather than subject. Includes treatment of patents, reports and periodicals. A useful guide to scientific literature, despite its emphasis on British sources.
R: *Choice* 7: 1014 (Oct. 1970); *CRL* 38: 430 (Jan. 1977); *WLB* 45: 316 (Nov. 1970); *ARBA* (1971, p. 470; 1974, p. 540; 1977, p. 624; 1983, p. 595); Sheehy (EA1).

Science Information Resources. **US Library of Congress Staff Members.** New York: Science Associates, 1976.

Volume 1, *Literature Guides*, is a collection of more than 50 literature guides issued by the LC since 1972 entitled *LC Science Tracer Bullets*. Each *Bullet* runs from 2 to 11 pages, begins with a definition of the scope of the subject, and provides basic sources of information. Subject and title indexes. Volume 2, entitled *Information Services*, 1969, included in the directory section of this book.

Scientific and Technical Information Sources. 2d ed. **Ching-chih Chen.** Cambridge, MA: MIT Press, 1986.

First edition, 1977; second printing, 1979.
A complete updated edition that covers information sources in all fields of science and technology, mostly published after 1977. Over 4,000 entries are included under 23 categories. All entries under each chapter are grouped by subject first, then alphabetically by title. In addition to full bibliographic information, both critical and descriptive annotations are given whenever possible. Review sources are provided as well. Complete indexes are available. Essential tool for all scientific and technical libraries.

Subject Collections. 5th ed. **Lee Ash, comp.** New York: Bowker, 1979.
Fourth edition, 1974.
Contains some 70,000 entries covering myriad subject headings. Serves as a guide to the special collections of universities, public and special libraries, and museums.
R: *BL* 76: 1387 (May 15, 1980).

ASTRONOMY

A Guide to the Literature of Astronomy. **Robert A. Seal**. Littleton, CO: Libraries Unlimited, 1977.
Selective, annotated list of the literature of astronomy. Consists of 4 sections: general astronomy; practical and spherical astronomy; theoretical astronomy; and descriptive astronomy. Includes nearly 600 citations to reference sources of scholarly and popular interest. Topical arrangement of material. Indexed by author, title, and subject. Basically, a guide for the beginner and nonspecialist.
R: *Sky and Telescope* 55: 430 (May 1978); 57: 73 (Jan. 1979); *RQ* 18: 217 (Winter 1978); *WLB* 52: 656 (Apr. 1978); *ARBA* (1978, p. 635); Sheehy (EB1).

MATHEMATICS

Statistics and Econometrics: A Guide to Information Sources. **Joseph Zaremba, ed.** Detroit: Gale Research, 1980.
R: *BL* 77: 350 (Oct. 15, 1980); *CRL* 42: 183 (Mar. 1981).

Use of Mathematical Literature. **Alison R. Dorling, ed.** Boston: Butterworth, 1977.
A well-rounded guide to the use of mathematical literature. First 3 chapters devoted to general reference sources. Subsequent chapters feature 14 bibliographic essays by specialists and cover specific aspects of mathematics. Includes author and subject indexes. Well-documented guide for graduate-level use.
R: *Journal of Documentation* 34: 91 (Mar. 1978); *RSR* 6: 19 (July/Sept. 1978); *ARBA* (1978, p. 627); Sheehy (EF1).

Using the Mathematical Literature: A Practical Guide. **Barbara Kirsch Schaefer**. New York: Dekker, 1979.
A selective guide to the vast and varied amount of mathematical literature. Emphasizes descriptions of different types of publications. Titles selected were in print in 1977. Chapter 2 features an excellent essay on historical perspectives.

R: *CRL* 40: 390 (July 1979); *Journal of Academic Librarianship* 5: 296 (Nov. 1979); *LJ* 104: 1330 (June 1979); *ARBA* (1980, p. 607).

PHYSICS

An Introductory Guide to Information Sources in Physics. **L. R. A. Melton.** London: Institute of Physics, 1978.

R: *New Scientist* 82: 293 (Apr. 26, 1979).

Use of Physics Literature. **Herbert Coblans, ed.** Reading, MA: Butterworth, 1975.

Intended for both librarians and scientists. Seventeen contributors present essay-type discussions that help readers to understand the control and documentation process in the physical sciences and its applications.
R: *RSR* 5: 18 (Apr./June 1977); *ARBA* (1977, p. 648).

CHEMISTRY

Guide to Basic Information Sources in Chemistry. **Arthur Antony.** New York: Halsted Press, 1979.

Focuses on techniques and tools available for locating chemical literature. Emphasis placed on pure chemistry with some additional references to related fields. Most entries reflect English-language works. Includes title and subject indexes. A handy tool for students and the public.
R: *American Scientist* 68: 83 (Jan./Feb. 1980); *Chemistry and Industry* 7: 274 (Apr. 5, 1980); *Journal of Chemical Education* 56: A382 (Dec. 1979); *Journal of the American Chemical Society* 102: 3306 (Apr. 23, 1980); *BL* 76: 1151 (Apr. 1, 1980); 76: 1159 (Apr. 15, 1980); 77: 69 (Sept. 1, 1980); *Choice* 16: 795 (Sept. 1979); *TBRI* 46: 81 (Mar. 1980); *ARBA* (1980, p. 614).

Guide to Gas Chromatography Literature. Vol. 4. **Austin V. Signeur.** New York: Plenum Press, 1979.

Easy reference tool to the literature of gas chromatography. Lists citations from journals and books, reports and papers presented at international technical meetings. Consists of over 16,000 entries arranged numerically as well as alphabetically by author or agency.
R: *ARBA* (1981, p. 633).

A Guide to the HPLC Literature. Vols. 1–. **Henri Colin et al.** New York: Wiley, 1984–.

Volume 1, 1966–1979; volume 2, 1980–1981; volume 3, 1982.

How to Find Chemical Information: A Guide for Practicing Chemists, Teachers, and Students. **Robert E. Maizell.** New York: Wiley, 1979.

A thorough, up-to-date guide to chemical information resources. Covers a broad range of topics: online searching, patents, government documents, etc. Ideal for all levels of chemists.

R: *Chemistry in Britain* 16: 280 (May 1980); *Journal of Academic Librarianship* 5: 304 (Nov. 1979); *CRL* 41: 81 (Jan. 1980); *RSR* 8: 30 (July/Sept. 1980); *ARBA* (1980, p. 615).

The Literature of Matrix Chemistry. **H. Skolnik**. New York: Wiley, 1982.

An up-to-date version of the status of the literature of chemistry and its use by chemists and chemical engineers. Beneficial to the chemist in need of chemical literature.

Use of Chemical Literature. 3d ed. **R. T. Bottle, ed.** London: Butterworth, 1979.

First edition, 1962; second edition, 1969.
Covers international resources on reference materials and divisions of chemistry. Well organized and informative.
R: *RSR* 8: 23 (July/Sept. 1980); Wal (p. 151).

BIOLOGICAL SCIENCES

Coping with the Biomedical Literature: A Primer for the Scientist and the Clinician. **Kenneth S. Warren, ed.** New York: Praeger, 1981.

Contributed articles from experts in biomedical information fields, arranged under 4 categories—the structure of the information system, production of biomedical information, utilizing biomedical information, and sources of biomedical information.

A Guide to Searching the Biological Literature. **Michael M. King and Linda S. King**. Boca Raton, FL: Science Media, 1978.

Discusses 5 significant reference works: *Excerpta Medica*, *Index Medicus*, *Biological Abstracts*, *Chemical Abstracts*, and *Science Citation Index*. Examines each work in the context of a common theme. The taped discussion is presented in 6 sections with intermission points provided for review. Developed for students and professionals active in the medical, biological, and biochemical fields.

Information Sources in Agriculture and Food Science. **G. P. Lilley**. Stoneham, MA: Butterworth, 1981.

A reference guide to the world literature of agriculture, horticulture, and food science in both general and specialized areas. International in scope and directed toward information and subject specialists.

Information Sources on Bioconversion of Agricultural Wastes. **United Nations**. New York: Unipub, 1979.

The Literature of the Life Sciences: Reading, Writing, Research. **David A. Kronick**. Philadelphia: ISI Press, 1985.

A guide to both print and online resources in the fields of biology, medicine, and chemistry.

North American Forest History: A Guide to Archives and Manuscripts in the United States and Canada. **Richard C. Davis, comp.** Santa Barbara, CA: ABC-Clio Books, 1977.

A companion volume to *North American Forest and Conservation History: A Bibliography*. Includes much archival information, some oral histories and photographs. Contains citations to over 3,800 manuscripts, topically arranged. Well-recommended.
R: *Choice* 14: 1338 (Dec. 1977); *ARBA* (1978, p. 750); Sheehy (EL7).

Smith's Guide to the Literature of the Life Sciences. 9th ed. **Roger C. Smith, W. Malcolm Reid, and Arlene E. Luchsinger.** Minneapolis: Burgess Publishing, 1980.

An authoritative concise guide to literature on the life sciences as well as on the career pursuit of a life scientist. Stresses library experience including discussions of topics such as proposal writing, fund requesting, preparation of research papers, theses and dissertation writing. Suitable as a self-study guide or biology literature text. For academic libraries and students.
R: *ARBA* (1981, p. 642).

Entomology: A Guide to Information Sources. **Pamela Gilbert and Chris J. Hamilton.** London: Mansell. Distr. New York: H. W. Wilson, 1983.

A list of 1,305 bibliographic entries on entomological information sources, and a directory of 316 entomological suppliers, photo sources, libraries with good collections on the subject, and societies. Indexed.
R: *Choice* 21: 1274 (May 1984); *ARBA* (1985, p. 532).

Taxonomic Literature: A Selective Guide to Botanical Publications with Dates, Commentaries, and Types. 2d ed. **R. S. Cowan and Frans Anthonie Stafleu.** Utrecht, Netherlands: Bohn, Scheltema and Holkema. 1979–.

First edition, 1967; second edition, Volume 2, H–Le, 1979 volume 3: Lh–O, 1981. A selective guide to botanical publications, with dates, commentaries, and types.
R: *Quarterly Bulletin of IAALD*, 24: 82 (Summer–Fall 1979): *ARBA* (1984, p. 645); Jenkins (G80); Wal (p. 217); Win (2EC10).

The Use of Biological Literature. 2d ed. **R. T. Bottle and H. V. Wyatt, eds.** London: Butterworth, 1972.

First edition, 1966.
A comprehensive survey of biological literature divided into subfields. Ample coverage of government publications, bibliographies, patents, and abstracts, though primary emphasis is on British sources. Includes chapters on library use and research methods.
R: *Assistant Librarian* 65: 68 (Apr. 1972); *Bibliographical Society of America Papers* 61: 290 (July/Sept. 1967); *WLB* 42: 220 (1967); *ARBA* (1973, p. 549); Jenkins (G1); Wal (p. 191); Win (2EC1).

Using the Biological Literature. **Elisabeth B. Davis.** New York: Dekker, 1981.

An outgrowth of handouts prepared for students using an academic biological library. Selective major information sources grouped in broad subject chapters, subdivided by form of publications. Annotations are given whenever possible. Useful as a quick guide.
R: *ARBA* (1983, p. 613).

EARTH SCIENCES

Geologic Reference Sources: A Subject and Regional Bibliography of Publications and Maps in the Geological Sciences. 2d ed. **Dederick C. Ward and Marjorie W. Wheeler**. Metuchen, NJ: Scarecrow Press, 1972.

First edition, 1967.
A bibliographic guide aimed at librarians and researchers. Primarily an unannotated approach to the literature of the geologic sciences, including maps and regional sources. Sources included range from fundamental texts to highly technical treatises and serials.
R: *Choice* 9: 1577 (Feb. 1973); *ARBA* (1974, p. 606); Jenkins (F7); Wal (p. 145); Win (2EE8).

Offshore Oil and Gas: A Guide to Sources of Information. **Richard Ardern, ed.** Edinburgh, Scotland: Capital Planning Information, 1978.

Focuses on the United Kingdom, Ireland, and Norway with some pertinent American references included. Provides a selective listing of material on many aspects of offshore oil and gas exploration: transport, licensing, environmental impact, health and safety, etc. Includes an address list, index and detailed map.
R: *NLW* 79: 239 (Dec. 1978).

Sources of Information in Water Resources: An Annotated Guide to Printed Materials. **Gerald J. Giefer and Water Resources Center Archives, University of California, Berkeley**. Port Washington, NY: Water Information Center, 1976.

A collection of references to current secondary sources in the field of water resources. Cites and annotates some 1,100 works, many reviewing federally sponsored programs. Tends to emphasize literature from the United States. Appears to be aimed at hydrologists and other specialists in the field.
R: *BL* 73: 280 (Oct. 1, 1976); *Choice* 13: 958 (Oct. 1976); *WLB* 51: 92 (Sept. 1976); *ARBA* (1977, p. 692); Sheehy (EJ30).

Use of Earth Sciences Literature. **D. N. Wood, ed.** London: Butterworth, 1973.

Emphasis is on British publications for this guide, which, in its 17 chapters, discusses both the nature of the literature (including library use, primary and secondary literature, search procedures, and translations) and specific bibliographic surveys in the geological sciences.
R: *ARBA* (1974, p. 604).

GENERAL ENGINEERING

Guide to Basic Information Sources in Engineering. **Ellis Mount**. New York: Wiley, 1976.

Arrangement of material under four general categories: "Technical Literature," "Books," "Periodicals and Technical Reports," and "Other sources of Information." Each category further subdivided by reference source: bibliographies, handbooks, encyclopedias, etc. Annotations are very brief. Includes author-title index. Intended primarily for engineering students and researchers.

R: *BL* 74: 232 (Sept. 1, 1977); *Choice* 14: 514 (June 1977); *LJ* 102: 1469 (July 1977); *RQ* p. 354 (Summer 1977); *ARBA* (1978, p. 758).

Information Sources in Biotechnology. **A. Crafts-Lighty.** New York: Nature Press/ Grove's Dictionaries of Music, 1983.

A general guide to the subject of literature and information sources in biotechnology. Subject index only.
R: *Choice* 21: 1584 (July/Aug. 1984); *ARBA* (1985, p. 580).

The Use of Engineering Literature. **K. W. Miloren, ed.** London: Butterworth, 1976.

Somewhat British emphasis in this guide to publications and literature. Searching in all aspects of engineering. Intended for engineers and librarians.
R: *ARBA* (1977, p. 752).

CHEMICAL ENGINEERING

Chemical Industries Information Sources. **Theodore P. Peck.** Detroit: Gale Research, 1979.

Guide to information sources in chemical engineering and related industries. Lists handbooks, dictionaries, directories, encyclopedias, indexing and abstracting services, etc. Also includes a list of specialized libraries and publishers and addresses. Intended for science, engineering and technology collections.
R: *Chemical Engineering* 87: 11 (Jan. 14, 1980); *Journal of the Electrochemical Society* 126: 439C (Oct. 1979); *TBRI* 45: 390 (Dec. 1979); *ARBA* (1980, p. 615); Mal (1980, p. 233).

Information Resources in Toxicology. **Philip Wexler.** New York: Elsevier, 1982.

The first guide available to the information sources of toxicology. Provides some answers to reference questions rather than leading to a source.
R: *Science and Technology Libraries* 3: 107 (Summer 1983).

Information Sources on the Natural and Synthetic Rubber Industry. **United Nations.** New York: Unipub, 1979.

Offshore Petroleum Engineering: A Bibliographic Guide to Publications and Information Sources. **Marjorie Chryssostomidis.** New York: Nichols Publishing, 1978.

This bibliography has a threefold purpose: to assist the neophyte, to provide the involved professional with additional, hard-to-locate references, and to identify the basic information sources in an offshore petroleum engineering collection. Cites some 2,600 books, articles, reports, and conference papers among other pertinent information. Subject arrangement of annotated citations. Indexed by author, title, and permuted topic.

Plastics and Rubber: World Sources of Information. **E. R. Yescombe.** London: Applied Science Publishers, 1976.

A comprehensive listing of literature and other information sources on rubber and plastics. Includes some 3,000 references to 1975. Features directory information on

related, worldwide organizations. Useful guide to students, researchers. and engineers connected with this subject and their industries.
R: *Composites* 8: 15 (Jan. 1977); *TBRI* 43: 323 (Nov. 1977).

CIVIL ENGINEERING

Construction Information Source and Reference Guide. 4th ed. **Jack W. Ward**. Phoenix: Construction Publications, 1981.

Six color-coded sections offer a variety of materials of interest to any student or professional concerned with construction or construction technology. Information provided relates to both standard reference tools and publications of professional institutes.
R: *ARBA* (1975, p. 787; 1982, p. 804).

Guide to Literature on Civil Engineering. **Rita McDonald**. Washington, DC: American Society for Engineering Education, 1972.

A small listing of civil engineering literature by types of sources, such as guides, bibliographies, dictionaries, etc. Unannotated.
R: *ARBA* (1974, p. 672).

Information Sources in Architecture. **Valerie J. Bradfield**. Stoneham, MA: Butterworth, 1983.

An invaluable guide to the full range of sources on architectural information for those interested in the many aspects of the construction process. Sources such as organizations and their libraries, books and bibliographies, databases and data banks, government literature, etc., are all discussed in relation to the stages involved in the construction project.

Sources of Construction Information: An Annotated Guide to Reports, Books, Periodicals, Standards and Codes. Vols. 1–. **Jules B. Godel**. Metuchen, NJ: Scarecrow Press, 1977–.

Contains annotated descriptions of books, reports, and standards dealing with the construction area. Entries include author, title, publisher, date, pages, and price. Useful to architects, planners, engineers, contractors, and building officials.
R: Sheehy (*E J*13).

ELECTRICAL AND ELECTRONICS ENGINEERING

COMPUTER TECHNOLOGY

ACM Guide to Computing Literature. Baltimore: Association for Computing Machinery, 1980. Annual.

1982, published in 1984.
An information source in a comprehensive index to the world's computing literature. Contains bibliographic listing, author listing, source index, subject indexes by keyword and category, and reviewer index in a well-organized form. For researchers, librarians, and educators. Has a thorough cross-referencing system for easy use.

Computer Science Resources: A Guide to Professional Literature. **Darlene Myers, comp.** White Plains, NY: Knowledge Industry Publications, 1981.

A comprehensive authoritative guide to computer science literature. Chapters are arranged by type of material and include journals, indexes and abstracts, books, newsletters, software, proceedings, etc. Includes a list of the names and addresses of 800 publishers. Helpful aid to researchers, managers, consultants, and systems analysts in data processing centers.

The Reader's Guide to Microcomputer Books. **Michael Nicita and Ronald Petrusha.** Brooklyn, NY: Golden–Lee Book, 1983.

Reviews and rates more than 400 microcomputer books according to content and quality of presentation. Geared to 4 audiences: novice, intermediate, advanced, or all.
R: *LJ* 108: 1698 (Sept. 1983); *ARBA* (1984, p. 623).

Robotics and Automation Today: A Guide to Information Sources. New York: Bowker, 1984.

A listing of over 4,000 reference materials, organizations, industrial research laboratories, manufacturers and scientific and technical specialists. Covers the technical aspects of automation as well as the social, economic, political, and human factors involved. For laypersons to professionals to find a source of answers to questions about the mechanization of work.

Robotics and CAD/CAM Marketplace 1985: A Worldwide Guide to Information Sources. New York: Bowker, 1985.

Provides a listing of some 4,000 reference materials, including handbooks, monographs, conference proceedings, online databases, and more. An essential guidebook for nearly everyone seeking information in this area.

INDUSTRIAL ENGINEERING

Guide to Information Services in Marine Technology. **Arnold Myers, comp.** Edinburgh, Scotland: Institute of Offshore Engineering, Heriot–Watt University, 1979.

Definite British emphasis. Provides details of United Kingdom sources of information in marine technology. Lists applicable library facilities, information retrieval and enquiry services, publications, institutions, etc. Includes section on foreign and international services. A self-service manual for UK users.
R: *NLW* 78: 56 (Mar. 1977); Wal (p. 211).

Information Sources on Industrial Maintenance and Repair. **United Nations.** New York: Unipub, 1979.

NUCLEAR ENGINEERING

Guide to Literature on Nuclear Engineering. **Harold N. Wiren.** Washington, DC: American Society for Engineering Education, 1972.

A brief guide to the literature, arranged by types of sources, including abstracting services, bibliographics, periodicals, directories, yearbooks, standards, etc.

The Nuclear Power Debate: A Guide to the Literature. **Jerry W. Mansfield**. New York: Garland, 1984.

An annotated bibliographic source on literature on nuclear power under 3 broad categories—pro-nuclear power, anti-nuclear power, and neutral treatment of nuclear power. Author, title, and subject indexes are provided.
R: *ARBA* (1985, p. 508).

Science Information Available from the Atomic Energy Commission. **US Atomic Energy Commission, Division of Technical Information**. Oak Ridge, TN: US Atomic Energy Commission, 1971 and revisions.

Lists and describes books, pamphlets, reports, translations, reference tools and services, educational films, and exhibits. Of interest to students and the general public, as well as nuclear scientists and engineers.
R: *ARBA* (1972, p. 543).

ENERGY

Energy Statistics: A Guide to Information Sources. **Sarojini Balachandran**. Detroit: Gale Research, 1980.

Expanded from *Energy Statistics: A Guide to Sources* and *Energy Statistics: An Update*. Provides a wealth of statistical sources on energy. Consists of 3 main sections: the first contains keywords/subject descriptors, which enable subject and geographic access to titles; the second consists of 40 annotated and alphabetically arranged sources; and the third includes a list of 600 annotated publications organized by subject. Also contains a directory of publishers, bibliography, personal and corporate author index, and a subject index. An invaluable aid for academic and public libraries.
R: *Choice* 18: 219 (Oct. 1980); *CRL* 41: 491 (Sept. 1980); *LJ* 105: 1720 (Sept. 1, 1980); *RQ* 20: 94 (Fall 1980); *ARBA* (1981, p. 679).

Energy Information Guide. 3 vols. **R. David Weber**. Santa Barbara, CA: ABC-Clio, 1982–1984.

Volume 1: *General and Alternative Energy Sources*, 1982; volume 2: *Nuclear and Electric Power*, 1983; volume 3: *Fossil Fuels*, 1984.
More than 2,000 items are included in this valuable 3-volume guide. Materials are arranged by type under 9 broad topics, all dealing in some way with the production, distribution, storage, and/or consumption of energy. Indexed by author, title, subject, and document number. A must for energy collection.
R: *BL* 80: 251 (Oct. 1, 1983); *Choice* 21:69 (Sept. 1983); *LJ* 107: 1451 (Aug. 1982); 108:994 (May 15, 1983); *ARBA* (1985, p. 507).

Federal Energy Information Sources and Data Bases. **Carolyn C. Bloch**. Park Ridge, NJ: Noyes Data, 1979.

Contains a listing of energy information available from the federal government. Four main sections cover cabinet departments, administrative agencies, quasi-government agencies, and congressional offices. Information on libraries, databases, projects, services, and publications is given for each department or agency; entries provide descriptions and addresses. Contains a general index and an index of information cen-

ters, retrieval systems, and libraries. Invaluable tool for scientists, engineers, and students in energy fields.
R: *RSR* 8: 24 (Jan./Mar. 1980); *ARBA* (1980, p. 649).

Information Sources in Power Engineering: A Guide to Energy Resources and Technology. **Karen S. Metz.** Westport, CT: Greenwood Press, 1976.
A guide to the literature in power engineering and energy. Describes both standard publications and information sources and systems. Intended for engineers, librarians, and management personnel.
R: *BL* 73: 1197 (Apr. 1, 1977); *ARBA* (1977, p. 679).

Information Sources on Non-Conventional Sources of Energy. UNIDO Guides to Information Sources, 30; ID/210 (UNIDO/LIB/Ser.D/30). **United Nations.** New York: United Nations, 1978.
Literature covers 4 principal forms of energy: geothermal, biomass, tidal, and solar. Lists directories; statistical sources; basic reference books; periodicals; monographic series; nonprint media; proceedings of conferences and congresses; professional, trade, and research organizations; etc.
R: *IBID* 7: 25 (Spring 1979).

ENVIRONMENTAL SCIENCES

Energy and Environment Information Resource Guide. **Lynne M. Neufeld and Martha Cornog, comps.** Philadelphia: National Federation of Abstracting and Information Services, 1982.
A guide to information sources in the field of energy and environment information. Very basic and general.
R: *Unesco Journal of Information Science, Librarianship and Archives Administration*: 4: 213 (July 1982).

Environmental Economics: A Guide to Information Sources. **Barry C. Field and Cleve E. Willis.** Detroit: Gale Research, 1979.
A bibliography of information sources on environmental economics. Sections cover conceptual foundations, empirical studies, and applications. Includes a major section on recreational aspects. Entries are annotated and consist primarily of current works. Appendixes contain a glossary and lists of journals, books, government agencies, newsletters, abstracts, etc. Author and subject indexes are provided.
R: *Choice* 16: 992 (Oct. 1979); *WLB* 54: 70 (Sept. 1979); *ARBA* (1980, p. 651).

Environmental Planning: A Guide to Information Sources. **Michael J. Meshenberg.** Detroit: Gale Research, 1976.
A selective annotated bibliography covering a full range of environmental quality topics. Topical arrangement of 13 selected fields with examples from book, periodical, and government document literature. Integrates historical material with current references. Access by author, title, and subject index. Intended as a tool for urban and regional planners.
R: *WLB* 51: 3623 (Dec. 1976); *ARBA* (1977, p. 681).

Environmental Toxicology: A Guide to Information Sources. **Robert L. Rudd.** Detroit: Gale Research, 1977.

Serves as an introductory reference tool for the neophyte and as an authoritative guide to very specific areas of environmental toxicology. Annotates some 1,000 sources in
4 main sections. Guide is based on the collection at the University of California at Davis. Literature included is mainly of a scholarly and research nature.
R: *Choice* 15: 672 (July/Aug. 1978); *WLB* 52: 585 (Mar. 1978); *ARBA* (1979, p. 695).

Guide to Ecology Information and Organizations. **John Gordon Burke and Jill Swanson Reddig.** New York: Wilson, 1976.

Excellent guide to nontechnical literature of ecology. Consists of 10 sections: citizen action guides, reference books, histories, monographs, government publications, indexes, nonprint media, periodicals, organizations, and government officials. Most valuable section lists 500 nontechnical monographs. Full bibliographic citations followed by concise annotations. Specifically aimed at public librarians and public library patrons.
R: *RQ* 16: 254 (Spring 1977); *WLB* 51: 362 (Dec. 1976); *ARBA* (1977, p. 681); Sheehy (EC14).

Human Ecology: A Guide to Information Sources. **Frederick Sargent, II.** Detroit: Gale Research, 1983.
Health Affairs Information Guide Series, Vol. 10.
A guide to the literature of human ecology, including sections on nature and scope, the setting, human-environment interactions, human manipulations of the environment, and more. Also includes author, title, and subject indexes, a list of journals, and a list of abstracts and indexes. Informative and well designed.
R: *ARBA* (1984, p. 679).

Man and the Environment Information Guide Series. Vols. 1–. **Seymour M. Gold, ed.** Detroit: Gale Research, 1975–.
A monographic series of guides to environmental information sources.
R: Sheehy (EJ16).

Noise Pollution: A Guide to Information Sources. **Clifford R. Bragdon.** Detroit: Gale Research, 1979.

A listing of more than 3,000 annotated references to literature on noise pollution. References taken from books, periodicals, nonprint sources, and documents. Entries are arranged under broad subject categories with a section on primary and secondary periodicals and abstracts and indexes. Includes author, title, and subject indexes. Useful for researchers in the field.
R: *CRL* 41: 93 (Jan. 1980); *ARBA* (1980, p. 651).

Societal Directions and Alternatives: A Critical Guide to the Literature. **Michael Marien.** La Fayette, NY: Information for Policy Design, 1976.

Sourcebook on the Environment: A Guide to the Literature. **Kenneth A. Hammond, George Macinko, and Wilma B. Fairchild, eds.** Chicago: University of Chicago Press, 1978.

An excellent sourcebook written by geographers interested in environmental health. Contains a broad range of information, including 24 bibliographic essays, list of government publications, etc. Informative, well researched.
R: *Environmental Research* 17: 481 (Dec. 1978); *Journal of Environmental Health* 41: 125 (Sept./Oct. 1978); *BL* 75: 577 (Nov. 1978); *RQ* 18: 205 (Winter 1978); *RSR* 8: 27 (Jan./Mar. 1980); *TBRI* 45: 35 (Jan. 1979); 45: 115 (Mar. 1979); *ARBA* (1980, p. 651).

Toxic Substances Sourcebook: The Professional's Guide to the Information Sources, Key Literature and Laws of a Critical New Field. **Steve Ross and Monica Pronen, eds.** New York: Environment Information Center, 1978.

A single-volume reference that contains detailed descriptions of the use of toxic substances in the United States. Takes into account National Institute of Occupational Safety and Health statistics, legislation, environmental concerns. Contains regular and keyword index, Standard Industrial Classification code terms, geography. Very highly recommended for special, academic, and public libraries.
R: *ARBA* (1979, p. 695).

Water Pollution: A Guide to Information Sources. **Allen W. Knight and Mary Ann Simmons, eds.** Detroit: Gale Research, 1980.

A guide to the location of information on water pollution. Lists reference books, articles, films, organizations, etc. Entries are organized by subject and type of material and include short descriptive annotations. Appendixes include a glossary of 100 terms and selected readings. Contains author, title, and subject indexes. Valuable reference aid for high school, college, and public libraries.
R: *BL* 77: 350 (Oct. 15, 1980); *Choice* 18: 1074 (Apr. 1981); *WLB* 55: 463 (Feb. 1981); *ARBA* (1981, p. 684).

TRANSPORTATION

Sources of Information in Transportation. 2d ed. **Ad Hoc Committee of Transportation Librarians, comp.** Washington, DC: Research and Special Programs Administration, US Dept. of Transportation. Distr. by National Technical Information Service, 1981.

A basic reference to all modes of transportation, including socioeconomic and technical aspects, especially US sources as well as major international references.
R: *Science and Technology Libraries* 3: 68 (Fall 1982).

Transguide: A Guide to Sources of Freight Transportation. **Reebie Associates, ed.** New York: Greenwich Press, 1980.

A noteworthy cross-referenced guide to various freight transportation information sources including guides, bibliographies, directories, and maps. Entries contain addresses, phone number, frequency, pages, price, etc. A valuable aid to graduate students and professionals in transportation studies as well as to managers, planners, and researchers.
R: *Choice* 18: 1240 (May 1981); *Science and Technology Libraries* 1: 63 (Spring 1981).

CHAPTER 3 BIBLIOGRAPHIES

GENERAL SCIENCE

American Science and Technology: A Bicentennial Bibliography. **George W. Black, Jr.** Carbondale, IL: Southern Illinois University Press, 1979.

Gathers some 1,000 references to journal literature dealing with biographical and historical aspects of American science and technology. The scope is the bicentennial year, 1976. All entries culled from 5 Wilson indexes. Citations arranged under subject headings with author, journal title, and proper name indexes.
R: *LJ* 104: 1239 (June 1, 1979); *WLB* 54: 65 (Sept. 1979); *ARBA* (1981, p. 617).

Annotated Bibliography of Technical and Specialized Dictionaries in Spanish–Spanish and Spanish– . . . : With Commentary. **Maria Luz Espinosa Elerick.** Translated by Charles Elerick and Richard V. Teschner. Troy, NY: Whitston, 1982.

An analysis of 86 dictionaries covering scientific, business, and technical fields such as engineering, economics, and earth sciences. Easy to understand.
R: *Choice* 20: 555 (Dec. 1982).

Bibliographic Guide to Technology. **New York Public Library, Research Libraries.** Boston: Hall, 1975–. Annual.

A multiple-access subject bibliography to publications cataloged by the Research Libraries of the New York Public Library, with additional citations from Library of Congress MARC tapes and conference publications cataloged by the Engineering Societies Library, New York.
R: Sheehy (EJ2); *ARBA* (1982, p. 693).

A Bibliography of the Philosophy of Science, 1945–1981. **Richard J. Blackwell, comp.** Westport, CT: Greenwood Press, 1983.

A bibliography of over 7,000 items including books, articles, papers from collections, and more. Though the size of the book is massive, it does have limitations: the time span is limited; the subject categories are narrowly defined; and it has no subject index, annotations, or commentaries.
R: *ARBA* (1984, p. 606).

British Natural History Books, 1495–1900: A Handlist. **R. B. Freeman.** Hamden, CT: Archon Books; Kent, England: Dawson, 1980.

A bibliography containing 60 general reference titles and over 4,000 entries of titles to date from 1495 through 1900. Worth the reading.
R: *Choice* 19: 1234 (May 1981); *ARBA* (1982, p. 719).

The Chronological Annotated Bibliography of Order Statistics: Pre-1950. Vol. I. **H. Leon Harter.** Columbus, OH: American Sciences Press, 1983.

Includes some 942 items published before 1950 with 50% from 1925 to 1949 and 77% from 1900 to 1949. Provides a full bibliographic citation, reviews of the book or article, a summary of the contents, and more for each item. For research collections

in mathematics and statistics as well as agriculture, demography, econometrics, and sociology.
R: *ARBA* (1984, p. 641).

Handbooks and Tables in Science and Technology. **Russell H. Powell, ed.** Phoenix: Oryx Press, 1979.
A list of over 2,000 scientific and technical handbooks used in chemistry, physics, biology, astronomy, geology, agriculture, etc. Composed of 2 sections: section 1 lists 1,500 titles alphabetically, providing full bibliographic information; section 2 lists compilations of standard data. Contains subject and author indexes. For large science libraries.
R: *BL* 76: 1151 (Apr. 1, 1980); 76: 1628 (July 1, 1980); *RQ* 19: 182 (Winter 1979); *RSR* 8: 28 (July/Sept. 1980); *WLB* 54: 196 (Nov. 1979); *ARBA* (1980, p. 604); Mal (1980, p. 41).

ISIS Cumulative Bibliography: A Bibliography of the History of Science Formed from ISIS Critical Bibliographies 1–90, 1913–1965. Author Index. Vol. 6. **Magda Whitrow, ed.** London: Mansell. Distr. New York: Wilson, 1980–.
Volume 5, Author Index, 1984.
The ISIS bibliographies are essential resources for the history of science. The author index includes about 75,000 entries.
R: *ARBA* (1985, p. 487).

Law and Science: A Selected Bibliography. **Morris L. Cohen, Naomi Ronen, and Jan Stepan.** Edited by Viven B. Shelanski and Marcel C. La Follette. Cambridge, MA: MIT Press, 1980.
Revised and updated edition that emphasizes the effect of new advancements in science and technology on law.
R: *CRL* 41: 396 (July 1980).

Pure and Applied Science Books, 1876–1982. New York: Bowker, 1982.
A comprehensive bibliography of all books in the areas of science and technology published or distributed in the United States during the past 107 years. Contains more than 170,000 Library of Congress cataloged entries indexed under some 25,000 Library of Congress subject headings.
R: *ARBA* (1983, p. 597).

Science for Society: A Bibliography. 6th ed. **Joseph M. Dasbach, prep.** Washington, DC: American Association for the Advancement of Science, Office of Science Education, 1976.
A selective, annotated bibliography of books and journal articles describing the interrelationships among society, science, and technology. Entries are current through 1975. Subject arrangement under 11 topics. Limited by lack of indexes. Highly recommended for undergraduate collections.
R: *ARBA* (1978, p. 616).

Sociology of Sciences: An Annotated Bibliography on Invisible Colleges, 1972–1981. **Daryl E. Chubin.** New York: Garland, 1983.
Invisible colleges, which are informal groups of scientists engaged in the exchange of information about their research, have been a major factor in the advancement of

various fields of science. Here is a review and collection of their findings presented in a most authoritative format.
R: *ARBA* (1984, p. 606).

Victorian Science and Religion: A Bibliography of Works on Ideas and Institutions, with Emphasis on Evolution, Belief, and Unbelief, Published from 1900–1975. **Sydney Eisen and Bernard Lightman, eds.** Hamden, CT: Shoe String Press, 1983.

A catalog of over 6,000 entries of secondary works dealing with ideas and institutions during the period. Categories include history, geology, biology, evolution and its effects on thought, and religion. Contains annotations and a subject index. A necessary reference tool.

ASTRONOMY

A Bibliography of Astronomy, 1970–1979. **Robert A. Seal and Sarah S. Martin.** Littleton, CO: Libraries Unlimited, 1982.

Updates 1970's *Astronomy and Astrophysics: A Bibliographical Guide* with over 2,000 sources such as journals, papers, monographs, catalogs, etc., in all areas of astronomy. Each entry contains bibliographic data, references, and date of reference.
R: *ARBA* (1983, p. 602).

Bibliography of Natural Radio Emissions from Astronomical Sources. Vols. 1–. **M. Stahr Carpenter, ed.** Ithaca, NY: Cornell University Press, 1962–.

An annual bibliography of radio astronomy containing approximately 800 items in each volume.

MATHEMATICS

Annotated Bibliography of Expository Writing in Mathematical Sciences. **Matthew P. Gaffney and Lynn Arthur Steen.** Washington, DC: Mathematical Association of America, 1976.

A Bibliography of Early Modern Algebra, 1500–1800. **Robin E. Rider.** Berkeley: University of California, Office for History of Science and Technology, 1982.

Seventh in a series of monographs.
Outlines the development of algebra from 1500 through 1799 with information on the history of the development of algebraic theory and an examination of publication practices and patterns during this period. Arranged chronologically with a full author index.
R: *ARBA* (1983, p. 611).

A Bibliography of Recreational Mathematics. Vol. 4. **W. L. Schaff.** Reston, VA: National Council of Teachers of Mathematics, 1978.

Contains up-to-date source materials on mathematical games. Arranged in 12 sections, with 3 appendixes. For math buffs.
R: *ARBA* (1979, p. 647).

Bibliography of Statistical Literature. 3 vols. with supps. **Maurice G. Kendall and Alison G. Doig.** Salem, NH: Arno, 1981.

Earlier edition, 1968.

Comprehensive listing of the significant contributions to statistics since the sixteenth century.

R: *Nature* 202: 330 (Apr. 25, 1964); 209: 750 (Feb. 19, 1966); Jenkins (B73); Wal (p. 75).

The High School Mathematics Library. Rev. ed. **William L. Schaaf.** Reston, VA: National Council of Teachers of Mathematics, 1982.

First edition, 1960; fifth edition, 1973; sixth edition, 1976.

The fifth edition arranges 950 book entries under 15 topical headings and 3 special sections; 200 titles in the list are starred as the core collection. Many entries briefly annotated. Useful to public and college libraries as well as school libraries.

R: *ARBA* (1974, p. 552).

Integer Programming and Related Areas: A Classified Bibliography, 1978–1981. **R. von Randow, ed.** New York: Springer-Verlag, 1982.

Reproduced from a computer output in all capital letters, this bibliography provides useful information on the applications of integer programming, complexity, dynamic programming, graph theoretic results, packing, and shortest paths. For researchers and graduate students in mathematical economics and related fields.

R: *ARBA* (1984, p. 642).

Japanese Mathematics: A Bibliography. **Shojo Honda, comp.** Washington, DC: Library of Congress, 1982.

A bibliography of printed books and manuscripts from the Asian Division of the Library of Congress. Covers pre-Meiji Japanese mathematics from the seventeenth century to 1867. Has 403 entries arranged alphabetically by romanized title using the Hepburn system and providing information on each entry.

R: *ARBA* (1984, p. 642).

PHYSICS

Acoustic Emission: A Bibliography with Abstracts. **Thomas F. Drouillard.** Edited by Frances J. Laner. New York: Plenum Press, 1979.

A comprehensive listing of the world literature on acoustic emission. Lists 1,994 references in a single volume. Also lists journal sources, an author index, and a handy subject index. All entries are annotated and translated into English. Includes plans to update by periodical supplements. Invaluable reference source for undergraduate and graduate students, engineers and scientists in the fields of mechanical engineering, stress analysis, pressure vessel design and maintenance, welding engineering, metallurgy, and quality control.

R: *Journal of Metals* 32: 73 (Feb. 1980); *Journal of the Acoustical Society of America* 66: 1907 (Dec. 1979); *NDT International* 12: 300 (Dec. 1979); *TBRI* 46: 72 (Feb. 1980); 46: 114 (Mar. 1980); *ARBA* (1980, p. 618).

Auger Electron Spectroscopy: A Bibliography, 1925–1975. **Donald T. Hawkins, comp.** New York: Plenum Press, 1977.

Various aspects of Auger electronic spectroscopy form the basis of this bibliography. Cites some 2,100 references under 6 separate sections. Two sections, "Theory, Physics of the Auger Effect" and "Surface Analysis by Auger Spectroscopy," account for nearly one-half of the work. Includes excellent permuted title index and an author-title index. Despite its almost exclusive reliance on *Chemical Abstracts* and *Physics Abstracts* for citations and its weak coverage of literature prior to 1967, this is a useful bibliography.
R: *Journal of Metals* 30: 6 (Jan. 1978); *ARBA* (1979, p. 652).

Bibliography on Atomic Energy Levels and Spectra, July 1975 through June 1979. National Bureau of Standards Special Publication no. 363, Supp. 2. **Romuald Zalubas and Arlene Albright.** Washington, DC: National Bureau of Standards. Distr. Washington, DC: US Government Printing Office, 1980.

Supplement 1, from July 1971 through June 1975, published in 1977.
Some 1,200 references are classified here by subject for individual atoms and atomic ions. First indexing is of spectra with the element, its spectrum, and page reference. Then lists reference numbers for each spectrum in relation to the final part, which is arranged numerically in ascending order. Also an author index. For scientists engaged in atomic physics research.
R: *ARBA* (1982, p. 711).

Bibliography of Microwave Optical Technology. **Arthur F. Harvey, ed.** New York: Plenum Press, 1976.

A selective listing of some 15,000 references dealing with microwave and optical technologies. Entries culled from over 700 technical and scientific journals. Time span covers the last decade. Supported by an author index and a comprehensive subject index. Useful addition to research libraries where interests delve into theoretical and applied principles of microwave optics.
R: *Optical Society of America Journal* 67: 410 (Mar. 1977); *TBRI* 43: 232 (June 1977); *ARBA* (1978, p. 648).

Bibliography on Atomic Line Shapes and Shifts (June 1975 through June 1978). **J. R. Fuhr, B. J. Miller, and G. A. Martin.** Washington, DC: National Bureau of Standards. Distr. Washington, DC: US Government Printing Office, 1978.

Bibliography consisting of 600 entries covering the period from 1975 to 1978. Divided into 5 sections: abbreviated references to general information; papers with numerical data; chronological history; author index; and errata for the second supplement.
R: *ARBA* (1980, p. 618).

Bibliography on Atomic Transition Probabilities (1914 through October 1977). National Bureau of Standards Special Publication no. 505. **J. R. Fuhr, B. J. Miller, and G. A. Martin.** Washington, DC: US National Bureau of Standards. Distr. Washington, DC: US Government Printing Office, 1978.

Updates an earlier (1974) publication on atomic transition probabilities. Lists 2,400 items in 4 separate sections; general interest, numerical data articles, articles arranged by publication date, and author index.
R: *ARBA* (1979, p. 659).

Crystal Growth Bibliography. 2 vols. **Anne M. Keesee, T. F. Connolly, and G. C. Battle, Jr., comps.** New York: Plenum Press, 1979.

A 2-volume, comprehensive bibliography of 5,022 entries covering the period from 1972 through 1977 and focusing on literature of the crystal growth of inorganic materials. References are taken from journals, technical reports, books, dissertations, symposia. Volume 1 lists the references in chronological order; volume 2 consists of author and permuted title indexes. For specialized audiences.
R: *Choice* 17: 202 (Apr. 1980).

Crystal Growth Bibliography Supplement. **A. M. Keesee, T. F. Connolly, and G. C. Battle, Jr., comps.** New York: IFI/Plenum, 1981.

Supplements *Crystal Growth Bibliography*, 1979, volumes 10A and 10B of *Solid-State Physics Literature Guides*.
Computer-produced bibliography of theoretical, review, experiment papers, technical reports, and books involved with crystal growth of inorganic materials.
R: *ARBA* (1983, p. 610).

Current Physics Bibliographies. New York: American Institute of Physics, 1973–

A series of specialized bibliographies in relatively narrow areas of physics and astronomy. Frequent but irregular updating from SPIN databases.

Heat Bibliography, 1948/52–. **National Engineering Laboratory.** Edinburgh: Statistical Office, National Engineering Laboratory, 1959–. Annual.

Each annual includes material noted in the National Engineering Library.
R: Wal (p. 108); Win (3EI32).

The History of Classical Physics: A Selected, Annotated Bibliography. Bibliographies of the History of Science and Technology, vol. 8. **R. W. Home.** With the assistance of Mark J. Gittins. New York: Garland, 1984.

A bibliographic tool that contains about 1,300 entries on writing on the history of classical physics, about 1700 to 1900, or from the Scientific Revolution to the beginning of modern physics.
R: *ARBA* (1985, p. 614).

The History of Modern Physics: An International Bibliography. Bibliographies of the History of Science and Technology, vol. 4. **Stephen G. Brush and Lanfrance Belloni.** New York: Garland, 1983.

An annotated bibliography of 2,073 entries, international in scope, that include books and articles on the history of physics after the discovery of x rays in 1895. Includes name, subject, and institutional indexes.
R: *Choice* 21: 1108 (Apr. 1984); *ARBA* (1985, p. 613).

An Inventory of Published Letters to and from Physicists, 1900–1950. **Bruce R. Wheaton and J. L. Heilbron.** Berkeley: University of California, Office for History of Science and Technology, 1982.

On microfiche, an inventory of almost 25,000 quotations—more than 40,000 references—from the correspondence of physicists appearing in the items listed in the bibliography *Literature on the History of Physics in the Twentieth Century*. Also, in book format, an alphabetical listing of 76 physicists with information about each. An important source.
R: *ARBA* (1984, p. 638).

Laser Crystals. **Alexander A. Kaminskii.** Berlin: Springer-Verlag, 1981.

The English version of the author's 1975 Russian-language work, supplemented with new data in its tables and figures.
R: *Laser Focus* 19: 127 (May 1983).

Solid State Physics Literature Guides. 10 vols. **Tom F. Connolly and Errett Turner.** New York: Plenum, 1970–1977.

Volume 1, *Ferroelectric Materials and Ferroelectricity*; volume 2, *Semiconductors—Preparation, Crystal Growth, and Selected Properties*; volume 3, *Groups IV, V, and VI Transition Metals and Compounds*; volume 4, *Electrical Properties of Solids*; volume 5, *Bibliography of Magnetic Materials and Tabulation of Magnetic Transition Temperatures*; volume 6, *Ferroelectrics Literature Index*; volume 9, *Laser Window and Mirror Materials*.
Mainly based on papers received by the Research Materials Information Center of the Oak Ridge National Laboratory. Most of the coverage is from 1960. These guides are well indexed.
R: *Choice* 8: 814 (Sept. 1971); *ARBA* (1972, p. 562); (1973, p. 546).

CHEMISTRY

Bibliographic Atlas of Protein Spectra in the Ultraviolet and Visible Regions. **Donald M. Kirschenbaum, ed.** New York: Plenum, 1983.

Covers all aspects of spectroscopic analysis of proteins researched in the last 10 years. Indexed for use as a handy tool for researchers in many areas.

Chembooks: New Books and Journals. Basel, Switzerland: Karger Libri, 1969–. Annual.

Volume 10, 1977–78.
Annual bibliography of new publications in the fields of pure and applied chemistry. Literature arranged under 22 subject areas. Brief annotations given. Author and publisher indexes.
R: Wal (p. 114).

Equilibrium Properties of Fluid Mixtures—2: A Bibliography of Experimental Data on Selected Fluids. **M. J. Hiza, A. J. Kidnay, and R. C. Miller.** New York: IFI/Plenum, 1982.

A current reference to available experimental phase equilibria and thermophysical properties data on mixtures of selected low molecular weight fluids. For the design engineer, data analyst, and the experimental chemist.
R: *ARBA* (1984, p. 635).

Literature Guide to the GLC of Body Fluids. **Austin V. Signeur.** New York: IFI/Plenum, 1982.

Over 4,500 references from the scientific literature on the determination of substances found in the analysis of human body fluids using the gas chromatographic method from the mid-1950s through 1981.
R: *ARBA* (1983, p. 606).

Physical and Chemical Properties of Water: A Bibliography: 1957–1974. **Donald T. Hawkins.** New York: Plenum Press, 1976.

Bibliography focuses on fundamental properties of water. Actually consists of 2 bibliographies: part 1 covers the literature from 1957 through 1968; part 2 covers 1969 to 1974. Cites nearly 3,600 references to monographs, government documents, patents, dissertations, and journal articles. Each section is arranged by subject. A separate author and permuted title (KWIC) index is provided for each section. Most references are drawn from *Chemical Abstracts*.
R: *RSR* 5: 13 (Apr./June 1977); *ARBA* (1978, p. 637).

BIOLOGICAL SCIENCES

A Bibliography of "Ab Initio" Molecular Wave Functions. **W. G. Richards et al.** New York: Oxford University Press, 1971. 3 supps. to 1981.

First published in 1971. Supplements for: 1970–73, published in 1974; 1974–77, published in 1978; 1978–80, published in 1981.
Arranged by diatomic, triatomic, and tetratomic molecules followed by polyatomic molecules. A useful assistant in laboratory work.
R: *Choice* 16: 1154 (Nov. 1979); *ARBA* (1983, p. 606).

Bibliography of Bioethics. 6 vols. **LeRoy Walters, ed.** Detroit: Gale Research, 1975–1980.

Volume 1, 1975; volume 2, 1976; volume 3, 1977; volume 4, 1978; volume 5, 1979; volume 6, 1980.
Contains over 1,600 references to documents in journals, monographs, bills, audiovisual materials, newspapers, and unpublished documents.
R: *CRL* 41: 191 (Mar. 1980); 42: 83 (Jan. 1981).

Bibliography on Zinc in Biological Systems. **John W. Gardner et al.** Provo, UT: Brigham Young University Press, 1976.
R: *TBRI* 43: 212 (June 1977).

Biological Sciences: A Bibliography of Bibliographies. **Theodore Besterman.** Totowa, NJ: Rowman & Littlefield, 1972.
R: *ARBA* (1973, p. 549).

Biotechnology: A Review and Annotated Bibliography. **Harry Rothman et al.** New York: Pergamon, 1981.

An introduction for the nonspecialist on the aspects of biotechnology, or biochemical engineering, through a bibliography of books and journal articles, reports, and companies involved in this area.
R: *Choice* 18: 1438 (June 1981); *ARBA* (1983, p. 596).

Chromosomal Variation in Man: A Catalog of Chromosomal Variants and Anomalies. 3d ed. **Digamber S. Borgaonkar.** New York: Liss, 1980.

First edition, 1975; second edition, 1977.
A helpful source for information on all of the 86 chromosome regions that have been reported up to this date, with descriptions of 63 syndromes.
R: *ARBA* (1982, p. 721).

Endangered Species: A Bibliography. **Oklahoma Cooperative Wildlife Research Unit.** Stillwater, OK: Oklahoma State University Press, 1977.

1,100 entries to literature covering the world's rare, endangered, and recently extinct plants and wildlife. Indexed by author-publisher, geographical location, and subject.
R: *Sci-Tech News* 31: 117 (Oct. 1977).

Exobiology: A Research Guide. **Martin H. Sable.** Brighton, MI: Green Oak Press, 1978.

Lists over 3,800 unannotated entries to much material not readily available elsewhere. Covers many forms of media from technical reports to TV and radio scripts and in some 2 dozen languages. The scope is 1648 to 1975. The entries are arranged under broad subjects and subdivided chronologically. Includes a useful directory listing of organizations and periodicals. Beneficial to large or highly specialized collections.
R: *BL* 76: 854 (Feb. 15, 1980); *LJ* 103: 1970 (Oct. 1, 1978); *ARBA* (1979, p. 663).

Food Science and Technology: A Bibliography of Recommended Materials. **Richard E. Wallace, ed.** Beltsville, MD: US Department of Agriculture, National Agricultural Library. New York: Special Libraries Association, Food and Nutrition Division, 1978.

Consists of 14 separate sections on food science and technology compiled by various members of the Food and Nutrition Division. Entries are alphabetically organized by title or author into one of 3 categories: serials, articles, and books. Contains 1,770 entries and an index to serials and monographs. Recommended reference for academic and special libraries.
R: *ARBA* (1980, p. 705).

Genetic Engineering, DNA and Cloning: A Bibliography in the Future of Genetics. **Joseph Menditto and Debbie Kirsch.** Troy, NY: Whitston Publishing, 1983.

More than 8,000 items on the scientific, social, ethical, economic, and legal aspects of genetic engineering, DNA and cloning. Complete, timely, and easy to read.
R: *ARBA* (1984, p. 643).

The Modified Nucleosides of Transfer RNA: A Bibliography of Biochemical and Biophysical Studies from 1970–1979. **Paul F. Agris.** New York: Liss, 1980.

A bibliography of over 1,000 citations divided into 16 categories, each with its own author and keyword indexes. Contains most of the pertinent material through 1979. Some citations are confusing, in that they are listed twice with varying information.
R: *ARBA*: (1982, p. 718).

Molecular Structures and Dimensions. Vols. 1–. **Olga Kennard et al., eds.** Utrecht, Bohn, Germany: Scheltema & Holkema. Distr. Pittsburgh: Polycrystal Book Service, 1970–. Annual.

Focuses on organic and organometallic crystal structures. Supported by author, formula (standard and permuted), and transition metal indexes. Volume 8 introduces a KWIC index.
R: Sheehy (ED18).

Key Works to the Fauna and Flora of the British Isles and Northwestern Europe. **G. J. Kerrich, D. L. Hawksworth, and R. W. Sims, eds.** New York: Academic Press, 1978.

A bibliography of literature pertaining to living organisms in the British Isles and northwestern Europe. Compiled by experts in the fields of botany, zoology, and entomology, the book is divided into broad classifications. An important guide to the literature for British research libraries with collections in natural history.
R: *ARBA* (1981, p. 642).

Agriculture

Agricultural Credit: Annotated Bibliography, Author and Subject Index. **Food and Agriculture Organization of the United Nations.** Rome, Italy: Food and Agriculture Organization of the United Nations, 1975. Distr. New York: Unipub, 1977.

A selective list of Food and Agriculture Organization publications and documents issued between 1967 and 1975. Represents worldwide literature on the agricultural and economic development of a country and its basis for agricultural credit. Arrangement of documents by accession number. Includes alphabetical author and analytical indexes.
R: *ARBA* (1978, p. 736).

Bibliography of Agricultural Bibliographies 1977: A Categorized Listing of Bibliographies Indexed in AGRICOLA. Bibliographies and Literature of Agriculture, BLA: 1/1977. **Charles N. Bebee, comp.** Beltsville, MD: US Department of Agriculture, Science and Education Administration, Technical Information Systems, 1978. Supps.

Annual bibliography of agricultural bibliographies extracted from AGRICOLA. Includes only English-language sources.
R: *Quarterly Bulletin of the IAALD* 24: 27 (Winter–Spring 1979).

Bibliography of Agricultural Residues: Fisheries, Forestry and Related Industries. FAO Agricultural Services Bulletin no. 35. **Food and Agriculture Organization of the United Nations.** Rome, Italy: Food and Agriculture Organization of the United Nations. Also supps.

Lists specialized journals and information services dealing with agricultural residues. Examples of categories include beverage industry residues, animal by-products, municipal and domestic wastes, and rubber tree products. Published in 3 languages: English, French, and Spanish.
R: IBID 6: 381 (Dec. 1978).

The Bibliography of Agriculture. Phoenix: Oryx Press. Monthly.
Volume 48, 1984.
Monthly index to the literature of agriculture and allied disciplines. Includes over 15,000 main entry citations of journal articles, pamphlets, government documents, special reports, and proceedings. Compiled by the National Agricultural Library; the Food and Nutrition Information and Education Reserach Center; the American Agricultural Economics Documentation Center; and Agriculture Canada. Main entry citation section provides full bibliographic data. Supported by subject, geographic, and author indexes. Annual cumulations. Retrospective bibliography (1970–1978) available on microfiche.
R: *ARBA* (1985, p. 495).

Current Bibliography of Agriculture in China. Vol. 1. Wageningen, Netherlands: Centre for Agricultural Publishing and Documentation (Pudoc), 1979.
Presents information sources on agricultural principles and practices in the People's Republic of China. Draws citations, some annotated, from books, reports, and journals. Estimates about 1,500–2,000 items per year.
R: *Quarterly Bulletin of the IAALD* 24: 54 (Spring–Summer 1979).

Forest Land Use: An Annotated Bibliography of Policy, Economic, and Management Issues, 1970–1980. **William E. Shands, comp.** With Barbara K. Rhodes and Noreen O'Meara. Washington, DC: Conservation Foundation, 1981.
Contains over 50 annotations of major books and reports on topics such as general land use. Each entry is introduced and described in relation to the topical area. Good as a summary of land-use policy in the 1970s; for undergraduates, graduates, and professionals.
R: *Choice* 19: 890 (March 1982); *ARBA* (1983, p. 708).

International Citrus Crops Bibliography. Philadelphia: BIOSIS, 1985.
This 1-volume publication covers the bibliographic records arranged by BIOSIS subject classification headings, with author and subject indexes. It covers the period 1974–1983 with over 5,000 entries.

International Corn Bibliography. 3 vols. Philadelphia: BIOSIS, 1985.
Volumes 1–2 consist of bibliographic records arranged by BIOSIS subject classification headings; volume 3 is composed of author and subject indexes.
The 3-volume set contains about 24,000 items and covers the period 1974–1983.

International Soybean Bibliography. Philadelphia: BIOSIS, 1983.

BOTANY

Bibliography of Plant Viruses and Index to Research. **Helen Purdy Beale, comp. and ed.** New York: Columbia University Press, 1976.

This international bibliography of plant-attacking viruses literature cites nearly 29,000 articles gathered from 6,500 periodicals. Particular emphasis is placed on Japanese and Soviet work. Covers an 80-year time span (1892–1970). Entries are arranged alphabetically by author. All titles are translated, except those appearing in English, French, or German. Monumental bibliographic effort aimed at plant pathologists and virologists.
R: *BioScience* 27: 288 (Apr. 1977); *LJ* 101: 1512 (July 1976); *TBRI* 43: 211 (June 1977); *ARBA* (1977, p. 729); Sheehy (EC21).

Edible Wild Plants: An Annotated List of References. **Diane Schwartz.** Bronx, NY: Council on Botanical and Horticultural Libraries, New York Botanical Garden, 1978.
R: *ARBA* (1980, p. 625).

Endangered Plant Species of the World and Their Endangered Habitats: A Selected Bibliography. **C. R. Long and M. A. Miasek, comps.** Monticello, IL: Council of Planning Librarians, Exchange Bibliographies, 1976.

Herbs: An Indexed Bibliography 1971–1980: The Scientific Literature on Selected Herbs, and Aromatic and Medicinal Plants of the Temperate Zone. **James E. Simon, Alena F. Chadwick, and Lyle E. Craker.** Hamden, CT: Archon Books/Shoe String Press, 1984.

Contains almost 8,000 entries of citations, most from scientific journals. Separate subject and author indexes, in reference to 63 major economical herbs. Gives a short description and detailed citations by 10 subject classifications. Valuable to research scientists with commercial interest in herbal plants.
R: *Choice* 22: 70 (Sept. 1984), *Library Professional Publications*: 27 (Spring 1983); *ARBA* (1985, p. 516).

Huntia: A Yearbook of Botanical and Horticultural Bibliography. Vols. 1–. Pittsburgh: Hunt Botanical Library, 1964–. Annual.

An eclectic annual, covering the literature on systematic botany and horticulture, medical botany, botanical exploration, etc.
R: *Science* 152: 916 (1966); *Journal of the Royal Horticultural Society* 89: 396 (Sept. 1964); Jenkins (G90); Wal (p. 418).

Photosynthesis Bibliography. 1975: References No. 21505–25161, Aar–Zur. Vol. 6. **Z. Sestak and J. Catsky, eds.** The Hague: Junk, 1980.
Volume 1, 1974.
An excellent bibliography comprising papers in all fields of photosynthesis research, arranged alphabetically by authors' names. Each volume indexed by author, subject, and plant. A comprehensive reference work for researchers in photosynthesis and related subjects.
R: *ARBA* (1981, p. 647).

Roses: A Bibliography of Botanical, Horticultural, and Other Works Related to the Genus Rosa. **Joanne Werger and Robert E. Burton.** Metuchen, NJ: Scarecrow Press, 1972.

Historical and bibliographic information from all areas relating to the rose, including science, art, and industry.
R: *ARBA* (1973, p. 560).

Trees and Shrubs of the United States: A Bibliography for Identification. **Elbert L. Little, Jr. and Barbara H. Honkala.** Washington, DC: US Forest Service, 1976. Distr. Washington, DC: US Government Printing Office, 1977.

A classified unannotated bibliography to sources of identification of woody plants. Lists some 470 books and journal articles currently in print along with some classic titles. Entries range from technical monographs to popular guides, including some state and federal publications. Appendix outlines special guides to winter identification, seeds and seedlings, and National Park habitats.
R: *ARBA* (1978, p. 667).

Water-in-Plants Bibliography, 1981. Vol. 7. **J. Pospíšilová and J. Solárová, eds.** Hingham, MA: Kluwer Boston, 1982.

An updated edition of an annual bibliography on the subject of water relations in plants. This volume covers papers that were published primarily in 1981. The 1,547 titles are international in coverage and complete in detail. A valuable annual source for crop scientists and plant physiologists.
R: *ARBA* (1984, p. 645).

NUTRITION

Food and Nutrition Bibliography. 11th ed. Phoenix: Oryx Press, 1984.

Ninth edition, 1980; tenth edition, 1982.
Compiled from data provided by the National Agricultural Library, this is a comprehensive annotated guide to 3,566 print and audiovisual materials covering all areas related to human nutrition. Indexed. Useful to researchers, educators, and librarians.
R: *ARBA* (1985, p. 496).

Human Food Uses: A Cross-Cultural, Comprehensive Annotated Bibliography. **Robert L. Freeman.** Westport, CT: Greenwood Press, 1981.

A catalog of over 9,000 monographs, articles, theses, etc., about food habit research and anthropology. Entries arranged alphabetically, with access through a keyword and key concept. Multilingual and uncritical, this work is good for medical and dietary professionals.
R: *ARBA* (1982, p. 798).

ZOOLOGY

A Bibliography of Birds. 4 vols. **Myron Reuben Strong.** Chicago: Natural History Museum, 1939–1959.

Comprehensive coverage through 1926 and selective to 1938. For later works, *Biological Abstracts* should be consulted.
R: *Bioscience* 18: 62 (Jan. 1968); Jenkins (G219); Win (EC107).

Bibliography of Fishes. 3 vols. **Bashford Dean, C. R. Eastman, ed.** New York: American Museum of Natural History, 1916–1923. Repr. New York: Hafner, 1972.

A 3-volume listing of 50,000 references to the structure, development, habitats, physiology, and distribution of fishes. The first 2 volumes are organized by author; the third volume consists of a subject index. Includes pre-Linnaean publications, a list of periodicals, and references to general bibliographies that cover ichthyology. For natural history libraries and marine biologists.

Bibliography of Reproduction: A Classified Monthly Title List Compiled from the World's Research Literature, Vertebrates, Including Man. Vol. 1. Cambridge, England: Reproduction Research Information Service, 1963–. Monthly.

Over 600 literature sources per issue for such subjects as biology, medicine, agriculture, and veterinary science. Includes author and animal indexes.
R: Wal (p. 226).

A Bibliography on Animal Rights and Related Matters. **Charles R. Magel.** Washington, DC: University Press of America, 1981.

Contains literature from biblical times to the present concerning animal rights and ethics. An authoritative and specialized tool for scholars seeking references to various concerns of animal experimentation or conservation.
R: *ARBA* (1983, p. 632).

Checklist of the Coleopterous Insects of Mexico, Central America, the West Indies, and South America. United States National Museum Bulletin, no. 185, pts. 1–6. **Richard E. Blackwelder.** Washington, DC: Smithsonian Institution, 1982.

Contains a list of species and their various aspects, a bibliography of the coleopterology of Latin America up to 1941, and an index of generic names. For those interested in entomology and zoology with an emphasis in systematics.
R: *ARBA* (1983, p. 640).

Primates of the World: Distribution, Abundance, and Conservation. **Jaclyn H. Wolfheim.** Seattle: University of Washington Press, 1983.

In 3 sections: a brief introduction; phylogenetically arranged species accounts, with distribution maps for each; and a concluding discussion that includes extensive data summaries in tabular format. For research libraries and those familiar with primate taxonomy and biology.
R: *ARBA* (1983, p. 667).

Zoobooks: A Bibliography of New and Forthcoming Books: Veterinary Medicine, Zoology . . . Books, Series, Proceedings, Journals. Basel, Switzerland: Karger Libri, 1969–. Annual.

A purchasing guide in the form of a classified list with author index. Text in English, German, French, Italian, and Spanish.
R: Wal (p. 222).

EARTH SCIENCES

American Geological Literature, 1669 to 1850. **Robert M. Hazen and Margaret Hindle Hazen.** Stroudsburg, PA: Dowden, Hutchinson & Ross, 1980.

Contains more than 11,000 citations relating to US literature on geology published from 1669 to 1850. Includes standard reference works as well as reviews, journal articles, pamphlets, and other nonnewspaper sources. Index is provided. Outstanding reference for geology and history of geology students and academic libraries.
R: *Choice* 18: 1072 (Apr. 1981).

Annotated Bibliographies of Mineral Deposits in Africa, Asia (Exclusive of the USSR), and Australia. New York: Pergamon Press, 1976.
R: *RSR* 5: 14 (July/Sept. 1977); *SL* 67: 5A (Apr. 1976); Sheehy (EE17).

Annotated Bibliographies of Mineral Deposits in the Western Hemisphere. GSA Memoir no. 131. **John D. Ridge.** Boulder: Geological Society of America, 1972.

Basic arrangement by continent and country, subdivided in some cases by state or province. Includes information on location of deposits, age of formations, Lindgren classification category, etc. Companion volume on deposits in Eastern Hemisphere in preparation.
R: *RSR* 5: 14 (July/Sept. 1977); *ARBA* (1973, p. 592); Sheehy (EE16).

Antarctic Bibliography. Vols. 1–. **US Library of Congress, Science and Technology Division.** Washington, DC: US Government Printing Office, 1962–.
Volume 8, 1976
International bibliography covering biological and geological sciences, atmospheric physics, expeditions, etc.
R: *ARBA* (1972, p. 594; 1978, p. 616).

Antarctic Bibliography: Indexes to Volumes 1–7. **Geza T. Thuronyi, ed.** Washington, DC: US Library of Congress. Distr. Washington, DC: US Government Printing Office, 1977.

Provides 4 indexes (author, subject, geographic, and grantee) to volumes 1–7 of the *Antarctic Bibliography*. Entries include volume number, letter abbreviations for subject category, accession number, and additional bibliographic information.
R: *ARBA* (1979, p. 645).

Arctic Bibliography. 26 vols. **Arctic Institute of North America.** Edited by Maret Martna and Maria Tremaine. Vols. 1–12: Washington, DC: US Government Printing Office. Vols. 13–21: Montreal: McGill-Queen's University Press. 1947–1975.

Features articles by international scholars; includes over 100,000 books and reports published in nearly 40 languages. Titles and their English abstracts have been culled from literature housed in Canadian and United States libraries. Has witnessed an increase in material on the medical, physiological, anthropological, and environmental aspects of the development of Arctic regions.
R: *Choice* 13: 641 (July/Aug. 1976); *ARBA* (1977, p. 651).

Bibliography and Index, 1845–1977, Great Basin Province, Paleozoic and Proterozoic Strata. **Maurice Kamen-Kaye.** Cambridge, MA: Maurice Kamen-Kaye, 1978.

Comprehensive listing of references to literature concerned with Great Basin Paleozoic and Proterozoic geology. Includes over 1,750 references. Indispensable tool for petroleum companies in their search for western North American petroleum reserves.
R: *American Association of Petroleum Geologists Bulletin* 63: 114 (Jan. 1979); *TBRI* 45: 166 (May 1979).

Bibliography of American Published Geology, 1669 to 1850. **Robert M. Hazen.** Boulder, CO: Geological Society of America, 1976.

A compilation of 13,700 entries of geological literature published in America. Entries culled from 89 secondary sources dating back to 1723.
R: *RSR* 5: 14 (July/Sept. 1977).

Bibliography of Fossil Vertebrates. New York: Geological Society of America. 1902–.

1959–1963 edition (pub. 1968); 1964–1968 edition (pub. 1972).
Five-year bibliographies published as issues of the society's memoirs, numbers 37, 57, 84, 92, 117, and 134. Arranged alphabetically by author, with subject index.
R: *Choice* 6: 991 (1969); *Scientific Information Notes* 7: 18 (June/July 1965); *ARBA* (1973, p. 585); Jenkins (F160); Win (3EE12).

Bibliography of New Mexico Paleontology. **Barry S. Kues and Stuart A. Northrop.** Albuquerque: University of New Mexico Press, 1981.

A collection of information from some 2,000 published papers from 1844 through 1979. Papers are limited to those that provided "information or interpretation concerning the nature, identity, distribution, age, ecology, or geologic context of New Mexico fossils."
R: *ARBA* (1983, p. 613).

A Bibliography of the Literature on North American Climates of the Past 13,000 Years. **Donald K. Grayson.** New York: Garland STPM Press, 1975.

Selective bibliography to pertinent literature. Arrangement of entries by author; access by geographical index.
R: Sheehy (EE12).

Bibliography on Cold Regions Science and Technology. Vols. 23–. Hanover, NH: US Army Cold Regions Research and Engineering Laboratory, 1969–.

Volume 26, 1972 (AD-752-083).
Former volumes issued under titles *Bibliography of Snow, Ice and Permafrost*, and *Bibliography on Snow, Ice and Frozen Ground*. Currently an unannotated, computer-produced list of citations.
R: *LCIB* 29: 656 (1970); Win (3EA4).

Catalogue and Index of Contributions to North American Geology, 1732–1891. **Nelson Horatio Darton.** Edited by I. Bernard Cohen. New York: Arno Press, 1980.

Reprint of 1896 edition.

The Earth Sciences: An Annotated Bibliography. **Roy Porter.** New York: Garland, 1983.

Refers to the history of the larger specialties, "cognate sciences," areal studies, and other areas of earth science.
R: *ARBA* (1984, p. 683).

Geologic Reference Sources: A Subject and Regional Bibliography of Publications and Maps in the Geological Sciences. 2d ed. **Dederick C. Ward, Marjorie W. Wheeler, and Robert A. Bier, Jr.** Metuchen, NJ: Scarecrow Press, 1981.

First edition, 1972.
New edition is 25 percent more extensive than the first. Divided into 3 sections—general, subject, and regional—which include indexes, books, articles and maps. Intended as an easy-to-use reference guide for the geology student and those in other disciplines as well.
R: *ARBA* (1982, p. 755).

Geologists and the History of Geology: An International Bibliography from the Origins to 1978. 5 vols. **William Antony S. Sarjeant.** New York: Arno Press, 1980.

Volume 1, *Introduction: Histories of Geology and Related Sciences*; volume 2, *The Individual Geologists, A–K*; volume 3, *The Individual Geologists, L–Z*; volume 4, *Geologists Indexed by Country and Specialty*; volume 5, *Index of Authors, Editors, and Translators*.
A monumental 5-volume set that covers geologists and the history of geology. Volume 1 deals with the important events in the history of geology; volumes 2 and 3 contain over 10,000 alphabetically organized biographical sketches of geologists and their works, as well as prospectors, and mining engineers. Volume 4 consists of nationality and country and specialty indexes, and volume 5 is a thorough index of authors, editors, and translators. The entire work comprises approximately 30,000 entries grouped by subject and unabbreviated. Indexes are cross-referenced. An outstanding reference tool essential to all history of science collections, geologists, and researchers.
R: *Choice* 18: 510 (Dec. 1980); *LJ* 105: 2191 (Oct. 15, 1980); *ARBA* (1981, p. 694).

Isotopes of Water: A Bibliography. **W. K. Summers and Carolyn J. Sittler.** Ann Arbor, MI: Ann Arbor Science, 1976.

Gathers together significant pre-1975 literature dealing with water isotopes. Includes some 2,300 references that promise to be only the tip of the iceberg of the literature in this field. References are drawn from international sources and arranged into subject categories. Highly recommended to hydrologists, geologists, geochemists, and water supply engineers.
R: *ARBA* (1977, p. 693).

Mount St. Helens: An Annotated Bibliography. **Caroline D. Harnly and David A. Tyckoson.** Metuchen, NJ: Scarecrow Press, 1984.

Includes some 1,700 entries of articles and other publications about Mount St. Helens even before the 1980 eruption.
R: *Choice* 21: 1444 (June 1984); *ARBA* (1985, p. 603).

World Palaeontological Collections. **R. J. Cleevely.** Distr. New York: Wilson, 1983.

Provides detailed bibliographies on general paleontology, collections, biography, and published museum catalogs as well as an index of collectors and institutions and their holdings. Geared to collectors.
R: *ARBA* (1984, p. 687).

Water Resources: A Bibliographic Guide to Reference Sources. **Valerie Ralston.** Storrs, CT: University of Connecticut Library, 1975.

Full bibliographic detail provided for various reference works on water resources. Includes significant numbers of state and federal government publications.
R: *ARBA* (1976, p. 705); Sheehy (EJ27).

OCEANOGRAPHY

Bibliography on Marine Geology and Geophysics. **US Department of Commerce.** Rockville, MD: National Oceanographic Data Center, 1972.

This annotated work also provides an appendix of references pertaining to bibliographies on the subject published prior to 1969.

A Guide to Publications and Subsequent Investigations of Deep Sea Drilling Project Materials. La Jolla, CA: University of California, Scripps Institute of Oceanography. Semiannual.

Ocean Engineering Information Series. 6 vols. **Evelyn Sinha.** La Jolla, CA: Ocean Engineering Information Service, 1967–1971.

Annotated bibliographies on topics in ocean engineering.

Oceans of the World: The Last Frontier. An Annotated Bibliography on the Law of the Sea. **B. Hurd and B. P. Compassero.** Cambridge, MA: MIT Sea Grant Program, 1974.

The Sea: A Select Bibliography on the Legal, Political, Economic, and Technological Aspects, 1978–1979. **United Nations.** New York: Unipub, 1980.

A bibliography of legal, political, scientific, technological, and economic aspects of the Law of the Sea. Contains a section on sources of marine pollution.
R: *IBID* 8: 88 (Summer 1980).

The Sea: Economic and Technological Aspects; A Select Bibliography. New York: United Nations, 1974.

The Sea: Legal and Political Aspects; A Select Bibliography. New York: United Nations, 1974.

GENERAL ENGINEERING

Barker Engineering Library Bulletin. MIT Libraries. Cambridge, MA: MIT Press, 1966–.

Bibliography and Index of Experimental Range and Stopping Power Data. The Stopping and Ranges of Ions in Matter. Vol. 2, **H. H. Andersen.** Elmsford, NY: Pergamon, 1977.

Five-volume series covering various aspects of stopping power data. Specialized reference set for the nuclear physicist.
R: *ARBA* (1979, p. 659).

Bibliography of Energy Conservation in Architecture: Keyword Searched. **Kaiman Lee.** Boston: Environmental Design and Research Center, 1977.

Lists journal articles, books, and reports from the 1970s which deal with energy conservation.
R: *ARBA* (1978, p. 414).

Engineering Eponyms. 2d ed. **C. P. Auger.** Phoenix: Oryx Press, 1975.

This useful annotated bibliography pertains to various aspects of mechanical engineering, such as selected elements, principles, and machines.

Technology Book Guide. Vols. 1–. **Gerald Swanson, ed.** Boston: Hall, 1974–. Monthly with annual cumulations.

English- and French-language books and serial titles are classified by the Library of Congress in all subfields of engineering technology.

AERONAUTICAL AND ASTRONAUTICAL ENGINEERING

Aeronautical Engineering: A Special Bibliography with Indexes (NASA-SP-7037). Washington, DC: National Aeronautics and Space Administration, 1970–.

Since 1971, monthly with annual cumulative index. Reports and articles on aerodynamics and aeronautics culled from STAR and IAA. Includes abstracts, and subject, author, and contract-number indexes. See also *NASA Continuing Bibliography Series.*

An Aerospace Bibliography. 2d ed. **Samuel Duncan Miller, comp.** Washington, DC: Office of Air Force History. Distr. Washington, DC: US Government Printing Office, 1979.

An updated and expanded version of *United States Air Force History: An Annotated Bibliography,* this selective but outstanding bibliography deals with various aspects of the US Air Force. Covers Air Force activities in war and peace, aircraft, space vehicles, museums, UFO's, etc. Articles are culled from books, periodicals, and official documents. Appendixes contain a bibliography of bibliographies and a listing of various places that provide information on aviation. Also includes excellent subject and author indexes. An invaluable aid for all aviation researchers.
R: *BL* 75: 1525 (June 15, 1979); *ARBA* (1980, p. 731).

Bibliography of Space Books and Articles from Non-Aerospace Journals, 1957–1977. **John J. Looney.** Washington, DC: History Office, National Aeronautics and Space Administration Headquarters. Distr. Washington, DC: US Government Printing Office, 1979.

A unique reference that lists nonspecialized and nontechnical literature relating to National Aeronautics and Space Administration and space flight from the pre-1970s.

Material covered does not appear in *STAR* or *International Aerospace Abstracts*. Entries are grouped into 14 subject categories and arranged alphabetically by author.
R: *BL* 76: 1497 (June 15, 1980); *ARBA* (1981, p. 752).

NASA Continuing Bibliography Series. Washington, DC: National Aeronautics and Space Administration.

Aerospace Medicine and Biology, NASA-SP-7011, monthly; *Aeronautical Engineering*, (NASA-SP-7037, monthly; *Patent Abstracts Bibliography*, NASA-SP-7039, semiannually; *Earth Resources*, NASA-SP-7041, quarterly; *Energy*, NASA-SP-7043, quarterly; *Management*, NASA-SP-7500, annually.
Annotated bibliographies of unclassified reports and articles. Most include accession numbers, price, and ordering information.

UFOs and Related Subjects, Annotated Bibliography. Repr. ed. **Lynn E. Cator, ed.** Detroit: Gale Research, 1979.

First edition, 1969.
Among its 1,600 items is the extensive UFO literature collection of the Library of Congress.
R: *ARBA* (1970, p.105).

CHEMICAL ENGINEERING

Chemical Vapor Deposition, 1960–1980: A Bibliography. **Donald T. Hawkins, ed.** New York: IFI/Plenum, 1981.

More than 5,000 citations pertaining to chemical vapor deposition and vapor transport processes are contained in this computer-produced bibliography of documents. For any technical library involved with research of chemical vapor deposition.
R: *ARBA* (1983, p. 723).

Coating Equipment and Processes. **Jack Weiner.** Appleton, WI: Institute of Paper Chemistry, 1975.

Gathers articles and patents on areas within the paper industry. Many originally appeared in the *Abstract Bulletin of the Institute of Paper Chemistry* and the *TAPPI Bibliography*. Arrangement is alphabetical by author; annotations range in length from 25 to 200 words. Supported by author, subject, and patent indexes. Of limited interest to specialized collections.
R: *ARBA* (1977, p. 642).

Guide to Gas Chromatography Literature. 4 vols. **Austin V. Signeur.** New York: Plenum, 1964–1979.

Volume 3 contains 15,741 citations arranged alphabetically by author and subject indexes.
R: *Journal of the American Chemical Society* 102: 5136 (July 16, 1980); *NTB* 50: 50 (1965); *ARBA* (1976, p. 649); Jenkins (D31); Wal (p. 131).

The History of Chemical Technology: An Annotated Bibliography. Bibliographies of the History of Science and Technology, vol. 5. **Robert P. Multhauf.** New York: Garland, 1984.

Contains about 1,500 bibliographic entries. Annotated and indexed by author and title. A valuable research tool.
R: *Choice* 21: 1590 (July/Aug. 1984); *ARBA* (1985, p. 543).

Ozone Chemistry and Technology: A Review of the Literature: 1961–1974. Philadelphia, PA: Science Information Services, 1975.

An overview by world-recognized experts in the field. Over 3,000 footnotes and citations.
R: *SL* 66: 8A (July 1975) (ad).

Plastics Book List. **George J. Patterson.** Westport, CT: Technomic, 1975.

Presents some 1,660 unannotated citations of monographs in the area of plastics. Material gathered from *Books in Print* and various technical journals. Arrangement by subject under 40 distinct categories. Entries current through 1974.
R: *ARBA* (1977, p. 772).

ELECTRICAL AND ELECTRONICS ENGINEERING

Digest of Literature on Dielectrics. Vols. 1–. **Prepared by the Committee on Digest of Literature of the Conference on Electrical Insulation, Division of Engineering and Industrial Research, US National Research Council.** Washington, DC: National Academy of Sciences, National Research Council, 1949–.

Volumes 1–10, 1939–1946 in one volume, 1949.
A critical bibliography of the previous year's literature in all phases of dielectrics.
R: Wal (p. 111; p. 317).

International Bibliography of Automatic Control. Vols. I–. Brussels: Presses Academiques Européenes; London: Butterworth, 1962–. Quarterly.

Covers articles and proceedings papers only.
R: Wal (p. 301).

Ion Implantations in Microelectronics: A Comprehensive Bibliography. **A. H. Agajanian.** New York: IFI/Plenum, 1981.

A bibliography of literature from 1976–1980, including journals, theses, etc. Over 2,000 citations are grouped under 52 subject headings arranged within 7 categories. Detailed author and subject indexes appear. Useful for technical library; prior knowledge helpful.
R: *ARBA* (1983, p. 723).

Literature in Digital Signal Processing: Author and Permuted Title Index. Rev. Ed. **Howard D. Helms, James F. Kaiser, and Lawrence R. Rabiner.** New York: Institute of Electrical and Electronics Engineers, 1975.

Comprehensive coverage of digital filtering and spectral analysis. Boasts over 1,800 citations to monographs, technical documents, journal articles, dissertations and conference reports. Covers literature through 1974. Consists of 3 distinct sections: author index, keyword title index, and list of references.
R: *ARBA* (1977, p. 758).

Literature Survey of Communication Satellite Systems and Technology. **J. H. W. Unger, comp.** New York: Institute of Electrical and Electronics Engineers Press, 1976.

Companion bibliography to *Communication Satellite Systems: An Overview of the Technology* (edited by Y. F. Lum and R. G. Gould). Cites over 3,600 references to literature published through 1974. Arranged in 3 sections: permuted title index, author index, and list of references. Highly recommended to researchers in the area of communication satellite systems.
R: *ARBA* (1978, p. 762).

Microelectronic Packaging: A Bibliography. **A. H. Agajanian, comp.** New York: Plenum Press, 1979.

Includes 3,000 references taken from various abstracts and indexes between 1976 and 1978 covering the field of microelectronic packaging. Entries are alphabetically organized by author and categorized into 27 subject headings and 42 subheadings. Contains subject index. A reference aid for workers in the field.
R: *ARBA* (1981, p. 752).

MOSFET Technologies: A Comprehensive Bibliography. **A. H. Agajanian, comp.** New York: Plenum Press, 1980.

Compilation of 4,499 references to MOSFET technologies literature spanning 1976 to 1980. Three chapters cover literature on technologies, properties, and characterization and dielectric and thin films. Designed to supplement *Semiconducting Devices: A Bibliography of Fabrication Technology, Properties, and Applications* by the same author. Suitable for beginners who need assistance in finding review articles.
R: *Choice* 18: 919 (Mar. 1981).

Computer Technology

Computer-Aided Design of Digital Systems: A Bibliography. **W. M. van Cleemput.** Woodland Hills, CA: Computer Science Press, 1960–1979.

Volume 1, 1960–1974; volume 2, 1975–1976; volume 3, 1976–1977; volume 4, 1977–1979.
Comprehensive work (over 3,000 citations) covering the entire field of digital systems hardware and software.
R: *SL* 67: 10A (July 1976); *ARBA* (1978, p. 770).

Computer Simulation 1951–1976: An Index to the Literature. **P. Holst.** Salem, NH: Mansell, 1979.

Lists over 6,000 entries covering a wide range of subjects in computer simulation. Compiled by the author, an expert in the field. Although not intended as the definitive source, it will be a valuable reference tool for those interested in computer simulations.
R: *Aslib Proceedings* 45: 1 (Jan. 1980).

Computer Technology: Logic, Memory, and Microprocessors; A Bibliography. **A. H. Agajanian.** New York: Plenum Press, 1978.

Considered a comprehensive international bibliography covering the literature from 1970 through 1977. Includes subject index. Comprehensive.

Computext Book Guides: Technology. Vols. 1–. Boston: Hall, 1974–. Monthly.
R: *ARBA* (1976, p. 632).

Microprocessor Applications in Science and Medicine: A Bibliography (1977–1978). **S. Deighton, ed.** London: Institute of Electrical Engineers, 1980.

In *Microprocessor Application Series.*
A bibliography bringing together some 200 references on the many new applications used in microprocessing from the IEE INSPEC database period of 1977 to 1978. A useful basis for retrospective searching.
R: ABL (Jan. 1981).

Robotics, 1960–1983: An Annotated Bibliography. **Andrew Garoogian.** Brooklyn, NY: CompuBibs/Vantage Information Consultants, 1984.

An annotated bibliography that includes information sources and citations under 14 broad categories. Appendixes of organizations, periodicals, and directories in the field. Nontechnical emphasis. More useful to public libraries, hobbyists, and students.
R: *Choice* 22: 62 (Sept. 1984); *ARBA* (1985, p. 599).

INDUSTRIAL ENGINEERING

The Blacksmith's Source Book: An Annotated Bibliography. **James Evans Fleming.** Carbondale, IL: Southern Illinois University, 1980.

An annotated bibliography containing 276 books, pamphlets, and journal articles on the areas of blacksmithing. Includes texts, manuals, reference works, specialized areas of the trade, historical background, and products used in the profession.
R: *Choice* 19: 220 (Oct. 1981); *ARBA* (1982, p. 803).

Vapor–Liquid Equilibrium Data Bibliography. Supp. 2. **I. Wichterle, J. Linek, and E. Hala.** New York: Elsevier, 1979.

This supplement covers the literature from January 1976 through December 1978. Information is arranged in formula-index style, with photo-reproduced computer output. No data given, only references.
R: *Journal of the American Chemical Society* 102: 3665 (May 7, 1980).

MATERIALS SCIENCE

A Bibliography on the Corrosion and Protection of Steel in Concrete. Washington, DC: US Government Printing Office, 1980.

A bibliography of papers, reports, and talks of steel in concrete and related subjects. Contains 394 references indexed by subject and author.
R: *Journal of Metals* 32: 65 (Apr. 1980).

Bibliography on the Fatigue of Materials, Components and Structures. Vol. 2. **J. Y. Mann, comp.** Elmsford, NY: Pergamon, 1978.

Volume 1, 1838–1950; volume 2, 1951–1960.
A chronologically and alphabetically organized bibliography on fatigue of materials. Publication titles are abbreviated and English-language translations accompany foreign-language article titles. Includes subject and author indexes. Of interest to engineering libraries.
R: *ARBA* (1980, p. 711).

MECHANICAL ENGINEERING

Bibliography on Engine Lubricating Oil: 1968–1983. Malcolm Fox et al. Brookfield, VT: Gower, 1985.

Engineering Eponyms: An Annotated Bibliography of Some Named Elements, Principles and Machines in Mechanical Engineering. 2d rev. ed. **C. P. Auger.** London: Library Association, 1975.

Covers mechanical eponyms, with the exception of the textile industry. Arranged alphabetically by eponym, including discussion and references.
R: *Chartered Mechanical Engineer* 00: 97 (Feb. 1966); *ARBA* (1975, p. 779); Wal (p. 286).

Heat Pipe Technology; A Bibliography with Abstracts. Albuquerque: University of New Mexico, Technology Application Center, 1971–. Quarterly updates.

Jet Pumps and Ejectors: A State-of-the-Art Review and Bibliography. 2d ed. **S. T. Bonnington and A. L. King.** Corning, NY: Air Science Co., 1976.

Features a chronological arrangement of some 400 references. Includes detailed subject and author indexes. Useful bibliography for plant designers and users.
R: *Mine and Quarry* 6: 23 (Feb. 1977); *TBRI* 43: 229 June 1977).

SAE Transactions and Literature Developed. Warrendale, PA: Society of Automotive Engineers. Annual. Vols. 1–, 1905–.

METALLURGY

Coal Mine Road Technology: An Assessment of References and Annotated Bibliography, January 1983. Washington, DC: Office of Surface Mining. Distr. Washington, DC: US Government Printing Office, 1983.

A 1-volume paperback that includes information on 128 articles on aspects of the design, construction, maintenance, and reclamation of coal mine roads. Additional 400 unannotated articles are included in appendix.
R: *ARBA* (1985, p. 544).

Multicomponent Alloy Construction Bibliography 1955–1973. **Alan Prince.** London: The Metals Society, 1978.

A bibliography containing more than 18,000 document references. Includes only those systems with more than 2 components. Entries are for ternary systems first,

then quaternary, etc. Arrangement within each type of system is based on the components.
R: *Journal of Metals* 31: 51 (Nov. 1979); *RSR* 8: 30 (July/Sept. 1980).

Researchers' Guide to Iron Ore: An Annotated Bibliography on the Economic Geography of Iron Ore. **Fillmore C. F. Earney.** Littleton, CO: Libraries Unlimited, 1974.

Covers the literature published in English since 1945. Divided into 3 main sections: general references, topical references, and regional references.
R: *LJ* 99: 3124 (Dec. 1, 1974).

Underwater Construction and Mining: A Bibliography with Abstracts, NTIS/PS-76/0365/7GA. **Guy E. Habercom, Jr.** Springfield, VA: National Technical Information Service, 1976.

Reported for 1964–May 1976.
Approximately 200 abstracts from government-sponsored research reports in areas of underwater minerals, mining techniques, legal implications, and construction techniques.

MILITARY SCIENCE

Chemical/Biological Warfare: A Selected Bibliography. 3d rev. ed. **Julian Perry Robinson, comp.** Los Angeles, CA: Center for the Study of Armament and Disarmament, California State University, 1979.

Second revised edition, 1974.
A listing of 355 references, most from the pre-1975 period. Arranged by subject, the bibliography provides international coverage, although most entries are of publications of US government agencies. Material deals with various aspects of chemical and biological warfare. Helpful to military libraries and researchers in this area.
R: *ARBA* (1980, p. 728).

NUCLEAR ENGINEERING

Deuterium and Heavy Water; A Selected Bibliography. **Gheorge Vasaru et al.** New York: American Elsevier, 1975.

Chronologically and then alphabetically by author, 3,763 publications from 1932 to May 1974 on the properties, analysis, and production of deuterium and heavy water, and on the behavior of heavy water as a moderator in nuclear reactors.
R: *ARBA* (1977, p. 650).

IAEA Bibliographical Series. Nos. 1–. Vienna: International Atomic Energy Agency, 1960–.

A multivolume series, each volume dealing with a facet of the peaceful uses of atomic energy. Includes introductory as well as bibliographic information.

Preparation of Nuclear Targets: A Comprehensive Bibliography. **Jozef Jaklovsky.** New York: IFI/Plenum, 1981.

A bibliography of over 6,000 numbered entries and patent entries labeled as a patent index, for the period from 1936 through June 1980. Some incomplete or inconsistent use of entries, citations, etc., may cause some difficulty for the reference librarian seeking to find some of the citations.
R: *ARBA* (1983, p. 609).

Public Regulation of Site Selection for Nuclear Power Plants: Present Procedures and Reform Proposals—An Annotated Bibliography. **Ernest D. Klema and Robert L. West.** Washington, DC: Resources for the Future, 1977.

Emphasis on alternative procedures to siting nuclear power plants. Lists some 40 extensively annotated citations in 4 main sections. In many cases, the annotation requires one-third of the page. Arrangement appears to be by importance. Excellent introduction puts the situation in its proper perspective.
R: *ARBA* (1979, p. 698).

Reactor Safety: A Literature Search. Oak Ridge, TN: Department of Energy Technical Information Center Service, 1978.

Earlier edition, 1972.
Several citations to references issued since 1971 on safety aspects of nuclear design, siting, etc.

ENERGY

Appropriate Energy Technology Library Bibliography. **Hannah R. Clark.** Berkeley: University of California, Energy and Environment Division, Lawrence Berkeley Laboratory, 1979.

Chinese Petroleum: An Annotated Bibliography. **Raymond Change.** Boston: Hall, 1982.

Contains nearly 900 entries. Lists reference works and articles and documents in English, Chinese, and Japanese. Chronological arrangement makes author, subject, and title indexes necessary for researchers needing verification of a citation. Good for social science and earth science libraries.
R: *ARBA* (1983, p. 650).

Coal Bibliography and Index. 2 vols. Houston: Gulf, Book Division, 1980–1981.

Focuses on "fugitive" coal literature published between 1880 and 1945. Also indexes recent and current coal documents. Wealth of information drawn from the massive energy collection of the Texas A & M University Library. Includes full range of bibliographic data, complete with brief abstract. Access to entries by 4 separate indexes: keyword-in-title, author, corporate author, and subject indexes.

Energy: A Continuing Bibliography with Indexes, NASA-SP-7043. Washington, DC: National Aeronautics and Space Administration, 1974–. Quarterly.

Emphasis on nuclear field. Covers regional, national, and international energy systems. Abstracts are culled from *STAR* and *IAA*.

Energy and Congress: An Annotated Bibliography of Congressional Hearings and Reports, 1974–1978. **Igor I. Kavass and Doris M. Bieber.** Buffalo, NY: Hein, 1980.

1971–1973 edition, published in 1974.
R: *RSR* 2: 146 (Oct./Dec. 1974).

Energy and the Social Sciences: A Bibliographic Guide to the Literature. **Ernest J. Yanarella and Ann-Marie Yanarella.** Boulder: Westview Press, 1982.

A bibliography divided into 9 parts, each concerned with an aspect of energy—its sources, its future, etc. Each section has an introduction and is divided into more detailed subsections. Good for social scientists, but literature searches are necessary for current sources.
R: *ARBA* (1983, p. 652).

Energy—A Scientific, Technical and Socioeconomic Bibliography. **Kitty Hsieh.** Corvallis, OR: Oregon State University Press, 1976.

Covers various aspects in the energy field. Part 1 lists general energy information sources (indexes, bibliographies, dictionaries). Parts 2–4 list basic information sources for specific forms of energy (coal, solar, nuclear). Includes concise subject index. Although emphasizes holdings in the Oregon State University Library, can serve as a guide for most academic libraries.
R: *ARBA* (1978, p. 684).

Energy Bibliography and Index. Vols. 1–. **Texas A & M University Libraries, comps.** Houston: Gulf, 1978–.

Most comprehensive source on energy research. Reflects the most exhaustive collections of Texas A & M University Library. Covers all types of energy in various forms of media: books, maps, monographs, periodicals, microforms, government documents, reports, and serials. Contains literature dating from the turn of the century to the present year. Over 25,000 items indexed by subject, keyword, personal and corporate author. Now available online from System Development Corporation.
R: *Choice* 17: 693 (July/Aug. 1980); *RSR* 6: 27 (Oct./Dec. 1978); *ARBA* (1980, p. 646).

Energy Costs and Costing: A Selected, Annotated Bibliography. **Emanuel Benjamin Ocran.** Metuchen, NJ: Scarecrow, 1983.

A compilation of sources from books, reports, and journal articles on the topics of energy costs and costing, including information on the accounting, costs, financing, pricing, and more for each area of concentration. Suitable for students, specialists, librarians, and others.
R: *ARBA* (1984, p. 673).

Energy Policy-Making: A Selected Bibliography. **Robert W. Rycroft et al.** Norman, OK: University of Oklahoma Press, 1977.

A 1,200-item unannotated bibliography dealing with energy policy research. Covers books, periodical articles, government documents, and doctoral dissertations written in English. Coverage limited to items published between 1970 and mid-1977. Emphasizes US energy policy. Geared toward social scientists working in the field.
R: *BL* 75: 324 (Oct. 1, 1978); *Choice* 15: 840 (Sept. 1978); *LJ* 103: 858 (Apr. 15, 1978); *ARBA* (1979, p. 687).

Energy: Sources of Print and Non-Print Materials. **Maureen Crowley, ed.** New York: Neal-Schuman, 1980.

Useful bibliographic listings of energy-involved organizations. Materials listed are inexpensive.

R: *LJ* 105: 29 (Jan. 1, 1980).

Heat and Power from the Sun—An Annotated Bibliography with a Survey of Available Products and Their Suppliers. 2d ed. **S. Loyd.** Bracknell, England: Building Services Research and Information Association, 1975.

Hydrogen Energy: A Bibliography with Abstracts. Albuquerque, NM: University of New Mexico, Technology Application Center, 1974–. Annual Supp.

Articles, conference proceedings, and research reports. Volume 1 was published as report TAC-H-74-501.

International Bibliography of Alternative Energy Sources. **G. Hutton and M. Rostron.** New York: Nichols Publishing, 1979.

International listing of books, papers, and journal articles covering various aspects of alternative energy research.

Liquified Natural Gas: A Bibliography. New York: American Gas Association, 1937/1967–.

Offshore Petroleum Engineering: A Bibliographic Guide to Publications and Information Sources. **Marjorie Chryssostomidis.** New York: Nichols Publishing, 1978.

A comprehensive bibliography on offshore petroleum engineering and related fields. Topics covered include offshore structures engineering, underwater and deep-sea operations, seafloor engineering, oceanography, etc. Lists over 2,700 citations from US and international books, reports, conference proceedings, and articles. Entries are grouped into 22 categories subdivided into 100 subsections. Subsections are further separated into books, conferences, or reports. All references are coded alphanumerically and contain full bibliographic data. Appendixes contain a list of sources, directory of publishers, and tips on locating and obtaining publications. Author, title, and permuted topics indexes are provided. A valuable reference work for petroleum engineers and academic libraries.

R: *Choice* 16: 1148 (Nov. 1979); *RQ* 19: 89 (Fall 1979); *ARBA* (1980, p. 720).

Solar Energy: A Bibliography. Oak Ridge, TN: Energy Research and Development Administration, 1973–. Annual.

R: *Sci-Tech News*, p. 57 (Apr. 1975).

Solar Energy Books. **US National Solar Energy Education Campaign.** Beltsville, MD: International Compendium, 1977.

Listing of government documents, technical reports, and commercial publications on aspects of solar and other alternative energy sources. Entries are arranged under 6 major headings and contain descriptive annotations. Intended as a catalog of materials available through the National Solar Energy Education Campaign.

R: *BL* 76: 1440 (June 1, 1980).

Solar Thermal Energy Utilization: A Bibliography with Abstracts. Albuquerque, NM: University of New Mexico, Technology Application Center, 1957/1974–. Quarterly supp.

Includes articles, research reports, and conference proceedings. Volumes 1–2 were published as reprint TAC-ST74-600.

Sun Power: A Bibliography of United States Government Documents on Solar Energy. **Sandra McAninch, comp.** Westport, CT: Greenwood Press, 1981.

A valuable reference tool for finding literature on solar and wind energy.

Synthetic Fuels Research: A Bibliography, 1945–1976. 3d ed. **Ruby L. Mathison, comp.** Arlington, VA: American Gas Association, 1977.

A compilation of references on coal gasification and liquefaction, plus synthetic fuels from other feedstocks such as shale oil and solid wastes. Includes references on hydrogen energy, methanol, and power generation from coal, solid wastes, etc.

ENVIRONMENTAL SCIENCES

GENERAL

Annotated Bibliography on the Ecology and Reclamation of Drastically Disturbed Areas. US Department of Agriculture Forest Service General Technical Report NE-21. **Miroslaw M. Czapowskyj.** Upper Darby, PA: Northeastern Forest Experiment Station, US Forest Service. Distr. Washington, DC: US Government Printing Office, 1976.

Lists nearly 600 succinctly annotated references to literature, dealing with mining effects and land reclamation in the US coal regions. Majority of citations encompass technical reports and conference papers. Represents a significant amount of basic and applied research material culled from university, government, and industrial projects. Entries arranged alphabetically by author. Includes geographical, material, and general subject index. Useful addition to environmental sciences collections.
R: *ARBA* (1978, p. 684).

A Bibliography of African Ecology: A Geographically and Topically Classified List of Books and Articles. **Dilwyn J. Rogers, comp.** Westport, CT: Greenwood Press, 1979.

Provides 7,800 references on African ecology categorized under 5 geographic regions. Includes a wide range of topics such as anthropology, history, and human health. For African and biology collections.
R: *Choice* 16: 998 (Oct. 1979); *ARBA* (1980, p. 652).

A Bibliography of Quantitative Ecology. **Vincent Schultz et al.** Stroudsburg, PA: Dowden, Hutchinson and Ross. Distr. New York: Wiley, 1976.

A computer-generated bibliography arranged alphabetically by subject. Covers such areas as population, models, and taxonomy. Entries contain keyword listing.
R: *ARBA* (1977, p. 683).

The Energy and Environment Bibliography: Access to Information. Rev ed. **Betty Warren.** San Francisco: Friends of the Earth Foundation, 1978.

First edition, 1977.
Incorporates some 400 references into this bibliography, which integrates energy with environmental concerns. Lists mainly books and pamphlet sources with a few graphic sources. Alphabetical arrangement of entries in 16 separate sections. Most entries give full bibliographic information. Includes some well-known authors and publishers in the energy field.
R: *BL* 74: 704 (Dec. 15, 1977); *Choice* 16: 208 (Apr. 1979); *ARBA* (1980, p. 648).

The Energy and Environment Checklist: An Annotated Bibliography of Resources. Rev. ed. **Betty Warren.** San Francisco: Friends of the Earth Foundation, 1980.

An expanded and revised edition that contains over 1,600 references. Provides coverage of books, journals, organizations, audiovisual materials, etc. Focuses on soft-energy options, although it covers energy and the environment. Bibliographic and ordering information as well as an annotation are included. Bibliography is organized by general categories such as energy conservation, fossil fuels, nuclear energy, solar energy, etc., but then subdivided within these areas. Contains an author index. An inexpensive and helpful aid for all libraries.
R: *Choice* 18: 930 (Mar. 1981).

Environment and Behavior: An International Multi-disciplinary Bibliography 1970–1981. 2 vols. **Lenelis Kruss and Reiner Arlt.** New York: Saur, 1984.

A useful bibliographic reference tool with over 6,000 title entries, representing monographs and journal articles from many different countries and disciplines. Consists of 2 separate volumes—an alphabetical listing by authors with a keyword index and a series of abstracts in English.

Environmental Impact Assessment: A Bibliography with Abstracts. **Brian D. Clark, Ronald Bisset, and Peter Wathern.** New York: Bowker, 1980.

A bibliographic reference to environmental impact assessment. Composed of 1,106 references divided into 5 sections: methods and manuals, general and critical reviews, social impact and public participation in planning, impact assessment in different countries, and information sources. Most entries are located in the fourth section, which focuses on involvement of United States, Canada, United Kingdom, Europe, and Australia. The fifth section lists bibliographies and periodicals to consult for further reference. Some entries are annotated, but subject and author indexes are provided. Mainly for academic and special libraries, private organizations, and government agencies.
R: *LJ* 105: 2315 (Nov. 1, 1980); *ARBA* (1981, p. 683).

Environmental Law: A Guide to Information Sources. **Mortimer D. Schwartz.** Detroit: Gale Research, 1977.

Fulfills its goal of being a useful introductory, bibliographic source. Concentrates on monographs and treatises dealing with environmental law as a field directed toward protecting, preserving, and rehabilitating the natural environment. Entries include occasional brief annotations. Recommended to all libraries.
R: *Choice* 14: 1027 (Oct. 1977); *WLB* 52: 189 (Oct. 1977); *ARBA* (1978, p. 685).

Environmental Planning: A Guide to Information Sources. Man and the Environment Information Guide Series, vol. 3. **Michael J. Meshenberg, ed.** Detroit: Gale Research, 1976.

A selective and annotated bibliography of literature on natural processes in the environment, as well as social and legal aspects of planning. Author and title index.

Environmental Values, 1860–1972: A Guide to Information Sources. Man and the Environment Information Guide Series, vol. 4. **Loren C. Owings, ed.** Detroit: Gale Research, 1976.

An annotated bibliography on the historical development in the United States of attitudes toward and concern for nature. Indexed.
R: *BL* 75: 74 (Sept. 1, 1978); *RQ* 16: 356 (Summer 1977); *ARBA* (1977, p. 683).

EPA Cumulative Bibliography, 1970–1976. PB-265920. 2 vols. Springfield, VA: National Technical Information Service, 1976.

A cumulative listing of all reports issued by the National Technical Information Service through the US Environmental Protection Agency. Contains full bibliographic information.
R: Sheehy (EJ15).

EPA Publications Bibliography: Quarterly Abstract Bulletin. Springfield, VA: National Technical Information Service, 1977–. Quarterly.

Includes abstracts, bibliography, title and subject index, as well as accession numbers, list of US Environmental Protection Agency libraries.
R: Sheehy (EJ18).

Freshwater and Terrestrial Radioecology: A Selected Bibliography. **Alfred W. Klement, Jr. and Vincent Schultz, eds.** New York: Van Nostrand Reinhold, 1980.

An outstanding bibliography of more than 20,000 references to radioactive nuclides in the ecology of freshwater and marine systems. References date back to 1898 with many culled from 13 previous bibliographies. All entries listed by surname of first author. Includes an appendix of other bibliographies. A worthy addition to any university library or primary investigator's collection.
R: *Choice* 18: 1395 (June 1981).

The Natural Environment: An Annotated Bibliography of Attitudes and Values. **Mary Anglemyer and Eleanor R. Seagraves, comps.** Washington, DC: Smithsonian Institution Press, 1984.

The first of a planned series formed in response to the *Global 2000 Report.* It includes 857 references from 1971–1983 in broad subject arrangement. A comprehensive index is available.
R: *ARBA* (1985, p. 512).

A Search for Environmental Ethics: An Initial Bibliography. **Mary Anglemyer et al., comps.** Washington, DC: Smithsonian Institution Press, 1980.

This bibliography is aimed at presenting the attitude of various groups toward the natural environment. Includes 450 books, articles and conference proceedings from

the natural sciences, social sciences, and humanities. Entries contain well-written annotations. Indexed by subject and name. For large library collections.
R: *ARBA* (1981, p. 683).

Selected References on Environmental Quality as It Relates to Health. **US National Library of Medicine, Bethesda, MD.** Washington, DC: US Government Printing Office, 1971–. Monthly.

Each issue contains approximately 1,000 citations of articles in journals covered by *Index Medicus*. Covers a variety of technical literature concerned with health hazards.
R: Katz (p. 26).

Wastewater Management: A Guide to Information Sources. Man and the Environment Information Guide Series, vol. 2. **George Tchobanoglous, ed.** Detroit: Gale Research, 1976.

A selected and annotated guide to periodical articles, technical reports, and collected papers on the engineering of wastewater collection, treatment, disposal, and reuse systems.
R: *ARBA* (1977, p. 689).

POLLUTION

Acid Rain: A Bibliography. Orlando: University of Central Florida Library, 1980.
R: *CRL* 41: 566 (Nov. 1980).

Air Pollution Publications: A Selected Bibliography with Abstracts; 1955–. Public Health Service Publication no. 979. **Science and Technology Division, US Library of Congress.** Prepared for the National Air Pollution Control Administration. Washington, DC: US Government Printing Office, 1964–. Annual.

References and abstracts of articles, books, reports, etc. Arranged under broad subject categories. Primarily literature generated from NAPCA.
R: *LCIB* 28: 358 (July 10, 1969); *ARBA* (1970, p. 164); Win (2EI18; 3EI11).

An Annotated Bibliography of Canadian Air Pollution Literature. **Christopher J. Sparrow and Leslie T. Foster, comps.** Ann Arbor: Ann Arbor Science, 1976.

Excellent annotations accompany over 1,000 references to Canadian air pollution literature. Includes articles from scholarly journals, books, conference proceedings, and government reports. Collected literature through 1973. Divided into 7 sections ranging from general articles to research. Supported by 3 indexes: subject, geographical, and author. A gem for all environmental collections in public and academic libraries.
R: *ARBA* (1977, p. 688).

Ground Water Pollution: A Bibliography. **Kelly W. Summers and Zane Spiegel.** Ann Arbor: Ann Arbor Science, 1974.

Includes over 400 entries to reference material dealing with groundwater contamination by pesticides, heavy metals and herbicides. Also explores the effects of urbanization and disposal of solid and animal wastes.
R: Sheehy (EJ28).

Motor Vehicle Emissions: A Bibliography with Abstracts. **US National Highway Traffic Safety Administration.** Springfield, VA: National Technical Information Service, 1974.

Toxicity of Chemicals and Pulping Wastes to Fish. **Louise Louden.** Appleton, WI: Institute of Paper Chemistry, 1979.

This bibliography is a collection of abstracts drawn from several abstracting and indexing services that cover the subject of toxic substances to fish. Lists some 1,600 books, articles, government reports, and conferences, with complete bibliographic information and source. Author and keyword indexes are provided.
R: *ARBA* (1981, p. 687).

TRANSPORTATION

Information Sources in Transportation, Material Management, and Physical Distribution: An Annotated Bibliography and Guide. **Bob J. Davis, ed. and comp.** Westport, CT: Greenwood Press, 1976.

Gathers together classic and hard-to-locate items in the areas of transportation, materials management, and physical distribution. Encompasses references to books, statistical publications, atlases, maps, government documents, and organizations. Excludes periodical articles and dissertations. Over 10,000 annotated citations are arranged under some 70 subject categories. Unique reference source for the educator, librarian, practitioner, and researcher.
R: *BL* 74: 1569 (June 1, 1978); *Choice* 14: 828 (Sept. 1977); *RQ* 17: 186 (Winter 1977); *ARBA* (1978, p. 777).

Traffic Noise: A Review and Bibliography on Surface Transportation Noise. **G. Vulkan and A. Gomersall.** Bedford, England: IFS Publications, 1979.

Selective bibliography of environmental noise. Compiled for students, consultants, and government employees.
R: *Journal of Sound and Vibration* 66: 641 (Oct. 22, 1979); *TBRI* 46: 119 (Mar. 1980).

Transportation System Management: Bibliography of Technical Reports. **Richard Oram.** Washington, DC: US Department of Transportation, 1976.

Focuses on 9 aspects of operational transportation improvements. Entries provide the following data: brief summary, report preparation information, date, and ordering details. Access to reports through a bleed index.
R: *ARBA* (1977, p. 774).

CHAPTER 4　　　　　　ENCYCLOPEDIAS

GENERAL SCIENCE

Growing Up with Science: The Illustrated Encyclopedia of Invention. **Michael Dempsey, ed.** Westport, CT: Stuttman. Distr. Freeport, NY: Marshall Cavendish, 1984.

Informative articles aimed at young readers to introduce the world of science, technology, and invention. The topics covered are astronomy, medicine, aviation, engineering, biology, physics, electronics, military science, chemistry, agriculture, and manufacturing. A major strength is the illustrations.
R: *ARBA* (1985, p. 489).

How It Works: The Illustrated Encyclopedia of Science and Technology. 20 vols. **Donald Clarke, ed.** New York: Marshall Cavendish, 1977.

A 20-volume encyclopedia intended mainly for younger readers. Articles cover such subjects as the computer, engine mechanics, photography, etc. Each article is supplemented by an illustration. For public libraries.
R: *BL* 75: 563 (Nov. 1978); *ARBA* (1978, p. 617).

McGraw-Hill Concise Encyclopedia of Science and Technology. **Sybil P. Parker, ed.-in-chief.** New York: McGraw-Hill, 1984.

Earlier edition, 1982.
Covers all major topics of science and technology in a smaller 1-volume edition. Replaces the earlier 15-volume set of the *McGraw-Hill Encyclopedia of Science and Technology* with the same accuracy, quality, and completeness found in this concise edition.
R: *Choice* 22: 66 (Sept. 1984); *LJ* 109: 1232 (June 15, 1984); *WLB* 59: 67 (Sept. 1984); *ARBA* (1985, p. 489).

McGraw-Hill Encyclopedia of Science and Technology. 5th ed. 15 vols. **Daniel N. Lapedes, ed.** New York: McGraw-Hill, 1982.

First edition, 1960; third edition, 1971; fourth edition, 1977.
Substantially revised and updated articles on a wide variety of topics in science and technology. Contains numerous illustrations; easy-to-use index. An important reference work.
R: *American Scientist* 59: 253 (Mar./Apr. 1971); *Chemical Engineering* 85: 11 (July 3, 1978); *Journal of Chemical Education* 48: 463 (July 1971); *Scientific American* 214: 138 (June 1966); *Australian Library Journal* 20: 325 (Mar. 1971); *BL* 74: 767 (Jan. 1, 1978); *Choice* 14: 1480 (Jan. 1978); *LJ* 92: 2146 (1967); 103: 723 (Apr. 1, 1978); *New Library World* 78: 176 (Sept. 1977); *RQ* 10: 274 (Spring 1971); *Subscription Books Bulletin* 64: 793 (1968) *TBRI* 44: 248 (Sept. 1978); *ARBA* (1978, p. 619); Wal (p. 17); Win (EA86; 1EA11).

McGraw-Hill Encyclopedia of Scientific and Technical Terms. 2d ed. **Daniel N. Lapedes, ed.** New York: McGraw-Hill, 1978.

An enlarged edition, which includes scientific and technical terms from a variety of fields.
R: *Earth Science* 31: 234 (Autumn 1978); *TBRI* 45: 127 (Apr. 1979).

New Encyclopedia of Science. 16 vols. **Tony Osman, ed.** Milwaukee, WI: Purnell Reference Books, 1979.

A 16-volume set covering the physical, natural, medical, and behavioral sciences. Entries are arranged alphabetically and cross-referenced. Includes definitions and biographical sketches of famous scientists. Contains more than 6,000 color photographs, diagrams, and charts. For high school libraries and secondary school students.
R: *ARBA* (1981, p. 621).

The Raintree Illustrated Science Encyclopedia. 20 vols. **Lawrence Urdang, ed.** Milwaukee: Raintree Publishers, 1984.

Earlier edition, 1979.
An adaptation of the *Encyclopedia of Nature and Science*, this 20-volume set contains over 3,000 entries and 4,000 illustrations covering facts, people, and principles from all fields of science and technology. Articles vary in length and range from broad to specific discussion. Contains an index of over 7,000 terms, a classified bibliography of books, and a unique science project section. Primarily for children in the upper elementary grades.
R: *BL* 77: 765 (Feb. 1, 1981); *ARBA* (1980, p. 602); (1985, p. 490).

Van Nostrand's Scientific Encyclopedia. 5th ed. **Douglas M. Considine, ed.** Princeton, NJ: Van Nostrand Reinhold, 1976.

First edition, 1938; second edition, 1947; third edition, 1958; fourth edition, 1968.
A well-received 1-volume work. Definitions provide both basic and technical information. Dated, however, in the area of topics of current interest.
R: *New Scientist* 41: 188 (Jan. 23, 1969); *SJ* 5: 117 (May 1969); *BL* 74: 1029 (Feb. 15, 1978); *BMLA* 65: 399 (July 1977); *Catholic Library World* 49: 358 (Mar. 1978); *Choice* 5: 1417 (1969); *CRL* 30: 84 (1969); *NLW* 78: 116 (June 1977); *RQ* 10: 28 (Fall 1970); *RSR* 5: 6 (Apr./June 1977); *WLB* 51: 600 (Mar. 1977); *ARBA* (1977, p. 629); Jenkins (A49); Sheehy (EA14); Wal (p. 18); Win (EA88; 2EA17).

ASTRONOMY

The Cambridge Encyclopedia of Astronomy. **Simon Mitton, ed.** New York: Crown, 1977.

Prepared by amateurs, this encyclopedia presents comprehensive information on astronomy and astrophysics. Well-illustrated and thorough. Recommended as an excellent reference volume for both public and academic libraries.
R: *American Scientist* 66: 490 (July/Aug. 1978); *Nature* 272: 786 (Apr. 27, 1978); *New Scientist* 76: 717 (Dec. 15, 1977); *Physics Bulletin* 29: 226 (May 1978); *Sky and Telescope* 55: 528 (June 1978); 56: 337 (Oct. 1978); *BL* 75: 1315 (Apr. 15, 1979); *Choice* 15: 666 (July/Aug. 1978); *LJ* 103: 721 (Apr. 1, 1978); 103: 762 (Apr. 1, 1978); *RSR* 6: 24 (July/Sept. 1978); *TBRI* 44: 87 (Mar. 1978); *ARBA* (1979, p. 649); Sheehy (EB4).

A Concise Encyclopedia of Astronomy. 2d ed. **Alfred Weigart and Helmut Zimmerman.** New York: Crane Russak, 1976.

First edition, 1968.
Translated from the German. Comprises a valuable single-volume reference for astronomers. Second edition is considerably revised and well-illustrated.

R: *Sky and Telescope* 36: 255, 314 (1968); *BL* 65: 56 (1969); *BL* 75: 246 (Sept. 15, 1978); *Choice* 14: 837 (Sept. 1977); Jenkins (E33); Sheehy (EB6).

Encyclopedia of Astronomy: A Comprehensive Survey of Our Solar System, Galaxy and Beyond. **Colin Ronan, ed.** London, New York: Hamlyn, 1979.

A useful text covering the solar system, galaxy and beyond. Includes star charts, diagrams, and illustrations. Well-written and illustrated.
R: *TBRI* 46: 92 (Mar. 1980). Wal (p. 145).

The Illustrated Encyclopedia of Astronomy and Space. Rev. ed. **Ian Ridpath, ed.** New York: Crowell, 1979.

First edition, 1976.
An updated reference that covers the fields of astronomy and space science. Describes in detail various instruments, launch vehicles, and satellites. Contains biographical sketches of astronomers from the Greeks to the present. Includes diagrams and black-and-white and color photographs. Indexed and cross-referenced. A valuable addition to high school, public, and academic libraries.
R: *Sky and Telescope* 59: 416 (May 1980); *BL* 77: 279 (Oct. 1, 1980); *RSR* 6: 21 (July/Sept. 1978); *SLJ* 26: 95 (May 1980); *TBRI* 43: 371 (Dec. 1977); *ARBA* (1978, p. 635; 1981, p. 629).

The Illustrated Encyclopedia of Space Technology: A Comprehensive History of Space Exploration. **Kenneth Gatland et al.** New York: Crown/Harmony, 1981.

A 1-volume encyclopedia covering past, present, and future of the space age. Includes photographs, drawings, diagrams, and maps. Contains a glossary, index, and chronology from 360 BC through 1980.
R: *BL* 77: 1427 (July 15, 1981); *WLB* 56: 382 (Jan. 1982).

The Illustrated Encyclopedia of the Universe. **Richard S. Lewis.** New York: Harmony Books, 1983.

An outstanding tool with specific emphasis on recent space exploration with the goal of explaining and assisting readers to understand the cosmos. Extensive color pictures, breathtaking double-page spreads, drawings, brief bibliographies for the 20 chapters, and specific subject index. A rare reference work with universal appeal.
R: *Choice* 21: 1158 (Apr. 1984); *ARBA* (1985, p. 601).

McGraw-Hill Encyclopedia of Astronomy. **Sybil P. Parker, ed.** New York: McGraw-Hill, 1983.

Authoritative reference tool with over 200 articles on theoretical, observational, and experimental aspects of astronomy. Extensive photographs and figures. Suitable for both technical and public libraries.
R: *WLB* 57: 883 (June 1983, p. 883); *ARBA* (1984, p. 631).

The UFO Encyclopedia. **Margaret Sachs.** New York: Putnam's, 1980.

Consists of 750 alphabetically organized entries relating to UFO phenomena. Provides information on people, locations of sightings, interest groups, technical terms, and publications. Extensively cross-referenced. Of interest to academic libraries.
R: *Choice* 18: 1241 (May 1981).

MATHEMATICS

Encyclopedia of Mathematics and Its Applications. Reading, MA: Addison-Wesley, 1976–. In progress.

Volume 1, *Integral Geometry and Geometric Probability, Section-Probability,* 1976; volume 2, *The Theory of Partitions, Section-Number Theory,* 1976; volume 3, *The Theory of Information and Coding, Section-Probability,* 1977; volume 4, *Symmetry and Separation of Variables, Section-Special Functions,* 1977; volume 7, 1979; volume 12, volume 14, part 2, *Extension and Applications,* 1981; volume 15, *The Logic of Quantum Mechanics,* 1981.
R: *ARBA* (1978, p. 627).

Encyclopedia of Statistical Sciences. Vols. 1–. **Samuel Katz and Norman L. Johnson, eds.** New York: Wiley-Interscience, 1982–.

Volume 2, 1982; volume 4, 1983; volume 5, 1985; volume 6, 1985.
Contains a wide range of statistical theory, methodology, and applications in the fields of natural sciences, engineering, and social sciences. Helpful to nonstatisticians as well as practicing statisticians.
R: *Science Technology Libraries* 5: 2 (Winter 1984).

Mathematics Encyclopedia. **Max S. Shapiro, ed.** Garden City, NY: Doubleday, 1977.

Covers all branches of mathematics. Includes tables, formulas, and symbols. Appropriate reference for high school students and teachers.
R: *School Science and Mathematics* 78: 362 (Apr. 1978); *LJ* 102: 1748 (Sept. 1977); *TBRI* 44: 212 (June 1978); *WLB* 52: 348 (Dec. 1977).

Universal Encyclopedia of Mathematics. Foreword by James R. Newman. New York: Simon and Schuster, 1969.

Based on *Meyers' Grossen Rechenduden,* 1964.
Topical and tabular coverage of fundamentals for secondary school and college students.
R: *Engineer* 217: 955 (May 1964); *Mathematics of Computation* 19: 164 (1965); *ARBA* (1970, p. 104); Jenkins (B39); Wal (p. 66); Win (EF19).

The VNR Concise Encyclopedia of Mathematics. **W. Gellert et al., eds.** New York: Van Nostrand Reinhold, 1977.

This comprehensive yet compact encyclopedia of mathematics covers all fields and includes historical biographies and photographs. A useful reference for libraries of all sizes.
R: *Science News* 111: 319 (May 1977); 114: 127 (Aug. 1978); *LJ* 103: 723 (Apr. 1, 1978); *WLB* 52: 84 (Sept. 1977).

PHYSICS

Concise Encyclopedia of Solid State Physics. **R. G. Lerner and G. L. Trigg, eds.** Reading, MA: Addison-Wesley, 1983.

This is a more concise version of 1981's *Encyclopedia of Physics*; it reduces the number of topics to 130 by selecting only articles dealing with solid-state physics.
R: *Applied Optics* 22: 1215 (Apr. 15, 1983).

Encyclopedia of Emulsion Technology. Basic Theory: **Paul Becher, ed.** New York: Dekker, 1983.

The Encyclopedia of Physics, 3d ed. **Robert M. Besancon, ed.** New York: Van Nostrand Reinhold, 1985.

First edition, 1966; second edition, 1974.
A major reference tool, containing over 350 articles, which has been designed to meet the needs of a broad range of users. Articles in the main divisions of physics are intended for readers with only general knowledge of the subject; those in the subdivision are aimed at a slightly more sophisticated audience; and articles in the finely divided areas presuppose considerable knowledge of physics and mathematics.
R: *Physics Today* 19: 97 (Oct. 1966); *Science* 152: 951 (1966); *Choice* 3: 495 (1966); *ARBA* (1975, p. 655); Jenkins (C48); Win (1EG1).

Encyclopedia of Physics. **Rita G. Lerner and George L. Trigg, eds.** Reading, MA: Addison-Wesley, 1980.

Articles cover a wide range of topics in physics and are written by authorities in their fields. Considered a major reference source.
R: *American Scientist* 69: 222 (Mar./Apr. 1981); *Nature* 290: 657 (Apr. 23, 1981).

Handbuch der Astrophysik. 7 vols. **A. von G. Eberhard et al.** Berlin: Springer-Verlag, 1928–1936.

Handbuch der Physik. **S. Flugge, ed.** Berlin: Springer-Verlag, 1974–76.

The major reference covering the entire field of physics. Each volume covers a specific aspect. Scholarly and highly technical articles intended for advanced researchers. Bibliographies at the end of each chapter are lengthy and helpful.
R: Sheehy (EG4).

Handbuch der Physik: Index to Titles of Volumes, Authors, and Titles of Articles and Subjects. Lexington, MA: Lincoln Laboratory Library, 1965.

McGraw-Hill Encyclopedia of Physics. **S. P. Parker, ed.** New York: McGraw-Hill, 1983.

An alphabetical arrangement of 760 articles from the *McGraw-Hill Encyclopedia of Science and Technology*. Includes the basic principles and recent advances from "Aberration (optics)" through "Zeeman effect." Drawings, graphs, charts, and photographs. For academic and larger public libraries.
R: *Science and Technology Libraries* 4: 120 (Winter 1983).

CHEMISTRY

Encyclopedia of Chemistry. 4th ed. **Douglas M. Considine, ed.** New York: Van Nostrand Reinhold, 1984.

Kingzett's Chemical Encyclopedia: A Digest of Chemistry and Its Industrial Applications. 9th ed. **Charles Thomas Kingzett.** New York: Van Nostrand Reinhold, 1966.

Eighth edition, 1952.
Standard 1-volume work gives brief digests of topics in various aspects of chemistry. Major headings are broken down into subheadings, facilitating access.
R: *LJ* 93: 1585 (1968); Jenkins (D69); Wal (p. 118); Win (ED24).

McGraw-Hill Encyclopedia of Chemistry. **Sybil P. Parker, ed.** New York: McGraw-Hill, 1980.

Contains 790 articles by 387 experts in fields of inorganic, organic, physical and analytical chemistry. There is a set of page references, cross references to topics, illustrations, photos, and line drawings in addition to equations. Comprehensive and current.
R: *Applied Optics* 22: 1534 (May 15, 1983).

Van Nostrand Reinhold Encyclopedia of Chemistry. 4th ed. **Douglas M. Considine, ed.** New York: Van Nostrand Reinhold, 1984.

First edition, 1956; second edition, 1966; third edition, 1973.
A comprehensive 1-volume tool with 1,300 alphabetically arranged entries with a detailed subject index. Intended for those readers with a general background in chemistry.
R: *WLB* 48: 264 (Nov. 1973); ARBA (1974, p. 557; 1985, p. 603).

SPECIFIC

Concise Encyclopedia of Biochemistry. New York: de Gruyter, 1983.

A translation and update of the German *Brockhaus ABC Biochemie* (1981). Dictionary arrangement of entries covering a wide range of the field of biochemistry. A valuable tool for all libraries interested in this subject.
R: *ARBA* (1985, p. 602).

The Encyclopedia of Biochemistry. **Roger John Williams and Edwin M. Lansford, Jr.** New York: Van Nostrand Reinhold, 1977.

Alphabetically arranged articles on broad topics.
R: *Bioscience* 18: 58 (1968); *Choice* 4: 644 (1967); *LJ* 92: 2553 (July 1967); *NTB* 52: 201 (1967); *RQ* 6: 200 (Summer 1967); Jenkins (D104); Wal (p. 201); Win (2EC8).

Encyclopedia of Electrochemistry of the Elements. Vols. 1–. **Allen J. Bard, ed.** New York: Dekker, 1973–.

Volume 13, *Organic Section*, 1979; volume 14, 1980.
An estimated 15 volumes will comprise this tool, which provides a methodical and critical review of the electrochemical behavior of chemical elements and their compounds.
R: *NTB* 60: 104 (Mar. 1975); *RSR* 3: 43 (Apr./June 1975); *Science* 192: 1256 (June 18, 1976); *ARBA* (1975, p. 651; 1976, p. 648; 1978, p. 641; 1979, p. 652).

Encyclopedia of the Alkaloids. 3 vols. **John Stephen Glasby.** New York: Plenum Press, 1975–1977.

Intended as a reference source for organic chemists.
R: *Chemistry in Britain* 14: 302 (June 1978); *Journal of the American Chemical Society* 98: 1062 (Feb. 18, 1976); 101: 1911 (Mar. 28, 1978); *TBRI* 42: 165 (May 1976); *ARBA* (1979, p. 653).

Merck Index: An Encyclopedia of Chemicals and Drugs. 9th ed. **Paul G. Stecher, ed.** Rahway, NJ: Merck, 1976.

Seventh edition, 1960; eighth edition, 1968.
Some 10,000 descriptions of individual substances, over 4,500 structural formulas, and 42,000 names of chemicals and drugs.
R: *Choice* 5: 1430 (1969); *RSR* 5: 11 (Apr./June 1977); Jenkins (D140); Sheehy (ED12); Wal (p. 476); Win (ED27; 2ED5).

BIOLOGICAL SCIENCES

The Encyclopedia of Microscopy and Microtechnique. **Peter Gray.** Melbourne, FL: Krieger, 1981.

Earlier edition, 1973.
Several articles on microscopy in the biological sciences. Liberal use of tables and formulas makes this a reliable source on microscopes and methods.
R: *American Scientist* 62: 490 (July/Aug. 1974); *Microchemical Journal* 19: 219 (June 1974); *Science* 184: 55–56 (Apr. 5, 1974); *TBRI* 40: 297 (Oct. 1974); *ARBA* (1974, p. 565).

The Encyclopedia of the Biological Sciences. 2d ed. **Peter Gray, ed.** Melbourne, FL: Krieger, 1981.

First edition, 1961; earlier edition, 1970.
Intended as a brief reference for biologists seeking information outside their field of specialty. Perhaps the only comprehensive 1-volume encyclopedia in the field. Contains several signal articles.
R: *Choice* 7: 818 (Sept. 1970); *Nature* 227: 208 (July 1970); *Science* 134: 93–94 (1961); *Subscription Books Bulletin* 58: 585 (1962); *ARBA* (1971, p. 493); Jenkins (G41); Wal (p. 192); Win (EC18; 3EC3).

ISI Atlas of Science: Biochemistry and Molecular Biology, 1978/80. Philadelphia: Institute for Scientific Information, 1981.

An encyclopedia collection of reviews, based on co-citation clustering, in which 1 author cites a pair of works by 2 other authors. Contains 102 sections, each with a summary of the review, a list of the core documents, a list of cited publications in *Science Citation Index*, and a chart of the research front. Previous knowledge helpful to students.
R: *ARBA* (1983, p. 616).

Reston Encyclopedia of Biomedical Engineering Terms. **Rudolf F. Graf and George J. Whalen.** Reston, VA: Reston Publishing, 1977.

Provides a combination of biomedical and engineering terms. Emphasis is on mechanical, electrical, and chemical engineering. Definitions are brief and nontechnical. For science libraries.

Botany

The Aquarium Encyclopedia. **Gunther Sterba and Dick Mills.** Translated by Susan Simpson. Cambridge, MA: MIT Press, 1982.

Provides information on all areas of aquaria including descriptions of the various orders and families of fish, invertebrates, and plants. Discusses the physiology and behavior of fish and includes other articles concerning filtering systems, fish diseases, etc. Comprehensive and illustrated enough for the novice as well as the professional to find useful.
R: *LJ* 109: 784 (Apr. 1984); *Choice* 20: 1570 (July/Aug. 1983).

Chilton's Encyclopedia of Gardening. **Martin Stangl.** Radnor, PA: Chilton, 1975.

An introductory reference that discusses the planning of a garden, choice of tools and plant foods, cures for diseases, and bug killers as well as the structure of plant tissues, fruits, and flowers.

Encyclopedia of American Forest and Conservation History. 2 vols. **Richard C. Davis, ed.** New York: Macmillan, 1983.

An authoritative reference tool with about 400 articles on the history of forestry conservation, forest industries, and other forest-related subjects in the United States. Illustrations and indexed. Useful appendixes. Recommended for school and academic libraries.
R: *Choice* 21: 1274 (May 1984); *WLB* 58: 594 (Apr. 1984); *ARBA* (1985, p. 514).

The Encyclopedia of Herbs and Herbalism. **Malcolm Stuart, ed.** New York: Grosset & Dunlap, 1979.

Written by authorities, this monumental work contains text and descriptions on individual herbs arranged alphabetically by botanical names. Also discusses the history of herbalism, medicinal uses, culinary uses and cultivation. Includes full-color photographs and glossary, bibliography, conversion tables, and indexes. Highly recommended for comprehensive library collections.
R: *Nature* 282: 649 (Dec. 6, 1979); *BL* 76: 1151 (Apr. 1, 1980); 76: 1565 (June 15, 1980); *ARBA* (1981, p. 657).

The Encyclopedia of Mushrooms. **Colin Dickinson and John Lucas, eds.** New York: Putnam's, 1979.

Text discusses habitat, biology, and life-styles of fungi and mushrooms and relates them to man and food. Numerous color illustrations and drawings add to the book's appeal. A second section is a guide for the identification of mushrooms. Contains an appended glossary, a brief bibliography, and indexes. Useful as a field guide and natural history reference for nonspecialists.
R: *New Scientist* 82: 745 (May 31, 1979); *RSR* 8: 16 (July/Sept. 1980); *ARBA* (1980, p. 634).

The Encyclopedia of Organic Gardening. Rev. ed. **Staff of *Organic Gardening Magazine*.** Emmaus, PA: Rodale Press, 1978.

First edition, 1959.

Updated edition of this standard work on organic gardening. Format includes extensive cross referencing, clearly presented line drawings, black-and-white photographs, and succinct textual material. Highly recommended.
R: *BL* 75: 1117 (Mar. 1, 1979); *LJ* 103: 1997 (Oct. 1, 1978); *ARBA* (1979, p. 758).

Encyclopedia of Plant Physiology. Vols. 1–. **A. Pirson and M. H. Zimmerman, eds.** New York: Springer-Verlag, 1975–.

Volume 1, *Transport in Plants 1: Phloem Transport*; volume 2, 1976.
R: *SL* 66: 2A (Mar. 1975).

Illustrated Encyclopedia of Indoor Plants. **Kenneth Beckett and Gillian Beckett.** Garden City, NY: Doubleday, 1976.

Concise information on 2,000 species of house plants. Describes growing conditions, methods of propagation. Includes line drawings and color illustrations. Arranged by scientific name with index by common name. A helpful reference.
R: *BL* 73: 790 (Feb. 1, 1977); *LJ* 102: 187 (Jan. 15, 1977); *ARBA* (1978, p. 752).

The Illustrated Encyclopedia of Succulents. **Gordon Rowley.** New York: Crown, 1978.

An outstanding work on succulent plants, including their ecology, evolution, systematics, and conservation. Well-referenced and illustrated with photographs. Recommended.
R: *LJ* 103: 2124 (Oct. 15, 1978).

The Illustrated Encyclopedia of Trees, Timbers, and Forests of the World. **Herbert L. Edlin et al.** New York: Crown, 1978.

Profusely illustrated color guide to over 250 tree species of the Northern Hemisphere, as well as some tropical and desert trees. Brings together information on the forest, as well as individual species of trees and their uses for timber. Contributions from experts. Each illustration includes tree, fruit and winter twigs. Recommended.
R: *BL* 77: 590 (Dec. 15, 1980); *LJ* 104: 638 (Mar. 1, 1979); *WLB* 53: 525 (Mar. 1979); *ARBA* (1981, p. 661).

The Illustrated Reference on Cacti and Other Succulents. Vol. 5. **Edgar Lamb and Brian Lamb.** Poole, England: Blandford Press. Distr. New York: Sterling Publishing, 1978.

An encyclopedic volume consisting of some 300 photographs of the world's succulents. Provides detailed growing instructions and a cumulative index of all genera. Suitable for hobbyists as well as botanists.
R: *ARBA* (1980, p. 627).

The Marshall Cavendish Illustrated Encyclopedia of Gardening. 20 vols. **Peter Hunt, ed; Edwin F. Steffek, American ed.** New York: Marshall Cavendish, 1977 [c1968–1970].

Arranged both alphabetically and seasonally. An illustrative directive encyclopedia geared toward the serious gardener. Clear color illustrations, cross-referenced.
R: *BL* 74: 1755 (July 15, 1978); *ARBA* (1979, p. 759).

The New York Botanical Garden Illustrated Encyclopedia of Horticulture. 10 vols. **Thomas H. Everett.** New York: Garland STPM Press, 1980.

Volume 1, *A–Be.*

A standard reference set on horticulture. Contains some 20,000 descriptions of plant species and varieties with 900 other entries on additional topics. Entries are arranged by scientific name and provide information on habitation, cultivation, landscaping, diseases, and pests. Includes 2,500 cross references and discussions of 260 plant families. Illustrated with more than 10,000 photographs. An indispensable work for botanists and public or academic libraries.

R: *Horticulture* 58: 9 (Dec. 1980); *ARBA* (1981, p. 744).

The Oxford Encyclopedia of Trees of the World. **Bayard Hora, ed.** New York: Oxford University Press, 1981.

A comprehensive work dealing with the types and life cycles of trees. Contains entries for 149 genera and covers over 350 species. Discusses tree botany, forest ecology, diseases, and identification. Well-illustrated. Includes glossary, bibliography, and common and scientific indexes. Valuable addition to public and academic libraries.

R: *BL* 77: 1324 (June 15, 1981); *ARBA* (1983, p. 631).

The Pocket Encyclopedia of Cacti and Succulents in Color. **Edgar Lamb and Brian Lamb.** England: Blandford, 1980.

Early edition, 1970.
Information only for the most inexperienced novices.
R: *ARBA* (1971, p. 501).

The Pocket Encyclopedia of Modern Roses. **Tony Gregory.** Poole, England: Blandford Press. Distr. New York: Sterling Publishing, 1984.

A popular tool that includes about 300 varieties of roses with color photographs and descriptions that include information on the raiser, the date of introduction, flower habitat, color, and cultivation requirements.
R: *ARBA* (1985, p. 502).

Popular Encyclopedia of Plants. **V. H. Heywood and S. R. Chant, eds.** New York: Cambridge University, 1982.

Over 2,200 articles describing ornamental and commercial plants. Also includes timber species, lower plants, and more. Many illustrations.
R: *Choice* 19: 849 (Mar. 1982); *ARBA* (1983, p. 621).

Reader's Digest Encyclopedia of Garden Plants and Flowers. **The Reader's Digest Association, ed.** London: Reader's Digest Association. Distr. New York: Norton, 1977 [c1975].

Arranged alphabetically by genus, this is an authoritative encyclopedia of gardening dealing with more than 3,000 plants. Color photographs contribute to making this a worthwhile encyclopedia.
R: *BL* 74: 640 (Dec. 1, 1977); *LJ* 102: 1004 (May 1, 1977); *ARBA* (1978, p. 755).

Rodale's Encyclopedia of Indoor Gardening. **Anne M. Halpin, ed.** Emmaus, PA: Rodale Press, 1980.

An encyclopedia of indoor gardening divided into 2 parts: the first part discusses the care of plants—diseases and pests, artificial lighting, and carnivorous plants; the second part describes 250 plants. Plants are listed alphabetically by scientific name and include care instructions. Provides extensive color photographs, black-and-white line drawings, and tables. Indexed. An invaluable reference for home use.
R: *ARBA* (1981, p. 745).

Secondary Plant Products. **E. A. Bell and B. V. Charlwood, eds.** New York: Springer-Verlag, 1980.

As part of the *Encyclopedia of Plant Physiology* series, this volume describes various secondary plant products. Discusses history, metabolism, and expression. Includes an extensive list of references as well as author, subject, and species indexes. Of interest to plant physiologists.
R: *ARBA* (1981, p. 647).

Tanaka's Cyclopedia of Edible Plants of the World. **Tyozaburo Tanaka.** Tokyo: Keigaku Publishing, 1976.

The largest compendium to date on edible plants, including information on more than 10,000 wild and cultivated plants. Provides insight into the biology and anthropology of plants and plant resources.
R: *RSR* 5: 8 (Apr./June 1977).

Wyman's Gardening Encyclopedia. Rev. ed. **Donald Wyman.** New York: Macmillan, 1977.

First edition, 1971.
An indispensable tool, giving information on nearly every aspect of gardening. Contains numerous illustrations, color photographs, most recent name changes of plants. Well-recommended.
R: *BL* 74: 238 (Sept. 15, 1977); *RSR* 7: 13 (July/Sept. 1979); *ARBA* (1978, p. 756); Sheehy (EL1).

ZOOLOGY

The Audubon Society Encyclopedia of Animal Life. **John Farrand, Jr.** New York: Crown, 1982.

Traces the evolution of the 33 phyla and 7 classes of the phylum Chordata with terms, charts, descriptions, and photos of each. Useful for reference or circulation in libraries.
R: *LJ* 107: 2329 (Dec. 1982); *ARBA* (1983, p. 632).

The Audubon Society Encyclopedia of North American Birds. **John K. Terres.** New York: Alfred A. Knopf, 1980.

A monumental work containing 6,000 alphabetized and cross-referenced entries, 1,675 illustrations and full-color photographs, 770 definitions, a bibliography of 4,000 articles and monographs, 126 biographies of famous naturalists. Profiles of about 850 birds provide nomenclature, physical descriptions, feeding habits, and other useful data. Highly recommended to all librareis.
R: *Nature* 290: 658 (Apr. 23, 1981); *Choice* 18: 1078 (Apr. 1981); *RSR* 9: 104 (Jan./Mar. 1981); *WLB* 55: 782 (June 1981).

Butterfly and Angelfishes of the World. Vol. 2. **G. R. Allen.** New York: Wiley, 1980.

Volume 2, *Atlantic Ocean, Caribbean Sea, Red Sea, Indo-Pacific.*
An illustrated reference to the butterfly and angelfishes of the world. In 2 volumes.

The Complete Encyclopedia of Horses. **M. E. Ensminger.** Cranbury, NJ: Barnes, 1977.

This superior encyclopedia of horses covers a broad range of topics including horses and the law, diseases, and nutrition. Useful tables, illustrations, and data are provided. Well-recommended.
R: *BL* 75: 824 (Jan. 1979); *WLB* 52: 648 (Apr. 1978); *ARBA* (1978, p. 742).

Dolphins, Whales, and Porpoises: An Encyclopedia of Sea Mammals. **David J. Coffey.** New York: Macmillan, 1977.

Material arranged by 3 major subject headings: Cetacea, Pinnipedia, and Sirenia. Discussions contain information on anatomy, physiology, and behavior. An illustrated source of ready reference.
R: *Limnology and Oceanography* 22: 1105 (Nov. 1977); *Oceans* 10: 70 (July 1977); *Sea Frontiers* 24: 57 (Jan.–Feb. 1978); *BL* 74: 403 (Oct. 15, 1977); *LJ* 102: 1002 (May 1, 1977); *TBRI* 43: 333 (Nov. 1977); 44: 52 (Feb. 1978); 44: 92 (Mar. 1978); *WLB* 52: 264 (Nov. 1977); *ARBA* (1978, p. 681).

Encyclopedia of Animal Care. 12th ed. **Geoffrey P. West, ed.** Baltimore: Williams & Wilkins, 1977.

A standard veterinary reference; updated and cross-referenced. Emphasizes first aid and preventive medicine.
R: *ARBA* (1978, p. 741).

Encyclopedia of Animals. **Jan Hanzak, Zdenek Veselovsky, and David Stephen.** Edited by Tom McCormick. New York: St. Martin's Press, 1979.

All living mammals arranged by 19 orders and subdivided by families. Aimed at the high school and small-college audience.
R: *BL* 71: 1133–34 (July 1, 1975); *Choice* 11: 1606 (Jan. 1975); *ARBA* (1976, p. 675).

The Encyclopedia of Aquarium Fishes in Color. **David J. Coffey.** New York: Arco Publishing, 1977.

A nicely illustrated, descriptive encyclopedia of fishes, aquarium equipment, fish diseases, etc. Covers over 200 species. Contains a wealth of information, excellent photographs. Well-recommended.
R: *BL* 74: 1135 (Mar. 1, 1978); *Choice* 14: 1621 (Feb. 1978); *LJ* 102: 1624 (Sept. 1977); *WLB* 52: 265 (Nov. 1977).

Encyclopedia of Aviculture: Keeping and Breeding Birds. **Richard Mark Martin.** New York: Arco Publishing, 1983.

Some 450 short entries related to the care and breeding of captive birds in a handbook-size backup source of reference.
R: *ARBA* (1984, p. 658).

The Encyclopedia of Mammals. **David Macdonald, ed.** New York: Facts on File, 1984.

An important guide to mammals in over 700 entries. Color illustrations, charts, maps, and bibliographies. Indexing is adequate. A useful reference for both general readers and zoologists.
R: *ARBA* (1985, p. 535).

Encyclopedia of Marine Invertebrates. **Jerry G. Walls, ed.** Neptune, NJ: T. F. H. Publications, 1982.

Fifteen chapters arranged phylogenetically, from Protozoa to Cephalochordata, of the various marine invertebrates suitable to be kept in aquariums. A bit unbalanced for the nonspecialist with little or no knowledge of zoology.
R: *ARBA* (1984, p. 669).

The Encyclopedia of North American Wildlife. **Stanley Klein.** New York: Facts on File, 1983.

Entries include mammals, birds, reptiles, amphibians, and fish with explanations and illustrations suitable for the general audience rather than the professional.
R: *ARBA* (1984, p. 655).

Encyclopedia of the Animal World. Sydney, Australia: Bay Books. Distr. Freeport, NY: Cavendish, 1980.

Has some 3,000 entries and many illustrations listing scientific names, family, order, class, physiology, and biochemistry aspects of each animal. Arranged alphabetically and written by specialists in this field as a guide to the animal kingdom.
R: *ARBA* (1982, p. 730).

The Encyclopedia of the Horse. **C. E. G. Hope and G. N. Jackson.** England: Ebury, 1981.

Early edition, 1973.
Popular account of breeding and sporting events. Many color illustrations.
R: *LJ* 98: 2560 (Sept. 15, 1973); *WLB* 48: 265 (Nov. 1973); *ARBA* (1974, p. 652).

Grzimek's Animal Life Encyclopedia. 13 vols. **Bernhard Grzimek.** New York: Van Nostrand Reinhold, 1972–1975.

Earlier edition in German, 1967.
The set is arranged by animal groups, with the material in each volume arranged by animal orders and families. Includes a systematic classification index and multilingual glossary.
R: *American Scientist* 61: 486–487 (July/Aug. 1973); 62: 484 (July/Aug. 1974); *Science* 177: 1184 (Sept. 29, 1972); *ARBA* (1976, p. 675; 1975, p. 689; 1973, p. 569); Sheehy (EC41).

Grzimek's Encyclopedia of Ethology. **Bernhard Grzimek, ed.** New York: Van Nostrand Reinhold, 1977.

An encyclopedia of animal behavior; information contributed by field experts is considered specialized and technical. Covers such topics as parental behavior, the language of bees, and stress in mammals. A comprehensive reference work.

R: *Condor* 79: 510 (Winter 1977); *BL* 75: 317 (Oct. 1, 1978); *LJ* 102: 1626 (Aug. 1977); *TBRI* 44: 176 (May 1978); *WLB* 52: 186 (Oct. 1977); 52: 503 (Feb. 1978); *ARBA* (1978, p. 670).

Grzimek's Encyclopedia of Evolution. **Bernhard Grzimek, ed.** New York: Van Nostrand Reinhold, 1977.

A companion volume to *Grzimek's Animal Life Encyclopedia.* Provides a historical background into biological evolution. Twenty-three articles by over 200 distinguished contributors. Includes illustrations and bibliographies. A standard reference for public, college, and university libraries.

R: *BioScience* 28: 124 (Feb. 1978); *BL* 74: 1128 (Mar. 1, 1978); *Choice* 14: 1480 (Jan. 1978); *LJ* 102: 1480 (July 1977); *TBRI* 44: 165 (May 1978); *WLB* 52: 186 (Oct. 1977); *ARBA* (1978, p. 652); Sheehy (EC18).

The Illustrated Encyclopedia of Birds: All the Birds of Britain and Europe in Color. 5 vols. **John Gooders, ed.** New York: Marshall Cavendish, 1979.

Volume 1, *Birds of Ocean and Estuary*; volume 2, *Birds of Mountain and Moorland*; volume 3, *Birds of Marsh and Shore*; volume 4, *Birds of Heath and Woodland*; volume 5, *Birds of Hedgerow and Garden.*

Translated from the Italian, this edition deals with 642 thoroughly discussed and illustrated species of birds. Species are listed by colloquial or common name in English, French, Spanish, Italian, German, and Latin. Details identification, habitats, food, reproduction, subspecies, and distribution. Includes distribution maps and color photographs and drawings depicting birds in flight. Species are categorized under families in 1 of 27 orders. Contains an index of bird names listing scientific and common names as well as a comprehensive index of all scientific names and topics. An indispensable reference tool for academic, research, special, and public libraries.

R: *BL* 77: 414 (Nov. 1, 1980); *LJ* 105: 188 (Jan. 15, 1980); *WLB* 54: 626 (Apr. 1980); *ARBA* (1981, p. 667).

Inside the Animal World: An Encyclopedia of Animal Behavior. **Maurice Burton and Robert Burton.** New York: Quadrangle Books/New York Times, 1977.

Written by 2 prominent zoologists. Discusses the behavioral responses of vertebrates.
R: *BL* 74: 637 (Dec. 1, 1977).

The Larousse Encyclopedia of the Animal World. New York: Larousse, 1975.

Earlier edition, 1967.

An illustrated reference that classifies animals from simple unicellular life to complex animals. Chapters examine a specific phylum and discuss structure, communication, reproduction, habitat, feeding, etc. A succinct source of information suitable for public or academic libraries. Indexed.

R: *Science* 158: 898 (1967); *LJ* 92: 3626 (Oct. 15, 1967); *ARBA* (1977, p. 665); Jenkins (G198); Wal (p. 224); Win (2EC15).

Macmillan Illustrated Animal Encyclopedia. **Philip Whitfield, ed.** New York: Macmillan, 1984.

A simple guide to the 1,925 species of the vertebrates. Worldwide coverage with color illustrations, paintings, and an index of English and scientific names. Intended for general readers.
R: *ARBA* (1985, p. 523).

Marine Life: An Illustrated Encyclopedia of Invertebrates in the Sea. **John David George and Jennifer J. George.** New York: Wiley, 1979.

A comprehensive encyclopedia of the biology and ecology of marine invertebrates. Species are arranged phylogenetically from sponges to lancelots; lists Latin names and scientific terms. Contains 1,300 color photographs and line drawings, as well as bibliography, glossary, and index to text and illustrations. Helpful to undergraduate zoology students and amateur naturalists.
R: *RSR* 8: 16 (July/Sept. 1980); 8: 86 (July/Sept. 1980); *ARBA* (1981, p. 675); Wal. p. 166.

The New Larousse Encyclopedia of Animal Life. rev. ed. New York: Larousse, 1980.

Earlier edition, 1967.
Contains new illustrations and information on living organisms in their natural habitats. Includes adequate but not in-depth accounts of invertebrate phyla and other types of invertebrates. Takes a somewhat conservative view on the classification of some groups.
R: *ARBA* (1982, p. 730).

The Ocean World of Jacques Cousteau. Rev. ed. 20 vols. Danbury, CT: World Publishing, 1975.

A multivolume set that deals with marine life and exploration. Includes color illustrations, maps, and diagrams.
R: *BL* 73: 742 (Jan. 15, 1977).

Wildlife in Danger. **James Fisher, Noel Simon, and Jack Vincent.** New York: Viking Press, 1969.

Encyclopedic coverage of plants and animals in danger of extinction. Emphasizes bird and mammal species.
R: Wal (p. 64); Win (3EC15).

EARTH SCIENCES

The Cambridge Encyclopedia of Earth Sciences. **David G. Smith, ed.** New York: Crown and New York: Cambridge University, 1982.
R: *ARBA* (1983, p. 662).

Clouds of the World: A Complete Color Encyclopedia. **Richard S. Scorer.** Harrisburg, PA: Stackpole Books, 1972.

An informative text combined with 337 color photographs make this tool a reliable and aesthetic reference.

R: *American Scientist* 61: 360–361 (May/June 1963); *WLB* 48: 613 (Mar. 1973); *ARBA* (1974, p. 607).

Color Encyclopedia of Gemstones. **Joel E. Arem.** New York: Van Nostrand Reinhold, 1977.

Arranged alphabetically. Lists mineral species and gemological data. Includes 210 entries, numerous illustrations, bibliography, and a refractive index. Recommended for public libraries.

R: *BL* 76: 1698 (July 15, 1980); *LJ* 103: 673 (Mar. 15, 1978); *WLB* 52: 651 (Apr. 1978); *ARBA* (1979, p. 703).

The Discoverers: An Encyclopedia of Explorers and Exploration. **Helen Delpar, ed.** New York: McGraw-Hill, 1980.

Contains information on geographical discoveries and explorers. Articles contain illustrations and bibliographies. Comprehensive, for school, public, and college libraries.

R: *WLB* 54: 396 (Feb. 1980).

Encyclopedia of Earth Sciences. Vols. 1–. **Rhodes Whitmore Fairbridge, ed.** New York: Van Nostrand Reinhold, 1966–.

Volume 1, *Encyclopedia of Oceanography*, 1966; volume 2, *Encyclopedia of Atmospheric Sciences and Astrogeology*, 1967; volume 3, *Encyclopedia of Geomorphology*, 1968; volume 4a, *Encyclopedia of Geochemistry and Environmental Sciences*, 1972; volume 6, *Encyclopedia of Sedimentology*, 1978; volume 7, *Encyclopedia of Paleontology*, 1979; volume 8a, *Encyclopedia of World Regional Geology, Part I: Western Hemisphere (Including Antarctica and Australia)*, 1975; volume 12, *Encyclopedia of Soil Science, Part I: Physics, Chemistry, Biology, Fertility, and Technology*, 1979; volume 13, *Encyclopedia of Applied Geology*, 1984.

Each volume of this encyclopedic set will be autonomous and will discuss a different aspect of earth science.

R: *Science* 165: 53 (July 4, 1969); *Choice* 4: 100 (1968); *ARBA* (1971, p. 518; 1985, p. 604); Jenkins (E25); Wal (p. 158); Win (2EE3; 2EE4). See also reviews under each volume entry.

The Encyclopedia of Geochemistry and Environmental Sciences. Encyclopedia of Earth Sciences Series, vol. 4a. **Rhodes Whitmore Fairbridge, ed.** New York: Academic Press, 1972.

Examines the important trends in environmental science. All entries include relevant chemical and mathematical notations, illustrations, etc.

R: *American Scientist* 63: 462 (July/Aug. 1975); *Science* 178: 1277 (Dec. 22, 1972); *ARBA* (1973, p. 587).

The Encyclopedia of Mineralogy. **Keith Frye, ed.** Stroudsburg, PA: Hutchinson Ross, 1981.

The Encyclopedia of Minerals and Gemstones. **Michael O'Donoghue, ed.** New York: Putnam, 1976.

Clearly written and well illustrated. Describes over 1,000 minerals. For the nonspecialist. Contains tables, bibliographies, glossary, index.

R: *Gems and Minerals* 473: 58 (Mar. 1977); *BL* 73: 1443 (May 15, 1977); *LJ* 101: 2359 (Nov. 15, 1976); *TBRI* 43: 168 (May 1977); *WLB* 51: 686 (Apr. 1977); *ARBA* (1977, p. 695).

The Encyclopedia of Paleontology. Encyclopedia of Earth Sciences Series, vol. 7. **Rhodes Whitmore Fairbridge and David Jablonski, eds.** Stroudsburg, PA: Dowden, Hutchinson & Ross. Distr. New York: Academic Press, 1979.

Alphabetically arranged entries cover over 100 topics pertaining to paleontology. Written by experts in the field, detailed summaries are enhanced by illustrations and references. Provides current ideas on animal and plant groups, plate tectonics, systematic philosophy. Valuable to students of paleontology. For public and university libraries.

R: *American Scientist* 68: 450 (July/Aug. 1980); *Nature* 284: 646 (Apr. 17, 1980); *Choice* 17: 696 (July/Aug. 1980); *ARBA* (1981, p. 643).

The Encyclopedia of Sedimentology. Encyclopedia of Earth Sciences Series, vol. 6. **Rhodes Whitmore Fairbridge and Joanne Bourgeois, eds.** Stroudsburg, PA: Dowden, Hutchinson & Ross. Distr. New York: Academic Press, 1978.

Comprehensive, well-written volume covering a wide range of topics in the field of sedimentology. Contains short but detailed articles on current issues such as carbonate sediments, canyons, crude-oil composition, and lunar sedimentology. Provides references, diagrams, and an index. Highly recommended to geologists, social scientists, archaeologists, and students. Belongs in academic and science library collections.

R: *Choice* 16: 202 (Apr. 1979); *ARBA* (1980, p. 655); Sheehy (EE4).

The Encyclopedia of Soil Science, Part I: Physics, Chemistry, Biology, Fertility, and Technology. Encyclopedia of Earth Sciences Series, vol. 12. **Rhodes Whitmore Fairbridge and Charles W. Finkl, Jr., eds.** Stroudsburg, PA: Dowden, Hutchinson & Ross. Distr. New York: Academic Press, 1979.

An interdisciplinary work that brings together a range of topics in soil science. Entries summarize current as well as historical material in physics, biology, chemistry, plant nutrition. Well illustrated, cross-referenced, and indexed. A bibliography of major references in soil science is also included. A thorough concise compilation recommended to students and practitioners.

R: *Choice* 17: 48 (Mar. 1980); *ARBA* (1981, p. 695).

The Encyclopedia of World Regional Geology, Part I: Western Hemisphere (Including Antarctica and Australia). Encyclopedia of Earth Sciences Series, vol. 8a. **Rhodes Whitmore Fairbridge, ed.** New York: Academic Press, 1975.

Consists of over 150 entries from specialists, including geological maps. Considers the regions of the world on a sliding scale of magnitude from continents to regions and countries and finally to overseas territories and isolated islands. Profusely illustrated and cross-indexed in 3 different ways: in the alphabetical sequence, at the end of each entry, and in the body of the text.

R: *American Scientist* 64: 443 (July/Aug. 1976); *Nature* 261: 174 (May 13, 1976); *ARBA* (1977, p. 691); Sheehy (EE5).

Encyclopedia of World Regional Geology. Part 3. **Rhodes Whitmore Fairbridge, ed.** New York: Van Nostrand Reinhold, 1985.

The Illustrated Encyclopedia of the Mineral Kingdom. **Alan Woolley, ed.** New York: Larousse, 1978.

An authoritative text, written for mineral collectors. Information is up-to-date and includes color photographs. A comprehensive reference for the layperson.
R: *Journal of Gemmology* 16: 281 (Oct. 1978); *Mineralogical Magazine* 42: 416 (Sept. 1978); *BL* 75: 1390 (May 1, 1979); *LJ* 103: 1497 (Aug. 1978); *TBRI* 45: 9 (Jan. 1979); 45: 94 (Mar. 1979); *ARBA* (1979, p. 705).

McGraw-Hill Encyclopedia of the Geological Sciences. **Daniel N. Lapedes, ed.** New York: McGraw-Hill, 1978.

A well-produced, illustrated and comprehensive reference on the geological sciences. Contains over 500 signed articles that include helpful bibliographies. Material included is drawn heavily from the *McGraw-Hill Encyclopedia of Science and Technology.* This volume is recommended, therefore, to large libraries with a strong subject collection.
R: *American Scientist* 67: 354 (May/June 1979); *Chemical Engineering* 86: 12 (Mar. 26, 1979); *Earth Science* 31: 234 (Autumn 1978); *Geophysics* 44: 1306 (July 1979); *BL* 76: 1083 (Mar. 15, 1980); *LJ* 104: 179 (Jan. 15, 1979); *TBRI* 45: 127 (Apr. 1979); 45: 247 (Sept. 1979); *WLB* 53: 473 (Feb. 1979); *ARBA* (1979, p. 703).

The Planet We Live On: Illustrated Encyclopedia of the Earth Sciences. **Cornelius S. Hurlbut, Jr., ed.** New York: Abrams, 1976.

An excellent single-volume encyclopedia. Articles contributed by authorities in their field reflect the high quality of this volume. Articles are illustrated with photographs and line drawings. Recommended highly to all libraries and to professionals on any level.
R: *School Science and Mathematics* 76: 452 (May–June 1977); *BL* 73: 1293 (Apr. 15, 1977): *LJ* 101: 2268 (Nov. 1, 1976); *RSR* 5: 15 (July/Sept. 1977); *TBRI* 43: 245 (Sept. 1977); *WLB* 51: 263 (Nov. 1976; *ARBA* (1977, p. 690); Sheehy (EE2).

Rainbow Prehistoric Life Encyclopedia. **Mark Lambert.** Edited by Adrian Sington. Chicago, IL: Rand McNally, c1981.

Presents a narration of the evolution of life on earth from the Big Bang to the end of the Mesolithic period with major emphases on reptiles and mammals. For children and young adults with some sense of science knowledge.
R: *ARBA* (1984, p. 688).

OCEANOGRAPHY

The Encyclopedia of Beaches and Coastal Environments. **Maurice L. Schwartz, ed.** Stroudsburg, PA: Hutchinson Ross Publishing. Distr. New York: Van Nostrand Reinhold, 1982.

Encyclopedia of Earth Sciences, Volume 15.
A valuable reference tool covering over 500 subject areas about coastal areas. Useful for all types of libraries and users.
R: *Choice* (July/Aug. 1983, pp. 1573–1574); *ARBA* (1984, p. 681).

McGraw-Hill Encyclopedia of Ocean and Atmospheric Sciences. **Sybil P. Parker, ed.** New York: McGraw-Hill, 1980.

An updated and revised work concerning various aspects of ocean and atmospheric sciences. Designed as a companion to the *McGraw-Hill Encyclopedia of the Geological Sciences*, this volume contains over 230 alphabetically arranged articles, many taken from the *McGraw-Hill Encyclopedia of Science and Technology*. Provides over 500 photographs, maps, and graphs to supplement the text. Includes a list of 200 contributors, a detailed subject index, and cross-references. Essential for academic and public libraries.
R: *Marine Engineering/Log* 84: 123 (Dec. 1979); *Physics Today* 33: 83 (Jan. 1980); *BL* 77: 410 (Nov. 1, 1980); *Choice* 17: 520 (June 1980); *LJ* 105: 395 (Feb. 1, 1980); *RQ* 19: 307 (Spring 1980); *TBRI* 46: 47 (Feb. 1980); *WLB* 54: 464 (Mar. 1980); *ARBA* (1981., p. 690).

Ocean World Encyclopedia. **Donald G. Groves and Lee M. Hunt.** New York: McGraw-Hill, 1980.

An encyclopedia of 425 alphabetically arranged articles dealing with aspects of oceanography and oceans. Contains 25 biographical entries, extensive cross-references, detailed index, and excellent black-and-white photographs. Nontechnical terminology makes the volume useful for high school, public, and academic libraries.
R: *BL* 77: 841 (Feb. 15, 1981); *LJ* 105: 1292 (June 1, 1980); *RQ* 19: 395 (Summer 1980); *WLB* 54: 670 (June 1980); *ARBA* (1981, p. 694).

Rand McNally Encyclopedia of World Rivers. Chicago, IL: Rand McNally, 1980.

Comprehensive coverage of 1,750 of the world's rivers. Information provided includes source, length, physical features, agriculture, flora and fauna, industrial activity, and history. Contains over 500 illustrations and color maps with more than 140 color photographs. Essential for all public academic libraries.
R: *BL* 77: 176 (Oct. 1, 1980).

GENERAL ENGINEERING

Encyclopedia of Engineering. **Sybil P. Parker, ed.-in-chief.** New York: McGraw-Hill, 1983.

A handy reference guide to engineering principles and practices.
R: *Mechanical Engineering* 105: 94 (Apr. 1983).

McGraw-Hill Encyclopedia of Engineering. New York: McGraw-Hill, 1983.

Contains about 700 articles that analyze and explain the major engineering disciplines and current technology.
R: *Petroleum Engineer International* 55: 158 (May 1983).

AERONAUTICAL AND ASTRONAUTICAL ENGINEERING

The Complete Illustrated Encyclopedia of the World's Aircraft. **David Mondey, ed.** New York: A & W Publishers, 1978.

Contains a history of aircraft and aviation. Also contains a list of manufacturers with photographs and descriptions of major aircraft. Highly recommended.
R: *LJ* 104: 1441 (July 1979); *ARBA* (1980, p. 724).

Encyclopedia of Aircraft. **Michael J. H. Taylor and John W. R. Taylor, eds.** New York: Putnam, 1978.

Alphabetically arranged articles on the history of aircraft from the Wright brothers forward. Illustrated by both black-and-white and color illustrations. Up to date through late 1977. Contains technical data, tables, glossary.
R: *BL* 76: 147 (Sept. 15, 1979); *Choice* 15: 1646 (Feb. 1979); *LJ* 104: 121 (Jan. 1, 1979); *ARBA* (1980, p. 725).

Encyclopedia of Aviation. New York: Scribner's, 1977.

A general encyclopedia with about 700 articles arranged alphabetically. Well-illustrated, serves as a ready reference for nontechnical aspects of aviation.
R: *LJ* 102: 2421, 2423 (Dec. 1, 1977); *RQ* 17: 358 (Summer 1978); *WLB* 52: 586 (Mar. 1978); *ARBA* (1978, p. 761).

The Illustrated Encyclopedia of Aviation. Reference ed. 20 vols. **Anthony Robinson, ed.** New York: Cavendish, 1979.

First edition, entitled *Wings*, 20 volumes, 1977.
Presented in magazine format, volumes cover aviation history and are extensively illustrated. Entries are organized alphabetically under broad subject categories such as war in the air, famous aeroplanes, and fighting airmen. Articles vary in length and contain technical and biographical data. Includes index. An outstanding compendium of information that is essential for all aviation collections.
R: *WLB* 54: 590 (May 1980); *ARBA* (1981, p. 765).

The Illustrated Encyclopedia of General Aviation. **Paul Garrison.** Blue Ridge Summit, PA: TAB Books, 1979.

A comprehensive encyclopedic work that provides explanations of terms relating to general aviation. Includes numerous drawings and photographs. For all reference and aviation collections.
R: *ARBA* (1981, p. 765).

The International Encyclopedia of Aviation. **David Mondey, ed.** New York: Crown, 1977.

Text divided into 6 major areas: origins and development; military aviation; rocketry and space; etc. Contains more than 1,200 illustrations. Considered an excellent publication, with contributions from experts in their fields.
R: *LCIB* 37: 264 (Apr. 21, 1978); *RQ* 17: 358 (Summer 1978).

CHEMICAL ENGINEERING

Chemical Technology: An Encyclopedic Treatment; The Economic Application of Modern Technological Development Based upon a Work Originally Devised by the Late **J. F. van Oss.** Vols. 1–. New York: Barnes & Noble, 1968–.

Volumes 7–8, 1975.
Published in Great Britain under the title *Materials and Technology*. Currently 7 of 8 volumes are completed on all aspects of chemical technology. Intended for both the layperson and the technologist.

R: *Journal of Geology* (Dec. 1971); *Choice* 7: 49 (1970); 10: 594 (June 1973); *LJ* 94: 2458 (June 15, 1969); *ARBA* (1970, p. 160; 1971, p. 544; 1974, p. 671; 1977, p. 755); Jenkins D63; Sheehy (EJ6); Win (3ED5).

The Encyclopedia of Chemical Electrode Potentials. **Marvin S. Antelman, ed.** New York: Plenum Press, 1981.

An encyclopedia divided into 5 parts covering chemical electrode potentials. Also contains complex formative EMF data calculated from Nernst equation thermodynamics.

Encyclopedia of Chemical Processing and Design. Vols. 1–. **John J. McKetta and William A. Cunningham, eds.** New York: Dekker, 1976–.
Volume 13, 1981, volume 24, 1985.
A multivolume encyclopedia that serves as a comprehensive reference source covering chemical processes, methods, practices, and standards in chemical industries. An excellent addition for both university and industrial libraries.
R: *Chemistry and Industry* 16: 624 (Aug. 19, 1978); *RSR* 6: 45 (Apr. 6, 1978); *TBRI* 43: 75 (Feb. 1977); 44: 31 (Jan. 1978); *ARBA* (1983, p. 718).

Encyclopedia of Emulsion Technology. Vol. 1. Basic Theory **Paul Becher, ed.** New York: Dekker, 1983.

A collection of various topics by leading authorities in the field of emulsion technology. Well presented.
R: *ARBA* (1984, p. 634).

Encyclopedia of Surfactants. Vols. 1–. **Michael Ash and Irene Ash.** New York: Chemical Publishing, 1980–.

First of a 3-volume set covering US and foreign manufacturers of this product. Useful for libraries serving this field.
R: *ARBA* (1982, p. 711).

Kirk-Othmer Concise Encyclopedia of Chemical Technology. New York: Wiley, 1985.

A self-contained encyclopedia of chemical technology in 1 volume with access to information contained in the 26-volume third edition of the *Kirk-Othmer Encyclopedia of Chemical Technology.* For both students and specialists.

Kirk-Othmer Encyclopedia of Chemical Technology. 3d ed. 26 vols. **Martin Grayson et al., eds.** New York: Wiley, 1978–1984.

Second revised edition, 1963–1971; third edition: volumes 1–4, *A - Cardiovascular Agents,* 1978; volumes 5–8, *Castor Oil—Emulsions,* 1979; volumes 9–12, *Enamels, Porcelain or Vitreous—Hydrogen Energy,* 1980; volumes 13–16, *Hydrogen-Ion Activity—Perfumes,* 1981; volumes 17–19, *Peroxides and Peroxy Compounds, Inorganic—Recycling (Rubber),* 1982; volumes 20–23, *Refractories—Vinyl Polymers,* 1983; volume 24, *Vitamin—Zone Refining,* 1984; volume 25, *Supplement,* 1984; volume 26, *Index,* 1984.
An authoritative multivolume encyclopedia completely revised and updated. Comprehensive information reflects most recent chemical technology to date. Signed articles; edition uses CAS Registry Numbers and SI units.
R: *British Chemical Engineering* 10: 50 (Jan. 1965); *Chemical Engineering* 85: 11 (Oct. 9, 1978); 86: 11 (Dec. 31, 1979); *Journal of the American Chemical Society* 101: 2255 (Apr.

11, 1979); 101: 7136 (Nov. 7, 1979); 102: 892 (Jan. 16, 1980); 102: 4284 (June 4, 1980): 102: 5135 (July 16, 1980); 102: 6391 (Sept. 24, 1980); 102: 2144 (Apr. 22, 1981); *Laboratory Practice* 104: 4035 (July, 1982); *Journal of the American Oil Chemists' Society* 56: 108A (Feb. 1979); 27: 729 (Sept. 1978); 28: 405 (Apr. 1979); 28: 465 (Apr. 1979); *New Scientist* 79: 213 (July 20, 1978); *New Library World* 79: 54 (Mar. 1978); *RSR* 6: 43 (Apr./June 1978); *TBRI* 45: 193 (May 1979); *ARBA* (1970, p. 160; 1980, p. 617; 1981, p. 635); Jenkins (D65); Sheehy (EJ7); Wal (p. 470); Win (EI56;1EI9; 2EI11; 3EI8).

The Pesticide Book. **George W. Ware.** San Francisco, CA: Freeman, 1978.

For the nonspecialist, this encyclopedia discusses pest control chemistry and its effect on invertebrates, plants, microorganisms. Also considers legal matters. Includes a complete glossary.

Pesticides Process Encyclopedia. **Marshall Sittig.** Park Ridge, NJ: Noyes Data, 1977.

Over 550 entries alphabetically arranged by common name or systematic name. Also contains valuable raw materials and trade name index.
R: *Chemistry and Industry* 19: 789 (Oct. 1, 1977); *Farm Chemicals* 140: 84 (Aug. 1977); *TBRI* 43: 358 (Nov. 1977); 44: 38 (Jan. 1978).

Ullmanns Encyklopadie der Technischen Chemie. 25 vols. New York: Verlag Chemie, 1979.

A 25-volume encyclopedia, arranged alphabetically. Considered a basic reference on chemical technology. Recommended particularly to technical and chemical engineering libraries. Considered an excellent resource.
R: *Chemistry and Industry* 9: 314 (May 5, 1979).

FOOD TECHNOLOGY

Alexis Lichine's New Encylopedia of Wines and Spirits. 3d ed. New York: Knopf, 1981.

Second.edition, 1974.
A fundamental reference. Main body is arranged alphabetically and includes information on history, cellars, vinification processes, and viticulture.
R: *BL* 71: 774 (Mar. 15, 1975); *LJ* 100: 45 (Jan. 1, 1975); *WLB* 50: 357–358 (Jan. 1975); *ARBA* (1976, p. 744); Jenkins (J123); Wal (p. 482); Win (2EK9).

Encyclopedia of Common Natural Ingredients Used in Food, Drugs and Cosmetics. **Albert Y. Leung.** New York: Wiley, 1980.

A practical reference for food and drug technologists. Describes over 300 naturally derived substances. Alphabetically arranged by common name. Information includes source, chemical composition, biological activity, etc. Includes a chemical name index and a subject index.
R: *Journal of the American Chemical Society* 102: 7628 (Dec. 3, 1980); *New Scientist* 88: 724 (Dec. 11, 1980); *Choice* 18: 508 (Dec. 1980); *ARBA* (1981, p. 636).

Encyclopedia of Food and Nutrition. **Catherine F. Adams.** New York: Drake Publishers, 1977.

A compendium of data on the nutritional value of foods. Includes water content; calories; and protein, fat, carbohydrate, and vitamin value per portion. Contains metric conversion table. Alphabetically arranged. For public libraries.
R: *ARBA* (1979, p. 754).

Encyclopedia of Food Science. Vol. 3. **Martin S. Peterson and Arnold H. Johnson.** Westport, CT: AVI, 1978.

Contains more than 250 articles on food science. Provides composition, attributes, manufacturing processes of various foodstuffs. Comprehensive, well arranged for ready reference. A valuable tool.
R: *Food Technology* 33: 99 (Apr. 1979); *TBRI* 45: 238 (June 1979); *ARBA* (1979, p. 756).

Encyclopedia of Food Technology. Encyclopedia of Food Technology and Food Science Series, vol. 3. **Martin S. Peterson and Arnold H. Johnson eds.** Westport, CT: AVI, 1978.
Volume 2, 1974.
Comprehensive treatment of food-processing methods and technology as well as useful additional information. Valuable source of biographic material on people in the field. Unanimously well received.
R: *BL* 71: 527–528 (Dec. 1, 1975); *ARBA* (1976, p. 749).

Food and Nutrition Encyclopedia. **A. H. Ensminger and others.** Clovis, CA: Pegus Press, 1983.
R: *LJ* (Nov. 1, 1982).

Foods and Food Production Encyclopedia. **Douglas M. Considine and Glen D. Considine, eds.** New York: Van Nostrand Reinhold, 1982.
R: *Science* 218: 402 (Oct. 22, 1982).

McGraw-Hill Encyclopedia of Food, Agriculture and Nutrition. **Daniel N. Lapedes, ed.** New York: McGraw-Hill, 1977.

Four hundred articles on a wide range of topics in nutrition. Alphabetically arranged, indexed, comprehensive. A signficant ready reference source in the area of food science for agronomists, nutritionists, and botanists. Some articles taken from the *McGraw-Hill Encyclopedia of Science and Technology.*
R: *American Scientist* 66: 500 (July/ Aug. 1978); *BL* 75: 1232 (Apr. 1, 1979); *LJ* 103: 674 (Mar. 15, 1978); *RSR* 7: 12 (July/Sept. 1979); *WLB* 52: 647 (Apr. 1978); *ARBA* (1978, p. 737).

Modern Encyclopedia of Wine. **Hugh Johnson.** New York: Simon and Schuster, 1983.

A useful tool that covers about 7,000 wine producers from 30 countries. For each type of wine, information is given on addresses, size of property, current winemaker, current production, etc.
R: *BL* 80: 659 (Jan. 1, 1984); *Choice* 21: 1448 (June 1984); *ARBA* (1985, p. 497).

The New Larousse Gastronomique: The Encyclopedia of Food, Wine and Cookery. **Prosper Montagne.** New York: Crown, 1977.

Substantially revised encyclopedia of gastronomics. Entries include names of food, equipment, and explanation of international cooking terms. Definitions are clear and comprehensive. A useful reference volume.
R: *BL* 74: 1760 (July 15, 1978).

The Winemaker's Encyclopedia. **Ben Turner and Roy Roycroft.** Salem, NH: Faber & Faber, 1979.

Provides 600 alphabetically arranged entries that provide a wealth of technical data. Includes a discussion of homemade wines and a bibliography. Suitable for students, amateur winemakers, and professionals.
R: *Choice* 17: 367 (May 1980); *ARBA* (1981, p. 735).

The World Encyclopedia of Food. **Patrick L. Coyle.** New York: Facts on File, 1982.

Contains over 4,000 international food and beverage entries alphabetically arranged. Uses 53 color plates and about 350 photos and drawings. Identifies and describes the food and where it comes from. Also gives nutritional information. Good for general reader.
R: *LJ* 107: 2329–2330 (Dec. 15, 1982); *ARBA* (1983, p. 705).

POLYMER TECHNOLOGY

Encyclopedia of Plastics, Polymers, and Resins. **Michael Ash and Irene Ash, comps.** New York: Chemical Publishing, 1982–1983.

Information provided in this encyclopedia is drawn from manufacturers' catalogs, brochures, and technical data sheets. For each product, the information provided includes chemical description, category/applications, form, general properties, toxicity/handling, and standard packages.
ARBA (1985, pp. 541–542).

Encyclopedia of Polymer Science and Engineering. 2d ed., New York: Wiley-Interscience, 1984–.

First edition, 1964–72; supplement, 1976–77.
Volume 1, *A to Amorphous Polymers*; volume 2, *Anionic Polymerization to Cellular Materials*.
A multivolume update of a comprehensive encyclopedia devoted to chemical substances, polymer properties, methods, processes, and uses of polymers. Each article has an extensive bibliography. Authoritative and for every library.

Encyclopedia of PVC. 3 vols. **Leonard I. Nass, ed.** New York: Dekker, 1976–1977.

In 3 volumes. Covers the basic technology of polyvinyl chloride. Articles contributed by authorities in their field. Detailed yet fundamental. Useful for industrial libraries.
R: *Chemistry and Industry* no. 19: 846 (Oct. 2, 1976); no. 15: 660 (Aug. 6, 1977); *Elastomerics* 110: 46 (Mar. 1978); *Journal of the Oil and Colour Chemists' Association* 59: 455 (Dec. 1976); *Polymer* 18: 640 (June 1977); *TBRI* 43: 75 (Feb. 1977); 43: 195 (May

1977); 43: 277 (Sept. 1977); 43: 356 (Nov. 1977); 44: 238 (Apr. 1978); *ARBA* (1979, p. 775).

Encyclopedia of Shampoo Ingredients. **Anthony Hunting.** Cranford, NJ: Micelle Press, 1983.

This informative reference tool containing shampoo formulas and ingredients is divided into 2 sections: the first section consists of shampoo names and their marketing position; the second section consists of offers and brief discussions of individual shampoo ingredients.
R: *Choice* 21: 952 (Mar. 1984); *ARBA* (1985, pp. 542–543).

Encyclopedia of the Terpenoids. **J. S. Glasby.** New York: Wiley, 1982.

This encyclopedia lists over 10,000 terpenoids covering the literature up to the end of 1979. Physical data, toxicity, medicinal uses or other uses, as well as many other topics are given for each compound.
R: *Chemistry in Britain* 18: 444 (June 1982).

TEXTILE TECHNOLOGY

Encyclopedia of Textiles. 3d ed. **American Fabrics Magazine.** Englewood Cliffs, NJ: Prentice-Hall, 1980.

Second edition, 1972.
Abundantly illustrated work containing simplified explorations of all processes and fibers. Production and consumption tables with flow-sheet diagrams included.
R: *SL* 58: 127 (1967); Jenkins (J129).

Encyclopedia of Textiles, Fibers and Non-Woven Fabrics. **M. Grayson, ed.** New York: Wiley, 1984.

A good reference for chemists, engineers, fiber physicists, and material scientists. Each chapter is written by a recognized authority on the subject.
R: *Polymer News* (Nov. 1984).

ARCHITECTURAL ENGINEERING

Encyclopedia of Architectural Technology. **Pedro Guedes, ed.** New York: McGraw-Hill, 1979.

An encyclopedia of structural, mechanical, and technical aspects of architecture, compiled by distinguished practicing architects.
R: *Civil Engineering-ASCE* 49: 42 (June 1979); *TBRI* 45: 276 (Sept. 1979).

Encyclopedia of Energy-Efficient Building Design: 391 Practical Case Studies. 2 vols. **Kaiman Lee.** Boston: Environmental Design and Research Center, 1977.

Abstracts of projects that supply innovative methods of natural and renewable energy.
R: *Mechanical Engineering* 99: 112 (Oct. 1977): *TBRI* 43: 397 (Dec. 1977).

Encyclopedia of Wood: Wood as an Engineering Material. New York: Sterling Publishing, 1980.

Encyclopedia of Wood: Wood as an Engineering Material. New York: Sterling Publishing, 1980.

A reprint of the *Wood Handbook*, this reference provides information on the mechanical and physical properties of wood. Of interest to engineers, builders, architects, hobbyists, and public and college libraries.
R: *ARBA* (1981, p. 758).

Macmillan Encyclopedia of Architects. **Adolf K. Placzek, ed.** New York: Macmillan, 1982.

Contains authoritative biographies and illustrations of 2,400 architects.
R: *Choice* 19: 1162 (May 1982).

Macmillan Encyclopedia of Architecture and Technological Change. **P. Guedes, ed.** New York: Macmillan, 1979.

Comprises a brief history of architecture; includes discussions of building types, structural design, and technological innovations.
R: *Aslib Proceedings* 45: 49 (Jan. 1980).

ELECTRICAL AND ELECTRONICS ENGINEERING

Encyclopedia of Integrated Circuits: A Practical Handbook of Essential Reference Data. **Walter H. Buchsbaum.** Englewood Cliffs, NJ: Prentice-Hall, 1981.

The purpose of the book is to describe how integrated circuits function. They are divided into 4 categories: analog, consumer, digital, and interface. Each entry has a description, logic diagram, and key parameters. Good for the hobbyist and engineer.
R: *EIII* Proceedings 70: 782 (July 1982); *ARBA* (1983, p. 723).

Encyclopedia of Semiconductor Technology. **Martin Grayson, ed.** New York: Wiley, 1984.

A useful reference work for most purposes on the subject. It provides a good general overview of semiconductor technology.
R: *ARBA* (1985, p. 547).

Illustrated Encyclopedic Dictionary of Electronic Circuits. **John Douglas-Young.** Englewood Cliffs, NJ: Prentice-Hall, 1983.

A reference tool with useful appendix that includes conversion factor tables, graphic symbols, and mathematical tables. A subject index is available. Though mostly for the beginning electronics buff, some sections are rather technical and complicated.
R: *ARBA* (1985, p. 546).

Illustrated Encyclopedia of Solid-State Circuits and Applications. **Donald R. Mackenroth and Leo G. Sands.** Englewood Cliffs, NJ: Prentice-Hall, 1984.

A book more suitable for vocational schools or young adult technology collections for its general descriptions of the circuits shown and their functions.
R: *ARBA* (1985, p. 547).

COMPUTER TECHNOLOGY

The Atrari User's Encyclopedia. **Gary Phillips and Jerry White.** Los Angeles: Book Company, 1984.

A useful compact volume that provides essential information on all Atari-related topics. Part 1 is a tutorial for Atari BASIC; part 2 includes sample programs; and part 3, the largest, is an A to Z listing of every word associated with Atari. Essential to all Atari users.

R: *SLJ* 30: 22 (May 1984); *ARBA* (1985, p. 590).

Concise Encyclopedia of Information Technology. **Adrian V. Stokes.** Englewood Cliffs, NJ: Prentice-Hall, 1983.

An encyclopedic dictionary of about 2,500 terms related to networking, computer, and information retrieval. British slant. A useful tool for novice computer users.

R: *ARBA* (1985, p. 584).

Encyclopedia of Computers and Data Processing. Vols. 1–. Detroit: International Electronics Information Services, 1978–.

Volume 1, *A–Besm.*

An illustrated, multivolume work that covers topics dealing with computer-related concepts and data processing. Articles are brief but clear and helpful; topics such as ARTS (automated radar airline reservation systems) and data storage are discussed in greater length and include references. Recommended for all reference collections.

R: *Choice* 15: 1498 (Jan. 1979); *ARBA* (1980, p. 717).

Encyclopedia of Computers and Electronics. Chicago, IL: Rand McNally, 1983.

A brief look at the fields of electronics and computers with explanations of microwaves, electrons, computer graphics, and more. This is only a general and brief guide, not necessarily the best tool for a beginner in the field.

R: *ARBA* (1984, p. 617).

Encyclopedia of Computer Science. **Anthony Ralston and Chester L. Meeks, ed.** New York: Petrocelli/Charter, 1976.

Contains more than 480 signed articles pertaining to computer science, information science, and data processing. Common abbreviations and acronyms as well as mathematical notations are collected in the appendix. Includes 3 useful numerical tables. Ideal for the nonspecialist and the layperson.

R: *Datamation* 22: 122 (Nov. 1976); *BL* 74: 699 (Dec. 15, 1977); *RQ* 17: 83 (Fall 1977); *RSR* 5: 26 (Oct./Dec. 1977); *TBRI* 43: 77 (Feb. 1977); Sheehy (EJ47).

Encyclopedia of Computer Science and Engineering. 2d ed. **Anthony Ralston and Edwin D. Reilly, Jr., eds.** New York: Van Nostrand Reinhold, 1983.

A 1-volume reference containing 550 entries and over 700 illustrations, tables, and charts for 9 broad subject areas.

R: *Measurements and Control* 17: 175 (Sept. 1983); *Choice* 20: 1108 (April 1983).

Encyclopedia of Computer Science and Technology. Vols. 1–. New York: Marcel Dekker, 1975–.

Volume 1. *Abstract Algebra*; volume 2, *AN/FSO-7 Computer*; volume 3, *Ballistics Calculations*; volume 4, *Brain Models*; volume 9, *Generative Epistemology to Laplace Transforms*; volume 10, *Linear and Matrix Algebra to Microorganisms*; volume 11, *Minicomputers to PASCAL*; volume 12, *Pattern Recognition to Reliability of Computer Systems*; volume 15, supplement, 1980.
A projected 20-volume set. Provides comprehensive articles on a broad scope of topics in computer science. Recommended for large reference collections and technical libraries.
R: *ARBA* (1976, p. 770).

Encyclopedia of Computer Terms. **Douglas Downing.** Woodbury, NY: Barron's Educational Series, 1983.
Presents 400 entries simplified and readable for the nonprofessional computer user interested in personal computing. A nice piece to add to one's collection, but information can be obtained easily elsewhere.
R: *ARBA* (1984, p. 616).

McGraw-Hill Encyclopedia of Electronics and Computers. **Sybil P. Parker, ed.** New York: McGraw-Hill, 1984.
A comprehensive volume with more than 470 articles, about 1,300 illustrations, and 5,500 index entries. Complementary text is written in understandable technical language with bibliographies and index provided. A must for both beginners and experts.
R: *Choice* 21: 806 (Feb. 1984); *ARBA* (1985, p. 547).

INDUSTRIAL ENGINEERING

Encyclopedia of North American Railroading: 150 Years of Railroading in the United States and Canada. **Freeman Hubbard.** New York: McGraw-Hill, 1981.
A well-organized encyclopedia covering 150 years of North American rail history. For hobbyists, railroad public relations personnel, and historians.
R: *BL* 78: 271 (Oct. 15, 1981); *WLB* 56: 302 (Dec. 1981).

Lyons' Encyclopedia of Valves. **Jerry L. Lyons and Carl L. Askland, Jr.** New York: Van Nostrand Reinhold, 1975.
An encyclopedia that details the design, manufacture, selection, installation, and use of valves. Text is accompanied by photographs and definitions. A major reference source for engineering and special libraries.
R: *Choice* 13: 960 (Oct. 1976); *ARBA* (1977, p. 769).

MATERIALS SCIENCE

Encyclopedia/Handbook of Materials, Parts and Finishes. **Henry R. Clauser, ed.** Westport, CT: Technomic, 1976.
Provides detailed information on materials, parts, and finishes used in industry. Information includes size and shape of parts, design capabilities, and production. A handy reference work.

Encyclopedia of Composite Materials and Components. **Martin Grayson, ed.** New York: Wiley, 1983.

Encyclopedia Reprint Series.

It groups reprints from the 25-volume of Kirk and Othmer's *Encyclopedia of Chemical Technology* (third edition, 1978) that are related to composite materials and components. Arranged by 50 main entries. A useful desk reference for specialists and students on the subject.

R: *ARBA* (1985, p. 553).

Encyclopedia of Materials Science and Engineering. **Michael B. Bever, ed.-in-chief.** Oxford, UK: Pergamon Press, 1984.

Features articles by over 1,200 distinguished authors and Nobel laureates. Includes articles, bibliographic references of key sources in the literature, diagrams, photographs, and many tables. 8 volumes.

R: *Microelectronics and Reliability* 23: 766 (1983).

Materials and Technology: A Systematic Encyclopedia. New York: Longman, 1968–.

Eventually will be an 8-volume set intended for laypeople and technologists.

MECHANICAL ENGINEERING

Encyclopedia of Fluid Mechanics. **Nicholas P. Cheremisinoff, ed.** 3 vols. New York: Wiley, 1985.

Volume 1, *Flow Phenomena and Measurement*, 1985; volume 2, *Dynamics of Single-Fluid Flows and Mixing*, 1985; volume 3, *Gas-Liquid Flows*, 1985.

ENERGY

Alternative Energy Sources: An Internatinal Compendium. **T. Nejat Veziroglu, ed.** New York: McGraw-Hill, 1979.

An easy-to-use guide to the development of alternative energy technology and applications with contributions from more than 400 energy experts around the world. Explores industrial, agricultural, and residential uses of alternative energy sources being tested and applied today in the Untied States, Europe, the Middle East, Asia, and Australia. For engineering and corporate libraries.

R: *RSR* 8: 27 (Jan./Mar. 1980).

Coal in America: An Encyclopedia of Reserves, Production, and Use. **Richard A. Schmidt.** New York: McGraw-Hill, 1979.

R: *Chemical Engineering* 87: 11 (Apr. 21, 1980).

Encyclopedia of Energy. **Daniel N. Lapedes, ed.** New York: McGraw-Hill, 1976.

In 2 sections: energy perspectives and energy technology. Presents an excellent overview for a broad audience. Comprehensive, clear, and well-illustrated. An important and useful reference.
R: *IEEE Spectrum* 14: 60 (July 1977); *TBRI* 43: 315 (Oct. 1977); *ARBA* (1977, p. 684).

International Petroleum Encyclopedia. Tulsa, OK: Petroleum Publishing, 1970–.
Latest edition, 1975.
A yearbook pinpointing on regional maps petroleum activity described in the previous petroleum trade journals.
R: *ARBA* (1972, p. 594); Wal (p. 492); Win (3EI40).

Kaiman's Encyclopedia of Energy Topics. 2 vols. **Kaiman Lee and Jacqueline Masloff.** Newtonville, MA: Environmental Design and Research Center, 1979.
A compilation of 621 articles relating to energy topics, with emphasis on housing aspects and solar power. Articles are arranged under 156 categories and vary in length. Useful for public or junior-college libraries and the layperson.
R: *ASHRAE Journal* 21: 94 (Aug. 1979); *TBRI* 45: 352 (Nov. 1979); *ARBA* (1981, p. 680).

McGraw-Hill Encyclopedia of Energy. 2d ed. **Sybil P. Parker, ed.** New York: McGraw-Hill, 1981.
First edition, 1976.
Provides authoritative and comprehensive coverage of energy issues, policies, and technologies. Contains 300 alphabetically organized articles concerning such topics as energy conservation, consumption, environmental protection, and future fuels. Many articles are taken from the *McGraw-Hill Encyclopedia of Science and Technology*. Includes over 800 illustrations as well as an index and cross references. An appendix gives conversion tables, publications, and a list of federal energy agencies. Highly recommended for scientific and technical libraries.
R: *BL* 74: 1136 (Mar. 1, 1978); *Choice* 18: 926 (Mar. 1981); *LJ* 102: 786 (Apr. 1, 1977); *RSR* 8: 63 (Oct./Dec. 1980); *WLB* 51: 600 (Mar. 1977); 55: 681 (Jan. 1981).

The World Energy Book: An A–Z Atlas and Statistical Source Book. **David Crabbe and Richard McBride, eds.** New York: Nichols Publishing, 1978.
Contains over 1,500 entries that relate to energy resources and are arranged alphabetically and cross-referenced. Definitions are clear and comprehensive. Tables, graphs, diagrams, and coversion charts appear in the appendix. A well-organized reference for undergraduate collections.
R: *Mechanical Engineering* 101: 108 (Aug. 1979); *New Scientist* 80: 955 (Dec. 21/28, 1978); *Science Digest* 86: 84 (Nov. 1979); *Scientific American* 240: 49 (June 1979); *Aslib Proceedings* 44: 80 (Feb. 1979); *New Library World* 80: 75 (Apr. 1979); *RQ* 18: 401 (Summer 1979); *TBRI* 45: 113 (Mar. 1979); 45: 311 (Oct. 1979); 45: 358 (Dec. 1979); *ARBA* (1979, p. 691).

ENVIRONMENTAL SCIENCES

Encyclopedia of Environmental Science and Engineering. 2 vols. **James R. Pfafflin and Edward N. Ziegler, eds.** New York: Gordon and Breach, 1976.

A basic encyclopedia of environmental matters, covering such areas as water handling, government regulations, pollution, and occupational health. Concisely written, a useful reference.
R: *Choice* 14: 830 (Sept. 1977); *RSR* 5: 34 (Oct./Dec. 1977); *ARBA* (1978, p. 686).

Grzimek's Encyclopedia of Ecology. **Bernhard Grzimek, ed.** New York: Van Nostrand Reinhold, 1977.

Chapters on various aspects of animal and human ecology. A reference for public, college, and school libraries. Contains well-written articles, bibliographies, and an index of English and scientific names.
R: *Science News* 111: 319 (May 1977); *Aslib Proceedings* 42: 471 (Sept. 1977); *BL* 74: 1128 (Mar. 1, 1978); *LJ* 102: 1626 (Aug. 1977); 103: 509 (Mar. 1, 1978); *WLB* 52: 186 (Oct. 1977); *ARBA* (1978, p. 686); Sheehy (EC15).

Hazardous Waste in America. **Samuel S. Epstein, Lester O. Brown, and Carl Pope.** San Francisco: Sierra Club Books. Distr. New York: Random House, 1982.

Information examining a wide range of occupational and environmental hazards resulting from the production and disposal of toxic wastes. Includes case studies, appendixes, and a few tables. Informative and useful.
R: *ARBA* (1984, p. 680).

McGraw-Hill Encyclopedia of Environmental Science. 2d ed. **Sybil P. Parker, ed.** New York: McGraw-Hill, 1980.

First edition, 1974.
An updated and revised edition of an outstanding reference on environmental science. Contains over 250 alphabetically arranged articles, many taken from the *McGraw-Hill Encyclopedia of Science and Technology*. Articles cover all aspects of environmental science from climate to waste management. Five feature articles on general subjects are provided. Includes over 650 photographs, charts, graphs, and line drawings to enhance the text. Indexed and cross-referenced. An indispensable and essential reference tool for science, business, and public libraries.
R: *American Scientist* 63: 236 (Mar./Apr. 1975); *BL* 77: 1171 (Apr. 15, 1981); *Choice* 18: 64 (Sept. 1980); *LJ* 105: 1501 (July 1980); *RQ* 20: 105 (Fall 1980); *RSR* 3: 80 (Apr./June 1975); *WLB* 55: 64 (Sept. 1980); *ARBA* (1981, p. 685).

Topics and Terms in Environmental Problems. **John R. Holum.** New York: Wiley, 1978.

Contains information on nearly 240 topics related to the environment, covering such areas as water pollution and air pollution. Comprises a quick reference volume for libraries with a heavy demand for information in this area.
R: *Choice* 15: 1350 (Dec. 1978); *LJ* 103: 859 (Apr. 15, 1978); *ARBA* (1979, p. 694).

TRANSPORTATION

Encyclopedia of Ships and Seafaring. **Peter Kemp, ed.** New York: Crown Publishers, 1980.

Authoritative, well-organized collection of material on marine technology and related subjects. Discusses the history of exploration, diving, salvage, ships and their develop-

ment, and important men of the sea. Generously illustrated; has a British orientation. Useful for school libraries.
R: *Choice* 18: 774 (Feb. 1981); *ARBA* (1981, p. 766).

The Illustrated Encyclopedia of Ships, Boats, Vessels, and Other Water-Borne Craft. **Graham Blackburn.** Woodstock, NY: Overlook Press. Distr. New York: Viking Press, 1978.

Over 750 descriptions of boats, each with a line-drawn illustration. Text is written in hand-lettered italics. Recommended for public libraries.
R: *ARBA* (1979, p. 778).

The Illustrated Encyclopedia of the World's Automobiles. **David Burgess Wise, ed.** New York: A & W Publishers, 1979.

Consists of more than 4,000 entries dealing with commercially produced automobiles for private use. Entries are alphabetically organized and include short technical descriptions and illustrations. Cross-referenced in a separate alphabet. Contains a historical summary and biographical sketches of automotive personalities. Includes a glossary and conversion tables. For home and public libraries.
R: *ARBA* (1981, p. 771).

The Oxford Companion to Ships and the Sea. **Peter Kemp, ed.** New York: Oxford University Press, 1976.

Nearly 4,000 articles arranged alphabetically. Provides information on ships, sailors, and the sea. Considered a definitive work.
R: *United States Naval Institute Proceedings* 103: 81 (Apr. 1977); *BL* 74: 228 (Sept. 15, 1977); *Choice* 14: 182 (Apr. 1977); *CRL* 37: 54 (Jan. 1977); *LJ* 102: 188 (Jan. 5, 1977); *RQ*, p. 356 (Summer 1977); *TBRI* 43: 234 (June 1977); *WLB* 51: 540 (Feb. 1977); *ARBA* (1978, p. 779).

The Rand McNally Encyclopedia of Transportation. Chicago, IL: Rand McNally, 1976.

Deals with all facets of land, sea, and air transportation including economics, demography, and international affairs. Information is comprehensive, complete, and accompanied by charts, diagrams, and photographs. Useful for school and public libraries.
R: *BL* 74: 577 (Nov. 15, 1977); *ARBA* (1977, p. 774).

MILITARY

Air Power: The World's Air Forces. **Anthony Robinson, ed.** New York: McGraw-Hill, 1980.

A well-illustrated volume on the air forces of the countries of the world, supplemented with charts and photographs. Includes British spelling and usage. Requires some background knowledge. For airplane and military buffs.
R: *WLB* 55: 682 (Jan. 1981).

The Encyclopedia of Air Warfare. **Christopher Chant et al.** New York: Crowell, 1976.

Presents a chronological development of the airplane and its role in war use in the past 10 years.
R: *BL* 72: 55 (Spet. 1, 1976); *WLB* 50: 808 (June 1976); *ARBA* (1977, p. 781).

Encyclopedia of US Air Force Aircraft and Missile Systems. Post-World War II Fighters, 1945–1973. Vol. 1. **Marcelle S. Knaack.** Washington, DC: Office of Air Force History. Distr. Washington, DC: US Government Printing Office, 1978.

Provides technical data on aircraft; arranged chronologically. Includes black-and-white photos and line drawings and bibliographies.
R: *BL* 75: 674 (Dec. 15, 1978); *ARBA* (1979, p. 783).

The Encyclopedia of World Air Power. **Bill Gunston, ed.** New York: Crescent Books, 1980.

An encyclopedic work of the air forces, aircraft, and air-launched missiles of the world. The air forces section is alphabetically organized by continent and contains photos of aircraft and insignia. The aircraft section consists of 400 aircraft alphabetically grouped by manufacturer. Provides excellent photographs, drawings, and brief sketches. Sixty-nine air-launched missiles are described. Includes number code and common names indexes.
R: *ARBA* (1981, p. 777).

The Illustrated Encyclopedia of the World's Rockets and Missiles: A Comprehensive Technical Directory and History of the Military Guided Missile Systems of the 20th Century. **B. Gunston.** London: Salamander Books, Ltd., 1979.

A comprehensive, illustrated encyclopedia on rockets and missiles of the world. Topics include surface-to-surface missiles; air-to-surface missiles; air-to-air missiles; and antisubmarine missiles. Includes descriptions, dimensions, launch weight, and range on each.
R: Wal (p. 208).

Military Small Arms of the 20th Century: A Comprehensive Illustrated Encyclopedia of the World's Small-Calibre Firearms, 1900–1977. **Ian V. Hogg and John Weeks.** New York: Hippocrene Books, 1977.

Illustrations and descriptions of military arms. Divided into 5 sections: guns, submachine guns, rifles, automatic rifles, antitank rifles. Includes an appendix of basic data on ammunition.
R: *ARBA* (1978, p. 784).

CHAPTER 5 DICTIONARIES

BIBLIOGRAPHICAL TOOL

World Dictionaries in Print: A Guide to General and Subject Dictionaries in World Languages. New York: Bowker, 1983.

Provides information on some 20,000 dictionaries from 3,500 publishers worldwide, many of which are in the areas of science and technology. A Directory of Publishers and Distributors and a Key to World Currency Symbols are included. Indexed by subject, title, language, and author/compiler/editor.

ABBREVIATIONS AND ACRONYMS

Abbreviations Dictionary. 5th ed. **Ralph DeSola**. New York: American Elsevier, 1978.

Fourth Edition, 1974.
Full identification of acronyms, initials, abbreviations, symbols, etc. Myriad features such as weather symbols, atomic numbers, astronomical constellations, etc.
R: *Journal of Metals* 31: 51 (Nov. 1979); *LJ* 93: 535 (1968); *TBRI* 45: 20 (June 1978); Jenkins (A58).

Acronyms, Initialisms, and Abbreviations Dictionary. 8th ed. 3 vols. **Ellen T. Crowley, ed.** Detroit: Gale Research, 1982.

Fifth edition, *Acronyms and Initialisms Dictionary: A Guide to Alphabetic Designations, Contractions, Acronyms, Initialisms, and Similar Condensed Appellations*, 1975; sixth edition, 1978; seventh edition, 1980.
Volume 1, *Acronyms, Initialisms, and Abbreviations Dictionary*; volume 2, *New Acronyms, Initialisms, and Abbreviations 1979 and 1980*; volume 3, *Reverse Acronyms, Initialisms, and Abbreviations Dictionary*.
Volume 1 contains alphabetic listings of acronyms used in the United States, Britain, France, Germany, and Russia. Volume 2 provides an updating supplement. Volume 3 first lists term/organization, then acronym. Alphabetically arranged.

Anglo-American and German Abbreviations in Data Processing. **Peter Wennrich.** New York: Saur, 1984.

Includes 35,000 technical expressions and abbreviations in data processing from over 100 international German and English language periodicals.

Reverse Acronyms, Initialisms, and Abbreviations Dictionary. 7th ed. **Ellen T. Crowley, ed.** Detroit: Gale Research, 1980.

Earlier edition, 1974.
R: *RSR* 8: 63 (Oct./Dec. 1980).

World Guide to Abbreviations of Organizations. 7th ed. **F. A. Buttress.** London: Leonard Hill. Dist. Detroit: Grand River Books/Gale Research, 1984.

First edition, 1954.

Contains 43,000 entries of acronyms and titles of organizations in such fields as science and technology, medicine, journalism, education, and industry. Includes 2 bibliographies. Comprehensive and international in scope.
R: *ARBA* (1985, p. 3).

Subject

Anglo-American and German Abbreviations in Environmental Protection. **Peter Wennrich.** New York: Saur, 1980.

An index containing terms of environmental protection and related areas. It includes precepts from the fields of biology, chemistry, agriculture, medicine, and physics pertaining to the environmental sciences. Suggest some previous knowledge of the context for interpretation purposes.
R: *ARBA* (1982, p. 748).

Anglo-American and German Abbreviations in Science and Technology: Anglo-Amerikanische und Deutsche Abkurzungen in Wissenschaft und Technik. 3 vols. **Peter Wennrich.** New York: Bowker. Munich: Verlag Dokumentation, 1976–1978. 6 Supp., 1980.

A 3-volume set that includes some 150,000 abbreviations and acronyms of scientific and technical terms. Bilingual in English and German. Clear format, recommended for special libraries.
R: *Journal of the American Chemical Society* 100: 7787 (Nov. 22, 1978); *BL* 75: 1112 (Mar. 1, 1979); *LJ* 102: 1264 (June 1, 1977); *RSR* 9: 105 (Jan./Mar. 1981); *ARBA* (1978, p. 614; 1979, p. 642); Sheehy (EA20).

Computer Acronyms, Abbreviations, Etc. **Claude P. Wrathall.** New York: Petrocelli Books, 1981.

Contains a listing of over 10,000 terms and definitions including national, international, and industry standards designations. Valuable for all levels of business activities.
R: *IEEE Proceedings* 70: 880 (Aug. 1982).

Dictionary of Biomedical Acronyms and Abbreviations. **J. Dupayrat.** New York: Wiley, 1985.

Dictionary of New Information Technology Acronyms. 1st ed. **Michael Gordon, Alan Singleton, and Clarence Rickards.** London: Kogan Page. Distr. Detroit: Gale Research, 1984.

An alphabetical listing of over 10,000 acronyms used in such disciplines of information technology as telecommunications, data processing, and microelectronics. Each entry is expanded and annotated if necessary. Useful for those in the field of information technology.
R: *Choice* 22: 400 (Nov. 1984); *ARBA* (1985, p. 582).

Pugh's Dictionary of Acronyms and Abbreviations: Abbreviations in Management, Technology and Information Science. **Eric Pugh, comp.** Phoenix: Oryx, 1982.

Contains 30,000 entries, including 5,000 new entries not found in the first edition. The author is inconsistent in choosing material, but does include some unusual references that would be useful in libraries.
R: *LJ* 1087: 107 (June, 1982).

NOMENCLATURES AND THESAURI

Agricultural Economics and Rural Sociology Multilingual Thesaurus. 5 vols. **Commission of the European Communities.** New York: Saur, 1979.

English, French, German, and Italian thesaurus follows the UNISIST/ISO guidelines. Covers agriculture and related fields including law, crops, and livestock. Each volume is in a different language, and a multilingual index of all descriptors comprises the fifth. An up-to-date tool that aids in searching bibliographic databases and enables easy information exchange. For agriculture economists.
R: *ARBA* (1981, p. 733).

ASM Thesaurus of Metallurgical Terms. 2d ed. **American Society for Metals.** Metals Park, OH: American Society for Metals, 1976.

First edition, 1968.
Contains the vocabulary for indexing and retrieving technical information in metallurgy, particularly the appropriate databases. Valuable for the information searcher.
R: *RSR* 5: 35 (Oct./Dec. 1977); Jenkins (K198); Wal (p. 501); Win (2EI34).

Biological Nomenclature. 2d ed. **Charles Jeffrey.** London: Edward Arnold, 1977.

First edition, 1973.
Explains the nomenclature of taxonomic systems in all fields of biology. A useful, lucid account.
R: *Endeavour* 2: 49 (1978); *Aslib Proceedings* 42: 567 (Dec. 1977); *TRBI* 44: 256 (Sept. 1978); Sheehy (EC6).

Common Plants: Botanical and Colloquial Nomenclature. **John J. Cunningham and Rosalie J. Cote.** New York: Garland STPM Press, 1977.

In 2 parts: traces chronological development of Linnaean binomial system and systematics and describes common botanical names and their folklore, medicinal, and religious derivation. Recommended to both researchers and laypeople.
R: *LJ* 102: 1002 (May 1, 1977); *RSR* 7: 9 (Jan./Mar. 1979); *ARBA* (1978, p. 654).

Compendium of Analytical Nomenclature; Definitive Rules 1977. **International Union of Pure and Applied Chemistry.** Edited by H. M. N. H. Irving, H. Freiser, and T. S. West. Elmsford, NY: Pergamon Press, 1978.

Surveys trends in analytical chemistry, dealing with such matters as automatic analysis, ion exchangers, and electrochemical data. Also includes over 1,500 definitions. Culls information from a variety of sources. Highly recommended.
R: *Analyst* 103: 1184 (Nov. 1978); *Association of Official Analytical Chemists Journal* 69: 968 (July 1979); *TBRI* 45: 88 (Mar. 1979); 45: 284 (Oct. 1979).

A Concordance to Darwin's "Origin of Species". 1st ed. **Paul H. Barrett, Donald J. Weinshank, and Timothy T. Gottleber, eds.** Ithaca, NY: Cornell University Press, 1981.

An alphabetical list of all the words used in Darwin's *On the Origin of Species by Means of Natural Selection* (1859) in a keyword-in-context index.
R: *ARBA* (1983, p. 614).

Enzyme Nomenclature: Recommendations (1972) of the Commission on Biochemical Nomenclature on the Nomenclature and Classification of Enzymes Together with Their Units and the Symbols of Enzyme Kinetics. **International Union of Biochemistry, Standing Committee on Enzymes.** New York: American Elsevier, 1973.

Provides Enzyme Code number, recommended name, systematic name, reaction catalyzed, and references for some 1,700 enzymes.
R: Sheehy (EC48).

Food: Multilingual Thesaurus. 5 vols. **Commission of the European Communities, Directorate-General for Research, Science and Education, ed.** New York: Saur, 1979.

Contains controlled vocabulary of food technology according to the UNISIST/ISO guidelines. Designed for indexing and retrieval in documentation systems. In English, French, German and Italian. Volume 5 serves as a multilingual index of all the descriptors appearing in the thesaurus. Useful to food science research personnel.
R: *CRL* 41: 396 (July 1980); *ARBA* (1981, p. 738).

GEOREF Thesaurus and Guide to Indexing. 3d ed. **Sharon J. Riley, ed.** Falls Church, VA: American Geological Institute, 1981.

This is an explanation of the GEOREF indexing system containing terms from the *Bibliography and Index of North American Geology*, an in-depth listing of geological terms. Contains many cross-references as well. Highly recommended for libraries with computer searching.
R: *ARBA* (1983, p. 665).

Handbook of Chemical Synonyms and Trade Names—A Dictionary and Commercial Handbook Containing Over 35,000 Definitions. 8th ed. **William Gardner et al.** West Palm Beach, FL: CRC Press, 1979.

R: *Chemical Engineering* 86: 12 (Mar. 12, 1979).

How to Name an Inorganic Substance. **W. C. Fernelius, ed.** New York: Pergamon Press, 1978.

A guide to the use of *Nomenclature of Inorganic Chemistry: Definitive Rules 1970*. Briefly defines 13 different types of names with samples, then outlines procedure in 2 pages of questions with answers. Includes tables of names for ions and radicals in formula-index order.
R: *Journal of the American Chemical Society* 101: 3419 (June 6, 1979); 102: 3664 (May 7, 1980); *RSR* 7: 19 (Apr./June 1979).

INIS: Thesaurus. **International Atomic Energy Agency.** New York: Unipub, 1980.

A revised thesaurus containing 15,974 accepted terms and 4,567 forbidden terms used in preparation of International Nuclear Information System (INIS) input in the areas of nuclear physics, reactor technology, and related topics.

INIS: Thesaurus. IAEA-INIS-13. 21st ed. New York: Unipub, 1982.

Tenth revised edition, 1977.
Contains terminology to be used for subject description for the preparation of International Nuclear Information System (INIS) input by national and regional centers.
R: *IBID* 4: 131 (June 1976).

NASA Thesaurus. 2 vols. Springfield, VA: National Technical Information Service, 1976.

An authorized subject listing of National Aeronautics and Space Administration scientific and technical information system documents in 2 volumes. Volume 1 consists of an alphabetical arrangement of subject terms; volume 2 lists all thesaurus entries.

Nomenclature of Organic chemistry. 4th ed. (sec. A–H). **International Union of Pure and Applied Chemistry.** New York: Pergamon Press, 1979.

Second edition (section C); third edition (sections A and B), 1971.
Rules for uniformity in terminological usage.
R: *Journal of the American Chemical Society* 102: 3665 (May 7, 1980); *NTB* 51: 264 (1966); *RSR* 8: 24 (July/Sept. 1980); Jenkins (D156); Wal (pp. 132, 136).

Organic Nomenclature: A Programmed Study Guide. **Carl R. Johnson.** New York: Worth Publishers, 1976.

Designed as a companion to Allinger's *Organic Chemistry*. Proceeds with fundamental concepts. Discusses specific rules of nomenclature. Recommended for librarians who teach science literature courses as a means of keeping up with chemical nomenclature.
R: *RSR* 6: 45 (Apr./June 1978).

SPINES Thesaurus: A Controlled and Structured Vocabulary of Science and Technology for Policy-Making Management and Development. 4 vols. **Unesco Secretariat and B. de Padirac, comps.** Paris: Unesco. Distr. New York: Unipub, 1977.

Four volumes that assist in codifying technical terms in bibliographical control. Presents helpful diagrams, abbreviation keys, introductory and historical material. Useful in international scientific communication, particularly in the UNISIST network.
R: *ARBA* (1978, p. 621).

Thesaurus of Agricultural Terms as Used in the Bibliography of Agriculture from Data Provided by the National Agricultural Library, US Department of Agriculture. Scottsdale, AZ: Oryx Press, 1976.

Helpful for users of the *Bibliography of Agriculture* database; contains over 25,000 terms. Well organized and cross-referenced. Considered a highly useful tool for librarians and computer searchers.
R: *ARBA* (1977, p. 730).

Thesaurus of Information Science Terminology. Rev. ed. **Claire K. Schultz.** Metuchen, NJ: Scarecrow Press, 1978.
Thesaurus with entries arranged alphabetically; indexed by multiword terms.

Thesaurus of Metallurgical Terms. 2d ed. Metals Park, OH: American Society for Metals, 1976.
Serves as the vocabulary authority for *Metals Abstracts Index*. Includes some 9,000 terms. Also useful in the *Alloys Index* classification scheme. A helpful reference.
R: *RSR* 5: 10 (Apr./June 1977).

Veterinary Multilingual Thesaurus. 5 vols. **Commission of the European Communities.** New York: Saur, 1979.
A multivolume set in 4 languages—English, French, German, and Italian. Fifth volume contains a multilanguage index. Coverage includes the study of animals in health and disease, and also microbiology, zoology, pharmacology, and pathology. Comprehensive work that veterinary researchers will find helpful.
R: *CRL* 41: 396 (July 1980); *Quarterly Bulletin of the IAALD* 25: 47 (1980); *ARBA* (1981,
p. 749).

MULTILINGUAL

BIBLIOGRAPHIES

Bibliography of Interlingual Scientific and Technical Dictionaries. 5th ed. Paris: Unesco, 1969. Also later ed.
Fourth edition, 1961.
Approximately 2,500 entries.
R: *NTB* 56: 4 (1971); Wal (p. 20); Win (EA96; 1EA14; 3EA15).

Dictionaries of English and Foreign Languages: A Bibliographical Guide to Both General and Technical Dictionaries with Historical and Explanatory Notes and References. Rev. and enl. **Robert L. Collison.** New York: Hafner, 1971.

Location Key to Foreign Language Dictionaries. **M. S. Davis.** England: Surrey and Sussex Libraries in cooperation, 1974.
R: *Aslib Proceedings* 29: 33 (Jan. 1975); *RSR* 3: 56 (Apr.–June 1975).

ENGLISH–CHINESE

Agricultural Terms: English—Chinese. **Food and Agriculture Organization of the United Nations.** Rome, Italy: Food and Agriculture Organization of the United Nations, 1977.

Bilingual English-Chinese dictionary of agricultural terms.
R: *IBID* 6: 378 (Dec. 1978).

Chinese–English, English–Chinese Astronomical Dictionary. **Hong-yee Chiu, ed.** New York: Consultants Bureau, 1966.
R: Jenkins (E34); Win (1EB2).

A Dictionary of Military Terms: Chinese–English, English–Chinese. **Joseph D. Lowe.** Boulder, CO: Westview Press, 1977.

Intended primarily for translators, a dictionary of about 2,500 terms in English and Chinese. Limited to primarily military terms, the book offers accurate and highly technical definitions.
R: *ARBA* (1979, p. 784).

An English–Chinese Dictionary of Engineering and Technology. **The Dictionary Editing Group, Zhong Wai Publishing.** New York: Wiley, 1981.

Over 173,000 clearly defined entries encompassing the entire spectrum of modern engineering and technology will help engineering and technical users read, understand, and translate with the precision demanded by their profession.
R: *Choice* 19: 1023 (Apr. 1982).

English–French

Dictionary of Civil Engineering and Construction Machinery and Equipment. 7th ed. 2 vols. **H. Bucksch.** New York: International Publications Service, 1979. Earlier edition, 1960.
Volume 1, *English–French*; volume 2, *French–English*.
R: Wal (p. 353).

Dictionary of Science and Technology: English–French. **A. F. Dorian, comp.** New York: American Elsevier, 1979.

Includes 150,000 terms covering topics from acoustics to zoology. Brief definitions are given in some cases. For university, technical, and public libraries.
R: *Laboratory Practice* 29: 747 (July 1980); *ARBA* (1980, p. 600).

Dictionary of Science and Technology: French–English **A. F. Dorian, comp.** New York: American Elsevier, 1980.

Companion to the English–French volume.
R: *ARBA* (1981, p. 620).

English-French Petroleum Dictionary. **Michael Arnould and Fabio Zubibi, eds.** Paris: Dunod, 1981.

A most useful translation of some 8,700 terms on petroleum technology from English to French.
R: *Science and Technology Libraries* 4: 121 (Winter 1983).

English/French Paints and Coatings Vocabulary. **L'Association Quebecoise des Industries de la Peinture.** Montreal, Quebec: L'Association Quebecoise des Industries de la Peinture, 1977.

A pocketsize bilingual dictionary of terms used in the coatings industry. For professionals.

R: *Journal of Coatings Technology* 49: 92 (Aug. 1977); *TBRI* 43: 350 (Nov. 1977).

French–English and English–French Dictionary of Technical Terms and Phrases. 2 vols. **J. O. Kettridge.** Boston: Routledge & Kegan Paul, 1980.

Translates 100,000 words and phrases from the fields of civil, mechanical, mining, and electrical engineering as well as other related subjects. Supplements update the work. Entries are coded by number. For large reference collections.

R: *ARBA* (1981, p. 621).

French–English Chemical Terminology. **Hans Fromherz and Alexander King.** Translated by Jack Jousset. New York: Verlag Chemie, 1969.

An introductory bilingual dictionary of chemical terminology.

French–English Science and Technology Dictionary. 4th ed. **L. DeVries and S. Hochman.** New York: McGraw-Hill, 1976.

First edition, 1940.

Fourth edition includes new terms and new translations and meanings of older terms. Comprises 4,500 new entries from the major modern technical fields. For scientists, translators, and businessmen.

R: *LJ* 101: 1516 (July 1976); *ARBA* (1977, p. 628); Sheehy (EA21); Wal (p. 26).

Glossary of Automotive Terminology: French–English, English–French. **Chrysler Corporation, comp.** Warrendale, PA: Society of Automotive Engineers, 1977.

Provides French–English automotive terminology for technical writers. Definitions recommended by the Society of Automotive Engineers. Recommended as a tool for industrial libraries.

R: *ARBA* (1979, p. 765).

ENGLISH–GERMAN

Dictionary of Electronics: English–German. **Alfred Oppermann, ed.** New York: Saur, 1980.

Includes about 100,000 concepts and 300,000 translations in the field of electronics and related subjects. Translations are arranged in alphabetical order.

Dictionary of Engineering and Technology. 4th ed. **Richard Ernst.** New York: Oxford University Press, 1980.

Third edition, 1974.

Volume 1, German-English; Volume 2, English-German.

Volume 1 contains over 150,000 entries of German terms and phrases used in engineering and technology. This is an essential source for libraries that have large re-

search collections in engineering and technology with extensive holdings in foreign languages.
R: *ARBA* (1982, p. 696).

Dictionary of Geosciences: English/German. 2d ed. **Adolf Watznauer, ed.** New York: Elsevier, 1982.
Includes over 35,000 terms from the first edition and adds over 3,800 new terms. Defines terms from mineralogy, pertography, and economic geology as well as words from other closely related sciences. Useful to the interpreter, researcher, and librarian.
R: *ARBA* (1983, p. 662).

Dictionary of Modern Engineering. 3d ed. 2 vols. **Alfred Oppermann.** New York: Saur, 1972–1974.
Volume 1, *English–German,* 1972; volume 2, *German–English,* 1974.
In 2 volumes: English–German and German–English dictionary of 450,000 terms relating to engineering.

Dictionary of Particle Technology, English–German, German–English: Worterbuch der Mechanischen Verfahrenstechnik, Englisch–Deutsch, Deutsch–Englisch. **K. Leschonski and F. T. C. Carter.** New York: American Elsevier, 1978.
Presents a list, in computer typeface, of German and English terminology in the new field of particle technology. Includes many general scientific and engineering words. For universities with specialized programs in particle technology.
R: *ARBA* (1980, p. 620).

Dictionary of Science and Technology: English–German. 2d ed. **A. F. Dorian, comp.** New York: American Elsevier, 1978.
Revised and expanded edition incorporates 16,000 terms, reflecting recent advances in the field. Provides rapid access. Well-recommended to scientists, librarians, and technologists.
R: *ARBA* (1979, p. 640); Sheehy (EA23).

Dictionary of Science and Technology: German–English. Handworterbuch der Naturwissenschaft und Technik; Deutsch–Englisch. 2d ed. **A. F. Dorian, comp.** Amsterdam: Elsevier. New York: American Elsevier, 1981.
First edition, 1970.
A comprehensive one-stop reference for the translator of German science and technology books. A companion volume to Dorian's *Dictionary of Science and Technology: English-German,* 1967. 120,000 words from 128 subject fields.
R: *Endeavour* 30: 156 (Sept. 1971); *Choice* 5: 749 (1968); *LJ* 93: 1125 (1968); 96: 1961 (1971); *ARBA* (1971, p. 475); Jenkins (A64); Wal (p. 24); Win (2EA21; 3EA17).

German–English Science Dictionary. 4th ed. **Louis DeVries.** New York: McGraw-Hill, 1978.
First edition, 1946; third edition, 1959.
Contains a total of over 65,000 terms. Fourth edition is much revised and expanded. Includes definitions from all branches of science. Helpful section on grammatical principles, syntax, constructions, and idioms. Also provides a list of abbreviations used

in scientific literature. Considered an indispensable reference tool for scientists, technicians, students, and translators.
R: *ARBA* (1979, p. 639); Mal (1980, p. 47); Sheehy (EA22).

Microelectronics Dictionary, English-German/German-English. **IWT Verlag GmbH, ed.** Vaterstetten: VDI International, 1980.

Worterbuch der Elektronik, Englisch-Deutsch. Dictionary of Electronics, English-German. **Alfred Oppermann, ed.** New York: Saur, 1980.

Includes exact equivalents and definitions of English words in German. Verbs are listed only if they are of strong terminological character. Grammatical information about German terms is limited. Intended for German specialists who want to read English.
R: *ARBA*: (1982, p. 810).

Wörterbuch Technischer Begriffe mit 4,300 Definitionen nach DIN: Deutsch und Englisch. 3d ed. **Henry G. Freeman.** Berlin: Beuth Verlag. Distr. Philadelphia: Heyden, 1983.

First edition, 1972.
An excellent technical dictionary providing accurate translations of engineering words and concepts from German to English. Definitions are derived from the written DIN standards. Includes an alphabetical index of English translations of the entries. Useful as a working tool for international communication in the fields of technology and engineering. For university and special libraries.
R: *ARBA* (1985, p. 537).

ENGLISH–RUSSIAN

The Concise Illustrated Russian-English Dictionary of Mechanical Engineering: 3795 Terms. **Vladimir V. Shvarts.** Moscow: Russian Language Publishers, 1981.

This dictionary contains a supplement to many of the terms with notes given in the internationally accepted symbols, formulas, diagrams, engineering drawings, sketches, etc. Arranged according to subject with an index of Russian terms.
R: *Science and Technology Libraries* 2: 115 (Summer 1982).

The Concise Russian–English Chemical Glossary: Acids, Esters, Ethers, and Salts. **James F. Shipp.** College Park, MD: Wychwood Press, 1983.

A specific guide in translating information concerning the 4 basic substances commonly found in chemical literature: acids, esters, ethers, and salts. For those specifically interested in this area.
R: *ARBA* (1984, p. 637).

English–Russian Dictionary of Applied Geophysics. **B. V. Gusev et al.** New York: Pergamon Press, 1984.

Contains over 30,000 alphabetized entries of terms used in geophysics and the petroleum industry. Includes a section on acronyms and their terms. Useful for petroleum engineers or businessmen who have dealings within the Soviet bloc.
R: *ARBA* (1985, p. 606).

English–Russian Dictionary of Refrigeration and Low-Temperature Technology. **Mikhail B. Rozenberg, ed.** Elmsford, NY: Pergamon, 1979.

Contains approximately 20,000 terms. For specialists in low-temperature physics and refrigeration.
R: *ASHRAE Journal* 21: 80 (Dec. 1979); *TBRI* 46: 78 (Feb. 1980).

English–Russian Physics Dictionary: About 60,000 Terms. **D. M. Tolstoi, ed.** Elmsford, NY: Pergamon, 1978.

Provides 60,000 terms from physics and related subjects. Russian equivalents are included.
R: *New Scientist* 81: 500 (Feb. 15, 1979).

English-Russian Polytechnical Dictionary: 80,000 Terms. 3d ed. **A. E. Chernukhin, ed.** Elmsford, NY: Pergamon, 1976.

First edition, 1962; second edition, 1971; reprinted edition, 1976.
One of the most comprehensive Russian-English technical dictionaries. Contains some 80,000 terms.
R: *ARBA* (1978, p. 617); Sheehy (EA24).

English-Russian Reliability and Quality Control Dictionary: 22,000 Terms. **E. G. Kovalenko.** Elmsford, NY: Pergamon, 1977.

For engineering libraries, a dictionary of some 22,000 Russian definitions. Also contains a list of English abbreviations and acronyms.
R: *ARBA* (1978, p. 759).

Russian–English Chemical and Polytechnical Dictionary. 3d ed. **Ludmilla Ignatiev Callaham.** New York: Wiley, 1975.

First edition, 1947; second edition, 1967.
Primarily intended for English-speaking engineers having a basic Russian scientific vocabulary. Extensive coverage of terms in organic, inorganic, analytical, physical, and nuclear chemistry as well as chemical technology.
R: *Chemistry and Industry* 18: 978 (June 15, 1963); *Science* 140: 654 (1963); *ARBA* (1976; p. 648); Jenkins (D85); Sheehy (ED7); Wal (pp. 29, 120); Win (ED44).

Russian–English Dictionary of the Mathematical Sciences. **A. J. Lohwater and S. H. Gould.** Providence, RI: American Mathematical Society, 1974 (c. 1961).

A joint venture of the US National Academy of Sciences, the Academy of Sciences of the USSR, and the AMS. Equivalents for more than 10,000 terms. Primarily of use to scientists in mathematics and theoretical physics.
R: *ARBA* (1976, p. 644); Wal (p. 68).

Russian–English Glossary of Fishing and Related Marine Terms. **M. Ben-Yomi, ed.** Forest Grove, OR: International Scholarly Book Services, 1975.

Russian–English Index to Scientific Apparatus Nomenclature. 2d ed. **James F. Shipp.** Philadelphia: Translation Research Institute/College Park, MD: Wychwood Press, 1983.

Includes terms used in the fields of physics, chemistry, medicine, electronics, and the geosciences. Limited in its applications and identifications but useful for scientific research libraries with large Russian-language collections.
R: *ARBA* (1984, p. 610).

Russian–English Oil-Field Dictionary. **D. E. Stoliarov, ed.** New York: Pergamon, 1983.

A 30,000-entry dictionary of oil-field terms used in the Soviet oil and gas industry covering various categories. Translations are from Russian to English and should be used in a special library with access to Russian language literature in the oil and gas industries.
R: *Science Technology Libraries* 5: 2 (Winter 1984); *ARBA* (1985, p. 509).

Russian–English Polytechnical Dictionary. **B. V. Kuznetsov, ed.** New York: Pergamon, 1981.

Russian–English Translators Dictionary: A Guide to Scientific and Technical Usage. 2d ed. **M. Zimmerman.** New York: Wiley-Interscience, 1984.

Emphasis is on more recent developments of laser techniques and space research, with various examples of usage taken from the latest English books and journals in several areas of science and technology.

Transliterated Dictionary of the Russian Language. **Eugene Garfield, ed.** Philadelphia: ISI Press, 1983.

A useful transliterated dictionary of Russian place names, cognates, scientific terms, and many other words.

English–Spanish

ARCO Motor Vehicle Dictionary: Spanish-English: English-Spanish. 1st paper ed. **Robert F. Lima, ed.** New York: Arco Publishing, 1980.

Early edition, 1969.
Over 40,000 entries in both Spanish and English are included. A useful technical dictionary of potential value to many different groups of people.
R: *ARBA* (1982, p. 813).

Dictionary of Materials Testing. 3 vols. **D. Werner Goedecke, comp.** Philadelphia: Heyden, 1980.

Volume 1, German–English–French; volume 2, English–German–French; volume 3, French–German–English.

Dictionary of Mathematics in Four Languages: English, German, French, Russian. **Gunther Eisenreich and Ralf Sube, comps.** New York: Elsevier Science Publishing, 1982.

A multilingual dictionary specifically for mathematic terms. Most useful for translators, interpreters, scholars, and generally anyone working in the field of mathematics.
R: *ARBA* (1984, p. 641).

A Dictionary of Statistical, Scientific, and Technical Terms: English-Spanish; Spanish-English. **Hardeo Sahai and Jose Berrios.** Belmont, CA: Wadsworth Publishing, 1981.

Lists more than 2,800 terms used in mathematics, biology, education, and other related fields. The first half uses English terms and their Spanish equivalents, the second half uses Spanish terms with their English meanings. Useful for statisticians.
R: *ARBA* (1983, p. 714).

English–Spanish, Spanish–English Encyclopedic Dictionary of Technical Terms. 3 vols.
Javier L. Collazo. New York: McGraw-Hill, 1980.

Comprehensive 3-volume dictionary containing 143,000 current entries from 240 engineering sciences, general science and technology. Volumes 1 and 2 are English terms translated into Spanish; volume 3 is Spanish–English. Includes standard abbreviations for both Spanish and English terms. Helpful to translators, engineers, techicians, instructors, and students in the fields of electrical engineering and telecommunications. For undergraduate and graduate libraries.
R: *Choice* 18: 220 (Oct. 1980); *RQ* 19: 402 (Summer 1980); *ARBA* (1981, p. 620).

A Glossary of Agricultural Terms: English–Spanish, Spanish–English. **ACTION/ Peace Corps, Information Collection and Exchange.** Washington, DC: ACTION/Peace Corps, 1976.

A glossary compiled by agricultural specialists.
R: *RSR* 5: 28 (July/Sept. 1977).

More Than Two Languages

Complete Multilingual Dictionary of Aviation and Aeronautical Terminology: English French, Spanish. **Henri Demaison, comp.** Lincolnwood, IL: Passport Books/ National Textbook, 1984.

Contains 13,000 terms arranged into 3 categories, with the major section in English. The French and Spanish sections refer to the English section for definitions or translation. Includes several useful appendixes. For aviation collections.
R: *ARBA* (1975, p. 616).

Complete Multilingual Dictionary of Computer Terminology: English, French, Italian, Spanish, Portuguese. **Georges Nania, comp.** Lincolnwood, IL: Passport Books/National Textbook, 1984.

Contains an alphabetized, numbered list of 12,300 computer terms in English, followed by their equivalents in French, Italian, Spanish, and Portuguese. Includes a 4-page list of acronyms in English. Useful only for translation because no definitions are provided.
R: *ARBA* (1985, p. 583).

Data Systems Dictionary: English–Russian–German. **Joachim Schulz, ed.** Wiesbaden, Germany: Oscar Bradstetter Verlag. Distr. Forest Grove, OR: International Scholarly Book Services, 1977.

An authoritative dictionary for engineering libraries. Entries collected by the Control Computers Institute and Siemens AG. Includes abbreviations, classifications, and major terms of the computer field.
R: *RSR* 6: 28 (Oct./Dec. 1978); *ARBA* (1979, p. 770).

Dictionary of Agriculture: German/English/French/Spanish/Russian. 4th rev. ed. **Gunther Haensch and Gisela Haberkamp de Antón.** New York: Elsevier Scientific, 1975.
Multilingual dictionary of agriculture and related fields. Arranged alphabetically and by section. Considered essential for agricultural libraries.
R: *ARBA* (1978, p. 737).

Dictionary for Automotive Engineering: English-French-German. 2d ed. **Jean De Coster.** New York: Saur, 1986.
Includes about 1,000 English terms defined with corresponding terms in German and French.

Dictionary of Biology: English/German/French/Spanish. **Gunther Haensch and Gisela Haberkamp de Antón.** New York: Elsevier Scientific, 1976.
Covers a wide scope of terms in biology with emphasis on more current issues such as environment and ecology. Highly recommended for all types of libraries.
R: *ARBA* (1978, p. 652).

Dictionary of Chemical Terminology: In Five Languages: English, German, French, Polish, and Russian. **Dobromila Kryt, ed.** New York: American Elsevier, 1980.
The first multilingual dictionary of modern chemistry. Contains 3,800 terms from standard fields of chemistry and related subjects. Entries culled from monographs, handbooks, journals, and IUPAC and ISO publications. Arranged in English followed by equivalents in other languages. Cross-referenced and indexed. Comprehensive work for scientists, engineers, researchers, and students.
R: *Choice* 18: 920 (Mar. 1981).

Dictionary of Dairy Terminology: In English, French, German and Spanish. **International Dairy Federation, comp.** New York: Elsevier, 1983.
A multilingual dictionary containing over 4,000 terms from the fields of dairy technology and economics, husbandry, biochemistry, immunology, microbiology, and nutrition. Terms are alphabetized in English and translated in French, Spanish, and German. For agricultural collections in special libraries.
R: *ARBA* (1985, p. 497).

Dictionary of Microprocessor Systems: In Four Languages: English, German, French, Russian. **Dieter Muller, ed.** New York: Elsevier, 1984.
Entries are listed in English, followed by the corresponding words or phrases in the other languages. An alphanumeric numbering system enables access to the English list from the other 3 word lists. Useful for researchers and engineers in the field. For large scientific and public libraries.
R: *ARBA* (1985, p. 582).

Dictionary of Physical Metallurgy: English, German, French, Polish, Russian. **Eugeniusz F. Tyrkiel.** New York: Elsevier Scientific, 1977.

Multilingual dictionary of 2,300 terms. Clear format allows for easy retrieval of information. Contains a wide scope of terms and definitions. Recommended for special libraries.
R: *Journal of Metals* 30: 25 (June 1978); *Metals Technology* 5: 176 (May 1978); *Choice* 15: 846 (Sept. 1978); *RSR* 6: 28 (Oct./Dec. 1978); *TBRI* 44: 319 (Oct. 1978); *ARBA* (1979, p. 776).

Dictionary of Plastics Technology in Four Languages: English, German, French, Russian. New York: Elsevier Scientific, 1982.

Some 8,700 terms on materials, production aids, processing, and applications concerning plastics technology are listed, first in English and then with parallel columns of the German, French, and Russian translations. Coded to allow for conversion from one of the non-English languages to another. For special libraries and those persons involved with foreign language works in plastics technology.
R: *Science and Technology Libraries* 4: 124 (Winter 1983).

Dictionary of Surface Active Agents, Cosmetics and Toiletries: English, French, German, Spanish, Italian, Dutch, Polish. **Gerardus Carriere.** New York: Elsevier Scientific, 1978.

Seven hundred entries in English, French, German, Spanish, Italian, Dutch, and Polish pertaining to the cosmetics industry. For chemists and those who deal with detergents and cosmetics.
R: *Chemistry in Australia* 46: 268 (June 1979); *Cosmetics and Toiletries* 93: 79 (Nov. 1978); *Drug & Cosmetic Industry* 123: 74 (Oct. 1978); *TBRI* 45: 32 (Jan. 1979); 45: 72 (Feb. 1979); 45: 348 (Nov. 1979); *ARBA* (1981, p. 634).

Elsevier's Dictionary of Automotive Engineering in Five Languages: English–German–French–Italian– and–Spanish. **Kohji Kondo, comp.** New York: Elsevier Scientific, 1977.

A multilingual dictionary divided into 3 parts: English terms with French, Italian and German definitions, separate language indexes, and illustrations of major components of automobiles. Excellent illustrations make this an unusual reference tool for engineering and technical libraries.
R: *RSR* 6: 26 (Oct./Dec. 1978); *ARBA* (1978, p. 764); Wal (p. 421).

Elsevier's Dictionary of Botany, I: Plant Names, in English, French, German, Latin and Russian. **P. Macura, comp.** New York: American Elsevier, 1979.

This volume of a multilingual dictionary consists of over 6,000 numbered entries pertaining to trees, plants, lichens, and mushrooms. Contains a Basic Table of alphabetical English entries with their multilingual equivalents, excluding Russian terms, which are collected in a separate numerical index. Each language provides its own index of terms. A well-designed volume for libraries serving users who need specialized dictionaries with multilingual equivalents.
R: *Nature* 284: 382 (Mar. 27, 1980); *RSR* 8: 18 (July/Sept. 1980); *ARBA* (1981, p. 649).

Elsevier's Dictionary of Botany, II: General Terms in English, French, German and Russian. **P. Macura, comp.** New York: Elsevier, 1982.

The Basic Table includes a list of nearly 10,000 English terms related to botany, and their French and German equivalents. Other sections of the book index words in French, German, and Russian. Well-organized, useful to specialists in the field.
R: *ARBA* (1983, p. 621).

Elsevier's Dictionary of Brewing: In English, French, German and Dutch. **European Brewery Convention, comp.** New York: Elsevier, 1983.

An up-to-date volume containing 4,000 terms. Entries are numbered and alphabetized in English and are followed by separate listings in French, German, and Dutch. Useful addition to business, scientific, and research libraries.
R: *ARBA* (1985, p. 497).

Elsevier's Dictionary of Food Science and Technology: In Five Languages: English–French–Spanish–German with an Index of Latin Names. **Ian D. Morton and Chloe Morton, eds.** New York: Elsevier Scientific, 1977.

A multilingual dictionary of terms relating to food science. Arranged alphabetically by English terms. French, Spanish, and German equivalents are provided. Cross-referenced, well recommended.
R: *Chemistry and Industry* 16: 624 (Aug. 19, 1978); *TBRI* 44: 356 (Nov. 1978); *ARBA* (1978, p. 744); Wal (p. 551).

Elsevier's Dictionary of Horticulture in Nine Languages: English, French, Dutch, German, Danish, Swedish, Spanish, Italian, Latin. **J. Nijdam, ed.** Compiled under the auspices of the Ministry of Agriculture and Fisheries at The Hague, The Netherlands. New York: American Elsevier, 1970.

Some 4,000 names and terms pertaining to many aspects of horticulture. Contains scholarly details such as authorities for Latin binomials of plants, diseases, and insects. Does not include highly specialized terms.
R: *LJ* 95: 2787 (Sept. 1, 1970); *WLB* 45: 89 (Sept. 1970); *ARBA* (1971, p. 503).

Elsevier's Dictionary of Measurement and Control in Six Languages: English/American–French–Spanish–Italian–Dutch–and–German. **W. E. Clason, comp.** New York: Elsevier Scientific, 1977.

Arranged alphabetically by English language term, with foreign equivalents provided. Definitions reflect the growth of the control field in such areas as engineering and mathematics. For special library reference collections.
R: *Choice* 15: 208 (Apr. 1978); *RSR* 6: 26 (Oct./Dec. 1978); *ARBA* (1979, p. 762); Wal (p. 340).

Elsevier's Dictionary of Metallurgy and Metal Working (English, French, Spanish, Italian, Dutch and German). **W. E. Clason.** Amsterdam: Elsevier, 1978.

Earlier edition entitled *Elsevier's Dictionary of Metallurgy in Six Languages: English/American, French, Spanish, Italian, Dutch and German*, 1967.
Equivalent terms without definitions.

Elsevier's Dictionary of Tools and Ironware in Six Languages: English/American, French, Spanish, Italian, Dutch and German. **W. E. Clason, comp.** New York: Elsevier, 1982.

Lists 2,576 English/American words relating to tools and ironware, followed by their equivalents in the other 5 languages. A list of the words in the other languages is given, with cross references to the English. Gives no definitions.
R: *ARBA* (1983, p. 725).

Elsevier's Nautical Dictionary in Six Languages: English/American, French, Spanish, Italian, Dutch, and German. 2d rev. ed. **J. P. Vandenberghe and L. Y. Chaballe, comps.** New York: American Elsevier, 1978.

Contains over 18,700 numbered entries from nautical and maritime subjects such as shipbuilding, vessels, navigation, and instruments. Main section of book is composed of English terms with equivalents in other 5 languages. Five indexes of terms follow the Basic Table, as well as an up-to-date bibliography. For business and special libraries with maritime collections.
R: *ARBA* (1980, p. 730).

Elsevier's Oil and Gas Field Dictionary: In Six Languages: English/American, French, Spanish, Italian, Dutch, and German. **L. Y. Chaballe, L. Masuy, and J. P. Vandenberghe, comps.** New York: American Elsevier, 1980.

A multilingual dictionary of 4,800 common terms from the natural gas and petroleum industry. Emphasizes geology, offshore conditions, drilling, and preparation of gas and oil for shipping. Arranged by English term; includes index in each language and cross-referenced Arabic supplement. For graduate or professional energy collections.
R: *Choice* 18: 1068 (Apr. 1981).

Elsevier's Dictionary of Personal and Office Computing: In English, German, French, Italian, and Portuguese. **O. Vollnhals.** New York: Elsevier North Holland, 1984.

A multilingual dictionary containing over 5,100 terms relating to personal office, mini-, micro-, and home computers. Areas covered include hardware, software, programming, applications, and allied fields. The main section is an alphabetized list of English terms followed by their equivalents in German, French, Italian, and Portuguese. An up-to-date, complete reference for computer manufacturers, vendors, translators, and users of office automation equipment and microcomputers.

Elsevier's Telecommunication Dictionary in Six Languages: English/American-French-Spanish-Italian-Dutch-German. 2d rev. ed. **W. E. Clason, comp.** New York: American Elsevier, 1976.

Divided into 33 areas of telecommunications, this language dictionary provides terms in English with the corresponding equivalents in each of the 5 other languages.

Energy Terminology: A Multilingual Glossary: A Glossary for Engineers, Research Workers, Industrialists and Economists Containing over 1,000 Standard Energy Terms in English, French, German and Spanish. Rev. and enl. ed. **The World Energy Conference.** New York: Pergamon, 1983.

Terms are arranged by topic, type of energy, or technological applications. With alphabetical indexes. Recommended for the special, reference, or larger public library.
R; *Choice* 21: 250 (Oct. 1983); *ARBA* (1985, p. 505).

Glossary of Soil Micromorphology: English, French, German, Spanish and Russian. **A. Jongerius and G. K. Rutherford, eds.** New York: Unipub, 1979.

An important glossary of soil micromorphology. Contains 661 terms classified under 51 headings. Definitions include quotations, sources, cross references, variations in different languages, and translations into each of the 4 languages. A complete, international tool that is invaluable to English-speaking soil micromorphologists translating from other languages. Subject indexes are included.
R: *Soil Science* 128: 254 (Oct. 1979); *TBRI* 46: 29 (Jan. 1980); *ARBA* (1980, p. 659).

Glossary of Transport: English, French, Italian, Dutch, German, Swedish. **Gordon Logie.** Amsterdam: Elsevier Scientific, 1980.

Presents 1,108 commonly used transportation terms and phrases beginning with English and following through with the other 5 languages. Covers 9 major categories—transport, transportation study, roads, road traffic, parking, road vehicles, railways, water-borne transport, and aviation—which are all further categorized. Because of the omission of Spanish entrants, this book is beneficial for libraries with special collections in regional planning or transportation.
R: *ARBA* (1982, p. 814).

IEC Multilingual Dictionary of Electricity. New York: Institute of Electrical and Electronics Engineers. Distr. New York: Wiley, 1984.

This volume is the English text section of a multilingual dictionary in both French and English with equivalent terms in 7 languages. An accompanying second volume contains an alphabetical index for each of the other 7 languages. It is an important work aimed at the internationalization of standards and terminology in science and technology.
R: *ARBA* (1985, p. 546).

Illustrated Glossary of Process Equipment. **Bernard H. Paruit, ed.** Houston: Gulf Publishing, 1984.

An illustrated dictionary covering the equipment and terms used in the chemical and hydrocarbon industries. In 3 languages: English, French, and Chinese. For special libraries.
R: *ARBA* (1985, p. 543).

International Maritime Dictionary: An Encyclopedic Dictionary of Useful Maritime Terms and Phrases, Together with Equivalents in French and German. 2d ed. **Rene de Kerchove.** New York: Van Nostrand Reinhold, 1984.

A reprint of the standard authoritative reference on terms relating to ships and all phases of maritime activity. Defines 10,000 words and phrases associated with such topics as seamanship, naval architecture, navigation, oceanography, and shipbuilding. Well illustrated and referenced. International in scope, with French and German equivalents provided for each term. A valuable addition to all library collections.
R: *ARBA* (1985, p. 621).

Multilingual Compendium of Plant Diseases. 2 vols. **Paul R. Miller and Hazel L. Pollard.** St. Paul, MN: American Phytopathological Society, 1976–1977.

In 2 volumes, this work covers diseases caused by bacteria and fungi and diseases caused by nematodes and viruses. Provides concise descriptions and definitions used in plant pathology. Includes indexes for each language and color photograph identification aids. Highly recommended for plant pathologists.
R: *RSR* 5: 8 (Apr./June 1977); 7: 12 (July/Sept. 1979).

The Multilingual Computer Dictionary. **Alan Isaacs, ed.** New York: Facts on File, 1981.

Presents a listing of programming and data-processing terms in English, German, Spanish, Italian, and Portuguese. It was carelessly prepared and could use some major corrections.
R: *ARBA* (1982, p. 807).

Multilingual Dictionary of Concrete. **Federation Internationale de la Precontrainte.** Amsterdam: Elsevier Scientific, 1976.

For engineering libraries, a dictionary of construction terms in English, French, German, Spanish, Dutch, and Russian.
R: *RSR* 5: 25 (Oct./Dec. 1977).

The Multilingual Energy Dictionary. **Alan Isaacs, ed.** New York: Facts on File, 1981.

Presents a listing of the basic terminology used in the production and manipulation of energy in English, French, German, Italian, Portuguese, and Spanish. Unorganized presentation of material.
R: *ARBA* (1982, p. 745).

New International Dictionary of Refrigeration. **International Institute of Refrigeration.** Elmsford, NY: Pergamon, 1977.

Multilingual English, French, German, Italian, Norwegian, Russian, and Spanish dictionary of refrigeration terms. Term presented in each language with English and French definitions. A comprehensive source in refrigeration technology.
R: *IBID* 6: 172 (June 1978).

Sanyo's Trilingual Glossary of Chemical Terms: English–Japanese–Chinese. **Hiroshi Yamada, comp.** Tokyo: Sanyo Shuppan Boeki. Distr. Philadelphia: Sadtler Research Laboratories, 1976.

One-volume glossary in 2 parts: alphabetical listing of English terms with Chinese and Japanese equivalents, and Chinese and Japanese indexes. Useful to chemists and engineers—English, Chinese and Japanese.
R: *Farm Chemicals* 140: 72 (Jan. 1977); *TBRI* 43: 133 (Apr. 1977); *ARBA* (1978, p. 641).

Science and Technology for Development: Terminology Bulletin. UN79/1/15 **United Nations.** New York: Unipub, 1979.

A compilations of over 2,800 terms in English alphabetical order. French, Russian, Spanish, Chinese, and Arabic equivalents are included. Indexed.

Standard Terms of the Energy Economy: A Glossary for Engineers, Research Workers, Industrialists and Economists Containing over 600 Standard Energy Terms in English, French, German and Spanish. **World Energy Conference, comp.** Elmsford, NY: Pergamon, 1978.

A multilingual dictionary of energy terms. Definitions are succinct. Includes multilingual index. Terms presented under 8 headings, including electricity industry, mining and processing of solid fuels, nuclear power technology, and water power.
R: *ARBA* (1979, p. 690).

SUBJECT DICTIONARIES

GENERAL—UNITS AND MEASUREMENTS

A Dictionary for Unit Conversion. **Yishu Chiu.** Washington, DC: George Washington University, School of Engineering and Applied Science, 1975.

A Dictionary of Scientific Units, Including Dimensionless Numbers and Scales. 4th ed. **H. G. Jerrerd and D. B. McNeill.** New York: Barnes & Noble, 1980.

Third edition, 1972.
Alphabetical arrangement for several units. Includes definitions and historical references. Emphasizes lesser-known units.
R: *Pharmaceutical Journal* 139: 68 (July 1964); *WLB* 47: 293 (Nov. 1972); *ARBA* (1973, p. 627); (1982, p. 696); Wal (p. 48).

Quantities and Units of Measurement: A Dictionary and Handbook. **J. V. Drazil.** London: Mansell. Distr. New York: Wilson, 1983.

A revised and expanded version of the *Dictionary of Quantities and Units*, 1971, by the same author. Part 1 is an alphabetical listing of units and provides symbols in SI, uses, and conversion factors to other SI units. Part 2 is an alphabetical list of quantities and selected constants. A comprehensive handbook suited for ready reference and essential for researchers, scientists, and librarians.
R: *ARBA* (1985, p. 610).

GENERAL SCIENCE

Chambers Dictionary of Science and Technology. 2 vols. Totowa, NJ: Littlefield, Adams, 1976.

R: *Nature* 236: 246 (Mar. 31, 1972); *New Scientist* 45: 577 (Mar. 19, 1970); *BL* 73: 1600 (June 15, 1977); *Choice* 9: 623 (July/Aug. 1972); *WLB* 47: 293 (Nov. 1972); *ARBA* (1973, p. 534; 1977, p. 628); Wal (p. 22).

Coming to Terms: From Alpha to X-ray: A Lexicon for the Science Watcher. **Wayne Biddle.** New York: Viking Press, 1981.

A listing of terms and concepts used in various scientific disciplines including biology, physics, astronomy, energy, and ecology. Illustrations are included as well as sources to consult for further reference. For academic and public libraries.
R: *LJ* 106: 1296 (June 15, 1981).

The Compact Dictionary of Exact Science and Technology. Vol. 1: English–German. **A. Rucera.** London: Oscar Brandstetter, 1980.

R: *Laboratory Practice* 29: 1293 (Dec. 1980).

Dictionary of Inventions and Discoveries. 2d rev. ed. **E. F. Carter, ed.** New York: Crane-Russak, 1976.

Earlier edition, 1966; first revised edition, 1969.
Primarily a student's or layperson's guide to major scientific and technological inventions and inventors. Alphabetical arrangement.
R: *NTB* 52: 88 (1967); *ARBA* (1970, p. 101); Jenkins (K64); Wal (p. 239).

A Dictionary of Named Effects and Laws in Chemistry, Physics, and Mathematics. 4th ed. **Denis William George Ballentyne and D. R. Lovett.** New York: Barnes & Noble, 1980.

First edition, 1958; second edition, 1961; third edition, 1971.
A number of laws and effects named for their discoverers. Intended for students and professionals.
R: *Journal of the American Chemical Society* 104: 2678 (May 1982); *Science and Technology Libraries* 2: 78 (Spring 1982); *ARBA* (1982, p. 696); Jenkins (A55); Wal (pp. 22, 67, 96, 119); Win (EA79).

Dictionary of Scientific and Technical Terms. 2d ed. **Daniel N. Lapedes, ed.** New York: McGraw-Hill, 1978.

First edition, 1974.
R: *American Scientist* 63: 484 (July/Aug. 1975); *Journal of the American Chemical Society* 101: 5866 (Sept. 12, 1979).

Gerrish's Technical Dictionary: Technical Terms Simplified. **Howard H. Gerrish.** South Holland, IL: Goodheart-Wilcox, 1982.

Provides easily understood definitions of technical terms used in trades such as electronics, graphics, and many more. Contains some line drawings and biographical information to clarify definitions. Useful for vocational schools and students.
R: *ARBA* (1983, p. 714).

Longman Dictionary of Scientific Usage. **A. Godman and E. M. Payne.** New York: Longman, 1980.

R: *Laboratory Practice* 30: 55 (Jan. 1981); *Nature* 286: 745 (Aug. 14, 1980).

Longman Illustrated Science Dictionary. **Arthur Godman.** New York: Longman, 1981.

An unconventional dictionary listing 1,500 scientific terms in subfields within the broad subjects of physics, biology, and chemistry. Extensive cross referencing makes it similar to an encyclopedia. Definitions are at the general reader's level. A backup tool.
R: *LJ* 107: 982 (May 1982).

McGraw-Hill Dictionary of Science and Engineering. **Sybil P. Parker, ed.** New York: McGraw-Hill, 1984.

A standard-size volume containing 35,000 terms from 102 fields of science and engineering. Terms selected from the *McGraw-Hill Dictionary of Scientific and Technical Terms*. Provides authoritative definitions and cross references. Suitable for anyone reading or writing about science and engineering as a nonprofessional. For public and school libraries.
R: *Choice* 22: 405 (Nov. 1984); *LJ* 109: 1315 (July 1984); *RQ* 24: 108 (Fall 1984); *WLB* 59: 67 (Sept. 1984); *ARBA* (1985, p. 490).

McGraw-Hill Dictionary of Scientific and Technical Terms. 3d ed. **Sybil P. Parker, ed.** New York: McGraw-Hill, 1984.

First edition, 1974; second edition, 1978.
Definition of terms from 100 general and specialized fields of science and technology. With emphasis in electronics and computer science in this edition. Broad in its uses.
R: *ARBA* (1984, p. 609).

The Penguin Dictionary of Science. 5th ed. **E. B. Uvarov, D. R. Chapman, and Alan Isaacs.** London: Harmondsworth, Penguin, 1979.

Third edition, 1971; fourth edition, 1977.
A book of several thousand scientific definitions from the fields of physics, chemistry, mathematics, and astronomy, with a few from biology as well. About 200 (mostly trivial) mistakes have been detected. Good for students and teachers.
R: *Physics Bulletin* 31: 175 (June 1980).

Physical Sciences Dictinary: Terms, Formulas, Data. **Cesare Emiliani.** New York: Oxford University Press, 1986.

A comprehensive 1-volume dictionary presents clear definitions of terms from physics, chemistry, the geological sciences, and cosmology. Symbols and abbreviations are spelled out. Features extensive sets of tables that complement the dictionary entries. Intended for students, researchers, and lay readers.

Science Policy: A Working Glossary. 3d ed. **US Congress, House of Representatives, Subcommittee on Science, Research, and Technology.** Science Policy Research Division, US Library of Congress, prep. Washington, DC: US Government Printing Office, 1976.

Alphabetical listing of 200 science terms used in determining scientific policy. Updated and revised edition contains information not available elsewhere. For librarians and scientists.
R: *ARBA* (1977, p. 629).

Scientific Words: Their Structure and Meaning. **Walter Edgar Flood.** Westport, CT: Greenwood Press, 1974.

Reprint of 1960 edition.
Alphabetical listing of roots, prefixes, and suffixes used in the formation of scientific terms. Considerable explication.
R: Wal (p. 23).

Astronomy

Astronomical Directory. Toronto, Ontario: Gall Publications, 1978.

Over 3,000 worldwide listings of planetariums, amateur groups, observatories, publications, and suppliers.
R: *Sky and Telescope* 57: 289 (Mar. 1979).

Astronomy: A Dictionary of Space and the Universe. **Iain Nicholson.** London: Arrow, 1977.

A comprehensive dictionary of space flight. Includes many diagrams and some color plates.
R: *Space Flight* 19: 452 (Dec. 1977); *TBRI* 44: 209 (June 1978).
Observatory 98: 71 (Apr. 1978).

The A–Z of Astronomy. **Patrick Moore.** New York: Scribner's, 1976.

Written for the amateur astronomer. Contains several hundred definitions and descriptions of celestial bodies, scientific instruments, and astronomical terms. For school and public libraries.
R: *BL* 74: 702 (Dec. 15, 1977); *RQ* 17: 275 (Spring 1978); *WLB* 52: 186 (Oct. 1977).

Dictionary of Astronomy, Space, and Atmospheric Phenomena. **David F. Tver, Lloyd Motz, and William K. Hartmann.** New York: Van Nostrand Reinhold, 1979.

Broad coverage of current advancements in meteorology, space exploration, and astronomy, as well as physics and mathematics. Contains about 2,300 brief entries; many are accompanied by tables of data and numerous line drawings. An appendix consists of data on the stars, the constellations, Messier objects, and other objects of the solar system. Cross-referenced. Addressed to the needs of amateur astronomers, students, and the general public.
R; *New Scientist* 87: 395 (July 31, 1980); *Sky and Telescope* 59: 235 (Mar. 1980); 59: 62 (Jan. 1980); *BL* 76: 1695 (July 15, 1980); *Choice* 17: 206 (Apr. 1980); *ARBA* (1980, p. 610).

Dictionary of Astronomy: Terms and Concepts of Space and the Universe. **Iain Nicolson.** New York: Barnes & Noble, 1980.

A comprehensive, reliable source of information on all aspects of astronomy including satellites, meteors, asteroids, and "intelligent life in the universe." Well-illustrated with useful diagrams. For astronomy students and enthusiasts.
R: *ARBA* (1981, p. 629).

The Facts on File Dictionary of Astronomy. **Valerie Illingworth, ed.** New York: Facts on File, 1979.

Up-to-date dictionary of astronomy which includes 1,700 basic terms found in astronomy, astrophysics, and space exploration. Contains numerous tables and 100 line drawings, which enhance the text. Alphabetical entries are brief but adequate. This book contains information not found in other dictionaries and is useful for students as well as specialists.

R: *Sky and Telescope* 58: 568 (Dec. 1979); 59: 235 (Mar. 1980); *BL* 76: 1151 (Apr. 1, 1980); 76: 1159 (Apr. 15, 1980); 76: 1240 (May 1, 1980); 76: 1565 (June 15, 1980); *LJ* 104: 2557 (Dec. 1, 1979); *ARBA* (1980, p. 609).

Glossary of Astronomy and Astrophysics. 2d ed. **Jeanne Hopkins.** Chicago: University of Chicago Press, 1980.

First edition, 1976.

Contains about 2,000 definitions, emphasizing terms used in current research. For astrophysics and science fiction collections.

R: *Astronomy* 5: 57 (May 1977); *Journal of the Royal Astronomical Society of Canada* 70: 206 (Aug. 1976); *Science* (June 18, 1976); *RSR* 5: 17 (Apr./June 1977); *TBRI* 43: 5 (Jan. 1977); 43: 164 (May 1977); 43: 204 (June 1977); *ARBA* (1977, p. 636); (1982, p. 706).

Key Definitions in Astronomy. **Jacqueline Mitton.** Totowa, NJ: Littlefield, Adams, 1982.

This work provides the most important and useful terms in the field of astronomy and covers those likely to be found in popular journals, newspapers, and on television. Limited in scope but intended for the educated layperson and student in this field.

R: *ARBA* (1983, p. 603).

MATHEMATICS

Dictionary of Gaming, Modelling and Simulation. **G. Ian Gibbs.** Beverly Hills, CA: Sage Publications, 1978.

Includes numerous statistical tests, mathematical formulas, and helpful illustrations that explain some terms. For public libraries as well as academic libraries with collections in electrical engineering, statistics, and computer science.

R: *Choice* 17: 48 (Mar. 1980).

Dictionary of Mathematics. **T. Alaric Millington and William Millington.** New York: Barnes & Noble, 1971 (c. 1966).

Conceptual and technical information for beginning students. Similar in scope to *The Crescent Dictionary of Mathematics*, New York: Macmillan, 1962.

R: *ABL* 31: entry 112 (Mar. 1966); *ARBA* (1972, p. 551); Jenkins (B35); Wal (p. 67); Win (1EF3).

A Dictionary of Statistical Terms. 4th rev. ed. **M. G. Kendall and W. R. Buckland.** New York: Longman, 1982.

Contains some 3,000 statistical terms, definitions, and many new concepts in a clear and easy-to-read format. A helpful reference aid for those in the statistical consultation field.

Dictionary of Symbols of Mathematical Logic. **R. Feys and F. B. Fitch.** Amsterdam: North Holland. Distr. New York: Humanities Press, 1969.

Explains and translates symbols currently used in mathematical logic. Covers many symbolic languages, with an emphasis on journalized deductive systems. Intended for both expert and novice.
R: *ABL* 34: entry 158 (Apr. 1969); *ARBA* (1971, p. 481); Wal (p. 71).

Encyclopedic Dictionary of Mathematics. 2 vols. **Mathematical Society of Japan.** Edited by Shokichi Iyanaga and Yukiyosi Kawada. Cambrige, MA: MIT Press, 1977.

An excellent dictionary, cumulated by international experts in mathematics. Entries from all fields of mathematics contain cross references. Recommended for specialists and advanced students.
R: *RSR* 6: 19 (July/Sept. 1978); *WLB* 52: 587 (Mar. 1978); Sheehy (EF5).

Encyclopedic Dictionary of Mathematics for Engineers and Applied Scientists. **Ian N. Sneddon, ed.** Elmsford, NY: Pergamon, 1976.

Alphabetical arrangment of terms used in math, math engineering, physics, and related fields. SI units used. A well-produced single-volume dictionary.
R: *Aeronautical Journal* 80: 371 (Aug. 1976); *Journal of Nuclear Energy* 4: 205 (1977); *RSR* 5: 24 (Oct./Dec. 1977); 6: 19 (July/Sept. 1978); *TBRI* 43: 79 (Feb. 1977); 43: 400 (Dec. 1977).

Facts on File Dictionary of Mathematics. **Carol Gibson, ed.** New York: Facts on File, 1981.
R: *BL* 77: 985 (Mar. 1, 1981).

A Handbook of Terms Used in Algebra and Analysis. **A. G. Howson, comp.** New York: Cambridge University Press, 1972.

Intended for undergraduate students. Nonalphabetic arrangement. Definitions given are mostly clear.
R: *ARBA* (1973, p. 539).

Mathematics Dictionary. 4th ed. **Glen James and R. C. James, eds.** Princeton, NJ: Van Nostrand Reinhold, 1976.

First edition, 1942; second edition, 1959, third edition, 1968.
Reliable general coverage for students, scientists, and engineers. Does presuppose a basic knowledge of the field. Includes logarithm tables, mathematical formulas, and a list of mathematical symbols. A multilingual edition is available.
R: *BL* 73: 452 (May 15, 1977); *Choice* 5: 1561 (1968); 13: 572 (Feb. 1977); *LJ* 93: 2847 (1968); *ARBA* (1977, p. 634); Jenkins (B33); Sheehy (EF4); Wal (p. 67); Win (EF16; 2EF2).

PHYSICS

Concise Dictionary of Physics. **J. Thewlis.** Elmsford, NY: Pergamon, 1979.
For a wide range of users, a truly concise dictionary of physical terms.
R: *Nature* 284: 82 (Mar. 6, 1980).

Concise Dictionary of Physics and Related Subjects. 2d rev. ed. **James Thewlis.** Elmsford, NY: Pergamon, 1979.

First edition, 1973.
Contains about 7,100 brief definitions pertaining to physics. Revised appendixes include complete listings of SI units and physical constants. Cross-referenced. For beginning students and the layperson. Recommended for academic, special, and public library collections.
R: *Physics Bulletin* 30: 532 (Dec. 1979); *Physics Education* 14: 456 (Nov. 1979); *TBRI* 46: 93 (Mar. 1980); *ARBA* (1981, p. 639); Wal (p. 149).

A Dictionary of Physical Sciences. **John Daintith, ed.** New York: Pica Press. Distr. New York: Universe Books, 1977.
Succinct volume of some 4,000 terms in astronomy, chemistry, and physics. Includes diagrams and drawings.
R: *Physics Bulletin* 29: 24 (Jan. 1978); *Physics Teacher* 15: 376 (Sept. 1977); *Aslib Proceedings* 42: 323 (June 1977); *BL* 74: 490 (Nov. 1, 1977); *Choice* 14: 1338 (Dec. 1977); *LJ* 102: 1623 (Aug. 1977); *TBRI* 43: 323 (Nov. 1977); *ARBA* (1978, p. 617); Sheehy (EA17).

Dictionary of Physics and Allied Sciences. 2 vols. **Charles J. Hyman and Ralph Idlin, eds.** New York: Ungar Publishing, 1978.
An essential dictionary of the physical sciences. Bilingual English-German/German-English.
R: *Choice* 15: 1192 (Nov. 1978); *ARBA* (1979, p. 659).

A Dictionary of Spectroscopy. 2d ed. **R. C. A. Denney.** New York: Wiley, 1982.
First edition, 1973.
Ready reference for interested students and general spectroscopists.
R: *Physics Today* 36: 74 (July 1983); *Choice* 21: 249 (Oct. 1983); *ARBA* (1975, p. 655).

Dictionary of Thermodynamics. **A. M. James.** New York: Halsted, 1976.
Provides sketches of important concepts, equations, and formulas. Comprises a ready reference for physicists and chemists. Also provides a bibliography and tabular data. Alphabetically arranged.
R: *Science News* 111: 319 (May 1977); *Choice* 13: 1572 (Feb. 1977); *RSR* 5: 37 (Oct./Dec. 1977); 6: 45 (Apr./June 1978); *ARBA* (1978, p. 639).

Encyclopaedic Dictionary of Physics. 9 vols. **J. Thewlis, ed.** Oxford and New York: Pergamon Press, 1961–1964. Also, supps. 1–5, 1966–1975.
Volume 8, subject and author indexes; volume 9, multilingual glossary. The nine basic volumes are updated via irregularly published supplements. Highly technical.
R: *Chemistry and Industry* 31: 1273–1275 (Aug. 3, 1963); *Choice* 4: 636 (1967); *Subscription Books Bulletin* 64: 1061 (1968); *ARBA* (1970, p. 111); Jenkins (C50); Sheehy, (EG2); Wal (p. 94); Win (EG14; 1EG3; 2EG3; 3EG3).

Facts on File Dictionary of Physics. **Eric Deeson, ed.** New York: Facts on File, 1981.
R: *BL* 77: 985 (Mar. 1, 1981).

Glossary for Radiologic Technologists. **Patricia A. Myers and Therese A. Martin.** New York: Praeger Publishers, 1981.

A glossary of current terminology used in the field of medical radiography. Includes brief descriptive definitions; numerous references, photos, and freehand drawings; a glossary; and a list of abbreviations. For community college and undergraduate libraries. Especially helpful to student radiographers.

R: *Choice* 18: 1396 (June 1981).

Illustrated Glossary for Solar and Solar-Terrestrial Physics. **A. Bruzek and C. J. Durrant, eds.** Dordrecht, Holland: Reidel, 1977.

An advanced dictionary of terms related to solar physics. Arranged by subject and chapter, well indexed, thorough definitions.

R: *Aslib Proceedings* 43: 263 (June 1978); *New Scientist* 78: 314 (May 4, 1978); *Observatory* 99: 99 (June 1979); *Physics Bulletin* 29: 527 (Nov. 1978); *Physics Today* 31: 60 (Sept. 1978); *RSR* 6: 21 (July/Sept. 1978); *Sky and Telescope* 56: 151 (Aug. 1978).

McGraw-Hill Dictionary of Physics and Mathematics. **Daniel N. Lapedes, ed.** New York: McGraw-Hill, 1978.

An offshoot of the *McGraw-Hill Dictionary of Scientific and Technical Terms.* Contains 20,000 terms and definitions from physics, mathematics, and related fields such as astronomy, electronics, and statistics. Provides synonyms, acronyms, abbreviations, and 700 illustrations. Appendix comprises helpful tables and diagrams. Cross-referenced; an authoritative, comprehensive dictionary recommended to all types of libraries, including public and special.

R: *American Scientist* 67: 226 (Mar./Apr. 1979); *Physics Bulletin* 30: 263 (May 1979); *Physics Teacher* 17: 138 (Feb. 1979); *Physics Today* 32: 59 (Apr. 1979); *BL* 76: 1083 (Mar. 15, 1980); *RQ* 18: 403 (Summer 1979); *TBRI* 45: 127 (Apr. 1979); *WLB* 53: 409 (Jan. 1979); *ARBA* (1979, p. 660).

The Penguin Dictionary of Physics. **Valerie H. Pitt, ed.** New York: Penguin Books, 1977.

Abridged edition of *New Dictionary of Physics.* Considered a standard reference, this book is of high quality and would be useful to students and teachers of physics.

R: *BL* 74: 640 (Dec. 1, 1977); *ARBA* (1978, p. 649).

CHEMISTRY

A–Z of Clinical Chemistry. **W. Hood.** New York: Halsted Press, 1980.

This dictionary provides information on techniques and terminologies of clinical chemistry as well as advances in the field. Contains references and presents 1,300 terms covering such topics as diagnostic methods, recent advances, and tests. Well-written, concise entries are cross-referenced. A list of general textbooks and additional readings are provided. For undergraduate students.

R: *Journal of the American Chemical Society* 103: 2912 (May 20, 1981); *Choice* 18: 1236 (May 1981).

Concise Etymological Dictionary of Chemistry. **Stanley C. Bevan et al.** Barking, Essex, England: Applied Science Publishers, 1976.

A fascinating and insightful dictionary of the meaning of chemical terminology.
R: *Analyst* 101: 830 (Oct. 1976); *Journal of the Oil and Colour Chemists' Association* 59: 296 (Aug. 1976); *TBRI* 43: 2 (Jan. 1977); 43: 82 (Mar. 1977).

The Condensed Chemical Dictionary. 9th ed. **Gessner G. Hawley, ed.** New York: Van Nostrand Reinhold, 1977.

First edition, 1919; seventh edition, 1966; eighth edition, 1971.
Continues format of previous editions, presenting both definitions and descriptions. Revised edition reflects new environmental concerns. Well-written and cross-referenced. A highly recommended reference for all chemists. Contains a wealth of information.
R: *AICHE J* 17: 1520 (Nov. 1971); *Chemical Engineering* 88: 105 (July 27, 1981); *Journal of the American Chemical Society* 93: 4637 (Sept. 8, 1971); *Choice* 3: 1003 (1966); *LJ* 93: 1585 (1968); *NTB* 52: 21 (1967); *RSR* 6: 42 (Apr./June 1978); *TBRI* 38: 11 (Jan. 1972); 43: 365 (Dec. 1977); *WLB* 52: 84 (Sept. 1977); *ARBA* (1972, p. 557; 1978, p. 640); Jenkins (D64); Sheehy (ED6); Wal (p. 118); Win (ED31; 1ED4).

Dictionary of Chemistry and Chemical Engineering (Wörterbuch der Chemie und der Chemischen Verfahrenstechnik). 2 vols. **Louis DeVries and Helga Kolb.** Weinheim, W Germany: Verlag Chemie, 1978.

Volume 1, *German/English*; volume 2, *English/German*.
Enlarged and revised bilingual dictionary. Contains excellent facts. Well-recommended.
R: *Journal of Chemical Education* 48: A786 (Dec. 1971); *Laboratory Practice* 28: 937 (Sept. 1979); *RSR* 7: 15 (Apr./June 1979).

A Dictionary of Chromatography. 2d ed. **R. C. Denney.** New York: Wiley, 1982.

For any scientist or spectroscopist who occasionally requires the use of spectroscopic terms.
R: *Librarians' Newsletter* 23: 31 (Feb. 1984).

Dictionary of Organic Compounds: The Constitution and Physical, Chemical and Other Properties of the Principal Carbon Compounds and Their Derivatives, Together with Relevant Literature References. 5th ed. 7 vols. **J. B. Buckingham, et al.** New York: Methuen, 1982. With a new supplement.

Fifth cumulative supplement, 1971; thirteenth supplement, 1977; fifteenth cumulative supplement, 1979.
Useful dictionary in 7 main volumes that supply physical data, structural formulas, and references for common organic compounds and natural products used in industry. Trade names and systematic nomenclature comprise the alphabetic listing of entries. A concise and convenient tool for university libraries.
R: *Journal of the American Chemical Society* 100: 6547 (Sept. 27, 1978); *RSR* 8: 86 (July/Sept. 1980); 19: 402 (Summer 1980); *ARBA* (1978, p. 639; 1981, p. 635).

Dictionary of Organometallic Compounds. **J. Buckingham.** New York: Methuen, 1985.

Facts on File Dictionary of Chemistry. **John Daintith, ed.** New York: Facts on File, 1981.
R: *BL* 77: 985 (Mar. 1, 1981).

Heilbron's Dictionary of Organic Compounds. 4th ed. **J. B. Thomson.** New York: Oxford University Press, 1979.
R: *Journal of the American Chemical Society* 102: 6907 (Oct. 22, 1980).

McGraw-Hill Dictionary of Chemistry. **Sybil P. Parker, ed.** New York: McGraw-Hill, 1984.
Contains 9,000 terms and definitions selected from the *McGraw-Hill Dictionary of Scientific and Technical Terms,* 3d ed., 1984. Alphabetical entries focus on the vocabulary of theoretical and applied chemistry. Includes specialized terms from the fields of atomic and nuclear physics. Cross-referenced. Valuable to chemists, chemical engineers, educators, and students in the field.
R: *ARBA* (1985, p. 603).

The Vocabulary of Organic Chemistry. **Milton Orchin et al.** New York: Wiley, 1980.
Contains definitions of over 1,200 of the most common and most important terms in organic chemistry. Entries in the 15 chapters are arranged in pedagogical order. A glossary that should be included in both graduate and undergraduate libraries.
R: *Choice* 18: 1293 (May 1981); *Journal of the American Chemical Society* 104: 1158 (Feb. 1982).

Chemical Names

Chemical Synonyms and Trade Names: A Dictionary and Commercial Handbook Containing Over 35,000 Definitions. 8th rev. ed. **Edward I. Cooke and Richard W. I. Cooke.** Cleveland: Chemical Rubber, 1978.
Seventh revised edition, 1971.
Revised edition contains 3,300 new terms and an additional 400 names in the manufacturer's index.
R: *ABL* 33: entry 334 (July 1968); *Choice* 8: 1317 (Dec. 1971); *RSR* 8: 24 (July/Sept. 1980); *ARBA* (1973, p. 543); Jenkins (D73); Sheehy (ED11); Wal (p. 472); Win (ED33).

Glossary of Chemical Terms. **Clifford A. Hampel and Gessner G. Hawley.** New York: Van Nostrand Reinhold, 1976.
Carefully selected glossary of chemical terms and acronyms. Cross-referenced. Covers all major chemical groups. Recommended for college and industrial libraries.
R: *Farm Chemicals* 140: 72 (Jan. 1977); *Food Technology* 31: 124 (Apr. 1977); *Journal of the American Chemical Society* 99: 1677 (Mar. 2, 1977); *RSR* 6: 45 (Apr./June 1978); *TBRI* 43: 124 (Apr. 1977); 43: 164 (May 1977); 43: 204 (June 1977); *WLB* 51: 445 (Jan. 1977); *ARBA* (1977, p. 645).

BIOLOGICAL SCIENCES

Barnes & Noble Thesaurus of Biology: The Principles of Biology Explained and Illustrated. **Anne C. Gutteridge.** New York: Barnes & Noble, 1983.

An excellent collection of 2,700 terms grouped by subject area. Includes helpful color illustrations, diagrams, charts, and chemical structures. Indexed. A handy reference for laypeople or students.
R: *ARBA* (1985, p. 516).

Dictionary of Life Sciences. 2d rev. ed. **E. A. Martin, ed.** New York: Pica Press. Distr. New York: Universe Books, 1983.

First edition, 1977.
Entries are encyclopedic and span all topics in the life sciences, with emphasis on genetics, molecular biology, microbiology, and immunology. Includes helpful diagrams and formulas. Useful for teachers and students of biology.
R: *BL* 74: 490 (Nov. 1, 1977); *Choice* 14: 1338 (Dec. 1977); *LJ* 102: 1625 (Aug. 1977); *ARBA* (1978, p. 653; 1985, p. 516).

Dictionary of Theoretical Concepts in Biology. **Keith E. Roe and Richard G. Frederick.** Metuchen, NJ: Scarecrow Press, 1981.

A useful volume that lists new biological concepts and the initial works and current literature in which they appear. A helpful reference for upper-level undergraduate and graduate biology students as well as professionals.
R: *Choice* 18: 1396 (June 1981); *WLB* 55: 702 (May 1981).

Facts on File Dictionary of Biology. **Elizabeth Tootill, ed.** New York: Facts on File, 1981.

R: *BL* 77: 985 (Mar. 1, 1981).

A Glossary of Genetics and Cytogenetics, Classical and Molecular. 4th ed. **R. Reiger.** New York: Springer-Verlag, 1976.

First edition, 1954: second edition, 1958; third edition, 1968.
Some 2,500 entries with list of literature citations.
R: *TBRI* 43: 337 (Nov. 1977); Wal (p. 197); Win (EC129; 3ED21).

Henderson's Dictionary of Biological Terms. 9th ed. **Sandra Holmes.** New York: Van Nostrand Reinhold, 1979.

Eighth edition, 1963.
Provides broad coverage in the enormous field of biology, which includes zoology, genetics, anatomy, and botany. Completely updates the previous edition by including terms from recent technical advancements and rapidly growing areas such as molecular biology, DNA research, and space medicine. A standard British work containing approximately 22,500 terms, which are clearly defined and cross-referenced. Features a list of abbreviations, tables, and 3 appendixes. For specialized and large reference collections.
R: *BL* 77: 1364 (June 15, 1981); *WLB* 55: 63 (Sept. 1980); *ARBA* (1981, p. 643).

Histological Methods and Terminology in Dictionary Form. **Frances M. Brimmer.** Tucson, AZ: Mosaic Press, 1980.

Includes entries dealing with enzymes, reagents, stains, and tissues that are important to histochemists as well as a variety of cell biology and biochemical phrases. Extensive cross-indexing is supplied. A supplementary reference source for upper-level undergraduate students.
R: *Choice* 18: 547 (Dec. 1980).

McGraw-Hill Dictionary of the Life Sciences. **Daniel N. Lapedes, ed.** New York: McGraw-Hill, 1976.
A dictionary of over 20,000 definitions and terms from 55 different fields of science including the biological sciences, chemistry, physics, and statistics. Synonyms are cross-referenced; over 800 illustrations are included. For both the specialist and the general public.
R: *BL* 74: 949 (Feb. 1, 1978); *RSR* 5: 7 (Apr./June 1977); *TBRI* 43: 37 (Apr. 1977); *WLB* 51: 538 (Feb. 1977); *ARBA* (1977, p. 653).

The Penguin Dictionary of Biology. 6th ed. **M. Abercrombie, C. J. Hickman, and M. L. Johnson.** New York: Penguin Books, 1980.
A compact dictionary of biology. Clear definitions are cross-referenced, and simple illustrations are included. For college and public libraries.
R: *Aslib Proceedings* 43: 151 (Mar. 1978).

The Reston Encyclopedia of Biomedical Engineering Terms. **Rudolf F. Graf and George J. Whalen.** Reston, VA: Reston Publishing, 1977.
Provides 6,500 brief medical and engineering terms. Also includes many terms of computer technology and other related disciplines. Comprehensive vital source.
R: *BL* 74: 954 (Feb. 1, 1978); *Choice* 14: 832 (Sept. 1977).

Botany

American Medical Ethnobotany: A Reference Dictionary. **Daniel E. Moerman.** New York: Garland STPM Press, 1977.
Over 3,500 entries on a broad scope of words pertaining to native American ethnobotany. Listed according to genera, uses, family, tribe, and source.
R: *LJ* 102: 1481 (July 1977); *ARBA* (1978, p. 655).

The Color Dictionary of Flowers and Plants for Home and Garden. Compact ed. **Roy Hay and Patrick M. Synge.** New York: Crown, 1982.
Earlier editions, 1969 and 1975.
Several color plates, alphabetical by botanical name within various sections and cross references to the photography sections. Supplements the Royal Horticultural Society's *Dictionary of Gardening*, which has no colored plates.
R: *LJ* 94: 2906 (1969); *ARBA* (1976, p. 754); Win (3EC8).

The Complete Dictionary of Wood. **Thomas Corkhill.** Briarcliff Manor, NY: Stein & Day, 1980.
A comprehensive dictionary of over 10,000 terms relating to trees, their properties, botanical names, etc. Includes definitions of woodworking, building, and finishing, as well.
R: *LJ* 105: 970 (Apr. 15, 1980); *ARBA* (1981, p. 757).

A Dictionary of Botany. **R. John Little and C. Eugene Jones.** New York: Van Nostrand Reinhold, 1980.

Provides up-to-date coverage with over 5,500 precisely defined entries pertaining to botany. Terms collected from over 100 sources, including texts, journals, and indexes. Well-illustrated with original line drawings; cross-referenced. A bibliography of 100 texts and glossaries concludes the work. For college and university libraries.
R: *BL* 77: 347 (Oct. 15, 1980); *LJ* 105: 1151 (May 15, 1980); *RSR* 8: 18 (July/Sept. 1980); *WLB* 54: 672 (June 1980); *ARBA* (1981, p. 648).

Dictionary of Mosses: An Alphabetical Listing of Genera Indicating Familial Disposition, Nomenclatural and Taxonomic Synonymy Together with a Systematic Arrangement of the Families of Mosses and a Catalogue of Family Names Used for Mosses. **Marshall R. Crosby and Robert E. Magill.** St. Louis: Missouri Botanical Garden, 1978.

Arranged alphabetically by moss genus. The most recent compilation of moss genera published in years. Taxonomic and nomenclatural information is provided, as well as a complete bibliography. An excellent reference for botanists.
R: *RSR* 7: 9 (Jan./Mar. 1979).

The Dictionary of Useful Plants. **Nelson Coon.** Emmaus, PA: Rodale Press, 1977.

Entries, primarily consisting of native and escaped plants of the United States are arranged alphabetically under plant families. Comprehensive data and useful illustrations for the layperson.
R: *WLB* 50: 530 (Mar. 1975); *ARBA* (1976, p. 663).

Elsevier's Dictionary of Weeds of Western Europe. **Gareth Williams, comp.** New York: Elsevier, 1982.

Lists 1,043 weeds by the names printed on labels, followed by the various common names of each weed in each of 12 European languages. The 12 other sections list the weed by the common name. Compiled to aid in choosing a herbicide to kill a specific weed.
R: *ARBA* (1983, p. 627).

The Facts on File Dictionary of Botany. **Elizabeth Tootill, ed.** New York: Facts on File, 1984.

Contains 3,000 entries in all areas of botany. Provides broad in-depth definitions of technical terms and general concepts in biochemistry, cell biology, and plant ecology. Useful reference for students and the general layperson.
R: *Choice* 22: 246 (Oct. 1984); *LJ* 109: 1438 (Aug. 1984); *ARBA* (1985, p. 517).

Glossary for Horticultural Crops. **J. Soule.** New York: Wiley, 1985.

A glossary of terms covering the entire field of horticulture and the related plant sciences, cross-referenced, indexed, and illustrated.

The Hillier Colour Dictionary of Trees and Shrubs. **Harold Hillier.** New York: Van Nostrand Reinhold, 1981.

Describes the many species and varieties of trees and shrubs that are readily found in Britain, with more than 600 color illustrations. Limited use in the United States.
R: *ARBA* (1983, p. 631).

Hortus Third: A Concise Dictionary of Plants Cultivated in the United States and Canada. **Liberty Hyde Bailey and Ethel Zoe Bailey, comps.** Revised and expanded by the staff of the Liberty Hyde Bailey Horotorium. New York: Macmillan, 1976.

First edition, *Hortus*, 1930; second edition, *Hortus Second*, 1941.
Third edition is substantially revised and considered an essential reference tool for anyone who deals with cultivated plants. Includes not only new scientific terminology but an index to 10,408 common plant names. Highly recommended for public, academic, and agricultural libraries.
R: *RSR* 5: 6 (Apr./June 1977); 7: 12 (July/Sept. 1979); *ARBA* (1978, p. 751); Sheehy (EC28).

Synopsis and Classification of Living Organisms. 2 vols. **S. Parker.** New York: McGraw-Hill, 1982.

A 2-volume comprehensive reference covering the classification and description of all living organisms. Includes an index of some 35,000 entries with scientific and common names. Useful for researchers, teachers, and students involved in comparative and taxonomic studies.

VNR Color Dictionary of Mushrooms. **Colin Dickinson and John Lucas, eds.** New York: Van Nostrand Reinhold, 1982.

Discusses 525 species of mushrooms, mainly from Europe and North America. Many descriptions of species include photographs or drawings done in color. Presented in an informal, enjoyable, and resourceful format.
R: *ARBA* (1983, p. 629).

Zoology

Black's Veterinary Dictionary. 13th ed. **Geoffrey Philip West, ed.** London: Adam & Charles Black. Distr. Totowa, NJ: Barnes & Noble, 1979.

First edition, 1928.
A reliable reference in the field of veterinary medicine. This up-to-date dictionary provides broad coverage in all aspects of animal care, such as physiology, diseases and fractures, rearing practices, treatment, genetics and anatomy, and medicines. Straightforward and easy-to-understand entries are illustrated with photographs and line drawings. Cross-referenced. A must for all medical collections.
R: *Choice* 18: 642 (Jan. 1981); *CRL* 42: 83 (Jan. 1981); *ARBA* (1981, p. 750).

The Dictionary of Birds in Color. **Bruce Campbell. Richard T. Holmes, ed.** New York: Viking Press, 1974.

A comprehensive and scholarly reference work on more than 1,000 species from 6 main geographical regions. Unusually high-quality color illustrations and informative text.
R: *BL* 71: 871–872 (Apr. 15, 1975); *Choice* 12: 367–368 (May 1975); *WLB* 50: 529 (Mar. 1975); *ARBA* (1976, p. 679).

The Dictionary of Butterflies and Moths in Color. **Allan Watson and Paul E. S. Whalley.** New York: McGraw-Hill, 1975.

Dictionary section of the work defines some 2,000 entries. Also contains 405 color photographs. Intended for anyone interested in entomology.
R: *ARBA* (1976, p. 690).

A Dictionary of Entomology. **A. W. A. Leftwich.** New York: Crane-Russak, 1976.

A dictionary for naturalists and entomologists. Defines terms relating to the study of insects. Includes over 3,000 definitions of species, in addition to biological terms. A choice dictionary for public libraries.
R: *Choice* 13: 496 (June 1976); *LJ* 101: 880 (Apr. 1, 1976); *WLB* 50: 746 (May 1976); *ARBA* (1977, p. 674).

A Dictionary of Immunology. 2d ed. **W. J. Herbert and P. C. Wilkinson, eds.** Philadelphia: J. B. Lippincott, 1977.

First edition, 1971.
A glossary of a wide variety of terms in immunology reflecting current terminology. Highly recommended for the undergraduate student.
R: *Nature* 273: 254 (May 1978); *TBRI* 45: 97 (Mar. 1979); *ARBA* (1973, p. 551); Sheehy (EC5).

Fishes of the World: An Illustrated Dictionary. **Alwyne Wheeler.** New York: Macmillan, 1975.

Fish species are entered under scientific names with common names, geographical range, commercial use, habitat, etc. Contains over 500 color photographs. Intended for both the expert and the novice ichthyologist.
R: *ARBA* (1976, p. 688); Sheehy (EC45).

Glossary of Inland Fishery Terms. European Inland Fisheries Advisory Commission Occasional Paper, 12; EIFAC/OP 12. **M. Leopold, comp.** Rome, Italy: Food and Agriculture Organization of the United Nations, 1978.

English-French list of inland fishery terms. Contains annotations and references.
R: *IBID* 7: 142 (Summer 1979).

The International Horseman's Dictionary. **Charles Stratton.** New York: Dial Press, 1975.

A companion volume for such works as Hope's *Encyclopedia of the Horse* and Taylor's *Harper's Encyclopedia for Horsemen*.
R: *LJ* 100: 2316 (Dec. 1975); *ARBA* (1976, p. 746).

The Illustrated Dinosaur Dictionary. **Helen Roney Sattler.** New York: Lothrop, Lee & Shepard Books/William Morrow, 1983.

A handy up-to-date reference on the 300 known dinosaurs. Entries include scientific classification, description, and information on any evidence available for the existence of each dinosaur. Enhanced by black-and-white line drawings and cross references. A scientifically accurate text for school and public libraries.
R: *SLJ* 30: 24 (May 1984); *ARBA* (1985, p. 609).

Nomina Anatomica Avium: An Annotated Dictionary of Birds. **J. J. Baumel and others, eds.** New York: Academic Press, 1979.

A source of anatomical information previously unpublished. The Latin language is used, one name of a bird is accepted, and new ones have been named. Gives an introduction to the Avian Anatomical Nomenclature Committee. A reference for ornithologists, anatomists, etc.
R: *RSR* 8: 18 (July/Sept. 1980).

Microbiology

A Dictionary of Microbial Taxonomy. **Samuel T. Cowan and L. R. Hill, eds.** England: Cambridge University Press, 1978.

Revised and expanded edition of *A Dictionary of Microbial Taxonomic Usage*, 1968. A comprehensive dictionary of microbial taxonomy, containing approximately 1,600 entries. A valuable reference source.
R: *Abstracts on Hygiene* 44: 85 (Jan. 1969); *American Scientist* 67: 479 (July/Aug. 1979); *Journal of Applied Bacteriology* 47: 194 (Aug. 1979); *Nature* 276: 741 (Dec. 14, 1978); *Aslib Proceedings* 43: 453 (Dec. 1978); *TBRI* 45: 53 (Feb. 1979); 45: 329 (Nov. 1979; Sheehy (EC3); Wal (p. 199); Win (3EC19a).

Dictionary of Microbiology. **Paul Singleton and Diana Sainsbury.** New York: Wiley, 1978.

Comprehensive dictionary covering concepts, terms, and techniques pertaining to microorganisms and their environment. Contains alphabetically arranged articles on microbiology and microbiologists. Includes an appendix of metabolic pathways and useful diagrammatic material. Recommended to students and professionals.
R: *Laboratory Practice* 28: 935 (Sept. 1979); *Nature* 278: 93 (Mar. 1, 1979); *Aslib Proceedings* 44: 384 (Sept. 1979); *Choice* 16: 1156 (Nov. 1979); *ARBA* (1980, p. 622).

Agriculture

An A–Z of Offshore Oil and Gas: An Illustrated International Glossary and Reference Guide to the Offshore Oil and Gas Industries and Their Technology. 2d ed. **Harry Whitehead.** Houston: Gulf Publishing, 1983.

A comprehensive reference containing terms associated with offshore oil and gas exploration, drilling, and production. Entries are short and cross-referenced. Includes 30 appendixes on offshore areas, national oil and gas agencies, and more. Useful for those involved in the petroleum industry.
R: *ARBA* (1985, p. 510).

Agricultural Terms: As Used in the Bibliography of Agriculture from Data Provided by the National Agricultural Library, US Department of Agriculture. 2d ed. Phoenix: Oryx Press, 1978.

For use in online searching. Contains a list of words used in the AGRICOLA data base. A widely revised and expanded edition of *The Thesaurus of Agricultural Terms*. Comprises over 37,000 terms used in finding articles indexed in the *Bibliography of Agriculture*.
R: *RQ* 19: 93 (Fall 1979); *ARBA* (1979, p. 747).

Black's Agricultural Dictionary. **D. B. Dalal-Clayton.** New York: Adam & Charles Black, 1982.

It is not nearly so comprehensive, with about 3,500 terms defined. Although the coverage is British, references to American usage are included if there are particular differences.
R: *Choice* 19: 1375 (June 1982); *ARBA* (1983, p. 703).

Dictionary of Agricultural and Food Engineering. 2d ed. **Arthur W. Farrall and J. A. Basselman, eds.** Danville, IL: Interstate Printers and Publishers, 1979.

An extensive selection of words, largely taken from the field of agricultural engineering. Recommended for general reference collections.
R: *Food Technology* 33: 124 (Oct. 1979): *TBRI* 45: 382 (Dec. 1979); *ARBA* (1981, p. 733).

Earth Sciences

A Dictionary of Earth Sciences. **Stella E. Stiegler, ed.** New York: PICA Press. Distr. New York: Universe Books, 1977.

A British dictionary of geology, paleontology, geophysics, and meteorology. Includes supplementary diagrams and illustrations.
R: *BL* 74: 490 (Nov. 1, 1977); *Choice* 14: 1338 (Dec. 1977); *LJ* 102: 1625 (Aug. 1977); *ARBA* (1978, p. 691); Sheehy (EE1).

Dictionary of Geography: Definitions and Explanations of Terms Used in Physical Geography. Rev. ed. **W. G. Moore.** New York: Barnes & Noble, 1978.

Provides definitions of terms in physical geography, earth science, astronomy, and climatology.

Dictionary of Geological Terms. 3d ed. **Robert L. Bates and Julia A. Jackson, eds.** Garden City, NY: Anchor Press/Doubleday, 1984.

First edition, 1957; second edition, 1960; second revised edition, 1976.
A revised and updated version of the previous edition. Usage is North American. Defines the most common terms used in the several earth sciences. For the layperson.
R: *ARBA* (1985, p. 604).

Dictionary of Geology. 5th ed. **John Challinor.** New York: Oxford University Press, 1978.

Third edition, 1967; fourth edition, 1973.
Though emphasis is on applications in Britain and Wales, this work is accurate and succinct enough to be appreciated in all collections. Contains a classified index that groups related words.
R: *BL* 71: 774 (Mar. 15, 1975); *Choice* 12: 44 (Mar. 1975); *RSR* 2: 123 (Oct.–Dec. 1974); *ARBA* (1976, p. 704); Jenkins (F42); Wal (p. 146); Win (EE44; 2EE10).

Dictionary of Geotechnics. **S. H. Somerville and M. A. Paul.** Boston: Butterworth, 1983.

Dictionary of soil and rock mechanics and engineering, suitable for use on both sides of the Atlantic. Focused for students and practitioners.
R: *Choice* 21: 68 (Sept. 1983); *ARBA* (1985, p. 545).

A Dictionary of Petroleum Terms. 3d ed. **Jodie Leecraft, ed.** Austin, TX: University of Texas, Petroleum Extension Service, 1983.

Provides definitions for 4,000 terms, including acronyms, units of measure, legal and economic terms, and slang. Four tables on SI units and metric equivalents complete the work. A good basic dictionary.
R: *ARBA* (1985, p. 509).

Dictionary of Petrology. **S. I. Tomkeieff, E. K. Walton, B. A. O. Randall, M. H. Battey, eds.** New York: Wiley, 1983.

An alphabetical listing of the terminology of sedimentary, metamorphic, and igneous petrology. Includes the definition, background, references, and any changes for each entry.

Geologic Names of the United States Through 1975. **Roger W. Swanson et al.** Washington, DC: US Government Printing Office, 1981.

A computer-produced listing of geologic names commonly in use in the United States as of 1975, with brief information on each. A useful reference tool for geologists and those in all science libraries. Updated annually in the *USGS Bulletin*.
R: *Science and Technology Libraries* 3: 73 (Winter 1982).

A Glossary of Geographical Terms; Based on a List Prepared by a Committee of the British Association for the Advancement of Science. 3d ed. **L. D. Stamp, ed.** New York: Longman, 1979.

First edition, 1961; second edition, 1966.
A valuable reference covering physical, human, economic, and political geography. Includes some foreign terms, and appendixes of special terms, and Greek and Latin roots.
R: *Geographical Magazine* (Jan. 1980, p. 316); Wal (p. 159).

Glossary of Marine Technology Terms. **Institute of Marine Engineers.** White Plains, NY: Sheridan House, 1980.

A British work covering a broad scope; definitions are supplied in the fields of physics, electricity and electronics, and marine and ship terminology. Contains helpful illustrations. For large university libraries and special collections.
R: *Choice* 18: 373 (Nov. 1980).

Glossary of Mineral Species 1983. 4th ed. **Michael Fleischer.** Tucson, AZ: Mineralogical Record, 1983.

First edition, 1971.
This alphabetical list of minerals provides chemical composition, polymorphism, and relationship to other minerals in each entry. For advanced mineral collectors who may need a quick reference to current and outdated mineral names.
R: *ARBA* (1985, p. 606).

The Illustrated Petroleum Reference Dictionary. 2d ed. **Robert D. Langenkamp, ed.** Tulsa: PennWell, 1982.

More than 3,000 entries and illustrations to aid in understanding the expressions used in the multinational petroleum industry. Complete and detailed.
R: *Science and Technology Libraries* 3: 81 (Spring 1983).

McGraw-Hill Dictionary of Earth Sciences. **Sybil P. Parker, ed.** New York: McGraw-Hill, 1984.
An easy-to-use comprehensive dictionary consisting of 15,000 terms used in all 18 disciplines of the earth sciences, including climatology, geology, mineralogy, and petrology. Includes alphabetized synonyms, acronyms, and abbreviations. Cross-referenced. An accurate reference for all libraries.
R: *LJ* 109: 1315 (July 1984); *ARBA* (1985, p. 604).

Mineral Names: What Do They Mean? **Richard Scott Mitchell.** New York: Van Nostrand Reinhold, 1979.
Well-organized book that contains the meaning of more than 2,600 mineral names. A valuable reference for students, collectors, and professionals. Recommended for public and academic libraries.
R: *Lapidary Journal* 33: 1800 (Nov. 1979); *TBRI* 46: 7 (Jan. 1980); *ARBA* (1980, p. 658).

Ocean and Marine Dictionary. **David F. Tver.** Centreville, MD: Cornell Maritime Press, 1979.
Provides concise, informative definitions on many aspects of the marine and ocean environment such as sailing nomenclature, currents, seashells, marine biology, and vegetation. Tables on area, volume, depth of oceans, velocity of sound in seawater, composition of seawater, and temperature are collected at the end of the book. Presents a wide biological and historical view of the marine world; for all oceanographic and marine libraries.
R: *Marine Engineering/Log* 85: 119 (Jan. 1980); *BL* 77: 71 (Sept. 1, 1980); *Choice* 18: 70 (Sept. 1980); *CRL* 41: 191 (Mar. 1980); *TBRI* 46: 94 (Mar. 1980); *WLB* 54: 529 (Apr. 1980); *ARBA* (1981, p. 695).

Ocran's Acronyms: A Dictionary of Abbreviations and Acronyms Used in Scientific and Technical Writing. **Emanuel Benjamin Ocran.** Boston: Routledge & Kegan Paul, 1978.
In 2 parts: alphabetical arrangement and listing by subject. Comprises an up-to-date reference of acronyms and abbreviations.
R: *BL* 76: 738 (Jan. 15, 1980); *ARBA* (1979, p. 641).

Oil Terms: A Dictionary of Terms Used in Oil Exploration and Development. **Leo Crook.** New York: International Publications Service, 1976.
Succinct definitions of terms used in oil exploration and production. Contains some 500 words; written for the beginning engineer. Also contains diagrams. Recommended.
R: *BL* 73: 426 (Nov. 1, 1976); *Choice* 13: 642 (July/Aug. 1976); *RQ* 16: 256 (Spring 1977); *ARBA* (1977, p. 684).

The Penguin Dictionary of Geology. **D. G. A. Whitten.** Baltimore: Penguin Books, 1976.

Alphabetically arranged; contains brief definitions, cross references, and up-to-date terminology.
R: *ARBA* (1974, p. 606; 1977, p. 692).

The Petroleum Dictionary. **David F. Tver and Richard W. Berry.** New York: Van Nostrand Reinhold, 1980.
Consists of 4,000 concise definitions of terms related to all aspects of the petroleum industry and related fields. Includes 150 tables and illustrations and some geological terminology. Recommended for geologic and chemical engineering collections as a supplemental volume.
R: *BL* 77: 479 (Nov. 15, 1980); *Choice* 18: 71 (Sept. 1980); *LJ* 105: 1070 (May 1, 1980); *WLB* 54: 673 (June 1980); *ARBA* (1980, p. 680).

VNR Color Dictionary of Minerals and Gemstones. **Michael O'Donoghue.** New York: Van Nostrand Reinhold, 1982.
More than 1,000 minerals are included, arranged according to the British Museum's Chemical Index of Minerals. Minerals are grouped according to elements and alloys. Useful to persons who wish to collect specimens and display them.
R: *ARBA* (1983, p. 665).

GENERAL ENGINEERING

McGraw-Hill Dictionary of Engineering. **Sybil P. Parker, ed.** New York: McGraw-Hill, 1984.
Consists of 16,000 entries selected from the *McGraw-Hill Dictionary of Scientific and Technical Terms*, 1983. Entries include concise definitions and indicate the main engineering field associated with the term. For public libraries.
R: *ARBA* (1985, p. 537).

AERONAUTICAL AND ASTRONAUTICAL ENGINEERING

The Dictionary of Space Technology. **Joseph A. Angelo.** New York: Facts on File, 1982.
Provides a thorough survey of most current US programs involved in space technology, especially the current Space Shuttle program and the Spacelab project. Illustrated. Geared toward the student as well as the scientist.
R: *ARBA* (1983, p. 717).

Jane's Aerospace Dictionary. **Bill Gunston.** New York: Jane's Publishing, 1981.
Over 15,000 entries from astronautics, civil and military aeronautics, electronics, meteorology and material science. Many acronyms are included, as are many of the field's obscure terms and phrases. Useful to libraries with substantial aerospace reference activity.
R: *ARBA* (1982, p. 802).

CHEMICAL ENGINEERING

A Consumer's Dictionary of Cosmetic Ingredients. Rev. ed. **Ruth Winter.** New York: Crown, 1984.

An alphabetical list, with definitions, of common ingredients found in cosmetics. Focuses on consumer interests such as harmful effects of products. Comprehensive and useful for both men and women.
R: *ARBA* (1985, p. 603).

Dictionary of Electrochemistry. 2d ed. **D. B. Hibbert and A. M. James.** New York: Wiley, 1985.

A revised dictionary containing 300 entries in fields such as fuel cells, corrosion, energy conversion, and bioelectrochemistry. Entries are cross-referenced and include illustrated line diagrams.

A Dictionary of Words About Alcohol. 2d ed. **Mark Keller, Mairi McCormick, and Vera Efron.** New Brunswick, NJ: Rutgers Center of Alcohol Studies, 1982.

Fairchild's Dictionary of Textiles. 6th ed. **Isabel B. Wingate.** New York: Fairchild, 1979.

Fifth edition, 1967.
Updated edition provides expanded coverage and includes some 14,000 definitions relating to textile technology, electronic equipment, methods of dyeing, printing, and finishing fabrics. Also contains an annotated list of trade associations and institutes. Recommended for all types of libraries.
R: *Textile Institute and Industry* 17: 437 (Dec. 1979); *LJ* 104: 1131 (May 15, 1979); *TBRI* 46: 120 (Mar. 1980); *WLB* 54: 196 (Nov. 1979); Wal (p. 247).

Glossary of Packaging Terms: Standard Definitions of Trade Terms Commonly Used in Packaging. New York: Packaging Institute, 1973–. Annual updates.

A Glossary of Wood: 10,000 Terms Relating to Timber and Its Use, Explained and Classified. **T. Corkhill.** London: Stobart, 1979.

A standard reference covering all aspects of timber, but mainly uses and fabrication of wood. Includes 10,000 terms, some 1,000 of them illustrated.
R: Wal (p. 245).

Industrial Engineering Terminology (ANSI Standard Z94. 0-1982). **American National Standards Institute, Z94 Committee.** Norcross, GA: Institute of Industrial Engineers, 1983.

The official compilation of the definitions adopted by the Z94 Committee of the American National Standards Institute (ANSI). Though there may be some difficulty in using it, all libraries with an interest in industrial engineering should consider it a necessity.
R: *Science Technology Libraries* 5: 2 (Winter 1984); *ARBA* (1985, p. 551).

International Glossary of Technical Terms for the Pulp and Paper Industry. **Paul D. Van Derveer and Leonard E. Haas, eds.** San Francisco: W. H. Freeman, 1976.

Interdisciplinary terms from biology, engineering, physics, and chemistry which relate to the paper industry. Terms in 5 languages: English, Swedish, German, French, and Spanish. Good format; for both the nonspecialist and the specialist.

Leather Technical Dictionary. **International Union of Leather Technologists' and Chemists' Societies.** Darmstadt, Germany: Roether Verlag, 1979.

Multilingual dictionary of German, French, Italian, Russian, and Spanish terms. Contains 5,429 entries. Recommended for its broad scope of the leather industry.
R: *Journal of the American Leather Chemists Association* 74: 385 (Oct. 1979); *TBRI* 46: 75 (Feb. 1980).

Paint/Coatings Dictionary. **Federation of Societies for Coatings Technology.** Philadelphia: Federation of Societies for Coatings Technology, 1978.

Highly recommended dictionary of paints and coatings.
R: *Metal Finishing* 77: 91 (Jan. 1979); *TBRI* 45: 113 (Mar. 1979).

Whittington's Dictionary of Plastics. 2d ed. **Lloyd R. Whittington, ed.** Westport, CT: Technomic, 1978.

Sponsored by the Society of Plastics Engineers, this new edition includes new developments in polymers, processing methods, and applications. Terms are mostly related to toxicity, ecological impacts, flammability, and other safety aspects.
R: *Science and Technology Libraries* 1: 135 (Fall 1980).

Nutrition and Food Technology

A Consumer's Dictionary of Food Additives. Rev. ed. **Ruth Winter.** New York: Crown, 1984.

Revised edition, 1978.
Provides information on recently approved or banned food additives and the recommendations of the Select Committee on GRAS Substances. Lists chemical ingredients alphabetically, with definition, function, and toxicity. Highly useful for public libraries and the layperson.
R: *Journal of the American Chemical Society* 94: 9287 (Dec. 27, 1972); *BL* 74: 1536 (June 1, 1978); *WLB* 53: 93 (Sept. 1978); *ARBA* (1979, p. 757; 1985, p. 497).

Dictionary of Food Ingredients. **Robert S. Igoe.** New York: Van Nostrand Reinhold, 1983.

An informative and applicable, alphabetical listing of some 1,000 substances found in foods. For the general reader.
R: *Choice* 20: 1110 (Apr. 1983).

Dictionary of Gastronomy. **Andre L. Simon and Robin Howe.** Woodstock, NY: Overlook Press, 1978.

This compilation of over 2,000 entries of culinary terms emphasizes definition and cross reference. Discusses foods, wines, and techniques common to Western Europe, as well as exotic dishes. Includes numerous color plates and illustrations.
R: *ARBA* (1980, p. 705).

Dictionary of Nutrition and Food Technology. 5th ed. **Arnold E. Bender.** New York: Chemical Publishing, 1977.

Third edition, 1969; fourth edition, 1975.
Approximately 3,000 briefly defined terms for food chemists and public health workers.

R: *Choice* 6: 991 (Oct. 1969); *ARBA* (1970, p. 152; 1978, p. 744); Jenkins (J113); Wal (p. 485).

Food: An Authoritative and Visual History and Dictionary of the Foods of the World. **Waverley Root.** New York: Simon & Schuster, 1980.

A substantial collection of information on food; the emphasis is more geographic, historical, or botanical than culinary. Includes numerous literary quotations and allusions, as well as personal anecdotes. The history of food is presented through an abundant supply of pictures and photographs. A reliable source of information.
R: *ARBA* (1981, p. 740).

Larousse Dictionary of Wines of the World. **Gerard Dubuigne.** New York: Larousse, 1976.

Alphabetically arranged, cross-referenced, and illustrated with color photographs and maps. Covers all major geographic areas that produce wine.
R: *BL* 73: 1297 (May 15, 1977); *LJ* 102: 107 (Jan. 1, 1977); *ARBA* (1978, p. 738).

CIVIL ENGINEERING

Architectural and Building Trades Dictionary. 2d ed. **R. E. Putnam and G. E. Carlson.** Chicago: American Technical Society, 1979.

Includes 7,500 definitions, as well as many illustrations and some tips on construction and installation.
R: *Plant Engineering* 33: 262 (Aug. 23, 1979); *TBRI* 45: 354 (Nov. 1979).

Dictionary of Building. 3d ed. **John S. Scott.** New York: Halsted Press/Wiley, 1984.

An updated edition of a British dictionary; it includes 100 new illustrations. Some terms are listed with US synonyms.
R: *ARBA* (1985, p. 545).

Construction Contract Dictionary. **Leonard Fletcher, Reginald Lee, and John A. Tackaberry.** New York: Spon, in association with Methuen, 1981.

A handbook of definitions for the many confusing terms used in legal contracts and the law; relates to construction contracts in particular. Technically accurate as of October 1, 1980. Intended to be updated from time to time.
R: *ARBA* (1983, P. 719).

Construction Glossary: An Encyclopedic Reference and Manual. **J. Stewart Stein.** New York: Wiley, 1980.

Extensive and varied definitions of terms in the construction industry. Includes code interpretations, standard uses, historic references, technical and scientific data, professional services, etc. Includes a section of reference data sources and an extensive index. A timely reference for engineers and contractors.
R: *ARBA* (1981, p. 758).

Construction Regulations Glossary. **J. Stewart Stein.** New York: Wiley, 1983.

Contains a variety of terminology and definitions associated with zoning ordinances, subdivision controls, and building codes. Ideal as a tool for code writers and for students in building codes and zoning ordinances fields.
R: *Choice* 21: 691 (Jan. 1984).

Dictionary of Civil Engineering. 3d ed. **J. S. Scott.** New York: Halsted Press, 1981.

Updated and metricated edition covers all aspects of civil engineering, including hydraulics, sewage disposal, structural design, modern techniques, etc. A table of civil engineering units is included. For engineers.

A Dictionary of Soil Mechanics and Foundation Engineering. **John A. Barker.** New York: Longman, 1981.

The dictionary consists of approximately 2,500 brief entries and is intended for use by a broad audience. Although the emphasis is British, there are definitions of some terms with American applications.
R: *Choice* 19: 888 (Mar. 1982); *ARBA* (1983, p. 726).

Illustrated Dictionary of Building. **Paul Marsh.** New York: Longman, 1982.

Some 4,000 definitions of terminology used in British building trades. Definitions are clear and concise, but the sketches prove to be very confusing. For libraries serving students dealing with British terminology.
R: *Choice* 20: 1578 (July/Aug. 1983).

Illustrated Encyclopedic Dictionary of Building and Construction Terms. **Hugh Brooks.** Englewood Cliffs, NJ: Prentice-Hall, 1976.

A dictionary of construction terms divided into 3 sections: alphabetical list, list by function, and commonly used formulas. Concise, understandable terms, usable by both the layperson and the specialist.
R: *RSR* 5: 24 (Oct./Dec. 1977); *WLB* 51: 187 (Oct. 1976); *ARBA* (1977, p. 756); Wal (p. 621).

The Road & Track Illustrated Dictionary. **John Dinkel.** New York: Norton, 1977.

For the auto enthusiast; contains illustrations and definitions that comprise a useful guide for definitions and explanations of auto mechanics. A recommended reference.
R: *BL* 74: 154 (Sept. 15, 1977); *LJ* 102: 1746 (Sept. 1, 1977); *ARBA* (1978, p. 763).

ELECTRICAL AND ELECTRONICS ENGINEERING

Dictionary of Electronics. **S. W. Amos.** Boston: Butterworth, 1981.

Encyclopedic Dictionary of Electronic Terms. **John E. Traister and Robert J. Traister.** Englewood Cliffs, NJ: Prentice-Hall, 1984.
R: *Choice* 22: 409 (Nov. 1984); *ARBA* (1985, p. 548).

Electronics Dictionary. 4th ed. **John Markus.** New York: McGraw-Hill, 1978.
Third edition, entitled *Electronics and Nucleonics Dictionary*, 1966.

A well-illustrated, comprehensive dictionary that includes approximately 17,000 terms in electronics. Presents current language of electronics with clear, concise terms accompanied by detailed illustrations. Also features the *Electronics Style Manual*, which simplifies troublesome spelling and grammatical problems.
R: *Marine Engineering/Log* 83: 110 (Oct. 1978); *LJ* 93: 1585 (1968); *NTB* 52; 60 (1967); *TBRI* 44: 396 (Dec. 1978); Jenkins (K97); Sheehy (EJ39); Wal (p. 322); Win (EI98; 2EI21).

IEEE Standard Dictionary of Electrical and Electronics Terms. 3d ed. **Frank Jay, ed.** New York: Institute of Electrical and Electronics Engineers. Distr. New York: Wiley, 1984.
First edition, 1972; second edition, 1977.
An expanded and revised edition that contains over 23,000 technical terms from all fields of electrical and electronic technology. Each definition is an official standard of IEEE. Includes 140 drawings and diagrams and an appendix of 15,000 common acronyms. For all science libraries.
Chemical Engineering 85: 16 (Aug. 14, 1978); *Plant Engineering* 32: 138 (Sept. 28, 1978); *Choice* 9: 1114 (Nov. 1972); *RSR* 6: 27 (Oct./Dec. 1978); *TBRI* 44: 394 (Dec. 1978); *ARBA* (1973, p. 652; 1979, p. 772; 1985, p. 546).

The Illustrated Dictionary of Electronics. **Rufus P. Turner.** Blue Ridge Summit, PA: TAB Books, 1980.
Covers a broad range of topics including computers, microcircuitry, cyclotrons, physics, chemistry, and audio electronics. Includes concise definitions, hundreds of abbreviations and acronyms, line drawings, circuit diagrams, tables and charts. A welcome addition to any library providing information on current technological developments and discoveries in electronics.
R: *ARBA* (1981, p. 760).

Modern Dictionary of Electronics. 5th ed. **Rudolf F. Graf.** Indianapolis: Sams, 1977.
First edition, 1962; third edition, 1968; fourth edition, 1972.
Defines approximately 20,000 terms. Includes appropriate tables and charts. Accurate, up-to-date, and useful to a broad spectrum of people in the field.
R: *SL* 59: 119 (1968); *ARBA* (1973, p. 651; 1979, p. 772); Jenkins (K95); Wal (p. 321); Win (2EI22).

The New Penguin Dictionary of Electronics. **Carol Young.** London: Penguin Books, 1979.
R: *Laboratory Practice* 29: 1081 (Oct. 1980).

Standard Dictionary of Electrical and Electronics Terms. 2d ed. Wiley, 1978.
R: *Librarians' Newsletter* (March 1983).

COMPUTER TECHNOLOGY

American National Dictionary for Information Processing Systems. **American National Standards Committee, X3, Information Processing Systems, dev.** Homewood, IL: Dow Jones–Irwin, 1984.

An easy-to-use dictionary providing terms associated with the language of computers and information processing. Includes International Standards Organization terms. Extensively cross-referenced and cross-indexed. For libraries with large standard collections.
R: *ARBA* (1985, p. 581).

The Beginner's Computer Dictionary. **Elizabeth S. Wall and Alexander C. Wall.** New York: Avon, 1984.

Well-illustrated text with cross-referenced words. Useful for educating children about computers and their underlying concepts and uses.
R: *ARBA* (1985, p. 584).

The Computer Alphabet Book. **Elizabeth S. Wall.** New York: Avon, 1984.

Uses letters of the alphabet to help the reader learn the definitions of 26 important computer words. Written at the grade-school level.
R: *ARBA* (1985, p. 584).

Computer Dictionary. **Patricia Conniffe.** New York: Scholastic Book Services, 1984.

Provides definitions for over 500 computer-related terms. Emphasizes hardware and software terms used in microcomputing. Suitable for children and adults with little knowledge of computers.
R: *ARBA* (1985, p. 582).

The Computer Dictionary. **John Prenis.** Philadelphia: Running Press, 1983.

Provides clear definitions of terms and jargon associated with computers. Illustrated with excellent line drawings. Intended for the novice.
R: *ARBA* (1985, p. 583).

Computer Dictionary. 3d ed. **Charles J. Sippl and Roger J. Sippl.** Indianapolis: Howard W. Sams, 1980.

A "browsing" dictionary to be used as a tutorial book for computer users rather than for computer specialists. Long and detailed definitions and explanations. Includes photographs and diagrams.
R: *Science and Technology Libraries* 1: 73 (Summer 1981).

Computer Dictionary. 2d ed. **Donald D. Spencer.** Ormond Beach, FL: Camelot Publishing, 1979.

First edition, 1977.
A selective dictionary containing 2,500 terms, phrases, and acronyms related to computer technology. Also includes some biographical information. Clear, concise information and format. Recommended for school, college, and public libraries.
R: *BL* 74: 1696 (July 1, 1978); 77: 531 (Dec. 1, 1980); *ARBA* (1980, p. 719).

Computer Dictionary and Handbook. 3d ed. **Charles J. Sippl and Charles P. Sippl.** Indianapolis: Sams, 1980.

First edition, 1966; second edition, 1972.
Some 22,000 lucid definitions. Much of the work is devoted to textual material on computer principles and applications.

R: *Choice* 3: 1006 (1966); *NTB* 51: 151 (1966); *Scientific Information Notes* 1: 292 (1969); *SL* 57: 669 (1966); *WLB* 48: 796 (May 1973); *ARBA* (1974, p. 677); Jenkins (B64); Wal (p. 545); Win (1EI28).

Computer Dictionary for Everyone. **Donald D. Spencer.** New York: Scribner's, 1979.

A nontechnical computer dictionary of 2,500 words, phrases, and acronyms. Related concepts and terms are cross-referenced. A quick-reference work useful to students of computer science and data processing, as well as to systems analysts and programmers.
R: *WLB* 54: 528 (Apr. 1980); *ARBA* (1981, p. 759).

The Computer Glossary: It's Not Just a Glossary! 2d ed. **Alan Freedman.** New York: Computer Language, 1981.

A beginner's guide to computer literacy in an easy-to-read and understandable context.
R: *ARBA* (1983, p. 721).

The Computer Graphics Glossary. **Stuart W. Hubbard.** Phoenix, AZ: Oryx Press, 1983.

A specialized dictionary of 750 computer graphics terms, including a glossary of product names, acronyms for professional organizations, and business terms. Useful for technical libraries.
R: *Choice* 21: 1446 (June 1984); *Science Technology Libraries* 5: 2 (Winter 1984); *ARBA* (1985, p. 588).

Dictionary of Computer Graphics. **John Vince.** White Plains, NY: Knowledge Industry Publications, 1984.

Presents comprehensive written technical information and illustrations of terms related to various aspects of computer graphics. Will be a useful tool.

Dictionary of Computing. **Valerie Illingworth, ed.** New York: Oxford University Press, 1983.

Developed by 50 people involved in the field of computers to explain over 3,750 terms used in this area. Covers all aspects of the field for the novice as well as the advanced computer user.
R: *ARBA* (1984, p. 617).

Dictionary of Computing: Data Communications Hardware and Software Basics Digital Electronics. **Frank J. Galland, ed.** New York: Wiley, 1982.

Lists 9,000 terms used in data communications, digital electronics, and basic hardware and software. Includes tables, pictures, examples, and cross references. The definitions are clear, precise, and useful to both British and American programmers, systems analysts, and computer engineers.
R: *ARBA* (1983, p. 721).

Dictionary of Computers, Data Processing, and Telecommunications. **Jerry M. Rosenberg.** New York: Wiley, 1984.

An extensive collection of over 10,000 terms and definitions associated with computers, data processing, and telecommunications. Terms are arranged alphabetically and are cross-referenced. Contains an appended French and Spanish glossary of equivalent terms. Recommended for public and academic libraries.
R: *Librarians' Newsletter* 23: 15 (Feb. 1984); *LJ* 108: 2326 (Dec. 15, 1983); *WLB* 58: 514–515 (Mar. 1984); *ARBA* (1985, p. 583).

Dictionary of Data Processing. 2d ed. **Jeff Maynard, ed.** Woburn, MA: Butterworth Scientific, 1982.

A compilation of over 4,000 definitions, acronyms, and abbreviations used as technical terms in computing, telecommunications, and office automation.
R: *ARBA* (1983, p. 722).

Dictionary of Data Processing: Including Applications in Industry, Administration, and Business; English–German–French. 3d rev. ed. **Alfred Wittman and Joel Klos.** New York: Elsevier/North-Holland, 1977.

Second revised edition, 1973.
A revised edition that contains more than 5,300 terms in English, German, and French. 150 new terms are included. Primarily for computer science collections.
R: *ARBA* (1979, p. 771).

Dictionary of Information Technology. **D. Longley and M. Shain, eds.** Wiley Interscience, 1982.

Defines more than 6,000 terms from the fields of computers and communications. Provides in-depth essays on specific topics, such as word processing and computer electronics. Easy to understand and useful for the novice.
R: *Librarians' Newsletter* (Nov. 1982).

Dictionary of Logical Terms and Symbols. **Carol Horn Greenstein.** New York: Van Nostrand Reinhold, 1978.

Concise, accessible dictionary of logical terms. Useful to engineers, computer scientists, logicians. Contains notation systems, flow-chart symbols, Venn diagrams, binary truth tables, and circuit diagrams. A good bibliography adds to the usefulness of this book.
R: *ARBA* (1979, p. 769).

A Dictionary of Minicomputing and Microcomputing. **Philip E. Burton.** New York: Garland, 1981.

The dictionary is organized as an alphabetically ordered dictionary, followed by 9 separate appendixes. Many of the terms do have unique definitions. All definitions that appear in the various appendixes are cross-referenced in the main dictionary.
R: *IEEE Proceedings.* 70: 880 (Aug. 1982); *ARBA* (1983, p. 720).

The Easy to Understand Computer Dictionary. **Irv Brechner.** Chicago: WIDL Video Publications. Distr. New York: Scribner's, 1983.

Provides definitions and illustrations for 75 computer terms associated with hardware, software, and computer basics. For young readers.
R: *ARBA* (1985, p. 589).

The Facts on File Dictionary of Microcomputers. **Anthony Chandor.** New York: Facts on File, 1981.

This dictionary defines approximately 2,500 terms in a strict alphabetical order. Oriented to word microcomputer owners. This dictionary is useful for public libraries.
R: *ARBA* (1982, p. 805).

The Hacker's Dictionary: A Guide to the World of Computer Wizards. **Guy L. Steele, Jr.** et al. New York: Harper & Row, 1983.

A guide to the world of computer hackers. Terms are arranged alphabetically; definitions are informative and anecdotal. Cross references provided.
R: *WLB* 58: 452 (Feb. 1984); *ARBA* (1985, p. 584).

The Illustrated Computer Dictionary. Rev. ed. **Donald D. Spencer.** Columbus, OH: Merrill, 1983.

R: *ARBA* (1985, p. 583).

Illustrated Dictionary of Microcomputer Terminology. **Michael Hordeski.** Blue Ridge Summit, PA: TAB Books, 1978.

Contains 4,000 terms related to the microcomputer industry. Examples are given to avoid confusion. Cross-referenced and alphabetized.
R: *Choice* 16: 1148 (Nov. 1979); *Journal of Academic Librarianship* 5: 114 (May 1979); *ARBA* (1980, p. 719).

Information Resource/Data Dictionary Systems. **Henry C. Lefkovits, Edgar H. Sibley, and Sandra L. Lefkovits.** Wellesley, MA: QED Information Sciences, 1983.

A technical survey of the information resource management field. Chapters 1 and 2 provide an in-depth analysis of 7 commercially available systems from IBM, MSP, Cullinet, Intel, and others.
R: *ARBA* (1985, p. 582).

McGraw-Hill Dictionary of Electronics and Computer Technology. **Sybil P. Parker, ed.** New York: McGraw-Hill, 1984.

Provides definitions for 10,000 terms that are derived from the *McGraw-Hill Dictionary of Scientific and Technical Terms*, 3d ed., 1984. A competently produced reference book.
R: *ARBA* (1985, p. 547).

Microcomputer Dictionary. 2d ed. **Charles J. Sippl.** Indianapolis: Howard W. Sams, 1981.

Includes over 5,000 computer terms and definitions, some from computer terminology other than microcomputer terminology. Some common microcomputer terms do not appear at all. Not useful for a novice, but rather for a hobbyist or technician.
R: *ARBA* (1983, p. 722).

The Penguin Dictionary of Microprocessors. **Anthony Chandor.** New York: Penguin Books, 1981.

Contains 2,800 up-to-date and clear definitions of microprocessing words. Some definitions are incomplete and not very helpful.
R: *ARBA* (1982, p. 806).

Personal Computers A–Z. **Joel Makower.** Garden City, NY: Quantum Press/Doubleday, 1984.

A collection of 350 terms and their definitions related to microcomputer hardware, software, and services. Definitions are up-to-date and comprehensible. Well-indexed. Useful for less experienced personal computer users.
R: *LJ* 109: 990 (May 15, 1984); *ARBA* (1985, p. 589).

The Personal Computer Glossary. **George Ledin, Jr.** Sherman Oaks, CA: Alfred Publishing, 1983.

A 2-part glossary containing 550 definitions of generic terms and 400 terms labeled *Computerisms* (jargon).
R: *ARBA* (1985, p. 589).

Robotics Sourcebook and Dictionary. **David F. Tver and Roger W. Bolz.** New York: Industrial Press, 1983.

A 4-part, handy reference focusing on the industrial applications of robotics. Part 1 describes the characteristics of basic robot styles. Part 2 describes tasks that robots can manage successfully; part 3 contains a robotics glossary and a dictionary of computer-control terminology. Part 4 is a product catalog that lists manufacturers, their products and addresses. A concise, well-illustrated reference for both industrial and library use.
R: *Choice* 21: 256 (Oct. 1983); *ARBA* (1985, p. 599).

Security Dictionary. **Richard A. Hofmeister and David J. Prince.** Indianapolis: Howard W. Sams, 1983.

Provides informative definitions of terms and equipment used in the security industry today. Emphasizes methods for protecting life and property from harm or loss. Includes appendixes listing security, fire protection, and electronic symbols.
R: *ARBA* (1985, p. 546).

The Sybex Personal Computer Dictionary. Berkeley: Sybex, 1984.

A good introduction to personal computing, including common terms, acronyms, and jargon. Includes a selection of microcomputer terms translated into 10 European languages, *EIA* and *IEEE* standards, and a list of microcomputer component suppliers.
R: *ARBA* (1985, p. 583).

User's Guide to Microcomputer Buzzwords. **David H. Dasenbrock.** Indianapolis: Howard W. Sams, 1983.

Words and phrases associated with microprocessors and their uses in communication are arranged alphabetically and clearly defined with the aid of drawings and diagrams. Terms are brief and understandable. Useful as a quick refresher or reference text on microcomputer terminology. For all professionals in the field.
R: *ARBA* (1985, p. 589).

COMMUNICATION

Communications Standard Dictionary. **Martin H. Weik.** New York: Van Nostrand Reinhold, 1982.

Contains well-clarified definitions of terms used by designers, developers, manufacturers, managers, etc., consistent with the latest standards. Alphabetically arranged.
R: *Proceedings of the IEEE* 71: 1118 (Sept. 1983); *Choice* 20: 1114 (April 1983).

Dictionary of Data Communications. **C. J. Sippl, ed.** 2d ed. New York: Wiley, 1985.

A dictionary of terms and jargon currently used in the field of data communications. The second edition has been updated and expanded to include the major areas of EMMS, remote databases, videotex, paging systems, intra- and inter-company LANs and low-cost direct service satellite systems.

Dictionary of Telecommunications. **Sidney John Aires.** Woburn, MA: Butterworth, 1981.

A dictionary of all the terms currently used in the United Kingdom and the United States. Definitions are detailed; illustrations are included to clarify.
R: Proceedings of the IEEE (1982, p. 784).

Electro-Optical Communications Dictionary. **Dennis Bodson and Dan Botez, eds.** Rochelle Park, NJ: Haden Books, 1983.

Provides technical definitions for 2,500 terms related to fiber-optic and light-wave communication technology. Definitions are consistent with standards used in the field. Entries provide cross references, additional uses and applications, and bibliographic references. Includes a list of 220 abbreviations and acronyms. Recommended for science libraries.
R: *ARBA* (1985, p. 545).

The Facts on File Dictionary of Telecommunications. **John Graham.** New York: Facts on File, 1983.

A more up-to-date and broader source of telecommunications terms. Suitable for smaller libraries as a fine source.
R: *Choice* 21: 250 (Oct. 1983).

Fiber Optics and Lightwave Communications Standard Dictionary. **Martin H. Weik.** New York: Van Nostrand Reinhold, 1981.

A complete collection of current terms used in the new fields of fiberoptronics and fiber-optic communications. Cross-referenced and illustrated. For upper-level undergraduate students and professionals.
R: *Choice* 18: 1244 (May 1981).

MATERIALS SCIENCE

Dictionary of Ceramic Science and Engineering. **Loran S. O'Bannon.** New York: Plenum, 1984.

Briefly describes 8,000 words, terms, materials, processes, products, and business terms used in ceramics and related industries. Its peculiar alphabetizing system may cause some confusion; otherwise, it will prove to be a valuable tool for libraries supporting research in ceramics.
R: *Choice* 22: 252 (Oct. 1984); *Science Technology Libraries* 5: 2 (Winter 1984); *ARBA* (1985, p. 554).

Dictionary of Gemology. **P. G. Read.** Woburn, MA: Butterworth, 1982.

Contains terms associated with gemology and certain other fields with explanations of the various types of gemological instruments. Brief, but complete enough for the specialist and the library.
R: *ARBA* (1984, p. 686).

Dictionary of Materials Testing. **Dipl-Ing Werner Goedecke, comp.**, VDI International, 1980.

MECHANICAL ENGINEERING

Air Conditioning, Heating and Refrigeration Dictionary. **Timothy Zurick.** Birmingham, MI: Business News Publishing, 1977.

Over 1,700 entries pertaining to the fields of air conditioning, heating, and refrigeration. Succinct and easy to use.
R: *ARBA* (1979, p. 765).

Dictionary of Drying. **Carl W. Hall.** New York: Marcel Dekker, 1979.

A compilation of about 4,000 terms on drying. Includes cross entries and a list of 75 English-language monographs published since 1950. Among key features are tables, charts, line drawings, evaluations of energy use, descriptions of drying processes, and conversion tables. For libraries serving engineers in the drying field.
R: *ARBA* (1981, p. 761).

Dictionary of Instrument Science. **T. Ramalingom.** New York: Wiley, 1982.
R: *Physics today* 35: (Oct. 1982).

Dictionary of Mechanical Engineering. 3d ed. **J. L. Nayler and G. H. F. Nayler.** Stoneham, MA: Butterworth, 1985.

First edition, 1967; second edition, 1975; second edition reprinted, 1978.
Precise definitions for approximately 6,000 terms. Amply illustrated. British emphasis.
R: *Choice* 5: 182 (1968); Jenkins (K106); Wal (p. 297).

Dictionary of Terms Used in the Safety Profession. **Stanley A. Abercrombie, comp. and ed.** Park Ridge, IL: American Society of Safety Engineers, 1981.

An alphabetical listing of terms used in the areas of construction safety, fire protection engineering, toxicology, industrial hygiene, etc. With addresses of professional and governmental agencies involved with the safety profession. A specialized dictionary for students in the occupational safety and health area.
R: *ARBA* (1983, p. 730).

The International Dictionary of Heating, Ventilating and Air Conditioning. **Documentation Committee of the Representatives of European Heating and Ventilating Associations, comp.** London: Spon, 1982.

Compiled by the representatives of an international association of heating engineers, this dictionary contains a list of terms in English, followed by the equivalent term in German, French, Hungarian, Italian, Dutch, Polish, Russian, Spanish, and Swedish. There is also an index for each of these languages, referring back to the English. Useful for engineers and libraries.
R: *ARBA* (1983, p. 716).

Plumbing Dictionary. 2d ed. **I. D. Jacobson, ed.** Cleveland: American Society of Sanitary Engineering, 1975.

Nontechnical but accurate definitions of terms used in plumbing. Covers 2,500 terms overall. Contains cross references and bibliographies. A useful reference tool.
R: *ARBA* (1977, p. 772).

AUTOMOTIVE

Automotive Dictionary. **W. Crouse and D. Anglin.** New York: McGraw-Hill, 1976.

Written by leading authorities on automotive mechanics, this handy dictionary contains about 1,000 common automotive terms defined in simple language. Includes helpful metric conversion tables. An easy reference tool for car owners.

Dictionary for Automotive Engineering. **Jean de Coster, ed.** New York: Saur, 1983.

Contains about 900 automotive terms in English, with the equivalents and definitions given in German and French. Terms are from a wide range of areas, including the engine and electronic ignition system. Useful for travelers and special libraries.

Technical Dictionary for Automotive Engineering. 2 vols. **Robert Bosch.** Brookfield, VT: Renouf USA, 1976.

A 2-volume, German–English, English–German dictionary containing 15,000 terms from automotive engineering. Features cross references for spelling variants, chemical formulas, synonyms, sources, and classifications symbols. A must for engineering and technical schools; helpful to translators.
R: *Choice* 18: 371 (Nov. 1980).

MILITARY SCIENCE

Dictionary of American Naval Fighting Ships. Vol. 6. US Department of the Navy, Naval History Division. Washington, DC: US Government Printing Office, 1976.

The sixth volume in an alphabetically arranged series. Covers the letters *R* and *S*. Provides historical sketches of ships and descriptions of different classes of ships.
R: *ARBA* (1977, p. 788).

Dictionary of Basic Military Terms: A Soviet View. Soviet Military Thought Series, no. 9. **US Department of the Air Force.** Washington, DC: US Government Printing Office, 1976.

Alphabetically arranged by romanized Soviet term with English equivalent. Contains Soviet military terms.
R: *ARBA* (1977, p. 778).

Jane's Dictionary of Naval Terms. **Joseph Palmer, comp.** London: Macdonald and Jane's. Distr. New York: Hippocrene Books, 1975.

This dictionary includes general terms of maritime use both from the United States and Great Britain. Contains an appendix of abbreviations. Recommended.
R: *BL* 73: 1604 (June 15, 1977).

Jane's Dictionary of Military Terms. **P. H. C. Hayward, comp.** London: Macdonald and Jane's. Distr. New York: Hippocrene Books, 1976.
R: *BL* 73: 1604 (June 15, 1977).

A Modern Military Dictionary: Ten Thousand Technical and Slang Terms of Military Usage. 2d repr. ed. **Max B. Garber and P. S. Bond.** Detroit: Gale Research, 1975.

First edition, 1936; second edition, Washington, DC: P. S. Bond, 1942.
Technical military dictionary that, because of the date published (1942), is most useful for historical study.
R: *ARBA* (1977, p. 778).

Naval Terms Dictionary. 4th ed. **John V. Noel, Jr. and Edward L. Beach.** Annapolis, MD: Naval Institute Press, 1978.

Comprises 5,000 alphabetical terms with definitions used by the American Navy, including both standard and slang terms. Covers such topics as maneuvers, equipment, weather, etc. Recommended.
R: *ARBA* (1979, p. 787).

ENERGY

Dictionary of Applied Energy Conservation. **David Kut.** New York: Nichols Publishing, 1982.

Contains over 1,500 definitions of both general and specific terms related to energy conservation. British approach and meanings. Useful to architects and engineers.
R: *ARBA* (1983, p. 653).

A Dictionary of Energy. **Martin Counihan.** Boston: Routledge & Kegan Paul, 1981.
R: *ARBA* (1983, p. 652).

Dictionary of Energy. **Carl W. Hall and George W. Hinman.** New York: Marcel Dekker, 1983.

Definitions of over 2,000 energy-related terms, diagrams, charts, and formulas.
R: *ARBA* (1984, p. 674).

Dictionary of Energy. **Malcolm Slesser, gen. ed.** London: Macmillan, 1983.

A comprehensive guide to the many terms used by professionals, scientists, and those involved in the various uses of energy. Presented first with a more elementary definition and then in greater detail for each term. Most complete and accurate.
R: *Nature* 302: 276 (March 1983).

Dictionary of Energy Technology. **Alan Gilpin in collaboration with Alan Williams.** Boston: Butterworth, 1982.

A Dictionary of the Environment. 2d ed. **Michael Allaby.** New York: New York University Press. Distr. New York. Columbia University Press, 1983.

First edition, 1977.

A revised and updated edition containing terms from the life and earth sciences, meteorology and climatology, and economics. Arranged alphabetically and extensively cross-referenced. Useful for scientists, conservationists, and students.
R: *ARBA* (1985, p. 511).

Energy Dictionary. **Daniel V. Hunt.** New York: Van Nostrand Reinhold, 1979.

Consists of some 4,000 terms dealing with all types of energy, environmental concerns, conservation and certain sciences. Over 300 illustrations (graphs, diagrams, charts, and photographs) enhance the text. Includes conversion tables and an extensive bibliography.
R: *Chemistry and Industry* 19: 789 (Oct. 4, 1980); *Electrical Apparatus* 32: 53 (June 1979); *JPT: Journal of Petroleum Technology* 31: 1591 (Dec. 1979); *BL* 76: 1159 (Apr. 15, 1980); 77: 641 (Jan. 1, 1981); *Choice* 16: 798 (Sept. 1979); *LJ* 104: 1127 (May 15, 1979); 105: 572 (Mar. 1, 1980); *RSR* 8: 26 (Jan./Mar. 1980); *TBRI* 45: 276 (Sept. 1979); 46: 74 (Feb. 1980); *WLB* 53: 728 (June 1979); *ARBA* (1980, p. 648); Wal (p. 197).

Solar Energy Dictionary. **Daniel V. Hunt.** New York: Industrial Press, 1982.

A thorough and up-to-date survey and basic guide to the current solar energy situation. For the energy specialist as well as the layperson.
R: *Mechanical Engineering*. 105: 107 (Feb. 1983); *ARBA* (1983, p. 653).

Integrated Energy Vocabulary. **US National Technical Information Service.** Springfield, VA: US National Technical Information Service, 1976.

Comprises some 30,000 terms from interdisciplinary fields related to science and technology.
R: *RSR* 5: 29 (July/Sept. 1977).

ENVIRONMENTAL SCIENCES

Dictionary of Dangerous Pollutants, Ecology, and Environment. **David F. Tver.** New York: Industrial Press, 1981.

Covers pollutions, energy, and environments. Useful for any type of library.
R: *ARBA* (1982, p. 748).

A Dictionary of Ecology, Evolution and Systematics. **R. J. Lincoln, G. A. Boxshall, and P. F. Clark.** New York: Cambridge University Press, 1982.

Contains some 10,000 terms of biology, chemistry, mathematics, and more, with an emphasis on their principles, processes, and classifications. For more advanced students working in this field.
R: *Nature* 299: 471 (Sept. 30, 1982); *ARBA* (1984, p. 680).

Dictionary of Environmental Engineering and Related Sciences: Diccionario de Ingenieria Ambiental Y Ciencias Afines. **Jose T. Villate.** Miami: Ediciones Universal, 1979.

A bilingual, English-Spanish, Spanish-English dictionary of 13,300 terms from environmental engineering and related sciences such as air pollution, ecology, agriculture, meteorology, oceanography, and water treatment. Recommended for all libraries.
R: *ARBA* (1981, p. 753).

Dictionary of Environmental Terms. **Alan Gilpin.** London: Routledge & Kegan Paul, 1976.

R: *Nature* 263: 803 (Oct. 28, 1976); *Aslib Proceedings* 43: 392 (Sept. 1978); *TBRI* 43: 4 (Jan. 1977).

A Dictionary of the Environment. **Michael Allaby.** New York: Van Nostrand Reinhold, 1977.

Some 6,000 terms from a broad range of fields that apply to the environmental sciences: botany, zoology, geology, etc. For both scientists and the layperson; a concise source of information.
R: *Chemical Engineering* 85: 11 (Nov. 20, 1978); *Nature* 272: 291 (Mar. 16, 1978); *New Scientist* 77: 377 (Feb. 9, 1978); *BL* 76: 146 (Sept. 15, 1979); *Choice* 15: 1191 (Nov. 1978); *RSR* 6: 26 (Oct./Dec. 1978); *TBRI* 44: 215 (June 1978); 45: 10 (Jan. 1979); *WLB* 53: 187 (Oct. 1978); *ARBA* (1979, p. 691); Sheehy (EJ19).

A Dictionary of the Natural Environment. **F. J. Monkhouse and J. Small.** New York: Wiley, 1978.

A dictionary of environmental and physical and geographical terms. Extensive definitions, maps, diagrams, and photographs. Cross-referenced.

Dictionary of Waste and Water Treatment. **John S. Scott and Paul G. Smith.** Woburn, MA: Butterworth, 1981.

Defines some 6,000 terms used in the British and United States associated with water treatment and supply, sewerage treatment and disposal, and other aspects of public health engineering. Brief yet complete definitions will be useful to biologists, engineers, and chemists.
R: *ARBA* (1983, p. 726).

Dictionary of Water and Sewage Engineering. 2d rev. and enl. ed. **R. Meinck and H. Mohle, eds.** Amsterdam, Oxford, and New York: Elsevier, 1977.

First edition, 1963.
Part 1 lists about 12,000 terms in German, followed by their equivalents in English, French, and Italian. Part 2 lists the terms in English, French, and Italian, linking them to part 1 with letters and numbers. Gives conversion tables and international units of measure.
R: *ARBA* (1979, p. 777); Wal (p. 410).

Ecology Field Glossary: A Naturalist's Vocabulary. **Walter H. Lewis.** Westport, CT: Greenwood Press, 1977.

Designed for the nonspecialist. Includes selected vocabulary from all fields of the environmental sciences. Recommended for reference use in public libraries.

R: *BL* 75: 497 (Nov. 1978); *LJ* 102: 2420 (Dec. 1, 1977); *RQ* 18: 98, 101 (Fall 1978); *RSR* 6: 21 (Oct./Dec. 1978); *ARBA* (1978, p. 687).

Environmental Glossary. 3d ed. **G. William Frick, ed.** Rockville, MD: Government Institutes, 1984.

First edition, 1980; second edition, 1982.

A revised and expanded edition listing over 3,000 terms, abbreviations, and acronyms relating to the environment, natural resources, and energy. Up-to-date and accurate. Useful to environmental lawyers.

R: *Science and Technology Libraries* 1: 67 (Spring 1981); *ARBA* (1985, p. 511).

Environmental Impact Statement Glossary: A Reference Source for EIS Writers, Reviewers, and Citizens. **Marc Landy, ed.** New York: Plenum Press, 1979.

Arranged in 16 sections or environmental impact themes, this useful glossary presents terms and phrases in the first 13, on such subjects as laws and regulations, water, transportation, and health. The last 3 themes comprise a collection of terms on population, geographic areas, and housing. Contains 4,000 agencies and over 85 documents. Includes a bibliography of documents and an alphabetized index of all entries. Helpful to anyone writing, studying, or reviewing an environmental impact statement.

R: *Choice* 17: 517 (June 1980); *ARBA* (1981, p. 685).

Glossary of the Environment with French and German Equivalents. **Conseil International de la Langue Française.** New York: Praeger Publishers, 1977.

A multidisciplined glossary of terms relating to environmental studies. Cross references are included, as is a listing of French and German equivalents. Up-to-date.

R: *ARBA* (1978, p. 685; 1979, p. 692).

Glossary on Air Pollution. Copenhagen, Denmark: The World Health Organization Regional Office for Europe, 1980.

Contains over 600 definitions, many taken from standard vocabularies of international organizations, such as the International Commission of Radiation Units.

R: *Water, Air, and Soil Pollution* 17: 337 (April 1982).

Lexicon of Terms Relating to the Assessment and Classification of Coal Resources. **Arthur H. J. Todd.** London: Graham & Trotman. Distr. New York: Nichols, 1982.

A compilation of many of the terms used in the description and assessment of coal deposits throughout the world. A useful source for those interested in the mining field.

R: *ARBA* (1984, p. 675).

TRANSPORTATION

Dictionary of Public Transport. Washington, DC: N. D. Lea Transportation Research Corporation, 1982.

A trilingual dictionary in English, French, and German containing some 2,000 public transportation and related terms and their definitions. Alphabetically sequenced.
R: *Science and Technology Libraries* 3: 76 (Winter 1982).

SAE Motor Vehicle, Safety and Environmental Terminology. **Society of Automotive Engineers.** Warrendale, PA: Society of Automotive Engineers, 1977.

Alphabetical listing of terms relating to motor vehicle safety. Also includes safety guidelines.
R: *ARBA* (1978, p. 764).

Transportation-Logistics Dictionary. 2d ed. **Joseph L. Cavinato, ed.** Washington, DC: Traffic Service, 1982.

An alphabetical listing of transportational terms covering a broad range of subjects, with a section of definitions of transportation rates and a list of standard abbreviations. For large reference collections and students of business administration involved with product distribution.
R: *ARBA* (1983, p. 738).

The VNR Dictionary of Ships and the Sea. **John V. Noel.** New York: Van Nostrand Reinhold, 1981.

A list of terms and phrases used in regard to the sea and ships. Entries are brief and may not be helpful to those unfamiliar with the sea.
R: *ARBA* (1982, p. 815).

CHAPTER 6   HANDBOOKS

GENERAL SCIENCE

The Amateur Naturalist's Handbook. Rev. ed. **Vinson Brown.** Englewood Cliffs, NJ: Prentice-Hall, 1980.

First edition, 1949.
The revised edition of a manual on nature study. New chapters on ecology and ethology are added. Includes drawings and an updated bibliography. Intended for beginning and advanced nonscientists.
R: *BL* 77: 1178 (May 1, 1981).

CRC Composite Index for CRC Handbooks. Cleveland: Chemical Rubber, 1977.

An index to the 49 CRC handbooks of the life, mathematical, and engineering sciences. Alphabetically arranged, cross-referenced. Helpful in accessing specific data. Includes over 250,000 entries.

Handbook of Chemistry and Physics: A Ready-Reference Book of Chemical and Physical Data. **Robert C. Weast, ed.** Boca Raton, FL: Chemical Rubber, 1913–. Annual.

Fifty-seventh edition, 1976–1977; sixty-third edition, 1982.
Annually revised handbook providing extensive data in physics and chemistry. Considered the "bible" of science students and researchers.
R: *Chemical Engineering* 86: 11 (Dec. 31, 1979); *Journal of the American Chemical Society* 93: 7122 (Dec. 15, 1971); 101: 5866 (Sept. 12, 1979); *Choice* 8: 46 (1971); *RSR* 5: 11 (Apr./June 1977); *ARBA* (1976, p. 649; 1972, p. 558); Jenkins (A79); Wal (p. 98); Win (ED46).

A Handbook of Public Speaking for Scientists and Engineers. **Peter Kenny.** Philadelphia: Heyden, 1982.

This handbook offers scientists and engineers some valuable guidance for speaking in public on technical matters. The only one of its kind, it is a practical and thorough guide for those in technical or specialists' fields seeking to become competent public speakers.
R: *New Scientist* 196: 49 (Oct. 7, 1982).

Handbook of Scientific and Technical Awards. Supp. no. 1. New York: Special Libraries Association, Science-Technology Division, 1981.

A general reference handbook to the recipients of various awards given by several scientific and technical societies. Provides useful information, but the way in which it is presented may cause confusion in some areas.
R: *ARBA* (1983, p. 599).

The Practicing Scientist's Handbook: A Guide for Physical and Terrestrial Scientists and Engineers. **Alfred J. Moses.** New York: Van Nostrand Reinhold, 1978.

Designed to provide a comprehensive source of materials' property data. Contains numerous charts, tables, and diagrams, which cover elements, organic compounds, alloys, glasses, polymers, ceramics, etc. A convenient source for scientists and engineers. Excellent for the practicing scientist.

R: *Analyst* 104: 586 (June 1979); *Chemical Engineering* 86: 14 (Jan. 29, 1979); *Chemistry and Industry* 12: 411 (June 16, 1979); *Mechanical Engineering* 101: 134 (Feb. 1979); *New Scientist* 81: 268 (Jan. 25, 1979); *Physics Bulletin* 30: 531 (Dec. 1979); *Physics Today* 32: 61 (June 1979); *TBRI* 45: 129 (Apr. 1979); 45: 248 (Sept. 1979); 45: 288 (Oct. 1979); *WLB* 53: 474 (Feb. 1979); *ARBA* (1979, p. 643).

Technical Writer's Handbook. **Harry E. Chandler.** Metals Park, OH: American Society for Metals, 1983.

Part 1, *Writing for Busy Readers*, explains how to write for technical and nontechnical people. Part 2, *Customs, Practices, and Standards for Technical Writing*, offers answers for writers who want to know "how do others do this?" Part 3, *An Anthology of Stylebooks*, contains rules and practices for spelling, punctuation, abbreviations, etc. Each part is one book in itself.
R: *Mechanical Engineering* 105: 113 (Nov. 1983).

ASTRONOMY

The Amateur Astronomer's Handbook. 4th rev. ed. **J. B. Sidgwick.** Hillside, NJ: Enslow, 1982.

Third edition, 1971; fourth edition, 1980.
Intended as a practical guide for the amateur who utilizes a small- or medium-sized telescope. Emphasizes technique, but does include a glossary, reading lists, and information on future phenomena.
R: *New Scientist* 87: 43 (July 3, 1980).

Astronomy and Telescopes: A Beginner's Handbook. **Robert J. Traister and Susan E. Harris.** Blue Ridge Summit, PA: TAB Books, 1983.

Emphasis is on theory, construction, and use of telescopes and auxiliary equipment for the new amateur astronomer. Presents a fine introduction as well as some valuable information for the buyer, user, or builder of telescopes.
R: *ARBA* (1984, p. 633).

Astronomy Handbook. **James Muirden.** New York: Arco, 1982.

Covers most topics in astronomy and is well-illustrated. Presumes no previous knowledge of astronomy. Recommended for lower-level students and general audiences.
R: *Choice* 20: 601 (Dec. 1982); *ARBA* (1983, p. 604).

Burnham's Celestial Handbook: An Observer's Guide to the Universe Beyond the Solar System. Rev. and enl. ed. **Robert Burnham, Jr.** New York: Dover, 1978.

Volume 3, *Pavo-Vulpecula*.
A complete, detailed multivolume set that contains over 7,000 descriptions of celestial objects. Entries are grouped alphabetically by constellation and include charts, diagrams, photographs, and descriptions. Volumes cover different galaxies, with volume 3 focusing on the galaxies from Pavo to Vulpecula. An invaluable reference for beginning astronomers.
R: *Astronomy* 7: 77 (July 1979); *Sky and Telescope* 56: 452 (Nov. 1978); 57: 389 (Apr. 1979); 57: 570 (June 1979); *RSR* 6: 24 (July/Sept. 1978); *TBRI* 45: 242 (Sept. 1979); *ARBA* (1979, p. 649; 1980, p. 611).

Handbook of Astronomy, Astrophysics and Geophysics. Vol. 1. **Charlotte W. Gordon, V. Canuto, and W. Ian Axford, eds.** London: Gordon & Breach, 1978.

Volume 1, *The Earth*; part 1, *The Upper Atmosphere, Ionosphere and Magnetosphere.* Comprehensive articles that cover a wide range of topics, including hydrogen in the upper atmosphere, the equatorial electrojet, etc.

R: *Nature* 277: 73 (Jan. 4, 1979); *Sky and Telescope* 57: 181 (Feb. 1979); *Aslib Proceedings* 43: 388 (Sept. 1978); *TBRI* 45: 86 (Mar. 1979).

The Handbook of the British Astronomical Association 1982. **Gordon E. Taylor, ed.** Piccadilly, London: Burlington House, 1981. Annual.

This handbook gives ephemerides of the sun, moon, planets, 92 double stars, and 13 periodic comets that were expected to be seen in 1982. Contains other information, positions, and visual magnitudes for several other astronomical aspects.

R: *Sky and Telescope* 63: 49 (Jan. 1982).

Handbook of Radio Sources. **A. G. Pacholczyk, ed.** Tucson: Pachart Publishing, 1978.

Contains data concerning quasars and galaxies.

R: *New Scientist* 79: 489 (Aug. 17, 1978); *Observatory* 99: 51 (Apr. 1979); *Sky and Telescope* 57: 176 (Feb. 1979); *TBRI* 44: 288 (Oct. 1978); 45: 129 (Apr. 1979).

Handbook of Solar Flare Monitoring and Propagation Forecasting. **Carl M. Cherman.** Blue Ridge Summit, PA: TAB Books, 1978.

For the beginning astronomer. Contains basic information on flare monitoring.

R: *Sky and Telescope* 57: 282 (Mar. 1979); *TBRI* 45: 163 (May 1979).

Handbook of Space Astronomy and Astrophysics. **M. V. Zombeck.** New York: Cambridge University Press, 1982.

A compilation of often-inaccessible information in tables, graphs, and formulas for use across a broad range of the physical sciences. Information is included on all aspects of astronomy and astrophysics, the Earth's atmosphere and environment, relativity and atomic physics.

R: *Nature* 300: 554 (Dec. 9, 1982).

Mysterious Universe: A Handbook of Astronomical Anomalies. **William R. Corliss.** Glen Arm, MD: Sourcebook Project, 1979.

A handbook of astronomical illustrations.

R: *Nature* 280: 92 (July 5, 1979); *New Scientist* 84: 888 (Dec. 13, 1979); *Sky and Telescope* 58: 166 (Aug. 1979); *TBRI* 46: 42 (Feb. 1980).

The Observer's Handbook 1982. 74th ed. **Roy L. Bishop, ed.** Royal Astronomical Society of Canada. Canada: Sky Publishing, 1981.

An in-depth continuation of the format introduced 2 years ago as an aid for amateur astronomers. All you need to know about astronomy, including occultation predictions for 15 standard stations in Canada and the United States and maps of North American grazing occultations.

R: *Sky and Telescope* 63: 159 (Feb. 1982).

The Planet Jupiter: The Observer's Handbook. 2d rev. ed. **Patrick Moore.** London: Faber and Faber, 1981.
First edition, 1958.
An up-to-date revision of the first edition by Bertrand M. Peek. Gives a complete description of Jupiter as observed from the Earth, with visual observations of color, position, rotation, and surface detail. Contains diagrams, tables, photos, a glossary, and an index. Recommended for Jupiter observers and college, community college, and general collections.
R: *Choice* 20: 115 (Sept. 1982).

Webb Society Deep-Sky Observer's Handbook. 5 vols. **Kenneth Glyn Jones, ed.** Hillside, NJ: Enslow, 1979–1982. Rev. edition, 1985–.

Volume 1, *Double Stars*; volume 2, *Planetary and Gaseous Nebulae*; volume 3, *Open and Globular Clusters*; volume 4, *Galaxies*; volume 5, *Clusters of Galaxies*. Volume 1. *Double Stars,* **revised edition by R. W. Argyle,** 1985.
Each volume covers 1 class of objects. It includes a review article, background information, positions, magnitudes, charts, and various other information. This series would be an excellent aid for any college and university library observatory collection.
R: *Astronomy* 8: 73 (Jan. 1980); *New Scientist* 87: 602 (Aug. 21, 1980); *Sky and Telescope* 58: 458 (Nov. 1979); 59: 147 (Feb. 1980); *Choice* 20: 601 (Dec. 1982); *TBRI* 46: 87 (Mar. 1980); *ARBA* (1980, p. 611; 1981, p. 630).

MATHEMATICS

ASM Handbook of Engineering Mathematics. **William G. Belding, ed.** Metals Park, OH: American Society for Metals, 1983.
Key equations of the basic mathematic equations and theorems used in the design and manufacturing environment of typical metalworking companies. For the engineer or engineering student with a basic knowledge of college-level mathematics.
R: *Science and Technology Libraries* 4: 121 (Winter 1983).

Engineering Mathematics Handbook: Definitions, Theorems, Formulas, Tables. 2d rev. ed. **Jan J. Tuma.** New York: McGraw-Hill, 1979.
First edition, 1971.
A comprehensive reference to various aspects of engineering mathematics. Volume is arranged in 5 sections according to the type of mathematics covered: (1) algebra, geometry, and trigonometry; (2) differential calculus; (3) ordinary and partial differential equations; (4) numerical methods, probability, and statistics; and (5) indefinite integrals. Invaluable reference tool for engineers, scientists, and architects.
R: *Chemical Engineering* 87: 11 (Apr. 7, 1980); *Instruments and Control Systems* 52: 18 (Sept. 1979); *Lighting Design and Application* 9: 52 (Nov. 1979); *Choice* 8: 366 (May 1971); 16: 1156 (Nov. 1979); *RQ* 19: 95 (Fall 1979); *TBRI* 45: 364 (Dec. 1979); 46: 40 (Jan. 1980); *ARBA* (1972, p. 636; 1971, p. 482; 1980, p. 608).

Green's Functions and Transfer Functions Handbook. **A. G. Butkovskiy.** New York: Wiley, 1982.
R: *Science* 218: 189 (Oct. 8, 1982).

Handbook for Linear Regression. **Mary Sue Younger.** N. Scituate, MA: Duxbury Press, 1979.

A reference handbook that examines simple and multiple linear regression methods. Contains a subject index and 4 appendixes that add to the usefulness of the book. Some background knowledge of statistics and matrix analysis is helpful but not required. Of interest to undergraduate libraries.
R: *Choice* 17: 111 (Mar. 1980).

Handbook of Applicable Mathematics. 6 vols. in 8 pts. **Walter Ledermann, ed.** New York: Wiley, 1980–.

Volume 1, *Algebra*, 1980; volume 2, *Probability*, 1980; volume 3, *Numerical Methods*, 1981; volume 4, *Analysis*, 1981; volume 5, *Geometry and Combinatorics*, 1981; volume 6, *Statistics*, in preparation.
A 6-volume reference set concerning the practical aspects and applications of mathematics. Articles do not require extensive knowledge of particular mathematical topics. Volumes cover algebra, geometry, probability and statistics, analysis, etc. An extensive bibliography is included at the end of each article.
R: *New Scientist* 88: 172 (Oct. 16, 1980); 90: 510 (May 21, 1981); *Choice* 18: 692 (Jan. 1981).

Handbook of Applied Mathematics. 5th ed. **Carl E. Pearson, ed.** New York: Van Nostrand Reinhold, 1974.

Third edition, 1955; fourth edition, 1966.
Covers methods of mathematical analysis, emphasizing technique. Contributions by experts; includes bibliographic references.
R: Jenkins (B55); Sheehy (EF8); Win (EF26; 1EF4).

Handbook of Applied Mathematics. Selected Results and Methods. 2d ed. **Carl E. Pearsen.** New York: Van Nostrand Reinhold, 1983.

A book of mathematical procedures and applications for solving scientific and engineering problems. Includes formulas for vector analysis, tensors, complex variables, etc.
R: *IEEE Spectrum* 20: 112A (May 1983).

Handbook of Hypergeometric Integrals; Theory, Applications, Tables, Computer Programs. **Harold Exton.** New York: Halsted Press, 1978.

An ordered list of analytical formulas. A useful reference for mathematicians, statisticians, physicists, computer analysts, etc.

Handbook of Mathematical Tables and Formulas. 5th ed. **R. Burington.** New York: McGraw-Hill, 1973.

A ready reference for the most important formulas and theorems from mathematics. Covers such material as linear algebra, Legendre polynomials, Laplace transforms, and Bessel functions. Recommended tool for all students and professionals who need quick reference.

Handbook of Measurement Science. **P. H. Sydenham.** 2 vols. New York: Wiley, 1982–1983.

Volume 1, *Theoretical Fundamentals*, 1982; volume 2, *Practical Fundamentals*, 1983.

Handbook of Numerical and Statistical Techniques: With Examples Mainly from the Life Sciences. **J. H. Pollard.** New York: Cambridge University Press, 1977.

A well-arranged handbook of statistics and statistical techniques. Easy to use and highly recommended.
R: *Nature* 270: 284 (Nov. 17, 1977); *Aslib Proceedings* 42: 488 (Oct. 1977); *TBRI* 44: 137 (Apr. 1978).

Handbook of Operations Research. 2 vols. **Joseph J. Moder and Salah E. Elmaghraby.** New York: Van Nostrand Reinhold, 1977–1978.

Volume 1, *Foundations and Fundamentals*; volume 2, *Models and Applications*.
Covers theory and concepts of operational research, as well as areas of application. Comprehensive.
R: *Quality Progress* 11: 32 (Nov. 1978); *RSR* 5: 26 (Oct./Dec. 1977); *TBRI* 45: 37 (Jan. 1979).

Handbook of Stochastic Methods. 2d ed. **Crispin W. Gardiner.** New York: Springer-Verlag, 1985.

Operations Research Handbook: Standard Algorithms and Methods with Examples. **H. A. Eiselt and H. von Frajer.** New York: De Gruyter, 1977.

Contains algorithms in the field of computer programming. All algorithms are presented in form principle, description, and example.
R: *Aslib Proceedings* 43: 87 (Feb. 1978).

Statistical Data Analysis Handbook. **F. J. Wall.** New York: McGraw-Hill, 1985.

PHYSICS

Advanced Oscilloscope Handbook for Technicians and Engineers. **Robert G. Middleton.** Reston, VA: Reston Publishing, 1977.

Written at an advanced high school level, a clear concise handbook of oscilloscope operation. Contains illustrations, examples of how to test and analyze wavelengths. A recommended reference for electrical engineers.
R: *Choice* 14: 897 (Sept. 1977).

American Institutes of Physics Handbook. 3d ed. 9 secs. **Dwight Gray, ed.** New York: McGraw-Hill, 1972.

First edition, 1957; second dition, 1963.
Collection of the fundamental principles of physics, including tables, graphs, and bibliographies. Each section edited by a specialist. Aimed at university and professional levels. Exceptionally well-received reference.
R: *Science Reference Notes* 10: 24 (1963); *ARBA* (1973, p. 547); Jenkins (C58); Win (EG33).

CRC Handbook of Laser Science and Technology. **Marvin J. Weber, ed.** Boca Raton, FL: CRC Press, 1982– .

Volume 1, *Lasers and Masers*, 1982; volume 2, *Gas Lasers*, 1982; volume 3, *Optical Materials–Nonlinear Optical Properties/Radiation Damage*, 1986; volume 4, *Optical Materials—Properties*, 1986.

Includes summaries and tabulations, provided by individuals who have been firsthand important participants in the past 20 years of laser development. Updated and expanded since the earlier edition of a decade ago.

R: *Laser Focus* 19; 99 (Mar. 1983); *Optical Engineering* 22: SR158 (Sept. 1983).

Handbook of Acoustical Enclosures and Barriers. **Richard K. Miller and W. V. Montone.** Atlanta: Fairmont, 1979.

For acoustical engineers. Provides information on design and includes tables and equations.

R: *Plant Engineering* 33: 156 (Dec. 13, 1979); *TBRI* 46: 116 (Mar. 1980).

Handbook of Chemical Lasers. **R. W. Gross and J. F. Bott.** New York: Wiley, 1976.

A comprehensive and highly technical review on chemical lasers. Includes bibliographies. A state-of-the-art book on theory.

R: *Chemistry and Industry* 19: 771 (Oct. 7, 1978); *Journal of the American Chemical Society* 99: 4868 (July 6, 1977); 100: 1329 (Feb. 15, 1978); *Nature* 268: 85 (July 7, 1977); *TBRI* 43: 275 (Sept. 1977); *ARBA* (1978, p. 644).

Handbook of Elementary Physics. 4th English ed. **N. I. Koshkin and M. G. Shirkevich.** Translated by G. Leib. Moscow: Mir, 1982.

A pocket-sized volume for quick reference or review. Includes 145 tables of data, 92 figures, and a good index, but lacks literature references. A translation of the eighth Russian edition.

R: *Science Technology Libraries* 5: 2 (Winter 1984).

Handbook of Fiber Optics: Theory and Applications. **Helmut F. Wolf, ed.** New York: Garland STPM Press, 1979.

A survey of all areas of fiber optics. Presents economic aspects and a summary of activities currently under way in the United States, Europe, and Japan. Includes illustrations and photographs. A graduate-level text for any fiber-optics collection.

R: *Choice* 17: 564 (June 1980).

Handbook of Flame Spectroscopy. **Michael L. Parsons et al.** New York: Plenum Press, 1976.

Discusses basic principles of flame spectroscopy. Culls useful data.

R: *Canadian Journal of Spectroscopy* 21: 29A (Nov./Dec. 1976); *TBRI* 43: 169 (May 1977).

Handbook of Fluids in Motion. **Nicholas P. Cheremisinoff and R. Gupta, eds.** Ann Arbor: Ann Arbor Science, 1983.

A reference to a broad spectrum of engineering research collection as a handbook-cum-encyclopedia on the area of applied fluid mechanics.

R: *Science and Technology Libraries* 4: 120 (Winter 1983).

Handbook of Multiphase Systems. **G. Hetsroni.** New York: McGraw-Hill, 1981.

A comprehensive survey of all aspects of multiphase systems. Discusses continuum physics laws, surface phenomena, sampling, averaging, condensation, fluidization, separation, etc. Includes tables, illustrations, and figures. Intended for chemical, mechanical, nuclear, and petroleum engineers.

Handbook of Optical Holography. **H. J. Caulfield, ed.** New York: Academic Press, 1979.
A practical guide to holography. For researchers, professionals, and artists.
R: *American Scientist* 68: 710 (Nov./Dec. 1980); *Physics Bulletin* 31: 248 (Aug. 1980).

Handbook of Optics. **Walter G. Driscoll and William Vaughan, eds.** New York: McGraw-Hill, 1978.
A major contribution to the field of applied physics. Intended for optical designers and engineers. An authoritative work that contains much graphic and tabular information. For undergraduate, graduate, and research libraries.
R: *American Journal of Physics* 47: 293 (Mar. 1979); *Applied Optics* 18: 416 (Feb. 15, 1979); *Chemical Engineering* 86: 14 (Mar. 26, 1979); *New Scientist* 80: 703 (Nov. 30, 1978); *Physics Today* 31: 69 (Oct. 1978); *Review of Scientific Instruments* 50: 266 (Feb. 1979); *TBRI* 45: 124 (Apr. 1979); 45: 203 (June 1979); *ARBA* (1979, p. 660); Sheehy (EG8).

Handbook of Physical Calculations. **Jan J. Tuma.** New York: McGraw-Hill, 1976.
A straightforward handbook of physical calculations. Presents in 1 volume a concise collection of major definitions, tables, formulas, and examples of elementary and intermediate technological physics. Emphasizes practical applications with simple, easy-to-understand fundamentals. A primer and refresher for engineers and technologists.
R: *Physics in Technology* 8: 32 (Jan. 1977); *RSR* 5: 24 (Oct./Dec. 1977); *TBRI* 43: 132 (Apr. 1977).

Handbook on Semiconductors. 4 vols. **T. S. Moss, gen. ed.** Amsterdam: North-Holland, 1980–1982.
Volume 1, *Bank Theory and Transport*, 1982; volume 2, *Optical Properties of Semiconductors*, 1980; volume 3, *Materials, Properties, and Preparation*, 1980; volume 4, *Device Physics*, 1981.
This 4-volume handbook presents a comprehensive view of the whole field of semiconductor knowledge. Each volume is edited by an internationally recognized leader in his field.
R: *Physics Today* 35: 84 (Oct. 1982).

Handbook of Spectroscopy. Vols. 1–. **J. W. Robinson, ed.** Cleveland: Chemical Rubber, 1974–.
A handy reference covering the main areas of spectroscopy, including IR, NMR, UV, Raman, mass spectrometry, etc. Provides a combined author and subject index. Of interest to chemists and physicists.

Handbook of Surfaces and Interfaces. 3 vols. **L. Dobrzynski, ed.** New York: Garland STPM Press, 1978.

A 3-volume handbook that extensively reviews the theory and techniques of surfaces and interfaces, particularly regarding atomic and electronic structures. For engineers, scientists, and graduate students of surface science.
R: *Advances in Colloid and Interface Science* 11: 93 (Mar. 1979); *TBRI* 45: 203 (June 1979).

Handbook of X-ray and Ultraviolet Photoelectron Spectroscopy. **D. Briggs, ed.** Philadelphia: Heyden, 1977.
Contains valuable references for both experienced and novice professionals. Concise, yet includes a broad scope of information in X-ray spectroscopy. Includes bibliographic references and index.
R: *Analytical Chemistry* 50: 1004A (Sept. 1978); *Physics Today* 31: 69 (Oct. 1978); *TBRI* 44: 282 (Oct. 1978).

Handbook on Plasma Instabilities. Vol. 2. **Ferdinand F. Cap.** New York: Academic Press, 1978.
Volume 1, 1976.
R: *Science* 193: 1042 (Sept. 10, 1976).

Infrared Handbook. **William L. Wolfe and George J. Zissis, eds.** Washington, DC: US Department of the Navy, Office of Naval Research, 1978.
A handbook prepared to replace 1968's *Handbook of Military Infrared Technology*. Provides ready-reference data, techniques and equations. Good for the "infrared community."
R: *Journal of Applied Photographic Engineering* 7: 91A (June 1982).

The Infrared Spectra Handbook of Priority Pollutants and Toxic Chemicals. Philadelphia: Sadtler Research Laboratories, 1982.
Contains spectra of over 500 compounds at least nominally falling into the category described by the title.
R: *Applied Spectroscopy* 37: 304 (May/June 1983).

The Laser Experimenter's Handbook. **Frank G. McAleese.** Blue Ridge Summit, PA: TAB Books, 1979.
A handbook containing topics such as matter, energy, and atomic nature; electromagnetic wave theory; light wave theory; and more. Provides an understanding of the physics of lasers needed to construct a functionally operating laser. Contains photographs and diagrams. For public libraries and personal purchase.

Laser Handbook. 3 vols. **F. T. Arecchi And E. D. Schulz-Dubois.** New York: American Elsevier, 1973–1979.
R: *American Scientist* 69: 109 (Jan./Feb. 1981); *Physics Today* 29: 53 (July 1969); Wal (p. 202).

Laser Safety Handbook. **Alex Mallow and Leon Chabot.** New York: Van Nostrand Reinhold, 1978.
A detailed handbook on laser safety. Well recommended.
R: *American Scientist* 68: 230 (Mar./Apr. 1980); *New Scientist* 81: 191 (Jan. 18, 1979); *TBRI* 45: 157 (Apr. 1979).

The Master Handbook of Acoustics. **F. Alton Everest.** Blue Ridge Summit, PA: TAB Books, 1981.

An attempt to popularize and simplify the complex subject of acoustics. It is not very useful for the acoustical engineer or architect, but recommended for the vocational school, community college, public library, and general public.

R: *Choice* 19: 1433 (June 1982).

Medical Physics Handbooks. 3 vols. **J. M. A. Lenihan, ed.** Bristol, England: Adam Hilger. Distr. Philadelphia: Heyden, 1979–1980.

Volume 1, *Ultrasonics.* John Woodcock. 1979; volume 2, *Computing Principles and Techniques.* Bruce Vickery. 1979; volume 3, *Physical Principles of Audiology.* Peter Haughton. 1980.

A comprehensive 3-volume handbook designed for use by scientists and engineers, as well as physicians seeking a better understanding of the physical principles behind bioengineering.

Modern Oscilloscope Handbook. **Douglas Bapton.** Englewood Cliffs, NJ: Prentice-Hall, 1979.

Contains clear and basic information on oscilloscopes and their applications. Useful for self-study. Recommended for physicists.

R: *Physics Teacher* 17: 606 (Dec. 1979); *Aslib Proceedings* 44: 528 (Dec. 1979); *TBRI* 46: 70 (Feb. 1980).

Neblette's Handbook of Photography and Reprography: Materials, Processes, and Systems. 7th ed. **John M. Sturge, ed.** New York: Van Nostrand Reinhold, 1977.

Sixth edition, 1962.

Expanded and revised handbook with contributions by experts. Covers all basic aspects of reprography including optics, chemistry, physics, manufacturing. Recommended.

R: *Optical Spectra* 11: 50 (Nov. 1977); *TBRI* 44: 39 (Jan. 1978); *WLB* 52: 189 (Oct. 1977).

Practical Instrumentation Handbook. **Bert Earley.** London: Scientific Era, 1976.

A highly useful book for plant engineers. Contains much information on instrumentation.

R: *Electronics and Power* 23: 71 (Jan. 1977); *TBRI* 43: 192 (May 1977).

Practical Oscilloscope Handbook. 2d ed. **Howard Bierman.** Rochelle Park, NJ: Hayden Book Co.

Provides the basics of oscilloscope operation as well as the most advanced and difficult methods with the latest scope designs. Will enable an engineer, technician, or serious hobbyist to set up, display, and interpret waveforms.

R: *Measurements and Control* 16: 169 (Feb. 1982).

Practical Oscilloscope Handbook. **John Douglas-Young.** Englewood Cliffs, NJ: Prentice-Hall, 1979.

A practical handbook on the selection and operation of oscilloscopes. Useful for technicians.

RF Radiometer Handbook. **G. Evans and C. W. McLeish.** Dedham, MA: Artech House, 1977.

For radio engineers and radio astronomers. Discusses the design and use of radiometers.

R: *Microwaves* 17: 118 (June 1978); *Proceedings of the IEEE* 66: 109 (Jan. 1978); *TBRI* 44: 83 (Mar. 1978); 44: 244 (Sept. 1978).

Refrigeration Processes: A Practical Handbook on the Physical Properties of Refrigerants and Their Applications. **H. M. Meacock.** Elmsford, NY: Pergamon, 1979.

A reference volume on the physical and thermodynamic properties of 29 refrigerants. Describes definitions, processes, and properties. Includes numerous tables, diagrams, and data sheets. Uses both English and metric units. Recommended for upper-level undergraduate students interested in thermodynamic properties.

R: *Choice* 17: 565 (June 1980).

Sadtler Handbook of Infrared Spectra. **William W. Simons, ed.** Philadelphia, PA: Sadtler Research Laboratories, 1978.

Comprises a small, convenient reference for student and practicing spectroscopist. Comprehensively covers spectra-structure correlation. Provides general information on organic spectra. Recommended.

R: *Applied Spectroscopy* 33: 420 (July/Aug. 1979); *Journal of Chromatographic Science* 16: 21A (Nov. 1978); *Journal of the American Chemical Society* 101: 4021 (July 4, 1979); *TBRI* 45: 50 (Feb. 1979); 45: 290 (Oct. 1979).

Safety with Lasers and Other Optical Sources: A Comprehensive Handbook. **David Sliney and Myron Wolbarsht.** New York: Plenum Press, 1980.

An authoritative handbook on laser safety and the control of other potential sources of optical radiation. Discusses practical problems and solutions. Useful to health and safety professionals, optical engineers, and others who work with optical equipment.

Semiconductor Laser Diodes; A User's Handbook. **M. E. Fabian.** Scotland: Electrochemical, 1981.

An attempt to familiarize readers with the laser diode and its associated terminology, and to show the laser as just another circuit element in system design. Written as an aid to engineers and managers.

R: *Proceedings of the IEEE* 70: 317 (Mar. 1982).

Shock and Vibration Handbook. 2d ed. 3 vols. **Cyril M. Harris and C. E. Crede, eds.** New York: McGraw-Hill, 1976.

First edition, 1961; volume 1, *Basic Theory and Measurements*; volume 2, *Data Analysis, Testing and Methods of Control*; volume 3, *Engineering Design and Environmental Conditions*.

Covers shock and vibration theory, instrumentation, and measurement techniques. Encompasses a broad scope of information. Contains much new material. Well-referenced and illustrated. Designed primarily for specialists.

R: *ASHRAE Journal* 19: 75 (Mar. 1977); *American Journal of Physics* 45: 689 (July 1977); *Automotive Engineering* 84: 15 (Dec. 1976); *Journal of the Acoustical Society of America* 33: 1435 (Oct. 1961); 61: 1658 (June 1977); *Naval Engineers Journal* 89: 30 (Aug. 1977); *Choice* 13: 1570 (Feb. 1977); *RSR* 5: 37 (Oct./Dec. 1977); *TBRI* 27: 186

(1961); 43: 116 (Mar. 1977); 43: 271 (Sept. 1977); 43: 313 (Oct. 1977); 43: 354 (Nov. 1977); *ARBA* (1977, p. 768); Sheehy (EJ56); Wal (p. 292); Win (EI149).

CHEMISTRY

The Beilstein Guide: A Manual for the Use of Beilstein's Handbuch der Organischen Chemie. **Oskar Weissbach.** New York: Springer-Verlag, 1976.

Beilstein's Handbuch der Organischen Chemie [Handbook of Organic Chemistry]. 4th ed. New York: Springer-Verlag, 1918–. Also, supp. 1, 1930–1949; supp. 2, 1920–1929; supp. 3, 1930–1949; supp. 4, 1956–.

First edition, 1881–1883; second edition, 1885–1889; third edition, 1893–1899; supplement, 1901–1906; volume 21, 1978–1979.

The main work of the fourth edition covers the literature on about 200,000 organic compounds to 1909. This set is the most extensive work in the field. Arrangements are based on classes of compounds.

R: Wal (p. 133).

R: *Journal of the American Chemical Society* 102: 7162 (Nov. 5, 1980); Wal (p. 133).

CRC Handbook of Clinical Chemistry. **Mario Werner, ed.** Boca Raton, FL: CRC Press, 1982–.

Volume 1, 1982; volume 2, 1985; volume 3, 1985.

A compendium of information, data, and methods for those whose work involves clinical chemistry. Extensive details and bibliographies for specific applications of various techniques. Abundant tables and figures.

CRC Handbook Series in Inorganic Electrochemistry. **Louis Meites, Petr Zuman, Elinore B. Rupp, and Ananthakrishnan Narayanan, eds.** Boca Raton, FL: CRC, 1983–.

Volume 5, *O–PD*, 1985; volume 6, *PM–SC*, 1986; volume 7, *SC–TM*, 1986.

A comprehensive reference set that contains information on the electrochemical behaviour of inorganic substances, including the complexes of metal ions with organic ligands. Includes a key to literature citations and an author index in each volume.

Chemical Technician's Ready Reference Handbook. 2d ed. **Gerston J. Shugar, Ronald A. Shugar, and Laurence Bauman.** New York: McGraw-Hill, 1981.

First edition, 1973.

A handbook aimed at technicians, students, and working chemists. A ready-reference guide that can be used as a student's laboratory tool.

R: *Chemistry and Industry* 10: 406 (May 18, 1974); *Journal of the American Chemical Society* 95: 8491 (Dec. 12, 1973); *Choice* 10: 1170 (Oct. 1973); *TBRI* 40: 270 (Sept. 1974); *ARBA* (1974, p. 558).

Chromatography: A Laboratory Handbook of Chromatographic and Electrophoretic Methods. 3d ed. **Erich Heftmann, ed.** New York: Van Nostrand Reinhold, 1975.

R: *Clinical Chemistry* 22: 128 (Jan. 1976); *Journal of the American Chemical Society* 98: 308 (Jan. 7, 1976); *TBRI* 42: 84 (Mar. 1976).

Gmelins Handbuch der Anorganischen Chemie [Handbook of Inorganic Chemistry]. 8th ed. 71 vols. **Gmelin-Institut.** New York: Springer-Verlag, 1922–. Also, *New Supplement Series.*

Eighth edition under the auspices of the German Chemical Society until 1945, then the Gmelin-Institut.

This is an authoritative and comprehensive compendium of all literature on the field. Contains systematic subject index. The *New Supplement Series* has more than 14 volumes already published on topics currently attracting interest in chemical research.

R: *Science Progress* (208): 678 (Oct. 1964); Wal (p. 131); (Win ED61).

R: *Journal of Metals* 30: 7 (Feb. 1978); *Journal of the American Chemical Society* 100: 1017 (Feb. 1, 1978); 100: 1643 (Mar. 1, 1978); 102: 3665 (May 7, 1980); *Science Progress* (208): 678 (Oct. 1964); Wal (p. 131); Win (ED61).

A Guide to Beilstein's Handbook. **Martha M. Vestling and Janice T. Liebe.** Washington, DC: AAAS Science Books and Films, 1977.

Discusses organization of the *Handbuch der Organischen Chemie.* Invaluable to students, librarians, and chemists.

R: *Journal of the American Chemical Society* 99: 5840 (Aug. 17, 1977); *TBRI* 43: 331 (Nov. 1977).

Handbook of Analytical Derivatization Reactions. **D. R. Knapp.** New York: Wiley, 1979.

An outstanding reference on the preparation of derivatives of organic compounds. Contains indexes that allow easy access to information. Helpful to scientists working with gas chromatography, mass spectrometry, or liquid chromatography. For scientists, teachers, and students in organic chemistry.

R: *Journal of Chemical Education,* 57: A196 (June 1980); *Journal of the American Chemical Society* 102: 4285 (June 4, 1980).

Handbook of Anion Determination. **W. John Williams.** London: Butterworth, 1979.

A useful handbook dealing with organic and inorganic anion determination and separation. Anions are treated separately with discussion of the accuracy of various methods. Theory, equations, and procedures are given. Includes ample reference lists. For all graduate and undergraduate chemistry libraries.

R: *Chemistry in Britain* 16: 280 (May 1980); *Journal of the American Chemical Society* 103: 2501 (May 6, 1981); *Laboratory Practice* 29: 35 (Jan. 1980); *Nature* 284: 648 (Apr. 17, 1980); *Choice* 17: 414 (May 1980); *RSR* 8: 32 (July/Sept. 1980).

Handbook of Automated Analysis, Continuous Flow Techniques. **W. A. Coakley.** New York: Dekker, 1981.

Helpful in setting up and using segmented stream continuous flow automatic analyzers. Main consideration is given to Technicon AutoAnalyzer equipment without mention of the SOLIDPrep sampler, the Continuous Digestor, or any other multichannel instruments. This book is more useful for newcomers to this field than to more experienced autoanalysts.

R: *Chemistry in Britain* 18: 511 (July 1982).

Handbook of Bimolecular and Termolecular Gas Reactions. Volumes 1–2. **Alistair Kerr and Stephen J. Moss, eds.** Boca Raton, FL: CRC Press, 1981.

This handbook provides a useful summary up to the end of 1977 of 4 previous collections of gas-phase rate constants. Volume 1 covers rate constants for metathetical reactions, reactions of radicals with carbon atoms, and inorganic radicals. Volume 2 covers termolecular reactions, atom and radical interactions, etc.
R: *Journal of the American Chemical Society* 104: 4734 (Aug. 1982).

Handbook of Chemical Equilibria in Analytical Chemistry. **S. Kotrly and L. Suncha.** New York: Halsted Press, 1985.

A concise introduction to the use of equilibrium data for solving practical chemical analysis problems. Reliable as a guide for selecting constants with which to solve these problems and to using different types of graphs with diagrams for solving problems. Well-organized and easy to follow.

Handbook of Chemical Microscopy. 4th ed. Vol. 1. **C. W. Mason.** New York: Wiley, 1983.

Includes material on scanning microscopy as well as microscopial qualitative tests from volume 2 of the third edition. All aspects of physical methods and chemical analysis are also covered.
R: *Chemistry in Britain* 18: 880 (Dec. 1982).

Handbook of Chemical Property Estimation Methods: Environmental Behavior of Organic Compounds. **Warren J. Lyman, William F. Reehl, and David H. Rosenblatt, eds.** New York: McGraw-Hill, 1982.

A comprehensive handbook on existing techniques for estimating physiochemical properties of organic chemicals of environmental concern. Discusses transport, properties, and importance of chemicals. Sections cover importance of properties, descriptions, and instructions for each method. Includes references and lists of symbols and sources of data. Recommended reference for environmental chemists, chemical engineers, and pharmacologists.
R: *ARBA* (1983, p. 607).

A Handbook of Computational Chemistry: A Practical Guide to Chemical Structure and Energy Calculations. **T. Clark, ed.** New York: Wiley, 1985.

A practical guide to performing chemical structure and energy calculations. Deals specifically with the 3 most commonly used programs—MM2, MOPAC, and GAUSSIAN 82.

Handbook of Data on Organic Compounds. **Robert C. Weast and Melvin J. Astle, eds.** Boca Raton, FL: CRC Press, 1985. 2 vols.

Alphabetical listing of about 24,000 organic compounds, giving information on common names and synonyms, melting and boiling points, molecular formula and weight, line formula, refractive index, density, color, crystalline form, specific rotation, and solubility. A must for every chemistry library.

A Handbook of Decomposition Methods in Analytical Chemistry. **Rudolf Bock and Iain L. Marr.** New York: Halsted Press/Wiley, 1979.

An advanced reference work for the analytic chemist. Covers common procedures, inorganic and organic solids, appendixes, and bibliographies.
R: *Analyst* 104: 1211 (Dec. 1979); *Journal of the American Chemical Society* 102: 4286 (June 4, 1980); *Laboratory Practice* 28: 517 (May 1979); *TBRI* 46: 82 (Mar. 1980).

Handbook of Environmental Data on Organic Chemicals. 2d ed. **Karel Verschueren.** New York: Van Nostrand Reinhold, 1983.

Earlier edition, 1977.
Expanded edition including mixtures and preparations as well as some 1,000 organic chemicals listed in alphabetical order with reference notes. For anyone seeking exact, compressed, or significant data in this field.
R: *ARBA* (1984, p. 637).

Handbook of Enzyme Inhibitors (1965–1977). **M. K. Jain.** New York: Wiley, 1982.

Contains almost 5,000 compounds that function as enzyme inhibitors in over 3,000 processes, listing the name, process it inhibits, type of inhibition, inhibition constant, and the reference.
R: *Nature* 298: 403 (July 22, 1982).

Handbook of Geochemistry. **K. H. Wedepohl, ed.** New York: Springer-Verlag, 1969–.

Volume 1 and volume 2, part 1, 1969; volume 2, part 2, 1970; volume 2, part 3, 1972; volume 2, part 4, 1975; volume 2, part 5, 1979.
R: *Journal of Geology* 79: 632 (Sept. 1971); *Nature* 229: 642 (Feb. 26, 1971); 257: 258 (Sept. 18, 1975); 281: 242 (Sept. 20, 1979); *Science* 167: 1115 (Feb. 20, 1970); *LCIB* 29: 219 (May 7, 1970); *ARBA* (1970, p. 127); Win (2EE5).

Handbook of Intermediary Metabolism of Aromatic Compounds. **Brian L. Goodwin.** London: Chapman & Hall, 1976.

An accurate informative reference on aromatic metabolism. Deals with enzymatic reactions. Provides alphabetical listing of compounds.
R: *Chemistry and Industry* no. 19: 848 (Oct. 2, 1976); *Chemistry in Britain* 12: 393 (Dec. 1976); *Phytochemistry* 16: 407 (1977); *TBRI* 43: 43 (Feb. 1977); 43: 94 (Mar. 1977); 43: 213 (June 1977).

Handbook of Laboratory Distillation. 2d ed. **Erich Krell.** New York: Elsevier Scientific, 1982.

First edition, 1958.
A standard reference manual in the field of distillation in the laboratory. Gives an introduction to practical work with apparatus and describes physical fundamentals and calculations of the separation process. Also beneficial for distillation on the industrial scale.
R: *Chemistry in Britain* 18: 728 (Apr. 1982).

Handbook of Lipid Research. 2 vols. New York: Plenum Press, 1978.

Volume 1, *Fatty Acids and Glycerides*, A. Kulsis, ed.; volume 2, *The Fat Soluble Vitamins*, H. F. DeLuca, ed.

Deals with the biochemistry of fat-soluble vitamins (A, D, E, and K). A well-produced reference volume.
R: *Journal of Chromatography* 170: 397 (Feb. 11, 1979); *Journal of the American Oil Chemists' Society* 56: 16A (Jan. 1979); *TBRI* 45: 151 (Apr. 1979); 45: 216 (June 1979).

Handbook of Metal Ligand Heats and Related Thermodynamic Quantities. 3d ed., rev. and exp. **James J. Christensen and Reed M. Izatt.** New York: Dekker, 1983.

This new edition contains data through 1980 for 1,547 separate ligands. To be used as a companion to various published tables.
R: *ARBA* (1984, p. 634).

Handbook of Modern Analytical Instruments. **Raghbir Singh Khandpur.** Blue Ridge Summit, PA: TAB Books, 1981.

Prepared for users of analytical instruments. It includes the principles of constructing and operating electrochemical, spectrometric, and chromatographic instruments. Would be more useful for the undergraduate library.

Handbook of Neurochemistry. 2d ed. Vol. 1. **Abel Lajtha, ed.** New York: Plenum, 1982.

A comprehensive and up-to-date reference for the area of neurochemistry.
R: *Nature* 296: 479 (Apr. 8, 1982).

Handbook of Organic Reagents in Inorganic Analysis. **Zavis Holzbecher et al.** Translated by Stanislav Kotrly. New York: Wiley, 1976.

Contains information on use of organic reagents, though it does not give specific procedures of analysis. Well-referenced. Useful in industrial libraries.
R: *Angewandte Chemie: International Edition in English* 15: 786 (Dec. 1976); *Chemistry in Britain* 12: 391 (Dec. 1976); *Choice* 13: 1456 (Jan. 1977); *RSR* 5: 13 (Apr./June 1977); *TBRI* 43: 85 (Mar. 1977); 43: 126 (Apr. 1977).

Handbook of Practical Organic Micro-Analysis: Recommended Methods for Determining Elements and Groups. **S. Bance.** New York: Halsted Press, 1980.

A clearly written presentation of organic microanalysis techniques. Provides detailed laboratory procedures, with emphasis on quantitative elemental analysis. Intended for graduate students in chemistry.
R: *Chemistry in Britain* 17: 250 (May 1981); *Choice* 18: 816 (Feb. 1981).

Handbook of Reactive Chemical Hazards. 2d ed. **L. Bretherick.** Woburn, MA: Butterworth, 1979.

First edition, 1975.
A compilation of data on dangerous reactions, with an emphasis on flammability and explosiveness. Intended for use in the research laboratory.
R: *Chemical Engineering* 87: 14 (Jan. 14, 1980); *Journal of the American Chemical Society* 98: 308 (Jan. 7, 1976); 102: 1474 (Feb. 13, 1980); *New Scientist* 81: 1054 (Mar. 29, 1979); *RSR* 5: 12 (Apr./June 1977); 8: 23 (July/Sept. 1980); *TBRI* 42: 111 (Mar. 1976); 46: 70 (Feb. 1980).

Handbook of Thermochemical Data for Compounds and Aqueous Species. **Herbert E. Barner and Ricard V. Scheuerman.** New York: Wiley, 1978.

Contains data concerning thermodynamic properties, methods of calculation, and bibliographies. A helpful reference for analytic laboratories.

R: *Journal of Metals* 32: 64 (Apr. 1980); *Journal of the Electrochemical Society* 126: 64C (Feb. 1979); *TBRI* 45: 121 (Apr. 1979).

Handbook of Thermodynamic Tables and Charts. **Kuzman Raznjevic.** New York: McGraw-Hill, 1976.

Contains charts and thermodynamic or transport data for a variety of substances in the solid, liquid, or gaseous states.

R: *Chemical Engineer* 318: 183 (Mar. 1977); *TBRI* 43: 207 (June 1977).

Handbook of Toxic and Hazardous Chemicals. **Marshall Sittig.** Park Ridge, NJ: Noyes, 1981.

Gives a concise form of chemical, health, and safety information on nearly 600 toxic and hazardous chemicals. A valuable addition to industrial, medical, and environmental libraries.

R: *Water, Air, and Soil Pollution* 17: 238 (Feb. 1982).

Handbook on the Physics and Chemistry of Rare Earths. Vol. 1. **Karl A. Gschneidner, Jr. and LeRoy Eyring, eds.** New York: Elsevier/North Holland, 1978.

Volume 1, *Metals.*

A well-written and organized volume that covers such aspects of rare earth metals as mechanical properties, temperatures, high-pressure studies, and electronic structure. Well-indexed. Considered an indispensable source for chemistry collections.

R: *Physics Bulletin* 31: 57 (Feb. 1980).

Hazards in the Chemical Laboratory. **G. D. Muir, ed.** Washington, DC: American Chemical Society, 1981.

The second edition of this valuable handbook covers various aspects of chemical hazards, protection, and precautions. Useful for laboratory employees.

Industrial Solvents Handbook. 2d ed. **Ibert Mellan.** Park Ridge, NJ: Noyes Data, 1977.

First edition, 1970.

Authoritative, well-referenced; comprises a complete guide to properties of industrial solvents.

R: *Journal of Coatings Technology* 50: 102 (July 1978); *Journal of the American Oil Chemists' Society* 55: 282A (Apr. 1978); *Metal Finishing* 75: 84 (July 1977); *TBRI* 43: 277 (Sept. 1977); 44: 238 (Apr. 1978); 44: 316 (Oct. 1978); Wal (p. 128).

Laboratory Handbook of Chromatographic and Allied Methods. **Otakar Mikes, ed.** New York: Halsted Press/Wiley, 1979.

Details chromatographic techniques in all fundamental types of chemistry. Useful reference for the beginning technician and all those who carry out chemical analysis.

R: *Analyst* 104: 892 (Sept. 1979); *Analytica Chimica Acta* 113: 200 (Jan. 1, 1980); *Chemistry and Industry* 2: 114 (Feb. 2, 1980); *Journal of Chromatography* 175: 224 (July 11,

1979); *Laboratory Practice* 28: 1087 (Oct. 1979); *Nature* 280: 706 (Aug. 23, 1979); *Pharmaceutical Journal* 223: 139 (Aug. 11, 1979); *TBRI* 45: 287 (Oct. 1979); 45: 361 (Dec. 1979); 46: 90 (Mar. 1980).

Laboratory Handbook of Paper and Thin-Layer Chromatography. **Jiri Gasparic and J. Churacek.** Translated by Z. Prochazka. New York: Wiley, 1978.

A convenient bench reference for the chemist. Illustrates theoretical principles and practical applications of organic compounds.

R: *Analytical Biochemistry* 92: 253 (Jan. 1, 1979); *Analytical Chemistry* 51: 399A (Mar. 1979); *Chemistry and Industry* 12: 410 (June 16, 1979); *Chemistry in Britain* 15: 151 (Mar. 1979); *Journal of Chromatography* 170: 298 (Feb. 11, 1979); *Journal of the American Chemical Society* 101: 4020 (July 4, 1979); *Laboratory Practice* 27: 876 (Oct. 1978); *Nature* 274: 828 (Aug. 24, 1978); *TBRI* 44: 284 (Oct. 1978); 45: 44 (Feb. 1979); 45: 164 (May 1979); 45: 205 (June 1979).

Lange's Handbook of Chemistry. 12th ed. **John A. Dean, ed.** New York: McGraw-Hill, 1979.

Tenth edition, 1961; tenth revised edition, 1967; eleventh edition, 1973.

A thorough updating, including expanded coverage on thermodynamics, organic solvents, etc. Includes tables. A handy reference.

R: *Chemical Engineering* 86: 11 (Mar. 12, 1979); 86: 11 (Dec. 31, 1979); *Journal of Pharmaceutical Sciences* 68: 805 (June 1979); *Journal of the American Chemical Society* 101: 6146 (Sept. 26, 1979); *WLB* 53: 654 (May 1979); *ARBA* (1974, p. 556; 1980, p. 617); Wal (pp. 123, 153); Win (ED49; 2ED9).

Official Methods of Analysis. 12th ed. Washington, DC: Association of Official Analytical Chemists, 1975.

Eleventh edition, 1970.

Quantitative and qualitative analytic methods for foods, fertilizers, pesticides, drugs, and cosmetics. Includes a section on preparation of standard solutions.

R: Jenkins (J46); Sheehy (ED9).

On-Line Process Analyzers. The Handbook. Vol. 1. **D. J. Huskins.** New York: Wiley, 1982.

A reference for chemists, physicists, chemical engineers, and analytical chemists. A step-by-step guide to the selection, installation, and application of analyzer systems and equipment.

R: *Choice* 19: 947 (Mar. 1982).

The Oxide Handbook. 2d ed. **G. V. Samsonov, ed.** Translated by Robert K. Johnson. New York: Plenum Press, 1981.

A convenient reference book providing an easy entrance into the Russian literature on oxides.

Parent Compound Handbook. 4 vols. **Chemical Abstracts Service.** Columbus, OH: American Chemical Society, 1976–.

Supersedes *The Ring Index* (first edition, 1940; second edition, 1960, with supplements 1–3, 1963–1965).

Considered a major reference work for those who use *Chemical Abstracts*. Each parent compound contains structural data, identifier, CAS registry number, Wiswesser Line Notation. Comprises a 4-volume reference of accessible information.
R: *Journal of the American Chemical Society* 100; 1328 (Feb. 15, 1978); Sheehy (ED3).

Practical Protein Biochemistry: A Handbook. **A. Darbre and M. D. Waterfield.** New York: Wiley, 1986.

Separation Procedures in Inorganic Analysis: A Practical Handbook. **Roland S. Young.** New York: Halsted Press, 1979.

A succinct handbook of separation procedures of the 57 inorganic elements. Useful reference in laboratory and plant settings.

Spectral and Chemical Characterization of Organic Compounds: A Laboratory Handbook. 2d ed. **W. J. Criddle and G. P. Ellis.** New York: Wiley, 1980.

First edition, 1976.

A convenient laboratory handbook that clearly presents a large amount of information. Comprises a clear guide to identification of organic compounds.
R: *Pharmaceutical Journal* 217: 524 (Dec. 4, 1976); *TBRI* 43: 85 (Mar. 1977).

Spectroscopic References to Polyatomic Molecules. **V. N. Verma.** New York: IFI/Plenum, 1980.

A listing of references to literature about the spectroscopy of almost 900 organic ring components arranged alphabetically by the name of the component. The citations do not give a title, but do indicate infrared, Raman, absorption, emission, etc.
R: *ARBA* (1982, p. 711).

BIOLOGICAL SCIENCES

Biochemical Engineering and Biotechnology Handbook. **Bernard Atkinson and Ferda Mavituna.** New York: Macmillan, 1982.

A handbook of authoritative facts and data concerning information such as quantities, temperature, coefficients, and the like about the field of biochemical engineering. For those involved in developing products based on biological processes.
R: *Nature* 298: 501 (Aug. 5, 1982).

Biological Handbooks. Vols. 1–. **Philip L. Altman and Dorothy D. Katz, eds. and comps.** Bethesda, MD: Federation of American Societies for Experimental Biology, 1976–.

Volume 1, *Cell Biology*.

A new series of handbooks published by the Federation of American Societies for Experimental Biology. Volume 1 contains much quantitative information on cell biology. Contributions from experts in the field. Clearly written and referenced.
R: *RSR* 5: 6 (Apr./June 1977).

CRC Handbook of Microbiology. 2d ed. Vols. 1–. Cleveland: Chemical Rubber, 1977–.

First edition, 1973; volume 1, Allen I. Laskin and Hubert A. Lechevalier, eds.

Multivolume series on microbiology provides information on microorganisms, taxonomic indexes, metabolism, etc.
R: Sheehy (EC10).

CRC Handbook Series in Zoonoses. Sect. A. 2 vols. **James H. Steele, ed.** Boca Raton, FL: Chemical Rubber, 1979–1980.

Section A, *Bacterial, Rickettsial, and Mycotic Diseases.*

Genetics. **M. W. Roberts.** Estover, Plymouth, England: Macdonald & Evans, 1977.

A reference account of basic genetics. Contains glossary, bibliography.
R: *Aslib Proceedings* 42: 571 (Nov./Dec. 1977).

Handbook of Biochemistry and Molecular Biology. 3d ed. **Gerald D. Fasman, ed.** Cleveland: Chemical Rubber, 1976–1977.

First edition, 1968; second edition, 1970.
The third edition appears in several sections: *Lipids, Carbohydrates, and Steroids Section,* volume 1, 1975; *Nucleic Acids Section,* 2 volumes, 1975; *Physical and Chemical Data Section,* 2 volumes, 1976; *Proteins Section,* 3 volumes, 1976; *Cumulative Index,* 1977.

A specialized supplement to the *Handbook of Chemistry and Physics* providing critical physical/chemical data, abbreviations and nomenclature for amino acids, peptides and proteins, purines, steroids, pyrimidines, nucleic acid, etc. Intended for use by researchers and graduate students.
R: *Journal of the American Chemical Society* 100: 1642 (Mar. 1, 1978); *Journal of Chemical Education* 48: 5563 (Sept. 1971); *ARBA* (1971, p. 491); Sheehy (EC49); Wal (p. 202); Win (2EC7; 3EC6).

Handbook of Ethological Methods. **Philip N. Lehner.** New York: Garland STPM Press, 1979.

A comprehensive reference to ethology. Material covered includes design of research, data collecting and observation techniques, and interpretation of results. Provides a bibliography and an appendix of statistical tables, lists of journals, and pointers on obtaining research funds. A valuable addition to college library collections and a helpful aid for upper-level undergraduate and guaduate students as well as educators.
R: *Choice* 17: 101 (Mar. 1980).

Handbook of Experimental Immunology. 3d ed. **Donald M. Weir, ed.** Philadelphia: Lippincott, 1978.

Second edition, 1973.
A classic reference containing a substantial amount of information. Includes an exhaustive list of references. Highly recommended for immunology labs.
R: *Annals of Allergy* 42: 112 (Feb. 1979); *TBRI* 45: 179 (May 1979).

Handbook of Genetics. 5 vols. **Robert C. King, ed.** New York: Plenum Press, 1976.

Volume 1, *Bacteria, Bacteriophages, and Fungi*; volume 2, *Plants, Plant Viruses, and Protists.*

Presents a great deal of worthwhile information; covers chromosomal proteins, DNA, and RNA.
R: *Nature* 256: 153 (July 10, 1975); 266: 570 (Apr. 7, 1977); *Science* 189: 283 (July 1975); *TBRI* 43: 214 (June 1977).

Handbook of Microbiology. 4 vols. **Allen I. Laskin and Hubert A. Lechevalier, eds.** Cleveland: Chemical Rubber, 1973–1974.

Volume 1, *Organismic Microbiology* 1973; volume 2, *Microbial Composition,* 1973; volume 3, *Microbial Products,* 1974; volume 4, *Microbial Metabolism, Genetics and Immunology,* 1974.

A comprehensive collection of data on bacteria, algae, protozoa, fungi, and viruses. Intended for both the working microbiologist and the university student.
R: *American Scientist* 62: 490–491 (July/Aug. 1974); *Journal of the Medical Society of New Jersey* 71: 241 (Mar. 1974); *TBRI* 40: 199 (May 1974).

Handbook of Physiology: A Critical, Comprehensive Presentation of Knowledge and Concepts. **J. M. Brookhart et al., eds.** Baltimore: Williams & Wilkins, 1977.

A 2-part handbook devoted to physiology and physiological research methods. Discusses such topics as the interaction of cells and membrane conduction. A highly recommended source of reference.
R: *Nature* 272: 654 (Apr. 1978).

Handbook of Protein Sequence Analysis: A Compilation of Amino Acid Sequences of Proteins with an Introduction to the Methodology. 2d ed. **L. R. Croft.** New York: Wiley, 1981.

Contains detailed descriptions of protein sequence determination methodology. An indispensable, up-to-date handbook.
R: *Nature* 292: 278 (July 16, 1981); *New Scientist* 91: 163 (July 16, 1981).

Handbook of Radioimmunoassay. **Guy E. Abraham, ed.** New York: Dekker, 1977.

A well-referenced handbook on a growing field in laboratory medicine.
R: *Laboratory Management* 16: 65 (Mar. 1978); *TBRI* 44: 215 (June 1978).

Handbook of Sensory Physiology. 10 vols. **Werner R. Loewenstein, ed.** New York: Springer-Verlag, 1971–.

Eight volumes currently completed on the physiology of nervous tissue in higher and lower life forms.
R: *American Scientist* 61: 225 (Mar./Apr. 1973); 63: 108 (Jan./Feb. 1975); 63: 582 (Sept./Oct. 1975); 63: 712 (Nov./Dec. 1975); *Nature* 254: 230 (Mar. 20, 1975); 275: 257 (Sept. 21, 1978).

Handbook of Toxic Fungal Metabolites. **Richard J. Cole and Richard H. Cox.** New York: Academic Press, 1981.

This handbook is presented in response to a growing awareness of the dangers from mycotoxins. It covers some 21 groups of mycotoxins with physical, chemical, spectral, and biological data. For libraries serving undergraduate and/or graduate students in

the fields of mycology, toxicology, medicine, food and feed sciences, natural products, and biochemistry.
R: *Choice* 19: 782 (Feb. 1982).

Utilization of Microorganisms in Meat Processing: Handbook for Meat Plant Operators. **J. Bacus.** New York: Research Studies Press, 1984.
A useful guide or reference combining common knowledge of microorganisms and meat fermentation with practical aspects and observations in meat processing. Based on information developed over the last 4 to 5 years.

AGRICULTURE

Agricultural Development Indicators: A Statistical Handbook. New York: International Agricultural Development Service, 1978.
Gathers important statistics helpful in solving agricultural problems in low-income countries.
R: *Quarterly Bulletin of the IAALD* 23: 65 (Fall–Winter 1978).

CRC Handbook of Natural Pesticides: Methods. **N. Bhushan Mandava, ed.** Boca Raton, FL: CRC, 1985.
Volume 1, *Theory, Practice, and Detection*; volume 2, *Isolation and Identification*.
A compendium of current methods for the detection and characterization of natural pesticides. Written for both nonspecialists and experienced researchers.

Farm Builder's Handbook. 3d ed. **R. J. Lytle.** Farmington, MI: Structures, 1981.
Second edition, 1973; third revised edition, 1978.
Intended for construction personnel working in agricultural structures. Information included on all types of farm buildings, silos, waste disposal systems, etc.
R: *ARBA* (1974, p. 673).

Field Crop Diseases Handbook. **Robert F. Nyvall.** Westport, CT: Avi Publishing, 1979.
Provides comprehensive treatment of more than 800 field-crop diseases for 25 important economic field crops. Diseases are examined under 4 main headings: cause, distribution, symptoms, and control. The diseases for each field crop are arranged by their cause. Contains 190 black-and-white photographs of diseases, a glossary of technical terms, references, and indexes of common names of diseases and Latin names of pathogens. Invaluable aid for all students, instructors, and professionals involved in agriculture or plant pathology.
R: *Choice* 17: 694 (July/Aug. 1980).

Insecticide and Fungicide Handbook for Crop Protection. 5th ed. **Hubert Martin and C. R. Worthing, eds.** Oxford, England: Blackwell Scientific, 1977.
Comprehensive discussions of fungicides and crop protection. Covers cereals, potatoes, beets, grass, fruits, and vegetables. Authoritative and up-to-date.
R: *Chemistry and Industry* 22: 917 (Nov. 19, 1977); *Farm Chemicals* 141: 140 (Sept. 1977); *TBRI* 43: 392 (Dec. 1977); 44: 71 (Feb. 1978).

Vegetable Growing Handbook. **Walter E. Splittstoesser.** Westport, CT: Avi Publishing, 1979.

A basic handbook on all phases of vegetable growing. Details conditions, nutrition, harvesting, storage, etc. The second half of the book, which consists of a list of individual vegetables and common problems, is organized in dictionary form. Appendixes include seed companies, nutritive values of foods, and metric/English measurements. Suitable for home use.
R: *ARBA* (1980, p. 709).

Botany

CRC Handbook of Flowering. 5 vols. **Abraham H. Halevy, ed.** Boca Raton, FL: CRC Press, 1985–.

Volumes 1–4, alphabetical listing, 1985; volume 5, special topics and an A–Z listing of sections that were unavailable for previous volumes, 1986.
A worldwide effort that brings together over 250 authorities to present topics on the control and regulation of flowering. For each plant, detailed information is given on environmental requirements for flower initiation and development, aspects of flower development, and chemical regulation of flowering. Tables, illustrations, and electron micrographs.

CRC Handbook of Fruit Set and Development. **Shaul P. Monselise, ed.** Boca Raton, FL: CRC, 1986.

A handbook that contains current case histories of development for 26 different fruits growing in widely different climates. These fruits include apple, banana, blueberry, citrus, pineapple, plum, strawberry, and tomato. Illustrations, micrographs, diagrams, tables, and bibliographies are included.

The Color Handbook of House Plants. **Elvin McDonald, Jacqueline Heriteau, and Francesca Morris, eds.** New York: Hawthorne Books/Wentworth Press, 1975.

Describes over 250 plants. Includes general discussions on soils, fertilizers, insects, lighting, etc. Well-written and indexed. A concise reference.
R: *ARBA* (1977, p. 748).

Ferns and Allied Plants: With Special Reference to Tropical America. **Rolla M. Tryon and Alice F. Tryon.** New York: Springer-Verlag, 1982.

A most comprehensive presentation of ferns and allied plants of the Pteridophyta represented in America. Includes photographs, maps, and charts.
R: *ARBA* (1984, p. 646).

Forestry Handbook. 2d ed. **Karl F. Wenger, ed.** New York: Wiley, 1984.

Presents the most up-to-date information and data on all phases of forestry and allied fields in the United States and Canada.

Grafter's Handbook. **Robert J. Garner.** New York: Oxford University Press, 1979.

A revised reference book for horticulturists. Well-illustrated and indexed.
R: *Garden* 3: 36 (Nov./Dec. 1979); *TBRI* 46: 28 (Jan. 1980).

Handbook of Phycological Methods. Vols. 1–. New York: Cambridge University Press, 1973–.

Volume 1, *Culture Methods and Growth Measurements*, edited by Janet R. Stein, 1973; volume 2, *Physiological and Biochemical Methods*, edited by Johan A. Hellebust and J. S. Craigie, 1978; volume 3, *Developmental and Crytological Methods*, edited by Elisabeth Gantt, 1980.

A beautifully produced series of handbooks containing contributions from numerous expert authors. Clearly written and highly technical. A valuable reference for graduate and professional-level phycologists.

R: *American Scientist* 62: 605 (Sept./Oct. 1974); *Phytochemistry* 18: 1422 (Aug. 1979); *South African Journal of Science* 75: 422 (Sept. 1979); *Aslib Proceedings* 44: 247 (June 1979); *Choice* 18: 683 (Jan. 1981); *TBRI* 45: 295 (Oct. 1979); 46: 53 (Feb. 1980).

Handbook of Plant Virus Infections. **E. Kurstak, ed.** Amsterdam: Elsevier Biomedical, 1981.

Covers all aspects in the field of plant virology research. Written by leading experts in the field.

R: *Nature* 296: 100 (Mar. 1982).

Handbook of Seagrass Biology: An Ecosystem Perspective. **Ronald C. Phillips and C. Peter McRoy, eds.** New York: Garland STPM Press, 1980.

A compilation of papers covering the state of the art of seagrasses. Sections discuss taxonomy, chemical composition, anatomy, and measurements. Intended for graduate students and professionals.

R: *Choice* 18: 116 (Sept. 1980).

Handbook of Tropical Forage Grasses. **B. Ira Judd.** New York: Garland STPM Press, 1979.

Written in a clear, concise fashion, this handbook discusses 36 tropical forage grasses in detail. Topics covered include morphology, importance, and uses. For each grass, the author gives Latin and English names, a nontechnical description, and instructions for fertilization, culture, pest control, and means of propagation. Includes up-to-date bibliographical references. An excellent reference tool for those in the field.

R: *ARBA* (1981, p. 746).

Handbook of Utilization of Aquatic Plants: A Review of World Literature. **E. C. S. Little.** Rome: Food and Agriculture Organization of the United Nations, 1979.

A handbook dealing with the utilization of aquatic plants for such objectives as fuel, building materials, feed, water purification, etc. Lists over 250 books, papers, and other publications. Of interest to ecology or agricultural collections.

R: *IBID* 8: 13 (Spring 1980).

Healing Plants: A Modern Herbal. **W. A. R. Thomson, ed.** London: Macmillan, 1978.

Contains alphabetical lists of healing plants and their properties. Also includes excellent botanical illustrations. Substantive information; a highly recommended reference.

R: *Nature* 279: 85 (May 3, 1979).

Horticultural Handbook. **W. W. Fletcher et al.** New York: Granada Publishing, 1981.
A clearly and concisely written reference book detailing all aspects of horticulture. Covers such topics as climate, soils, fertilizers, irrigation, crops, machinery, etc. An excellent aid for all students and professional horticulturists.

An Integrated System of Classification of Flowering Plants. **Arthur Cronquist.** Frenchtown, NJ: Columbia, 1981.
A collection of 383 families in 83 orders, each described in detail. They are arranged in order based on the strobilar theory of evolution. The major families are illustrated with line drawings. A welcome addition to taxonomic literature.
R: *Choice* 19: 647 (Jan. 1982).

Knott's Handbook for Vegetable Growers. 2d ed. **O. A. Lorenz and D. N. Maynard.** New York: Wiley, 1980.
A completely revised and updated reference examining all aspects of vegetable production. Details such topics as seeding, pest and weed control, fertilizing, soil analyses, etc. Contains charts, lists, and graphs for quick reference. Fully indexed. A valuable reference for farmers, students, or home gardeners.

Medicinal Plants of North Africa. **Loutfy Boulos.** Algonac, MI: Reference Publications, 1983.
Presents the basic information on plants used in folk medicine in North Africa. Not organized in its presentation, which may bring confusion to the nonbotanist.
R: *ARBA* (1984, p. 648).

Mycotoxic Fungi, Mycotoxins, Mycotoxicoses: An Encyclopedic Handbook. 3 vols. **Thomas D. Wyllie and L. G. Morehouse, eds.** New York: Dekker, 1977–1978.
Volume 2, *Mycotoxicoses of Domestic and Laboratory Animals, Poultry, and Aquatic Invertebrates and Vertebrates.*
A handbook of 3 volumes with contributions from field experts. Most helpful to toxicologists.
R: *Journal of the American Oil Chemists' Society* 55: 868A (Dec. 1978); *Journal of the American Veterinary Medical Association* 173: 896 (Oct. 1, 1978); *RSR* 7: 11 (Jan./Mar. 1979); *TBRI* 44: 377 (Dec. 1978); 45: 102 (Mar. 1979).

The Concise New British Flora. **W. Keble Martin.** Nomenclature edited and revised by Douglas H. Kent. London: Michael Joseph and Ebury Press. Distr. Lawrence, MA: Merrimack Book Service, 1982.
First edition, 1965; second edition, 1969; both entitled *The Concise British Flora.* Includes beautiful illustrations of some 1,400 wildflowers of Britain, with brief, authoritative descriptions of size, form, habitat, and time of blooming. Revised and updated in accordance with the current Botanical Society of the British Isles listing.
R: *ARBA* (1984, p. 647).

North American Trees (Exclusive of Mexico and Tropical United States): A Handbook Designed for Field Use, with Plates and Distribution Maps. 3d ed. **Richard J. Preston, Jr.** Ames, IA: Iowa State University Press, 1976.

First edition, 1961.
Contains logically arranged identification keys to the diverse trees of North America. Contains extensive description of trees, which are grouped by genera, and includes over 160 full-page illustrations. Well-recommended for students, scientists, and the interested layperson.
R: *BL* 74: 774 (Jan. 1, 1978); *ARBA* (1977, p. 664).

Weed Control Handbook. 8th ed. 2 vols. **J. D. Fryer and R. J. Makepeace, eds.** Oxford, England: Blackwell Scientific, 1978.

Volume 1, *Principles*; volume 2, *Recommendations*.
A weed-control handbook in 2 volumes, both of which are much revised. Provides the user with the latest information on the continuing development of herbicides, with notes on control of some individual weeds. Highly recommended.

Westcott's Plant Disease Handbook. 4th ed. **Cynthia Westcott.** New York: Van Nostrand Reinhold, 1979.

Third edition, 1971.
Helps to identify organisms that attack plants and maintain their control. Also contains a list of host plants and their organisms. Arranged by common names of plant diseases. Glossary of terms and extensive bibliography. A useful reference manual intended for both novices and professionals.
R: *American Horticulturist* 58: 12 (Dec. 1979/Jan. 1980); *TBRI* 46: 68 (Feb. 1980); *ARBA* (1980, p. 709); Jenkins (J56).

ZOOLOGY

The Audubon Illustrated Handbook of American Birds. **Edgar M. Reilly, Jr.** New York: McGraw-Hill, 1968.
R: *American Scientist* 57: 147 (Summer 1969); *ARBA* (1970, p. 124).

The Audubon Society Handbook for Birders. **Stephen W. Kress.** New York: Scribner's, 1981.

A detailed and well-illustrated handbook on all aspects of birdwatching. Contains material on equipment, observation methods, sketching, photography, and recording of bird sounds. Also provides extensive publication sources about birds and birdwatching, college courses, and educational programs. Appendixes include sources of bird watching equipment and publications. Valuable reference for naturalists and bird enthusiasts.
R: *BL* 77: 1131 (Apr. 15, 1981).

CRC Handbook of Laboratory Animal Sciences. 3 vols. **Edward C. Melby and Norman H. Altman.** Cleveland: Chemical Rubber, 1974–1976.

Deals with the legalities, management, diseases, and nutrition of laboratory animals. In 3 volumes, with cumulated index in last.
R: *Journal of the American Veterinary Medical Association* 166: 924 (May 1, 1975); *Journal of Veterinary Research* 36: 715 (May 1975); *TBRI* 41: 260 (Sept. 1975).

CRC Handbook of Tapeworm Identification. **Gerald D. Schmidt.** Boca Raton, FL: CRC, 1985.

An one-volume comprehensive guide to world tapeworms. Over 3,800 species within 601 genera and 66 families are described, and for each genus there is at least one line drawing of a sample species. Extensive bibliography of over 4,000 references.

Eastern Birds of Prey: A Guide to the Private Lives of Eastern Raptors. **Neal Clark.** Thorndike, ME: Thorndike Press, 1983.

A limited study on birds of prey that are most commonly found east of the Mississippi and are both diurnal and nocturnal. Includes black-and-white photographs of each.
R: *ARBA* (1984, p. 656).

The Guinness Book of Animal Facts and Feats. 3d ed. **Gerald L. Wood.** Enfield, England: Guinness Superlatives. Distr. New York: Sterling, 1982.

Gives superlatives and answers questions about the largest, the fastest, or the rarest mammals, birds, or invertebrates and more. Arranged by taxonomic groups with an extensively detailed index. Most records are for the British Isles, with a few for North America. Includes photographs.
R: *ARBA* (1984, p. 655).

Handbook of Canadian Mammals. Marsupials and Insectivores. Vol. 1. **C. G. Van Zyll de Jong.** Ottawa: National Museum of Natural Sciences/National Museums of Canada. Distr. Chicago: University of Chicago Press, 1983.

A small but attractive volume (one of 7 that deals with Canadian mammals: marsupials and insectivores). Includes color plates, black-and-white drawings, and shell renditions. For the naturalist and amateur mammalogist.
R: *ARBA* (1984, p. 667).

Handbook of Freshwater Fishery Biology. Life History Data on Freshwater Fishes of the United States and Canada, Exclusive of the Perciformers. 4th ed. **Kenneth D. Carlander.** Ames, IA: Iowa State University Press, 1979.

First edition, 1953; third edition, 1969.
A tabular reference book on fish growth. Also covers feeding habits, habitat, and more.
R: *ARBA* (1970, p. 125; 1978, p. 676).

Handbook of North American Birds. 3 vols. **Ralph S. Palmer, ed.** New Haven, CT: Yale University Press, 1962–1976.

Volume 1, *Loons Through Flamingos*; volume 2, *Waterfowl* (part 1); volume 3, *Waterfowl* (concluding part).
An extensively researched and accurate 3-volume handbook. Contains all data necessary for identification in the field. Considered a definitive source for biologists and environmentalists.
R: *Choice* 13: 1276 (Dec. 1976); *LJ* 101: 999 (Apr. 15, 1976); *ARBA* (1977, p. 671); Sheehy (EC44).

Handbook of the Birds of Europe, the Middle East and North Africa: The Birds of the Western Palearctic. Vols. 1–. **Stanley Cramp, ed.** New York: Oxford University Press, 1977–.

Volume 1, *Ostrich to Ducks*, 1977; volume 2, *Hawks to Bustards*, 1980.

A 7-volume set on birds of the Western Palearctic. Contains detailed descriptions of 105 species of birds from 4 orders: Accipitriformes, Falconiformes, Galliformes, and Guiformes. Includes numerous excellent color paintings and black-and-white drawings. An indispensable reference for all bird experts and enthusiasts.

R: *American Scientist* 66: 752 (Nov./Dec. 1978); *Nature* 272: 652 (Apr. 13, 1978); 287: 568 (Oct. 9, 1980); *ARBA* (1981, p. 665).

Handbook of the Birds of India and Pakistan: Together with Those of Bangladesh, Nepal, Bhutan, and Ceylon. Stone Curlews to Owls. Vol. 3. 2d ed. **Salim Ali and S. Dillon Ripley.** New York: Oxford University Press, 1981.

First edition, 1968–74; second edition was begun in 1978.

Covers the latter part of the Charadriiformes, and all of the Columbiformes, Psittaciformes, Cuculiformes, and Strigiformes. Provides only minor changes from the first edition.

R: *ARBA* (1983, p. 634).

Insect Pests of Farm, Garden, and Orchard. **R. H. Davidson and W. L. Lyon.** New York: Wiley, 1979.

A comprehensive, well-written reference covering the major insect pests on farms of North America. Details their control, classification, and habits and provides applications and remedies for insect bites. Includes ample illustrations and photographs and an excellent index. An invaluable aid for college or high school students taking an introductory entomology course.

Mammal Species of the World: A Taxonomic and Geographic Reference. **James H. Honacki, Kenneth E. Kinman, and James W. Koeppl.** Lawrence, KS: the Association of Systematics Collections and the Allen Press, 1982.

This taxonomic and geographic listing of the mammal species of the world attempts to serve as a standard reference to mammalian nomenclature. This exhaustive reference will undoubtedly need periodic updates.

R: *ARBA* (1984, p. 665).

Rodale's Color Handbook of Garden Insects. **Anna Carr.** Emmaus, PA: Rodale Press, 1979.

A comprehensive illustrated reference to garden insects. Insects are arranged according to habits and superficial structure. Contains color photographs of over 200 insects accompanied by information on life cycles, feeding, predators, and physical characteristics. Discusses development and plant damage. Includes 2 pictorial keys to common garden insects as well as an appendix and index of common and scientific names. Invaluable reference tool for insect experts.

R: *Blair and Ketchum's Country Journal* 4: 26 (Apr. 1980); *BL* 77: 1057 (Mar. 15, 1981).

The Sierra Club Handbook of Whales and Dolphins. **Stephen Leatherwood and Randall R. Reeves.** San Francisco: Sierra Club Books. Distr. New York: Random House, 1983.

An excellent field guide and up-to-date source of information on these animals.
R: *ARBA* (1984, p. 666).

Venomous Arthropod Handbook: Envenomization Symptoms/Treatment, Identification, Biology, and Control. **Terry L. Biery.** Brooks Air Force Base, TX: US Air Force School of Aerospace Medicine, Disease Surveillance Branch. Distr. Washington, DC: US Government Printing Office, 1977.

Describes venomous arthropods in the United States and the severity of their envenomizations. Includes information on 17 species, such as color photograph for identification, description of behavior, methods of avoidance and control, etc. Provides extensive bibliographies. Appendix contains technical information on treatment. Highly recommended.
R: *ARBA* (1978, p. 677).

Whale Manual '78. **Friends of the Earth.** San Francisco: Friends of the Earth Foundation, 1978.

Covers many aspects of whales and whaling, including history, ecology, and hunting. Contains much useful data. Recommended for natural history collections.
R: *Nature* 275: 348 (Sept. 28, 1978); *LJ* 104: 201 (Jan. 15, 1979); *TBRI* 44: 374 (Dec. 1978).

EARTH SCIENCES

The Earthquake Handbook. **Peter Verney.** New York: Paddington, 1979.

Fully describes earthquake phenomena in clear understandable language. Provides historical and mythological accounts. Discusses plate tectonics, geography, and prediction. A well-written book for the nonspecialist.
R: *Geological Magazine* 116: 412 (Sept. 1979); *Nature* 281: 612 (Oct. 18, 1979); *New Scientist* 82: 480 (May 10, 1979); *TBRI* 46: 10 (Jan. 1980).

Estuarine Hydrography and Sedimentation: A Handbook. **K. R. Dyer, ed.** New York: Cambridge University Press, 1980.

The first of 3 handbooks dealing with the physical and geological aspects of estuaries. Covers tidal measurement methods, water currents, hydrographic surveying, sampling and analysis of bottom sediment. Helpful reference for researchers, graduate and upper-level undergraduate students, biologists, chemists, and geologists.
R: *Choice* 18: 276 (Oct. 1980).

Gemstone and Mineral Data Book: A Compilation of Data, Recipes, Formulas, and Instructions for the Mineralogist, Gemologist, Lapidary, Jeweler, Craftsman and Collector. **John Sinkankas.** New York: Van Nostrand Reinhold, 1981.

A typical handbook with tables, lists, and discussions. There are 12 sections, each covering a different aspect of minerals. Physical properties, testing, cleaning, and identifying are also discussed. Good as a lab manual on all levels.
R: *ARBA* (1982, p. 754).

Handbook in Applied Meterology. **David D. Houghton.** New York: Wiley, 1985.

Handbook of Engineering Geomorphology. **P. G. Fookes and P. R. Vaughan, eds.** New York: Methuen, 1985.

Handbook of Iron Meteorites: Their History, Distribution, Composition and Structure. 3 vols. **Vagn F. Buchwald.** Berkeley: University of California Press, 1976, for the Center for Meteorite Studies, Arizona State University. Also supps.
Considered to be a standard reference for anyone studying iron meteorites.
R: *Science* 194: 313 (1976); *RSR* 5: 14 (July/Sept. 1977); *ARBA* (1977, p. 693).

Minerals Handbook, 1982–83. **Phillip Crowson.** New York: Van Nostrand Reinhold, 1982.
Introduces the international supply-and-demand characteristics of 37 minerals and metals in a most organized manner. Intended for the nonspecialist but recommended for libraries dealing with geology and gemology, and definitely where economic analysts hold forth.
R: *Science and Technology Libraries* 4: 117 (Winter 1983).

The Weather Handbook. Rev. ed. **H. McKinley Conway and Linda Liston.** Atlanta: Conway Research, 1977.
Earlier edition, 1963.
A comparative guide to weather phenomena. Contains data on temperature, precipitation, snowfall, humidity, hurricane tracks, etc. for over 200 US cities and some 300 others.
R: Jenkins (F107); Win (EE96).

OCEANOGRAPHY

Handbook of Marine Science. 2 vols. **F. G. Walton Smith and Frederick A. Kalber.** Cleveland: Chemical Rubber, 1974.
A comprehensive handbook for oceanographers. Reflecting the intedisciplinary nature of the field, the first volume covers chemical oceanography, marine geology, atmospheric science, and ocean engineering. The second volume includes biological and fisheries data, fisheries statistics, zooplankton population data, etc.
R: *Choice* 12: 984 (Oct. 1975); *ARBA* (1976, p. 703); Sheehy (EE23).

Handbook of Marine Science: Compounds from Marine Organisms. Sec. 2, vol. 1. **Joseph T. Baker and Vreni Murphy, eds.** Cleveland: Chemical Rubber, 1976.

GENERAL ENGINEERING

Engineering Formulas. 4th ed. **Kurt Gieck.** New York: McGraw-Hill, 1983.
A collection of technical formulas that provide a brief, clear, and useful tool to the more important technical and mathematical formulas used in this field.
R: *IEEE Proceedings* 72: 239 (Feb. 1984).

Handbook of Engineering Fundamentals. 3d ed. Wiley Engineering Handbook Series. **Ovid Wallace Eshbach and Mott Souders.** New York: Wiley, 1975.

Second edition, 1952.
A ready-reference tool to the basic concepts in engineering. Primarily of interest to science and engineering students. Includes tables.
R: *RSR* 4: 51 (Jan.–Mar. 1976); *SL* 67: 10A (Apr. 1976); *TBRI* 42: (May 1976); *ARBA* (1977, p. 751); Sheehy (EJ3); Wal (p. 289); Win (E13).

Handbook of Engineering Management. **J. E. Ullman et. al.** New York: Wiley, 1986.

Handbook of the Engineering Sciences. 2 vols. **James H. Potter, ed.** Princeton: Van Nostrand Reinhold, 1967.
Volume 1 provides an understanding of the basic sciences on which engineering principles are contingent; volume 2 is concerned with the applications of such knowledge to various engineering fields.
R: *Chemistry and Industry* 3: 90 (Jan. 20, 1968); *Electronics* (Aug. 7, 1967); *Mechanical Engineering* 90: 84 (June 1968); *ABL* 33: entry 294 (June 1968); *Choice* 5: 464 (1968); *TBRI* 33: 185 (Oct. 1967); Jenkins (K126); Wal (p. 285); Win (2E15).

Practical Inventor's Handbook. **Orville Green et al.** New York: McGraw-Hill, 1979.
Explains how to develop, protect, and sell inventions.
R: *Chemical Engineering* 87: 16 (Jan. 14, 1980); *Science Digest* 86: 86 (Oct. 1979); *RSR* 8: 25 (Jan./Mar. 1980); *TBRI* 45; 350 (Nov. 1979).

Standard Handbook of Engineering Calculations. **Tyler G. Hicks, ed.** New York: McGraw-Hill, 1972.
Over 2,000 step-by-step procedures for solving specific engineering problems. Covers 12 fields of engineering. Has been recommended as an excellent guide for persons studying for professional engineering licenses.
R: *IEEE Spectrum* p. 104 (Jan. 1973); *J SMPTE* p. 950 (Dec. 1972); *TBRI* 39: 104 (Mar. 1973); *ARBA* (1973, p. 629).

AERONAUTICAL AND ASTRONAUTICAL ENGINEERING

CRC Handbook of Space Technology: Status and Projections. **Michael Hord.** Boca Raton, FL: CRC Press, 1985.
A compact volume that defines in quantitative terms the opportunities and limits for future space system capabilities. Figures are accompanied by descriptive narrative.

The Instrument Pilot Handbook: Reference Manual and Exam Guide. **Courtney L. Flatau and Jerome F. Mitchell.** New York: Van Nostrand Reinhold, 1980.
Helpful preparation for the aviation pilot's written exam. More than 400 sample questions and 350 illustrations cover such subjects as navigation aids and calculations, basic flight instruments, and flight procedures. Useful references for all libraries with materials used by aviation pilots.
R: *Choice* 18: 688 (Jan. 1981).

International Handbook of Aerospace Awards and Trophies. **US National Air and Space Museum Library, Smithsonian Institution Libraries, ed.** Washington, DC: Smithsonian Institution Press, 1978.

A detailed compilation of aerospace awards. Includes awards of organizations such as the American Astronautical Society, the Aerospace Medical Association, the Aero Club de France, and the Franklin Institute. For public and academic libraries.
R: *ARBA* (1980, p. 715).

The Observer's Book of Aircraft. **William Green, comp.** New York: Warne, Frederick, & Co., 1951–. Annual.

Twenty-fourth edition, 1975; thirtieth edition, 1981.
In essence, a state-of-the-aircraft-industry offering an illustrated guide to military, commercial, and private airplanes either in production or nearing production. Not so comprehensive as Jane's *All the World's Aircraft*, but still of use to business and engineering personnel.
R: *ARBA* (1976, p. 781).

CHEMICAL ENGINEERING

Chemical Engineers' Handbook. 6th ed. **Robert H. Perry and Don W. Green, eds.** New York: McGraw-Hill, 1984.

Fourth edition, 1963; fifth edition, 1973.
Classic handbook for chemical engineers. In addition to the updated original 25 sections, 2 new sections dealing with waste management and biochemical engineering have been added. More than 100 experts contributed to the new edition. More than 1,800 illustrations.
R: *ARBA* (1985, p. 543).

Chemical Technicians' Ready Reference Handbook. 2d ed. **G. Shugar et al.** New York: McGraw-Hill, 1981.

A thoroughly expanded and revised edition of a reference handbook for chemical technicians. Contains formulas, equations, principles, and methods needed for all laboratory procedures. Provides new materials and information on toxic vapors, compressed gases, cleaning glassware, etc. A recommended reference for all chemists and chemical laboratory technicians.
R: *Journal of the American Chemical Society* 104: 3548 (June 1982).

Equipment Design Handbook for Refineries and Chemical Plants. 2d ed. 2 vols. **Frank L. Evans, Jr.** Houston: Gulf Publishing, 1980.

First edition, 1971–1974.
Practical information for the design engineer interested in oil refining and chemical industries. Broadly divided into 5 sections—on drivers, compressors, ejectors, pumps, and refrigeration. Volume 2 covers nonrotating equipment.
R: *ARBA* (1973, p. 660).

European Chemical Industry Handbook. **Stuart H. Wamsley.** London: Hedderwick, Stirling, Grumbar, 1980.

A guide to the west European chemical industry covering the years 1973 to 1978. Provides a wide range of statistical and other data compiled from national, international, and trade sources. Useful to all technical libraries.
R: *Chemical Engineer* 361: 647 (Oct. 1980).

Fluid Flow Pocket Handbook. **Nicholas Cheremisinoff, ed.** Houston: Gulf Publishing, 1984.

A useful tool for engineers, technicians, and students on the flow of gas and liquids. Ten sections on physical properties, governing equations of flow, pipe flow calculations, gas flow calculations, discharge through variable head meters, flow through coils and around tubes, channel flows, two-phase flows, gas-solid flows, and pump calculations. Tables, figures, references, glossary, and a brief index.
R: *ARBA* (1985, p. 46).

Handbook of Aerosol Technology. 2d ed. Paul A. Sanders. New York: Van Nostrand Reinhold, 1979.

In 3 sections: homogeneous systems, emulsions and foams, suspensions and miscellaneous. A basic text for paint chemists.
R: *Chemical Engineering* 87: 307 (Nov. 17, 1980); *Journal of Coatings Technology* 51: 100 (Dec. 1979); *TBRI* 46: 118 (Mar. 1980).

Handbook of Chemical Engineering Calculations. **Nicholas P. Chapey, ed.** New York: McGraw-Hill, 1984.

An extension of the *Standard Handbook of Engineering Calculations*, this handbook is a helpful aid in solving the main process-related problems that often come into play in chemical engineering practice.
R: *Science Technology Libraries* 5: 2 (Winter 1984).

Handbook of Essential Formulae and Data on Heat Transfer for Engineers. **H. Y. Wong.** Harlow, Essex, England: Longman, 1977.

A summary handbook of heat-transfer data. Highly recommended.
R: *Chemical Engineering* 86: 11 (Mar. 26, 1979); *International Journal of Heat and Mass Transfer* 21: 1007 (July 1978); *Aslib Proceedings* 43: 26 (Jan. 1978); *TBRI* 44: 320 (Oct. 1978).

Handbook of Industrial Water Conditioning. 7th ed. **Betz Laboratories.** Trevose, PA: Betz Laboratories, 1976.

For chemical engineers involved in water conditioning. A useful reference.
R: *Chemical Engineering* 84: 11 (Feb. 14, 1977); *TBRI* 43: 151 (Apr. 1977).

Handbook of Oil Industry Terms and Phrases. **R. D. Langenkamp, ed.** Tulsa: Pennwell Publishing, 1981.

A listing of nearly 2,500 terms from personnel, equipment, and techniques, along with other terms from other areas; defined in layperson's terms. Provides a basic, limited knowledge of the oil industry.
R: *ARBA* (1983, p. 729).

Handbook of Organic Waste Conversion. **Michael W. M. Bewick, comp.** New York: Van Nostrand Reinhold, 1981.

A collection of 15 papers dealing with the utilization and conversion of organic wastes. Topics cover food processing, sewage sledge, and antibiotic production. Contains praiseworthy illustrations and tables, as well as extensive references. A graduate-level text for engineers and industrial technologists.
R: *Chemical Engineering* 88: 119 (Aug. 10, 1981); *Choice* 18: 277 (Oct. 1980).

Handbook of Separation Techniques for Chemical Engineers. **Philip A. Schweitzer, ed.** New York: McGraw-Hill, 1979.

Contains descriptive data on industrial methods of chemical separation. Includes separation techniques for every major type of mixture; presents basic theories and numerous examples. A helpful reference for plant engineers and project managers.
R: *Chemical Engineering* 87: 11 (June 30, 1980); *Choice* 17: 105 (Mar. 1980); *RQ* 19: 306 (Spring 1980).

Handbook of Silicone Rubber Fabrication. **Wilfred Lynch.** New York: Van Nostrand Reinhold, 1978.

In 3 parts: properties, application, and fabrication of silicone rubber. Serves as an introductory source of information.
R: *Rubber Chemistry and Technology* 51: G91 (Sept.–Oct. 1978); *TBRI* 44: 396 (Dec. 1978).

Handbook of US Colorants for Foods, Drugs, and Cosmetics. 2d ed. **Daniel M. Marmion.** New York: Wiley, 1984.

First edition, 1979.
A 3-part handbook covering colorants for foods, drugs, and cosmetics. Part A deals with color additives, their background, use regulations, suppliers, and specifications. A short glossary in the section enhances its usefulness. Part B discusses colorant analysis. Subjects are described in technical language and include strength, identification, impurities, etc. Part C is concerned with the resolution of mixtures and the analysis of commercial products. Contains a thorough subject index and annotated bibliographies. Excellent addition to libraries or laboratories.
R: *Chemical Engineering* 87: 105 (Dec. 15, 1980); *Journal of the American Chemical Society* 102: 2136 (Mar. 12, 1980); *Laboratory Practice* 29: 645 (June 1980); *Pharmaceutical Journal* 223: 385 (Oct. 13, 1979); *Choice* 16: 1562 (Feb. 1980); *TBRI* 46: 36 (Jan. 1980); *ARBA* (1981, p. 638).

Pesticide Handbook—Entoma. 26th ed.–. College Park, MD: Entomological Society of America, 1975–. Biennial.

Twenty-seventh edition, 1977/1978.
Revised edition of a classic general reference on pesticides and pesticide regulations.
R: Sheehy (EL).

Food Technology

CRC Handbook of Food Additives. 2d ed. **Thomas E. Furia, ed.** Cleveland: Chemical Rubber, 1979.

First edition, 1968.
Covers various data on additives based on FDA standards. Each chapter, written by chemists, includes definitions, chemical properties and reactions, representative uses,

and safety inhibitors. Also includes a guide to current regulations for frequently used additives.
R: *Food Technology* p. 108 (Apr. 1973); *Journal of Food Technology* 9: 125 (Mar. 1974); *TBRI* 40: 260 (Sept. 1974); *ARBA* (1974, p. 655); Wal (p. 485).

CRC Handbook of Nutritive Value of Processed Food. 2 vols. **Miloslav Rechcigl, Jr., ed.** Boca Raton, FL: CRC Press, 1982.
Volume 1, *Food for Human Use*; volume 2, *Animal Feedstuffs*.
These 2 volumes explore various processing methods and their effects on the nutritive value of food for humans and animal feedstuffs. Extensive illustrations and tables.

CRC Handbook of Tropical Food Crops. **Franklin W. Martin, ed.** Boca Raton, FL: CRC Press, 1984.
A comprehensive yet compact volume of tropical food-crop information. Nomenclature, taxonomy, the principal adaptations or limits to adaptation, cultural factors, and cultivators are discussed for each crop. Intended for a wide range of readers.

Cane Sugar Handbook: A Manual for Cane Sugar Manufacturers and Their Chemists. 10th ed. **George Peterkin Meade and J. C. P. Chen.** New York: Wiley, 1977.
A handbook for sugar manufacturers. Updated edition contains definitive information. Includes 36 reference tables. Recommended for industrial libraries.
R: *Analyst* 103: 671 (June 1978); *Food Technology* 32: 125 (Feb. 1978); *International Sugar Journal* 79: 325 (Nov. 1977); *Journal of the American Chemical Society* 100: 4639 (May 24, 1978); *Journal of the Association of Official Analytical Chemists* 60: 1440 (Nov. 1977); *TBRI* 44: 77 (Feb. 1978); 44: 116 (Mar. 1978); 44: 157 (Apr. 1978); 44: 357 (Nov. 1978).

The Complete Food Handbook. 1980 ed. **Rodger P. Doyle and James L. Redding.** New York: Grove Press. Distr. New York: Random House, 1979.
First edition, 1976.
An authoritative guide to purchasing foods while focusing on nutrition and economy. Chapters cover all the different food groups and types of foods. Contains information on nutritional supplements and fiber, additives, labeling practices, processing, etc. Provides appendixes on food colors and flavors, dietary goals, and nutritive ratings. Includes bibliography, glossary, and index. An invaluable addition for all libraries.
R: *WLB* 53: 525 (Mar. 1979); *ARBA* (1978, p. 744; 1981, p. 739).

Fenaroli's Handbook of Flavor Ingredients. 2d ed. 2 vols. **Thomas E. Furia and Nicolo Bellance, trans. and eds.** Cleveland: Chemical Rubber, 1975.
First edition, 1971.
Translated and adapted from Geovanny Fenaroli's Italian books *Sostanze Aromatiche*, volumes 1–3, and *Aromatizzazione*.
A handbook to the chemical and physical properties of flavoring ingredients. Entries provide formulas, synthesis stability, solubility, etc. Of special interest to frozen-food processors.
R: *ARBA* (1972, p. 558).

Handbook of Food and Nutrition. **M. Swaminathan.** Madras, India: Ganesh, 1977.

A handbook of food technology for a wide range of users.
R: *Food Technology* 32: 119 (Mar. 1978); *TBRI* 44: 278 (Sept. 1978).

Handbook of Sugars. 2d ed. **Harry M. Pancoast and W. Ray Junk.** Westport, CT: AVI Publishing, 1980.

A handbook dealing with various aspects of the sugar industry. Covers consumption, processing, and nutritive values. For specialized industrial libraries and those involved in the food manufacturing industry.
R: *ARBA* (1981, p. 740).

Handbook of Tropical Foods. **Harvey T. Chan, Jr., ed.** New York: Dekker, 1983.

A heavily documented source of the most recent and available information on selected tropical foods.
R: *Science Technology Libraries* 5: 2 (Winter 1984).

POLYMER TECHNOLOGY

Adhesive Handbook. 2d ed. **J. Shields.** Woburn, MA: Butterworth, 1976.

First edition, 1970.
Fills a need for locating information on adhesives. Describes processes of bonding. Provides a glossary and a comprehensive reference.
R: *Composites* 7: 227 (Oct. 1976); *Proceedings of the Royal Australian Chemical Institute* 43: iii (Sept. 1976); *Structural Engineer* 5: 229 (May 1977); *RSR* 5: 36 (Dec. 1977); *TBRI* 43: 39 (Jan. 1977); 43: 78 (Feb. 1977); 43: 279 (Sept. 1977); Wal (p. 499).

Handbook of Analysis of Synthetic Polymers and Plastics. **Jerzy Urbanski et al.** New York: Wiley, 1977.

Details the analysis of synthetic polymers and plastics. In 2 parts: Instrumental and conventional methods of analysis, and analysis of individual polymers. For academic and research libraries.
R: *Analytica Chimica Acta* 96: 440 (Feb. 1, 1978); *Chemistry and Industry* 6: 202 (Mar. 18, 1978); *Chemistry in Britain* 14: 408 (Aug. 1978); *Journal of Chromatography* 152: 597 (May 21, 1978); *Choice* 15: 260 (Apr. 1978); *TBRI* 44: 199 (May 1978); 44: 279 (Sept. 1978).

Handbook of Fillers and Reinforcements for Plastics. **Harry S. Katz and J. V. Milewski, eds.** New York: Van Nostrand Reinhold, 1979.

A major compilation of useful and valuable information. Contributions from 30 experts.
R: *Chemical Engineering* 86: 11 (June 4, 1979); *Journal of Polymer Science; Polymer Letters Edition* 16: 551 (Oct. 1978); *Rubber World* 178: 54 (Aug. 1978); *TBRI* 44: 354 (Nov. 1978); 44: 394 (Dec. 1978); 45: 277 (Sept. 1979).

Handbook of Plastic Product Design Engineering. **Sidney Levy and Harry DuBois.** New York: Van Nostrand Reinhold, 1977.

Provides information on plastic products engineering.
R: *Automotive Engineering* 86: 138 (Feb. 1978); *TBRI* 44: 196 (May 1978).

Handbook of Plastics Testing Technology. **Vishu Shah.** New York: Wiley, 1984.

Handbook of Water-Soluble Gums and Resins. **Robert L. Davidson, ed.** New York: McGraw-Hill, 1980.

A comprehensive volume written by 29 leading experts in the field of hydrogels. The handbook was designed as a supplement to the *Encyclopedia of Polymer Science and Technology*; each chapter covers one type of synthetic or natural gum or resin. Chapters include general information, uses, formulations, trade names, laboratory techniques, and supplementary reading lists. Contains graphs, tables, diagrams, black-and-white photographs, and a praiseworthy index. Recommended for academic libraries whose users include chemists, chemical engineers, and food scientists.
R: *ARBA* (1981, p. 636).

Plastics Engineering Handbook. 4th ed. **Society of the Plastics Industry.** New York: Van Nostrand Reinhold, 1976.

First edition, 1960.
Revised edition covers all major plastics processing methods. Details properties, processing methods, and performance data. Considered one of the best available sources.
R: *Chemical Engineering* 84: 12, 14 (Jan. 31, 1977); *Plastics Engineering* 32: 70 (Nov. 1976); *RSR* 5: 36 (Oct./Dec. 1977); *TBRI* 43: 120 (Mar. 1977); 43: 158 (Apr. 1977).

Polymer Handbook. 2d ed. **J. Brandup and E. H. Immergut, eds.** New York: Wiley-Interscience, 1975.

First edition, 1966.
Gives basic data and references on the physical and chemical properties of polymers. Includes nomenclature, polymerization, solutions, and solid-state properties and constants. Highly regarded as a basic reference on polymers for the polymer scientist.
R: *ABL* 31: entry 343 (July 1966); *Choice* 3: 891 (Dec. 1966); Sheehy (EJ8); Wal (p. 530); Win (1E I12).

Spray Drying Handbook. **K. Masters.** 4th ed. New York: Wiley, 1985.

Technician's Handbook of Plastics. **Peter A. Grandilli.** New York: Van Nostrand Reinhold, 1981.

A well-written, technically accurate handbook containing the theory, practical application, and problems of the basic plastic processes of injection molding, blow molding, extrusion, compression, transfer molding, and a number of thermoset processes. For technical institutes and community colleges.
R: *Choice* 19: 264 (Oct. 1981).

TEXTILE TECHNOLOGY

Handbook of Chemistry Specialties: Textile Fiber Processing, Preparation and Bleaching. **J. E. Nettles.** New York: Wiley, 1983.

This reference source contains a review of the chemistry and technology of compounds used in textiles processing. Very logical and informative in its presentation.

Handbook of Textile Fibres. 5th ed. 2 vols. **J. Gordon Cook.** Watford, England: Merrow, 1975.

Fourth edition, 1968.
Volume 1, *Natural Fibres*; volume 2, *Man-Made Fibres.*
Emphasis on production and use. Discusses natural polymer fibers and fibers from synthetic polymers in detail. Includes a directory of man-made fibers in which generic class and manufacturer are given.
R: *ABL* 34: entry 100 (Feb. 1969); Wal (p. 522).

Handbook of Textile Fibers, Dyes, and Finishes. **Howard L. Needles.** New York: Garland STPM Press, 1980.

An introductory reference to basic properties of textile fibers, dyes, and finishes. Suitable for lower-level undergraduate and community college students.
R: *Choice* 18: 982 (Mar. 1981).

The Standard Handbook of Textiles. 8th ed. **A. J. Hall.** London: Iliffe, 1975.

Seventh edition, 1969.
Basic, nontechnical reference that includes information from the viewpoint of both the manufacturer and the user. Many photographs and illustrations bearing on the conversion of textile fibers into textile materials.
R: *TBRI* 42: 73 (Feb. 1976); *Textile Research Journal* 45: 874 (Dec. 1975); Wal (p. 522).

CIVIL ENGINEERING

General

Civil Engineer's Reference Book. 3d ed. **L. S. Blake.** London: Butterworth, 1975.

Second edition, 1969.
Designed specifically to meet the needs of the recently graduated professional, and the experienced engineer, geologist, and mining engineer.
R: *Australian Mining* 67: 89 (Oct. 1975); *TBRI* 42: 110 (Mar. 1976); Sheehy (EJ12).

Handbook of Applied Hydraulics. 3d ed. **C. Davis and K. Sorenson.** New York: McGraw-Hill, 1969.

An authoritative handbook on the design and construction of hydraulic structures. Covers basic principles of hydrology and hydraulics, dams, water use, control, and disposal. Contains practical examples and applications of interest to civil engineers.

Handbook of Geology in Civil Engineering. 3d ed. **Robert F. Legget and Paul F. Karrow.** New York: McGraw-Hill, 1982.

A book focusing on the geological interrelation with civil engineering and construction. Deals with the major forms of engineering and special problems and environmental concerns. Excellent for civil engineers and architects, as well as geologists.
R: *Civil Engineering* 53: 89 (Mar. 1983).

Standard Handbook for Civil Engineers. 2d ed. **Frederick S. Merritt.** New York: McGraw-Hill, 1976.

First edition, 1968.
Revised edition includes a wealth of new material, reflecting advances in computer operations and structural engineering. Contains important descriptive and numerical data on bridges, railroads, highways, tunnels, harbor construction, and environmental engineering. Comprehensive yet concise.
R: *Plant Engineering* 31: 172 (June 23, 1977); *Choice* 6: 342 (May 1969); *RSR* 5: 25 (Oct./Dec. 1977); *TBRI* 43: 277 (Sept. 1977); *ARBA* (1977, p. 756); Jenkins (K153); Sheehy (EJ14); Wal (p. 352); Win (2EI16).

CONCRETE AND CONSTRUCTION ENGINEERING

American Metric Construction Handbook. **Robert J. Lytle.** Farmington, MI: Structures Publishing, 1976.

Guides the user to the metric system as it applies to the construction industry. Provides a series of appendixes that provide data such as conversion tables with highly accurate interpolation. Considered a generally useful reference. For engineers, contractors, and materials suppliers.
R: *ARBA* (1977, p. 752).

Architects' Data: The Handbook of Building Types. 2d ed. **Ernst Neufert.** New York: Halsted Press, 1980.

A completely updated and modernized edition of a classic handbook on guidelines in the planning and design of building projects. Provides data on user requirements, site selection, and planning criteria. Each section is composed of illustrations and layouts. SI units are used, but conversion tables are included. An outstanding reference source for any architect, designer, contractor, or student.

Architectural Handbook: Environmental Analysis; Architectural Programming, Design and Technology; and Construction. **Alfred M. Kemper.** New York: Wiley, 1979.

Describes a complete range of architectural processes in building design, construction, environmental architecture, etc.
R: *Civil Engineering* 49: 39 (Nov. 1979); *Military Engineer* 71: 375 (Sept.–Oct. 1979); *TBRI* 46: 35 (Jan. 1980); 46: 75 (Feb. 1980).

Building Design and Construction Handbook. 4th ed. **Frederick S. Merritt, ed.** New York: McGraw-Hill, 1982.

Third edition, 1975, entitled: *Building Construction Handbook.*
A completely rewritten and revised edition of a classic handbook on all aspects of modern building construction and design. Contains exhaustive illustrations. Intended for nonspecialists.
R: *Heating/Piping/Air Conditioning* 54: 109 (July 1982).

Building Systems Integration Handbook. **Richard D. Rush.** New York: Wiley, 1985.

Organized in 4 building systems: Structure, envelope, mechanical, interior; and 5 levels: remote, touching, connected, meshed, and unique. This reference sets standards for the building professions. Contributions by over 100 building and design experts.

Concise Soil Mechanics. **M. J. Smith.** Estover, Plymouth, England: Macdonald & Evans, 1977.
A good introductory handbook that provides classification procedures and standard definitions in soil mechanics.
R: *Aslib Proceedings* 43: 209 (Apr. 1978).

Component and Modular Techniques: A Builder's Handbook. 2d ed. **Robert Lytle and Robert C. Reschke.** New York: McGraw-Hill, 1981.
A reference handbook on construction techniques for 3 popular types of woodframe residential and light commercial buildings. Covers panelized home packages, structural framing components, and construction ideas. Includes over 250 illustrations and diagrams. Contains a manufacturing list and a glossary. A welcome addition to civil engineers' and builders' collections.

The Complete Concrete, Masonry, and Brick Handbook. Reprint. **J. T. Adams.** New York: Van Nostrand Reinhold, 1983.
A great tool, written in a comprehensive manner, for anyone who wants to use tools and materials in a productive way. For personal or public libraries.
R: *Science Technology Libraries* 5: 2 (Winter 1984).

Cutting for Construction: A Handbook of Methods and Applications of Hard Cutting and Breaking on Site. **David W. Lazenby and Paul Phillips.** New York: Halsted Press, 1978.
A highly detailed handbook on methods of on-site construction demolition which elimate hazards to personnel and environment. Details necessary materials, safety precautions, and specifications, as well as various techniques such as cutting, explosives, fire, etc. Case studies are included. An important reference for engineers and construction workers.
R: *Scientific American* 242: 44 (Mar. 1980).

Environmental Science Handbook for Architects and Builders. **S. V. Szokolay.** New York: Halsted Press, 1980.
Deals with design systems that consider man-environment relationships. Takes into account the physical sciences and social sciences, in order that buildings meet human needs.

Excavation Handbook. **Horace K. Church.** New York: McGraw-Hill, 1980.
Gives thorough treatment of the engineering aspects of soil and rock excavation methods and equipment. Areas examined include geological aspects of excavation, equipment, costs, rock fragmentation methods, open-cut mining, etc. A glossary of excavation terms and extensive illustrations complement the text. Valuable aid for all engineering students and professionals, particularly those involved in construction or mining.
R: *Choice* 18: 981 (Mar. 1981).

Farm Builder's Handbook. 3d ed. **Robert Lytle.** New York: McGraw-Hill, 1981.
An indispensable volume for those involved in the building and planning of farm or pole buildings. Includes more than 400 illustrations, tables, and graphs. Also contains a helpful index.

Handbook of Architectural and Civil Drafting. **John A. Nelson.** New York: Van Nostrand Reinhold, 1983.

A guide for beginners using a nontechnical approach in introducing the world of drafting. For personal, public, or college libraries.
R: *Science Technology Libraries* 5: 2 (Winter 1984).

Handbook of Architectural Details for Commercial Buildings. **Joseph DeChiara, ed.** New York: McGraw-Hill, 1980.

Contains a full collection of detailed architectural drawings of commercial buildings throughout the United States. Contains clear solutions and excellent illustrations. For architects, drafters, and contractors.

Handbook of Composite Construction Engineering. **Gajanan M. Sabnis, ed.** New York: Van Nostrand Reinhold, 1979.

Provides a survey of composite construction including fundamentals, uses, and applications. Covers such materials as steel, concrete, wood, and timber. Contains many examples as well as a list of references. A helpful guide to structural engineers.
R: *ARBA* (1980, p. 713).

Handbook of Construction Equipment Maintenance. **Lindley R. Higgins, ed.** New York: McGraw-Hill, 1979.

A collection of 55 articles that examine various aspects of equipment maintenance. Subjects covered include engines, pumps, compactors, etc. Contains information concerning personnel, facilities, budgeting, and basic tools. Also provides over 7,500 practical applications as well as 1,000 illustrations. Valuable to contractors and construction engineers.
R: *Chemical Engineering* 87: 12 (Jan. 14, 1980).

Handbook of Construction Management and Organization. 2d ed. **J. B. Bonny and Joseph B. Frein.** New York: Van Nostrand Reinhold, 1980.
First edition, 1973.

Handbook of Structural Concrete. **F. K. Kong et al.** New York, McGraw-Hill, 1983.

An authoritative reference work on the state of the art and science of structural concrete. The volume contains 41 chapters in 6 parts, with extensive references, illustrations, and index. Invaluable for practicing professionals, researchers, and students.
R: *ARBA* (1985, p. 544).

PCI Design Handbook: Precast and Prestressed Concrete. 2d ed. Chicago: Prestressed Concrete Institute, 1978.

Professional Handbook of Building Construction. **E. Allen.** New York: Wiley, 1985.

A basic reference that deals with the common materials of building and the methods by which they are incorporated into buildings. Covers all of the most common construction systems and materials. Well illustrated with over 400 line drawings and 300 photographs.

Reinforced Concrete Designer's Handbook. 9th ed. **Charles E. Reynolds.** London: Cement and Concrete Association, 1981.

Seventh edition, 1971.
R: *Civil Engineering* 41: 14 (Oct. 1971).

Road and Bridge Construction Handbook. **Michael Lapinski.** New York: Van Nostrand Reinhold, 1978.

Explains road-building technology in simple everyday terms. Examines each phase of construction. Recommended for all members of the construction team.

OTHER

Domestic Water Treatment. **Jay H. Lehr et al.** New York: McGraw-Hill, 1980.

A detailed examination of domestic water sources, contaminants, treatments, and construction and maintenance of water supply systems. Include coverage of analysis and treatment techniques, special types of water for specific uses, and allowable constituents in water. Appendixes provide addresses of state agencies, handling advice of specific chemicals used in treatment, and a glossary. Of interest to city administrators, home owners, and construction specialists.
R: *BL* 76: 1241 (May 1, 1980).

Handbook of Dam Engineering. **Alfred R. Golez, ed.** New York: Van Nostrand Reinhold, 1977.

Provides detailed guidance for dam construction. Includes information on earthquake hazards, construction materials, and public safety controls. Up-to-date, valuable reference.
R: *New Scientist* 78: xviii (Apr. 27, 1978); *RSR* 5: 25 (Oct./Dec. 1977).

Handbook of Highway Engineering. **Robert F. Baker, ed.** Melbourne, FL: Krieger, 1982.

Reprint of 1975 edition.
Designed for the highway engineer, the reference provides principles, processes, and data necessary for the application of relevant technology. Twenty-six sections complete with text, data, diagrams, and photographs. Among the topics covered are highway policies, administration, urban transportation planning, geometric design standards, etc.
R: *ARBA* (1976, p. 782).

Handbook of Hydraulics for the Solution of Hydraulic Engineering Problems. 6th ed. **Ernest F. Brater and Horace Williams King.** New York: McGraw-Hill, 1976.

Fifth edition, 1963.
An illustrated handbook on the principles of hydraulics. Revised edition contains much current information including numerous tables, charts, graphs. A handy reference for civil engineers and practicing engineers who seek solutions to on-the-job problems.
R: *Marine Engineering/Log* 81: 124 (Dec. 1976); *RSR* 5: 24 (Oct./Dec. 1977); *TBRI* 43: 113 (Mar. 1977); *ARBA* (1977, p. 766); Sheehy (EJ32; EJ141).

Handbook of Soil Mechanics. Vol. 2. **Arpad Kezdi.** Translated by P. Szoke. New York: American Elsevier, 1980.

Volume 2, *Soil Testing.*

A well-presented handbook of field and laboratory soil tests. Includes tests such as groundwater, porepressure, load-bearing capacity, and earthwork strength. Contains sample sheets, photographs, and line drawings. Volume uses SI units. Worthwhile reference for engineers involved in soil mechanics and foundation engineering.

R: *Choice* 18: 124 (Sept. 1980).

Port Development: A Handbook for Planners in Developing Countries. **Secretariat of UNCTAD, prep.** New York: United Nations, 1978.

Deals with port-planning policy and with procedures for forecasting traffic, productivity, and studying problems. Discusses mathematical techniques and financial aspects.

R: *IBID* 7: 60 (Spring 1979).

Structural Engineering Handbook. 2d ed. **Edwin H. Gaylord, Jr. and Charles N. Gaylord, eds.** New York; McGraw-Hill, 1979.

First edition, 1968.

A thoroughly updated and revised edition that examines all phases of structural engineering. Includes new coverage of finite element analysis, steel poles for transmission lines, and reinforced concrete structures. Contains practical examples. A valuable addition to engineering libraries and a worthwhile purchase for structural engineers, architects, and construction managers.

R: *Chemical Engineering* 87: 11 (Jan. 14, 1980); *RQ* 19: 195 (Winter 1979); *ARBA* (1980, p. 721).

Tunnel Engineering Handbook. **John O. Bickel and T. R. Kuesel, eds.** New York: Van Nostrand Reinhold, 1982.

Contains valuable information of vital areas in the field of tunnel engineering. Includes topics such as geological and geophysical investigations, materials handling, ventilation, lighting, and the like.

R: *Civil Engineering* 52: 32 (May 1982).

Water Treatment Handbook. 5th ed. **Degremont Company Editors.** New York: Halsted Press, 1979.

A comprehensive source covering a wide range of water treatment situations such as drinking water, swimming pool water, industrial situations, etc. Covers analytical aspects, construction and design, maintenance. Contributions by field experts. Highly recommended to civil engineering libraries.

ELECTRICAL AND ELECTRONICS ENGINEERING

Electronics Engineer's Handbook. 2d ed. **Donald G. Fink, ed.-in-chief.** New York: McGraw-Hill, 1982.

First edition, 1975.

A companion handbook to the *Standard Handbook for Electrical Engineers.* An applications-oriented approach to "the essential principles, data, and design information on

the components, circuits, equipment, and systems of electronics engineering as a whole."
R: *IEEE Proceedings* 63: 1376 (Sept. 1975); *TBRI* 41: 397 (Dec. 1975); *ARBA* (1976, p. 772); Sheehy (EJ40).

Electronics Engineers' Reference Book. 4th ed. **L. W. Turner, ed.** London: Butterworth, 1976.

First edition, 1958; second edition, 1962; third edition, 1967 (published in the United States as *Handbook of Electronic Engineering.* Cleveland: CRC Press, 1967).
Up-to-date and thorough reference covering all basic fields in electronics.
R: *Choice* 7: 1016 (1970); Win (EI117; 3EI18).

Electronic Imaging Techniques: A Handbook of Conventional and Computer-Controlled Animation, Optical and Editing Processes. **Eli L. Levitan.** New York: Van Nostrand Reinhold, 1977.

A handbook devoted to electronic imaging techniques. Discusses computer animation, digital, on-line/off-line systems, optical effects, video, etc. Extremely well-illustrated. Highly recommended for engineering and media libraries.
R: *Choice* 14: 1390 (Dec. 1977).

Electronics Designers' Handbook. 2d ed. **L. J. Giacoletto, ed.** New York: Mc-Graw-Hill, 1977.

First edition, 1957.
A revised edition of this standard reference. Extensive coverage of computer design, electronic components, communication systems, circuitry, etc. Highly recommended for technical libraries.
R: *Proceedings of the IEEE* 66: 988 (Aug. 1978); *RSR* 5: 27 (Oct./Dec. 1977); *TBRI* 44: 353 (Nov. 1978).

Fundamentals Handbook of Electrical and Computer Engineering. Circuits, Fields and Electronics. Vol. 1. **S. Chang.** New York: Wiley, 1982.

A handbook covering the field of electrical and computer engineering in 3 volumes. Emphasizes active applications of engineering design and the ever-increasing tie between electrical engineering, mathematics, and physics.

Handbook of Modern Electronic and Electrical Engineering. **C. Belove.** New York: Wiley, 1985.

Standard Handbook for Electrical Engineers. 11th ed. **Donald G. Fink, ed.** New York: McGraw-Hill, 1978.

First edition, 1907; ninth edition, 1957; tenth edition, 1968.
A well-received reference that aims to incorporate into a single volume all pertinent data. Comprehensive treatment of technical data that bears on the economic aspects of engineering practice. Ample discussion of quantities, units, conversion factors, circuits, writing designs, codes, etc.
R: *Radio Electronics* 54: 106 (Feb. 1983); *ARBA* (1976, p. 770; 1979, p. 763); Jenkins (K160); Wal (p. 313); Win (E1114, 3E117).

Electronics

American Electrician's Handbook. 10th ed. **Terrell Croft et al., eds.** New York: McGraw-Hill 1981.

A complete revision of a classic handbook on the design, operation, selection, installation, and maintenance of electrical equipment. Contains new coverage of installation and splicing techniques, generators and motors, medium-voltage cables, etc. Recommended for electricians, contractors, inspectors, electrical equipment manufacturers, and electrical engineers.
R: *RQ* 20: 324 (Spring 1981); *RSR* 9: 104 (Jan./Mar. 1981).

The Beginner's Handbook of Electronics. **George H. Olsen and Forrest M. Mims, III.** Englewood Cliffs, NJ: Prentice-Hall, 1980.

Describes the use of capacitors, transistors, resistors, and other basics of modern electronics such as circuitry, components, semiconductors. A useful reference for both the amateur and the professional.

Electric Cable Handbook. **D. McAllister, ed.** London: Granada Publishing, 1982.

A valuable reference and technical book for those involved with power cable. Includes tables, charts, photographs, and 250 pages of appendixes that prove to be useful sources of information.
R: *IEEE Spectrum* 20: 14 (Oct. 1983).

Electronic Components Handbook. **Thomas H. Jones.** Reston, VA: Reston Publishing, 1978.

An introductory reference to military and commercial electronic components including capacitors, relays, resistors, and semiconductors. Provides definitions, charts, photographs, component specifications and data, list of manufacturers. Useful to engineers and technicians.
R: *Electrical Apparatus* 30: 33 (Aug. 1978); *IEEE Spectrum* 16: 95 (Jan. 1979); *Choice* 15: 1549 (Jan. 1979); *TBRI* 44: 394 (Dec. 1978); 45: 116 (Mar. 1979); *ARBA* (1980, p. 719).

Electronics Designer's Handbook. 2d rev. and enl. ed. **L. J. Giacoletto.** New York: McGraw-Hill, 1977.

An excellent reference aimed at the professional engineer. Discusses solid-state devices, amplifiers, information design, and computers.
R: *Choice* 13: 1339 (Dec. 1977); *ARBA* (1978, p. 771).

Electronic Designer's Handbook: A Practical Guide to Transistor Circuit Design. 3d ed. **T. K. Hemingway.** London: Business Books, 1979.
Second edition, 1970.

Electronic Filter Design Handbook. **Arthur Bernard Williams.** New York: McGraw-Hill, 1980.

Provides treatment of numerous aspects of filter design. Intended for upper-level undergraduate students interested in communications hardware and communications.
R: *IEEE Proceedings* 70: 317 (Mar. 1982); *Choice* 18: 821 (Feb. 1981).

Electronics Engineers' Handbook. 2d ed. **Donald G. Fink and Donald Christiansen, eds.** New York: McGraw-Hill, 1981.

A thoroughly revised edition of an outstanding handbook dealing with the data, properties, processs, standards, and applications of electronics engineering. Includes information about principles, devices, circuits and components, functions, and applications (audio and visual engineering, telecommunications, electronic navigation, etc.). Also contains new material on microprocessors and computer-aided design of circuits as well as other digital-related areas. An indispensable reference for all electronics engineers.
R: *IEEE Proceedings* 70: 1376 (Mar. 1982).

The Giant Handbook of Electronic Circuits. **Rayond A. Collins, ed.** Blue Ridge Summit, PA: TAB Books, 1980.

This handbook for designers and hobbyist provides schematic diagrams for a number of electronic circuits. Contents include AM and FM broadcast receivers, automatic circuits, computer-related circuits, music-related circuits, and more.

Handbook for Electronics Engineering Technicians. 2d ed. **Milton Kaufman and Arthur H. Seidman, eds.** New York: McGraw-Hill, 1984.

A convenient 1-volume reference work that provides practical information on discrete circuits and analog and digital integrated circuits for electronics engineers and technicians. It does not require technical knowledge to use this tool. Extensive index is available.
R: *ARBA* (1985, p. 548).

A Handbook of Active Filters. **David E. Johnson, Johnny Ray, and Harry P. Moore.** Englewood Cliffs, NJ: Prentice-Hall, 1980.

A thorough presentation of practical active filter designs. Contains numerous design examples and tables. Provides extensive bibliography. Of interest to academic libraries.
R: *IEEE Proceedings* 70: 314 (Mar. 1982); *Choice* 17: 698 (July/Aug. 1980).

Handbook of Basic Electronic Troubleshooting. **John D. Lenk.** Englewood Cliffs, NJ: Prentice-Hall, 1978.

Provides both an introduction for beginners and a complete reference for advanced technicians to basic electronic troubleshooting. Outlines procedures for locating faulty circuits and defective components in equipment. Contains more than 160 wiring diagrams and voltage and resistance data.
R: *RSR* 5: 29 (Oct./Dec. 1977).

Handbook of Batteries and Fuel Cells. **David Linden, ed.** New York: McGraw-Hill, 1984.

A useful volume for general public and academic libraries. Text with tables and illustrations. References are not extensive.
R: *ARBA* (1985, p. 549).

Handbook of Components for Electronics. **Charles A. Harper, ed.** New York: McGraw-Hill, 1977.

A detailed guide to electronic components. Examines all types of electronic circuitry and devices ranging from integrated circuits to semiconductors, resistors and capacitors to transformers and relays. Provides information on performance, reliability, standards, dimensions, and configurations. Also includes tables, illustrations, glossary of terms, and a cross-referenced index. For engineering libraries.
R: *Choice* 14: 832 (Sept. 1977); *RSR* 5: 28 (Oct./Dec. 1977).

Handbook of Electronic Circuit Designs. **John D. Lenk.** Englewood Cliffs, NJ: Prentice-Hall, 1976.

Covers a wide range of electronic circuit designs, including filters, attenuators, photo transistors, amplifiers, etc. For public libraries.
R: *Choice* 11: 392 (May 1976); *ARBA* (1977, p. 763).

Handbook of Electronic Communications. **Gary M. Miller.** Englewood Cliffs, NJ: Prentice-Hall, 1979.

Written for working technicians. Contains extensive information on electronic communication systems. Well-illustrated, contains helpful data sheets. Recommended.

Handbook of Electronic Formulas, Symbols and Definitions. **John R. Brand.** New York: Van Nostrand Reinhold, 1979.

A pocket reference of definitions of electrical formulas and symbols useful in the electronics area. Terms are organized alphabetically by symbol into specific subject categories including passive circuits, transistors, and operational amplifiers. Contains schematic diagrams and an appendix of electronic terms and their symbols. Assumes some electronics background. A helpful reference for engineers, students, and hobbyists.
R: *Choice* 17: 415 (May 1980); *ARBA* (1980, p. 712).

Handbook of Electronics Calculations for Engineers and Technicians. **Milton Kaufman and Arthur H. Seidman, eds.** New York: McGraw-Hill, 1979.

A practical guide of about 500 electronics problems and solutions. Solves problems in DC and AC circuit analysis, power supplies, microprocessors, tuned amplifiers, and computer-aided circuit design. A helpful reference to practicing electrical engineers and technicians. Should be in every academic library.
R: *JSMPTE* 88: 800 (Nov. 1979); *Choice* 17: 105 (Mar. 1980); *RQ* 19: 193 (Winter 1979); *TBRI* 46: 75 (Feb. 1980).

Handbook of Electronics Packaging Design and Engineering. **Bernard S. Matisoff.** New York: Van Nostrand Reinhold, 1982.

An easy reference source on choosing the most effective packaging technique for on-the-job applications. Also contains formulas and information on thermal control, enclosures, etc. Good for making decisions in the packaging field.
R: *IEEE Proceedings* 70: 1468 (Dec. 1982).

Handbook of Mechanical and Electrical Systems for Buildings. **H. Bovay.** New York: McGraw-Hill, 1981.

A survey of the latest methods used in the design of electrical and mechanical building systems. Chapters discuss air-conditioning systems, plumbing, alarm systems, communications systems, etc. A handy reference for engineers, architects, and builders.
R: *Choice* 19: 265 (Oct. 1981).

Handbook of Modern Electrical and Electronic Engineering. **C. Belove.** New York: Wiley, 1986.

Handbook of Modern Electrical Wiring. **John E. Traister.** Reston: VA: Reston Publishing, 1976.

A complete practical guide that outlines the basic principles and actual installation of electrical wiring. Provides many applications and praiseworthy illustrations. A reference for beginning workers in the electrical construction field, but also for students, contractors, and the general public.

Handbook of Practical Electrical Design. **J. F. McPartland, ed.** New York: McGraw-Hill, 1984.

A thorough reference work on the design aspects of electrical systems for commercial, industrial, and residential buildings. Specific emphasis on the National Electrical Code (NEC), but also includes Occupational Safety and Health Administration regulations and other codes, such as those of the National Fire Protection Association. Suitable for both academic libraries with electrical engineering programs and libraries serving the general public.
R: *Choice* (Jan. 1984, p. 729); *ARBA* (1985, p. 549).

Handbook of Power Generation: Transformers and Generators. **John E. Traister.** Englewood Cliffs, NJ: Prentice-Hall, 1983.

Focus is on the practical on-the-job dealings for a broad spectrum of applications in electrical power generation and distribution, with a slight mention of theories. A helpful aid for electrical engineering texts.
R: *Science and Technology Libraries* 4: 125 (Winter 1983).

Handbook of Simplified Electrical Wiring Design. **John Lenk.** Englewood Cliffs, NJ: Prentice-Hall, 1978.

A practical basic guide on solving specific wiring problems. Gives shortcuts while not requiring advanced knowledge of mathematics. Book emphasizes designing and planning of electrical wiring. Focuses on the fundamentals of power distribution, grounding problems, applications of transformers, etc. For electricians and home owners.

Handbook of Simplified Solid-State Circuit Design. 2d ed. **John D. Lenk.** Englewood Cliffs, NJ: Prentice-Hall, 1978.

First edition, 1971.
Contains step-by-step procedures in solid-state design.
R: *Mechanical Engineering* 100: 119 (Aug. 1978); *TBRI* 44: 315 (Oct. 1978).

Handbook of Transformer Design and Applications. **W. M. Flanagan.** New York: McGraw-Hill, 1985.

The International Countermeasures Handbook: 1978–79. 4th ed. **Harry F. Eustace, ed.** Palo Alto, CA: EW Communications. Distr. New York: Franklin Watts, 1978.

A thorough presentation on electronics warfare systems including microwave and infrared/electro-optical technology, tactical and airborne systems, flares and aerosols, etc. Contains information on US expenditures, Soviet systems, and companies that produce electronic warfare equipment. Provides tables, diagrams, charts, and an index. More technical language makes this reference suitable for those involved in the electronic warfare industry.
R: *ARBA* (1980, p. 734).

Lighting Handbook. Rev. ed. **Westinghouse Electric Corporation.** Bloomfield, NJ: Westinghouse Electric, 1977.

Contains technical and illustrative information on lighting principles.
R: *Plant Engineering* 32: 252 (Feb. 16, 1978); *TBRI* 44: 200 (May 1978).

The Lineman's and Cableman's Handbook. 6th ed. **Edwin B. Kurtz and Thomas M. Shoemaker.** New York: McGraw-Hill, 1981.

Fifth edition, 1976.
An updated edition of a standard handbook on the fundamentals and procedures for the maintenance, operation, and construction of electric distribution and transmission lines. Includes new material on transmission circuits, construction specifications, insulators, grounding, etc. Contains safety applications and self-testing questions. Extensively illustrated by photographs and line drawings. Valuable reference for linemen, electricians, cablemen, and field supervisors.
R: *RSR* 5: 28 (Oct./Dec. 1977); *ARBA* (1977, p. 763).

McGraw-Hill's National Electrical Code Handbook. 18th ed. J. F. McPartland et al. New York: McGraw-Hill, 1984.

Sixteenth edition, 1979; seventeenth edition, 1981.
A popular handbook that explains various provisions of the National Electrical Code. Contains new specifications and changes in design, sizing, grounding, etc. A useful reference for electricians, electrical engineers, and technicians.
R: *Chemical Engineering* 87: 14 (Apr. 7, 1980); *Electrical World* 192: 146 (Oct. 15, 1979); *RQ* 19: 194 (Winter 1979); *TBRI* 45: 388 (Dec. 1979); *ARBA* (1985, p. 550).

NFPA Handbook of the National Electrical Code. **Wilford L. Summers.** Boston: National Fire Protection Agency, 1978.

Third edition, 1972; fourth edition, 1975.
Reproduction of the most recent National Electrical Code with diagrams, illustrations, and commentary. A necessity for electrical engineers, contractors, librarians, and legalists.
R: *ARBA* (1973, p. 656; 1976, p. 773); Sheehy (EJ43).

Practical Electrical Wiring. 12th ed. **Herbert P. Richter and W. Creighton Schwann.** New York: McGraw-Hill, 1981.

A simple presentation on electrical wiring. Gives details on theory and basic principles and wiring for residential, nonresidential, and farm areas. Provides tables, examples, and a bibliography. For both professional and novice electricians.

Printed Circuits Handbook. 2d ed. **Clyde F. Coombs, Jr., ed.** New York: McGraw-Hill, 1978.

In-depth coverage of the subject. Includes illustrations and provides data, guidelines, and proven troubleshooting techniques. New edition contains detailed information on environmental protection and computer-assisted design processes. Considered a must for the practicing electrician.
R: *Metal Finishing* 77: 110 (Apr. 1979); *Solid State Technology* 22: 95 (July 1979); *TBRI* 45: 233 (June 1979); 45: 311 (Oct. 1979); 45: 385 (Dec. 1979).

RC Active Filter Design Handbook. Wiley Electrical and Electronics Technology Handbook Series. **F. W. Stephenson.** New York: Wiley, 1985.

Switchgear and Control Handbook. **Robert W. Smeaton, ed.** New York: McGraw-Hill, 1977.

A comprehensive, up-to-date, and authoritative guide to switchgear design. For engineers involved in electical power applications.
R: *Electronics and Power* 23: 654 (Aug. 1977); *IEEE Spectrum* 14: 85, 87 (Sept. 1977); *RSR* 5: 29 (Oct./Dec. 1977); *TBRI* 43: 358 (Nov. 1977).

Transistor Substitution Handbook. 23d ed. **Sams Engineering Staff.** Indianapolis: Sams, 1984.

Thirteenth edition, 1973; fourteenth edition, 1974; seventeenth edition, 1978. Computer-generated handbook of American and foreign transistors, listed in numerical and alphabetical order.
R: *ARBA* (1973, p. 655).

Tube Substitution Handbook. 28th ed. **Sams Engineering Staff.** Indianapolis: Sams, 1985.

Sixteenth edition, 1973; twentieth edition, 1977.

Radio, TV, Etc.

Complete Broadcast Antenna Handbook: Design, Installation, Operation, and Maintenance. **John E. Cunningham.** Blue Ridge Summit, PA: TAB Books, 1977.

Complete information on design, installation, and modes of transmission of antennae. Succinct reference source for beginners.
R: *Telecommunication Journal* 45: 419 (July 1978); *TBRI* 44: 352 (Nov. 1978).

The Complete Handbook of Magnetic Recording. **Finn Jorgensen.** Blue Ridge Summit, PA: TAB Books, 1980.

A well-written handbook on various areas of magnetic recording. Covers history, recording and playback theory, magnetic heads, tapes and disks, recording techniques, and equipment. Illustrations enhance usefulness of text. Valuable to anyone interested in magnetic recording.
R: *Choice* 18: 820 (Feb. 1981).

The Complete Handbook of Radio Transmitters. **Joseph J. Carr.** Blue Ridge Summit, PA: TAB Books, 1980.

The Complete Handbook of Radio Receivers. **Joseph J. Carr.** Blue Ridge Summit, PA: TAB Books, 1980.

A clear understandable presentation of material dealing with radio transmitters and receivers. Two complementary volumes include safety precautions and circuit diagrams. Intended for both experts and novices.
R: *Choice* 18: 1127 (Apr. 1981).

Handbook for Radio Engineering Managers. **J. F. Ross.** Woburn, MA: Butterworth, 1980.

A worthwhile reference on various aspects of radio engineering. Sections deal with: management and organization, engineering economy, safety practices, fires, environmental aspects, and contract administration. Contains references, tables, illustrations, and an index. For upper-level undergraduate and graduate students in electronics.
R: *Choice* 18: 1296 (May 1981).

Handbook of Modern Solid-State Amplifiers. **John D. Lenk.** Englewood Cliffs, NJ: Prentice-Hall, 1979.

Emphasis on theory, analysis, design, and test procedures. Includes troubleshooting techniques for solid-state amplifiers. Should be of use to both students and electronics technicians and designers.
R: *Choice* 11: 1662 (Jan. 1975).

Newnes Radio and Electronics Engineer's Pocket Book. 15th ed. **Editorial Staff of Electronics Today International.** Rev. ed. Woburn, MA: Butterworth, 1978.

A pocket-sized handbook that will serve electrical engineers as a quick source of reference. Contains such information as metric weights and measures, wiring sizes, and radio-wave frequencies.

The Radio Amateur's Handbook. 58th ed. **American Radio Relay League, Headquarters Staff.** Newington, CT: American Radio Relay League, 1980.
First edition, 1926.
Theoretical and popular approach to radio fundamentals. The basic information on current practice is presented in how-to-do-it terms. Excellent use of diagrams and photographic aids.
R: *ARBA* (1970, p. 163; 1971, p. 562; 1972, p. 655); Jenkins (K163); Win (EI121).

Radio Control Handbook. 4th rev. ed. **Howard G. McEntee.** Blue Ridge Summit, PA: TAB Books, 1979.
First edition, 1954; second edition, 1961; third edition, 1971.
Technical information geared to the hobbyist who flies radio-controlled model planes. Considerable electronics and workshop knowledge is presupposed. Includes numerous illustrations and ciruit drawings.
R: *ARBA* (1972, p. 655).

The Radio Handbook. 22d ed. **W. I. Orr.** New Augusta, IN: Editors and Engineers, 1982.
First edition, 1935; eighteenth edition, 1970; nineteenth edition, 1972; twenty-first edition, 1978.

Provides authoritative information on high-frequency and very high-frequency radio communication.
R: *Radio Electronics* 54: 106 (Feb. 1983); Jenkins (K168); Wal (p. 326).

Radio Propagation Handbook. **Peter N. Saveskie.** Blue Ridge Summit, PA: TAB Books, 1980.
A comprehensive reference on all aspects of radio-wave propagation. The book is divided into text and appendixes. Chapters cover numerous types of radio propagation with sample problems and solutions. The appendixes provide data such as Rayleigh fading, K-factor data, antenna height, outage, etc. Recommended for graduate students or engineers involved in telecommunication and radio-wave propagation.
R: *Choice* 18: 982 (Mar. 1981).

Radio, TV and Audio Technical Reference Book. 5th ed. **S. W. Amos, ed.** Boston: Newnes-Butterworth. Distr. Levittown, NY: Transatlantic Arts, 1977.
First edition, 1954; fourth edition, 1963.
Prepared with contributions by experts. Includes detailed information on the most recent advances in radio, TV, and electronic technology. Fully illustrated and of high standard. Recommended to engineers, designers, and technical libraries.
R: *Choice* 14: 1531 (Jan. 1978); *ARBA* (1979, p. 771).

Telecommunication Transmission Handbook. 2d ed. **Roger L. Freeman.** Somerset, NJ: Wiley, 1981.
Provides practical and real-world information used in telecommunication design for single links or for complete networks. For practicing telecommunications engineers and advanced students.
R: *Radio Electronics* 53: 107 (May 1982).

World Radio TV Handbook. Vol. 33. **J. M. Frost, ed.** New York: Watson-Guptill, 1979.
First edition, 1947.
This well-established annual includes technical data on world radio and television stations, listed under continents and countries. Includes world time-charts.
R: Wal (p. 203).

COMPUTER TECHNOLOGY

Active Filter Design: Handbook for Use with Programmable Pocket Calculators and Mini Computers. **G. S. Moschytz and P. Horn.** New York: Wiley, 1981.
Details the design of 20 low-cost active filters that can be used for most general applications. Characteristics include minimum power, low RC component count, easy tunability, low sensitivities to active and passive components, and good stability. For both the professional active filter designer and the nonspecialist.

The BASIC Handbook: Encyclopedia of BASIC Computer Language. 2d ed. **David A. Lien.** San Diego: Compusoft Publishing, 1981.
More of a reference of 500 entries related to BASIC commands and any operands. Each command includes a sample, plus alternate spellings, variations in usage in

other languages, and a cross-reference. A standard and essential dictionaric guide for anyone coding with BASIC on any machine.
R: *Computer* 16: 127 (Jan. 1983); *ARBA* (1983, p. 719).

CRC Handbook of Digital System Design for Scientists and Engineers: Designed with Analog, Digital, and LSI. **Wen C. Lin.** Boca Raton, FL: CRC Press, 1981.
For those scientists and engineers who wish to do practical design of digital systems for their applications, this handbook will be a useful source of information.
R: *Proceedings of the IEEE* 71: 911 (July 1983).

Computer Performance Modeling Handbook. **Stephen S. Lavenberg, ed.** New York: Academic Press, 1983.
This guide assists in the formulation and application of performance models and provides analytical results from various models and guidance in the simulation of performance models. A superb reference manual and handbook for those in performance modeling in industry, goverment, or academia.
R: *Computer* 15: 8 (Dec. 1982).

Designer's Handbook of Integrated Circuits. **Arthur B. Williams, ed.** New York: McGraw-Hill, 1984.
An application-oriented reference volume with emphasis on the selection of integrated circuits and IC design. Valuable tool for both practicing engineers and individuals involved in the design and use of integrated circuits.
R: *ARBA* (1985, p. 550).

Digital Logic and Computer Design. **M. Morris Mano.** Englewood Cliffs, NJ: Prentice-Hall, 1979.
A comprehensive handbook on computer design. Discusses both newer and more traditional aspects of computer design.

Electronic Communications Applications Handbook. **Cheshier.** New York: Wiley, 1986.

Ferromagnetic-Core Design and Application Handbook. **M. F. "Doug" DeMaw.** Englewood Cliffs, NJ: Prentice-Hall, 1981.
Provides detailed coverage of magnetic core devices and their uses in modern electronic circuits. Includes such topics as ferrite and powdered-iron components, narrow- and broad-band transformers and inductors, RF chokes, and ferrite loop antennas, ferrite beads. Utilizes numerous practical circuit examples. Contains an extensive bibliography and reference data. A helpful reference for technicians, engineers, and graduate and upper-level undergraduate students.
R: *Choice* 18: 1442 (June 1981).

Fundamentals Handbook of Electrical Computer Engineering. Vols. 1–3. **Sheldon S. L. Chang.** New York: Wiley, 1982.
A treatment of electrical and computer engineering with emphasis on design applications in the engineering industries. Shows the shifts from analog to digital systems and from discrete to integrated components.
R: *IEEE Spectrum* 20: 13 (January 1983); *LJ* 107: 1360 (Aug. 1982).

The Handbook of Artificial Intelligence. 2 vols. **Avron Barr and Edward Feigenbaum, eds.** Los Altos, CA: William Kaufman, 1981–82.

Covers search routines, knowledge representation, and understanding the natural language as it is written and spoken, relative to artificial intelligence. At times the material is presented in an unrelated or unequal manner that proves to be somewhat too sophisticated for the novice. Overall, a useful source of information in this field.
R: *Byte* 8: 450 (July 1983); 8: 486 (Sept. 1983).

Handbook of Computers and Computing. **Arthus H. Seidman and Ivan Flores, eds.** New York: Van Nostrand Reinhold, 1984.

The handbook consists of 50 articles aranged under 6 subject areas—components, devices, hardware systems, languages, software systems, and procedures. Subject index is excellent. Intended for professionals and students with technical background.
R: *ARBA* (1985, p. 587).

Handbook of Digital IC Applications. **David L. Heiserman.** Englewood Cliffs, NJ: Prentice-Hall, 1980.

This handbook covers such topics as the role of integrated circuits in digital electronics; principles of digital counters; digital measuring systems; basic memory systems, among others. For advanced undergraduate or graduate engineering students and academic libraries.

Handbook of Electronic Systems Design. **Charles A. Harper.** New York: McGraw-Hill, 1980.

A helpful volume on aspects of system design to consider when designing electronic systems in business and industry. Includes such systems as computer, radar, communications, navigation, and digital. Of interest to systems engineers, technical managers, and technical libraries.
R: *Choice* 17: 698 (July/Aug. 1980).

Handbook of Electropainting Technology. **Willibald Machu.** Translated by Peter Neufeld. Ayr, Scotland: Electrochemical Publications, 1978.

A thorough reference volume on applying organic coatings by electrophoresis. Considered an excellent reference.
R: *Metal Finishing* 77: 92 (Jan. 1979); *Plating and Surface Finishing* 66: 62 (Feb. 1979); *TBRI* 45: 117 (Mar. 1979).

Handbook of Microcircuit Design and Application. **David F. Stout.** New York: McGraw-Hill, 1980.

This handbook of microcircuit design is grouped into 27 chapters that cover broad topics. Index for quick reference is included. Of interest to specialists, research and special libraries. Includes design procedures and examples of applications in a concise but explicit format.
R: *IEEE Proceedings* 70: 686 (June 1982); *Choice* 17: 698 (July/Aug. 1980); *Computer* 15: 134 (April 1982).

Handbook of Microcomputer-Based Instrumentation and Controls. **John D. Lenk.** New York: Prentice-Hall, 1984.

A handbook that first covers the basic elements and theory of instrumentation/control equipment and later describes actual control and instrumentation systems used in industry. For engineers, technicians, programmers/analysts, and students.
R: *Science Technology Libraries* 5: 2 (1984).

Handbook of Microprocessors, Microcomputers and Minicomputers. **John D. Lenk.** Englewood Cliffs, NJ: Prentice-Hall, 1979.

Nontechnical, state-of-the-art handbook on microprocessor equipment. Compares major systems and provides detailed descriptions of all major lines. For computer specialists.
R: *Computer* 15: 166 (Jan. 1982); *Aslib Proceedings* 44: 538 (Dec. 1979).

Handbook of Practical Microcomputer Troubleshooting. **John D. Lenk.** Reston, VA: Reston Publishing, 1979.

A practical handbook that reviews the fundamentals of microcomputer systems, test equipment, and the troubleshooting of such systems. Although designed for computer technicians or hobbyists with a knowledge of microcomputers, it is also useful to undergraduates.
R: *Choice* 17: 106 (Mar. 1980).

Handbook of Semiconductor and Bubble Memories. **Walter A. Treibel and Alfred E. Chu.** Englewood Cliffs, NJ: Prentice-Hall, 1982.

A thorough overview of both old and new memory technologies, including read-only memory, random-access memory, first-in-first-out, and programmable logic array, as well as charge-coupled devices and magnetic bubble memories. Includes a number of excellent figures and tables, a useful index and bibliography, and a problem and question set with answers for each of the 10 chapters. Beneficial for graduate, professional, and upper-division undergraduate levels; for those with an assumed background in digital systems and circuit theory.
R: *Choice* 19: 788 (Feb. 1982).

Handbook of Software Maintenance. **G. Parikh.** New York: Wiley, 1985.

A useful reference tool to effective and economical software management, containing technical and managerial tips, techniques, guidelines, sources, and case studies.

Handbook of Solid-State Devices: Characteristics and Applications. **Michael Thomason.** Reston, VA: Reston Publishing, 1979.

A state-of-the-art reference on solid-state devices. Discusses semiconductor tests, recent developments in technology, etc. For technicians and engineers.

Integrated Circuits Application Handbook. **A. H. Seidman.** New York: Wiley, 1983.

A reference covering digital and linear integral circuits and integrated circuit fabrication. Each chapter is written by an expert and includes examples and calculation how-to's. Gives figures and tables.
R: *Librarian's Newsletter* (Oct. 1982).

The Master Handbook of IC Circuits. **Thomas K. Powers.** Blue Ridge Summit, PA: TAB Books, 1981.

Contains schematics for linear integrated circuits, voltage regulators, CMOS integrated circuits, TTL/LS integrated circuits, radio and TV circuits and special-purpose devices. Also has an index for specific circuits.
R: *IEEE Proceedings* 70: 1376 (Nov. 1982).

Master Handbook of Microprocessor Chips. **Charles K. Adams.** Blue Ridge Summit, PA: TAB Books, 1981.
A helpful guide to easy identification of all support integrated circuits applicable to specific microprocessors. Gives full details on support integrated circuits and their functions.
R: *Radio Electronics* 53: 106 (July 1982).

The McGraw-Hill Computer Handbook. **Harry Helms, ed.** New York: McGraw-Hill, 1983.
Provides practical details on how a computer, its peripherals, and its software work together. Covers computer theory, number systems, operating systems, graphics, programming languages, and more. For the beginner in computer science.
R: *ARBA* (1984, p. 619).

Microcomputers in Education: A Handbook of Resources. **Katherine Clay, ed.** Phoenix: Oryx Press, 1982.
A compilation of references to materials pertaining to computers and their use in education. Cites such topics as computer literacy, philosophy, classroom application, research studies, and more. A useful source, but a bit dated.
R: *ARBA* (1984, p. 621).

The Microcomputer Users' Handbook, 1983–84. **Dennis Longley and Michael Shain.** New York: Wiley. Annual.
A step-by-step reference covering the functions and reviews of various microcomputers. For the microcomputer user in business or for the potential buyer of a computer system.
R: *Librarians' Newsletter* 22: 1 (Dec. 1983).

Microprocessor Applications Handbook. **David F. Stout, ed.** New York: McGraw-Hill, 1981.
A quick reference to applications of microprocessor systems. Discusses both hardware and software topics, including interfacing, error correction in memory systems, keyboard scanning methods, voice recognition, telephony, video games, etc. Useful to systems and circuit design engineers, physicists, and chemists.

Microprocessor Handbook. **J. D. Greenfield.** New York: Wiley, 1985.
A handy reference on the features and capabilities of the most popular microprocessors currently available, including six 8-bit microprocessors and three of the most popular 16-bit microprocessors in the Intel, Zilog, and Motorola lines.

Mini- and Microcomputer Control in Industrial Processes: Handbook of Systems and Application Strategies. **M. Robert Skrokov, ed.** New York: Van Nostrand Reinhold, 1980.

Examines the administration and management of an industrial computer control project. Includes a worthwhile index and glossary. Useful for process control project managers.
R: *Choice* 18: (Jan. 1981).

Systems Troubleshooting Handbook. **Faulkenberry.** New York: Wiley, 1985.

Tower's International Microprocessor Selector. **T. D. Towers.** Blue Ridge Summit, PA: TAB Books, 1982.

A book containing data on 7,000 US, British, Japanese, and European microprocessors. Data includes operating temperatures, voltages, types, functions, related families, manufacturers, and training devices. Good for designers and repairers of digital equipment.
R: *ARBA* (1983, p. 722).

A User's Handbook of Semiconductor Memories. **E. R. Hnatek.** New York: Wiley, 1977.

A comprehensive overview of semiconductor memories and their applications. An up-to-date 1-volume reference.
R: *Choice* 15: 264 (Apr. 1978).

INDUSTRIAL ENGINEERING

Automotive Technician's Handbook. **William H. Crouse and Donald L. Anglin.** New York: McGraw-Hill, 1979.

A concise and clear handbook that focuses on aspects of automotive technology, including engine repair, manual and automatic transmissions, brakes, tune-ups, electrical equipment, heating, and air conditioning. Contains detailed explanations of repair procedures and inspection techniques. Provides numerous charts and illustrations. The reference is directed toward a technician but is suitable for both professional and home use.
R: *Choice* 17: 247 (Apr. 1980).

Cement Manufacturer's Handbook. **Kurt E. Peray.** New York: Chemical Publishing, 1979.

A valuable handbook for discussing cement plant operations. Mainly for research and special libraries in industry.

Consulting Engineer. 2d ed. **Claude Maxwell Stanley.** New York: Wiley, 1982.

Covers all aspects of the practice and functions of a consulting engineer, including relationships with clients; the organization; mangement and problems; and other information. Useful for engineers, architects, managers, and anyone involved in this field.
R: *Proceedings of the IEEE* 70: 1376 (Nov. 1982).

Electroplating Engineering Handbook. 4th ed. **Lawrence J. Durney, ed.** New York: Van Nostrand Reinhold, 1984.

Third edition, 1971.

A useful handbook with 44 chapters and subchapters in basically 2 parts: "General Processing Data" and "Engineering Fundamentals and Practice." Illustrated and indexed.
R: *ARBA* (1985, p. 541).

Engineering and Industrial Graphics Handbook. **George E. Rowbotham, ed.** New York: McGraw-Hill, 1981.

A comprehensive handbook on the numerous aspects of graphics. Covers utilization, techniques, and tools of graphics technology. Suitable for both technical and nontechnical managers.
R: *Civil Engineering* 52: 28 (May 1982).

Finishing Handbook and Directory. 30th ed. London: Sawell Publications, 1979.

A useful handbook and directory dealing with varied aspects of metal, wood, and plastic finishing; electroplating and processes; paint application; and drying equipment.
R: *Wal* (p. 205).

Handbook of Dimensional Measurement. 2d ed. **Francis T. Farago.** New York: Industrial Press, 1982.

First edition, 1969.
Practical information on the methods and equipment of metrology for modern industrial production. Numerous tables with diagrammatic illustrations facilitating ready reference.
R: *ARBA* (1970, p. 144).

Handbook of Industrial Engineering. **Gavriel Salvendy, ed.** New York: Wiley-Interscience, 1982.

Covers various aspects of a systemwide approach to all segments of services, manufacturing industries, and corporations in order to reach their optimal productivity. Well-organized and carefully written. This book will be useful to any professional and graduate library.
R: *Choice* 20: 298 (Oct. 1982).

Handbook of Laboratory Waste Disposal: A Practical Manual. **M. J. Pitt and E. Pitt.** New York: Wiley, 1985.

A practical reference tool on waste disposal in the laboratory. Covers a wide range of laboratory disciplines, including laboratories in educational institutions, with an emphasis on safety. Includes illustrations, bibliographies and appendixes.

Handbook of Package Engineering. 2d ed. **Joseph F. Hanlon.** New York: McGraw-Hill, 1984.

A comprehensive 1-volume handbook with extensive information on all types of materials used in packaging. Well-indexed and illustrated. Recommended to all types of libraries.
R: *ARBA* (1985, p. 541).

Industrial Hazard and Safety Handbook. **Ralph W. King and J. Magid.** London: Newnes-Butterworth, 1979.

Contains concise information and useful tables. A practical up-to-date reference work for industrial and safety engineers.
R: *Chemical Engineer* 350: 803 (Nov. 1979); *Chemistry in Britain* 15: 538 (Oct. 1979); *TBRI* 46: 35 (Jan. 1980); 46: 75 (Feb. 1980).

Industrial Safety Handbook. 2d ed. **William Handley, ed.** New York: McGraw-Hill, 1977.

The second edition of this handbook contains important information on the prevention of accidents in the workplace. Covers all pertinent issues. An informative guide for anyone concerned with the health and safety of workers.
R: *Chemical Engineer* 327: 882 (Dec. 1977); *National Safety News* 116: 102 (Dec. 1977); *Professional Safety* 22: 15 (Sept. 1977); *TBRI* 44: 114 (Mar. 1978); *ARBA* (1978, p. 774).

Maintenance Engineering Handbook. 3d ed. **Lindley R. Higgins and L. C. Morrow, eds.** New York: McGraw-Hill, 1977.

An updated and revised edition of a basic reference on general maintenance of industrial plants. Includes new material on pollution control equipment, fuel conservation, and plant security. Contains over 1,000 tables, charts, and graphs as well as topical glossaries and a detailed index. Useful for industrial engineers and plant managers.
R: *Industrial Engineering* 9: 64 (Nov. 1977); *RSR* 5: 37 (Oct./Dec. 1977); *TBRI* 44: 34 (Jan. 1978); *ARBA* (1978, p. 776).

Paint Handbook. **G. Weismantel.** New York: McGraw-Hill, 1980.

A comprehensive reference dealing with selection of paints and finishes for different surfaces. Arranged by surfaces painted, rather than by coatings used. A helpful volume for paint contractors, corrosion engineers, and architects.

Paints and Coatings Handbook. 2d ed. **Abel Banov.** New York: McGraw-Hill, 1980.

A basic reference source on paints and coatings. Includes many recent developments in the field. Contains illustrations and an index. For paint contractors and architects.

Patty's Industrial Hygiene and Toxicology. 3d rev. ed. Vols. 1–. **G. D. Clayton and F. E. Clayton, eds.** New York: Wiley, 1978–.

First edition, 1948; second edition, 1958; third edition, volume 1, *General Principles*, 1978; volume 2A, *Toxicology*, 1980; volume 2B, *Toxicology*, 1981; volume 3, *Theory and Rationale of Industrial Hygiene Practice*, edited by L. V. Cralley and L. J. Cralley, 1979.
A classic reference source in the area of occupational health. Covers all aspects of toxicology and industrial hygiene. For chemists, administrators, and laypeople.
R: *Chemical Engineering* 86: 11 (June 4, 1979); *Chemistry and Industry* 9: 314 (May 5, 1979); *New Scientist* 79: 957 (Sept. 28, 1978).

Product Safety and Liability: A Desk Reference. **John Kolb and Steven S. Ross.** New York: McGraw-Hill, 1980.

An introductory volume on product safety. Discusses such subjects as the regulatory process, marketing, insurance, production design and control. Ten appendixes list tables of exposure limits for toxic substances, handling regulations, ANSI standards,

etc. Includes a bibliography, glossary, and general and standards indexes. A well-presented reference for public, academic, and special libraries.
R: *Chemical Engineering* 87: 127 (Dec. 1, 1980); *ARBA* (1981, p. 761).

Quality Technology Handbook. 3d ed. **R. S. Sharpe et al., eds.** Guildford, Surrey, England: IPC Science and Technology, 1978.

For those involved with quality control. Considered a well-organized reference, especially for industrial libraries.
R: *British Journal of Non-Destructive Testing* 20: 254 (Sept. 1978); *Physics in Technology* 10: 36 (Jan. 1979); *TBRI* 44: 399 (Dec. 1978); 45: 160 (Apr. 1979).

Sensor and Analyzer Handbook. **Harry A. Norton.** Englewood Cliffs, NJ: Prentice-Hall, 1982.

Brings a beginning for an in-depth understanding of individual sensors. Acts as a good supplementary text for course work on sensor designs and uses, and it is also helpful for the practicing engineer.
R: *IEEE Spectrum* 19: 86 (Sept. 1982).

Standard Handbook of Plant Engineering. **Robert C. Rosalen and James O. Rice.** New York: McGraw-Hill, 1982.

Contains proven methods for achieving efficient, cost-effective operations for planning, construction, and maintenance of plant facilities. Includes the latest techniques and uses in energy conservation, pollution control, and data processing.
R: *Civil Engineering* 52: 24 (Dec. 1982); *ARBA* (1983, p. 727).

MATERIALS SCIENCE

ASM Metals Reference Book: A Handbook of Data about Metals and Metal-Working. Metals Park, OH: American Society for Metals, 1981.

A collection of tabular numerical and graphic data. Alloy phase diagrams are included, introduced by a list of general references and sources. Most of the data is on the composition and properties of metals.
R: *ARBA* (1983, p. 727).

CRC Handbook of Electrical Resistivities of Binary Metallic Alloys. **Klaus Schroder, ed.** Boca Raton, FL: CRC Press, 1983.

After an introductory chapter, tabulations for the binary alloys of each metal follow. Includes references for each section.
R: *Science Technology Libraries* 5: 2 (Winter 1984).

CRC Handbook of Materials Science. Boca Raton, FL: CRC Press, 1974–.

Volume 1, *General Properties*, 1974; volume 2, *Metals, Composites, and Refractory Materials*, 1975; volume 3, *Nonmetallic Materials and Applications*, 1975; volume 4, *Wood*, 1980. Volumes 1–3, edited by Charles T. Lynch; volume 4, edited by Robert Summitt.
A readily accessible guide to the physical properties of solid-state and structural materials. Interdisciplinary in approach and content, it covers a broad variety of types of materials.

CRC Handbook of Physical Properties of Rocks. 2 vols. **Robert S. Carmichael, ed.** Boca Raton, FL: CRC Press, 1981–1982.

Consists of a conglomeration of individual reports with limited data and massive assemblies of data. It is an up-to-date guide to physical properties for reference and comparison of various properties or types of materials.

R: *Civil Engineering* 52: 29 (Sept. 1982); *Science and Technology Libraries* 3: 72 (Winter 1982).

Composite Materials Handbook. **M. M. Schwartz.** New York: McGraw-Hill, 1984.

A practical handbook on the topics of resin and metal matrix materials, mostly at the macrostructure level. Extensive illustrations and charts, with useful glossary and index as well. Valuable to working engineers.

R: *ARBA* (1985, p. 554).

Construction Materials: Types, Uses and Applications. **Caleb Hornbostel.** New York: Wiley: 1978.

Alphabetically arranged, this volume contains much information on construction materials. Discussed under such topics as physical and chemical properties, history, and manufacturing.

Corrosion Resistant Materials Handbook. 2d ed. **I. Mellan, ed.** Park Ridge, NJ: Noyes Data, 1971.

Second edition, 1971.
A valuable reference work with extensive cross-referencing. Useful to those concerned with corrosion.

R: *Anti-Corrosion Methods and Materials* 23: 26 (Dec. 1976); *Journal of the Society of Environmental Engineers* 15: 22 (Dec. 1976); *Plant Engineering* 31: 316 (Oct. 13, 1977); *TBRI* 43: 156 (Apr. 1977); 43: 397 (Dec. 1977).

Data Handbook for Clay Materials and Other Non-metallic Minerals. **H. van Olphen and J. J. Fripiat, eds.** Elmsford, NY: Pergamon, 1979.

A comprehensive listing of chemical, physical, and mineralogical data on various clays and nonmetallic materials. Covers properties, origin of clays, methods of analysis, etc. An indispensable reference for graduate-level researchers in the field.

R: *Choice* 17: 103 (Mar. 1980).

Encyclopedia/Handbook of Materials, Parts and Finishes. Henry R. Clauser, ed. Westport, CT: Technomic, 1976.

Concise source of information on thousands of industrial materials. Lists generic name and describes chemical and engineering properties. Brings together much information in 1 volume. For engineers, chemists, designers, etc.

R: *ASTM Standardization News* 5: 43, 46 (Aug. 1977); *Chemtech* 7: 301 (June 1977); *Elastomerics* 109: 56 (Mar. 1977); *Choice* 14: 830 (Sept. 1977); *TBRI* 43: 192 (May 1977); 43: 273 (Sept. 1977); 43: 311 (Oct. 1977).

Fiberglass-Reinforced Plastics Deskbook. **Nicholas P. Cheremisinoff and Paul N. Cheremisinoff.** Ann Arbor, MI: Ann Arbor Science, 1978.

A handbook on modern uses and applications of fiberglasses. Contains graphs, tabulated data, diagrams, and ASTM standards. Of interest to engineers and technologists.
R: *Chemical Engineering* 86: 11 (Apr. 23, 1979).

Handbook of Adhesives. 2d ed. **Irving Skeist, ed.** New York: Van Nostrand Reinhold, 1977.

First edition, 1962.
For chemists and engineers, a classic handbook on adhesives. Covers most recent technology; provides glossary and bibliography.
R: *Engineering Materials and Design* 6: 131 (Feb. 1963); *Industrial Chemist* 39: 327 (June 1963); *Journal of the American Chemical Society* 100: 1977 (Mar. 15, 1978); *Scientific American* 237: 37 (Nov. 1977); *Tappi* 60: 48 (Aug. 1977); *TBRI* 43: 318 (Oct. 1977); 44: 277 (Sept. 1978); *ARBA* (1978, p. 775); Wal (p. 499).

Handbook of Composites. **George Lubin, ed.** New York: Van Nostrand Reinhold, 1982.

This comprehensive handbook has been compiled to help the reader understand composite meterials and their characteristics, design, and applications. The book is arranged into 4 sections: raw materials, processing methods, design, and application.
R: *ARBA* (1983, p. 715).

Handbook of Materials Handling. **R. G. T. Lindkvist.** New York: Wiley, 1985.

Handbook on Mechanical Properties of Rock. 4 vols. **R. D. Lama and V. S. Vutukuri.** Cleveland, OH: Trans Tech, 1976–1978.

Comprises a digest of facts of mechanical properties of rocks. Well-referenced and informative. A recommended reference work for teachers, researchers, and engineers.
R: *Ground Engineering* 9: 44 (Sept. 1976); *International Journal of Rock Mechanics and Mining Sciences* 16: 269 (Aug. 1979); *Mining Magazine* 140: 49 (Jan. 1979); *TBRI* 43: 80 (Mar. 1977); 45: 352 (Nov. 1979): 45: 156 (Apr. 1979).

Handbook of Oceanographic Engineering Materials. **S. C. Dexter.** New York: Wiley, 1979.

A succinct handbook that contains information on a variety of oceanographic engineering materials.

Handbook of Powder Science and Technology. **M. E. Rayed and L. Otten, eds.** New York: Van Nostrand Reinhold, 1984.

A comprehensvie 1-volume handbook with 19 chapters that cover every aspect of powder science and technology. Extensive illustrations, charts, tables, and references. Useful for all research and academic libraries with an interest in the subject.
R: *ARBA* (1985, p. 553).

Handbook of Refractory Compounds. **Grigorii Valentinovich Samsonov and I. M. Vinitskii.** Translated by Kenneth Shaw. New York: Plenum Press, 1980.

Translated from Russian.

An excellent reference providing data on refractory compounds. Topics include stoichiometry and thermodynamic, electrical, mechanical, and chemical properties. A lengthy appendix of binary systems is included, as well as an extensive and outstanding reference list of 845 sources. An invaluable addition to graduate and professional libraries.
R: *Physics Today* 33: 73 (Nov. 1980); *Choice* 18: 779 (Feb. 1981).

Materials Handling Handbook. 2d ed. **C. R. Asfahl.** New York: Wiley, 1985.

Materials Handbook. 11th ed. **George S. Brady and Henry R. Clauser.** New York: McGraw-Hill, 1978.

First edition, 1929; ninth edition, 1963; tenth edition, 1971.
A standard handbook listing over 13,000 industrial materials and substances. Includes chemicals, fuels, minerals, textiles, foodstuffs, synthetics, etc.; provides details on composition, charateristics, uses, and trade names. Recommended reference for engineers, technicians, and purchasing managers.
R: *Chemical Engineering* 85: 14 (May 8, 1978); *Materials and Methods* 43: 256 (May 1956); *TBRI* 7: 103 (June 1956); *WLB* 52: 84 (Sept. 1977); *ARBA* (1978, p. 774); Jenkins (A77); Wal (p. 292); Win (EI130).

Permanent Magnet Design and Application Handbook. **Lester R. Moskowitz.** Boston: Cahners Books, 1976.

Contains charts and tables on a wide variety of magnetic materials. Includes histories, classifications, design, and environmental effects.
R: *Choice* 13: 1460 (Jan. 1977); *RSR* 5: 29 (Oct./Dec. 1977); *ARBA* (1977, p. 764).

MECHANICAL ENGINEERING

GENERAL

ASME Handbook. 2d ed. 4 vols. **American Society of Mechanical Engineers.** New York: McGraw-Hill, 1965–.

First edition, 1953–1958; *Metals Engineering–Design,* 2d ed., 1965; *Metals Properties,* 1954; *Engineering Tables,* 1956; *Metals Engineering–Processes,* 2d ed., 1965.
A standard handbook for the mechanical engineer. Multi-authored, including systematic data and tables. Includes much hard to find data. Eventually all volumes will be updated.
R: *American Machinist* pp. 277, 279 (May 7, 1956); Jenkins (K172); Wal (pp. 298, 299, 355); Win (EI143; 1EI34).

Handbook of Measurement Science. **Peter H. Sydenham.** New York: Wiley, 1982.

A catalog and description of the principles underlying the measurement process and measurement systems. This volume is mainly a summary of the basic theory of the nature and behavior of this design and system. Written by 13 experts.
R: *Applied Optics* 22: 540 (Feb. 15, 1983).

Handbook of Mechanics, Materials, and Structures. **A. Blake and M. D. Martin.** New York: Wiley, 1985.

Handbook of Precision Engineering. Vols. 1–. **A. Davidson.** New York: McGraw-Hill, 1971–.

Volume 1, *Fundamentals*, 1971; volume 2, *Materials*, 1971; volume 3, *Fabrication of Non-Metals*, 1972; volume 4, *Physical and Chemical Fabrication Techniques*, 1972; volume 5, *Joining Techniques*, 1972; volume 6, *Mechanical Design Applications*, 1972; volume 8, *Surface Treatment*, 1973.

A series of volumes covering all aspects of precision engineering. Covers fundamentals, materials, and techniques. Of interest to engineers in the field.

R: *American Scientist* 60: 267 (Mar./Apr. 1972); 60: 391 (May/June 1972); *Engineer* (Dec. 23, 1970); *ARBA* (1972, p. 658; 1973, p. 659; 1974, p. 688; 1975, pp. 802, 803), Sheehy (EJ55); Wal (p. 301).

Marks' Standard Handbook for Mechanical Engineers. 8th ed. **Theodore Baumeister and Eugene Avallone.** New York: McGraw-Hill, 1978.

Sixth edition, 1958; seventh edition, 1967, entitled *Standard Handbook for Mechanical Engineers.*

Updated edition; considered one of the most thorough references. Includes material on Occupational Safety and Health Administration, the Environmental Protection Agency, and industrial engineers. Arranged by subject; includes line drawings and graphs.

R: *Choice* 5: 182 (Apr. 1968); *ARBA* (1979, p. 776); Wal (p. 296); Win (EI154; 2EI31a).

Mechanical Engineers' Handbook. **M. P. Kutz.** New York: Wiley, 1986.

Standard Handbook for Mechanical Engineers. 8th ed. **Theodore Baumeister, ed.** New York: American Society for Mechanical Engineers, 1978.

Covers such topics as refrigeration, strength, and handling of materials, cryogenics, etc. Contains an extensive analytic index. Valuable to mechanical engineers.

SPECIFIC

CRC Handbook of Lubrication (Theory and Practice of Tribology). **E. R. Booser, ed.** Boca Raton, FL: CRC Press, 1984.

Volume 1, *Applications and Maintenance*, 1983; volume 2, *Theory and Design*, 1984. The handbook covers the general area of lubrication and tribology in all its facets. Both metric and English units are provided throughout both volumes.

Die Casting Handbook. River Grove, IL: The Society of Die Casting Engineers, 1983.

Provides all necessary information about die casting, including safety tips, a glossary, and many tables and graphs. Its comprehensive form makes it very useful for both beginners and professionals of product designing, die casting, etc.

R: *Mechanical Engineering* 105: 103 (July 1983).

Flow Measurement Engineering Handbook. **R. W. Miller.** New York: McGraw-Hill, 1982.

A well-written and organized update on the latest flow-measuring devices and on the most recently developed equations, analysis techniques, and standards. Most helpful

to the practicing engineer interested in a working tool and reference guide to aid in the selection, design, and application of flowmeters and flow-measurement systems.
R: *Mechanical Engineering* 105: 100 (Aug. 1983).

Handbook of Advanced Robotics. **Edward L. Safford, Jr.** Blue Ridge Summit, PA: TAB, 1983.

A clear description of robots, with hundreds of drawings and photos. Good general descriptions in layperson's terms for all but the circuit designs, which are more complicated. Good for medium-sized and large collections.
R: *LJ* (March 1983, p. 446).

Handbook of Air Conditioning, Heating and Ventilating. 3d ed. **Eugene Stamper and Richard L. Koral, eds.** New York: Industrial Press, 1979.

A ready-reference data handbook. Useful to those involved in temperature engineering.
R: *ASHRAE J.* 21: 94 (Aug. 1979); *Heating/Piping/Air Conditioning* 51: 190 (Oct. 1979); *Refrigeration and Air Conditioning* 82: 67 (Oct. 1979); *TBRI* 45: 354 (Nov. 1979); 45: 392 (Dec. 1979); 46: 80 (Feb. 1980).

Handbook of Industrial Robotics. **Shimon Y. Nof.** New York: Wiley, 1985.

Handbook of Machinery Adhesives. **Haviland.** New York: Dekker, 1985.

Handbook of Mechanics, Materials and Structures. **A. Blake.** New York: Wiley, 1985.

A handy reference guide encompassing various facets of engineering design. Presents information on mechanics, materials and structures, finite-element modeling, developments in fracture control, and basic regulatory aspects of pressure vessel design. Intended for design engineers.

Handbook of Mining and Tunnelling Machinery. **Barbara Stack.** New York: Wiley, 1982.

Gives the evolutionary development of tunnelling, drilling, and mining machinery, from the earliest inception to the latest operating system.
R: *Civil Engineering* 52: 35 (June 1982).

Handbook of Noise and Vibration Control. 4th ed. **R. H. Waring, ed.** Morden, Surrey, England: Trade and Technical Press, 1979.
R: Wal (p. 103).

Handbook of Practical Gear Design. **Darle W. Dudley.** New York; McGraw-Hill, 1984.

Revised edition of *Practical Gear Design*, 1954.
A definitive reference work on gear design. Each chapter is augmented by useful charts, diagrams, tables, etc., with extensive bibliography and useful index. Useful to any scientific and engineering library interested in this subject.
R: *ARBA* (1985, p. 551).

Handbook of Precision Sheet, Strip and Foil. **Hamilton B. Bowman.** Metals Park, OH: American Society for Metals, 1980.

A handbook of data on thin, flat-rolled metal which is produced on specialized equipment to close tolerances. Among the topics included are terminology; thickness and its control; heat treating; coil size; corrosion; and conditions and defects. For research libraries and special libraries in industry.

Handbook of Pressure-Sensitive Adhesive Technology. **Donatas Satas, ed.** New York: Van Nostrand Reinhold, 1982.

Encompasses all aspects of pressure-sensitive adhesive products used in industry, the home, or the office. Geared for the technologist in the adhesive fields and in other related fields.
R: *ARBA* (1983, p. 715).

Handbook of Remote Control and Automation Techniques. **John E. Cunningham.** Blue Ridge Summit, PA: TAB Books, 1978.

A handbook for the installation of automated and remote-controlled systems. Systems have been pretested and built.

Handbook of Valves, Piping and Pipelines. **R. H. Warring.** Houston: Gulf Publishing, 1982.

Describes in detail more than 20 different types of valves and thoroughly examines all aspects of piping and pipelining.
R: *Civil Engineering* 52:34 (Nov. 1982).

Handbook on Industrial Robotics. **S. Y. Nof.** New York: Wiley, 1985.

Heat Exchanger Design Handbook. 5 vols. New York: Hemisphere Publishing Corp. 1983.

A presentation of the basic theory and mathematical foundations of heat transfer, giving specific examples of calculational procedures for a wide range of problems.
R: *Power* 127: 124 (July 1983).

Heat Trasnfer Pocket Handbook. **Nicholas P. Cheremisinoff.** Houston: Gulf Publishing, 1984.

A compact reference work that provides a short and concise guide to solving heat transfer problems. It is not a reference text, but is a guide to calculations, formulas, data, and design practices. Glossary, unit conversion factors, and a simple index are included.
R: *ARBA* (1985, p. 542).

Industrial Lubrication: A Practical Handbook for Lubrication and Production Engineers. **Michael Billett.** Elmsford, NY: Pergamon, 1979.

Contains practical information on lubrication. Stresses safety factors, selection, care, storage, and evaluation. Provides ample illustrations. Suitable for engineers and practitioners.
R: *Choice* 17: 415 (May 1980).

Industrial Robotics Handbook. **V. Daniel Hunt.** New York: Industrial Press, 1983.
Incorporates a broad overview of industrial robots with information on the applications for robots, procedures for system selection, technical characteristics for some 80 systems, and various discussions on the related topics of robots. For all levels.
R: *Science and Technology Libraries* 4: 123 (Winter 1983).

Iron Castings Handbook. 3d ed. **Charles F. Walton, ed.** Des Plaines, IL: Iron Castings Society, 1981.
Provides helpful and authoritative information for those involved in component design, materials specification, purchasing, and casting processing. Primarily for castings users, but may also serve as a useful reference book for engineering schools and foundries.
R: *Science and Technology Libraries* 3: 80 (Spring 1983).

Machinery's Handbook: A Reference Book for the Mechanical Engineer, Designer, Manufacturing Engineer, Draftsman, Toolmaker, and Machinist. 22d ed. **Erik Oberg, Franklin D. Jones, and Holbrook L. Horton.** New York: Industrial Press, 1984.
First edition, 1914; eighteenth edition, 1970; nineteenth edition, 1972; twentieth edition, 1975.
A standard reference work on machines and mechanical products, frequently revised and updated. Comprehensive coverage includes tables, standards, text, illustrations, etc. Some 155 topics appear under broad rubrics—friction and wear, twist drills, jigs and fixtures, soldering, etc. A must for libraries that need information on machinery and manufacturing.
R: *ARBA* (1972, p. 659; 1976, p. 779; 1985, p. 554).

Mechanical Technician's Handbook. **Maurice J. Webb, ed.** New York: McGraw-Hill, 1982.
This book covers building models and apparatus from instructions; selecting components and instruments for the assembly of apparatus; and safely conducting tests. Also contains tables and illustrations. Good for technicians who make and test new products.
R: *ARBA* (1983, p. 726).

Motor Trade Handbook. **W. A. Livesey.** Stoneham, MA: Butterworth, 1985.

Pressure Vessel Design Handbook. **Henry H. Bednar.** New York: Van Nostrand Reinhold, 1985.

Pressure Vessel Handbook. 4th ed. **E. F. Kegyesy.** Tulsa: Pressure Vessel Handbook, 1977.
First edition, 1972; second edition, 1973; third edition, 1975.

Pump Handbook. **Igor J. Karassik et al., eds.** New York: McGraw-Hill, 1976.
A multi-authored handbook written by authorities in their fields. Deals with theory, construction, and performance characteristics of all types of pumps. The handbook

also provides information on selection and testing of pumps. Considered a timely and invaluable source of reference for engineering libraries.
R: *AIChE Journal* 22: 1156 (Nov. 1976); *Chemical Engineering* 83: 11 (Nov. 22, 1976); *Farm Chemicals* 140: 72 (Jan. 1977); *Journal of the American Oil Chemists' Society* 54: 51A (Feb. 1977); *Naval Engineers Journal* 89: 32 (Aug. 1977); *TBRI* 43: 36 (Jan. 1977); 43: 116 (Mar. 1977); 43: 154 (Apr. 1977); 43: 355 (Nov. 1977); *ARBA* (1977, p. 769).

SAE Handbook. New York: Society of Automotive Engineers, 1924–. Annual.
Collection of reports and standards pertaining to the automotive industry.
R: *ARBA* (1970, p. 153; 1976, p. 763); Jenkins (K178); Wal (p. 373); Win (EI53).

Spring Designer's Handbook. **Harold Carlson.** New York: Dekker, 1978.
Covers practical application of spring-design principles. Includes step-by-step design procedures, data, and graphic material. For engineers at all levels of training.
R: *Mechanical Engineering* 100: 119 (Aug. 1978); *TBRI* 44: 351 (Nov. 1978).

Standard Handbook of Fastening and Joining. **Robert O. Parmley, ed.** New York: McGraw-Hill, 1977.
Up-to-date, accurate source of information on fastenings and joining. Covers such topics as standard fasteners and pins; pipe, tube, and hose connections; metal welding; and adhesive bonding. Includes illustrations and current standards presented in tabular form. Invaluable to all disciplines in engineering, but especially design engineers, technicians, machinists, and students.
R: *Plant Engineering* 31: 130 (Aug. 4, 1977); *TBRI* 43: 317 (Oct. 1977).

Standard Handbook of Machine Design. **J. E. Shigley and C. R. Misckhe.** New York: McGraw-Hill, 1986.

Surface Vehicle Sound Measurement Procedures. **Society of Automotive Engineers.** Warrendale, PA: Society of Automotive Engineers, 1979.
Contains data and practical information on vehicular noise such as road vehicles, motorboats, and snowmobiles. For acoustical engineers.
R: *Journal of Sound and Vibration* 66: 641 (Oct. 22, 1979); *TBRI* 46: 78 (Feb. 1980).

Thermal Insulation Handbook. **William C. Turner and John F. Malloy.** New York: McGraw-Hill, 1981.
A practical reference source on the techniques and information needed to solve insulation problems. Describes heat transfer, equipment, proper insulation, weather barriers, inspection, and maintenance. Of interest to engineers and contractors.
R: *Civil Engineering—ASCE* 52: 26 (Jan. 1982).

Tool and Manufacturing Engineers Handbook: A Reference Work for Manufacturing Engineers. 3d ed. **Society of Manufacturing Engineers.** Edited by Daniel B. Dallas. New York: McGraw-Hill, 1976.
Second edition, entitled *Tool Engineers Handbook*, 1959. ·
A new edition of this classic handbook, which covers all phases of manufacturing, planning, control, design, and tooling. Presents detailed information and practical data and techniques. Also contains metric conversion charts and information on new technology. For all engineering libraries.

R: *Industrial Lubrication and Tribology* 28: 120 (July–Aug. 1976); *National Safety News* 114: 107 (Oct. 1976); *Choice* 13: 1280 (Dec. 1976); *TBRI* 43: 32 (Jan. 1977); 43: 39 (Jan. 1977); *ARBA* (1977, p. 770); Sheehy (EJ58).

Weather Data Handbook for HVAC and Cooling Equipment Design. Ecodyne Corporation, Cooling Products Division. New York: McGraw-Hill, 1980.

Provides detailed tabulations of wet bulb temperature frequencies for the summer months; tabulations in 3 frequency categories for winter and summer conditions; and tabulations of various weather parameters. Contains material from the Fluor Products Company's *Evaluated Weather Data for Cooling Equipment Design* (1958) and Addendum no. 1, 1964, which are no longer in print. For use if a particular design problem should require more detailed or specific analysis.
R: *Science and Technology Libraries* 1: 75 (Summer 1981).

Welding Handbook. 7th ed. 2 vols. Miami: American Welding Society, 1976–1978.

Volume 1, 1976; volume 2, *Welding Processes—Arc and Gas Welding and Cutting, Brazing, and Soldering*, 1978.
Well-arranged, informative books on welding. Well-illustrated; published in numerous sections with specialized topics.
R: *National Safety News* 116: 98 (Oct. 1977); *TBRI* 44: 40 (Jan. 1978); Wal (p. 205).

METALLURGY

Engineering and Mining Journal Operating Handbook of Mineral Surface Mining and Exploration. **Engineering and Mining Journal.** New York: McGraw-Hill, 1978.

A comprehensive reference handbook of the methods, applications, tools, discoveries, and problems of surface mining. Recommended to mining engineers.

Handbook of Analytical Control of Iron and Steel Production. **T. S. Harrison.** New York: Halsted Press, 1979.

A well-documented research handbook. Contains analytical techniques, integrated iron and steel works. For reference libraries.
R: *Journal of Metals* 32: 66 (Apr. 1980).

Handbook of International Alloy Compositions and Designations. **H. Hucek and M. Wahll.** Columbus, OH: Metals and Ceramics Information Center, Batelle Columbus Laboratories, 1976–.

Volume 1, *Titanium.*
A handbook providing reference books to establish the chemical composition and similarities of the wide variety of alloy designations for metals and alloys in use throughout the modern world. Shows the relationship of alloys and standards of one country with similar alloys and standards of other countries.
R: *Journal of Metals* 30: 40 (May 1978); *RSR* 8: 28 (July/Sept. 1980).

Handbook of Metal Forming Processes. Betzalel Avitzur. New York: Wiley, 1983.

Handbook of Metal Treatments and Testing. **Robert B. Ross.** New York: Wiley, 1977.

Describes and defines some 1,000 processes and testing methods in metallurgy. Recommended for practicing engineers and industrial libraries.
R: *Endeavour* no. 2: 50 (no. 1, 1978); *Journal of Metals* 30: 40 (Mar. 1978); *Metal Finishing* 76: 124 (Jan. 1978); *Aslib Proceedings* 43: 60 (Jan. 1978); *Choice* 15: 379 (May 1978); *TBRI* 44: 118 (Mar. 1978); 44: 276 (Sept. 1978).

Handbook of Stainless Steels. **Donald Peckner and I. M. Bernstein.** New York: McGraw-Hill, 1977.

Comprehensively presents data on noncorrosive metals, including areas such as metallurgy, physical characteristics, machining, etc. Extensive references are provided, as well as over 740 illustrations. For every engineering library.
R: *Chemical Engineering* 85: 11 (May 8. 1978); *Journal of Metals* 30: 7 (Feb. 1978); *Metal Science* 12: 312 (July 1978); *RSR* 5: 36 (Oct./Dec. 1977); *TBRI* 44: 397 (Dec. 1978); *ARBA* (1978, p. 776).

Handbook of Superalloys. **M. J. Wahll, D. J. Maykuth, and H. J. Hucek, eds.** Columbus, OH: Battelle Press, 1980.

A handbook of superalloys containing information sources, a list of tables, appendixes, and alloy indexes. For professionals involved with international specifications for these alloys.

Induction Heating Handbook. **Evan John Davies and Peter Simpson.** New York: McGraw-Hill, 1979.

A comprehensive and well-written presentation of the theory and industrial applications of induction heatings. Also covers the electrical design of induction heating systems. Cites numerous examples. Assumes prior knowledge of certain electrical engineering areas. Invaluable for graduate and upper-level undergraduate students in metallurgy, as well as for metallurgical engineers, technicians, and quality control staff.
R: *Foundry Trade Journal* 147: 790 (Oct. 11, 1979); *Choice* 18: 277 (Oct. 1980); *TBRI* 46: 72 (Feb. 1980).

Metal Bulletin Handbook. 10th ed. **Ruby Packard, ed.** Surrey, England: Metal Bulletin, 1977.

This updated edition contains new statistical data, and the memoranda section has been reviewed and extended.
R: *Journal of Metals* 30: 40 (May 1978).

Metals Handbook. 9th ed. Vols. 1–. Metals Park, OH: American Society for Metals, 1978–.

First edition, 1927; eighth edition, 1961–1973. Volume 1, *Properties and Selection: Irons and Steels,* 1978; volume 7, *Powder Metallurgy,* 1984.
Covers all aspects of metals through the use of essays, graphs, tables, and illustrations.
R: *Foundry Trade Journal* 145: 1207 (Nov. 9, 1978); *NTB* 50: 97 (1965); *TBRI* 45: 111 (Mar. 1979); Jenkins (K186); *ARBA* (1985, p. 554); Sheehy (EJ61; EJ62); Wal (p. 334); Win (EI191; 2EI38; 3EI38).

Mining Chemicals Handbook. **American Cyanamid Company.** Wayne, NJ: American Cyanamid, 1976.

Contains information on chemical mining processes. Also includes information on dewatering aids, metallics ore flotation. Includes tables.
R: *Mining Magazine* 136: 505 (June 1977); *TBRI* 43: 350 (Nov. 1977).

Operating Handbook of Mineral Processing. **Richard Thomas, ed.** New York: McGraw-Hill, 1977.

Up-to-date, detailed source of data on mineral processing. Compiled from source material gathered during visits to research establishments and processing plants. Includes many articles contributed by leading authorities in the field.
R: *California Geology* 31: 168 (July 1978); *TBRI* 44: 319 (Oct. 1978).

Seawater Corrosion Handbook. **M. Schumacher, ed.** Park Ridge, NJ: Noyes Data, 1979.

A handbook that details the corrosion of metals and nonmetals in marine environments. A valuable reference for marine engineers.
R: *Journal of the South African Institute of Mining and Metallurgy* 79: 361 (July 1979); *TBRI* 45: 392 (Dec. 1979).

Steel Castings Handbook. 5th ed. **Peter F. Wieser, ed.** Rocky River, OH: Steel Founders Society of America, 1980.

Sections include designing and manufacturing castings, material selection, specifying steel castings, and buying steel castings. Four appendixes giving the composition limits and ranges of various aspects of steel castings as stated by the American Iron and Steel Institute, Society of Automotive Engineers, and the Alloy Casting Institute. Also, a glossary of foundry terms.
R: *Science and Technology Libraries* 2: 89 (Winter 1981).

Welding Handbook. 6th and 7th eds. 5 vols. **A. L. Phillips, ed.** New York: American Welding Society, 1973–1982.

First edition, 1938; fifth edition, 1962–1967; sixth edition, 1968.
Volume 1, *Fundamentals of Welding*, 7th ed., 1976; volume 2, *Welding Processes—Arc and Gas Welding and Cutting, Brazing, and Soldering*, 7th ed., 1978; volume 3, *Welding Processes—Resistance and Solid-State Welding and Other Joining Processes*, 7th ed., 1980; volume 4, *Metals and Their Weldability*, 7th ed., 1982; volume 5, *Applications of Welding*, 6th ed., 1973.
Information gleaned from company files, technical reports, and articles. Authoritative publication in 5 distinct sections.
R: *Engineer* 233: 60 (Aug. 12, 1971); 233: 57 (Oct. 21, 1971); Jenkins (K188); Wal (p. 335); Win (EI145, 1EI145).

Zinc and Its Alloys and Compounds. **S. W. K. Morgan.** New York: Wiley, 1985.

A basic reference book on the physical and chemical properties of zinc and its industrial applications, particularly in the field of corrosion protection.

NUCLEAR ENGINEERING

The Anti-Nuclear Handbook. **Stephen Croall.** New York: Pantheon Books, 1979.

A readable handbook on nuclear power. Well-illustrated; a recommended reference for public and university libraries.
R: *LJ* 104: 1125 (May 15, 1979).

Atomic Handbook. Vols. 1–. **J. W. Shortall, ed.** London: Morgan, 1965–.

Directory and statistical information on atomic energy developments throughout Europe. Addresses of 4,000 individuals and data on 18 international organizations and 1,000 firms provided.
R: Jenkins: (C73); Wal (p. 308); Win (1EI41).

CRC Handbook of Nuclear Reactors Calculations. **Yigal Ronen, ed.** Boca Raton, FL: CRC, 1986.

Volume 1, 1986; volume 2, 1986; volume 3, 1986.
A comprehensive handbook that breaks down the complex field of nuclear reactor calculations into major steps. Each step presents a detailed analysis of the problems to be solved, the parameters involved, and the elaborate computer programs developed to perform the calculations. It bridges the gap between nuclear reactor theory and the implementation of that theory. Intended primarily for use by nuclear engineers. Indexed.

Engineering Compendium on Radiation Shielding. **R. C. Jaeger et al.** New York: Springer-Verlag, 1968–.

Volume 1, *Shielding Fundmentals and Methods*; volume 2, *Shielding Materials and Design*; volume 3, *Shield Design and Engineering*, 1970.
R: *Civil Engineering and Public Works Review* (Aug. 1971).

Reactor Handbook. 2d ed. 4 vols. **US Atomic Energy Commission.** New York: Wiley-Interscience, 1960–1964.

Volume 1, *Materials;* volume 2, *Fuel Processing;* volume 3a, Nuclear Physics; volume 3b, *Shielding;* volume 4, *Nuclear Engineering.*
An encyclopedic compilation of information on reactor theory, data, equipment, and processes.
R: *NTB* 53: 33 (1968); Jenkins (K196); Win (EI211).

ENERGY

AIA Energy Notebook. Washington, DC: American Institute of Architects, 1979.

A valuable source for information on energy conservation. Stresses the economic and aesthetic use of solar energy. A bibliography and glossary are included. Updated through monthly newsletters and a quarterly supplement.
R: *AIA Journal* 69: 38 (Apr. 1980).

The Complete Energy Saving Handbook for Homeowners. **James W. Morrison.** New York: Harper & Row, 1979.

Covers various topics of concern to the homeowner; use of insulation, fireplaces, heating stoves, solar hot-water heaters, energy audits, tax credits, etc. Extensively illustrated; easily understood text.
R: *LJ* 105: 572 (Mar. 1, 1980).

The Consumer's Energy Handbook. **Peter Norback and Craig Norback.** New York: Van Nostrand Reinhold, 1981.
Part 1, "Energy Facts and Information" contains information on how a nuclear plant works, solar legislation of the states, etc. Part 2: "Energy-Saving Tips" looks at automatic thermostats, buying insulation, choosing firewood, etc. Part 3 is a directory of associations and organizations, offices, contractors, and other sources. Not all information is useful for the consumer.
R: *ARBA* (1983, p. 658).

Efficient Electricity Use: A Reference Book on Energy Management for Engineers, Architects, Planners, and Managers. 2d ed. **Craig B. Smith, ed.** Elmsford, NY: Pergamon, 1978.
A timely handbook that outlines current technology for improving the efficiency of electrical energy use. Also discusses the impact of future improvements. Contains well-done illustrations and tables. Language is technical but understandable by nontechnical persons. Of interest to engineers, architects, managers, the general public, and libraries concerned with energy.

The Energy Conservation Idea Handbook: A Compendium of Imaginative and Innovative Examples of Ideas and Practices at Colleges and Universities Today. Washington, DC: Academy for Educational Development, 1980.

Energy Deskbook. **Samuel Glasstone.** US Department of Energy, Technical Information Center, 1982.
A reference book of definitions of energy-related terms and descriptions of current and potential energy sources and their uses. Its emphasis is on general principles, not technology.
R: *Civil Engineering* 52: 27 (Dec. 1982).

Energy Handbook. 2d ed. **Robert L. Loftness.** New York: Van Nostrand Reinhold, 1984.
First edition, 1979.
A handy reference source for extensive information on energy-related subjects, ranging from energy resources to energy consumption and technology, etc. Thoroughly descriptive and referenced. Recommended for all types of libraries.
R: *ARBA* (1980, p. 650; 1985, p. 100).

Energy Management Handbook. **Wayne C. Turner et al., eds.** New York: Wiley, 1982.
A reference guide to all of the data and information needed to design and conduct a successful energy-management program. Includes areas of energy auditing, boilers and fired systems, waste heat recovery, electrical energy, etc.
R: *Heating/Piping/Air Conditioning* 54: 117 (Nov. 1982); *LJ* 107: 1360 (Aug. 1982).

Energy Managers' Handbook. **Gordon A. Payne.** Guildford, Surrey, England: IPC Science and Technology, 1977.

Designed for middle-management professionals who deal with effective ways to manage energy use and conservation. Includes checklists and tabular data, as well as illustrations.

R: *Building Services Engineer* 45: A10 (Aug. 1977); *Energy World* no. 43: 13 (Dec. 1977); *Food Processing Industry* 46: 50 (Oct. 1977); *Fuel* 56: 463 (Oct. 1977); *Mechanical Enginering* 99: 113 (Oct. 1977); *New Scientist* 77: 312 (Feb. 2, 1978); *Aslib Proceedings* 42: 445 (Aug. 1977); *TBRI* 43: 398 (Dec. 1977); 44: 36 (Jan. 1978); 44: 177 (Mar. 1978); 44: 197 (May 1978).

Energy Reference Handbook: A Glossary with Abbreviations and Conversion Tables. 2d ed. Thomas F. P. Sullivan, ed. Washington, DC: Government Institutes, 1977.

First edition, 1975.

Contains conversion tables, current references, and technical facts about energy.

R: *Plant Engineering* 32: 208 (Apr. 13, 1978); *TBRI* 44: 278 (Sept. 1978).

The Energy Saver's Handbook for Town and City People. The Scientific Staff of the Massachusetts Audubon Society. Emmaus, PA: Rodale Press, 1982.

A book dealing with saving energy costs in residences. Contains well-selected and well-presented information with hundreds of sketches to describe topics. Some data is geared for climate in Massachusetts; most information is applicable anywhere.

R: *ARBA* (1983, p. 658).

Energy Saving Handbook for Homes, Businesses and Institutions. **Edwin B. Feldman.** New York: Fell, 1979.

Comprehensive handbook on all aspects of energy use and control in both the home and workplace. Covers lighting, maintenance, equipment, etc.

R: *Heating/Piping/Air Conditioning* 51: 77 (Dec. 1979); *TBRI* 46: 72 (Feb. 1980).

Energy Technology Handbook. **Douglas M. Considine, ed.** New York: McGraw-Hill, 1977.

Contributions from over 140 energy technology experts comprise an exhaustive survey handbook. Well-written and arranged. Contains numerous charts and graphs. Considered one of the most comprehensive references available. Useful source for legislators, economists, planners, librarians, and community groups.

R: *Chemical Engineering* 85: 11 (Jan. 30, 1978); *Naval Engineers Journal* 90: 93 (Feb. 1978); *Choice* 14: 1478 (Jan. 1978); *LJ* 102: 1626 (Sept. 1977); *LJ* 103: 722 (Apr. 1, 1978); *RSR* 5: 31 (Oct./Dec. 1977); *TBRI* 44: 73 (Feb. 1978); *TBRI* 44: 193 (May 1978); *ARBA* (1978, p. 688); Sheehy (EJ52); Wal (p. 332).

Handbook of Energy Conservation for Mechanical Systems in Buildings. **Robert W. Roose, ed.** New York: Van Nostrand Reinhold, 1977.

Up-to-date guidelines for energy conservation.

R: *ASHRAE J* 20: 70 (Aug. 1978); *TBRI* 44: 358 (Nov. 1978).

Handbook of Energy Technology and Economics. **Robert A. Meyers.** New York: Wiley-Interscience, 1983.

A reference book containing all the engineering and economic data to evaluate and prepare research proposals for plausibility. Also helps users to assess the impact of technology on cities and to plan stategies. Good for engineers and research scientists.
R: *IEEE Spectrum* 20: 13 (Jan. 1983).

Handbook of Experiences in the Design and Installation of Solar Heating and Cooling Systems. Atlanta: ASHRAE, 1982.
Presents information on the experiences of designing, installing, operating, testing, and maintaining solar heating and cooling systems.
R: *Heating/Piping/Air Conditioning* 54: 148 (Mar. 1982).

Handbook of Industrial Energy Analysis. **I. Boustead and G. F. Hancock.** New York: Halsted Press/Wiley, 1979.
Aids in energy management in industrial operations. Includes introductory material on energy analysis and environmental sciences. Helpful in solving energy problems.
R: *Chemistry and Industry* 23: 850 (Dec. 1, 1979); *Chemistry in Britain* 15: 538 (Oct. 1979); *Journal of the American Chemical Society* 102: 1474 (Feb. 13, 1980); *New Scientist* 83: 46 (July 5, 1979); *TBRI* 46: 30 (Jan. 1980); 46: 112 (Mar. 1980).

Handbook of Synfuels Technology. **Robert A. Meyers.** New York: McGraw-Hill, 1984.
A useful 1-volume handbook that presents in 6 parts all the relevant topics related to synthetic fuel technologies. Much of the information is presented in charts, tables, diagrams, maps, and photos. In addition to data on US development, those in the Federal Republic of Germany, The Netherlands, Canada, and South Africa are also included. A general index is available. Useful to scientists, engineers, public policy makers, and planners.

Industrial Energy Conservation: A Handbook for Engineers and Managers. **David A. Reay.** Elmsford, NY: Pergamon, 1977.
Discusses specific energy resources and alternative sources of energy. A helpful source of information for engineers.
R: *Energy World* no. 49: 7 (June 1978); *TBRI* 44: 276 (Sept. 1978).

Solar Energy Handbook. **Jan F. Kreider and Frank Kreith.** New York: McGraw-Hill, 1981.
A practical handbook dealing with the data and procedures available for the design of solar collector systems and their economic assessments. Chapters discuss applications, principles, solar conversion, architecture, and techniques of harnessing the sun. Assumes some knowledge of classical physics. Of interest to public and technical libraries that serve graduate and upper-level undergraduate students.
R: *Choice* 18: 1444 (June 1981).

The Solar Energy Handbook. 5th rev. ed. **Henry C. Landa et al.** Milwaukee: Film Instruction Company of America, 1977.
A clear, practical reference to applications of solar energy. Outlines theory and stresses utilization. Suitable for quick reference.

Solar Energy Technology Handbook. 2 pts. **William C. Dickinson and Paul N. Cheremisinoff, eds.** New York: Dekker, 1980.

Part A, *Engineering Fundamentals*; part B, *Applications, Systems Design, and Economics*.

Synthetic Fuels Data Handbook. 2d ed. **Gary L. Baughman, ed.** Denver: Cameron Engineers, 1979.

R: *Chemical Engineering* 86: 11 (Dec. 17, 1979).

ENVIRONMENTAL SCIENCES

Agricultural Chemicals and Pesticides: A Handbook of the Toxic Effects. **E. J. Fairchild, ed.** Tunbridge Wells, Kent, England: Castle House Publications, 1978.

All toxicity data on agricultural pesticides are included in this National Institute of Occupational Safety and Health registry. A valuable reference for safe working environments.

R: *Chemistry in Britain* 15: 400 (Aug. 1979); *TBRI* 45: 346 (Nov. 1979).

CRC Coastal Processes and Erosion. **Paul D. Komar, ed.** Boca Raton, FL: CRC Press, 1983.

A timely source book on coastal-zone planning. A first scientific text on aspects of coastal erosion problems. Includes studies of both the east and west coasts of the United States, Australia, and the barrier islands.

CRC Handbook of Environmental Control. 5 vols. **Richard G. Bond and Conrad P. Straub, eds.** Cleveland: Chemical Rubber, 1973–1978.

Volume 1, *Air Pollution*, 1972; volume 2, *Solid Waste*, 1973; volume 3, *Water Supply and Treatment*, 1973; volume 4, *Waste Water Treatment and Disposal*, 1973; volume 5, *Hospital and Health Facilities*, 1975; *Series Index*, 1978.

A knowledge of chemistry is necessary for use of this specialists' tool. The large quantity of tabular data provides searching and analyses of several categories of pollutants.

R: *American Scientist* 65: 235 (Mar./Apr. 1977); *ARBA* (1974, p. 601; 1975, p. 696); Sheehy (EJ23).

CRC Handbook of Environmental Radiation. **Alfred W. Klement, Jr., ed.** Boca Raton, FL: CRC Press, 1982.

A handy handbook on the subject of environmental radiation, covering principles of estimating and measuring exposures to radioactivity, real radiological protection problems, etc.

CRC Handbook of Mass Spectra of Environmental Contaminants. **Ronald A. Hites.** Boca Raton, FL: CRC, 1985.

A collection of the electron impact mass spectra of 394 commonly encountered environmental pollutants. Indexed by common chemical names. CAS Registry Number, exact molecular weight, and intensity peaks. Intended for specialists in the field.

Clay's Handbook of Environmental Health. 14th ed. **H. M. Clay.** Edited by F. G. Davies and W. H. Bassett. London: Lewis, 1977.

A basic handbook that covers the topics of administration; construction technology; housing and urban development; health and safety; pollution control; and food safety and hygiene.
R: Wal (p. 409).

Environmental Assessment and Impact Statement Handbook. **Paul N. Cheremisinoff and A. C. Morresi.** Ann Arbor, MI: Ann Arbor Science, 1977.
Serves as a guide to environmental assessment of industrial problems. Provides thorough practical guidelines. Valuable for all engineering libraries.
R: *Journal of the New England Water Works Association* 91: 331 (Dec. 1977); *Plant Engineering* 31: 182 (Aug. 18, 1977); *RSR* 5: 33 (Oct./Dec. 1977); *TBRI* 43: 351 (Nov. 1977); 44: 193 (May 1978).

Environmental Impact Analysis Handbook. **John G. Rau and David C. Wooten, eds.** New York: McGraw-Hill, 1979.
An authoritative handbook that provides detailed coverage on the preparation of environmental impact assessments or statements. Contains example problems, illustrations, and extensive references, which enhance the usefulness of the volume. An invaluable aid to professionals, students, and educators involved in the field.
R: *Choice* 17: 406 (May 1980); *RQ* 19: 306 (Spring 1980).

Environmental Law Handbook. 7th ed. **J. Gordon Arbuckle et al., eds.** Rockville, MD: Government Institutes, 1983.
Fifth edition, 1978.
A comprehensive, up-to-date, and authoritative reference text in the environmental law field. The basic approach in each chapter is to analyze the topic by reference to the applicable federal statutes involved. Useful both to laypersons and attorneys.
R: *Journal of Environmental Health* 41: 62 (July/Aug. 1978); *ARBA* (1985, p. 513).

Environmental Management Handbook for the Hydrocarbon Processing Industries. **James D. Wall.** Houston: Gulf Publishing, 1980.
A compilation of presentations and articles on the practices and technologies for managers, designers, and environmental specialists. Contains articles on wastewater, air pollution control, odor, noise, spills, etc. Because many articles are dated before 1974, this handbook cannot be recommended as either a practical manual or as a status repot.
R: *Journal of the American Chemical Society* 104: 3783 (June 1982).

Environmental Regulation Handbook, 1980. **Steve Ross, ed.** New York: Environment Information Center, 1979.
Includes details on the federal laws and regulations concerning all environmental areas, such as air pollution, toxic substances, radioactive materials, pesticides, etc. Flow charts, a master index, and chapter headings are provided for easy reference. For those interested in environmental law.

Handbook of Advanced Wastewater Treatment. 2d ed. **Russell L. Culp et al.** New York: Van Nostrand Reinhold, 1977.
Describes fundamental processes of wastewater management. New technologies are discussed. Information provided is detailed.

R: *Chemical Engineering* 85: 12 (Dec. 4, 1978); *Plant Engineering* 32: 282 (Oct. 12, 1978); *TBRI* 44: 392 (Dec. 1978); 45: 33 (Jan. 1979).

Handbook of Environmental Control. 6 vols. **Richard G. Bond and Conrad P. Straub, eds.** Cleveland: Chemical Rubber, 1973–1978.

Volume 1, *Air Pollution*, 1973; volume 2, *Solid Waste*, 1973; volume 3, *Water Supply and Treatment*, 1973; volume 4, *Wastewater Treatment and Disposal*, 1974; volume 5, *Hospital and Health-Care Facilities*, 1975; volume 6, *Series Index, 1978*.
A 6-volume handbook series that examines different areas of environmental control. The *Series Index* is organized by subject and chemical substance. Assumes a background in chemistry and is intended mainly for specialists.

Handbook of Environmental Data and Ecological Parameters. **S. E. Jorgensen, ed.** Elmsford, NY: Pergamon, 1979.
A massive compilation of thousands of tables and diagrams relating to biology, ecology, and the environment. Intended as a counterpart to the *CRC Handbook of Chemistry and Physics*. Contains basic facts and short 1-line summaries with keywords and references. A useful reference for graduate students, faculty and professionals, government agencies, and consulting firms.
R: *Choice* 17: 242 (Apr. 1980).

Handbook of Environmental Data on Organic Chemicals. **Karel Verschueren.** New York: Van Nostrand Reinhold, 1978.
Includes 1,000 organic sustances alphabetically arranged. Provides data on their biological effects in physical and chemical terms. Recommended to those who work with volatile organic chemicals. Also includes definitions of terms and sampling and measurement techniques.
R: *Chemical Engineering* 85: 16 (Oct. 9, 1978); *Journal of the American Oil Chemists' Society* 55: 510A (July 1978); *Choice* 15: 1032 (Oct. 1978); *RSR* 6: 28 (Oct./Dec. 1978); *TBRI* 44: 279 (Sept. 1978); *ARBA* (1979, p. 655).

Handbook of Environmental Engineering. 2 vols. **Lawrence K. Wang and Norman C. Pereira, eds.** Clifton, NJ: Humana Press, 1979–1980.
Volume 1, *Air and Noise Pollution Control*, 1979; volume 2, *Solid Waste Processing and Resource Recovery*, 1980.
R: *Journal of the American Chemical Society* 102: 3668 (May 7, 1980).

Handbook of Environmental Health and Safety: Principles and Practices. **Herman Koren.** Elmsford, NY: Pergamon, 1980.
An encyclopedic reference treating environmental and ecological problems, including food technology, insect and rodent control, solid waste management, and air and water pollution. Contains numerous bibliographic references. Helpful to educators and administrators.
R: *American Scientist* 68: 700 (Nov./Dec. 1980); *Choice* 18: 426 (Nov. 1980).

Handbook of Industrial Waste Disposal. **Richard A. Conway and Richard D. Ross.** New York: Van Nostrand Reinhold, 1980.
A useful guide on the problems of industrial waste disposal. Covers wastewater equalization; activated carbon adsorption; incineration of solid waste; disposal of hazardous

materials; and more. Presents technology needed to meet environmental legislation. For research and special libraries.
R: *Chemical Engineering* 88: 137 (Aug. 24, 1981).

Handbook of Industrial Wastes Pretreatment. **Jon C. Dyer, Arnold S. Vernick, and Howard D. Feiler.** New York: Garland STPM Press, 1981.
A detailed reference on the pretreatment of industrial wastes. Covers regulations, implementation of pretreatment programs, as well as other considerations. Praiseworthy bibliography and illustrations. Recommended to practicing engineers, graduate students, and regulatory agencies.
R: *Choice* 18: 1442 (June 1981).

Handbook of Municipal Administration and Engineering. **W. Foster.** New York: McGraw-Hill, 1978.
A handy, how-to reference on municipal administration problems. Discusses sewage control, wastewater flow monitoring, asphalt repair, etc. A time-saving handbook for city managers and engineers as well as directors of public works.

Handbook of Noise Assessment. **Daryl N. May, ed.** New York: Van Nostrand Reinhold, 1978.
Describes psychological and physiological effects of noise. Presents data of acceptable noise levels. Well-organized, up-to-date reference.
R: *Audiology* 18: 263 (May/June 1979); *Journal of the Acoustical Society of America* 65: 554 (Feb. 1979); *Science News* 114: 82 (Aug. 1978); *TBRI* 45: 195 (May 1979); 45: 278 (Sept. 1979).

Handbook of Noise Control. 2d ed. **Cyril M. Harris, ed.** New York: McGraw-Hill, 1979.
Well-written and referenced handbook on noise control. Comprises an excellent compendium of up-to-date information.
R: *Chemical Engineering* 86: 11 (Apr. 9, 1979); *Journal of the Acoustical Society of America* 66: 611 (Aug. 1979); *National Safety News* 120: 101 (July 1979); *TBRI* 45: 235 (June 1979); 45: 313 (Oct. 1979); 45: 350 (Nov. 1979).

Handbook of Solid Waste Management. **David G. Wilson.** New York: Van Nostrand Reinhold, 1978.
A quick reference for all those involved in waste management.
R: *American Scientist* 66: 508 (July/Aug. 1978); *Chemical Engineering* 85: 11 (June 5, 1978); *Plant Engineering* 32: 132, 134 (July 6, 1978); *TBRI* 44: 280 (Sept. 1978).

Handbook of Stack Sampling and Analysis. **Richard Powals et al.** Westport, CT: Technomic, 1978.
Describes collection systems and analytical methods for air pollution listing. Contains glossary and bibliography.
R: *Plant Engineering* 33: 436 (Apr. 19, 1979); *TBRI* 45: 238 (June 1979).

Handbook of Wastewater Collection and Treatment. **Muhammad Anis H. Al-Layla, Shamin Ahmad, and E. Joe Middlebrooks.** New York: Garland STPM Press, 1980.

Discusses various aspects of wastewater treatment and collection. Topics covered include short-circuiting, plug flow, adsorption, etc. For those involved in the field.
R: *Choice* 17: 697 (July/Aug. 1980).

Hazardous Materials Spills Handbook. **Gary F. Bennett.** New York: McGraw-Hill, 1982.
Provides a vast range of background information and practical techniques on the problems of and the solutions to spills of hazardous chemicals.
R: *Civil Engineering* 53: 99 (Jan. 1983); *ARBA* (1983, p. 660).

Industrial Noise Control Handbook. **Paul N. Cheremisinoff and P. P. Cheremisinoff.** Ann Arbor, MI: Ann Arbor Science, 1977.
A comprehensive source on noise control in manufacturing and industrial settings. Provides information valuable in setting up a safe and productive work environment.
R: *Engineers' Digest* 39: 69 (Sept. 1978); *National Safety News* 118: 133 (Aug. 1978); *Plant Engineering* 31: 172 (June 23, 1977); *TBRI* 43: 273 (Sept. 1977); 44: 312 (Oct. 1978); 45: 32 (Jan. 1979).

Industrial Source Sampling. **David L. Brenchley et al.** Ann Arbor, MI: Ann Arbor Science, 1979.
Focuses on air pollution tests prescribed by the Environmental Protection Agency. Contains useful information on source sampling. For industrial managers and engineers.
R: *Plant Engineering* 33: 202 (Mar. 8, 1979); *TBRI* 45: 191 (May 1979).

Industrial Wastewater Management Handbook. **Hardam S. Azad, ed.** New York: McGraw-Hill, 1976.
In 2 sections: (1) fundamentals of wastewater management such as legislation, standards, etc.; (2) data and techniques of wastewater management. Deals with pollution control. Contains a wealth of information, tables, and illustrations, Highly recommended. For management personnel and engineers involved with environmental conservation.
R: *Chemical Engineering* 84: 11 (Apr. 11, 1977); *RSR* 5: 33 (Oct./Dec. 1977); *Journal of the American Chemical Society* 100: 6800 (Oct. 11, 1978); *TBRI* 43: 227 (June 1977); *ARBA* (1977, p. 767).

The NALCO Water Treatment Handbook. **NALCO Chemical Company.** Edited by Frank N. Kemmer. New York: McGraw-Hill, 1979.
A reference handbook arranged in 4 sections that describe the needs and uses of water. Sections cover the nature of water; unit operations and treatment; industrial and municipal use of water; and special technology in water treatment. Provides practical information in easy-to-understand terms. Of interest to undergraduate science libraries; also suitable for plant managers, project engineers, municipal supervisors, and layperson.
R: *Chemical Engineering* 87: 12 (July 14, 1980); *Public Works* 110: 32 (Dec. 1979); *Choice* 17: 106 (Mar. 1980); *TBRI* 46: 44 (Feb. 1980).

Noise Control: Handbook of Principles and Practices. **David M. Lipscomb and A. C. Taylor, Jr., eds.** New York: Van Nostrand Reinhold, 1978.

Helpful in solving community noise problems. Discusses aircraft noise; highway, industrial, and construction noise; and noise legislation.
R: *National Safety News* 118: 116 (Nov. 1978); *Sound and Vibration* 12: 11 (Nov. 1978); *TBRI* 45: 37 (Jan. 1979); 45: 117 (Mar. 1979); 45: 194 (May 1979).

Solid Waste Handbook—A Practical Guide. **W. Robinson.** New York: Wiley-Interscience, 1986.

POLLUTION

Air Pollution Control and Design Handbook. **Paul N. Cheremisinoff and R. A. Young, eds.** New York: Dekker, 1977.
For engineering and monitoring air pollution control. A helpful up-to-date reference.
R: *Bulletin of the American Meteorological Society* 58: 1219 (Nov. 1977); *Chemical Engineering* 85: 11 (Feb. 27, 1978); *TBRI* 44: 73 (Feb. 1978); 44: 152 (Apr. 1978).

Environment Regulation Handbook: Air Pollution, Land Use, Mobile Sources, NEPA, Noise, Pesticides, Radioactive Materials, Solid Wastes, Water Pollution.
New York: Environmental Information Center, 1973–.
Eclectic information on environmental topics in looseleaf format. Includes comprehensive regulatory sections and directory information. Duplicates some material in *Environmental Reporter* and the *Environmental Law Reporter*.
R: *ARBA* (1975, p. 697).

Examination of Water for Pollution Control: A Reference Handbook in Three Volumes. **Michael J. Suess, ed.** New York: Pergamon Press, 1982).
An encyclopedia of information on water pollutants. Reference procedures, alternates and critical evaluations are included for all substances. Represents efforts of 250 scientists to develop a unified system applicable to fresh-water and wastewater.
R: *Choice* 20: 290 (October 1982).

Handbook of Air Pollution Analysis. **Roger Perry and Robert J. Young, eds.** New York: Halsted Press, 1978.
A comprehensive source of information on atmospheric pollution; compiled by experts in the field.
R: *American Scientist* 67: 236 (Mar./Apr. 1979); *Analyst* 103: 671 (June 1978); *Chemistry and Industry* 13: 471 (July 1, 1978); *New Scientist* 78: 842 (June 22, 1978); *TBRI* 44: 317 (Oct. 1978); 44: 275 (Sept. 1978).

Handbook of Air Pollution Technology. **Seymour Calvert and Harold M. Englund, eds.** New York: Wiley-Interscience, 1984.
An up-to-date and comprehensive guide for engineers designing cost-effective air pollution systems.
R: *Chemical and Engineering News* 62: 39 (Feb. 1984).

Handbook of Pollution Control Management. **Herbert F. Lund.** Englewood Cliffs, NJ: Prentice-Hall, 1978.
This guidebook to alternate systems of pollution control includes 70 examples of multimillion-dollar savings in actual applications. Topics covered include forecasting costs,

recycling waste products, training equipment operators, government financial assistance, and more.
R: *Journal of Metals* 30: 43 (Oct. 1978).

Handbook of Water Resources and Pollution Control. **Harry W. Gehm and Jacob I. Bregman, eds.** New York: Van Nostrand Reinhold, 1976.

Discusses all major innovations in water pollution technology. Also presents data on nearly 200 public water systems in the United States. Contributions from a wide range of experts. Recommended for water engineers.
R: *Plant Engineering* 30: 232 (Sept. 16, 1976); *Public Works* 107: 144 (Sept. 1976); *LJ* 101: 1843 (Sept. 15, 1976); *WLB* 51: 444 (Jan. 1977); *TBRI* 43: 34 (Jan. 1977); *ARBA* (1977, p. 689).

Industrial Air Pollution Handbook. **Albert Parker, ed.** New York: McGraw-Hill, 1978.

Provides comprehensive coverage of common air pollution problems, such as pollutants in gaseous discharge, sulphur oxides, grit, and dust. Presents information on methods of surveying and reducing air pollution by giving solutions for specific industries. Also discusses air pollution's effects on plants, animals, and health. Contains a list of references. Useful to government inspectors, plant managers, and engineering consultants.
R: *Chemistry and Industry* 13: 471 (July 1, 1978); *Chemical Engineering* 85: 12 (Nov. 20, 1978); *Chemistry in Britain* 15: 208 (Apr. 1979); *Plant Engineering* 33: 442 (Apr. 19, 1979); *TBRI* 45: 238 (June 1979).

International Operations Handbook for Measurement of Background Atmospheric Pollution. **World Meterological Organization.** New York: World Meterological Organization, 1978.

Describes air pollution monitoring, including measurement of carbon dioxide, sulphur, carbon monoxide, nitrogen oxides, etc. Well-referenced.
R: *IBID* 7: 29 (Spring 1979).

Waste Recycling and Pollution Control Handbook. **A. V. Bridgewater and C. J. Mumford.** New York: Van Nostrand Reinhold, 1981.

A handbook on the management of liquid, solid, gaseous, and noise wastes. It is concerned with successful recovery techniques, recycling economics, and legislative practices in the United Kingdom. Includes a comprehensive bibliography. Useful for engineers involved in environmental control.
R: *Chemical Engineering* 88: 115 (Apr. 6, 1981).

CHAPTER 7 TABLES, ALMANACS, DATABOOKS, STATISTICAL SOURCES

TABLES

GENERAL SCIENCE

International Critical Tables of Numerical Data, Physics, Chemistry and Technology. 7 vols. **US National Research Council.** New York: McGraw-Hill, 1926–1930.

"Critical" in the sense that each contributing specialist was requested to give in each case the "best" value that could be derived from information available. Vast compilation of quantitative information.
R: Jenkins (A83); Win (EA139).

Statistical Tables for the Social, Biological, and Physical Sciences. **F. C. Powell.** New York: Cambridge University Press, 1982.

A collection of tables taken from books, journals, and other sources. Includes an introduction, tests, mathematical derivations, and examples. Recommended for professionals and for graduate students familiar with introductory work.
R: *Choice* 20: 562 (Dec. 1982).

Tables of Physical and Chemical Constants and Some Mathematical Functions. 14 rev. ed. **George William Clarkson Kaye and T. H. Laby.** New York: Longman, 1973.

A completely revised and updated edition of physical and chemical constants. Values are in SI units. Includes footnote references to the literature. Valuable to all scientists and engineers.

SI Units

Basic Electrical and Electronic Tests and Measurements. **Derek Cameron.** Englewood Cliffs, NJ: Prentice-Hall, 1978.

Provides an introduction to electrical tests; contains standardized measurements. Well-illustrated and written.

Introduction to the International System of Units with Conversion Tables and Normal Ranges. **Herbert Lippert and H. P. Lehrman.** Baltimore: Urban & Schwarzenberg, 1978.

Comprises an introduction to the legislation covering the International System of Units. A reference for scientists and physicians. Contains bibliography and index.
R: *Journal of the Indiana State Medical Association* 71: 713 (July 1978); *TBRI* 44: 248 (Sept. 1978).

The Metric System and Metric Conversion: A Checklist of References. **M. Michaelson, ed.** Mankato, MN: Minnesota Scholarly Press, 1978.

A detailed list of references concerning the metric system and conversion. Entries are grouped under 8 main headings: business, automotive engineering, construction and architecture, general, education, foreign countries, SI units, and agriculture. For public libraries.
R: *Choice* 16: 650 (July/Aug. 1979); *ARBA* (1980, p. 600).

SI: The International System of Units. 3d ed. **Chester Hall Page and P. Vigoureux, ed.** London: Her Majesty's Stationery Office, 1977.

An official translation of the French International Bureau of Weights and Measurements' publication. Well-organized, convenient, 1-volume reference.
R: *Journal of Physics E: Scientific Instruments* 10: 1303 (Dec. 1977); *TBRI* 44: 129 (Apr. 1978).

Steam and Air Tables in SI Units. **Thomas F. Irvine, Jr. and J. P. Hartnett, eds.** Washington, DC: Hemisphere Publishing, 1976.

A compilation of useful data.
R: *Chemical Engineering Science* 31: 857 (1976); *TBRI* 43: 35 (Jan. 1977).

Technical Data on Fuel. 7th ed. **J. W. Rose and J. R. Cooper, eds.** New York: Halsted Press, 1978.

Revised edition is in SI units. Contains excellent tables and graphs. Considered a valuable reference for engineers in the fuel industry.
R: *Electronics and Power* 24: 62 (Jan. 1978); *Fuel* 56: 463 (Oct. 1977); *TBRI* 44: 37 (Jan. 1978); 44: 158 (Apr. 1978); 45: 77 (Feb. 1979).

Units of Weight and Measure, International (Metric), and US Customary. **J. Chisholm.** Detroit, MI: Gale Research, 1975.

Conversion tables are simplified for laypersons, with some tables providing instant conversion answers.

The Wiley Metric Guide. **P. J. O'Neill.** New York: Wiley, 1976.

Contains compact data, containing both imperial equivalents and price conversions. Also provides a list of units named after famous people and an industry conversion plan.

ASTRONOMY

A Master List of Nonstellar Optical Astronomical Objects. **Robert S. Dixon and George Sonneborn, comps.** Columbus, OH: Ohio State University Press, 1980.

A listing of nonstellar objects based on 270 catalogs. Refers astronomers to the original catalogs and decreases chances of pseudodiscoveries. A standard reference for professional astronomers.
R: *Choice* 18: 813 (Feb. 1981).

Mathematics

Basic Mathematics. **Joseph P. Lerro.** Boston: Cahners Books, 1976.
Contains a useful compilation of formulas, tables, and definitions in algebra; trigonometry; and plane, solid, and analytic geometry. Extensive use of charts and monograms. Writing is succinct; includes bibliography. Useful in explaining basic mathematical concepts.
R: *ARBA* (1977, p. 634).

CRC Handbook of Tables for Mathematics. 4th rev. ed. **Samuel M. Selby, ed.** Boca Raton, FL: Chemical Rubber, 1975.
Fourth edition, 1970.
Originally published as a supplement to the *Handbook of Chemistry and Physics*. A comprehensive reference for all science students and professionals. In addition to the tables, brief explanations of formulae and notes on their use, etc. are also given.
R: *Mathematics of Computation* 19: 680 (1965); *Quarterly Journal of Experimental Physiology* 61: 75 (Jan. 1976); *TBRI* 42: 170 (May 1976); Jenkins (B54).

CRC Standard Mathematical Tables. 27th ed. **William H. Beyer, ed.** Boca Raton, FL: CRC Press, 1984.
First edition, 1964; twenty-third edition, 1975; twenty-fourth edition, 1976.
Math tables selected to aid scientific, engineering, industrial, and educational personnel. Includes numerical tables, logarithmic tables, trigonometry tables, binomial functions, differential equations, etc.
R: *Mathematical Gazette* 55: 470 (Dec. 1971); Jenkins (B52); Wal (p. 73).

Handbook of Mathematical Functions with Formulas, Graphs, and Mathematical Tables. **US National Bureau of Standards.** New York: Wiley-Interscience, 1984.
Incorporates numerical tables of mathematical functions. For those without easy access to computers, the tables serve as preliminary surveys of problems before the actual programming is attempted on the computer.

Handbook of Mathematical Tables and Formulas. 5th ed. **Richard S. Burington.** New York: McGraw-Hill, 1973.
First edition, 1933; fourth edition, 1965.
A collection of frequently used tables intended to meet the needs of students and professionals in mathematics, engineering, physics, chemistry, etc. Part 1 includes main formulas and theorems of algebra, geometry, trigonometry, calculus, etc. Part 2 includes tables of logarithms, trigonometry, powers and roots, probability distributions, etc.
R: *Mathematics of Computation* 19: 503 (1965); 27: 998 (Oct. 1973); *TBRI* 40: 84 (Feb. 1974); *WLB* 47: 613 (Mar. 1973); *ARBA* (1974, p. 552); Jenkins (B51); Mal (1976, p. 69); Wal (p. 72); Win (EF42).

An Index of Mathematical Tables. 2d ed. 2 vols. **Alan Fletcher.** Reading, MA: Addison-Wesley, 1962.
First edition, 1946.

An index to practical tables in mathematics dating from the sixteenth century to 1961. Includes within its 4 sections a historical introduction, bibliography of sources of tables, and errors in published tables.
R: *Nature* 201: 537 (Feb. 8, 1964); *Science* 137: 332 (1962); Jenkins (B46); Wal (p. 74); Win (EF32).

Mathematical Tables. Vols. 1–. **UK National Physical Laboratory.** London: Her Majesty's Stationery Office, 1956–.

Mathematical Tables Project. 40 vols. **Work Projects Administration of the Federal Works Agency.** Conducted under the sponsorship of the National Bureau of Standards. New York: Columbia University Press, 1939–1944.
R: Win (EF46).

Mathematics of Computation. Vols. 1–. Providence: American Mathematical Society, 1943–. Quarterly.
Volumes 1–13, *Mathematical Tables and Other Aids to Computation.*
The "Reviews and Descriptions of Tables and Books" section also lists unpublished mathematical tables.
R: *Product Engineering* 28: 194 (July 1957); *Scientific Information Notes* 10: 23 (Apr./May 1968); Jenkins (B49); Mal (1976, p. 70); Wal (p. 73).

The Penguin Book of Tables. Rev. and short. ed. New York: Penguin, 1974.
Earlier edition, 1968.
An inexpensive student reference. Includes fundamental mathematical tables and useful data such as metric equivalents and *systeme internationale* (SI) symbols and units.
R: *ARBA* (1975, p. 644).

Physical and Mathematical Tables. **T. M. Yarwood and F. Castle.** New York: Macmillan, 1970.
R: *Mathematical Gazette* 55: 461 (Dec. 1971).

Royal Society Mathematical Tables. Vols. 1-11-. **Royal Society of London.** Cambridge, England: University Press, 1950–.
Volume 9, 1968; volume 10, 1964; volume 11, 1964.
Each volume on a specific subject as, for example, *Coulomb wave functions*, volume 11. A continuation of *Mathematical Tables* (British Association for the Advancement of Science, 1931–1952).
R: Jenkins (B57); Wal (p. 73); Win (EF48; 1EF6; 3EF6).

A Table of Series and Products. **Eldon R. Hansen.** Englewood Cliffs, NJ: Prentice-Hall, 1975.
A listing of series that can be used to express a mathematical function as a sum of more elementary functions. Also gives the closed form for a given series. For mathematicians.

Tables of Integrals, Series, and Products. **Iosif Moiseevich Ryzhik and I. S. Gradshteyn, trans.** New York: Academic Press, 1980.

Russian edition, Moscow, 1962; entitled *Tablitsy Intergralov, Summ, Riadov i Proizvedenii.*
R: Win (1EF7).

Tables of Physical and Chemical Constants and Some Mathematical Functions. 14th rev. ed. **G. W. C. Kaye and T. H. Laby.** Edited by A. E. Bailey. New York: Longman, 1973.

Thirteenth edition, 1966.
Updated information in the fields of general physics, chemistry, and mathematical functions. Footnote references throughout. All tabulated values are in SI units.
R: *NTB* 52: 72 (1967); Jenkins (A80); Wal (pp. 98, 123); Win (EG38).

Tables of the F–E Related Distribution Algorithms. **K. Mardia and Zemroch.** New York: Academic Press, 1979.

US National Bureau of Standards. Applied Math Tables. Vols. 1–. Washington, DC: US Government Printing Office, 1948–.
Presently over 60 volumes. An authoritative series.

Statistics

Biometrika Tables for Statisticians Rep. ed. 2 vols. **Egon Sharpe Pearson and H. O. Hartley, eds.** London: Biometrika Trust, 1976.

Reprinted with corrections.
Two volumes of revised and expanded tables, both common and more specialized.
R: Sheehy (EF11).

CRC Handbook of Tables for Probability and Statistics. 2d ed. **William H. Beyer.** Chemical Rubber, 1968.

First Edition, 1966.
A collection of statistical and mathematical tables used in the fields of probability and statistics. Brief explanations of each table.
R: *Science* 154: 1316 (1966); *ABL* 34: entry 161 (Apr. 1969); Jenkins (B82); Win (1EF5).

Outliers in Statistical Data. 2d ed. Wiley Series in Probability and Mathematical Statistics. **V. Barnett and T. Lewis.** New York: Wiley, 1985.

An updated edition of a standard reference work on outliers. Relevant illustration and tabulation as well as suggestions for further research are provided.

Selected Tables in Mathematical Statistics. Vols. 1–. **Harman Leon Harter and Donald B. Owen.** Chicago: Markham, 1970–.

Volume 5, 1977.
A collection of statistics assembled by the Institute of Mathematical Statistics. All tables are selected from papers submitted to the Institute's committee on mathematical tables.
R: *Choice* 7: 1493 (Jan. 1971); *ARBA* (1972, p. 552); Win (3EF5).

Selected Tables in Mathematical Statistics. The Distribution of the Size of the Maximum Cluster of Points on a Line. Vol. 6. **Norman D. Neff and Joseph I. Naus.** Providence: American Mathematical Society, 1980.

Deals with the probabilities of large clusters under various models. Helpful to quality-control engineers, mathematicians, and ecologists.
R: *ARBA* (1981, p. 626).

Statistical Tables. 2d ed. **F. James Rohlf and Robert R. Sokal.** San Francisco: W. H. Freeman, 1981.
First edition, 1969.

Statistical Tables for Biological, Agricultural and Medical Research. 6th ed. **Ronald Aylmer Fisher and Frank Yates.** New York: Hafner, 1978.
Comprehensive collection of tables and bibliography on sources on statistical method.
R: *Choice* 6: 994 (Oct. 1969); Jenkins (C62); Wal (pp. 76, 195); Win (EC20).

Statistics Tables: For Mathematicians, Engineers, Economists and the Behavioral and Management Sciences. **H. R. Neave.** Edison, NJ: Allen & Unwin, 1978.
An up-to-date collection of statistical tables; helpful to a broad range of students.
R: *Aslib Proceedings* 43: 255 (June 1978).

PHYSICS

Master Handbook of Electronic Tables and Formulas. 3d ed. **Martin Clifford.** Blue Ridge Summit, PA: TAB Books, 1980.
R: *RSR* 9: 101 (Jan./Mar. 1981).

Master Tables for Electromagnetic Depth Sounding Interpretation. **Rajni K. Verna.** New York: Plenum Press, 1980.
Contains tables that assist in interpreting electromagnetic depth sounding. Provides extensive and accurate data on a wide range of parameters.
R: *RQ* 20: 105 (Fall 1980).

MIT Wavelength Tables. Wavelengths by Element. Vol. 2. **Frederick M. Phelps III, prep.** Cambridge, MA: MIT Press, 1982.
This is a table containing an account of 109,325 atomic emission lines. Each is arranged by element, including wavelengths in vacuum and corresponding wavenumbers.
R: *Laser Focus* 19: 99.

Neutron Activation Tables. **Gerhard Erdtmann.** Weinheim, Germany: Verlag Chemie, 1976.
A valuable collection of tables. Useful as an analytical tool for the production of radioisotopes and the activation of reactors.
R: *Journal of the American Chemical Society* 100: 7788 (Nov. 22, 1978); *Nuclear Science and Engineering* 63: 514 (Aug. 1977); *TBRI* 43: 353 (Nov. 1977).

Noise Evaluation Tables in Digitalized Form. **J. P. Seller, ed.** Hemel Hempstead, England: Seller, 1977.
Brings together acoustical reference quantities, criteria, and other factors in numerical form. Contains tables, computer printouts, and photographs.
R: *Journal of Sound and Vibration* 54: 626 (Oct. 22, 1977); *TBRI* 44: 78 (Feb. 1978).

Nuclear Tables. 7 vols. **W. Kunz and J. Schintlmeister.** Oxford: Pergamon Press, 1958–1968.

A compilation of critical data, including tables of nuclear reaction data with citations to sources of further information.
R: *Physics Today* 21: 137 (Jan. 1968).

The RAE Table of Earth Satellites, 1957–1980. **D. G. King-Hele et al., comps.** New York: Facts on File, 1981.

Includes a chronological list of the 2,145 satellites and space vehicles launched since 1957's Sputnik 1, through 1980's Cosmos 1236. For each satellite and rocket, the name, international designation, launch date, lifetime, mass, shape, size, and orbital parameters are given.
R: *Sky and Telescope* 63: 268 (Mar. 1982).

Smithsonian Physical Tables. 9th ed. **William E. Forsythe.** Washington, DC: Smithsonian Institution, 1964.

Contains 900 tables of general interest to the scientist and engineer interested in physics in a broad sense.

Spectroscopic Data. 2 vols. S. N. Suchard and J. E. Melzer, eds. New York: Plenum Press, 1976.

Volume 2, *Homonuclear Diatomic Molecules.*
An alphabetically arranged table of electronic spectroscopic data.
R: *Journal of the American Chemical Society* 99: 311 (Jan. 5, 1977); *TBRI* 43: 89 (Mar. 1977).

Sunrise and Sunset Tables for Key Cities and Weather Stations of the US: A Complete Collection of the United States Naval Observatory's Comprehensive Tables Providing the Hour and Minute of Sunrise and Sunset for Every Day of the Year for Each of 369 Key Locations in the United States, and Having Validity for the Entire Twentieth Century. **U.S. Nautical Almanac Office.** Washington, DC: US Government Printing Office. Repr. Detroit: Gale Research, 1977.

Easy-to-use tables of sunrise/sunset times, valuable any year in the 20th century for 369 cities in the United States. Provides data in standard time.
R: *BL* 75: 502 (Nov. 1978); *WLB* 52: 264 (Nov. 1977); *ARBA* (1978, p. 692).

Table of Isotopes. 7th ed. **Charles Michael Lederer and Virginia Shirley, eds.** New York: Wiley-Interscience, 1978.

This new edition presents data in one large table, preceded by an isotope index, arranged according to mass number. The table includes mass-chain decay schemes, nuclear level schemes, thermal neutron cross sections, and references. Appendixes give constants and conversions factors, nuclear spectroscopy standards, atomic levels, and more. For libraries beyond the secondary school level.
R: *Journal of the American Chemical Society* 102: 892 (Jan. 16, 1980); *Laboratory Practice* 28: 631 (June 1979); *Choice* 4: 1224 (Jan. 1968); *RSR* 8: 31 (July/Sept. 1980); Sheehy (EG10); Win (2EG7).

Table of Laser Lines in Gases and Vapors. 2d ed. **Rasmus Beck, W. Englisch, and K. Gurs.** New York: Springer Publishing, 1978.

First edition, 1976.
A catalog of 5,193 laser lines. Consists of a variety of tables that comprise a handy source of data. A clearly produced reference for chemists.
R: *Applied Optics* 17: 3854 (Dec. 15, 1978); *Atmospheric Environment* 12: 1811 (1978); *Optics and Laser Technology* 9: 143 (June 1977); *RSR* 6: 44 (Apr./June 1978); 8: 23 (July/Sept. 1980); *TBRI* 43: 241 (Sept. 1977); 44: 361 (Dec. 1978); 45: 82 (Mar. 1979).

Table of Molecular Weights: A Companion Volume to the Merck Index. 9th ed. **Martha Windholz et al., eds.** Rahway, NJ: Merck, 1978.

Contains a list of high-resolution molecular weights arranged in ascending order. Provides full chemical information. To be used in conjunction with *The Merck Index*.
R: Sheehy (ED13).

Tables of Standard Electrode Potentials. **Guilio Milazzo, S. Caroli, and V. K. Sharma.** New York: Wiley, 1978.

A series of reference tables of standard reduction potentials. Contains an extensive compilation of factual and numerical data. Useful to university and research libraries.
R: *Journal of Molecular Structure* 54: 307 (July 1979); *Journal of the American Chemical Society* 100: 7787 (Nov. 22, 1978); *Journal of the Electrochemical Society* 125: 261C (June 1978); *Laboratory Practice* 27: 973 (Nov. 1978); *TBRI* 44: 250 (Sept. 1978); 45: 47 (Feb. 1979); 45: 324 (Nov. 1979).

Tables of Wavenumbers for the Calibration of Infrared Spectrometers. 2d ed. **Andrew R. H. Cole, ed.** Elmsford, NY: Pergamon, 1977.

A greatly revised compilation of data. Should prove valuable for spectroscopists who work in infrared laboratories.
R: *Applied Optics* 16: 2598 (Oct. 1977); *Applied Spectroscopy* 32: 117 (Jan./Feb. 1978); *Journal of the American Chemical Society* 100: 4640 (May 24, 1978); *Journal of the Optical Society of America* 68: 144 (Jan. 1978); *RSR* 6: 43 (Apr./June 1978); *TBRI* 43: 363 (Dec. 1977); 44: 123 (Apr. 1978); 44: 204 (June 1978).

Tables on the Thermophysical Properties of Liquids and Gases: In Normal and Dissociated States. 2d ed. **N. B. Vargaftik.** New York: Halsted Press, 1975.

Tabulation of Infrared Spectral Data. **David Dolphin and Alexander E. Wick.** New York: Wiley, 1977.

A notable compilation of data based on model compounds. Tables illustrate characteristic frequencies of organic groups. Data is presented in tabular form.
R: *Applied Spectroscopy* 32: 333 (May/June 1978); *Journal of the American Chemical Society* 100: 1329 (Feb. 15, 1978); *RSR* 6: 44 (Apr./June 1978); *TBRI* 44: 124 (Apr. 1978); 44: 283 (Oct. 1978); *ARBA* (1979, p. 654).

CHEMISTRY

Basic Tables in Chemistry. **R. A. Keller, ed.** New York: McGraw-Hill, 1967.

A fundamental student handbook.
R: *Choice* 5: 34 (Mar. 1968); Wal (p. 123); Win (2ED8).

CRC Handbook of Tables for Organic Compound Identification. 3d ed. **Zvi Rappoport.** Cleveland: Chemical Rubber, 1967.

First edition, 1960; second edition, 1964.
Originally entitled *Tables for Identification of Organic Compounds.*
Aids chemists in the identification of organic compounds. More than 8,100 parent compounds arranged by functional groups.
R: Wal (p. 135); Win (2ED18).

Chemical Tables. **Bela A. Nemeth.** Translated by Istvan Finlay. New York: Wiley, 1976.

A handy reference book of alphabetically arranged tables.
R: *ARBA* (1977, p. 646).

Chemical Tables for Laboratory and Industry. **W. Helbing and A. Burkart.** New York: Halsted Press, 1980.

Translation of *Chemie-Tabellen für Labor und Betrieb.*
Contains tables handy in research work, including 1,300 organic and inorganic chemical types. Also includes a glossary of 5,000 keywords. In 3 sections: matter, number, and process.

Correlation Tables for the Structural Determination of Organic Compounds. **M. Pestemer.** New York: Verlag Chemie, 1975.

Provides a systematic compilation of spectral data of nearly 2,300 organic compounds. Helpful as an identification tool for chemists, particularly when used with other documents.

Geochemical Tables. **H. J. Rosler and H. Lange.** New York: American Elsevier, 1973.

Information tables in chemistry and math that have a bearing on earth science. The short introductions and definitions of special terms that accompany the tables make this a suitable tool for librarians and students.

JANAF Thermochemical Tables. 2d ed. **Dow Chemical Company, Thermal Research Laboratory.** Washington, DC: National Bureau of Standards, 1971.
R: Sheehy (ED10).

Selected Values of Chemical Thermodynamic Properties: Tables for Elements 35–55 in Standard Order of Arrangement. **US National Bureau of Standards.** Washington, DC: US Government Printing Office, 1969.

Selected Values of Chemical Thermodynamic Properties: Tables for Elements 54–61 in the Standard Order of Arrangement. **US National Bureau of Standards.** Washington, DC: US Government Printing Office, 1971.

BIOLOGICAL SCIENCES

Catalogue of Type Invertebrate, Plant, and Trace Fossils in the Royal Ontario Museum. **Janet Waddington et al.** Toronto: Royal Ontario Museum, 1978.

Lists 2,000 fossil invertebrates, plants, and trace fossils. Alphabetically arranged by genus.
R: *Journal of Paleontology* 53: 1262 (Sept. 1979): *TBRI* 45: 327 (Nov. 1979).

Cell Biology. **Philip L. Altman and Dorothy D. Katz, eds.** Bethesda, MD: Federation of American Societies for Experimental Biology, 1976.

Contains 102 detailed tables of data on cell biology. Includes appendix of animal and plant names.
R: Sheehy (EC8).

EARTH SCIENCES

Climates of the States: National Oceanic and Atmospheric Administration Narrative Summaries, Tables, and Maps for Each State. 2d ed. 2 vols. **James A. Ruffner, ed.** Detroit: Gale Research, 1980.

First edition, 1960.
Volume 1, *Climates of the States (Tables, Graphs, Maps, and Bibliographies);* volume 2, *Federal and State Public Services in Climate and Weather.*
Provides coverage of all aspects of the climate in the United States and Puerto Rico. Tables, graphs, maps, and figures complement the text. A valuable reference work for travelers, farmers, transportation officials, and construction engineers.
R: *ARBA* (1981, p. 691).

Climates of the States, with Current Tables of Normals 1941–1970 and Extremes to 1975. 2 vols. **Officials of the National Oceanic and Atmospheric Administration.** Detroit: Gale Research, 1978.

In 2 volumes: a handy compilation of data originally issued by the US Weather Bureau from 1941–1970. Relies on charts, maps, tabulations. Provides physical descriptions, seasonal temperatures, precipitation, snowfall, wind, etc.
R: *BL* 76: 852 (Feb. 15, 1980); *WLB* 53: 525 (Mar. 1979).

Discharge of Selected Rivers of the World. Mean Monthly and Extreme Discharges (1972–1975): A Contribution to the International Hydrological Programme. Vol. 3, pt. 3. Paris: Unesco. Distr. New York: Unipub, 1979.

A collection of tables showing means and extremes of the flows of the major rivers of the world. Rivers are organized successively by continent, country, reporting stations, river, and location. Certain sections are in English, French, Russian, and Spanish. Primarily for academic and special libraries.
R: *ARBA* (1981, p. 694).

Handbook of Oceanographic Tables. **US Naval Oceanographic Office.** Washington, DC: US Government Printing Office, 1966–.

International Oceanographic Tables. 2 vols. New York: Unipub, 1966–1974.
Volume 1, 1966; volume 2, 1974.

Mineral Tables: Hand-Specimen Properties of 1,500 Minerals. **Richard V. Dietrich.** New York: McGraw-Hill, 1969.

A computer-produced list of properties, culled mostly from Dana's *System of Mineralogy*. Tables are divided by minerals with metallic and nonmetallic luster and further subdivided according to color and hardness. Includes a table listing minerals according to chemical composition.

R: *ARBA* (1970, p. 128).

Smithsonian Meteorological Tables. Smithsonian Miscellaneous Collections. 114. 6th rev. ed. Washington, DC: Smithsonian Institution, 1951.

Intended for the meteorologist, this handbook includes the information necessary for compiling meteorological reports. Various conversion tables.

R: *Nature* 170. 175 (1952); Jenkins (F108).

General Engineering

CRC Handbook of Tables for Applied Engineering Science. 2d ed. **Ray E. Bolz and George L. Tuve, eds.** Cleveland: Chemical Rubber, 1973.

First edition, 1970.

Since earlier edition, tables have been added covering such subjects as lasers, radiation, cryogenics, etc. Well-documented tables cover a wide variety of disciplines.

R: *Engineering* p. 217 (May 1971); *ARBA* (1972, p. 636; 1975, p. 773); Wal (p. 288).

Engineering Tables and Data. **A. M. Howatson, P. G. Lund, and J. D. Todd.** London: Chapman & Hall. Distr. New York: Halsted Press, 1972.

Tables of interest to students arranged in 6 major sections. Lack of index hinders easy access.

R: *ARBA* (1974, p. 664).

Subject Engineering

Chemical Equilibria in Carbon-Hydrogen-Oxygen Systems. **Robert E. Baron, James H. Porter, and Ogden H. Hammond, Jr.** Cambridge, MA: MIT Press, 1976.

Tables useful to those interested in coal combustion or gasification. A thorough digest of thermodynamic data. Well-arranged for easy retrieval of information.

R: *RSR* 5: 12 (Apr./June 1977).

Design Tables for Beams on Elastic Foundations and Related Structural Problems. **K. T. S. R. Iyengar and S. A. Ramu.** Barking, Essex, England: Applied Science Publishers, 1979.

A book for structural engineers. Provides data on elastically supported beams. Consists of 2 parts: theoretical background and guide to the use of tables.

R: *Aslib Proceedings* 45: 30 (Jan. 1980).

Energy Information Handbook. **US House of Representatives, Committee on Interstate and Foreign Commerce, Subcommittee on Energy and Power.** Washington, DC: US Government Printing Office, 1977.

Energy data published by the federal government; includes 769 charts and tables that deal with various aspects of energy, beginning with 1960. Includes forecasts through the year 2000.
R: *RSR* 6: 28 (Oct./Dec. 1978).

Gas Tables: Thermodynamic Properties of Air Products of Combustion and Component Gases Compressible Flow Functions, Including Those of Ascher H. Shapiro and Gilbert M. Edelman. 2d ed. **Joseph Henry Keenan, Jing Chao, and Joseph Kaye.** New York: Wiley, 1980.
First edition, entitled: *Thermodynamic Properties of Air*, 1948.
A completely updated expanded edition containing 62 tables of thermodynamic properties of gases and air. Covers combustion products of hydrocarbon fuels, compressible flow tables for Rayleigh and Fanno-lines, isentropic flow, etc. Valuable aid for graduate and upper-level undergraduate engineering students and professionals in the field.
R: *Journal of the American Chemical Society* 103: 2911 (May 20, 1981); *Choice* 18: 426 (Nov. 1980); *ARBA* (1981, p. 640).

Handbook of Electronic Tables and Formulas. 5th ed. **Stanley Meacham and Donald Herrington.** Indianapolis: Sams 1979.
Fourth edition, 1973.
Formulas, laws, constants, symbols, codes and service, and installation data.

Mechanical Engineer's Reference Tables. **Z. Elizanowski.** New York: Chemical, 1966.
Basic collection of tables, values, and symbols. Includes 67 tables dealing with details of mechanical engineering.

Mineral Tables. **R. Dietrich.** New York: McGraw-Hill, 1969.
Provides broad tabular coverage of 1,500 minerals. Arranged by cationic elemental compounds. Useful to students, geologists, mining engineers, and mineral collectors.

A New Table of Indefinite Integrals, Computer Processed. **Melvin Klerer and Fred Grossman.** New York: Dover, 1971.
Over 2,000 indefinite integrals.

Plant Engineer's Handbook of Formulas, Charts, and Tables. 2d ed. **Donald W. Moffat.** Englewood Cliffs, NJ: Prentice-Hall, 1982.
First edition, 1974.
A single-volume source combining the important formulas and tables needed by the practicing plant engineer. Data is divided according to type of materials, such as concrete, blocks, wood, steel, aluminum, etc.
R: *ARBA* (1975, p. 805).

The RAE Table of Earth Satellites 1957–1982. 2d ed. **D. G. King-Hele et al., comps.** New York: Wiley, 1983.

Chronologically lists data on the 2,389 worldwide launches of satellites and space vehicles between 1957 and 1982. A useful library tool for those interested in the international space program.
R: *Science Technology Libraries* 5: 2 (Winter 1984).

Solar Age Catalog: A Guide to Solar Energy Knowledge and Materials from the Editors of "Solar Age". Port Jervis, NY: SolarVision, 1977.
For the general public. Provides data tables on a range of components, products, services, engineers, etc. involved with solar energy. Also contains a glossary of terms, and a bibliography.
R: *BL* 74: 1328 (Apr. 15, 1978); *ARBA* (1979, p. 699).

ALMANACS, DATABOOKS

GENERAL SCIENCE

NSF Factbook. US National Science Foundation. Orange, NJ: Academic Media, 1976. Annual.

Scientific Quotations: The Harvest of a Quiet Eye. **Alan L. Mackay.** Edited by Maurice Ebison. New York: Crane, Russak, 1977.
A fine collection of quotations by famous scientists and philosophers, from Aristotle through Neil Armstrong. Illustrates the influence of science on society, for a wide audience.
R: *ARBA* (1979, p. 643).

ASTRONOMY

Air Almanac. **US Nautical Almanac Office.** Washington, DC: US Government Printing Office, 1941–.
Compendium of astronomical data required for air navigation. Includes navigational star chart and sky diagrams as well as necessary charts and tables.
R: Jenkins (E48).

American Ephemeris and Nautical Almanac. **US Nautical Almanac Office.** Washington, DC: US Government Printing Office, 1855–. Annual.
Latest edition, 1977.
Standard navigational source giving celestial positions from major reference points. The almanac covers the sun, moon, universal and sidereal time, eclipses, ephemerides of major planets, mean places of 1,078 stars, etc.
R: Jenkins (E49).

The Astronomical Almanac for the Year 1981. Washington, DC: US Government Printing Office, 1980.
A revised reference book that includes many new topics, including expanded lists of stars; the introduction of radio and x-ray sources, quasars, and pulsars; orbital elements of 138 minor planets; and more attention to time, calendar, and coordinate systems.
R: *Sky & Telescope* 60: 518 (Dec. 1980).

Astronomical Ephemeris. London: Her Majesty's Stationery Office, 1767–. Annual.
See *American Ephemeris and Nautical Almanac.*
R: Wal (p. 84); Win (EB29).

The Astronomy Data Book. 2d ed. **J. Hedley Robinson and James Muirden.** New York: Wiley, 1979.
First edition, 1972.
Handy reference tool containing a concise assortment of significant astronomical data. Includes numerous tables, star charts, and a limited glossary. Complements more complete reference tools. For the student or amateur astronomer.
R: *Nature* 282: 660 (Dec. 6, 1979); *LJ* 104: 1241 (June 1, 1979); 105: 569 (Mar. 1, 1980); *ARBA* (1980, p. 612).

The Compilation, Critical Evaluation, and Distribution of Stellar Data. **Carlos Jaschek and G. A. Wilkins, ed.** Boston: D. Reidel, 1977.
Discusses many different fields of astronomy. Emphasizes standardization of stellar identification.

Ephemeris of the Sun, Polaris, and Other Selected Stars, with Companion Data and Tables. **US Bureau of Land Management.** Washington, DC: US Government Printing Office, 1910–.

The Guinness Book of Astronomy: Facts and Feats. 2d ed. **Patrick Moore.** Enfield, England: Guinness Superlatives. Distr. New York: Sterling, 1983.
First edition, 1979.
An improved and expanded edition of the famous first facts and tables of data on our universe. Aranged by subject for easy reference to students or librarians. An excellent source.
R: *ARBA* (1984, p. 632).

Landolf-Bornstein Numerical Data and Functional Relationships in Science and Technology: New Series. 7th ed. Astronomy and Astrophysics. Grp. 6, vol. 1. Berlin: Springer-Verlag, 1965.
Similar information to the *American Ephemeris,* but also includes chapters on optical instruments, interstellar space, radio-astronomical devices, telescopes, galaxies, etc. English and German languages.

The Nautical Almanac. **US Nautical Almanac Office.** Produced in cooperation with the Royal Greenwich Observatory. Washington, DC: US Government Printing Office, 1909–.
This almanac provides the data necessary for successful astronomical navigation at sea.

Sky Catalogue 2000.0. Stars to Magnitude 8.0. Vol. 1. **Alan Hirshfeld and Roger W. Sinnott, eds.** New York: Cambridge University Press, 1982.

Using the magnetic tapes of the US National Aeronautics and Space Administration Goddard SKYMAP project, the editors list all 45,269 stars brighter than visual magnitude 8.05. Data includes proper motion, magnitude, and distance.

MATHEMATICS

International Compendium on Numerical Data Projects: A Survey and Analysis. **International Council of Scientific Unions, Committee on Data for Science and Technology.** New York: Springer-Verlag, 1969.

Sourcebook on data centers; lists availability of literature and data.

R: *UB L* 24: entry 190 (May/June 1970); Wal (p. 48).

PHYSICS

Astrophysical Formulae: A Compendium for the Physicist and Astrophysicist. **K. R. Lang.** New York: Springer-Verlag, 1980.

Earlier edition, 1974.

A comprehensive reference of over 2,100 fundamental formulas of astrophysics. Includes tables of basic physical data on astronomical objects.

R: Sheehy (EG9).

Atomic Energy-Level and Grotian Diagrams. **Stanley Bashkin and John O. Stoner, Jr.** Amsterdam: North-Holland, 1978.

Volume 1, *Hydrogen I-Phosphorous XV, Addenda*; volume 2, *Sulfur I-Titanium XXII*. An updated databook in 2 volumes. Includes higher atomic numbers than previously given.

R: *RSR* 8: 23 (July/Sept. 1980).

Atomic Energy Levels; Data for Parametric Calculations. **Serafin Fraga, K. M. S. Saxena, and Jacek Korwowski.** Amsterdam: Elsevier, 1979.

Information on parameters for various electronic configurations of atomic species with 5 to 102 electrons.

R: *RSR* 8: 27 (July/Sept. 1980).

CINDA 81: An Index to the Literature on Microscopic Neutron Data. Vienna: International Atomic Energy Agency. Distr. New York: Unipub, 1981. Annual or biennial.

CINDA 72 superseded all earlier issues. Worldwide cooperation has produced this index of over 100,000 bibliographic extracts culled from journals, proceedings, and unpublished material. Entries arranged by element and mass number, then by cross section of other quantities.

R: *ARBA* (1972, p. 562; 1974, p. 560); Wal (p. 112).

Color Science: Concepts and Methods, Quantitative Data and Formulas. 2d ed. **Gunter Wyszecki and W. S. Stiles.** New York: Wiley, 1982.

First edition, 1967.

R: *Physics Today* 21: 83 (July 1968); 35: 59 (Apr. 1982); Jenkins (C70); Wal (p. 107); Win (2EG8).

Datensammlungen in der Physik: Data Compilations in Physics. 3 vols. Karlsruhe, W Germany: Zentralstelle für Atomkernenergie-Dokumentation, 1976–1978.

An index to about 2,400 data compilations in physics. Arranged under topical heading in German and English; indexed in English.
R: Sheehy (EG11).

For Good Measure: A Complete Compendium of International Weights Measures. **William D. Johnstone.** New York: Holt, Rinehart and Winston, 1976.

A handy reference book consisting of 8 chapters on different units of measurement found in various published sources. Information can be found through use of the index to unit names and a general index.
R: *LJ* 100: 2233–2235 (Dec. 1, 1975); *ARBA* (1976, p. 634).

Handbook of Atomic Data. **Serafin Fraga et al.** New York: Elsevier Scientific, 1976.

Includes tables of data on atoms and ions.
R: *Theoretica Chimica Acta* 45: 73 (1977); *TBRI* 43: 323 (Nov. 1977).

Handbook of Nuclear Data for Neutron Activation Analysis. **A. I. Aliev et al.** Translated by Baruch Benny. Jerusalem: Program for Scientific Translation, 1970; New York: Halsted Press, 1970.

Handbook on Nuclear Activation Cross-Sections: Neutron, Proton, and Charged Particle Nuclear Reaction Cross-Section Data. New York: Unipub, 1974.

Heat & Mass Transfer Data Book. 3d ed. **C. P. Kothandaraman and S. Subramanyan.** New York: Wiley, 1977.

A handy collection of essential formulas. Helpful to students and engineers involved in heat transfer.

Mossbauer Effect Data Index: Covering the 1976 Literature. **J. G. Stevens and V. E. Stevens, eds.** New York: Plenum Press, 1978. Annual, vols. 1–, 1971–.

Original edition entitled *MEDI 1958–1965*.
Offers bibliographic control of the ever-increasing output of Mossbauer literature and data.
R: *Physics Today* 29: 77 (May 1976); *ARBA* (1973, p. 547).

Phase Diagrams: A Literature Sourcebook. **J. Wisniak.** Amsterdam: Elsevier Scientific Publishing, 1981.

This alphabetic listing of elements presented in published phase diagrams allows an engineer or scientist to see if a phase diagram for a certain binary multicomponent mixture has been published.
R: *Journal of the American Chemist Society* 104: 2950 (May 1982).

Reference Data for Acoustic Noise Control. **W. L. Ghering, ed.** Ann Arbor: Ann Arbor Science, 1978.

A ready reference concerning acoustic noise control problems. Covers such topics as noise description, sound level meter weightings, noise standards, noise transmission loss, statistical energy analysis, etc. Includes over 80 figures, tables, graphs, and line

drawings. Contains an appendix of tables for combining decibels, references to literature, a bibliography, and a praiseworthy subject index. An excellent aid for acoustic engineers.
R: *RSR* 8: 25 (Jan./Mar. 1980); *ARBA* (1980, p. 619).

Thermophysical Properties of Matter. 13 vols. **Y. S. Touloukian and C. Y. Ho, eds.** New York: Plenum, 1970–1977.

Volumes 1–3, *Thermal Conductivity*; volumes 4–6, *Specific Heat*; volumes 7–9, *Thermal Radiative Properties*; volume 10, *Thermal Diffusivity*; volume 11, *Viscosity*; volumes 12–13, *Thermal Expansion*.
This comprehensive tool constitutes the revision and expansion of the *TPRC Data Book* originally released in 1960. Revision of each volume is expected every 5 years. Every volume contains a text, numerical data with source references, and an index.
R: *ABL* 36: entry 378 (Sept. 1971); *RSR* 5: 12 (Apr./June 1977); *ARBA* (1973, p. 548).

Thermophysical Properties Research Literature Retrieval Guide 1900–1980. 7 vols. **J. R. Chaney and V. Ramdas, eds.** New York: IFI/Plenum, 1982.

Volume 1, *Elements*; volume 2, *Inorganic Compounds*; volume 3, *Organic Compounds and Polymeric Materials;* volume 4, *Alloys, Intermetallic Compounds, and Ceramics;* volume 5, *Oxide Mixtures and Minerals;* volume 6, *Mixtures and Solutions*; volume 7, *Coatings, Systems, Composites, Foods, Animal and Vegetable Products.*
Earlier edition edited by Y. S. Touloukian, 1964–1972.
Most comprehensive source of information available for interested engineers, polymer scientists, metallurgists, and solid-state physicists. Each of the 7 volumes treats a specific class of material and contains a materials directory, search parameters, bibliography, and author index.
R: *ABL* 33: entry 440 (Sept. 1968); *ARBA* (1974, p. 563); Wal (p. 108).

Thermophysical Properties Research Literature Retrieval Guide. Supp. 2. 6 vols. **J. Koolhaas, V. Ramdas, and T. M. Putnam, eds.** New York: Plenum Press, 1979.

Comprehensive coverage of the world literature from 1971 to 1977. Cites some 15,300 references on 14 thermophysical properties of 12,700 materials. Cross-indexed by an additional 8,000 synonyms and trade names.

CHEMISTRY

The Aldrich Library of Infrared Spectra. 2d ed. **Charles J. Pouchert.** Milwaukee: Aldrich Chemical, 1976.

First edition, 1970.
An updated edition that presents 10,000 spectra of organic functional groups. Contains illustration and description.
R: *RSR* 5: 11 (Apr./June 1977); Sheehy (ED19).

Alphabetical List of Compound Names Formulae, and References to Published Infrared Spectra: An Index to 92,000 Published Infrared Spectra. **American Society for Testing and Materials.** Philadelphia: American Society for Testing and Materials, 1969.
R: Wal (p. 105).

The Chemist's Companion: A Handbook of Practical Data, Techniques, and References. **Arnold J. Gordon and Richard A. Ford.** New York: Wiley, 1973.

A ready-reference tool for the research scientist, classroom teacher, and working chemist. Includes information on the properties of atoms and molecules, spectroscopy, photochemistry, chromatography, experimental techniques, etc. Valuable directory of suppliers and commercial sources of chemical apparatus.
R: *Choice* 10: 1163 (Oct. 1973); *Journal of Chemical Education* 5: A50 (Jan. 1974); *TBRI* 40: 46 (Feb. 1974); *ARBA* (1974, p. 557); Mal (1976, p. 120).

DECHEMA Chemistry Data Series. **J. Gmiehling, U. Onken, and W. Arlt, eds.** Frankfurt, W Germany: DECHEMA Deutsche Gesellschraft für Chemisches Apparatswesen, 1977–.

Volume 1, *Vapor-Liquid Equilibrium Data Collection,* part 1, *Aqueous-Organic Systems;* part 2, *Organic Hydroxyl Compounds*; part 3/4, *Aldehydes, Ketones, Ethers*; part 5, *Esters and Carboxylic Acids*; part 6, *Aliphatic Hydrocarbons*; part 7, *Aromatic Hydrocarbons*; part 8, *Halogen, Nitrogen, Sulfur, and Other Compounds.*
A series providing vapor-liquid equilibrium data. Includes binary and ternary systems. For libraries supporting components of the organic chemistry industry.
R: *RSR* 8: 27 (July/Sept. 1980).

Fiesers' Reagents for Organic Synthesis. Vol. 9. **M. Fieser, R. L. Danheiser, and W. Roush.** New York: Wiley, 1981.

Volume 9 of a monumental series that provides data on hundreds of reagents used in organic synthesis. Indispensable aid for organic chemists.

Handbook of Proton Ionization Heats and Related Thermodynamic Quantities. Brigham Young University Center for Thermochemical Studies, contrib. no. 85. **James J. Christensen et al.** New York: Wiley-Interscience, 1976.

Hazardous Chemicals Data Book. Park Ridge, NJ: Noyes Data, 1980.

Identification of Molecular Spectra. 4th ed. **Reginald W. B. Pearse and A. G. Gaydon.** London: Chapman & Hall, 1976.

A standard reference that contains summary information of molecular spectra, mainly in the region 2,000 to 10,000.
R: *Journal of the Chemical Society: Faraday Transactions II* 73: 1151 (1977); *TBRI* 43: 370 (Dec. 1977).

Index of Vibrational Spectra of Inorganic and Organometallic Compounds (1964–1966). Vol. 3. **N. N. Greenwood and E. J. F. Ross.** Woburn, MA; Butterworth, 1977.

Lists compounds alphabetically by chemical formula, providing the following data: physical state, spectrum, wave number, document number, and article citations from which the data were taken. A major reference tool for spectroscopists.
R: *ARBA* (1979, p. 655).

International Data Series B; Thermodynamic Properties of Aqueous Organic Systems. London, England: Engineering Sciences Data Unit, 1979–.

A newly published series focusing on thermodynamic and other properties of dilute aqueous solutions of organic compounds.
R: *RSR* 8: 29 (July/Sept. 1980).

Organic Electronic Spectral Data, 1977. Vol. 19. **John P. Phillips et al., eds.** New York: Wiley, 1983.
Volume 16, 1980.
An abstract from some 151 journals of data published in 1977. Approximately 20,000 entries are arranged according to the molecular formula index system used by *Chemical Abstracts*.
R: *Science Technology Libraries* 5: 2 (Winter 1984).

Pesticide Book. **George W. Ware.** San Francisco: W. H. Freeman, 1978.
A reference on the chemical structure and formulation, as well as the hazards and legalities of pesticides. Contains a glossary and index.
R: *Farm Chemicals* 142: 122 (June 1979).

Physical Properties: A Guide to the Physical, Thermodynamic, and Transport Property Data of Industrially Important Chemical Compounds. **Carl L. Yaws, ed.** New York: McGraw-Hill, 1979.
A book of graphs covering substances from certain elements and simple inorganic compounds to simple hydrocarbons. Graphs show plots versus termperature for such properties as heat of vaporization, vapor pressure, surface tension, etc. For chemists and engineers in the chemical process industries.
R: *Journal of the American Chemical Society* 102: 892 (Jan. 16, 1980); *RQ* 19: 96 (Fall 1979).

Powder Diffraction File: Inorganic. Rev. sets. Philadelphia: Joint Committee on Powder Diffraction Standards, 1960–.

Powder Diffraction File: Organic. Rev. sets. Philadelphia: Joint Committee on Powder Diffraction Standards, 1960–.

Reagents for Organic Synthesis. 9 vols. **Louis Frederick Fieser and Mary Fieser, eds.** New York: Wiley, 1967–1981.
Alphabetical listing of reagents, with structural formulas, molecular weight, physical constants, preferred methods of preparation, and suppliers.
R: *Choice* 4: 966 (Nov. 1967); *Journal of the American Chemical Society* 100: 1328 (Feb. 15, 1978); 102: 1475 (Feb. 13, 1980); Sheehy (ED17); Win (2ED13).

Thermochemical Data of Organic Compounds. **J. B. Pedley et al.** New York: Methuen, 1985.

Thermochemical Properties of Inorganic Substances. **I. Barin and O. Knacke.** New York: Springer-Verlag, 1973.

Thermochemical Properties of Inorganic Substances. Supp. **Ihsan O. K. Barin and O. Kubaschewski.** Berlin: Springer-Verlag, 1977.
Supplemental data on 800 substances. An extremely valuable reference.
R: *RSR* 7: 15 (Apr./June 1979).

Biological Sciences

Biological Laboratory Data. 2d ed. **Leslie J. Hale.** New York: Wiley, 1965.

Covers only the most basic data and supplements with reference to sources containing more specialized information. Includes data from math, physics, and chemistry.

Biology Data Book. 2d ed. 3 vols. **Philip L. Altman and Dorothy S. Dittmer, comps. and eds.** Bethesda, MD: Federation of American Societies for Experimental Biology, 1972–1974.

Presents basic, established data in the biological and medical sciences. Each volume covers different aspects and subject areas of biology, and is independently indexed. Tables include contributor's name and a list of references. Intended as a comprehensive laboratory reference.
R: *Nature* 206: 971 (1965); *ARBA* (1974, p. 566); Jenkins (G49); Wal (p. 194); Win (EC19).

Bowes and Church's Food Values of Portions Commonly Used. 13th rev. ed. **Jean A. T. Pennington and Helen Nichols Church.** New York: Harper & Row, 1980.

First edition, 1937.
Provides a statistical survey of food values of single-portion servings. Food is organized under categories, and data is given in household units and grams. Lists calories, protein, vitamins, carbohydrates, fat, amino acids, iron, etc. Includes charts on recommended daily dietary allowances by age and sex and tables of caffeine and cholesterol content. Also contains a section on fast-food chains, bibliography, and index. Represents a wide assortment of foods. Invaluable aid to nutritionists, dieticians, and the general public.
R: *ARBA* (1981, p. 740).

A Complete Checklist of the Birds of the World. **Richard Howard and Alick Moore.** New York: Oxford University Press, 1980.

Covers birds of the world down to the subspecies level. Includes geographical distribution and common name. Cross-referenced and indexed. Useful to professional ornithologists as well as amateur birdwatchers. For science libraries.
R: *Choice* 18: 1072 (Apr. 1981); *Nature* 289: 829 (Feb. 26, 1981).

Dangerous Plants. **John Tampion.** New York: Universe Books, 1977.

An identification book for the general botanist; discusses toxic plants.
R: *Science News* 111: 319 (May 1977).

Dictionary of Tropical American Crops and Their Diseases. **Federick L. Wellman.** Metuchen, NJ: Scarecrow Press, 1977.

This highly recommended reference lists alphabetically crops endemic to the Americas. It discusses diseases and their causes and provides an index to scientific and common names.
R: *ARBA* (1978, p. 738).

Factual Data Banks in Agriculture. New York: Unipub, 1978.

Examines agriculture-related data banks in Europe, such as food science, dairy, animal disease, oil science, forestry, genetics, etc. Also outlines users' needs.

Flowering Plants of the World. **V. H. Heywood, ed.** Oxford and London: Oxford University, 1978.

Contains a description of over 300 of the major flowering plant families in the world. An excellent and interesting book for biologists and others interested in plants.
R: *Nature* 278: 795 (April 26, 1979).

1980 Handbook of Agricultural Charts. Agriculture Handbook no. 574. Washington, DC: US Department of Agriculture, 1980.

A compilation of charts that detail agricultural trends and data. Surveys farms, population and rural development, world production, nutrition programs, and trade and commodity trends. For agricultural libraries.
R: *BL* 77: 801 (Feb. 15, 1981).

Nutritional Quality Index of Foods. **R. Gaurth Hansen, Bonita W. Wyse, and Ann W. Sorenson.** Westport, CT: AVI Publishing, 1979.

A full description of nutritional qualities of food and standardized, recommended dietary allowances are provided in this index of precise data.

Nutritive Value of American Foods: In Common Units. **Catherine F. Adams.** Washington, DC: US Government Printing Office, 1975.

A list of caloric and nutritive value of approximately 1,500 foods. Food composition and weight-volume relationships are analyzed. Useful for those involved in planning diets.
R: *ARBA* (1977, p. 743).

Orders and Families of Recent Mammals of the World. **S. Anderson and J. K. Jones.** New York: Wiley-Interscience, 1984.

A quick reference containing information on diagnostic features, general characteristics, habits, habitats, distribution, and genera of the orders and families living and recently extinct mammals of the world.

World Wheat Statistics, 1979. New York: Unipub, 1979.

A quadralingual (English/French/Russian/Spanish) compilation of data on various aspects of wheat production. Covers imports and exports, supplies, prices, usage, and freight rates. Also provides statistics for durum wheat. For agricultural libraries.
R: *IBID* 7: 327 (Winter 1979).

EARTH SCIENCES

Catalogue of Meteorological Data for Research. Geneva: World Meteorological Organization. Distr. New York: Unipub, 1965–1980.

Part 1, 1965; part 2, 1970; part 3, 1972; part 4, 1980.
A specialized reference for meteorological research centers.
R: *ARBA* (1973, p. 591).

Climates of the States. **James A. Ruffner.** Detroit: Gale Research, 1978.

Consists of a reprint of a series originally published by the US Government, as well as additional new material. Contains maps, statistics on temperature, humidity, wind, etc. A highly recommended reference.
R: *Choice* 12: 1759 (Feb. 1975); *ARBA* (1976, p. 706; 1979, p. 701); Sheehy (EE14).

Climatological Data. **US National Oceanic and Atmospheric Administration.** Washington, DC: US Government Printing Office, 1959–.

Summarizes the reports of US weather stations for each state.

The Coastal Almanac: For 1980—The Year of the Coast. **Paul L. Ringold and John Clark.** San Francisco: W. H. Freeman, 1980.

A collection of data concerning the uses and characteristics of the US coastal zone. Covers a wide variety of topics and includes illustrations. Contains an index. Of interest to all libraries.
R: *Choice* 18: 1077 (Apr. 1981).

Engineering Properties of Soils and Rocks. 2d ed. **F. G. Bell.** London: Butterworth, 1983.

This useful reference tool gives properties of individual soil and rock types. Also includes information on ways they vary according to their geological setting and the influence of sorting, packing, and grain shape. Recommended for use by undergraduate engineering students.
R: *Science Technology Libraries* 5: 2 (Winter 1984).

Farwell's Rules of the Nautical Road. 5th ed. **Frank E. Bassett and Richard A. Smith, preps.** Annapolis: Naval Institute Press, 1977.

First edition, 1941; second revised edition, 1954; third edition, 1959; fourth edition, 1968.
Prepared by the US Navy, takes note of all technological data. An expanded and updated edition. Contains historical, technical, and legal facts.
R: *ARBA* (1978, p. 791).

The Great International Disaster Book. 3d ed. **James Cornell.** New York: Scribner's, 1982.

An updated record of the most recently occurring disasters, with information about the causes and effects of each.
R: *ARBA* (1984, p. 682).

Ocean Data Resources. **US Congress, Senate, Committee on Commerce.** Washington, DC: US Government Printing Office, 1975.

The Ocean Almanac. **Robert Hendrickson.** Garden City, NY: Doubleday, 1984.

A compendium on sea creatures, nautical lore and legend, master mariners, naval disasters, etc. Extensive indexing.
R: *LJ* 109: 75 (Jan. 1984); *ARBA* (1985, p. 610).

Statistical Analysis of Geological Data. **J. S. Koch and R. F. Link.** New York: Wiley, 1981.

Earlier edition, 1970.
R: *American Association of Petroleum Geologists Bulletin* (Aug. 1971); *Economic Geology* (Aug. 1971); *Journal of Geology* (July 12, 1971).

Surface Water Data. 8 vols. Canada: Inland Waters Branch, 1971–1973. Methodical coverage of the provinces and territories.

The Weather Almanac. 3d ed. **James A. Ruffner and Frank E. Bair, eds.** Detroit: Gale Research, 1981.

First edition, 1974; second edition, 1977.
Comprises an extensive collection of data on weather and climate through the United States. Contains numerous charts, tables, maps, and diagrams. Revised edition contains many newer facts: temperature-humidity index, livestock safety index, heating and cooling degree data, glossary, and index.
R: *ARBA* (1982, p. 1517).

Weather of the United States Cities. 2 vols. **James A. Ruffner and Frank E. Bair, eds.** Detroit: Gale Research, 1980.

R: *BL* 77: 350 (Oct. 15, 1980).

World Catalogue of Very Large Floods: A Contribution to the International Hydrological Programme. Paris: Unesco Press, 1976. Distr. New York: Unipub, 1977.

Provides representative data on floods occurring in a variety of condition. Aids in water management and flood control. Data is given for flood basins in 35 countries. Recommended for libraries with water management interests.
R: *ARBA* (1978, p. 693).

GENERAL ENGINEERING

Engineering Data Book. 9th ed. Tulsa: Natural Gas Processors Supplies Association, 1972.

Engineering Formulas. 3d ed. **Kurt Gieck.** New York: McGraw–Hill, 1979.

First edition, 1971; second edition, 1977.
A thoroughly revised edition that contains more than 1,800 formulas and 400 diagrams. Covers topics in electrical engineering, optics, geometry, dynamics, hydraulics, heat, and chemistry. Contains new material on Fourier series, and permutations and combinations. Volume uses SI units, although conversion information is provided. Includes an index and identification of symbols for variables. A handy, pocket-sized book recommended for academic and public libraries; for use by engineers, scientists, technicians, and students.
R: *RQ* 19: 402 (Summer 1980); *RSR* 5: 23 (Oct./Dec. 1977); *ARBA* (1978, p. 761; 1981, p. 754).

Chemical Engineering

Chemical Engineering Drawing Symbols. **D. G. Austin.** New York: Wiley, 1979.

Contains over 1,150 drawing symbols, both British and American. Comprises a comprehensive reference; helpful to those who interpret engineering diagrams. Includes numerous examples, symbols, applications.
R: *Journal of the American Chemical Society* 102: 1474 (Feb. 13, 1980); *RSR* 8: 26 (July/Sept. 1980).

Comparative Properties of Plastics. 4 vols. San Diego: International Plastics Selector, 1978–.

Guides the engineer to the appropriate selection of plastics, providing specification data on thousands of different types. Includes information on density, strength, flex, etc. A worthwhile reference.
R: *RSR* 6: 26 (Oct./Dec. 1978).

Dangerous Properties of Industrial Materials. 5th ed. **Newton Irving Sax.** New York: Van Nostrand Reinhold, 1979.

Updated edition containing data on 15,000 common industrial and laboratory materials. Provides human and experimental toxicological data as well as incompatibilities and warnings against other hazards. Includes detailed glossary and bibliographic references. For anyone interested in industrial and laboratory safety.
R: *TBRI* 46: 118 (Mar. 1980); *WLB* 54: 140 (Oct. 1979); *ARBA* (1980, p. 713); Wal (p. 188).

Fabric Almanac. 2d ed. **M. Klapper.** New York: Fairchild, 1971.

Contains an introduction to textiles in the 1970s. Includes key industry statistics as well as fabric glossary, principal US man-made fibers and their uses, natural fibers and their sources, etc.
R: Jenkins (J132).

A Formulary of Paints and Other Coatings. Vol. 1. **Michael Ash and Irene Ash, comps.** New York: Chemical Publishing, 1978.

Deals with paints and finishes, providing specific lists of ingredients and formulas. Well-arranged for easy access to information; considered an important addition to the chemical literature. For all special, engineering, and research libraries.
R: *TBRI* 45: 148 (Apr. 1980); *ARBA* (1979, p. 654).

Civil Engineering

Civil Engineer's Reference Book. 3d ed. **Leslie S. Blake, ed.** London: Butterworth. Distr. Levittown, NY: Transatlantic Arts, 1975.

Second edition, entitled *Civil Engineering Reference Book*, 1969.
A completely revised reference. Includes data on civil engineering site management. Contains extensive bibliographies.
R: Sheehy (EJ12).

ELECTRONICS ENGINEERING

Almanac for Computers, 1982. **US Nautical Almanac Office.** Washington, DC: US Naval Observatory, 1981.

Contains chebyshev expansions for use with small computers and calculators. Also includes data on solar, lunar, and planetary positions.
R: *Sky and Telescope* 63: 49 (Jan. 1982).

Buchsbaum's Complete Handbook of Practical Electronic Reference Data. 2d ed. **Walter H. Buchsbaum.** Englewood Cliffs, NJ: Prentice-Hall, 1978.

First edition, 1973.
Excellent use of graphs, conversion tables, etc. makes this a reliable basic-information source. Fundamental engineering knowledge is sufficient for use.

Digital Integrated Circuit DATA Book. 17th ed. **Derivation and Tabulation Associates.** Orange, NJ: DATA, 1975. Annual.

Computerist's Handy Databook/Dictionary. **Clayton L. Hallmark.** Blue Ridge Summit, PA: TAB Books, 1979.

A ready reference to computers. Includes definitions, tables, formulas, charts, symbols. Contains a section on microprocessors. Useful to students.
R: *ARBA* (1980, p. 718).

Electronic Circuits Notebook: Proven Designs for Systems Applications. **Samuel Weber, ed.** New York: McGraw-Hill, 1981.

A companion volume to *Circuits for Electronics Engineers*, this popular reference covers the most common engineering problems and their solutions. Contains 268 circuits, organized alphabetically by function. An invaluable aid for working engineers.

Electronic Properties Research Literature Retrieval Guide: 1972–1976. 4 vols. **J. F. Chaney and T. M. Putnam, eds.** New York: Plenum Press, 1979.

A 4-volume reference work on all aspects of electronic properties. Includes nearly 20,000 references and covers nearly 10,000 materials. Comprehensive and well-arranged for easy use.

Handbook of Electronics Industry Cost Estimating Data. T. Taylor. New York: Wiley, 1985.

Illustrated Handbook of Electronic Tables, Symbols, Measurements and Values. **Raymond H. Ludwig.** West Nyack, NY: Parker Publishing, 1977.

An illustrated handbook that covers a broad range of useful reference data and formulas of electronics. Thorough and clear. Comprises a useful reference source for engineering libraries.
R: *RSR* 6: 27 (Oct./Dec. 1978).

Linear Integrated Circuit DATA Book. 12th ed. **Derivation and Tabulation Associates.** Orange, NJ: DATA, 1975. Annual.

Microprocessor Data Book. **S. A. Money.** New York: McGraw-Hill, 1981.

Describes a microprocessor chip, a memory chip, and other devices. Aids in choosing a microprocessor and gives a list of manufacturers. A valuable reference tool for the design engineer.
R: *Radio Electronics* 54: 106 (Feb. 1983).

MATERIALS SCIENCE

Chemical Materials for Construction. **Philip Maslow.** New York: McGraw-Hill, 1981.

A reference containing information on chemical materials used in construction. Includes bibliography, appendix, and index. Useful to architects, civil engineers, contractors, and technologists.

Data Handbook for Clay Materials and Other Nonmetallic Minerals. **H. Van Olphen and J. J. Fripiat.** New York: Pergamon Press, 1979.

A well-documented reference book divided into 2 parts covering the chemical, physical, and mineralogical properties of clays and other minerals such as chrysotile, talc, gibbsite, magnesite, calcite, and gypsum. For materials engineers.
R: *RSR* 8: 27 (Jan./Mar. 1980).

Desk-Top Data Bank: Elastomeric Materials. **International Plastics Selector.** San Diego: International Plastics Selector, 1977.

Extensive information on 15,000 gum stocks, elastomers, and liquid systems. A comprehensive databook.
R: *Elastomerics* 110: 46 (Dec. 1977); *TBRI* 44: 114 (Mar. 1978).

MECHANICAL ENGINEERING

Data Book for Pipe Fitters and Pipe Welders. **Edward H. Williamson.** Folkestone, Kent, England: Bailey Brothers and Swinfen, 1977.

A handy pocket-sized reference that contains data on fitting and welding pipes. Includes tables and diagrams.
R: *International Journal of Pressure Vessels and Piping* 5: 321 (Oct. 1977); *Welding and Metal Fabrication* 45: 511 (Oct. 1977). *TBRI* 44: 80 (Feb. 1978); 44: 160 (Apr. 1978).

Mechanisms and Dynamics of Machinery. 3d ed. **H. H. Mabie and F. W. Ocvirk.** New York: Wiley, 1978.
SI Version.
Provides data on force, gravitational preference of machinery, mass, etc. Expresses all dimensions in SI units.

Engineering Sciences Data; Mechanical Engineering Series: Fluid Mechanics, Internal Flow. 5 vols. London: Engineering Sciences Data Unit, 1972–1976.

National Service Data-Advance. San Diego: National Automotive Service. Annual.

Thermodynamic and Thermophysial Properties of Combustion Products. Vols. 1–. Jerusalem: Program for Scientific Translations, 1975–.

METALLURGY

Engineering Properties of Steel. **Philip D. Harvey, ed.** Metals Park, OH: American Society for Metals, 1982.

Presents data for carbon, alloy, stainless and heat resisting, tool, ASTM structural, and maraging steels. Also, the chemical composition, general characteristics and uses, mechanical properties, and machining data for many steels used in industrial applications.
R: *Science and Technology Libraries* 4: 123 (Winter 1983).

Magnetism Diagrams for Transition Metal Ions. **E. Koenig and S. Kremer.** New York: Plenum Press, 1979.

This book includes diagrams showing magnetic moment as a function of temperature and other parameters. A companion to *Ligand Field Energy Diagrams*.
R: *RSR* 8: 29 (July/Sept. 1980).

Metal Progress Databook. Little Rock, AR: American Society for Metals, 1976. Annual.

Metals Reference Book. 5th ed. **Colin J. Smithells, ed.** Woburn, MA: Butterworth, 1976.

Fourth edition, 1967.
Considered one of the most authoritative references relating to metallurgy. Data is easily accessible through tables and charts. Highly recommended.
R: *Chemistry and Industry* 00: 1694 (Aug. 7, 1967); *Iron and Steel International* 50: 51 (Apr. 1977); *TBRI* 43: 238 (June 1977); *ARBA* (1976, p. 663; 1977, p. 771); Jenkins (K192); Sheehy (EJ63); Wal (p. 503).

Welding Design and Fabrication Data Book. Cleveland: Industrial, 1960/1961–. Annual.

Title on spine: *Welding Data Book*. Section A, directory of products and manufacturers; section B, trade names; section C, manufacturers' catalogs/outlets; section D, engineering and application data; section E, where to buy locally.
R: Win (E1158).

NUCLEAR ENGINEERING

Neutron Nuclear Data Evaluation. **International Atomic Energy Agency.** New York: Unipub, 1973.

Neutron Standard Reference Data. **International Atomic Energy Agency.** New York: Unipub, 1975.

Nuclear Data in Science and Technology. 2 vols. **International Atomic Energy Agency.** New York: Unipub, 1974.

Nuclear Proliferation Factbook. **US Library of Congress, Environment and National Resources Policy Division.** Washington, DC: US Government Printing Office, 1977.

Includes a number of basic documents concerning the proliferation of nuclear documents. Covers statistics, such as uranium supply, reactors, etc., taken from over 40 sources. Contains a bibliography taken from *Arms Control Today* 1976–1977. Contains a wealth of statistical information.
R: *ARBA* (1979, p. 697).

ENERGY

Coal Data: A Reference. **Energy Information Administration.** Washington, DC: US Government Printing Office, 1980.

Presents data on coal resources, mining, production, consumption, prices, and transportation in the United States. Includes a bibliography. For personnel in the coal industry.
R: *BL* 77: 448 (Nov. 15, 1980).

Coal Data Book. **President's Commission on Coal.** Washington, DC: US Government Printing Office, 1980.

Contains data on various aspects of coal production and consumption. Provides background knowledge on coal technology, transportation, and regulation. For those in the coal industry.
R: *BL* 76: 1352 (May 15, 1980).

Energy Crisis. vol. 3. **Lester A. Sobel.** New York: Facts on File, 1975–1977.
Volume 1, 1969–1973; volume 2, 1974–1975.
Compilation of relevant material gleaned from the weekly *Facts on File*.

The Energy Factbook. **Richard Dorf.** New York: McGraw-Hill, 1980.

Up-to-date energy almanac discussing important energy facts such as sources of supply, future use, conservation, and economic outlook for the United States. Accurate energy information for professionals, engineers, business leaders, and government officials.
R: *WLB* 55: 620 (Apr. 1981).

Oil and Gas Resources: Worldwide Petroleum Supply Limits: Future for World Natural Gas Supply. Full report to the Conservation Commission of the World Energy Conference. **P. Desprairies et al.** Guildford, Surrey, England: IPC Science and Technology, 1978.

A classified listing of petroleum resources, including oil, shale, tar, deep-sea fuels, etc. Includes tables, graphs, and diagrams.
R: *Aslib Proceedings* 43: 465 (Dec. 1978).

The Solar Energy Almanac. **Martin McPhillips, ed.** New York: Facts on File, 1983.

Provides the basic groundwork for research in solar energy at an elementary level. Lacks bibliographic references and detailed information. A good starter book for those interested in solar energy research.
R: *ARBA* (1984, p. 672).

World Energy Supplies, 1973–1978. **United Nations.** New York: Unipub, 1979.

Expanded edition covers production, consumption, import, and export of solid fuels, petroleum, nuclear fuels, and other energy sources by 193 countries.

Environmental Sciences

Air Pollution Sampling and Analysis Deskbook. **Paul N. Cherimisinoff and A. C. Morressi.** Ann Arbor: Ann Arbor Science, 1978.
R: *TBRI* 45: 150 (Apr. 1979).

The Cousteau Almanac of the Environment: An Inventory of Life on a Water Planet. **Jacques Yves Cousteau and Staff of the Cousteau Society.** Garden City, NY: Doubleday, 1981.

A popular compendium of facts and articles dealing with the environment. Focuses on the protection of the environment, mismanagement, and achievements. Includes illustration, references, and an index.
R: *BL* 77: 1318 (June 15, 1981).

Data for Radioactive Waste Management and Nuclear Applications. **D. C. Stewart.** New York: Wiley, 1985.

A specialized guide on the management of radioactive wastes, including information applicable to related applications. Covers physical data, chemical data, types of radioactive wastes, and data for different operations.

Environmental Impact Data Book. **Jack Golden et al.** Ann Arbor, MI: Ann Arbor Science, 1979.

Provides tabular data for usage in the preparation of environmental impact statements. Covers topics such as air quality, water resources, ecosystems, toxic chemicals, models, etc. Contains tables, charts, and an extensive list of references. Recommended reference for those involved in environmental impact statement preparation.
R: *Journal of Water Pollution Control Federation* 51: 863A (Nov. 1979); *TBRI* 46: 73 (Feb. 1980); *ARBA* (1980, p. 653). Mal (1980, p. 211).

Mass Spectrometry of Priority Pollutants. **B. S. Middleditch et al.** New York: Plenum, 1981.

A collaboration of information on the mass spectra for some 100 or more hazard-causing toxic pollutants found in effluents, drinking water, or fish. A helpful aid for those mass spectroscopists in the pollution field.
R: *ABL* 46: (Sept. 1981).

Storm Data for the United States, 1970–1974 and 1975–1979: A Quinquennial Compilation of the US Environmental Data Service's Official Monthly Reports. **US Weather Bureau.** Detroit: Gale Research, 1982.

Includes detailed data on climatological disasters and ordinary weather disturbances since 1970. Data is arranged chronologically by month and alphabetically by state. Good for research in meteorology and agriculture.
R: *ARBA* (1983, p. 664).

Water and Waste Treatment Data Book. **Permutit Company.** Paramus, NJ: Permutit Company, 1981.

A revised edition containing 91 sections of information, including charts, tables, and formulas. Intended for engineers and workers in waste-water treatment plants.
R: *Civil Engineering* 52: 41 (April 1982).

Where You Live May Be Hazardous to Your Health: A Health Index to Over 200 American Communities. **Robert A. Shakman.** Briarcliff Manor, NY: Stein & Day, 1979.

Contains details of factors relating to the environmental safety of more than 200 communities throughout the United States. Presents information useful to public and academic libraries.
R: *WLB* 54: 257 (Dec. 1979).

TRANSPORTATION

Air Facts and Feats. Rev. ed. **John W. R. Taylor, Michael J. H. Taylor, and David Mondey.** New York: Sterling Publishing, 1978.

A compendium of facts on aeronautical events. Contains appendixes that list speed, height, distance, and disasters. Good bibliography and index; well-illustrated.
R: *BL* 75: 540 (Nov. 15, 1978); *ARBA* (1979, p. 781).

Guinness Aircraft Facts and Feats. **Michael Taylor and David Mondey, eds.** Enfield, England: Guinness Superlatives. Distr. New York: Sterling Publishing, 1984.

Revision of the third edition of *Air Facts and Feats*, 1977.

Facts about specific individuals and planes, records of various types, etc. General and aircraft indexes.
R: *BL* 81: 468 (Nov. 15, 1984); *SLJ* 31: 179–180 (Oct. 1984); *ARBA* (1985, p. 617).

The Guinness Book of Ships and Shipping: Facts and Feats. **Tom Hartman, ed.** Enfield, England: Guinness Superlatives. Distr. New York: Sterling Publishing, 1983.

A compendium of brief facts concentrating on ships and shipping. An interesting collection of more than trivia. A handy book for naval collection.
R: *BL* 80: 778 (Feb. 1, 1984); *ARBA* (1985, p. 621).

World Motor Vehicle Data Book. Detroit, MI: Motor Vehicle Manual, 1986.

STATISTICAL SOURCES

GUIDE TO SOURCES OF STATISTICAL INFORMATION

American Statistics Index; A Comprehensive Guide and Index to the Statistical Publications of the United States Government. Washington, DC: Congressional Information Service, 1974–. Monthly.

The 1974 edition contains retrospective information. Identifies statistical data published by the federal government. Index and abstract sections.

Bibliography of Statistical Literature. 3 vols. Rep. ed. **Maurice G. Kendall and Alison G. Doig.** Salem, NH: Ayer, 1981.

Earlier edition, 1968.
Volume 1, 1950–1958; volume 2, 1940–1949; volume 3, pre-1940.
Retrospective guide to sources dating back to the sixteenth century.

Federal Statistical Directory: A Personnel Directory and Guide to Information Sources. 27th ed. **Richard J. D'Aleo and William R. Evinger.** Springfield, VA: ICUC Press, 1983.

Originally intended primarily for internal use by statistical agencies within the federal government, this directory has become a standard reference guide. This edition is the first to be produced privately. It contains a great deal of federal collected statistics. A must for all types of libraries.
R: *ARBA* (1985, p. 264).

Guide to US Government Statistics. 4th ed. **John Andriot.** McLean, VA: Documents Index, 1973 and new editions.

Third edition, 1961.
An annotated guide to recurring statistical publications of US agencies. Arranged by Sudocs classification scheme. Although many publications of interest to scientists are included, primary emphasis is on economic and social statistics.

Statistical Sources. 7th ed. **Paul Wasserman and Jacqueline O'Brien.** Detroit: Gale Research, 1982.

First edition, 1971; fourth edition, 1974; fifth edition, 1977.
Sources of numeric data under specific alphabetical subject headings. Presents about 22,000 citations on 12,000 subjects. Arranged in dictionary style. Sources include annuals, yearbooks, directories, and other publications.

GOVERNMENTAL STATISTICAL SOURCES

Government publications from both international and national organizations are an ever-increasing source of statistical information. At present nearly all government agencies generate some sort of numerical audit. In the United States, the Office of Management and Budget's Statistical Policy Division exists solely for the purpose of coordinating the statistical collection and dissemination functions of the federal government. Other government agencies involved primarily in statistical activities are the Department of Agriculture's Statistical Reporting Service, the Department of Commerce's Bureau of the

Census, and HEW's National Center for Health Statistics and National Center for Educational Statistics. The following is a sample list of important governmental statistical publications of interest to scientists, engineers, and technicians:

Agricultural Finance Statistics. US Department of Agriculture, Statistical Bulletin no. 706. **US Department of Agriculture, Economic Research Service.** Washington, DC: US Government Printing Office, 1984.

Agricultural Statistics. **US Department of Agriculture.** Al.47. Washington, DC: US Government Printing Office, 1936–. Annual.

Formerly contained in the *Yearbook of Agriculture*, this tool consists of statistics on yield, acreage, farm resources, etc.
R: *ARBA* (1977, p. 726).

Annual Report of the Council on Environmental Quality. EX 14.1: date. Washington, DC: US Government Printing Office. Annual.

Annual Report of the Federal Power Commission. FP 1.1: date. Washington, DC: US Government Printing Office, 1921–. Annual.

Overviews of the electric power and national gas industries.

Climatological Data: National Summary. **US National Oceanic and Atmospheric Administration.** C 55.214: V. Washington, DC: US Government Printing Office, 1950–. Monthly.

General summary of weather conditions and condensed climatological summary for the states.

Crop Production Reports. **US Department of Agriculture. Crop Reporting Board.** A94.24. Washington, DC: US Government Printing Office. Annual.

Annual summaries on acreage, yield, and production for both the national and individual states.

Directory of Federal Statistics for Local Areas. **US Department of Commerce Bureau of the Census.** Washington, DC: US Government Printing Office, 1966–.

Directory of Federal Statistics for States. **US Department of Commerce, Bureau of the Census.** C 3.6/2: St2/3, Washington, DC: US Government Printing Office, 1967–.

Arranged by subject.

Directory of Non-Federal Statistics for States and Local Areas. **US Department of Commerce Bureau of the Census.** Washington, DC: US Government Printing Office, 1970–.

Energy Resources of the US Geological Survey. I 19.4/2: 650. Washington, DC: US Government Printing Office, 1972–.

Estimates total known and undiscovered US energy resources.

Estimated Federal Expenditures on Domestic Transportation Capital Improvement and Operating Programs by State for Fiscal Years. 1957–1971. **US Office of Transportation Planning Analysis.** Washington, DC: US Department of Transportation, 1974.

Fatal and Injury Accident Rates on Federal-Aid and Other Highway Systems/1975. **US Federal Highway Administration.** Washington, DC: US Department of Transportation, 1976. Annual.

Handbook of Airline Statistics. **US Civil Aeronautics Board.** Washington, DC: US Government Printing Office. Annual.
R: *ARBA* (1973, p. 663).

Highway Statistics. **US Federal Highway Administration.** Washington, DC: US Department of Transportation. Annual.

Highway Statistics: Summary to 1975. **US Federal Highway Administration.** Washington, DC: US Government Printing Office, 1977.

A retrospective collection of statistical data related to all aspects of motor and highway matters, including taxes, construction, revenues, and usage.
R: *ARBA* (1978, p. 780).

Monthly Climatic Data for the World. **World Meteorological Organization in Cooperation with the National Oceanic and Atmospheric Administration.** C 55.211: Vol 2. Washington, DC: NOAA. 1948–. Monthly.

Data for selected world cities.

National Air Monitoring Program: Air Quality and Emission Trends, Annual Report. **Environmental Protection Agency.** EP 1.47. Washington, DC: US Government Printing Office. 1973–. Annual.

Status of air pollution measurement control from federal, state, and local monitoring stations.

National Transportation Statistics. **US Research and Special Programs Administration, Transportation Systems Center.** Washington, DC: US Government Printing Office. Annual.

A compilation of transportation and energy statistics. Uses 4 formats: tree displays, modal profiles, performance indicators, and transportation trends. Includes a glossary of terms and a bibliography. For reference libraries with transportation collections.
R: *ARBA* (1981, p. 769).

1979 National Survey of Compensation Paid Scientists and Engineers Engaged in Research and Development Activities. DOE/AD/7405-3. Washington, DC: US Department of Energy, 1979.

A statistical survey of the compensation paid to scientists and engineers who are engaged in research and development. Covers information including age, highest degree earned, and years since first degree.
R: *BL* 76: 1188 (Apr. 15, 1980).

Research Reports. **US Department of Agriculture, Farmer's Cooperative Service.** Washington, DC: US Government Printing Office, 1968–. Irregular.

Emphasizes production in farm cooperatives.

Statistical Abstract of the United States. **US Bureau of the Census.** Washington, DC: US Government Printing Office, 1878–. Annual.

A comprehensive, quantitative compendium of statistics on the political, social, and economic organization of the United States. Excellent ready-reference tool for statistics on information of national importance. References given to sources of all tables.

USSR Agricultural Atlas. **US Central Intelligence Agency, comp.** Washington, DC: US Government Printing Office, 1976.

Compilation of agricultural statistics and data on Soviet agricultural practices and policies. Fully illustrated with colored maps and charts and some black-and-white photographs. Analyzes Soviet technology, irrigation, land use, and agricultural products. Includes superficial comparison between US and Soviet systems. All information contributed by Central Intelligence Agency.
R: *ARBA* (1977, p. 727).

World Power Data. **US Federal Power Commission.** FP 1.12: date. Washinton, DC: US Government Printing Office, 1962–. Annual.

Geographically arranged statistics on electric and generating capacity and electric-energy production.

Nongovernmental Statistical Sources

Besides government publications, statistical information is also available from many commercial publishers: international, national, and local professional and trade organizations; private interest groups; and so on. Much of the information on these types of sources can be found in the chapters on yearbooks, directories, guides, databooks, and so forth, in this work. The following is a sample list of statistical sources for important subjects that are not produced by government agencies:

Aeorspace Facts and Figures. Washington, DC: Aerospace Industries Association of America, 1960–. Annual.

Latest edition, 1975/76.
An annual publication of statistical tables, providing information on production, trade, research and development, employment, etc., in the aerospace industry.

American Bureau of Metal Statistics Yearbook. New York: American Bureau of Metal Statistics. 1921–. Annual.

Databook on worldwide mine and smelter production, imports, exports, standard specifications, prices, etc. Most comprehensive source for statistics on nonferrous metals. Intended for use in research and market analysis.
R: *ARBA* (1970, p. 165; 1973, p. 661).

Annual Bulletin of Electric Energy Statistics for Europe. **United Nations, Economic Commission for Europe.** New York: Unipub. Annual.
Volume 22, 1976.
R: *IBID* 6: 36 (Mar. 1978).

Annual Bulletin of General Energy Statistics for Europe. Vols. 1–. New York: Unipub, 1968–. Annual.
Volume 9, 1976.
Trilingual (English/French/Russian). Arranges data on the energy situation as a whole in Europe and the United States by country. Production of energy form and balance sheets of individual forms of energy.
R: *IBID* 4: 220 (Sept. 1976); 6: 279 (Sept. 1978); 7: 235 (Fall 1979).

Annual Bulletin of Transport Statistics for Europe 1978. **United Nations.** New York: Unipub, 1979.

Annual Summary of Information on Natural Disasters: Earthquakes, Tsunamis, Volcanic Eruptions, Landslides, Avalanches, 1975. No. 10. Paris: Unesco. Distr. New York: Unipub, 1979.
A bilingual (English/French) survey of natural disasters. Information given includes location, time, dimensions, type, casualties, causes, etc. Earthquake section contains a map indicating magnitude and foci. Lists institutions that supply similar data. For earth scientists.
R: *ARBA* (1981, p. 690).

Automobile Facts and Figures. **Motor Vehicle Manufacturers Association of the United States.** Detroit: Motor Vehicle Manufacturers Association of the United States. Annual.

Building Construction Cost Data—1978. 36th annual ed. **Robert S. Godfrey.** Duxbury, MA: Robert Snow Means, 1978. Annual.
R: *Civil Engineering* (Dec. 1971).

Electronic News Financial Factbook and Directory. New York: Fairchild, 1970–. Annual.
In-depth profiles of major electronics companies, including financial data, addresses, statistical information, etc.

Energy in the World Economy: A Statistical Review of Trends in Output, Trade, and Consumption Since 1925. **Joel Darmstadter.** Baltimore, MD: Johns Hopkins University Press, for Resources for the Future, 1972.
R: Wal (p. 479).

Energy Statistics Yearbook. New York: United Nations. Annual.
Twenty-sixth edition, 1984. Formerly the *Yearbook of World Energy Statistics.*
A comprehensive volume of international energy statistics. In both English and French, it provides a global comparable data on long-term trends in the energy supply. A basic source of energy statistics for most larger libraries. No index.
R: *ARBA* (1985, p. 506).

Energy Use in the United States by State and Region: A Statistical Compendium of 1972 Consumption, Prices, and Expenditures. **Irving Hoch.** Washington, DC: Resources for the Future, 1978.
An exhaustive compilation of data on energy prices, expenditures, and consumption. Tabular data is organized by source, function, and sector. The 409 tables are distributed into 2 main sections: substantive results and appendixes, and supporting materials. Contains lists of references. For libraries with energy collections.
R: *ARBA* (1980, p. 649).

International List of Selected, Supplementary, and Auxiliary Ships, 1979. Geneva: World Meteorological Organization. Distr. New York: Unipub, 1979.
A list of ships that participate in the Voluntary Observation Scheme of the World Meteorological Organization. Data include a listing of ships by country, weather observation equipment carried, radio call signs, and geographical routes served. A specialized reference for meteorologists.
R: *ARBA* (1981, p. 691).

International Rayon and Synthetic Fibres Statistical Yearbook. Paris: International Rayon and Synthetic Fibres. Annual.
Trilingual statistics on production, consumption, and foreign trade.

Lloyd's Register of Shipping Statistical Tables. London: Lloyd's Register of Shipping. Annual.

Motor Truck Facts. **Motor Vehicle Manufacturers Association of the United States.** Detroit: Motor Vehicle Manufacturers Association of the United States. Annual.

National Construction Estimator, 1981. **Gary Moselle and Albert S. Paxton, eds.** Los Angeles: Craftsman, 1981.
Lists building costs for both light and heavy construction.
R: *ARBA* (1974, p. 673).

Natural Resource Commodities: A Century of Statistics. **Robert S. Manthy.** Baltimore: Johns Hopkins University Press, 1978.
An updated text covering historical statistical data on natural resources to the year 1973. Includes all major agricultural, forest, and mineral products produced and consumed in the United States. For engineering libraries.
R; *RSR* 8: 26 (Jan./Mar. 1980).

Non-Ferrous Metal Data, 1978. New York: American Bureau of Metal Statistics, 1979.

Original title: *Year Book of the American Bureau of Metal Statistics.*
Consists of more than 180 statistical tables relating to the production, consumption, imports, exports, and prices of nonferrous metals. Also includes miscellaneous tables on exchange rates, equivalents, stockpile materials, etc. Valuable addition for business and economics libraries.
R: *ARBA* (1980, p. 658).

Nuclear Resources: Contribution of Nuclear Power to World Energy Supply 1975–2020. Full Report to the Conservation Commission of the World Energy Conference. **J. S. Foster et al.** Guildford, Surrey, England: IPC Science and Technology, 1978.
Provides statistical information on the potential growth of nuclear power, linking this with uranium supply. Also details the technology of fuel fabrication.
R: *Aslib Proceedings* 43: 466 (Dec. 1978).

Projected Pulp and Paper Mills in the World 1979–1989. Food and Agriculture Organization of the United Nations. New York: Unipub, 1980.
Statistical tables that list pulp and paper expansions from 1979–1989, including those that are tentative and under construction. Entries cover details such as name of company, location and type of mill, product, and raw materials.

Quarterly Oil Statistics. Organization for Economic Cooperation and Development. Washington, DC: Organization for Economic Cooperation and Development, 1977–.
Provides current, accurate data on production, trade, refinery intake, and consumption of oil. For all research libraries.
R: *RSR* 4: 31 (Oct./Dec. 1977).

Tropical Cyclones of the North Atlantic Ocean, 1871–1977. **Charles J. Neumann et al.** Washington, DC: US Government Printing Office, 1978.
A compilation of statistical data on 871 North Atlantic tropical cyclones recorded from 1871 through 1977. Contains information on intensity, date of occurrence, and movement. Also provides general knowledge on cyclones. Includes an extensive bibliography. Recommended for any reference library.
R: *ARBA* (1980, p. 655).

Unesco Statistical Yearbook. New York: Unipub. Annual.
Statistics from over 200 countries and territories.

United Kingdom Mineral Statistics. Great Britain. **Institute of Geological Sciences.** London: Her Majesty's Stationery Office. Annual.
First edition, 1973.
A statistical source covering minerals in the economy, mineral production, and commodity reviews and summaries from 1971–77. Includes 11 maps and diagrams.
R: Wal (p. 244).

Water Resources of the World: Selected Statistics. **Frits van der Leeden, ed.** Port Washington, NY: Water Information Center, 1975.

A compilation of worldwide water resource statistics; presents data in tabular form. Also includes numerous maps and diagrams.
R: *ARBA* (1976, p. 705); Sheehy (EJ31).

World Energy Supplies, 1964–1978. New York: Unipub, 1980.
Covering the principal elements of production, import, export, bunkers, stock change, and apparent consumption of commercial energy for 202 countries and areas (statistics compiled as of March 1976).

The World Food Book: An A–Z Atlas and Statistical Source Book. **David Crabbe and Simon Lawson, eds.** New York: Nichols Publishing, 1981.
A 3-part reference tool containing 800 food terms in alphabetical order, atlases, statistical charts and tables of food production. This book has a British bias in spelling and is not as extensive as others.
R: *Choice* 19: 751 (Feb. 1982).

World Road Statistics, 1971–1975. **International Road Federation.** Washington, DC: International Road Federation, 1976.
R: Wal (p. 362).

Yearbook of Fishery Statistics. **Food and Agriculture Organization of the United Nations.** New York: Unipub. Annual.
Volume 51, *Fishery Commodities,* 1981.
R: *IBID* 4: 127 (June 1976); 6: 171 (June 1978); 7: 141 (Summer 1979).

Yearbook of Forest Products. New York: Unipub. Annual.
Former title: *Yearbook of Forest Products Statistics.*
R: *IBID* 4: 128 (June 1976); 6: 292 (Sept. 1978); 7: 144 (Summer 1979); Jenkins (J92).

CHAPTER 8 MANUALS, SOURCE BOOKS, LABORATORY MANUALS AND WORKBOOKS, AND HOW-TO-DO-IT MANUALS

MANUALS

SCIENCE

Auger Electron Spectroscopy Reference Manual. **G. E. McGuire.** New York: Plenum Press, 1979.
RSR 8: 87 (July/Sept. 1980).

First Aid Manual for Chemical Accidents: For Use in Cases of Poisoning by Non-Pharmaceutical Chemicals. **Marc J. Lefevre, ed.** English language edited by Ernest I. Becker. New York: Academic Press, 1980.

A valuable reference source for the treatment of chemical accidents. Includes a listing of chemicals used today in homes and laboratories; emergency procedures for accidents; and an index of products. Outlines symptoms and treatment of poisoning. Chapters are color-coded for easy access. English translation of French edition. Useful for university and industrial laboratories, research facilities, emergency medical personnel, police, and physicians.

The Flammarion Book of Astronomy. **Translated by Annabel and Bernard Pagel.** London: Allen & Unwin; New York: Simon and Schuster, 1964.
First published as *Astronomie Populaire*, Paris, Flammarion, 1880.
While regarded as a popular work, this classic remains a reliable source covering the exposition of the universe.
R: *Nature* 204: 815–816 (Nov. 28, 1964); *New Scientist* 43: 121–122 (Oct. 1964); *Science* 146: 1153 (1964); Jenkins (E26); Wal (p. 78); Win (EB18).

Laboratory Manual for Photographic Science. **E. N. Mitchell and T. W. Haywood.** New York: Wiley, 1984.
An introductory manual that addresses what, how, and why photographic instruments and photographic processes work. Examples included.

Manual for the Organization of Scientific Congresses. **Helena B. Lemp.** Basel: S. Karger, 1979.
R: *Physics Today* 32: 60 (Nov. 1979); *TBRI* 46: 6 (Jan. 1980).

Manual of Symbols and Terminology for Physiochemical Quantities and Units. 2d ed. **International Union of Pure and Applied Chemistry, Physical Chemis-**

try Divison, Commission on Symbols, Terminology, and Units. Edited by David Hardy Whiffen. Elmsford, NY: Pergamon, 1979.

First edition, 1973.

A slightly revised manual of symbols and terminology used by physical scientists. For academic and industry libraries.

R: *Journal of the American Chemical Society* 102: 3665 (May 7, 1980); *RSR* 8: 25 (July/Sept. 1980).

A Manual of Underwater Photography. **T. Glover, G. E. Harwood, and J. N. Lythgoe.** New York: Academic Press, 1977.

Most useful to those interested in fabricating their own equipment for underwater photography.

R: *American Scientist* 66: 500 (July/Aug. 1978).

Mathematics into Type: Copyediting and Proofreading of Mathematics for Editorial Assistants and Authors. **Ellen Swanson.** Providence: American Mathematical Society, 1979.

Earlier edition, 1972.

This guide to the preparation of mathematical copy for publication was originally a procedures manual for editors at the American Mathematical Society. Primarily a style manual.

R: *ARBA* (1972, p. 551).

Metric Manual. **Lawrence D. Pedde et al.** Detroit: Gale Research, 1980.

An authoritative, thorough, and comprehensive manual on the metric system. Covers a wide variety of engineering problems. Useful addition in most libraries and in the personal collections of scientists and engineers.

R: *ARBA* (1981, p. 756).

Metric System Guide. Rev ed. 5 vols. **J. J. Keller.** Neenah, WI: Keller, 1978.

Volume 1, *Metrication in the United States*; volume 2, *Legislation and Regulatory Activities*; volume 3, *Metric Units Edition*; volume 4, *References Sources*; volume 5, *Definitions and Terminology.*

BIOLOGICAL SCIENCES

Atlas and Manual of Plant Pathology. **Ervin H. Barnes.** New York: Plenum Press, 1979.

An excellent reference volume that, because of its graphs, data charts, experiments, equipment lists, and discussion questions, can be used as a botany text. Well-illustrated and referenced. For college and university libraries.

R: *ARBA* (1980, p. 625).

Bergey's Manual of Determinative Bacteriology. 8th ed. **Society of American Bacteriologists.** Edited by Robert E. Buchanan and Norman E. Gibbons. Baltimore: Williams & Wilkins, 1974.

Descriptive classification of bacteria. Companion volume *Index Bergeyana* is an alphabetical listing of the bacterial taxa.

R: *American Scientist* 63: 472 (July/Aug. 1975); Win (EC120).

Council of Biology Editors Style Manual. 4th ed. **Conference of Biological Editors Style Manual Committee.** Washington, DC: American Institute of Biological Sciences, 1978.
Original title: *Style Manual for Biological Journals.* Second edition, 1964.
A commonly used guide to acceptable forms of abbreviations and citation formats in the biological sciences. For students and researchers.

Manual of Clinical Biology. 3d ed. Bethesda, MD: American Society for Microbiology, 1980.

Manual of Clinical Laboratory Methods. 4th ed. **Opal Hepler.** Springfield, IL: Thomas, 1977.

Manual of Clinical Microbiology. 3d ed. **Edwin H. Lennette et al., eds.** Washington, DC: American Society for Microbiology, 1980.
Second edition, 1974.
Standard handbook covering a wide range of topics in this discipline, including mycology, bacteriology, virology, etc.
R: *Choice* 12: 1798 (Feb. 1975); *ARBA* (1976, p. 658).

Botany

Atlas and Manual of Plant Pathology. 2d ed. **Ervin H. Barnes.** New York: Plenum Press, 1979.
Simple-language text devoted to observational and experimental techniques.
R: Mal (1976, p. 188).

The Concise Herbal Encyclopedia. **Donald Law.** New York: St. Martin's Press, 1981.
Earlier edition, 1974.
Emphasis on the medicinal properties of herbs. Although several chapters have an alphabetical arrangement, this book is more a history and manual than an encyclopedia.
R: *AL* 5: 367 (July/Aug. 1974); *LJ* 99: 1376 (May 15, 1974); *ARBA* (1975, p. 668).

Exotic Plant Manual: Exotic Plants to Live With. 5th ed. **Alfred Byrd Graf, ed.** East Rutherford, NJ: Roehrs. Distr. New York: Scribner's, 1978.
Much material has been culled from Graf's scholarly *Exotica III* to provide amateur and professional gardeners with a sound guide to exotic plants. Includes a number of photographs with descriptions. Most comprehensive work of its kind.
R: *Choice* 8: 808 (Sept. 1971); *LJ* 95: 1597 (May 1, 1971); *ARBA* (1972, p. 575).

Fern Growers Manual. **Barbara J. Hoshizaki.** New York: Knopf, 1976.
Considered the best book on fern horticulture. A classic reference.
R: *Garden* 26: 202 (Dec. 1976); *TBRI* 43: 149 (Apr. 1977).

The Macmillan Book of Natural Herbs. **Marie-Luise Kreuter.** New York: Macmillan, 1985.

Phytoplankton Manual. **A. Sournia, ed.** New York: Unipub, 1979.

Provides a quantitative study of phytoplankton at all stages of research and collecting, including sea sampling; preserving and storing samples; and identification. Articles by 30 authors cover reasons for studying phytoplankton, how to interpret results, etc. Includes an outstanding bibliography and a list of manufacturers of plankton study equipment. An indispensable reference tool for marine biologists and undergraduate and graduate science libraries.
R: *Nature* 282: 886 (Dec. 20, 1979); *Choice* 17: 102 (Mar. 1980); *IBID* 7: 139 (Summer 1979).

Research Experience in Plant Physiology: A Laboratory Manual. 2d ed. **T. C. Moore.** New York: Springer-Verlag, 1981.

First edition, 1973.
This laboratory manual is intended for an upper-division undergraduate or graduate course in plant physiology. Includes exercises pertaining to most major areas of plant physiology.

World Vegetables: Principles, Production and Nutritive Values. **Mas Yamaguchi.** Westport, CT: AVI Publishing, 1983.

A convenient book on the subject in 3 parts—information on the evolution of vegetables, the principles of growing vegetables, and the world's vegetables in 17 chapters. Extensive tables, figures, and illustrations. Glossary and index are available.
R: *ARBA* (1985, p. 505).

Zoology

Manual of Methods for Fisheries Resource Survey and Appraisal. 5 vols. Rome: Food and Agriculture Organization of the United Nations, 1971–1975.
R: *IBID* 4: 126 (June 1976).

Manual of Methods for Fish Stock Assessment. 3 vols. Rome: Food and Agriculture Organization of the United Nations, 1966–1980.

Volume 1, *Fish Population Analysis,* 1969; volume 2, *Tables of Yield Functions,* revised edition, 1980, first edition, 1966; volume 3, *Selectivity of Fishing Gear,* 1976.
R: *IBID* 4: 126 (June 1976).

A Manual of Mammalogy: With Keys to Families of the World. 2d ed. **Anthony F. DeBlase and Robert E. Martin.** Dubuque, IA: Brown, 1981.

Earth Sciences

Ground Water Manual. New York: Wiley-Interscience, 1983.

Covers all aspects of groundwater resources in their investigation, development, and management. With tables, graphs, illustrations, and bibliography.

Manual of Applied Geology for Engineers. **Institution of Civil Engineers.** London: Institution of Civil Engineers, 1976.

A manual useful to both geologists and civil engineers. Covers soil and rock mechanics. Contains input from governmental and industrial sources.
R: *Geotechnique* 27: 274 (June 1977); *TBRI* 43: 314 (Oct. 1977).

Manual of Mineralogy. 20th ed. **C. Klein and C. Hurlbut.** New York: Wiley, 1985.

A comprehensive manual of mineralogy with descriptive discussions of about 200 of the most common minerals. Illustrated with color plates.

Manual on the Global Data-Processing System. 2 vols. **World Meteorological Organization.** New York: World Meteorological Organization, 1977.

Discusses standardized procedures of the World Weather Watch and Global Data-Processing System.
R: *IBID* 6: 166 (June 1978).

General Engineering

Consulting Engineering Practice Manual. **American Consulting Engineers Council.** Edited by Stanley Cohen. New York: McGraw-Hill, 1981.

A detailed examination of the field of consulting engineering. Covers various fields of discipline, clientele, promotion of services, and interaction with other professionals. A useful reference for engineering students faced with career decisions.

Engineering Manual: A Practical Reference of Design Methods and Data in Building Systems, Chemical, Civil, Electrical, Mechanical, and Environmental Engineering and Energy Conversion. 3d ed. **Robert H. Perry, ed.** New York: McGraw-Hill, 1976.

First edition, 1959; second edition, 1967.

Includes expanded and updated information providing essential working concepts in all special fields of engineering. Contains numerous formulas and tables pertaining to the daily needs of the engineer.
R: *NTB* 52: 268 (1968); *RSR* 5: 24 (Oct./Dec. 1977); *SL* 58: 733 (1967); *ARBA* (1977, p. 753); Jenkins (K125); Wal (p. 289); Win (EI7; 2EI4).

Engineer-in-Training Review Manual. **Michael R. Lindeburg.** San Carlos, CA: Professional Engineering Institute, 1979.

Designed for engineers preparing for the engineer-in-training examination. Contains information on all relevant subjects.
R: *Lighting Design & Application* 9: 52 (Nov. 1979); *TBRI* 46: 35 (Jan. 1980).

Handbook of Technical Writing Practices. Vols. 1 and 2. **S. Jordan.** New York: Wiley, 1971.

R: *Engineer* (Dec. 23, 1971).

Aviation

Airman's Information Manual. **Aeronautical Staff of Aero Publishers.** Fallbrook, CA: Aero Publishers, 1981.

This manual for pilots contains standard information and provides pilots with most of the material they need. Also provides supplementary information on airport facilities, an appropriate directory, etc.
R: *BL* 74: 1414 (May 1, 1978); 77: 909 (Mar. 1, 1981); *ARBA* (1979, p. 778).

Airman's Information Manual. **Walter P. Winner.** Glendale, CA: Aviation, 1982.

Airman's Information Manual. Washington, DC: US Government Printing Office.

Part 1, *Basic Flight Manual and ATC Procedures,* quarterly; part 2, *Airport Directory,* semiannually; part 3, *Operational Data,* quarterly; part 3a, *Notices to Airmen,* biweekly; part 4, *Graphic Notices,* quarterly.

Airport Services Manual. 2d ed. 3 vols. **International Civil Aviation Organization.** New York: Unipub, 1978.

First edition, 1975.
Provides assistance in adopting measures for safe airport operation.
R: *IBID* 4: 148 (June 1976); 7: 58 (Spring 1979).

Aviation Weather and Weather Services. Detroit: Gale Research, 1980.

Originally published by the Federal Aviation Administration and the National Oceanic and Atmospheric Administration, 1975.
An official government manual that contains all aspects of weather as it affects pilots. Provides numerous illustrations and describes latest aviation services.

CHEMICAL ENGINEERING

Cold and Freezer Storage Manual. 2d ed. **Willis R. Woolrich and Elliot R. Hallowell.** Westport, CT: AVI, 1980.

First edition, 1970.
Reference manual to cold-storage warehousing. Sections discuss construction design and materials, equipment, recommended management practices, etc.

Food Industries Manual. 20th ed. **Anthony Woolen, ed.** New York: Chemical, 1980.

A comprehensive source of practical food technology, applicable to all branches of the food industry.

Manual of Economic Analysis of Chemical Processes: Feasibility Studies in Refinery and Petrochemical Processes. **Alain Chaiwel et al.** Translated by Ryle Miller and Ethel B. Miller. New York: McGraw-Hill, 1980.

Translated from the French. Presents explanations, data, and methods needed to calculate the profit in chemical manufacturing. Discusses cost engineering, business forecasting, and market research. For chemical engineers, consultants, and investors.

Manual of Food Quality Control. Vols. 1–. Rome: Food and Agriculture Organization of the United Nations, 1979–.

Volume 1, *Food Control Laboratory*, 1979; volume 2, *Additives, Contaminants, Techniques*, 1979; volume 3, *Commodities*, 1979; volume 6, *Food for Export*, 1979.

A comprehensive series of manuals focusing on all aspects of food quality control. Covers exports, additives, commodities, chemical analysis, etc. Recommended reference for all nutritionists, dietitians, and academic libraries.

R: *IBID* 8: 16 (Spring 1980); 8: 91 (Summer 1980).

Pesticide Manual. 6th ed. **Charles R. Worthing, ed.** Croydon, England: British Crop Protection Council, 1979.

This updated manual includes 47 new entries that cover active components currently in use or under development in the pesticide industry. Essential for workers in pesticides and related fields.

R: *Chemistry and Industry* 11: 463 (June 7, 1980).

Textile Manual. Montreal: Canadian Textile Journal, 1975. Annual.

CIVIL ENGINEERING

ACI Manual of Concrete Inspection. 7th ed. **American Concrete Institute.** Detroit: American Concrete Institute, 1981.

Sixth edition, 1975.

Concrete Bridge Designer's Manual. **E. Pennells.** London: Viewpoint Publications, 1978.

A generously illustrated manual on concrete bridge design. A handy guide for the engineer.

Concrete Manual. 8th ed. **USDI.** New York: Wiley, 1983.

Design Mix Manual for Concrete Construction. **Leslie "Doc" Long et al.** New York: McGraw-Hill, 1981.

A self-teaching manual on the basics of concrete design. Examines the workability, durability, consistency, strength, and density of concrete mixes. Includes 270 tables, information sheets, procedures, computations, a glossary, and a bibliography. Valuable tool for engineers, contractors, lab technicians, and students.

Field Engineer's Manual. **R. O. Parmley.** New York: McGraw-Hill, 1981.

A comprehensive, ready reference manual to construction problems and planning. Discusses construction materials, surveying, mechanical systems, electrical construction, drainage, hydraulics, etc. For engineers and construction personnel.

Geotechnical Engineering Investigation Manual. **Roy E. Hunt.** New York: McGraw-Hill, 1983.

Divided into investigation methods, characteristics of geologic materials and formations, and geologic hazards in such a way as to merge geology with civil engineering.

R: *Civil Engineering* 53: 92 (Nov. 1983).

Manual for Maintenance Inspection of Bridges. Washington, DC: American Association for State Highway and Transportation Officials, Operating Subcommittee on Bridges and Structures, 1978.
Second edition, 1974.

Manual of Concrete Practice. 3 vols. Detroit: American Concrete Institute, 1972–1975.
ACI standards.

Manual of Energy Saving in Existing Buildings and Plants. 2 vols. **Stephen P. E. Baron.** Englewood Cliffs, NJ: Prentice-Hall, 1978.
Volume 1, *Operation and Maintenance*; volume 2, *Facility Modifications*.
A 2-volume handbook that illustrates energy-saving methods in commercial, industrial, and residential buildings. Shows how to reduce costs; contains detailed checklists.

A Manual of Geology for Civil Engineers. **J. Pitts.** New York: Wiley, 1985.
A useful manual that treats the major aspects of descriptive geology, especially rock types and structural studies.

Manual of Highway Road Materials and Design. **Harold Atkins.** Englewood Cliffs, NJ: Prentice-Hall, 1979.
Focuses on field techniques used in testing highway and road materials. Comprises a basic guide to testing and sampling soils, aggregates, asphalts, and concretes. Contains helpful diagrams, photographs, and illustrations.

Manual of Precast Concrete Construction. 3 vols. **T. Koncz.** New York: International Publications Service, 1971–1978.
Volume 1: *Principles, Roof and Floor Units, Wall Panels,* 2d ed., 1976; volume 2: *Industrial Shed-Type and Low-Rise Buildings,* 1971; volume 3: *Multi-Story Industrial and Administrative Buildings; School, University and Residential Buildings,* 1978.

Manual of Soil Laboratory Testing. Vols. 1–. **K. H. Head.** New York: Halsted Press, 1980–.
Volume 1, *Soil Classification and Compaction Tests,* 1980; volume 2, *Permeability, Quick Shear Strength and Compressibility Tests,* 1982; volume 3; *Effective Stress Testing Principles, Theory and Applications,* 1986.
A practical manual on the technology of soil testing for civil engineering purposes. Topics covered include equipment; description of soils; testing and techniques; and safety. Chapters contain introduction; definitions; tests and procedures; and basic concepts. Suitable for lower-level undergraduate and community college students interested in soil testing. Useful to laboratory technicians and civil engineers.
R: *Choice* 18: 817 (Feb. 1981).

Manual of Water Utility Operations. 7th ed. **Clayton H. Billings, ed.** Austin: Texas Water Utilities Association, 1979.

A standard reference for municipal water supply operations. Updated information complies to US Environmental Protection Agency regulations.
R: *Public Works* 110: 32 (Dec. 1979); *TBRI* 46: 70 (Feb. 1980).

Metric Architectural Drawing: A Manual for Designers and Draftsmen. **B. L. Frishman, L. Loshak, and C. S. Strelka.** New York: Wiley, 1981.
A manual for architects for conversion into the SI metric world.
R: *AIA Journal* (Sept. 1981).

Structural and Construction Design Manual. **James M. Gere and Helmut Krawinkler.** Palo Alto, CA: Equipment GuideBook, 1978.
R: *Plant Engineering* 32: 282 (Oct. 12, 1978); *TBRI* 44: 392 (Dec. 1978).

Timber Construction Manual. **American Institute of Timber Construction.** 3d ed. New York: Wiley, 1985.
An updated guide to the design of structural timber members and their fastenings. Emphasizes the design of single members such as columns, beams, arches, and trusses and light repetitive members. Features tables, examples, and national standards.

ELECTRICAL AND ELECTRONICS ENGINEERING

AIM 65 Laboratory Manual and Study Guide. **L. Scanlon.** New York: Wiley, 1981.
Gives a basic step-by-step introduction to assembly language for 6502 microprocessors through a series of experiments. Emphasizes practical applications programming. For beginning programmers.

Automotive Electrical Reference Manual. **D. A. Westlund.** Englewood Cliffs, NJ: Prentice-Hall, 1983.

Data Processing Documentation and Procedures Manual. **Larry Long.** Englewood Cliffs, NJ: Prentice-Hall, 1979.
An essential comprehensive guide to development form and procedures of information systems. Thoroughly explains all functions.

Electrical Engineers' Reference Book. 13th ed. **M. G. Say.** London: Butterworth, 1973. Distr. Levittown, NY: Transatlantic Arts, 1975.
Twelfth edition, 1968.
Actually the thirteenth edition of *Newnes Electrical Engineers' Reference Book.* Contains 24 sections grouped under 4 broad subject areas: basics, energy supply, power plant, and applications. Comprehensive up-to-date handbook for electrical engineers.
R: *Choice* 12: 374 (May 1975); *ARBA* (1976, p. 772); Wal (p. 313).

Electronic Data Reference Manual. **Matthew Mandl.** Englewood Cliffs, NJ: Prentice-Hall, 1979.
Provides essential information on current developments in electrical communications. Covers integrated circuits, AM, FM. For reference on service and construction design.

Contains concise illustrations. A manual designed for engineers and repair persons. Suitable as a reference book in an engineering or technical library.
R: *Choice* 17: 106 (Mar. 1980); *ARBA* (1981, p. 759).

Electronic Design with Off-the-Shelf Integrated Circuits. **Z. H. Meiksin and Philip C. Thackray.** Englewood Cliffs, NJ: Prentice-Hall, 1979.
A well-illustrated manual that provides statistical data on how to arrange circuitry. A handy reference.

Microprocessor-Microcomputer Technology. **Frederick F. Driscoll.** New York: Van Nostrand Reinhold, 1983.
A helpful tool in using 8080A, 6800, and 6502 microprocessors as well as a wide range of other microcomputer systems.
R: *IEEE Spectrum* 20: 16H (Oct. 1983).

Modern Electronic Circuits Reference Manual. **John Markus.** New York: McGraw-Hill, 1980.
Includes over 3,600 circuits utilizing a range of different transistors and integrated circuits. Circuits are referenced to original publication sources. Lists of addresses of publications, abbreviations, and an index are provided. A worthwhile purchase for any technical library.
R: *Choice* 18: 820 (Feb. 1981); *RQ* 20: 325 (Spring 1981); *RSR* 8: 63 (Oct./Dec. 1980).

Opto-Electronics/Fiber Optics Applications Manual. 2d ed. **Hewlett-Packard Corporation.** New York: McGraw-Hill, 1982.
First edition, 1977.
Supplies technical information needed to design circuits, solve problems, and specify components. Contains software and circuit examples and solutions to common problems. Important topics includes photometry/radiometry, and reliability of opto-electronic components. For engineering collections.

Programmer's ANSI COBOL Reference Manual. **Donald A. Sordillo.** Englewood Cliffs, NJ: Prentice-Hall, 1978.
A highly recommended reference for those involved with COBOL operations, which include planning and programming. Arranged both alphabetically and consequentially, this work contains over 500 cross-referenced entries, as well as charts and diagrams.
R: *Datamation* 24: 310 (Sept. 1978); *TBRI* 44: 332 (Nov. 1978).

Quick Reference Manual for Silicon Integrated Circuit Technology. **W. E. Beadle, J. C. C. Tsai, and R. D. Plummer.** New York: Wiley, 1985.
A quick reference tool on all questions related to the design, development, processing, and manufacture of reliable semiconductors.

Radio Control Manual—Systems, Circuits, Construction. 3d ed. **Edward L. Safford, Jr.** Blue Ridge Summit, PA: TAB Books, 1979.
A basic reference on electronic systems and instruments used in radio control systems. Includes recent innovations as well as diagrams and photographs. Suitable for students with knowledge of electronic instrumentation.

Reference Manual for Telecommunications. **R. L. Freeman.** New York: Wiley, 1985.

A source book for scientists, engineers, and technicians involved in system and subsystem design, telecommunications planning and plant extension.

Solid-State Circuit Design Users' Manual. **Matthew Mandle.** Reston, VA: Reston Publishing, 1977.

A reference for electricians; provides data on design programs, parameters, and graphs.
R: *IEEE Proceedings* 65: 1632 (Nov. 1977); *TBRI* 44: 77 (Feb. 1978).

VHF/UHF Manual. 3d ed. **D. S. Evans and G. R. Jessop.** London: Radio Society of Great Britain, 1976.

An informative manual for radio technicians and engineers.
R: *Telecommunications Journal* 43: 744 (Dec. 1976); *TBRI* 43: 114 (Mar. 1977).

Visual Display Terminals: A Manual Covering Ergonomics, Workplace Design, Health and Safety, Task Organization. **A. Cakir, D. J. Hart, and T. F. M. Stewart.** New York: Wiley, 1980.

A manual on the design of visual display terminals and workplaces. Useful for designers and planners and computer systems.

INDUSTRIAL ENGINEERING

Industrial New Product Development: A Manual for the 1980s. **J. W. Carson and T. Rickards.** New York: Halsted Press, 1979.

A systematic approach to developing new industrial products. Presents information in a step-by-step manner. Contains checklists and diagrams.

Standard Plant Operators' Manual. 3d ed. **Stephen Michael Elonka.** New York: McGraw-Hill, 1980.

Second edition, 1975.
Emphasizes energy conservation in the power plant. Reprint of many popular articles on the subject, with ample photos and illustrations.
R: *ASHRAE J* 17: 96 (Aug. 1975); *TBRI* 41: 396 (Dec. 1975); *ARBA* (1976, p. 777).

MECHANICAL ENGINEERING

Design Manual for High Temperature Hot Water and Steam Systems. **R. E. Cofield.** New York: Wiley-Interscience, 1984.

A single-volume technical engineering source demonstrating a most practical hands-on approach for solving problems involved in the design and analysis of high-temperature hot water and also steam energy systems.
R: *Librarians' Newsletter* 23: 5 (Feb. 1984).

Fowler's Mechanical Engineer's Pocket Book. **W. H. Fowler, comp.** Manchester, England: Science. Annual.

Seventy-sixth edition, 1976.

Mechanical Engineer's Reference Book. 11th ed. **A. Parrish.** Cleveland: Chemical Rubber, 1973.

Metals Joining Manual. **Mel M. Schwartz.** New York: McGraw-Hill, 1979.

A reference volume describing various metals-joining processes. Includes both accepted processes and those still being evaluated. Contains illustrations and concise descriptions. For engineering libraries.

Welding Manual for Engineering Steel Forgings. **H. F. Tremlett.** London: Pentech, 1977.

For the staff of engineering firms; a reference in welding engineering.
R: *Welding and Metal Fabrication* 45: 511 (Oct. 1977); *TBRI* 44: 79 (Feb. 1978).

METALLURGY

Canadian Mining Journal's Reference Manual and Buyer's Guide. Montreal: Canadian Mining Journal, 1974. Annual.

Manual of Mineralogy. 19th ed. **C. S. Hurlbut and C. Klein.** New York: Wiley, 1977.

First edition, 1848.
A widely read introduction to mineralogy. Invaluable to geologists and amateur mineralogists.
R: *Journal of Metals* 30: 140 (Aug. 1978).

A Manual of New Mineral Names, 1892–1978. **Peter G. Embrey and John P. Fuller.** New York: Oxford University Press, 1980.

An alphabetical compilation of over 5,500 minerals. Author listing also included. A worthwhile reference for college libraries.
R: *Choice* 18: 926 (Mar. 1981).

Manual of Steel Construction. 7th ed. New York: American Institute of Steel Construction, 1973.

A basic tool for structural engineers.

Mineral Processing Technology: An Introduction to the Practical Aspects of Ore Treatment and Mineral Recovery. **B. A. Wills.** Oxford: Pergamon, 1979.

Outlines treatments of ore and mineral recovery, including size reduction, physical separation, and flotation. A valuable reference for students of mineral processing.
R: *Aslib Proceedings* 44: 493 (Nov. 1979).

Techniques in Mineral Exploration. **J. H. Reedman.** Barking, Essex, England: Applied Science Publishers, 1979.

Deals with main aspects of mineral exploration, including surveying, mapping, photogeology, remote sensing, prospecting, deep sampling, and drilling. Intended to serve as a single-volume reference for working geologists.
R: *Aslib Proceedings* 44: 531 (Dec. 1979).

Environmental Sciences

Air Pollution Sampling and Analysis Deskbook. **Paul N. Cheremisinoff and Angelo C. Morres.** Ann Arbor, MI: Ann Arbor Science, 1978.

A compilation of several books on air pollution analysis and sampling. Manuals by several agencies and societies have been included. Contains the pros and cons of several methods of sampling, as well as tables.
R: *RSR* 8: 3; 26 (July/Sept. 1980).

Earthscape: A Manual of Environmental Planning. **John O. Simonds.** New York: McGraw-Hill, 1977.

A massive, comprehensive book on maintaining and improving environmental quality. Profusely illustrated with photographs, drawings, and diagrams. Contains the best ideas from current literature. Highly recommended.
R: *Blair & Ketchum's Country Journal* 6: 107 (Jan. 1979); *TBRI* 45: 78 (Feb. 1979).

Laboratory and Field Manual of Ecology. **Richard Brewer and Margaret T. McCann.** Philadelphia: Saunders, 1982.
R: *Science* 217: 465 (July 30, 1982).

Manual for Environmental Impact Evaluation. **Sherman J. Rosen.** Englewood Cliffs, NJ: Prentice-Hall, 1976.

Covers various aspects of environmental impact evaluation such as the National Environmental Policy Act and techniques and procedures for those involved in environmental assessment. Useful to those in the field of environmental evaluation.

Manual of Analytical Quality Control for Pesticides and Related Compounds in Human and Environmental Samples. **Lafayette College, Department of Chemistry.** Easton, PA: Lafayette College Press, 1976.

A reference for the pesticide chemist. Aids in the analysis of pesticide substances.
R: *RSR* 5: 31 (July/Sept. 1977).

Manual of Grey Water Treatment Practice. **J. H. T. Winneberger, ed.** Ann Arbor, MI: Ann Arbor Science, 1976.

Provides detailed information on grey water and land disposal systems.
R: *Journal of the Institution of Water Engineers and Scientists* 30: 401 (Nov. 1976); *TBRI* 43: 120 (Apr. 1977).

Manual of Decontamination of Surfaces. **International Atomic Energy Agency.** New York: International Atomic Energy Agency, 1979.

Provides guidelines on methods of decontaminating work spaces, laboratory equipment, and clothing, and on the removal of loose skin contamination from personnel.
R: *IBID* 7: 161 (Summer 1979).

Manual on Oil Pollution. 5 vols. **Intergovernmental Maritime Consultative Organization.** New York: Unipub, 1976–.

Section 1, *Prevention*, 1976; section 2, *Contingency Planning*, 1978; section 3, *Salvage*, in preparation; section 4, *Methods of Dealing with Spillages*, 1977; section 5, *Legal Aspects*, will be developed in the future.
Especially helpful for developing countries; provides guidelines on marine pollution and related contingency plans in oil pollution emergencies.
R: *IBID* 7: 139 (Summer 1979).

Manual on Urban Air Quality Management. WHO Regional Publications European Series, No. 1. **M. J. Suess and S. R. Crawford, eds.** Geneva: World Health Organization, 1976. Distr. Albany, NY: Q Corporation.

Orbital Remote Sensing of Coastal and Offshore Environments: A Manual of Interpretation. **H. G. Gierloff-Emden.** Hawthorne, NY: De Gruyter, 1977.
Contains a useful summary of sensor and image characteristics relating to the American space program; includes 13 color images of the North and South American coasts. Comprises a thorough interpretation guide.
R: *Aslib Proceedings* 43: 29 (Jan. 1978).

Standard Methods for the Examination of Water and Wastewater. 15th ed. **Arnold Greenberg et al., eds.** Washington, DC: American Public Health Association, 1981.
First edition, 1905; thirteenth edition, 1971; fourteenth edition, 1975.
Covers all major areas of determining constitution of water. Includes bibliography and index.
R: Sheehy (ED8).

Toxic and Hazardous Industrial Chemical Safety Manual for Handling and Disposal; With Toxicity and Hazard Data. **Gijutsu S. K. Kaigai.** Tokyo: International Technical Information Institute, 1976.
Contains safety information on 702 commercial substances. Provides information about use, toxicity, properties, synonyms, disposal, handling and storage, emergency treatment, etc. A detailed manual on the treatment and handling of chemicals for those involved with toxic chemical disposal.
R: *RSR* 5: 13 (Apr./June 1977).

NAVIGATION

Navigation Afloat: A Manual for the Seaman. **Alton B. Moody.** New York: Van Nostrand Reinhold, 1981.
A detailed discussion of the basics of charting and pilotage. Material is not simplified and uses American terminology, symbols, and charting methods. Includes an extensive bibliography and glossary, as well as a listing of symbols and abbreviations. A comprehensive reference for seamen.
R: *BL* 77: 1180 (May 1, 1981).

SOURCE BOOKS

SCIENCE

Archaeological Chemistry: A Sourcebook on the Applications of Chemistry to Archaeology. **Zvi Coffer.** New York: Wiley-Interscience, 1980.
Contains several topics focusing on the area where chemistry and archaeology intersect. It is not deep enough to be used for a text or as a sourcebook for the archaeologist-chemist.
R: *American Scientist* 68: 685 (Nov./Dec. 1980).

Gary Null's Nutrition Source Book for the '80's. **Gary Null, ed.** New York: Collier Books/Macmillan, 1983.
A food guide for people on restricted diets and for those who wish to develop better eating habits. More suitable for home use.
R: *ARBA* (1985, p. 500).

Genetic Engineering/Biotechnology Source Book. **Robert G. Pergolizzi.** New York: McGraw-Hill, 1982.
A listing of over 1,500 genetic engineering research projects funded by the US government and private sources since 1978. Lists the type of project, money received, etc. Good for researchers and executives.
R: *Nature* 299: 765 (Oct. 28, 1982).

Industrial Enzymology: The Application of Enzymes in Industry. **Tony Godfrey and Jon Reichelt.** New York: Macmillan, 1982.
Covers the sources of enzymes and their functions in industry. Includes suppliers of enzymes throughout the world, and a tabulated list of enzymes permitted in many countries. Unique cross-referencing.
R: *Nature* 298: 501 (Aug. 5, 1982).

Laboratory Text for Organic Chemistry: A Source Book of Chemical and Physical Techniques. **D. J. Pasto and C. R. Johnson.** Englewood Cliffs, NJ: Prentice-Hall, 1979.
In 3 parts: organic laboratory techniques, absorption spectroscopy, and mass spectroscopy. Intended for use in conjunction with an organic text book. A useful reference.
R: *Chemistry in Britain* 16: 432 (Aug. 1980); *Aslib Proceedings* 44: 513 (Dec. 1979).

Light, Life, and Chemical Change: A Source Book in Photochemistry. **J. Coyle, R. Hill, and D. Roberts, eds.** Milton Keynes, England: Open University Press, 1982.
This collection of source books shows the relevance of light-induced changes to life and the environment. Examines biological and technical applications. Useful to those interested in chemistry and technology.
R: *Chemistry in Britain* 18: 728 (Apr. 1982).

1001 Questions Answered About Astronomy. Rev. ed. **James S. Pickering.** New York: Dodd, Mead, 1975.

Provides information on various astronomical phenomena, including the sun, earth, planets, stars, comets, and galaxies.

Source Book for Food Scientists. **Herbert W. Ockerman.** Westport, CT: AVI, 1978.

An invaluable compilation of facts and data relevant to food science and technology. In 2 parts: alphabetic descriptions of foods and food terms, and tables illustrating food composition and properties. Highly recommended.
R: *Chemical Engineering* 85: 11 (Sept. 11, 1978); *ARBA* (1979, p. 755).

Source Book for Programmable Calculators. **Texas Instruments.** New York: McGraw-Hill, 1979.

Explores the multiple uses of programmable calculators. Helpful in many scientific applications.
R: *Mechanical Engineering* 101: 109 (Dec. 1979); *TBRI* 46: 49 (Feb. 1980).

A Source Book in Astronomy and Astrophysics, 1900–1975. **Kenneth R. Lang and Owen Gingerich, eds.** Cambridge, MA: Harvard University Press, 1979.

Presents original articles and papers dealing with advances in astronomy. Includes 132 articles from popular magazines as well as scientific journals. Divided into 8 chapters, which discuss such topics as the solar system, stellar evolution, variable stars, galaxies and quasars, and relativity. Each article consists of an introduction, citations to earlier works, background information, and the main body, which is abridged and annotated. Provides references and subject and author indexes. For science reference collections in observatory and academic libraries. Of interest to astronomers and science historians.
R: *American Scientist* 68: 448 (July/Aug. 1980); *Nature* 284: 707 (Apr. 24, 1980); *New Scientist* 87: 799 (Sept. 11, 1980); *Physics Today* 33: 58 (Apr. 1980); *Scientific American* 242: 30 (May 1980); *Sky and Telescope* 59: 154 (Feb. 1980); 60: 53 (July 1980); *ARBA* (1981, p. 628).

Source Book of Experiments for the Teaching of Microbiology. **S. B. Primrose and A. C. Warlaw, eds.** New York: Academic Press, 1982.

A listing of experiments to stimulate student interest. Over 100 experiments were chosen from the 1,000 submitted. Information about apparatus needed and the level and subject is included. Excellent for teachers.
R: *Nature* (Nov. 11, 1982).

Sources of Invention. 2d ed. **J. Jewkes.** London: Macmillan, 1971.
R: Wal (p. 239).

Strange Universe: A Source Book of Curious Astronomical Observations. 2 vols. **William R. Corliss, comp.** Glen Arm, MD: Source Book Project, 1975.

Covers mysteries of the earth's atmosphere. Maintains classified arrangement and is well indexed, making information easy to retrieve. Recommended for use by public and academic libraries.
R: *BL* 74: 1764 (July 15, 1978).

A Source Book of Titanium Alloy Superconductivity. **E. W. Collings.** Columbus, OH: Battelle Memorial Institute. New York: Plenum, 1983.

The first volume to present, systematically and exclusively, information regarding titanium alloys as superconductors. A complete manual of historical, scientific, and technical data on binary and multicomponent superconducting alloys.
R: *Physics Today* 36: 92 (Oct. 1983).

ENGINEERING

The Aviator's Catalog: A Source Book of Aeronautica. **Timothy R. V. Foster.** New York: Van Nostrand Reinhold, 1981.

This book, with an extensive listing of information on the subject matter, will answer many questions about flying. Useful for potential and already trained pilots and for anyone with a general question about flying.
R: *ARBA* (1982, p. 811).

The Blue Book for the Atari Computer: The Complete "Where to Find It" Book of Software, Hardware, and Accessories for the Atari 400, 800, and XL Series Computers. Chicago: WIDL Video Publications. Distr. New York: Scribner's, 1983.
R: *ARBA* (1985, p. 591).

The Blue Book for the Apple Computer: The Complete "Where to Find It" Book of Software, Hardware, and Accessories for the Apple II, Apple IIe, Franklin, and Other Apple-Compatible Computers. 3d ed. Chicago: WIDL Video Publications. Distr. New York: Scribner's, 1983.
R: *ARBA* (1985, p. 590).

Bowker's Complete Sourcebook of Personal Computing 1985. New York: Bowker, 1984. Annual update.

A guide that includes reviews and descriptions of microcomputer software, and a directory of 5,000 companies, plus software/hardware checklists, user groups, online database information, etc.

Directory of Industrial Heat Processing and Combustion Equipment, US Manufactures: 1981–1982. 3d ed. New York: Information Clearinghouse, 1981.

Up-to-date source of technical information on all kinds of equipment, including furnaces, ovens, kilns, etc. Covers industrial heating and combustion systems.

A Guidebook to Nuclear Reactors. **Anthony V. Nero, Jr.** Berkeley, CA: University of California Press, 1979.

A clear and accurate source on nuclear reactors. Explains operation, fuel supply, and safety problems. Intended for the layperson or college student. A thorough source of reference. Highly recommended.
R: *American Scientist* 67: 472 (July–Aug. 1979); *IEEE Spectrum* 16: 97 (Oct. 1979); *Nature* 283: 231 (Jan. 10, 1980); *Physics Teacher* 17: 607 (Dec. 1979); *LJ* 104: 813 (Apr. 1, 1979); *TBRI* 45: 315 (Oct. 1979); 45: 390 (Dec. 1979); 46: 76 (Feb. 1980).

Heat Exchanger Sourcebook. **J. W. Palen, ed.** New York: Hemisphere, 1985.

IC Schematic Sourcemaster. **Kendall W. Sessions.** New York: Wiley, 1978.

Provides 1,500 schematic diagrams for electronic circuits. Information is clear and concise. Covers the following circuit classes: power sources, signal conditioners, and voltage regulators. All schematics contributed by integrated circuit manufacturer. Recommended.

1984–1985 International Micrographics Source Book. White Plains, NY: Knowledge Industry Publications, 1984.

Completely revised and updated, this single-volume compendium covers more than 20 categories of worldwide sources of micrographics information. Also includes an easy-to-use alphabetical index.

Metalcutting: Today's Techniques for Engineers and Shop Personnel. **American Machinist editors.** New York: McGraw-Hill, 1979.

A collection of special reports covering basic metalworking processes such as laser cutting and metalcutting. Ample illustrations of tools are included, as are tables, charts, checklists, and glossaries. Of interest to both students and industrial engineers.

Parent/Teacher's Microcomputing Sourcebook for Children 1985. New York: Bowker, 1985. Annual update.

A sourcebook that taps into the specific world of children's computing with information for hardware and software; indexed by age group and subject, books and magazines, etc.

Source Book of Flavors. **Henry B. Heath.** Westport, CT: AVI, 1981.

Describes all aspects of the flavor industry, its formulations, and related organizations. Specifically for use by flavor chemists, students, and apprentices in this area.
R: *ARBA* (1983, p. 607).

Source Book of Food Enzymology. **Sigmund Schwimmer.** Westport, CT: AVI, 1981.

Examines and evaluates food enzymes and their applications. Includes in-depth sections on enzyme production, control, and management. An index allows easy access to data and references. Good for food researchers and students.
R: *Choice* 19: 1266 (Dec. 1982).

Source Book on Brazing and Brazing Technology. **Melvin M. Schwartz, ed.** Metals Park, Ohio: American Society for Metals, 1980.

A source book of articles on topics such as introduction to brazing processes, selected brazing processes, special brazing applications, and more. For research and special libraries.

Source Book on Food and Nutrition. 2d ed. **Ioannis S. Scarpa et al., eds.** Chicago: Marquis Academic Media; Marquis Who's Who, 1980.

First edition, 1978.
A compilation of articles in nutrition and food science taken from journals and government documents. Topics discussed include consumer information, dietary goals, food technology, dieting, and food safety. For dieticians and nutritionists.
R: *RQ* 19: 97 (Fall 1979); 20: 106 (Fall 1980); *ARBA* (1981, p. 741).

Source Book on Industrial Alloy and Engineering Data. **American Society of Metals.** Metals Park, OH: American Society of Metals, 1978.
A source book divided into 10 sections covering such topics as carbon and alloy steels, tool steels, aluminum and aluminum alloys, and powder metal parts. Data arranged in tabular and graphical form. Includes an additional section of conversion tables.
R: *RSR* 8: 26 (July/Sept. 1980).

Source Book on Materials for Elevated-Temperature Applications. **Elihu F. Bradley, ed.** Metals Park, OH: American Society of Metals, 1980.
An outstanding reference guide to the selection of materials that are useful at elevated temperatures. Covers stainless steels, heat-resistant alloys, and major superalloys. Topics discussed in detail include protective coatings, welding techniques, structure and property control. Valuable for all engineers and designers involved in the elevated-temperature field.
R: *Journal of Metals* 32: 67 (Apr. 1980).

Source Book on Powder Metallurgy. **American Society for Metals.** Metals Park, OH: American Society for Metals, 1979.

Source Book: Small Systems Software and Services Source Book. **Ruth K. Koolish, ed.** Glenview, IL: Information Sources, 1982.

Source Book I: Small Systems Software and Services Source Book. Supp. 1983.
A directory for the small business computer user with consultants and relevant companies cited. Both volumes together contain 3,000 entries of software, but the supplement is much more comprehensive overall.
R: *ARBA* (1984, p. 626).

Standard Boiler Room Questions and Answers. 3d ed. **Stephen M. Elonka and Alex Higgins.** New York: McGraw-Hill, 1981.
An authoritative reference to various aspects of steam generation. Utilizes a question-and-answer approach in covering equipment, safety, and types of boilers. A standard reference for experienced and beginning boiler operators.

Standard Plant Operator's Questions and Answers. 2d ed. 2 vols. **Stephen M. Elonka and Joseph F. Robinson.** New York: McGraw-Hill, 1981.
A 2-volume practical reference to the construction, maintenance, and operation of power plant equipment. Includes over 1,400 questions and answers for preparation for the license examination. For plant operators.

Technology Policy and Development: A Third World Perspective. **Pradip K. Ghosh, ed.** Westport, CT: Greenwood Press, 1984.

A handy guide to analyzing current trends in technology policy and development in the Third World. Good source book for both libraries and individuals.
R: *ARBA* (1985, p. 492).

Transmission Line Reference Book: 345KV and Above. 2d ed. Palo Alto, CA: Electric Power Research Institute, 1982.

A useful source of technology and data for the electrical design of EHV and UHV transmission lines up to 1500KV.
R: *IEEE Proceedings* 71: 477 (Mar. 1983).

Underground Coal Gasification. **George H. Lamb.** Park Ridge, NJ: Noyes Data, 1977.

Provides an overview of underground gasification, relying mainly on US government studies. A useful reference.
R: *Fuel* 57: 61 (Jan. 1978); *TBRI* 44: 195 (May 1978).

ENERGY AND ENVIRONMENTAL SCIENCES

Energy Management: A Source Book of Current Practices. Atlanta: Fairmont Press, 1982.

Contains an extensive listing of the latest technology relating to industrial and building utilization. Some topics covered are waste heat recovery, products and applications, and energy use analysis.
R: *Heating/Piping/Air Conditioning* 54: 150 (Mar. 1982).

The Energy Source Book. **Alexander McRae and Janice L. Dudas, eds.** Germantown, MD: Aspen Systems Corp., 1978.

Data culled from a wide variety of sources. Covers all forms of energy and provides quantitative description. Well-referenced, thorough, and profusely illustrated with charts, diagrams, and maps. A reliable reference for academic and public libraries.
R: *Chemical Engineering* 86: 12 (Jan. 15, 1979); *Plant Engineering* 33: 218 (Mar. 22, 1979); *L J* 103: 962 (May 1, 1978); *TBRI* 45: 117 (Mar. 1979); 45: 168 (May 1979); *ARBA* (1979, p. 687).

Environmental Engineering and Sanitation. 3d ed. **Joseph A. Salvato.** New York: Wiley, 1982.

A source covering problems in sanitary and environmental engineering and administration with state-of-the-art tools and methods in waste management, disease control, etc. Good for health officers, planners, consultants, and engineers.
R: *Science* 215: 7 (Jan. 1, 1982).

Lagoon Information Source Book. **E. Joe Middlebrooks et al.** Ann Arbor, MI: Ann Arbor Science, 1978.

A reference on wastewater sanitation lagoons, including information on performance, operation, and historical development of lagoons. Also provides an annotated bibliography. Considered a useful and detailed reference.
R: *ARBA* (1979, p. 777).

The Passive Solar Energy Book: A Complete Guide to Passive Solar Home, Greenhouse, and Building Design. **Edward Mazria.** Emmaus, PA: Rodale Press, 1979.

Extensive technical data and narrative information concerning passive systems is organized into short entries. Serves as a primer, source book, and workbook. Considered an essential purchase.
R: *LJ* 104: 586 (Mar. 1, 1979); *TBRI* 46: 77 (Feb. 1980).

Solar Energy Source Book: For the Home Owner, Commercial Builder and Manufacturer. 2d ed. **Christopher W. Martz, ed.** Washington, DC: Solar Energy Institute of America, 1978.

In 2 major sections, products and services, each arranged by broad subject categories. Provides detailed information, including a section on solar energy legislation, bibliography, glossary, and directory.
R: *Choice* 15: 380 (May 1978); *ARBA* (1979, p. 699).

Source Book for Farm Energy Alternatives. **James D. Ritchie.** New York: McGraw-Hill, 1982.

Intended for farmers and farm builders, this book addresses such areas as controlling temperatures to save animals, using energy-efficient cropping, promoting farm-grown fuels, etc. Lets farmers evaluate energy options.

Source Book on Asbestos Diseases: Medical, Legal, and Engineering Aspects. **George A. Peters and Barbara J. Peters.** New York: Garland Publishing, 1980.

A useful guide for anyone interested in defending a lawsuit involving the medical consequences of asbestos. Contains mostly a chronological bibliography and selected references, as well as medical aspects and controls and protection for worker and consumer.
R: *American Scientist* 68: 699 (Nov./Dec. 1980).

Source Book on the Production of Electricity from Geothermal Energy. **Joseph Kestin et al., eds.** Washington, DC: US Department of Energy, 1980.

Three Mile Island Source Book: Annotations of a Disaster. **Philip Starr and William Pearman.** New York: Garland, 1983.

Includes news coverage, government documents (both federal and state), and professional papers of Three Mile Island from November 18, 1966 to March 29, 1981. For both advocates and opponents of nuclear power and its further development.
R: *ARBA* (1984, p. 679).

Toxic Substances Control Source Book. **Alexander McRae and Leslie Whelchel, eds.** Germantown, MD: Aspen Systems, 1978.

A well-written reference focusing on various aspects of toxic substances control. Information includes material on federal laws and regulation; impact and management; and specific cases involving toxic substances. Includes 9 appendixes containing the full text of the Toxic Substances Control Act of 1976 and details on latest research. Recommended for those who are involved with toxic substances control in industry, government, and business.
R: *Choice* 15: 1541 (Jan. 1979); *ARBA* (1980, p. 654).

LABORATORY MANUALS AND WORKBOOKS

SCIENCE

Laboratory Manual for Schools and Colleges. **John Creedy.** London: Heinemann Educational, 1977.

Provides detailed information for school laboratory technicians. Covers all disciplines.
R: *Education in Chemistry* 14: 186 (Nov. 1977); *Physics Education* 12: 460 (Nov. 1977); *Physics Teacher* 17: 400 (Sept. 1979); *TBRI* 44: 2 (Jan. 1978); 44: 83 (Mar. 1978); 45: 320 (Nov. 1979).

MATHEMATICS

Mathematics for Technical Occupations. **Dennis Bila et al.** Englewood Cliffs, NJ: Prentice-Hall, 1978.

A workbook format that presents math problems relevant to technical occupations. Instructional material is clearly presented; covers a wide range of topics.

PHYSICS

Experiments in College Physics. 6th ed. **Bernard Cioffari.** Lexington, MA: Heath, 1978.

Sixth edition contains over 50 lab experiments. Experiments are well-structured and relate to all aspects of physics. Each experiment includes theory, data form, calculation, and question sections.
R: *Physics Teacher* 16: 494 (Oct. 1978); *TBRI* 44: 363 (Dec. 1978).

Experiments in Physics: A Laboratory Manual for Scientists and Engineers. **D. W. Preston.** New York: Wiley, 1985.

This laboratory manual is geared to beginning college students taking calculus-based physics courses.

Exploring Laser Light: Laboratory Exercises and Lecture Demonstrations Performed with Low-Power Helium-Neon Lasers. **Thomas Kallard.** New York: Optosonic, 1977.
R: *Optical Society of America Journal* 67: 1129 (Aug. 1977); *TBRI* 43: 325 (Nov. 1977).

Laboratory Manual for Physics in the Modern World. 2d ed. **Jerry Marion, Jean P. Hatheway, and Stephen M. Burroughs.** New York: Academic Press, 1981.

A collection of topics, including mechanics; electricity and magnetism; and waves, light and optics in the study of physics.

Landmark Experiments in Twentieth-Century Physics. **George L. Tigg.** New York: Crane-Russak, 1975.
R: *RSR* 4: 53 (Jan.–Mar. 1976).

Physics Demonstration Experiments. 2 vols. **H. F. Mieners, ed.** New York: Ronald Press, 1970.

A definitive reference on classic demonstration experiments in physics. Contains more than 1,100 entries and twice as many photographs and line drawings. Covers such topics as mechanics, magnetism, electricity, and atomic and nuclear physics. An effective teaching aid.

Physics Laboratory Manual. 3d ed. **Clifford N. Wall, Raphael B. Levine, and Fritjof E. Christensen.** Englewood Cliffs, NJ: Prentice-Hall, 1972.

Designed as a comprehensive beginning course in physics. Contains questions, problems, and numerous experiments.

CHEMISTRY

How to Solve General Chemistry Problems. 6th ed. **C. H. Sorum.** Englewood Cliffs, NJ: Prentice-Hall, 1981.

Fourth edition, 1969; fifth edition, 1976.

Meant for self-teaching. Problems are put forth and step-by-step solutions are given.

Identification of Molecular Spectra. **Reginald W. B. Pearse and A. G. Gaydon.** New York: Halsted Press, 1976.

A comprehensive manual devoted solely to the identification of molecular spectra. Provides ample data. A useful reference.

R: *Journal of the Optical Society of America* 67: 1129 (Aug. 1977); *TBRI* 43: 328 (Nov. 1977).

Instrumental Liquid Chromatography: A Practical Manual on High-Performance Liquid Chromatographic Methods. **N. A. Parris.** New York: Elsevier Scientific, 1976.

A laboratory manual on liquid chromatography. Practical, useful.

R: *Journal of Chromatography* 125: 451 (Oct. 13, 1976); *TBRI* 43: 8 (Jan. 1977).

Laboratory Manual for Fundamentals of Chemistry. 2d ed. **J. A. Beran.** New York: Wiley, 1984.

First edition, 1981.

For undergraduates majoring in science; emphasizes developing good laboratory techniques and skills.

R: *Librarians' Newsletter* 22: 40 (Dec. 1983).

Laboratory Manual of Physical Chemistry. 2d ed. **H. D. Crockford, J. W. Nowell, H. W. Baird, and F. W. Getzen.** New York: Wiley, 1976.

R: *Journal of the American Chemical Society* 100: 7787 (Nov. 22, 1978).

Quantitative Chemical Analysis: A Laboratory Manual. 4th ed. **R. A. Chalmers and M. S. Cresser.** New York: Halsted Press, 1982.

A practical manual featuring experiments illustrating the theory, and showing why procedural details are necessary. Contains historical background material and reference to the basic literature. For chemists at all levels.

Systematic Identification of Organic Compounds: A Laboratory Manual. 6th ed. **Ralph L. Shriner et al.** New York: Wiley, 1980.

Discusses chemical and spectral techniques of organic compound identification. Describes laboratory procedures, wet tests, and other useful data.

BIOLOGICAL SCIENCES

Basic Anatomy: A Laboratory Manual: The Human Skeleton, The Cat. **B. L. Allen.** San Francisco: W. H. Freeman, 1970.
Designed for basic laboratory courses. Combines the study of both feline and human anatomy.

Biological Systems: A Laboratory Manual. 4th ed. **Shelby D. Gerking.** Minneapolis: Burgess, 1974.

A Coded Workbook of Birds of the World. Non-passerines. Vol. 1. 2d ed. **Ernest Preston Edwards.** Sweet Briar, VA: Ernest P. Edwards, 1982.
First edition, 1974.
A listing of the living and recently extinct (since 1680) species of birds of the world. Each includes a unique alphanumeric identification code, its scientific name, English name, and a reference to the maps in the introduction. For most academic and larger public libraries.
R: *ARBA* (1984, p. 657).

Diagnostic Manual for the Identification of Insect Pathogens. **George O. Poinar, Jr. and G. M. Thomas.** New York: Plenum Press, 1978.
An illustrated manual of insect pathogens. Contains over 100 figures of disease organisms. A useful classroom tool.
R: *Quarterly Review of Biology* 53: 469 (Dec. 1978); *TBRI* 45: 190 (May 1979).

Exercise Manual in Immunology. **Lazar M. and Paula Schwartz.** Baltimore: Williams & Wilkins, 1975.

Experiments in Physiology. 3d ed. **Gerald D. Tharp.** Minneapolis: Burgess, 1976.

Freeze-Drying Biological Specimens: A Laboratory Manual. **Rolland O. Hower.** Washington, DC: Smithsonian Institution Press, 1979.
Comprises a comprehensive introduction to all aspects of freeze-drying. Well-illustrated. Covers freeze-drying of whole organisms and preparations of specimens for scanning electron microscopy. Includes line drawings, photographic illustrations, references, a glossary, and an index. Also contains charts of drying rates. Recommended for upper-level undergraduate and graduate students, as well as professionals.
R: *Choice* 17: 243 (Apr. 1980).

Fundamental Experiments in Microbiology. **Koby T. Crabtree and Ronald D. Hinsdill.** Philadelphia: Saunders, 1974.
Laboratory manual for students. Emphasizes soil and water microbiology.

Laboratory Exercises in Microbiology. 4th ed. **Michael J. Pelczar and E. C. S. Chen.** New York: McGraw-Hill, 1977.

Fourth edition contains new and additional information. Provides bibliography, appendixes, several new experiments.

A Laboratory Manual and Study Guide for Anatomy and Physiology. 3d ed. **Kenneth G. Neal and Barbara Kalbus.** Minneapolis: Burgess, 1976.

A Laboratory Manual for General Botany. **Margaret Balbach and Lawrence C. Bliss.** Philadelphia: Saunders, 1982.

R: *Science* 217: 465 (July 30, 1982).

Laboratory Manual of Cell Biology. **David C. Hall and S. E. Hawkins.** New York: Crane-Russak, 1974.

Laboratory Manual of General Ecology. 4th ed. **George W. Cox.** Dubuque, IA: Brown, 1980.

Laboratory Training Manual on the Use of Nuclear Techniques in Animal Research. Vienna: International Atomic Energy Agency. Distr. New York: Unipub, 1979.

Membrane Proteins: A Laboratory Manual. **A. Azzi, U. Brobeck, and P. Zahler.** New York: Springer-Verlag, 1981.

Describes many techniques useful in studying biological membranes first presented in an International Advanced Course on Membrane Proteins in Switzerland. Good as a text for an advanced course in methodology.

R: *Choice* 19: 950 (Mar. 1982).

Methods of Enzymatic Analysis. 2d ed. 4 vols. **H. U. Bergmeyer, ed.** Translated by D. H. Williamson. New York: Academic Press, 1974.

Expanded edition contains up-to-date information on analytical biochemistry procedures. Comprises a unique, easy-to-use laboratory manual. Fully descriptive. Recommended for analytical chemists, biochemists, clinical chemists, and physicians.

Molecules, Measurements, Meanings: A Laboratory Manual in Biochemistry. **David Krogmann.** San Francisco: Freeman, 1971.

R: *Journal of Chemical Education* (Dec. 1971).

Practical Invertebrate Zoology: A Laboratory Manual for the Study of the Major Groups of Invertebrates, Excluding Protochordates. 2d ed. **R. P. Dales.** New York: Halsted Press, 1981.

A comprehensive manual on invertebrate morphology and systematics. Provides a thorough background in animal construction and functional morphology. For students, lectureres, and researchers.

EARTH SCIENCES

Exercises in Physical Geology. 5th ed. **W. K. Hamblin and J. D. Howard.** Minneapolis: Burgess, 1980.

Fourth edition, 1975.

Geomorphology Laboratory Manual with Report Forms. **Marie Morisawa.** New York: Wiley, 1976.

Historical Geology: Manual of Laboratory Exercises. 3d ed. **Forbes S. Robertson and Frederick C. Marshall.** Minneapolis: Burgess, 1975.

ENGINEERING

Basic Electricity: A Text-Lab Manual. 5th ed. **Paul B. Zbar.** New York: McGraw-Hill, 1981.

A lab manual dealing with basic electrical concepts, with emphasis on solid-state devices and publication. Intended for introductory electrical and electronics courses at a community college or vocational school.

Building Science Laboratory Manual. **Henry J. Cowan and John Dixon.** London: Applied Science Publishers, 1978.

Presents material on the establishment of a building science laboratory. Includes bibliography, index, and references, as well as a variety of laboratory experiments. For university libraries.

Digital Electronics: A Hands-On Learning Approach. **George Young.** Rochelle Park, NJ: Hayden Book, 1980.

A laboratory manual that covers the basic concepts of digital electronics through experimental procedures. Includes praiseworthy diagrams. A helpful tool for technical and community college students for use as a laboratory guide in a basic electrical engineering course.
R: *Choice* 18: 821 (Feb. 1981).

Electronics Quizbook. **Rudolf F. Graf and George J. Whalen.** Indianapolis: Sams, 1975.

Exercises in Computer Systems Analysis. **W. Everling.** New York: Springer-Verlag, 1975.

Experimental Foods Laboratory Manual. **M. McWilliams.** Fullerton, CA: Plycon Press, 1977.

Considered a much-needed manual. Covers a wide range of foods.
R: *Food Technology* 32: 125 (Feb. 1978); *TBRI* 44: 156 (Apr. 1978).

FORTRAN 77: A Practical Approach. 4th ed. **Wilfred Rule.** New York: Van Nostrand Reinhold, 1983.

Provides FORTRAN grammar, information on writing effective programs, use of the many special features of the FORTRAN 77, etc. in the most advanced form of the FORTRAN programming language. Many sample programs.
R: *IEEE Spectrum* 20: 16H (Oct. 1983).

Introductory Soil Science: A Study Guide and Laboratory Manual. **Leon J. Johnson.** New York: Macmillan, 1979.

Presents laboratory exercises on concepts in soil science. Contains helpful illustrations and useful data.

Intuitive IC Electronics: A Sophisticated Primer for Engineers and Technicians. **Thomas M. Frederiksen.** New York: McGraw-Hill, 1981.
Intended as an advanced-level supplement to other texts, this primer explains the internal structures and mechanism of semiconductor devices. Covers such integrated circuit devices as the PN junction, bipolar transistor, MOS integrated circuit, multiple transistor integrated circuits, dynamic random-access memory, nonvolatile memories, etc. An excellent reference and lab guide for those involved in the semiconductor field: design engineers, technicians, and sales personnel.

Laboratory Experiments for Microprocessor Systems. **John Crane.** Englewood Cliffs, NJ: Prentice-Hall, 1981.
A lab manual intended to supplement a text in microprocessor fundamentals. The manual is composed of 21 experiments on hardware and software and covers basic signals, structures, and procedures applicable to any microprocessor. A basic understanding of digital electronics is expected, although computer experience is not required.

Laboratory Manual in Food Chemistry. **Alvin E. Woods and L. Aurand.** Westport, CT: AVI Publishing, 1977.
Written for advanced students of food chemistry. Covers a wide variety of topics as well as laboratory experiments.
R: *Food Technology* 31: 96 (Oct. 1977); *TBRI* 43: 360 (Nov. 1977).

Laboratory Manual on the Use of Radiotracer Techniques in Industry and Environmental Pollution. **International Atomic Energy Agency.** New York: Unipub, 1976.

Laboratory Training Manual on the Use of Nuclear Techniques in Animal Research. Technical Reports Series, 193. **International Atomic Energy Agency.** New York: International Atomic Energy Agency, 1979.
Provides animal science researchers with a basic understanding of ionizing radiation and measurement, its hazards, and applications. Covers a wide range of relevant topics. Includes bibliography.
R: *IBID* 7: 255 (Fall 1979).

Laboratory Waste Disposal Manual. 3d ed. Washington, DC: Manufacturing Chemists Association, 1976.
Second edition, 1969.
Provides concise instructions for waste disposal, arranged by chemical class. For both small laboratories and plants.
R: *Chemtech* 7: 15 (Jan. 1977); *TBRI* 43: 235 (June 1977).

Moving the Earth: The Workbook of Excavation. 3d ed. **Herbert L. Nichols, Jr.** Greenwich, CT: North Castle Books, 1976.

A revised guide to the excavation industry. Discusses land clearing, survey, measurement, soil and mud cellars, landscaping, conveyor machines, tractors, bulldozers, etc. A valuable reference for foremen, operators, engineers, and architects.
R: *RSR* 5: 25 (Oct./Dec. 1977).

Solar Energy in Buildings for Engineering, Architecture, and Construction. **Charles Chauliaguet, Pierre Baratsabal, and Jean-Pierre Batellier.** New York: Wiley, 1979.
Designed for engineers and architects involved in solar design. Covers such areas as solar economy, thermal design, applications, and examples.

Solar Energy Experiments. **Thomas W. Norton.** Emmaus, PA: Rodale Press, 1977.
Experiments ranging from simple to complex contain methods of measuring the motion of the sun throughout the day. Serves as a self-study course for the advanced student.
R: *Physics Teacher* 16: 182, 184 (Mar. 1978); *TBRI* 44: 238 (Apr. 1978).

VLSI System Design: When and How to Design Very-Large-Scale Integrated Circuits. **Saburo Muroga.** New York: Wiley, 1982.
A demonstration and explanation of the design and use of integrated-circuit chips. For engineers working directly with chip design and production, as well as for executives involved in the business aspects of the product.
R: *IEEE Proceedings* 71: 784 (June 1983).

HOW-TO-DO-IT AND POPULAR MANUALS

GENERAL

Technical Editor's and Secretary's Desk Guide. **George Freedman and Deborah A. Freedman.** New York: McGraw-Hill, 1984.
First edition, *A Handbook for the Technical and Scientific Secretary*, 1974.
Includes sections on basic-science knowledge for the secretary, and techniques for the technical secretary.
R: Jenkins (A148); Wal (p. 2EA18).

How to Write and Publish a Scientific Paper. **Robert A. Day.** Philadelphia: ISI Press, 1979.
A clearly written guide to writing, organizing, revising, and publishing a scientific paper. Covers all types of papers from abstracts to theses. Appendixes include lists of abbreviations, symbols, spelling errors, and terminology to avoid. A worthwhile review for graduate students and professionals.
R: *Choice* 17: 96 (Mar. 1980).

Science

ABC's of Calculus. **Rufus P. Turner.** Indianapolis: Sams, 1975.

The Calculator Handbook. **A. N. Feldzamen and Faye Herde.** New York: Berkeley, 1973.

A practical guide in the use of the smaller calculators, for businessmen, students, and housewives. Discusses both selection and basic requirements.
R: *ARBA* (1975, p. 790).

Teaching Manual on Food and Nutrition for Non-Science Majors. 3d ed. **Daniel Melnick.** Washington, DC: Nutrition Foundation, 1979.

Outlines a series of nutrition lectures for college students with little or no science background.
R: *Journal of the American Dietetic Association* 75: 389 (Sept. 1979); *TBRI* 45: 352 (Nov. 1979).

Biological Sciences

How to Know the Aquatic Plants. Pictured-Key Nature Series. 2d ed. **George W. Prescott.** Dubuque, IA: William C. Brown, 1980.

First edition, 1969.

An introductory text to the identification of common plants growing in moist habitats in the United States. Discusses the nature and ecology of aquatic plants such as algae, flowering plants, bryophytes, and pteridophytes. Descriptions accompanied by line drawings; indexed. For interested amateurs.
R: *RSR* 8: 14 (July/Sept. 1980); *ARBA* (1981, p. 650).

How to Know the Ferns and Fern Allies. Pictured-Key Nature Series. **John T. Mickel.** Dubuque, IA: William C. Brown, 1979.

A practical guide that presents information on fern structure and life history, hybridization, nomenclature, etc. Provides index and glossary. For beginning and intermediate fern students.
R: *ARBA* (1979, p. 669; 1981, p. 651).

How to Know the Insects. Pictured-Key Nature Series. 3d ed. **Roger C. Bland and Harry E. Jaques.** Dubuque, IA: William C. Brown, 1978.

First edition, 1947; second edition, 1974.
R: *Nature* 284: 102 (Mar. 6, 1980); *ARBA* (1979, p. 683).

How to Know the Lichens. Pictured-Key Nature Series. 2d ed. **Mason E. Hale.** Dubuque, IA: William C. Brown, 1979.

First edition, 1969.
Completely updated edition describes 427 lichen species. Entries accompanied by illustrations. An authoritative, nontechnical guide for lay collectors.
R: *ARBA* (1979, p. 669; 1980, p. 635).

How to Know the Seed Plants. Pictured-Key Nature Series. **Arthur Cronquist.** Dubuque, IA: William C. Brown, 1979.

An informative book on the identification of seed plants. Contains an illustrated key to the angiosperms and gymnosperms of the United States; outline of classification by family, index, and glossary. Includes line drawings. For amateur naturalists.
R: *ARBA* (1979, p. 669; 1980, p. 630).

How to Know the Trees. Pictured-Key Nature Series. 3d ed. **Howard A. Miller and Harry E. Jaques.** Dubuque, IA: William C. Brown, 1978.

First edition, entitled *How to Know the Trees: An Illustrated Key to the Most Common Species of Trees Found East of the Rocky Mountains,* 1946; second edition, 1972. Covers 312 species of native and exotic trees.

How to Know the Weeds. Pictured-Key Nature Series. 3d ed. **R. E. Wilkinson and Harry E. Jaques.** Dubuque, IA: William C. Brown, 1979.

Second edition, 1972.
An informative, spiral-bound volume on the identification of common weeds, which include fungi, pteridophytes, monocots, and dicots. Contains taxonomic descriptions, line drawings, geographic range, index, and pictured glossary. Provides an introductory section on weed control, as well as a listing of the weeds under appropriate families and orders. For college and university libraries, but also helpful to amateurs and professionals.
R: *ARBA* (1980, p. 631).

The Illustrated Science and Invention Encyclopedia: How It Works. 21 vols. Westport, CT: Webster's Unified, 1976.

Over 3,400 entries describe "how it works." Examples include the jukebox, violin, videotape, pacemaker, zipper, etc. A handy reference on inventions. For public libraries.

The Indoor Water Gardener's How-to Handbook. **Peter Loewer.** New York: Popular Library, 1976.

A valuable gardening reference book.
R: *L J* 98: 3275 (Nov. 1, 1973); *ARBA* (1974, p. 661).

ENGINEERING

Construction Engineering

Design Manual for Solar Heating of Buildings and Domestic Hot Water. **Richard L. Field.** Gaithersburg, MD: Solpub, 1977.

A step-by-step approach to the various stages in solar design; takes into account all aspects, including cost-effectiveness and year-round performance.
R: *Building Services Engineer* 45: A12 (Feb. 1978); *TBRI* 44: 235 (June 1978).

Encyclopedia of Home Construction. **M. Kreiger.** New York: McGraw-Hill, 1978.

Discusses all aspects of home construction and care and problem prevention. Details roof replacement, electrical wiring, and care of wooden floors. A useful book for homeowners.

The Encyclopedia of How It's Built. **Donald Clarke, ed.** New York: A & W Publishers, 1979.

A companion volume to the *Encyclopedia of How It Works*. International in coverage; 27 chapters pertain to such topics as arches, vaults, wells, and construction machinery. Illustrations and diagrams enhance the readable text. Recommended for public libraries.
R: *LJ* 105: 188 (Jan. 15, 1980); *TBRI* (1981, p. 753).

5,000 Questions Answered About Maintaining, Repairing, and Improving Your Home. **Stanley Schuler.** New York: Macmillan, 1976.

A manual of home repair, arranged in alphabetical order, describing over 60 subjects. Writing style is clear and specific. Does not include the repair of electrical appliances. Well-indexed. A suitable reference for home and public libraries.
R: *BL* 73: 925 (Feb. 15, 1977).

How to Build a Solar Heater. 2d ed. **Ted Lucas.** New York: Crown, 1981.

How to Save Energy and Cut Costs in Existing Industrial and Commercial Buildings: An Energy Conservation Manual. **Fred S. Dubin, H. L. Mindell, and S. Bloome.** Park Ridge, NJ: Noyes Data, 1976.

A practical manual on energy conservation. Provides descriptive details on modifying old systems and building new ones. Very valuable for energy technologists.
R: *LJ* 103: 722 (Apr. 1, 1978); *RSR* 5: 30 (Oct./Dec. 1977).

ELECTRICAL AND ELECTRONICS ENGINEERING

American Electricians Handbook: A Reference Book for the Practical Electrical Man. 10th ed. **John H. Watt, ed.** New York: McGraw-Hill, 1980.

First edition, 1913; eighth edition, 1961, Terrell Williams Croft, ed.; ninth edition, 1970.
A well-received basic electrician's source. Includes fundamentals of electrical work, equipment, batteries, interior wiring, transformers, etc.
R: *ARBA* (1971, p. 561); Jenkins (K155); Win (3EI16).

Electrical Engineering License Review. 5th ed. **Lincoln D. Jones and Donald G. Newnan.** New York: Engineering Press, 1980.

Third edition, 1972; fourth edition, 1975.

Practical Electrical Wiring: Residential, Farm, and Industrial. 13th ed. **H. P. Richter and W. Creighton Schwan.** New York: McGraw-Hill, 1984.

Eleventh edition, 1978.
An up-to-date manual that clearly explains proper electrical wiring installation. Scope is limited to voltages of 600 or less; all instructions are in compliance with the National Electrical Code.
R: *ARBA* (1985, p. 550).

Computers

Apple II User's Guide. **Lon Poole, Martin McNiff, and Steve Cook.** Berkeley: Osborne/McGraw-Hill, 1981.

An easy reference manual with detailed coverage of DOS, disk formats, memory usage, graphics data structures, and BASIC. Beneficial for the serious programmer.
R: *IEEE Proceedings* 70: 1246 (Oct. 1982).

The Basic Apple IIc: A Self-Teaching Guide. **G. Cornell and W. Abikoff.** New York: Wiley, 1985.

CBASIC User Guide. **Adam Osborne, Gordon Eubanks, Jr., and Martin McNiff.** New York: McGraw-Hill, 1981.

Illustrates, discusses, and explains various commands and programs used with the CBASIC language. An essential tool for any CBASIC programmer to have.
R: *Byte* 8: 430 (Aug. 1983).

Commodore 64 LOGO: A Learning and Teaching Guide. **P. Goodyear.** New York: Wiley, 1984.

Data Communication for Microcomputers: With Practical Applications and Experiments. **Joseph C. Nichols et al.** New York: McGraw-Hill, 1982.

Covers common data-communications problems that arise in the microcomputer-based area. This easy-to-follow guide will be an essential aid to the basic principles needed to implement the hardware and software necessary in such matters.
R: *IEEE Proceedings* 71: 448 (Mar. 1983).

How to Debug Your Personal Computer. **Robert Bruce and Jim Huffman.** Revised by Charlie Buffington. Reston, VA: Reston Publishing, 1980.

This revised edition shows how to locate and correct problems in software and hardware. Includes sections on the operation of personal computers and simple repair data for personal computer components, as well as system components and peripherals.

IBM PC PASCAL: A Self-Teaching Guide. **J. Conlan.** New York: Wiley, 1984.

The Microcomputer Users Handbook—1984. **D. Longley and M. Shain.** New York: Wiley, 1983.

A step-by-step guide to aid the microcomputer user in business by providing detailed information on various systems, dealers, and suppliers of microcomputers worldwide.

OP AMP Network Design Manual. **J. R. Hufault.** New York: Wiley, 1985.

PCjr Data File Programming: A Self-Teaching Guide. **J. Brown and L. Finkel.** New York: Wiley, 1985.

Practical PASCAL for Microcomputers. **Roger Graham.** New York: Wiley, 1983.

A detailed guide on how PASCAL works and how to use it most effectively. Includes subroutines useful in math, graphics, and text- and data-processing programs, and all standard PASCAL statements and operations.
R: *Librarians' Newsletter* 22: 26 (Dec. 1983).

Quick Reference Guides: 10-Unit Sets—(Timex Sinclair 1000; VIC-20 BASIC; Commodore 64 BASIC; TI 99/4 BASIC; TI 99/2 BASIC). **G. Held.** New York: Wiley, 1983.

Each guide provides easy access to the micro's myriad of programming symbols and statements, commands, and controls. A quick and easy reference to be kept beside the computer for instant use.

Troubleshooting Microprocessors and Digital Logic. **Robert L. Goodman.** Blue Ridge Summit, PA: TAB Books, 1980.

A well-illustrated digital/logic troubleshooting manual covering topics such as microcomputer applications and operations, the microprocessor and its support device, troubleshooting microprocessor systems, and more. For the electronics hobbyist, available to public libraries and personal purchase.

Use Guide to COBOL 85. **P. R. Brown.** New York: Wiley, 1985.

Using DEC Personal Computers: A Self-Teaching Guide. **R. Skvarcius.** New York: Wiley, 1985.

Using Personal Computers in Public Agencies. **J. R. Ottensmann.** New York: Wiley, 1985.

Z8000 CPU User's Reference Manual. Englewood Cliffs, NJ: Prentice-Hall, 1982.

Applicable to the Z8001 and Z8002 microprocessors with hardware information, Z8000 family specifications, programmers' quick reference guide, and a glossary of terms.

R: *Measurements and Control* 17: 146 (Apr. 1983).

Electronics

ABC's of Electronics. 3d ed. **Farl J. Waters.** Indianapolis: Sams, 1978.

Second edition, 1972.

Electronic Troubleshooting; A Manual for Engineers and Technicians. 2d ed. **Clyde N. Herrick.** Reston, VA: Reston, 1977.

First edition, 1974.

Troubleshooting procedures for electronic appliances, radio receivers, hi-fi and stereo equipment, color TV, etc. Minimizes theory and emphasizes application.

How to Design, Build, and Use Electronic Control Systems. **Frank P. Tedeschi.** Blue Ridge Summit, PA: TAB Books, 1981.

A survey of control system theory and design techniques. Concentrates on basic concepts and component descriptions. Discusses theoretical topics such as Laplace transforms, Bode plots, and root-locus techniques. Contains helpful glossary. Mainly for engineering students with a suitable background in math.

R: *Choice* 18: 1444 (June 1981).

Power Electronics: Problems Manual. **F. Csbaki et al.** Translated by B. Balkay. Wellingborough, UK: Collet's Holdings, 1979.

Radio and Stereo

ABC's of Hi-Fi and Stereo. 3d ed. **Hans Fantel.** Indianapolis: Sams, 1974.

Amateur Radio Advanced-Class License Study Guide. 2d ed., rev. **James Kyle and Ken Sessions.** Blue Ridge Summit, PA: TAB Books, 1975.
First edition, 1970.

Amateur Radio General-Class License Study Guide. 2d ed., rev. **James Kyle and Ken Sessions.** Blue Ridge Summit, PA: TAB Books, 1975.
First edition, 1971.

Manual for Operational Amplifier Users. **John D. Lenk.** Reston, VA: Reston, 1976.

Television

Basic Television. 6 vols. 2d ed., rev. **Alexander Schure.** New York: Hayden, 1975.

Television Interference Manual. 2d ed. **B. Priestley.** London: Radio Society of Great Britain, 1979.
A valuable reference for radio amateurs. Helps to solve the social and technical problems of radio interference.
R: *Telecommunications Journal* 46: 653 (Oct. 1979); *TBRI* 46: 78 (Feb. 1980).

Mechanics

Design Manual for Structural Tubing. **General Analytics, Inc.** Washington, DC: American Iron and Steel Institute, 1978.
Guides engineers in selecting best structural components in tubing design.
R: *Plant Engineering* 32: 132 (July 6, 1978); *TBRI* 44: 273 (Sept. 1978).

Dictionary of Tools Used in the Woodworking and Allied Trades, c. 1700–1970. New York: Scribner's, 1976.
This work defines approximately 60 types of axes and about 100 hammers. Information was obtained mostly from manufacturers' catalogs.
R: *LJ* 101: 2471 (Dec. 1, 1976).

Tools and How to Use Them: An Illustrated Encyclopedia. **Albert Jackson and David Day.** New York: Knopf, 1978.
A useful manual that presents information on the use of tools. Arranged by type of tool. Indexed and cross-referenced. Highly recommended; includes good illustrations.
R: *LJ* 104: 414 (Feb. 1, 1979); *WLB* 53: 523 (Mar. 1979).

The Way Things Work: An Encyclopedia of Modern Technology. 2 vols. New York: Simon & Schuster, 1971.

Directed at the general public. Describes the mechanics involved in over 200 machines and processes from colored television to nuclear reactors.
R: Sheehy (EA15).

Automobiles

Auto Electronics Simplified: Complete Guide to Service-Repair of Automotive Electronic Systems. **Clayton Hallmark.** Blue Ridge Summit, PA: TAB Books, 1975.

Basic Auto Repair Manual. 8th rev. ed. **Spence Murray.** Peterson, 1977. Annual.

Seventh edition, 1975.

Complete Auto Electric Handbook. **Clayton Hallmark.** Blue Ridge Summit, PA: TAB Books, 1975.

Complete Auto Electric Handbook: A Practical Guide to Diagnosis and Repair. **Clayton Hallmark.** Blue Ridge Summit, PA: TAB Books, 1974.

CHAPTER 9   GUIDES AND FIELD GUIDES

GUIDES

GENERAL SCIENCE

Asimov's Guide to Science. Rev. ed. Isaac Asimov. New York: Basic Books, 1972.

First edition, 1960, entitled *Intelligent Man's Guide to Science*; revised edition, 1965. The 1972 revision consists of physical sciences and biological sciences. Intended for the intelligent layperson.
R: *WLB* 47:95 (Sept. 1975); *ARBA* (1973, p. 535).

Guide to Scientific Instruments. **American Association for the Advancement of Science.** Washington, DC: AAAS. Annual.

Annual issue lists laboratory instruments and equipment with names and addresses of manufacturers.

A Guide to Writing Better Technical Papers. **C. Harkins and D. L. Plung.** New York: Wiley, 1982.

A guide containing articles designed to suggest more effective ways of writing technical papers for scientists and engineers.

Illustrated Fact Book of Science. **Michael W. Dempsey, ed.** New York: Arco Publishing, 1983.

Originally published as *Rainbow Fact Book of Science*.
A handy guide that contains over 35 sections describing various branches of science. Color photographs and drawings. Suitable for high-school libraries.
R: *ARBA* (1985, p. 492).

MATHEMATICS

Essentials of Mathematics. 4th ed. **Russell V. Person.** New York: Wiley, 1980.

Describes basic principles of arithmetic, algebra, geometry, calculus, and trigonometry. Includes several examples with step-by-step explanations. Contains expanded discussion of the metric system.

Vector Analysis: A Physicist's Guide to the Mathematics of Fields in Three Dimensions. **N. Kemmer.** New York: Cambridge University Press, 1976.

A beautifully illustrated guide that discusses methods of vector analysis.

PHYSICS

Applied Optics: A Guide to Optical System Design. 2 vols. **Leo Levi.** New York: Wiley, 1969–1980.

Volume 1, 1969; volume 2, 1980.
A comprehensive, 2-volume sourcebook that provides data and details on such topics as atmospheric optics, integrated optics to photoelectric detectors, vision, and photog-

raphy. Presentation consists of 125 tables and more than 350 diagrams and graphs. Cross-referenced and indexed. Recommended for all university and college libraries.
R: *IEEE Spectrum* 19: 66 (July 1982); *Choice* 18: 284 (Oct. 1980).

Atom-Molecule Collision Theory: A Guide for the Experimentalist. Physics of Atoms and Molecules. **Richard B. Bernstein, ed.** New York: Plenum Press, 1979.

A guide containing theory and computational methods required to describe the experimentally observable behavior of a system. A volume in the *Physics of Atoms and Molecules* series. For researchers in the field of atom-molecule collisions.
R: *Physics Bulletin* 30: 263 (May 1979).

Colour Index. 3d ed. 5 vols. Yorkshire, England: The Society of Dyers and Colourists, 1971.

Second edition, 1956–1958; supplement, 1963.
A reference for the professional dealing with elements and aspects of color. Developers, direct dyes, etc., amply discussed. Includes tables of comparison, chemical formulas, and reference lists.
R: Wal (p. 497).

Molecular and Crystal Structure Models. **Anne Walton.** New York: Halsted Press, 1978.

A photographically illustrated guide to molecular and crystalline structure. Discusses theoretical background and covers models used in a variety of situations. Contains appendixes and index.

Permanent Magnets. **Wilfred Wright and M. McCaig.** Engineering Design Guides, no. 20. New York: Oxford University Press, 1977.

A concise guide to concepts of permanent magnetism, theory, and design of magnets. Clear explanations and sample calculations. Highly recommended.
R: *New Scientist* 76: 719 (Dec. 15, 1977); *TBRI* 44: 132 (Apr. 1978).

Physics for Engineers and Scientists. 2d ed. **D. Elwell and A. J. Pointon.** New York: Wiley, 1979.

Deals with the fundamentals of basic physics. Covers atomic theory, themodynamics, and solid-state physics. Includes helpful references and an international temperature scale.

A User's Guide to Vacuum Technology. **John F. O'Hanlon.** New York: Wiley, 1980.

A practical guide for users of vacuum systems. Discusses new technology, utilizing turbomolecular, getter, ion, and cryogenic pumps. Examines operating procedures and economic factors. Useful for industrial workers.
R: *Nature* 289: 333 (Jan. 22, 1981); *Physics Today* 34: 64 (Apr. 1981); *Choice* 18: 982 (Mar. 1981).

CHEMISTRY

ACS Laboratory Guide. Easton, PA: American Chemical Society. Annual.
Published every August.

Contains information on supply houses, instruments, equipment, manufacturers, research services, new books, trade names, and companies.

Guide for Safety in the Chemical Laboratory. 2d ed. **Manufacturing Chemists Association**. New York: Van Nostrand Reinhold, 1972.

First edition, 1954.

Completely revised and updated reference on laboratory safety; includes new chemicals and materials such as cryogenic fluids and radioactive substances. Both comprehensive and practical, it is a needed item in high school, university, and industrial laboratories.

R: *ARBA* (1973, p. 657).

A Guidebook to Biochemistry. 4th ed. **Michael Yudkin and Robin Offerd.** New York: Cambridge University Press, 1980.

A 3-part introduction to the basic concepts of biochemistry and macromolecular chemistry. Sections cover structure and function of macromolecules, cellular metabolic reactions, and molecular genetics. Useful to biochemistry students.

A Guidebook to Mechanism in Organic Chemistry. 5th ed. **Peter Sykes**. New York: Longman, 1981.

Third edition, 1970; fourth edition, 1975.

This updated version of the third edition adds 2 chapters. The first is "Symmetry Controlled Reactions," and the second is "Linear Free Energy Relationships." Good for undergraduate and graduate students in chemistry.

R: *American Scientist* 59: 108 (Jan./Feb. 1971); *Journal of the American Chemical Society* 104: 7392 (Dec. 1982).

Guidebook to Organic Synthesis. **Raymond K. Mackie and David M. Smith**. New York: Longman, 1982.

A presentation of reactive groups showing, with diagrams and words, behavior with different reactions and conditions. Topics include ring closure and opening and carbon-to-carbon bonds. Good for chemistry students; college level.

R: *Choice* 20: 458 (Nov. 1982).

BIOLOGICAL SCIENCES

See also under "Field Guides."

The Complete Family Nature Guide. **Jean Reese Worthley**. Illustrated by Chris Fastie. Garden City, NY: Doubleday, 1976.

Five Kingdoms: An Illustrated Guide to the Phyla of Life on Earth. **Lynn Margulis and Karlene V. Schwartz**. San Francisco: W. H. Freeman, 1982.

The research, funded by the National Aeronautic and Space Administration in an attempt to understand the diversity of living things on earth, has resulted in this catalog of some 89 phyla from the five kingdoms of life on earth. Clearly presented amd thoroughly explained for any level.

R: *ARBA* (1983, p. 617).

Getting Food from Water: A Guide to Backyard Aquaculture. **Gene Logsdon**. Emmaus, PA: Rodale Press, 1978.

Information on how to grow food in water regions. Contains guidelines to proper growing conditions and technical information. Includes illustrations, photographs, and diagrams. Recommended.

A Guide to the Identification of the Genera of Bacteria. 2d ed. **V. B. D. Skerman**. Baltimore: Williams & Wilkins, 1973.

Methods used for the identification of the genera are stressed.
R: *ARBA* (1974, p. 569).

A Guide to Identifying and Classifying Yeasts. **J. A. Barnett, R. W. Payne, and D. Yarrow**. Cambridge: Cambridge University Press, 1979.

Condensed description of yeasts, including physiological and morphological attributes. Provides identification keys for small and large groups. Recommended for experts.
R: *Nature* 283: 799 (Feb. 21, 1980); *RSR* 8: 16 (July/Sept. 1980).

Guide to Sources for Agricultural and Biological Research. **J. Richard Blanchard and Lois Farrell, eds.** Berkeley: University of California Press, 1981.

A volume dealing with research resources concerned with the production of food, and resources on wildlife management and pollution. Each chapter has literature guides, abstracts and indexes reviews, etc.
R: *ARBA* (1983, p. 703).

Index Bergeyana. **Robert E. Buchanan et al., eds.** Baltimore: Williams & Wilkins, 1966.

An annotated alphabetical listing of names of the taxa of the bacteria. The *Index* evaluates more than 20,000 names of bacteria taxa.
R: Wal (p. 200); Win (1EC11).

Industrial Microbiology. **Gerald Reed**. New York: Macmillan, 1982.

A guide to the industrial technology and applications in the microbiology of foods and their ingredients.
R: *Nature* 299: 765 (Oct. 28, 1982).

Introduction and Guide to the Marine Bluegreen Algae. **Harold Judson Humm and Susanne R. Wicks.** New York: Wiley, 1980.

An authoritative taxonomic guide to marine blue green algae. The first section consists of a general introduction; the second section contains taxonomic keys. Includes schematic drawings and species descriptions. Valuable for marine scientists and graduate and upper-level undergraduate students.
R: *Choice* 18: 815 (Feb. 1981).

The Microtomist's Formulary and Guide. **Peter Gray**. London: Constable, 1954. Reprint. Huntington, NY: Krieger, 1975.

Guide to methods, techniques, and stains bearing on the use of microscopic slides in biological research.
R: Wal (p. 204).

Naturalist's Color Guide. **Frank B. Smithe**. New York: American Museum of Natural History, 1975.

In loose-leaf format. Provides identification of the major families of color successions. Separate appendix provides examples for the use of each color included in this guide. Useful to larger libraries and to the teaching of biology.
R: *Choice* 12: 1294 (Dec. 1975); *ARBA* (1976, p. 659); Sheehy (EG12).

A Practical Guide to Molecular Cloning. **Bernard Perbal**. New York: Wiley, 1984.

A description of the enzymes, vectors, techniques, mapping, etc., of all phases in molecular cloning experiments.

Scientific Illustrations: A Guide to Biological, Zoological, and Medical Rendering Techniques, Design, Printing, and Display. **Phyllis Wood**. New York: Van Nostrand Reinhold, 1979.

Techniques of scientific illustration are clearly and simply presented. Sample illustrations in black and white and color.
R: *BioScience* 30: 54 (Jan. 1980); *TBRI* 46: 102 (Mar. 1980).

The World's Worst Weeds: Distribution and Biology. **LeRoy G. Holm et al**. Honolulu, HI: University Press of Hawaii, 1977.

Contains information on 300 weeds, emphasizing their distribution and biology. Arranged by weed. Includes appendixes on control, as well as a bibliography. Comprehensive.
R: Sheehy (EL5).

Agriculture

Beuscher's Law and the Farmer. 4th ed. **Revised by Harold W. Hannah**. New York: Springer-Verlag, 1975.

Third edition, 1960.
The fourth edition has been revised and expanded to include the most recent legal developments in regard to farming. A valuable reference tool for farmers and students of agriculture.
R: *ARBA* (1976, p. 742).

Guide to the Chemicals Used in Crop Production. **E. Y. Spencer**. Ottawa: Agriculture Canada Branch, 1981.
R: *Science* 218: 189 (Oct. 8, 1982).

Handbook of Agricultural Occupations. 3d ed. **Norman K. Hoover**. Danville, IL: Interstate, 1977.

Second edition, 1969.
A vocational guide outlining occupations in agriculture and related areas.

Review of Agricultural Policies in OECD Member Countries. Paris: Organization for Economic Cooperation and Development, 1975. Annual.

Annual review, based on individual-country reports prepared by the organization's Working Party on Agricultural Policies of the Committee for Agriculture. Provides a comprehensive description of agriculture policy within the organization's oribit and gives special attention to actions taken to ease current problems. Individual-country reports in this series analyze the problems of agriculture in each member country and describe policies followed in the fields of markets, prices, structure, etc.
R: *IBID* 4: 14 (Mar. 1976).

Using Commercial Fertilizers. 4th ed. **Malcolm H. McVickar and William Walker.** Danville, IL: Interstate, 1978.

Third edition, 1970.

A practical guide describing the application of fertilizers. Both a reference manual and a text.

Botany

Aquarium Plants. **Niels Jacobsen.** Poole, Dorset, England: Blandford Press. Distr. New York: Sterling Publishing, 1979.

Concise book containing 200 excellent illustrations of plants suitable for home aquaria. Discusses heating, nutrients, filtration, and planting. Indexed. A highly recommended guide for all libraries.
R: *ARBA* (1980, p. 701).

Basic Plant Pathology Methods. **Onkar D. Dhangra and James B. Sinclair.** Boca Raton, FL: CRC Press, 1985.

A comprehensive up-to-date guide to the various methods and techniques for basic plant pathology. Over 200 formulas and methods of preparation of various culture media, salt solutions, buffers, and cleaning solutions. References, appendixes, and index.

A Checklist of Names for 3,000 Vascular Plants of Economic Importance. Agricultural Handbook, no. 505. **Edward E. Terrell.** Washington, DC: US Department of Agriculture, Agricultural Research Service Distr. Washington, DC: US Government Printing Office, 1977.

Consists of 2 checklists: scientific names (genus and species) and common names. Emphasizes North American plants. Contains much useful information.
R: *ARBA* (1979, p. 665).

Checklist of North American Plants for Wildlife Biologists. **Thomas G. Scott and Clinton H. Wasser.** Washington, DC: Wildlife Society, 1980.

Compilation of 16,000 scientific and common names of vascular plants of North America published in the *Journal of Wildlife Management, Wildlife Monographs, Wildlife Society Bulletin.* Lists are arranged alphabetically. For university and college libraries serving botanists.
R: *RSR* 18: 18 (July/Sept. 1980); *ARBA* (1981, p. 655).

Chemical and Botanical Guide to Lichen Products. **Chicita F. Culberson.** Chapel Hill, NC: University of North Carolina Press, 1969.

Second supplement: 1976–77, published by St. Louis, MO: The Bryologist, 1977.
A quick guide to more than 300 lichen, with physical data, chemical structures, and annotated references. A list of about 2,000 species is also given.
R: *RSR* 7: 7 (Jan./Mar. 1979).

The Encyclopedia Botanica: The Definitive Guide to Indoor Blossoming and Foliage Plants. **Dennis A. Brown.** New York: Dial Press, 1978.
More a guide than an encyclopedia. Features information on plant care, terrariums, hydroponics, pests, and diseases. Includes color and black-and-white photographs.
R: BL 75: 1322 (Apr. 15, 1979): *LJ* 104: 201 (Jan. 15, 1979); *ARBA* (1980, p. 706).

Freshwater Wetlands: A Guide to Common Indicator Plants of the Northeast. **Dennis W. Magee.** Amherst, MA: University of Massachusetts Press, 1981.
Provides a key and some 621 complete and informative descriptions of wetland plant species found in the northeast. Explains the range, habitat, and individual morphological characteristics of each species. For the nonbotanist, conservation group, or other interested parties, this guide will prove most helpful.
R: *ARBA* (1983, p. 622).

An Illustrated Guide to Pollen Analysis. **Peter D. Moore and J. A. Webb.** New York: Wiley, 1978.
Useful as a text, lab manual, and field guide. Contains numerous photomicrographs, which provide a workable key to pollen identification. Well-executed and thorough. For a wide variety of users.
R: *Nature* 298: 102 (Mar. 1, 1979); *Scientific American* 241: 35 (Aug. 1979); *TBRI* 45: 257 (Sept. 1979).

Jungles. **Edward S. Ayensu, ed.** New York: Crown Publishers, 1980.
An amply illustrated volume dealing with various aspects of tropical rain forests. Discusses flora, fauna, and evolution of the environment. Includes lists of products, photographs, drawings, and diagrams. Of interest to botanists and specialists.
R: *BL* 77: 907 (Mar. 1, 1981).

North American Range Plants. 2d ed. **J. Stubbendieck, Stephan L. Hatch, and Kathie J. Kjar.** Lincoln, Nebraska: University of Nebraska Press, 1982.
A guide to the 200 most important species of range plants that live on prairies and grasslands in North America. For the livestock owner or range manager.
R: *ARBA* (1983, p. 626).

Plant Propagation for the Amateur Gardener. **John I. Wright, ed.** Dorset, England: Blandford Press. Distr. New York: Sterling, 1983.
A comprehensive guidebook covering a wide range of plant propagation methods: extensive line drawings depict specific horticultural techniques and operations.
R: *ARBA* (1985, p. 505).

Seashore Plants of California. **E. Yale Dawson and Michael S. Foster.** Berkeley: University of California Press, 1982.
A unique introduction to the identification of marine algae (seaweeds) and flowering (terrestrial) plants found along the California coasts. Arranged by a dichotomous key

for each major group of algae, with brief descriptions. Limited in its description of terrestrial plants.
R: *ARBA* (1983, p. 622).

Seaweeds: A Color-Coded, Illustrated Guide to Common Marine Plants of the East Coast of the United States. **Charles J. Hillson.** University Park, PA: Pennsylvania State University Press, 1977.

A color-keyed guide to algae of the East Coast. Breaks seaweeds into brown, green, and red groups. Thorough and helpful. Well-recommended.
R: *LJ* 103: 511 (Mar. 1, 1978).

Succulent Flora of Southern Africa: A Comprehensive and Authoritative Guide to the Indigenous Succulents of South Africa, Botswana, South-West Africa/Namibia, Angola, Zambia, Zimbabwe/Rhodesia and Mozambique. **Doreen Court.** Distr. Salem, NH: Merrimack Book Service, 1981.

Nine major groups of southern African succulents and over 100 genera and thousands of species are arranged by family, genus, and then locality. Contains approximately 300 clear and beautiful color photographs, maps, and a glossary. For amateurs and botanists.
R: *ARBA* (1983, p. 630).

Tom Brown's Guide to Wild Edible and Medicinal Plants. **Tom Brown, Jr.** New York: Berkley Books, 1985.

The Tree Key: A Guide to Identification in Garden, Field and Forest. **Herbert Edlin, comp.** New York: Scribner's, 1978.

Covers 235 species and 77 genera of trees in western Europe and North America. Detailed, well-written, and illustrated. Recommended for the professional botanist or the amateur naturalist.
R: *Choice* 16: 101 (Mar. 1979); *WLB* 53: 473 (Feb. 1979); *ARBA* (1980, p. 637).

Trees and Shrubs Hardy in the British Isles. 8th ed. 4 vols. **W. J. Bean.** London: Murray, 1970–80.

Seventh edition, 1950–51; volumes 1–3, 1970–75; volume 4, 1980.
This guide to trees and shrubs in the British Isles lists entries under Latin name. Includes description of species and varieties, common names, synonyms, and location of illustration. Notes date of introduction and economic use.
R: *Journal of the Royal Horticultural Society* 45: 463 (Oct. 1970); Wal (p. 222).

Nutrition

The Complete Book of Herbs and Spices. **Claire Loewenfeld and Philippa Back.** North Pomfret, VT: David & Charles, 1980.

Comprehensive coverage of herbs and spices in all areas of the world. Gives background information covering history and tradition, and the use of herbs and spices in health, cooking, and cosmetics. Indexed.
R: *BL* 70: 350 (Nov. 15, 1974); *ARBA* (1975, p. 669).

Everyone's Guide to Better Food and Nutrition. **Barbara B. Deskins**. Middle Village, NY: David, 1975.
Intended for the layperson, written by an experienced nutrition educator.
R: *ARBA* (1976, p. 747).

The Food Service Market for Frozen Foods. Hershey, PA: National Frozen Food Association, 1975. Annual.
A guide for anyone in the food service market. Provides details on population shifts, consumer spending patterns, food volume sales, etc.

Guide to the Safe Use of Food Additives. 2d series. **Food and Agriculture Organization of the United Nations**. Rome: Food and Agriculture Organization of the United Nations, 1979.
Compilation of material from Sessions of the Joint Food and Agriculture Organization and World Health Organization Expert Committee on Food Additives and the Codex Alimentarius Commission. Concerns the use of chemicals in food preparation.
R: *IBID* 7: 345 (Winter 1979).

Guide to the Wines of the United States. **Dominick Abel**. Ney York: Cornerstone Library, 1979.
A survey of wines made in the United States. Covers the California, Pacific Northwest, and New York areas. For wine experts.
R: *LJ* 105: 613 (May 1, 1980); *ARBA* (1981, p. 734).

Health, Food, and Nutrition in Third-World Development. **Pradip K. Ghosh, ed**. Westport, CT: Greenwood Press, 1984.
Arranged in 4 parts: part 1 contains reprints of 18 essays; part 2 contains a 56-item bibliography and statistical tables and figures; part 3 contains 838 citations of source materials, most of them dated and with inconsistent format; and part 4 is a directory of information sources.
R: *ARBA* (1985, p. 499).

An Illustrated Guide to Wine. **George Rainbird, ed.** New York: Harmony Books/Crown, 1983.
A 166-page illustrated guide that arranges wine entries geographically by country and area. Exceptional glossy color photographs. Indexed.
R: *ARBA* (1985, p. 501).

The No-Nonsense Guide to Food and Nutrition. **Marion McGill and Orrea Pye**. New York: Norton, 1978.
An extremely comprehensive guide to basic concepts in nutrition, with a health and consumer-related focus. Clear, concise. For laypeople, dietitians, nutritionists.
R: *Journal of the American Dietetic Association* 74: 396 (Mar. 1979); *TBRI* 45: 195 (May 1979).

Nutrition Almanac. **J. D. Kirschman**. New York: McGraw-Hill, 1979.

The Pocket Guide to Spirits and Liqueurs. **Emanuel Greenberg and Madeline Greenberg, eds.** New York: Perigee/Putnam, 1983.

A handy guide that organizes major types and brands of spirits in 6 major categories—whiskey, brandy, clear spirits, rum, tequila, and liqueurs. Each section begins with an organoleptic survey and concludes with an application section.
R: *ARBA* (1985, p. 499).

The Signet Book of American Wine. 3d rev. ed. **Peter Quimme.** New York: New American Library, 1980.

First edition, 1975.
A 2-part guide to American wines. Part 1 contains a history of US wines. Part 2 consists of a region-by-region guide focusing on climate, wineries, vinification practices, and styles of wines. Of interest to public libraries.
R: *ARBA* (1981, p. 734).

Terry Robard's New Book of Wine: The Ultimate Guide to Wines Throughout the World. **Terry Robard, ed.** New York: Putnam, 1984.

A guide intended for a wide audience. It covers about 1,000 entries arranged in dictionary style.
R: *BL* 80: 1217 (May 1, 1984); *LJ* 109: 890 (May 1, 1984); *ARBA* (1985, p. 501).

The Wines of America. 2d ed. **Leon D. Adams.** New York: McGraw–Hill, 1978.

An authoritative reference on the wines of North America. Up-to-date and recommended.
R: *WLB* 53: 472 (Feb. 1979).

Zoology

The Audubon Society Book of Wild Animals. **Les Line and Edward Ricciuti.** New York: Harry N. Abrams, 1977.

Recommended for its pictorial qualities rather than its verbal information. A companion volume to the Audubon Society field guides. Includes some 200 high-quality color plates.
R: *American Scientist* 66: 624 (Sept./Oct. 1978); *WLB* 52: 503 (Feb. 1978).

Butterfly and Angelfishes of the World. 2 vols. New York: Wiley, 1978–1980.

Volume 1, *Australia*, 1978; volume 2, *Atlantic Ocean, Caribbean Sea, Red Sea, Indo-Pacific*, 1980.
International coverage of classification, natural history, and aquarium maintenance of butterfly and angelfishes. Contains numerous photographs. Includes indexes of common and scientific names, and a brief bibliography. For amateur naturalists and aquarists. Highly recommended.
R: *ARBA* (1980, p. 641; 1981, p. 670).

Comparative Anatomy of Domestic Animals: A Guide. **Bonnie Beaver.** Ames, IA: Iowa State University Press, 1980.

R: *RSR* 9: 101 (Jan./Mar. 1981).

A Guide to Bird Behavior. In the Wild and at Your Feeder. Vol. 2 **Donald W. Stokes and Lillian Q. Stokes, eds.** Boston: Little, Brown, 1983.

A compact guide with information on each species; includes a behavior calendar charting the months in which each of 7 activities takes place, a display guide describing both visual and auditory displays, and behavior descriptions. Illustrated with black-and-white portraits of the birds and line drawings.
R: *BL* 80: 930 (Mar. 1, 1984); *ARBA* (1985, p. 528).

A Guide to Observing Insect Lives. **Donald W. Stokes.** Boston: Little, Brown, 1983.

Arranged by seasonal aspects of each insect with a location guide and description of their lives and behavior. Includes drawings and descriptions of each insect. It is not an essential guide for every reference collection.
R: *ARBA* (1984, p. 663).

Guide to Study of Animal Populations. **James T. Tanner.** Knoxville, TN: University of Tennessee Press, 1978.

Contains introductory statistical methods for measuring the characteristics of animal populations. Intended for researchers and graduate students in applied biology. Recommended.
R: *American Scientist* 67: 106 (Jan.–Feb. 1979); *TBRI* 45: 178 (May 1979).

EARTH SCIENCES

Annual Summary of Information on Natural Disasters. Nos. 1–. Paris: Unesco. Distr. New York: Unipub. 1966–. Annual.
No. 10, 1979.

Basic annual information on earthquakes, tsunami, volcanic eruptions, landslides, and avalanches. Intended for technical, academic earth-science libraries, and large public libraries.
R: *IBID* 7: 238 (Fall 1979); *ARBA* (1976, p. 703).

Applied Water Resource Systems Planning. **David Major and Roberto Lenton.** Englewood Cliffs, NJ: Prentice-Hall, 1979.

Fully explains water resource planning procedures developed over the past 20 years. Includes examples from all over the world.

Catalogue of the Active Volcanoes of the World, Including Solfatara Fields. Pts. 1–. Naples: International Association of Volcanology, 1951–.
Part 21, 1967.
R: Jenkins (F175); Wal (p. 153); Win (EE149; 1EE24; 3EE15).

Classic Mineral Localities of the World: Asia and Australia. **Philip Scalisi and David Cook.** New York: Van Nostrand Reinhold, 1983.

Arranged geographically, covering Asia country by country and Australia state by state. For each, there is a description of the classic mines, dates of discovery, activity, and types of minerals mined. With maps, crystal diagrams, and some photos. For serious mineral collectors, and all geological libraries.
R: *ARBA* (1984, p. 687).

A Color Illustrated Guide to Carbonate Rock Constituents, Textures, Cements, and Porosities of Sandstones and Associated Rocks. **Peter A. Scholle.** Tulsa, OK: American Association of Petroleum Geologists, 1978.

Contains identifying photomicrographs that aid in the interpretation of sandstones. A well-produced volume for geologists.

R: *American Association of Petroleum Geologists Bulletin* 63: 1981 (Oct. 1979); *Geological Magazine* 115: 473 (Nov. 1978); *Geology* 6: 625 (Oct. 1978); *Journal of Sedimentary Petrology 48: 1380 (Dec. 1978); TBRI* 44: 370 (Dec. 1978); 45: 92 (Mar. 1979); 46: 8 (Jan. 1980).

Earthquakes, Tides, Unidentified Sounds and Related Phenomena: A Catalog of Geophysical Anomalies. **William R. Corliss, comp.** Glen Arm, MD: Sourcebook Project, 1983.

The author reports on various unexplained phenomena, such as geysers at sea, chemical differences in ocean waters, dogs howling before earthquakes, and more, in an attempt to stimulate investigation in these areas.

R: *ARBA* (1984, p 683).

The Ecology of Fossils: An Illustrated Guide. **W. S. McKerrow, ed.** Cambridge, MA: MIT Press, 1978.

Describes 125 distinctly different fossils with illustrative diagrams. British geological concentration, covers fossils from various periods since the Precambrian. Detailed, recommended.

R: *American Scientist* 67: 100 (Jan.–Feb. 1979); *Nature* 275: 777 (Oct. 26, 1978); *Physics Teacher* 16: 496 (Oct. 1978); *Science* 203: 999 (Mar. 9, 1979); *Scientific American* 237: 42 (Nov. 1978); *TBRI* 44: 367 (Dec. 1978); 45: 5 (Jan. 1979); 45: 208 (June 1979).

A Guide to Classification in Geology. **J. W. Murray.** New York: Halsted Press, 1981.

Provides information relating to the classification of rocks and details the classifications used in engineering and geology. Valuable for earth scientists.

The Hamlyn Nature Guide to Fossils. **Richard Moody.** London: Hamlyn, 1979.

A guide to fossil collection and identification, providing generic and geographical distribution. Includes high-quality illustrations and labeled diagrams. A useful pocket-sized guide.

R: *Nature* 282: 652 (Dec. 6, 1979).

Index of Generic Names of Fossil Plants, 1820–1965. Prepared for the US Geological Survey. **Henry N. Andrews, Jr.** Washington, DC: US Government Printing Office, 1970.

Updated US Geological Survey Bulletin no. 1013, which covered the period 1850–1950. Attempts to compile a comprehensive list of generic names of fossil plants.

R: *ARBA* (1971, p. 520).

Meteorites, Classification and Properties. **J. T. Wasson.** New York: Springer-Verlag, 1974.

This book provides a complete discussion of scientific research methods applied to meteorites and an up-to-date review of their classification with computer-generated listings of all meteorites for which accurate classification data are available. Intended for scientists and graduate students involved in earth science research.

Paleobiology of the Invertebrates: Data Retrieval from the Fossil Record. 2d ed. **P. Tasch.** New York: Wiley, 1980.

A guide to fossil data of several invertebrate phyla. Includes taxonomy, genetics, and details of skeletogenesis.

A Pictorial Guide to Fossils. **Gerard R. Case.** New York: Van Nostrand Reinhold, 1982.

A useful tool in identifying fossils worldwide, from protozoa to large mammals and rodents. Most illustrations are of those fossils in the United States and are not of the highest quality. Limited in its use of recognizable details, which may prove to be a disadvantage for its use by professionals rather than hobbyists.
R: *ARBA* (1983, p. 615).

The Water Naturalist. **Heather Angel and Pat Wolseley.** New York: Facts on File, 1982.

Provides information and do-it-yourself projects on many aspects of water, such as still water, running water, the edge of the sea, and the open sea. Photographs, drawings, and diagrams help make this a most unique reference guide.
R: *ARBA* (1983, p. 614).

Engineering

A Guidebook for Technology Assessment and Impact Analysis. **Alan L. Porter et al.** New York: American Elsevier, 1980.

Details the process of technology assessment and impact analysis. Includes an annotated bibliography. Helpful for academic libraries.
R: *Choice* 18: 628 (June 1981).

The Wiley Engineer's Desk Reference: A Concise Guide for the Professional Engineer. **S. I. Heisler.** New York: Wiley-Interscience, 1984.

Addresses practical problems as well as references to theoretical and derivative data. A practical and essential guide for engineers working in mechanics, materials, hydraulics, electronics, energy sources, and process control.

Aeronautical and Astronautical Engineering

Aeronautical Chart Catalogue. 23d ed. Montreal, Canada: International Civil Aviation Organization. 1978.

Nineteenth edition, 1976.
Aeronautical charts or chart series available in civil aviation.
R: *IBID* 4: 148 (June 1976); 7: 58 (Spring 1979).

Civil Aircraft of the World. Rev. ed. **John W. R. Taylor and Gordon Swanborough.** New York: Scribner's, 1980.

Comprehensive guide to civil aircraft. Includes model number and manufacturer indexes. Discusses major airliners, air taxis, private aircraft, and helicopters.
R: *BL* 71: 197 (Sept. 22 1975); *WLB* 47: 365 (Dec. 1972; 50: 126 (Oct. 1975); *ARBA* (1973, p. 663; 1976, p. 782).

Combat Aircraft of the World—1909 to the Present. **John W. R. Taylor, ed.** New York: Putnam, 1969.

A unique and massive guide to military aircraft. Includes approximately 2,000 photographs and line drawings. Statistics and performance data are geared to use by professionals involved in aviation.
R: *LJ* 94: 3994 (Nov. 1, 1969); *ARBA* (1970, p. 150); Wal (p. 350).

Federal Aviation Regulations for Aviation Mechanics. 6th ed. **Federal Aviation Administration, ed.** Glendale, CA: Aviation, 1980.

Earlier edition, 1975.

Jane's All the World's Aircraft. New York: Arco. 1909–. Annual.

Comprehensive listing of aircraft. Entries are amply described and illustrated.
R: *Aircraft Engineering* 35: 83 (Mar. 1963); *Interavia* 25: 1397 (Nov. 1970); *ARBA* (1979, p. 781); Jenkins (K181); Wal (p. 385; p. 393); Win (EI46).

Military Aircraft of the World. **John W. R. Taylor and Gordon Swanborough.** New York: Scribner's, 1979.

Revised edition contains information on newest aircraft. Includes photographs and technical data on more than 300 types of aircraft. Index by type of aircraft and manufacturer. Well-recommended.
R: *BL* 73: 204 (Sept. 15, 1976); *WLB* 50: 809 (June 1976); *ARBA* (1977, p. 782; 1980, p. 732).

Rockets, Missiles and Spacecraft of the National Air and Space Museum. Rev. ed. Washington, DC: Smithsonian Institution Press, 1983.

This museum catalog guide for visitors contains brief descriptive articles about each of the major space artifacts on display in the National Air and Space Museum. Black-and-white photos. Useful for lay public.
R: *ARBA* (1985, p. 616).

CHEMICAL ENGINEERING

Adhesives: Guidebook and Directory. Park Ridge, NJ: Noyes Data. Annual.

Almanac of the Canning, Freezing, Preserving Industries. Westminster, MD: Judge. Annual.

Sixtieth edition, 1975.
A ready-reference tool for anyone involved in the food industry.
R: Jenkins (J112).

The Book of American Trade Marks. Vols. 1–. **David E. Carter, ed.** New York: Art Direction, 1976–.

Volumes 1–3, reprinted edition, 1978; volume 4, 1976; volume 5, 1977.

Chemical Engineering Drawing Symbols. **D. G. Austin.** New York: Halsted Press, 1979.

Contains drawing symbols for a wide variety of equipment types and instrumentation in chemical engineering. Well-indexed and referenced.

Pesticides Guide: Registration, Classification and Applications. Neenah, WI: J. J. Keller, 1979.

A loose-leaf handbook concerning pesticides with 3 main sections: federal regulations, compliance, and references. Useful in industry and agriculture.
R: *ARBA* (1981, p. 688).

Plasticizers: Guidebook and Directory. Park Ridge, NJ: Noyes Data, 1976. Annual.

Polymer Additives; Guidebook and Directory. Park Ridge, NJ: Noyes Data. Annual.

Lists alphabetically by manufacturer name. Comprehensive listing of "commercially available protective additives."

Polymer Melt Rheology: A Guide for Industrial Practice. **F. N. Cogswell.** New York: Halsted Press, 1981.

A simplified mathematical treatment of polymer melt rheology with applications to industry. Discusses methods for measuring polymer flows, dependence on chemical structure, and advice on setting up an industrial rheology laboratory. Useful to chemical engineers and materials scientists.

Selection and Use of Thermoplastics. Engineering Design Guides, no. 19. **P. C. Powell.** New York: Oxford University Press, 1977.

Discusses significant aspects of thermoplastics, including methods of testing, service requirements, etc. Presents a great deal of useful information.
R: *Aslib Proceedings* 42: 588 (Nov.–Dec. 1977).

Textile Technology

Davison's Textile Catalog and Buyers' Guide. Ridgewood, NJ: Davison, Annual.

Identification of Textile Materials. 7th ed. Metuchen, NJ: Textile Book, 1975.

Sixth edition, 1970.
A working tool for the purpose of laboratory tests leading to identification of unknown fibers. Also provides a glossary and introductory description of the textile fibers available commercially.
R: Wal (p. 525).

Illustrated Guide to Textiles. 3d ed. **Marjory Joseph and Audrey Gieseking.** Plycon Press, 1981.

Second edition, 1972.

Textile Industries Buyers Guide. Atlanta, GA: Textile Industries. Annual.

A special issue of *Textile Industries* designed to provide textile executives with a guide to sources of equipment, supplies, and services required today for the manufacturing of textiles. Includes 2,500 headings, 1,800 suppliers of one or more products, and 170 advertisements. Indexed.

CIVIL ENGINEERING

AIA Metric Building and Construction Guide. **American Institute of Architects.** Edited by S. Braybrooke. New York: Wiley, 1980.

Contains essential information on the International System of Measurement. Includes charts, tables, drawings, and conversion factors.

The Builder's Complete Guide to Construction Business Success. **L. Reiner.** New York: McGraw-Hill, 1978.

This comprehensive volume details the fundamentals of building small residential and office buildings. Explains business organization; reading plans and specifications; estimates and purchasing decisions; and cost and progress control. A reference on the building industry for those interested in starting a construction business.

Building Construction: Materials and Types of Construction. 5th ed. **W. C. Huntington and R. E. Mickadeit.** New York: Wiley, 1981.

An updated and expanded edition that covers materials in building construction. Includes a new chapter on acoustics.

Cement and Concrete Reference Catalog. Skokie, IL: Portland Cement Association, 1975. Annual.

Construction Dewatering: A Guide to Theory and Practice. **J. P. Powers.** New York: Wiley, 1980.

Helpful guide to problems in groundwater control, using knowledge from geology, chemistry, and soil mechanics. Covers such topics as soil and water interrelationships, permeability of soils, and design of pumping tests. For civil engineers.

Construction Inspection: A Field Guide to Practice. **James E. Clyde.** New York: Wiley, 1979.

A field guide that deals with ground construction. Features a detailed list of nearly 500 matters related to civil engineering projects. Helpful data included.

Construction Materials for Architecture. Wiley Series of Practical Construction Guides. **W. J. Rosen.** New York: Wiley, 1985.

A single reference source for new innovations in building materials. Features excellent, architectural-quality illustrations, photographs, tables, and a comprehensive bibliography.

Designer's Guide to OSHA: A Practical Design Guide to the Occupational Safety and Health Act for Architects, Engineers, and Builders. 2d ed. **Peter S. Hopf.** New York: McGraw-Hill, 1981.

First edition, 1975.

Comprehensive working tool on the Occupational Safety and Health Act. Uses a unique graphic format to illustrate the law; provides hundreds of sketches, tables, and diagrams of common design problems to aid architects and engineers in conforming to the law. Discusses the inspection process, safe working environments, and all OSHA changes since the early 1970s. Essential for design professionals.

The Garden Book. **John Brookes, ed.** New York: Crown, 1984.

A detailed guide to various types, shapes, and sizes of gardens. Emphasizes planning and design for construction of gardens. Illustrated and well-indexed. Good for both amateurs and professionals.

R: *LJ* 109:1458 (Aug. 1984; *ARBA* (1985, p. 502).

The Hawkweed Passive Solar House Book. **Hawkweed Group.** Chicago: Rand McNally, 1980.

A practical guide to the design and construction of solar-heated homes and offices. Examines techniques and heating and cooling needs. Contains floor plans, drawings, and photographs. Includes bibliography, glossary, and index. For architects and civil engineers.

R: *BL* 77: 908 (Mar. 1, 1981).

Know Your Woods: The Complete Up-to-Date Guide to Woods. Rev. ed. **Albert Constantine, Jr.** New York: Scribner's, 1975.

A thorough reference tool. Describes more than 300 principal woods. With detailed *General Index.*

R: *ARBA* (1973, p. 566).

Principles of Home Inspection; A Guide to Residential Construction, Inspection, and Maintenance. **Jospeh G. McNeill.** New York: Van Nostrand Reinhold, 1979.

A practical guide to residential home construction, inspection, and maintenance. Includes appendixes and an index. For engineers, city planners, and home owners.

The Standard Forms of Building Contract. **D. Walker-Smith and H. A. Close.** London: Knight, June 1971.

R: *Civil Engineering and Public Works Review* (Oct. 1971).

Structural Engineering for Professional Engineers' Examination, Including Statics, Mechanics of Materials and Civil Engineering. 3d ed. **Max Kurtz.** New York: McGraw-Hill, 1978.

Completely revised and updated. Aims to help engineers pass the structural and civil engineering licensing examinations. Reviews basic theory, applications, and recent design codes.

US Plastics in Building and Construction; Marketing Guide and Company Directory. Stamford, CT: Technomic. Annual.

Wood Structures: A Design Guide and Commentary. **American Society of Civil Engineers, Structural Division, Task Committee on Status-of-the-Art: Wood; Committee on Wood.** New York: ASCE, 1975.

ELECTRICAL AND ELECTRONICS ENGINEERING

Advanced Stero System Equipment. **Lloyd Hardin.** Englewood Cliffs, NJ: Prentice-Hall, 1980.

Discusses state-of-the-art devices in stereo system equipment. Contains photographs, drawings, and graphs. Covers latest technological advances. Highly recommended.

A Basic Guide to Power Electronics. **Albert Kloss.** New York Wiley, 1984.

Complete Guide to Compact Disc Player (CD) Troubleshooting and Repair. **John D. Lenk.** Englewood Cliffs, NJ: Prentice-Hall, 1986.

D.C. Power Supplies: Application and Theory. **Robert Traister.** Englewood Cliffs, NJ: Prentice-Hall, 1979.

A basic guide to power supplies, especially those involving design. Includes more than 80 illustrations; discusses theory as well as practical matters in the electronic circuitry of power supplies.

Electrical Distribution Engineering. **Anthony J. Pansini, ed.** New York: McGraw-Hill, 1983.

A textbook with handy guide section of more than 100 pages on materials and equipment used in electric power distribution, such as conductors, insulators, transformers. Recommended for special libraries interested in electric power engineering.
R: *ARBA* (1985, p. 550).

Electronic Displays. **Sol Sherr.** New York: Wiley, 1979.

Explains the mechanisms of electronic displays and readouts. Covers every type of system and device, such as watches, calculators, digital TV. A convenient source of guidelines.

Guide to the National Electrial Code. **Thomas L. Harman and Charles E. Allen.** Englewood Cliffs, NJ: Prentice-Hall, 1979.

Details rules necessary for installing electrical wiring, but emphasizes questions that appear on the master electrician's examination. Covers wiring systems, installation, and theory. Includes extensive, illustrative tables and circuit diagrams, which support the text and are easily understood. Supplemented by an adequate index and a short bibliography. For practicing engineers and contractors, as well as and students.
R: *Choice* 16: 1610 (Feb. 1980); *RSR* 8: 25 (Jan./Mar. 1980); *ARBA* (1981, p. 755).

Illustrated Guidebook to Electronic Devices and Circuits. **Frederick W. Hughes.** Englewood Cliffs, NJ: Prentice-Hall, 1983.

International Frequency List. 7th ed. Geneva: International Frequency Registration Board, 1974. Also supps.

Japan EBG. Japan Electronics Buyers' Guide. Tokyo: Dempa. Distr. New York: International Publications. Annual.

Popular Circuits Ready-Reference. **John Markus.** New York: McGraw-Hill, 1982.

A useful compendium of circuits for most purposes.

Practical Applications of Data Communications: A User's Guide. **H. Karp.** New York: McGraw-Hill, 1980.

A collection of 58 articles dealing with advances in data-communications technology. Surveys data-link performance, data security, distributed data processing, and digitized voice. Valuable aid for system engineers, programmers, and network designers.

The Radio-Electronic Master. Garden City, NY: United Technical. Annual.

Annual catalog of standard electronic components and equipment. Indexes to products, manufacturers, and trade names.

A Reference Guide to Practical Electronics. **R. Krieger.** New York: McGraw-Hill, 1980.

A comprehensive reference guide to college-level electronics. Includes 100 equations ranging from LC circuits to communications technology. Examples illustrate each equation. Invaluable aid to students, hobbyists, and in-service technicians.

Tube Caddy: Tube Substitution Guide. **H. A. Middleton.** New York: Hayden, 1960-. Annual.

Lists almost 6,000 tubes used in radio, television, and other electronic equipment. Ideal for the smaller collection.
R: *ARBA* (1974, p. 680).

A User's Guide to Selecting Electronic Components. **G. L. Ginsberg.** New York: Wiley, 1981.

Provides detailed information on the structure and performance of resistors, capacitors, power sources, and more. Collected from some 4,000 articles and catalogs.
R: *IEEE Proceedings* 70: 416 (Apr. 1982).

Video and Digital Electronic Displays: A User's Guide. **S. Sherr.** New York: Wiley, 1982.

Provides a simple exposition of the main features of electronic display systems, from the digital watch to the color television set. Covering 10 sections, this book will prove to be a useful introduction to the world of electronic displays for many people.
R: *Applied Optics* 22: 1438 (May 1983).

Wire and Wire Products. Buyer's Guide, 1976. Perth Amboy, NJ: Syncro Machine, 1976 and updates.

Workbench Guide to Practical Solid-State Electronics. **Fredrick Hughes.** Englewood Cliffs, NJ: Prentice-Hall, 1978.

A valuable book for technicians and amateurs. Well-illustrated with diagrams and photographs. Includes example projects.

Computer Technology

Microcomputer Hardware

Bowker/Bantam 1984 Complete Sourcebook of Personal Computing. New York: Bantam Books/Bowker, 1983.

Choose Your Own Computer: A Guide to Buying the Best Personal Computer for Your Needs. **Peter Rodwell, ed.** Woodbury, NY: Barron's Educational Series, 1983.
Guide information is given for each model, with several paragraphs of general description and a photograph of the model, followed by technical specifications. Includes 67 models.
R: *ARBA* (1985, p. 593).

The Complete Book of Home Computers. **Van Waterford.** Blue Ridge Summit, PA: TAB Books, 1982.
Provides basic facts and photographs of various home computer systems. Models are not rated in any way, and there is a heavy devotion to Ohio Scientific home computers. Will be a useful reference guide to most collections.
R: *ARBA* (1984, p. 624).

Family Computers under $200. **Doug Mosher, ed.** Berkeley: Sybex, 1984.
A practical guide to computers such as the Timex/Sinclair 1000 and Commodore VIC 20.
R: *ARBA* (1985, p. 592).

Here Come the Clones!: The Complete Guide to IBM PC Compatible Computers. **Melody Newrock, ed.** New York: Micro Text Publications/McGraw-Hill, 1984.
A guide with chapters that focus on an individual line of computers and highlight software/hardware compatibility, keyboards, circuit-board comments, displays, disk drives, and other features. Includes a directory of manufacturers and a glossary.
R: *ARBA* (1985, p. 593).

Microcomputer Buyer's Guide. **Tony Webster.** New York: McGraw-Hill, 1983. Annual.
Earlier edition, 1981.
A summary of more than 500 microcomputers and microcomputer systems from over 180 major suppliers, with information on each supplier. Includes photographs and a glossary of computer terms. A valuable guide in comparing computer systems.
R: *ARBA* (1984, p. 625).

The Personal Computer Book. 4th ed. **Peter A. McWilliams.** New York: Ballantine/Random House, 1982.
R: *ARBA* (1985, p. 592).

Personal Computer Buyers' Guide: With Exclusive Product Reference Guide. **Dennis J. Grimes and Brian W. Kelly.** Cambridge, MA: Ballinger, 1983.

Includes articles, product descriptions, and a handy product reference guide to personal computer products. Provides the beginner with basic information on personal computers.
R: *ARBA* (1984, p. 621).

Small Business Computers: A Guide to Evaluation and Selection. **Koichiro R. Isshiki.** Englewood Cliffs, NJ: Prentice-Hall, 1982.

An introduction to computers and their applications, with a step-by-step guide to deciding whether to employ a computer service or to buy a computer. Illustrated with pictures, diagrams, and examples of analyses. For the presumably uninitiated small-business owner or manager considering the use or purchase of a computer.
R: IEEE Spectrum 20: 80 (Feb. 1983).

Microcomputer Software

The Best of IBM Pc Software. **Stanley R. Trost, ed.** Berkeley: Sybex, 1984.
R: *ARBA* (1985, p. 596).

The Best of CP/M Software. **John Halamka, ed.** Berkeley: Sybex, 1984.

Busy Person's Guide to Selecting the Right Word Processor: A Visual Shortcut to Understanding and Buying. **Alan Gadney, ed.** Glendale, CA: Festival Publications, 1984.

A crash course in word processing, with a product guide complete with addresses and current prices. Illustrations included.
R: *ARBA* (1985, p. 586).

Bowker's Software in Print, 1984. New York: Bowker, 1984. Annual.

An essential reference guide for anyone using software in business, research, education, or the home. Includes more than 23,000 software systems used worldwide, covering programs, products, vendor profiles, and more. Supplements are printed 3 times a year.

A Critic's Guide to Software for Apple and Apple-Compatible Computers. **Phillip I. Good, ed.** Radnor, PA: Chilton, 1983.
R: *BL*: 941 (Mar. 1, 1984; *ARBA* (1985, p. 597).

A Critic's Guide to Software for CP/M Computers. **Phillip I. Good, ed.** Radnor, PA: Chilton, 1983.
R: *BL* 80: 941 (Mar. 1, 1984; *ARBA* (1985, p. 597).

A Critic's Guide to Software for IBM-Pc and Pc-Compatible Computers. **Phillip I. Good, ed.** Radnor, PA: Chilton, 1983.
R: *BL* 80: 941 (Mar. 1, 1984; *ARBA* (1985, p. 597).

Datapro/McGraw-Hill Guide to IBM Personal Computer Software. New York: Datapro/McGraw-Hill, 1983.
R: *ARBA* (1985, p. 596).

Datapro/McGraw-Hill Guide to CP/M Software. New York: Datapro/McGraw-Hill, 1983.
R: *ARBA* (1985, p. 596).

Datapro/McGraw-Hill Guide to Apple Software. New York: Datapro/McGraw-Hill, 1983.
R: *ARBA* (1985, p. 596).

Legal Care for Your Software. **Daniel Remer.** Berkeley: Nolo Press, 1982.
Provides examples, explanations, and procedures concerning the legal methods of protecting one's software products. Simply written for those who write or publish commercial computer software or for anyone else interested in knowing the legal details behind doing so. A useful guide in this area.
R: *Byte* 9: 478 (Feb. 1984).

Microcomputer Software Buyer's Guide. **Tony Webster and Richard Champion, eds.** New York: Computer Reference Guide/McGraw-Hill, 1984.
A guide that tells what major software packages are for what computer systems. Useful for most libraries.
R: *WLB* 59: 292 (Dec. 1984; *ARBA* (1985, p. 598).

Chen, Ching-chih. *MicroUse Directory: Software.* Newton, MA: MicroUse Information, 1984. Annual updates.
A comprehensive guide to 1,500 selected micro-based software, 1,100 of which are general-purpose packages. Entries are arranged alphabetically with 5 functional indexes—type of software, vendor/developer, hardware, DOS, and random-access memory requirement.

The Software Finder: A Guide to Educational Microcomputer Software for Apple II; ATARI 400/800; Commodore PET, CBM; VIC-20, 64; RADIO SHACK TRS-80 Models I–III, Color Computer, and CP/M. Vol. 3, no. 2. Dresden, ME: Dresden Associates, 1983. Semiannal.
Earlier title: *School Microware Directory.*
Describes some 2,800 software packages from 320 suppliers for school- and college-oriented software for 11 microcomputers. Informative and useful.
R: *ARBA* (1984, p. 629).

Software Maintenance Guidebook. **Robert L. Glass and Ronald A. Noiseux.** Englewood Cliffs, NJ: Prentice-Hall, 1981.
A comprehensive survey of the tools and methodologies of software maintenance. Contains practical examples and bibliographies. Valuable for software maintenance personnel.

Software Reliability Guidebook. **Robert L. Glass.** Englewood Cliffs, NJ: Prentice-Hall, 1980.
Discusses techniques used in achieving software reliability. Techniques are presented and evaluated, and references are provided. Can be used to supplement university-

level courses in software engineering. For college students and others related to or in the software industry.

The Unix System Guidebook. **Peter P. Silvester.** New York: Springer-Verlag, 1984.

For beginners interested in learning about the Unix (a Bell Laboratories trademark) time-sharing operating system. Emphasis is on Unix Version 7.
R: *Science Technology Libraries* 5: 2 (Winter 1984).

Programming

The BASIC Book: A Cross-Referenced Guide to the BASIC Language. **Harry L. Helms.** New York: McGraw-Hill, 1983.

A guide to the conversion of BASIC programs from one computer to another—such as Apple, Atari, Commodore, IBM, Radio Shack, and Texas Instruments. A useful "Key Word Ready Reference" identifies the various commands used. For the experienced microcomputer user or those with a working knowledge of BASIC.
R: *ARBA* (1984, p. 620).

A Basic Programmers' Guide to PASCAL. **Mark J. Borgerson.** New York: Wiley, 1983.

A comprehensive guide to the basic elements of the PASCAL language up to the more complex and advanced features. Written in simple, clear language for any level of previous knowledge in this area. An outstanding guide for any technical library or motivated individual.
R: *Computer* 16: 117 (Aug. 1983).

Guide to Good Programming Practice. 2d ed. **B. L. Meek, N. J. Rushby, and P. Heath.** New York: Wiley, 1983.

First edition, 1980.
A guide to the various aspects of programming, including strategy and design; program writing; and development. Useful section on person-machine interface; checklists.
R: *Data Processing Digest* 26: 29 (Oct. 1980).

Program Design and Construction. **David Higgins.** Englewood Cliffs, NJ: Prentice-Hall, 1979.

A practical guide to computer programming. Divided into 2 sections: problem-solving design and techniques used to construct a computer program.

Other Related Topics

The Complete Guide to Computer Camps and Workshops. **Mike Benton.** New York: Bobbs Merrill, 1984.

A complete, useful guide to information on some 250 computer-oriented day programs, summer camps, and workshops geared for the 6–16-year-old age group. Includes with some adult programs. For parents seeking an up-to-date source of answers for their questions about sending their child to camp.
R: *LJ* 9: 109 (May 1984).

The Computer Resource Guide. **Kenneth L. Gilman, ed.** Detroit: Gale Research, 1982.

This 2-part guide covers some 100 software programs with information for each, and then lists foundations and corporations with a history of grant-making for computer software and hardware, in alphabetical order. A convenient and practical source for universities and large public libraries.
R: *Choice* 21: 399 (Nov. 1983).

Computer-Aided Data Analysis: A Practical Guide. **W. R. Green.** New York: Wiley, 1985.

Designed for nontechnical computer users, this is an introduction to the data-analysis techniques in common use in the applied sciences.

Computers for Everybody. 2d ed. **Jerry Willis and Merl Miller.** Beaverton, OR: Dilithium Press, 1983.

A specific, comprehensive guide to the use and purchase of a home computer for the beginner. Somewhat personalized tour through the micro world gives a fine introduction.
R: *ARBA* (1984, p. 625).

The Design of Operating Systems for Small Computer Systems. **Stephen Kaisler.** New York: Wiley-Interscience, 1982.

A fine guide to current techniques, procedures, and shortcuts for minicomputer and microcomputer operating system design. Most descriptive.
R: *IEEE Spectrum* 20: 24 (Apr. 1983).

A Practical Guide to Computer Communications and Networking. **R. J. Deasington.** New York: Wiley, 1983.

A Practical Guide to Computer Methods for Engineers. **Terry E. Shoup.** Englewood Cliffs, NJ: Prentice-Hall, 1979.

Written specifically for engineers and includes example algorithms with engineering applications. Presents succinct discussions of fundamental algorithms with figures and logic diagrams.
R: *Materials Engineering* 90: 112 (Sept. 1979); *Mechanical Engineering* 101: 111 (Sept. 1979); *TBRI* 45: 392 (Dec. 1979).

A Practical Introduction to Computer Graphics. **Ian O. Angell.** New York: Halsted Press, 1981.

In an easy-to-understand format, this work looks at the various aspects of computer graphics with insight into the theory and mathematics behind their creation. Familiarity with basic mathematical concepts is essential to using this book.
R: *Byte* 8: 374 (Feb. 1983).

A User's Guide to Computer Peripherals. **Donald Eadie.** Englewood Cliffs, NJ: Prentice-Hall, 1982.

An introduction emphasizing minicomputers and microcomputers. Excellent figures, tables, and charts help explain all sorts of old and new peripheral devices. Especially good for vocational and technical schools.
R: *Choice* 20: 129 (Sept. 1982).

User's Guidebook to Digital CMOS Integrated Circuits. **Eugene R. Hnatek.** New York: McGraw-Hill, 1981.

A brief, generally unorganized guide book covering a wide range of topics on CMOS. Useful as a quick reference to CMOS for a practicing engineer who is familiar with the field.
R: *Computer* 15: 118 (Nov. 1982); *IEEE Proceedings* 71: 686 (May 1983).

INDUSTRIAL ENGINEERING

Employer's Guide to Labor Relations. **James W. Hunt.** Washington, DC: Bureau of National Affairs, 1979.

Flowmeters: A Basic Guide and Sourcebook for Users. **A. T. J. Hayward.** New York: Halsted Press, 1980.

Sets guidelines for engineers to help them choose the appropriate flowmeter. Comprehensive, including flowmeter tables. Contains selective bibliography.

Guide to Quality Control. 2d rev. ed. **Kaoru Ishikawa.** Tokyo: Asian Productivity Organization, 1976.

Designed as a guide to quality-control techniques for department and section chiefs. Includes practice problems.
R: *IBID* 4: 129 (June 1976).

Industrial Heat Exchangers: A Basic Guide. **G. Walker.** New York: Hemisphere Publishing, 1982.

Provides a discussion of the many types of heat exchangers available commercially. A basic elementary guide for the consumer of heat exchangers seeking to define and select the particular equipment best suited for their individual needs.
R: *Power* 127: 130 (June 1983).

Ion Exchange in Water Treatment—A Practical Guide for Plant Engineers and Chemists. Cleveland: Duolite International, 1983.

Provides a step-by-step diagnosis for concurrent regenerated units, countercurrent regenerated units, and mixed beds. A practical, effective guide for the user of ion-exchange equpiment in water treatment.
R: *Power* 127: 96 (Dec. 1983).

Water Treatment Plant Design for the Practicing Engineer. **Robert L. Sanks, ed.** Ann Arbor MI: Ann Arbor Science, 1978.

This reference guide for engineers emphasizes the practical aspects of plant design. Includes numerous diagrams and much helpful information not readily available from other sources.

Marine Engineering

Bulk Carriers in the World Fleet; Oceangoing Merchant-Type Ships of 1,000 Gross Tons and Over (Excludes Vessels of the Great Lakes). Washington, DC: US Government Printing Office. Annual.

Guide to Port Entry, 1984/1985. London: Shipping Guides, 1984.

Mariner's Annual Ordering Guide. New Hope, PA: Charles Kerr Enterprises. Annual.

A universal marine-materials identification guide of items commonly used to build and operate merchant ships. Indexed.

Safety and Operational Guidelines for Undersea Vehicles. **Marine Technology Society, Undersea Vehicle Committee.** New York: MTS, 1974.

Materials Science

Aluminum: Properties and Physical Metallurgy. **John E. Hatch, ed.** Metals Park, OH: American Society for Metals, 1984.

Chemical and Process Plant: A Guide to the Selection of Engineering Materials. 2d ed. **L. S. Evans.** New York: Halsted Press, 1981.

A comprehensive guide to materials useful in the construction and maintenance of process plants.

Dangerous Properties of Industrial Materials. 6th ed. **N. Irving Sax.** New York: Van Nostrand Reinhold, 1984.

Second edition, 1963; third edition, 1968; fourth edition, 1975; fifth edition, 1979. Provides a single source of quick, up-to-date, concise, hazard-analysis information about 18,000 common industrial and laboratory materials, including a 40,000-synonym index. Indispensable tool.
R: *Chemical Engineering* 86: 11 (Dec. 17, 1979); *New Scientist* 82: 1109 (June 28, 1979); *RSR* 5: 12 (Apr./June 1977); *WLB* 50: 756 (June 1975); *ARBA* (1976, p. 779; 1985, p. 514); Jenkins (D144); Wal (p. 476).

Fire Protection Guide on Hazardous Materials. 7th ed. **Amy E. Dean and Keith Tower.** Boston: National Fire Protection Association, 1978.

Fifth edition, 1973.

This handbook is actually a combination of standards and manuals of the National Fire Protection Association. The first manual lists 8,800 trade-name products. The second manual describes the fire hazard properties of more than 1,300 flammable substances, and the third manual gives dangerous hazards of 388 chemicals. The fourth covers 2,350 mixtures of chemicals. Finally, there is a guide that determines the degree of hazard of various materials.
R: *Marine Engineering Log* 83: 109 (Dec. 1978); *TBRI* 45: 118 (Mar. 1979); *ARBA* (1975, p. 774).

Mechanical Fastening of Plastics: An Engineering Handbook. **Brayton Lincoln, Kenneth J. Gomes, and James F. Braden.** New York: Dekker, 1984.

An authoritative guide to the fabrication and use of plastic fasteners. Nontechnical in content, illustrated, and indexed.
R: *ARBA* (1985, p. 545).

Plastics Materials Guide. London: Engineering, Chemical, and Marine Press. Annual.

METALLURGY

Guide to Non-Ferrous Metals and Their Markets. 3d ed. **Peter Robbins.** New York: Nichols, 1982.

A discussion of the production, development, and marketing aspects of nonferrous metals, with a reference guide to follow. Includes addresses and names of associations, information centers, and their members.
R: *ARBA* (1983, p. 728).

Smithells Metals Reference Book. **Eric A. Brandes, ed.** 6th ed. Woburn, MA: Butterworth, 1983.

First edition, 1949; fifth edition, 1976.
A convenient summary of data relating to metallurgy. One volume, packed with 36 chapters (over 1,600 pages). Indexed. A must for all libraries with an interest in metals.
R: *ARBA* (1985, p. 552).

Steel Selection: A Guide for Improving Performance and Profits. **R. F. Kern and M. E. Suess.** New York: Willey, 1979.

Guides the materials engineer to the correct selection of steel. Bridges a gap between theory and applications.

Worldwide Guide to Equivalent Nonferrous Metals and Alloys. Vol. 2. **Paul M. Unterweiser, ed.** Metals Park, OH: American Society for Metals, 1981.

Volume 1: *Worldwide Guide to Equivalent Irons and Steels*, 1979.
A reference guide of the nonferrous metals and alloys that are the applicable specifications and designations of equivalents produced in the United States, Germany, France, the United Kingdom, Canada, Japan, Italy, Sweden, Mexico, various South American countries, and other lesser-known countries. Contains 8 major families of nonferrous metals and alloys, as well as other metals and their alloys.
R: *Science and Technology Libraries* 2: 89 (Winter 1981).

MECHANICAL ENGINEERING

ASME Guide for Gas Transmission and Distribution Piping Systems. American Society of Mechanical Engineers. New York: American Society of Mechanical Engineers, 1980.

Second edition, 1973.

The Basic Book of Metalworking. **Richard Little and Sherry Little.** Chicago: American Technical Society, 1979.

Covers techniques of metalworking, including welding and forging. Contains a glossary and index. For high school students.

Boiler Operators Guide. 2d ed. **Anthony L. Kohan and Harry M. Spring.** New York: McGraw-Hill, 1981.

A popular guide to solving problems related to the installation, operation, maintenance, and repair of boilers. Discusses such topics as the interrelationship of controls and safety devices, the practical aspects of boiler construction, and state and American Society of Mechanical Engineers' code requirements for repair to boilers.

Control of Noise in Ventilation Systems: A Designer's Guide. **M. Iqbal et al.** New York: Wiley, 1977.

Written for practicing air-conditioning engineers. Offers a step-by-step approach to all aspects of acoustical control of ventilation systems.
R: *ASHRAE J* 19: 68 (Aug. 1977); *Plant Engineering* 31: 142 (July 7, 1977); *TBRI* 43: 275 (Sept. 1977): 43: 396 (Dec. 1977).

Diesel Engine Reference Book. **L. C. R. Lilly, ed.** Woburn, MA: Butterworth, 1984.

A standard reference guide in the field. It covers all types of diesel engines manufactured in Europe, the United States, and Japan. Each chapter concludes with a list of references. Useful appendixes, illustrations and drawings. Indexed.
R: *ARBA* (1985, p. 620).

Diesel Spotter's Guide Update. **Jerry A. Pinkepank and Louis A. Marre.** Milwaukee, WI: Kalmbach Publishing, 1979.

Continues the *Second Diesel Spotter's Guide,* a standard volume on the identification of North American diesel locomotives. Descriptions of each model contain technical detail and statistics on production. Arranged by type of locomotive, this update includes discussions of lightweight train power cars, electric locomotives, and "slugs." Enhanced by numerous photographs and index. For the railroad buff.
R: *ARBA* (1981, p. 768).

Electric Motors. 3d ed. **Edwin P. Anderson.** Indianapolis: Sams, 1977.

Second edition, 1968.
A comprehensive guide covering all the fundamental principles of electric motors.

Hydraulic Pumps and Motors: Selection and Application for Hydraulic Power Control Systems. **Raymond P. Lambeck.** New York: Dekker, 1983.

A concise reference guide that provides a general overview of various types of pumps, motors, and controls. Recommended to all libraries with engineering collections.
R: *ARBA* (1985, p. 552).

Machinery Noise. **A. H. Middleton.** New York: Oxford University Press, 1978.

A guide to noise control in industrial settings. SI units used throughout. A useful addition to the literature.
R: *Journal of Sound and Vibration* 61: 67 (Dec. 22, 1978); *TBRI* 45: 118 (Mar. 1979).

Mechanical Engineering for Professional Engineers' Examinations: Including Questions and Answers for Engineer-in-Training Review. 3d ed. **John D. Constance.** New York: McGraw-Hill, 1978.

Provides a quick review of theory for mechanical engineers. Includes self-testing questions. A handy guide.
R: *Mechanical Engineering* 100: 78 (Sept. 1978); *TBRI* 44: 352 (Nov. 1978).

The Mechanic's Guide to Electronic Emission Control and Tune-Up. **Larry W. Carley and Robert Freudenberger.** Englewood Cliffs, NJ: Prentice-Hall, 1986.

Metric Guide to Mechanical Design and Drafting. **Fredrick T. Gutman.** New York: Industrial Press, 1978.

For draftspersons and designers, this guide covers basic features of SI units, units of metric length, volume, mass. Contains clear diagrams and a wealth of practical information.
R: *Automotive Engineering* 87: 117 (Mar. 1979); *General Engineer* 90: 241 (Oct. 1979); *TBRI* 45: 194 (May 1979); 46: 73 (Feb. 1980).

Turbomachines: A Guide to Design, Selection, and Theory. **O. E. Balje.** New York: Wiley, 1981.

An overview of the state-of-the-art of turbomachine design. Discusses performance, theory, and parameters. Of interest to mechanical engineers.

TRANSPORTATION

The Train-Watcher's Guide to North American Railroads: Significant Facts, Figures, and Features of over 140 Railroads in the United States, Canada, and Mexico. **George H. Drury, comp.** Milwaukee, WI: Kalmbach Publishing, 1984.

A basic guide to railroads in North America. Names of railroads are alphabetically arranged in the main section. Glossary and index are available. For any library with an interest in the subject.
R: *ARBA* (1985, p. 618).

World Guide to Battery-Powered Road Transportation: Comparative Technical and Performance Specifications. **Jeffrey M. Christian, comp.** Edited by Gary G. Reibsamen. New York: McGraw-Hill, 1980.

A listing of more than 100 electric vehicles. Vehicles are arranged alphabetically by manufacturer. Entries contain description of vehicle, service data, history, technical data, and performance and operation statistics. Also includes lists of dealers and manufacturers. Indexed by vehicle manufacturer, vehicle name, type, and country. For public and academic libraries, for use by engineers and laypersons.
R: *Choice* 18: 122 (Sept. 1980); *RQ* 19: 401 (Summer 1980); *ARBA* (1981, p. 763).

Nuclear Engineering

Agreements Registered with the International Atomic Energy Agency. New York: Unipub, 1978.

Sixth edition, 1976.
Lists chronologically all agreements registered with IAEA up to December 31, 1973, to which IAEA registration numbers have been allocated; and agreements registered between January 1, 1974 and July 31, 1975, without registration numbers, as this list is provisional; and tabulates the agreements by state and subject.
R: *IBID* 4: 143 (June 1976).

Commissioning Procedures for Nuclear Power Plants: A Safety Guide. **International Atomic Energy Agency.** New York: Unipub, 1980.

A safety guide to the commissioning of land-based stationary thermal neutron power plants.

External Man-Induced Events in Relation to Nuclear Power Plant Siting: A Safety Guide. **International Atomic Energy Agency.** New York: Unipub, 1981.

A safety guide examining the region and discussing the hazardous phenomena associated with external human-induced events.

Fire Protection in Nuclear Power Plants: A Safety Guide. Safety Series, 50-SG-D2. **International Atomic Energy Agency.** Vienna: International Atomic Energy Agency, 1979.

Discusses fire-protection requirements and operational guidance for protection from fire in nuclear power plants. Detailed. Provides bibliography.
R: *IBID* 7: 255 (Fall 1979).

Governmental Organization for the Regulation of Nuclear Power Plants: A Code of Practice. Safety Series, 50-C-G; IAEA Safety Standards. **International Atomic Energy Agency.** Vienna: International Atomic Energy Agency, 1978.

Regulations for the operation of nuclear power plants.
R: *IBID* 7: 50 (Spring 1979).

A Guide to Nuclear Power Technology: A Resource for Decision-Making. **Frank J. Rahn et al.** New York: Wiley, 1984.

A comprehensive source on all aspects of nuclear technology. Glossary and index are available. Indispensable tool for practicing engineers, physicists, government planners, and legal experts.

Guide to the Safe Handling of Radioactive Wastes at Nuclear Power Plants. **International Atomic Energy Agency.** New York: Unipub, 1980.

A guide summarizing present techniques of treating, conditioning, storing, and disposing of gaseous, liquid and solid wastes on and from nuclear power plant sites.

Manpower Development for Nuclear Power: A Guidebook. **International Atomic Energy Agency.** New York: Unipub, 1980.

A guide to the role, requirements, planning, and implementation of manpower-development programs associated with nuclear power. Examines national participation in the activities of a nuclear power program. For policy-makers and managers of nuclear power programs.

Operational Limits and Conditions for Nuclear Power Plants: A Safety Guide. IAEA Safety Series, 50-SG-03. **International Atomic Energy Agency.** Vienna: International Atomic Energy Agency, 1979.

Covers operational limits for nuclear power plants. Comprehensive and detailed.
R: *IBID* 7: 256 (Fall 1979).

Protection System and Related Features in Nuclear Power Plants: A Safety Guide. **International Atomic Energy Agency.** New York: Unipub, 1980.

A safety guide containing specific design principles and many specific design requirements for the protection system, which detects departures from acceptable conditions in nuclear power plants.

Quality Assurance in the Manufacture of Items for Nuclear Power Plants: A Safety Guide. **International Atomic Energy Agency.** New York: Unipub, 1981.

A safety guide of requirements related to the establishment of a quality-assurance program to ensure the safety of items manufactured by nuclear power plants.

Safety in Nuclear Power Plant Siting: A Code of Practice. IAEA Safety Series, 50-C-S; IAEA Safety Standards. **International Atomic Energy Agency.** Vienna: International Atomic Energy Agency, 1978.

Presents criteria for selecting nuclear power plant sites. Includes specific data, safety requirements, etc.
R: *IBID* 7: 51 (Spring 1979).

Technical Evaluation of Bids for Nuclear Power Plants: A Guidebook. **International Atomic Energy Agency.** New York: Unipub, 1981.

A guidebook on the organization and supervision of the technical evaluation of bids for a nuclear power plant. Focuses on the importance of evaluating technical bids.

ENERGY

Applied Solar Energy: A Guide to the Design, Installation, and Maintenance of Heating and Hot Water Services. **D. Kut and G. Hare.** New York: Halsted Press, 1979.

Covers the application of solar energy to the supply of hot water. Details practical aspects of assembling and maintaining systems. Includes numerous illustrations and a table of conversion factors.

The Architect's Guide to Energy Conservation. **S. Jarmul.** New York: McGraw-Hill, 1980.

Provides up-to-date information on the latest techniques for energy conservation in buildings. Case studies of new and existing buildings show the project cost and energy savings. New equipment, infrared scanning of buildings, and recycling energy are also discussed. For architects and energy specialists.

A Comprehensive Guide to Solar Water Heaters. **R. Montgomery.** New York: Wiley, 1985.

Intended for lower-level courses in solar systems applications or solar heating installation.

Consumer Handbook of Solar Energy for the United States and Canada. **John H. Keyes.** Dobbs Ferry, NY: Morgan & Morgan, 1979.

A practical guide to the purchase of solar energy heating equipment—collectors, storage devices, pumps, and solar equipment. Contains extensive charts and tables, worksheets, and monograms. For public libraries.
R: *LJ* 104: 1581 (Aug. 1979).

Efficient Energy Use: A Reference Book on Energy Management for Engineers, Architects, Planners, and Managers. 2d ed. **Craig B. Smith, ed.** Elmsford, NY: Pergamon, 1978.

First edition, entitled *Efficient Electricity Use: A Practical Handbook for an Energy-Constrained World,* 1976.
In 4 parts: general introduction, description of specific uses of energy, political constraints, specific technologies. Includes tables, charts, and graphs.
R: *ARBA* (1979, p. 689).

Energy Conservation in Buildings 1973–1983. **Penny Farmer.** New York: Scholium International, 1983.

Discusses methods of energy conservation, such as insulation, design, construction, heat recovery, and costs, in a variety of building types.
R: *Civil Engineering* 54: 79 (Feb. 1984).

Energy Policy and Third World Development. **Pradip K. Ghosh, ed.** Westport, CT: Greenwood Press, 1984.

A useful guide that includes 12 essays on energy and Third World Developments in part 1, and tables of statistical information and a bibliography of information sources in part 2. Indexed. Recommended for academic and public libraries.
R: *ARBA* (1985, p. 506).

Energy Products Specification Guide: Conservation Solar, Wind and Photovoltaics. Harrisville, NH: SolarVision. Distr. Atlanta: Fairmont Press, 1984.

An excellent guide to sources of alternative energy products. Organized in sections on solar components and subsystems for buildings, complete solar energy systems for buildings, high-efficiency building products, etc.
R: *ARBA* (1985, p. 506).

Energy Resources. **J. T. McMullan, R. Morgan, and R. B. Murray.** New York: Halsted Press, 1978.

Deals with techniques of energy conversion with natural sources, fossil fuels, and nuclear power. Presentation is descriptive. Deals also with waste management.
R: *Aslib Proceedings* 43: 242 (May 1978).

Engineer's Guide to Solar Energy. **Yvonne Howell and J. A. Bereny.** San Mateo, CA: Solar Energy Information Services, 1979.
A working reference guide and learning aid to solar devices.
R: *Plant Engineering* 33: 182 (Sept. 6, 1979); *TBRI* 45: 350 (Nov. 1979).

Fuel from Farms: A Guide to Small-Scale Ethanol Production. **US Department of Energy.** Washington, DC: US Government Printing Office, 1980.
Surveys the present development of on-farm fermentation, ethanol production, and the economic and technical aspects involved.
R: *BL* 76: 1352 (May 15, 1980).

Guide to the Energy Industries. Cambridge, MA: Harfax/Ballinger, 1983.
Includes marketing and financial data on various energy industries, such as coal, natural gas, nuclear energy, petroleum, and others. The publisher listing is often outdated, and the indexes are poor.
R: *ARBA* (1984, p. 676).

Home Wind Power. Charlotte, VT: Garden Way Publishing, 1981.
Useful, practical guide that includes an annotated directory of equipment suppliers; monthly average wind power charts; and lists of state energy offices, organizations, and publications. Contains a glossary and a bibliography. For public and academic libraries.
R: *BL* 77: 1325 (June 15, 1981).

The Illustrated Guide to Home Retrofit for Energy Savings. **The Energy Resources Center.** New York: McGraw-Hill, 1981.
A detailed guide to energy conservation. Step-by-step directions cover materials, description, preparation, and installation. For home use.

Oil: A Plain Man's Guide to the World Energy Crisis. **Philip Windsor.** Ipswich, MA: Gambit, 1976.
Intended for the layperson.

Oilfields of the World. 2d ed. **E. N. Tiratsoo.** Houston: Gulf Publishers, 1976.
First edition, 1973.
Provides maps and tables of world petroleum fields.

The Passive Solar Energy Book: A Complete Guide to Passive Solar Home, Greenhouse, and Building Design. **Edward Mazria.** Emmaus, PA: Rodale Press, 1979.
For the layperson. Illustrates home conversion to solar energy. Cross-referenced. Recommended for people all over the country.
R: *Science Digest* 86: 88 (Dec. 1979); *Choice* 17: 565 (June 1980); *TBRI* 46: 76 (Feb. 1980).

Plant Engineers' and Managers' Guide to Energy Conservation. **Albert Thumann.** New York: Van Nostrand Reinhold, 1977.
A comprehensive, though not overly detailed, guide to energy management in plants. Contains much practical advice; suggests means of energy cost reduction.

R: *Chemical Engineering* 85: 15 (May 22, 1978); *Heating/Piping /Air Conditioning* 49: 126 (Nov. 1977); *IEEE Spectrum* 15: 62 (Aug. 1978); *Industrial Engineering* 10: 68 (Aug. 1978); *TBRI* 44: 79 (Feb. 1978); 44: 319 (Oct. 1978); 44: 359 (Nov. 1978).

The Solar Age Resource Book: The Complete Guidebook to the Dramatic Power of Solar Energy. **Martin McPhillips, ed.** New York: Everest House, 1979.

Consists of 16 articles dealing with aspects of solar energy. Main section of book contains buyers' guides. Solar products buyers' guide comprises solar packages and components of systems sections. Includes information on trade name, manufacturer, installation, warranty, services, etc. Of interest to architects, engineers, designers, and consultants.
R: *WLB* 54: 193 (Nov. 1979); *ARBA* (1981, p. 679).

The Solar Decision Book: Your Guide to Making a Sound Investment. **Richard H. Montogomery and Jim Budnick.** New York: Wiley, 1979.

A step-by-step guide for installing solar heating. Contains numerous illustrations; discusses technicalities, planning. Recommended for both academic and public libraries.
R: *LJ* 104: 1581 (Aug. 1979).

Solar Energy Application in Buildings. **A. A. M. Sayigh, ed.** New York: Academic Press, 1979.

An up-to-date guide that presents basic principles of solar conversion for various climates throughout the world. Expertly written; contains over 300 references.

Solar Energy in Developing Countries: An Overview and Buyers' Guide for Solar Scientists and Engineers. **A. Eggers-Lura.** Oxford: Pergamon, 1979.

A basic outline of aspects of solar energy. Includes a bibliography over 2,000 entries, as well as information on solar hardware and equipment suppliers. For architectural libraries.
R: *Aslib Proceedings* 44: 7 (July 1979).

The Solar Greenhouse Book. **James C. McCullagh.** Emmaus, PA: Rodale Press, 1978.

Provides useful advice on using solar energy in greenhouses. Useful in practically all climates.
R: *Conservation* 33: 46 (Nov.–Dec. 1978); *TBRI* 45: 30 (Jan. 1979).

Solar Products Specifications Guide: A Technical Specifications Guide that Continuously Monitors the Development of Solar Products. Harrisville, NH: SolarVision, 1979.

This valuable guide to solar equipment provides extensive technical information on products and manufacturers. For architects.
R: *ASHRAE* 21: 96 (Nov. 1979); *TBRI* 46: 79 (Feb. 1980).

Windpower: A Handbook on Wind Energy Conversion Systems. **V. Daniel Hunt.** New York: Van Nostrand Reinhold, 1981.

A historical and technological overview of wind power. Discusses current and future programs and objectives. Of interest to those in the field.
R: *BL* 77: 1325 (June 15, 1981).

Environmental Sciences

Atmospheric Dispersion in Nuclear Power Plant Siting: A Safety Guide. **International Atomic Energy Agency.** New York: Unipub, 1980.

A guide discussing transport of radioactive releases from a nuclear power plant into the environment. Also covers other related topics.

Environmental Biology for Engineers: A Guide to Environmental Assessment. **G. Camougls.** New York: McGraw-Hill, 1980.

Details how to relate environmental biology to engineering projects and environmental assessments. Contains environmental biology terminology. Useful for scientists and engineers.

A Guide to Site and Environmental Planning. 2d ed. **H. M. Rubenstein.** New York: Wiley, 1980.

A fully illustrated guide to environmental planning and design, including land use, storm drainage, bicycle paths, sample charts, etc.
R: *RIBA Journal* (Oct. 1980); *Urban Design Newsletter* (May 1980).

How to Remove Pollutants and Toxic Materials from Air and Water: A Practical Guide. **Marshall Sittig.** Park Ridge, NJ: Noyes Data, 1978.

Alphabetically arranged discussions of methods of removing toxic material from air and water. Helpful to industrial professionals, public health officials, etc.
R: *Plant Engineering* 32: 210 (Dec. 7, 1978); *TBRI* 45: 78 (Feb. 1979).

Noise Control for Engineers. **Harold W. Lord, Harold A. Evensen, and William S. Gatley.** New York: McGraw-Hill, 1980.

Discusses vibrations and acoustics. Consists of 3 sections: noise control engineering, practical aspects of noise, and applications and case studies. Useful for graduate students in environmental engineering, and industrial health and safety.

Pocket Guide to Chemical Hazards. Publication no. 78–210. **US National Institute for Occupational Safety and Health.** Cincinnati: US National Institute for Occupational Safety and Health, 1979.

Culls information gathered from the National Institute for Occupational Safety and Health Standards Completion Program. Health hazard information is succinct and up-to-date.
R: *National Safety News* 120: 108 (Nov. 1979); *TBRI* 46: 76 (Feb. 1980).

Shallow Ground Disposal of Radioactive Wastes: A Guidebook. **International Atomic Energy Agency.** New York: Unipub, 1981.

A useful guide to shallow-ground disposal of selected radioactive wastes. Includes guidelines to regulating and implementing organizations.

Site Selection and Evaluation for Nuclear Power Plants with Respect to Population Distribution: A Safety Guide. **International Atomic Energy Agency.** New York: Unipub, 1980.

Examines factors such as population distribution in determining safe sites for nuclear power plants.

FIELD GUIDES

ASTRONOMY

Catalog of Cometary Orbits. **Brian G. Marsden, ed.** Hillside, NJ: Enslow Publishers, 1983.

A useful catalog that provides comprehensive orbital data on 1,109 cometary apparitions. Extensive bibliographic sources. Useful for academic libraries with an astronomy interest.
R: *ARBA* (1985, p. 601).

Edmund Sky Guide. **Terence Dickinson and Sam Brown.** Barrington, NJ: Edmund Scientific, 1977.

A guide to the night sky. Highly recommended to anyone interested in science and astronomy.
R: *Journal of the Royal Astronomical Society of Canada* 71: 473 (Dec. 1977); *TBRI* 44: 83 (Mar. 1978).

Field Book of the Skies. 4th ed. **William Tyler Olcott and R. Newton Mayall.** New York: Putnam, 1954.

Considered one of the best such books for the more experienced amateur astronomer.

Field Guide to the Stars and Planets, Including the Moon, Satellites, Comets, and Other Features of the Universe. **Donald Howard Menzel.** Boston: Houghton Mifflin, 1975.

Earlier edition, 1986.
Star charts and photographic atlas of the sky, for amateur astronomers.
R: *Sky and Telescope* 29: 36 (1965); *BL* 60: 944 (1964); Jenkins (E56); Mal (5–32); Wal (p. 82).

The Larousse Guide to Astronomy. **David Baker.** New York: Larousse, 1978.

A profusely illustrated guide to the solar system. Provides detailed descriptions, color maps, tables, bibliography, and index. Discusses 85 constellations, cosmology theories, and black holes. For the amateur and the professional.
R: *BL* 76: 574 (Dec. 1, 1979); *LJ* 103: 1757 (Sept. 15, 1978); *ARBA* (1980, p. 610).

The Messier Catalogue. **Charles Messier.** Translated and edited by P. H. Niles. Clifton Park, NY: Auriga, 1981.

A quick guide to the French astronomer's famous 1787 list of star clusters, galaxies, and nebulae. For historians and amateurs.
R: *ARBA* (1983, p. 604).

The New Guide to the Planets. 3d ed. **Patrick Moore.** New York: Norton, 1972.

Second edition, 1971.
This is the newest edition of *Guide to the Planets*, published 20 years ago, and *The Planets*, published in 1962. A quick ready-reference book for the amateur astronomer. It covers the wandering stars, birth and movement of the planets, rocket exploration, and particular planets.
R: *LJ* 97: 2174 (June 15, 1972); *New Scientist and Science Journal* (Dec. 23, 1971); *ARBA* (1973, p. 540).

New Guide to the Stars. **Patrick Moore.** New York: Norton, 1976.
For the novice, a guide to stellar and galactic astronomy. Highly recommended for its nontechnical simplicity.
R: *Astronomy* 5: 58 (Oct. 1977); *TBRI* 43: 369 (Dec. 1977).

Pictorial Guide to the Moon. 3d ed. **Dinsmore Alter.** Edited by Joseph H. Jackson. New York: Crowell, 1979.
Earlier edition, 1973.
Latest maps of the whole moon, and more than 50 new photographs of the moon.
R: *LJ* 92: 4425 (1967); *ARBA* (1974, p. 554); Jenkins (E65).

Pictorial Guide to the Planets. 3d ed. **Joseph H. Jackson.** New York: Harper & Row, 1981.
First edition, 1965; second edition, 1973.
The pictures are current with the publication date. Several tables follow the chapters that give special emphasis to the earth and moon.
R: *Choice* 2: 404 (1965); Win (1EB1).

Planet Guidebook. Vols 1 and 2. **Japan Lunar and Planetary Observers Network, ed.** Tokyo: Scioundo Shinkosha Publishing, 1981.
A 2-volume manual in Japanese with sketches of the planets by artists and the space probes. The first volume includes Mercury, Venus, and Mars; the second deals with Saturn, Uranus, Neptune, and Pluto. Useful to observers.
R: *Sky and Telescope* 63: 481 (May 1982).

The Pocket Guide to Astronomy. **Patrick Moore.** New York: Simon and Schuster, 1980.
A basic guide to astronomy. Includes star charts and data about various celestial bodies. Contains excellent color plates and includes moon features and general indexes. Helpful to amateur stargazers.
R: *Sky and Telescope* 60: 321 (Oct. 1980); 60: 413 (Nov. 1980); *ARBA* (1981, p. 631).

Skyguide: A Field Guide for Amateur Astronomers. **Mark R. Chartrand III.** Racine, WI: Western Publishing, 1982.
Discusses and illustrates such topics as the sun, moon, planets, and stars. With maps, charts, and tables. For the educated layperson interested in learning about the heavens and basic observations.
R: *ARBA* (1984, p. 631).

BOTANY

All the Plants of the Bible. **Winifred Walker.** New York: Doubleday, 1979.

Includes descriptions of 100 plants mentioned in the Bible. Discusses uses of the plant in biblical times and supplies color illustrations. Eight plants that appear in the apocryphal books comprise a supplement.
R: *BL* 76: 928 (Mar. 1, 1980); 76: 1562 (June 15, 1980); *ARBA* (1981, p. 650).

Alternative Foods: A World Guide to Lesser-Known Edible Plants. **James Sholto Douglas.** London: Pelham Books. Distr. Levittown, NY: Transatlantic Arts, 1978.

Fourteen chapters with introductions provide lists of uncommon edible plants of the world that serve as alternative food sources. Headings include edible flowers, mushroom substitutes, and energy providers. Information such as geographical distribution, growth season, and preparation are provided for each plant. Special features are line drawings, tables, a short bibliography, indexes, and a directory of companies where seeds can be purchased. For public, college, and university library collections.
R: *ARBA* (1980, p. 703).

American Medicinal Plants: An Illustrated and Descriptive Guide to Plants Indigenous to and Naturalized in the United States which Are Used in Medicine. **Charles F. Millspaugh.** Philadelphia: Yorston, 1892. Reprint. New York: Dover, 1974.

Contains 180 full-page plates accompanying the descriptions. Includes a new table of revised classification and nomenclature.
R: *ARBA* (1976, p. 664).

The Audubon Society Field Guide to North American Trees: Eastern Region. **Elbert L. Little.** New York: Knopf, 1980.

A pictorial guide to the trees of North America; features color photographs, helpful descriptions, clear drawings, and easy-to-use identification keys classified according to leaves, flowers, fruit, and autumn leaves. Information on each tree includes physical data, notes on use of the tree, products, history, and cultivation. Volume covers 315 native species from the North American continent east of the Rocky Mountains and south of the Arctic tree line. Supplies index and glossary. Highly recommended for all academic, public, and school libraries.
R: *BL* 77: 186 (Oct. 1, 1980); *LJ* 105: 2220 (Oct. 15, 1980); *RSR* 8: 87 (July/Sept. 1980); *SLJ* 27: 96 (Nov. 1980); *ARBA* (1981, p. 662).

The Audubon Society Field Guide to North American Trees: Western Region. **Elbert L. Little.** New York: Knopf, 1980.

A companion volume to *The Audubon Society Field Guide to North American Trees: Eastern Region.* Contains similar photographs, identification keys, glossary, index, and format. Volume describes 314 species from the Rocky Mountains to the Pacific Coast. For all academic and school libraries.
R: *BL* 77: 186 (Oct. 1, 1980); *LJ* 105: 2220 (Oct. 15, 1980); *RSR* 8; 87 (July/Sept. 1980); *SLJ* 27: 96 (Nov. 1980); *ARBA* (1981, p. 663).

Checklist of United States Trees (Native and Naturalized). Agriculture Handbook, no. 541. **Elbert L. Little, Jr.** Washington, DC: Forest Service, US Department of Agriculture. Distr. Washington, DC: US Government Printing Office, 1979.

A revised reference volume that serves as the official standard for tree names in the Forest Service. Alphabetically arranged entries contain 679 native species and 169 naturalized species. Presents scientific and common names of trees and lists them by geographic range. Includes complete references, a series of informative appendixes, and an index of common names. Recommended for college and university libraries. Helpful to foresters and botanists.
R: *American Forests* 86: 63 (Jan. 1980); *TBRI* 46: 111 (Mar. 1980); *ARBA* (1981, p. 663).

Chinese Herbs: Their Botany, Chemistry and Pharmacodynamics. **John D. Keys.** Rutland, VT: C. E. Tuttle, 1976.

Comprehensive listing of more than 250 herbal plants arranged by botanical classification. Provides pharmacological information and thorough descriptions. Includes several appendixes of Chinese medicine, references, and a glossary. Well-recommended guide to medical lore.
R: *BL* 74: 633 (Dec. 1, 1977).

Common Plants of the Mid-Atlantic Coast: A Field Guide. **Gene M. Silberhorn.** Baltimore: Johns Hopkins University, 1982.

A general introduction to shoreline ecology with emphasis on the description and illustration of individual species of plants with a dichotomous key to each plant described. Intended for general purposes.
R: *ARBA* (1983, p. 626).

The Complete Book of Herbs and Herb Growing. **Roy Genders.** New York: Sterling Publishing, 1980.

A comprehensive volume on the horticulture and harvesting of herbs. Also provides an alphabetical listing of herbs.
R: *LJ* 105: 2095 (Oct. 1, 1980).

Complete Book of Herbs and Spices. **Sarah Garland.** New York: Viking, 1979.
Contains history, gardening, and cooking of herbs. Well-illustrated.
R: *American Horticulturist* 59: 41 (Feb.–Mar. 1980).

Complete Book of Mushrooms: Over 1,000 Species and Varieties of American, European, and Asiatic Mushrooms with 460 Illustrations in Black and White and in Color. **Augusto Rinaldi and Vassili Tyndalo.** Translated by Italia and Alberto Mancinelli. New York: Crown Publishers, 1974.

Complete guide in 2 parts. The first section discusses 1,000 species of mushrooms and fungi according to their odor, flavor, physical features, and range. The second part is devoted to a general overview of mushrooms and provides information on hunting, cultivation, edibility, history, and identification. The book features a section on cooking, preserving, and nutrients of edible mushrooms. Provides a glossary and index. For amateur mushroom hunters.

The Complete Guide to Water Plants: A Reference Book. Rev. ed. **Helmut Muhlberg.** East Ardsley, England: EP Publishing. Distr. New York: Sterling, 1982.

Describes 200 species by scientific name, distribution, characteristics, etc. For those with a serious interest in aquatic plants.

R: *ARBA* (1983, p. 623).

The Complete Handbook of Cacti and Succulents: A Comprehensive Guide to Cacti and Succulents in Their Habitats, to Their Care and Cultivation in House and Greenhouse and to the Genera and Their Species. **Clive Innes.** New York: Van Nostrand Reinhold, 1977.

Covers a wide-ranging and diverse number of cacti and succulents. Part 1 deals with wild cacti; and part 2 deals with cultivated species. Includes brief information and photographs. An authoritative information source. Recommended for library collections with botanical interests.

R: *The Garden* 103: 81 (Feb. 1978); *TBRI* 44: 218 (June 1978); *ARBA* (1978, p. 658).

The Complete Trees of North America: Field Guide and Natural History. **Thomas S. Elias.** New York: Van Nostrand Reinhold, 1980.

Covers over 750 evergreen and deciduous trees native to North America. Each entry comprises general description, habitat, distribution map, and practical uses. Includes 1,600 helpful line drawings, easy-to-understand text, and a list of wildlife associations. A useful compilation for amateur naturalists.

R: *American Scientist* 69: 232 (Mar./Apr. 1981); *BL* 77: 292 (Oct. 15, 1980).

Conifers. **David Carr.** North Pomfret, VT: Batsford, 1979.

Provides useful information on culture and detailed descriptions of 50 species. Includes lists of American and British names of trees; growth charts; climatic zones maps; line drawings; habitat sketches; glossary of botanical terms; and a current reference list.

R: *ARBA* (1981, p. 743).

Culpeper's Color Herbal. **David Potterton, ed.** New York: Sterling Publishing, 1983.

Original publication, 1649.

The work of Nicholas Culpeper, a seventeenth-century British astrologer-physician, provides the Latin and common name of some 400 herbs wth colorful illustrations, flowering time, astrological character, and medicinal values of each. This is followed by Potterton's discussion of the modern uses of each plant. An interesting way to follow the history of medicine to the modern uses of herbs.

R: *ARBA* (1984, p. 649).

Dangerous Plants. **John Tampion.** New York: Universe Books, 1977.

A timely guide to 100 harmful plants. Authoritatively written for a wide audience. Well-recommended.

R: *BL* 74: 323 (Sept. 15, 1977).

Diseases and Pests of Ornamental Plants. 5th ed. **Pascal P. Pirone.** New York: Wiley, 1978.

First edition, 1943; second edition, 1948; third edition, 1960; fourth edition, 1970. Discusses the diseases and pests that affect 500 genera of ornamental plants grown in the home, under glass, or outside. Explains how to use the most effective fungicides, insecticides, control materials, and pesticides that are the least harmful to the environment. Highly recommended for amateur and professional growers of ornamental plants.
R: *ARBA* (1980, p. 708).

Dwarf Rhododendrons. **Peter Cox.** New York: Macmillan, 1973.

A comprehensive reference guide to dwarf rhododendrons. A list of rhododendron collectors and societies, a glossary, and a bibliography are also provided.
R: *ARBA* (1974, p. 578).

Endangered and Threatened Plants of the United States. **Edward S. Ayensu and Robert A. DeFilipps.** Washington, DC: Smithsonian Institution and the World Wildlife Fund, 1978.

Offers a wealth of information on all areas of endangered and threatened plant species of the United States. Contents include definitions, methods, and participants used in compilation of data. Main sections consist of lists of endangered and extinct plants (organized by plant family and state), as well as helpful bibliographies. Of interest of conservation commissions.
R: *Choice* 16: 1196 (Nov. 1979); *ARBA* (1980, p. 628).

A Field Guide to Berries and Berrylike Fruits. **Madeline Angell.** New York: Bobbs-Merrill, 1981.

Describes and explains the habitats, families, and uses of the many small, pulpy fruits that are known as berries. A complete and inclusive guide for nature lovers in the United States and Canada.
R: *ARBA* (1983, p. 623).

A Field Guide to Edible Wild Plants of Eastern and Central North America. Peterson Field Guide Series, vol. 23. **Lee Peterson.** Boston: Houghton Mifflin, 1978.

Identifies edible wild plants and provides symbols pertaining to plants' gastronomic use. Helps to distinguish poisonous plants. A well-recommended field guide.
R: *American Forests* 85: 63 (Apr. 1979); *BL* 75: 324 (Oct. 1978); *TBRI* 45: 217 (June 1979).

Field Guide to Medicinal Wild Plants. **Bradford Angier.** Harrisburg, PA: Stackpole Books, 1978.

Describes 150 plant groups that are native to Europe and North America and that have been used for their healing properties. Accounts are detailed and precise. Contains useful information; for home and college libraries.
R: *BL* 76: 305 (Oct. 1, 1979); *ARBA* (1980, p. 633).

Field Guide to North American Edible Wild Plants. **Thomas S. Elias and Peter A. Dykeman.** New York: Outdoor Life Books. Distr. New York: Van Nostrand Reinhold, 1982.

A guide to locating, identifying, harvesting, and preparing wild plants. Arranged by edible season and then by plant type. For general and professional use.
R: *ARBA* (1984, p. 649).

A Field Guide to Poisonous Plants and Mushrooms of North America. **Charles Kinsley Levy and Richard B. Primack, eds.** Brattleboro, VT: Stephen Greene Press, 1984.
A compact guide that provides clear, concise information about the physical characteristics and toxicity of over 250 species. Detailed line drawings accompany each entry, with supplemental color photographs of 36 species. Indexed by popular and scientific name. A practical tool for both the general public and physicians.
R: *ARBA* (1985, p. 522).

A Field Guide to Tropical and Subtropical Plants. **Frances Perry and Roy Hay.** New York: Van Nostrand Reinhold, 1983.
This guide to more than 200 tropical plants offers information about habitat, flower and leaf pattern, and use. Includes a color photo of each. Good for libraries serving a clientele of travelers.

Field and Laboratory Guide to Tree Pathology. **Robert O. Blanchard and Terry A. Tattar.** New York: Academic Press, 1981.
An overall coverage of the terminology. Covers infectious and noninfectious diseases, and field and laboratory exercises in the field of tree pathology.

Grass Systematics. 2d ed. **Frank W. Gould and Robert B. Shaw, eds.** College Station, TX: Texas A & M University Press, 1983.
A good general introduction to grass biology, together with keys for the identification of the North American Genera of grasses. Fine glossary and a good index.
R: *ARBA* (1985, p. 522).

Grasses: An Identification Guide. **Lauren Brown.** Boston: Houghton Mifflin, 1979.
A field guide to members of the grass family. Emphasis on grasses found primarily in the Northeast. A welcome contribution to the botany literature.
R: *Scientific American* 240: 48 (June 1979): *LJ* 104: 1127 (May 15, 1979); 105: 573 (Mar. 1, 1980).

Herbs: Their Cultivation and Usage. **John Hemphill and Rosemary Hemphill, eds.** Poole, England: Blandford Press. Distr. New York: Sterling Publishing, 1984.
A simple guide that includes 31 common culinary herbs with 140 full-color photographs.
R: *ARBA* (1985, p. 503).

Index to the Gray Herbarium of Harvard University. 10 vols. Boston: Hall, 1968.
Consists of approximately 259,000 photo-reproduced cards from the Gray Herbarium catalog. Covers plants of the Western Hemisphere from 1886 to the present, providing name and literature citations. Provides extensive cross-references.

Medicines from the Earth: A Guide to Healing Plants. **William A. R. Thomson, ed.** New York: McGraw-Hill, 1978.

Describes nearly 250 medical herbs that can be cultivated for medical purposes. Excellent color illustrations, brief but informative entries. Highly recommended.
R: *ARBA* (1979, p. 671).

A Modern American Herbal: Useful Herbaceous Plants. Vol. 2. **Chester B. Dugdale.** New York: Barnes, 1978.

Arranged in 2 sections, mono-and dicots; subarranged by family. For each species included, a description of habitat, common and scientific name, medical use, toxicity, etc., are provided. Well-indexed. Contains glossary and bibliography. Highly recommended for general collections.
R: *ARBA* (1979, p. 670).

Mushrooms & Toadstools: A Color Field Guide. **U. Nonis.** New York: Hippocrene Books, 1982.

Describes 168 common species of mushrooms by comparison, pictograms, and common characteristics of each. Accuracy in descriptions and organization of each species are lacking. Limited in its usefulness.
R: *ARBA* (1984, p. 649).

The Mushroom Trail Guide. **Phyllis G. Glick.** New York: Holt, Rinehart and Winston, 1979.

Handy guide to the several hundred species of North American mushrooms. Clear concise drawings and identification keys. Paperback format encourages its outdoor use. Of value to many libraries.
R: *LJ* 104: 501 (Feb. 15, 1979); 105: 573 (Mar. 1, 1980); *ARBA* (1980, p. 635).

Mycology Guidebook. Rev. ed. **Mycological Society of America.** Edited by Russell B. Stevens. Seattle: University of Washington Press, 1981.

Earlier edition, 1974.
The definitive manual for teaching assistants and beginning specialists. Includes sections on field observations, taxonomic groups, ecological aspects, and fungus physiology and genetics.
R: *Choice* 11: 972 (Sept. 1974); *ARBA* (1975, p. 670).

The Natural Vegetation of North America: An Introduction. **J. L. Vankat.** New York: Wiley, 1979.

Details major types of vegetation of North America, including such factors as soil, climate, and topography. In 2 parts: part 1 covers basic facts of vegetation science; part 2 discusses background material. Outlines major plant species and communities, while emphasizing adaptation and environment. Includes numerous illustrations and suggested readings. For biologists and botanists.

The New York Times Book of Vegetable Gardening. **Joan Lee Faust.** New York: Quadrangle/New York Times, 1982.

Reproduction of 1975 edition.
Detailed guide to planting and cultivating 55 different vegetables and herbs. Numerous illustrations.
R: *RQ* 15: 354 (Summer 1975); *ARBA* (1976, p. 752).

Outline of Plant Classification. **Sandra Holmes.** London: Longman, 1983.

Gives a summary of each taxonomic kingdom division, class, and order listed in hierarchial order. Lots of information for such a thin volume.
R: *STL* 5: 2 (Winter 1984).

Scented Flora of the World. **Roy Genders.** New York: St. Martin's Press, 1977.

Arranged alphabetically by genera, this work discusses history, identification, and use of scented flora. Contains appendixes, classified lists, and photographs. Informative; one of the few references of its kind. For special libraries.
R: *BL* 74: 709 (Dec. 15, 1977).

Shrubs in the Landscape. **Joseph Hudak, ed.** New York: McGraw-Hill, 1984.

Over 1,000 species, varieties, and cultivars of ornamental shrubs covering a range from the southernmost regions of the United States to the lower areas of Canada. Extensive black-and-white photos and 80 color plates. Highly recommended.
R: *BL* 81: 546 (Dec. 15, 1984); *LJ* 109: 2059 (Nov. 1, 1984); *ARBA* (1985, p. 504).

A Synonymized Checklist of the Vascular Flora of the United States, Canada, and Greenland. Vol. 2. **John T. Kartesz and Rosemarie Kartesz.** Chapel Hill: University of North Carolina Press, 1980.

A comprehensive reference that lists about 57,000 names and 20,000 synonyms of North American flora. Entries are divided into 3 sections: Pteridophyta, Gymnospermae, and Angiospermae. Names are alphabetically organized by hierarchy and contain subspecies, varieties, and hybrids. Includes an index of family and genus names. Recommended for botanists.
R: *ARBA* (1981, p. 653).

Trees of North America and Europe. **Roger Phillips and Sheila Grant.** Edited by Tom Wellsted. New York: Random House, 1978.

A color photographic guide to more than 500 trees. Entries are organized alphabetically by Latin name. Contains a leaf key, which arranges leaves by shape into conifers, simple broad-leaves, and compound broad-leaves. Includes information about size, fruit, height, and origin. A glossary, abbreviation key, and index of common names are provided. Necessary addition to public and academic libraries for use by botanists and laypersons.
R: *BL* 77: 146 (Sept. 15, 1980); *LJ* 103: 2435 (Dec. 1, 1978); *ARBA* (1979, p. 675).

Trees of the World. **Scott Leathart.** New York: A & W Publishers, 1977.

Profusely illustrated in both color and black and white, a guide to trees of the world. For each tree illustrated, accompanying data are presented on family, genus, habitat, distribution, and general characteristics. Well-indexed. Contains a wealth of information. Highly recommended to school, public, and academic libraries.
R: *BL* 74: 1764 (July 15, 1978); *Choice* 15: 1397 (Dec. 1978); *ARBA* (1979, p. 674).

Flowers

Field Guide to Orchids of North America: From Alaska, Greenland, and the Arctic South to the Mexican Border. **John G. Williams and Andrew E. Williams, eds.** New York: Universe Books, 1983.

A handy field guide that covers over 200 species, subspecies, and varieties of North American orchids. Contains more than 100 plates and is well-indexed. A complete glossary of botanical terms is available.
R: *ARBA* (1985, p. 521).

Identification of Flowering Plant Families. 2d ed. **P. H. Davis and J. Cullen.** Cambridge: Cambridge University Press, 1979.

First edition, 1965. Published by Oliver and Boyd.
A taxonomic guide to flowering plants. Useful to students.
R: *Nature* 282: 651 (Dec. 6, 1979).

The "New York Times" Book of Annuals and Perennials. **Joan Lee Faust.** New York: Times Books, 1980.

A useful guide to the cultivation of 100 popular flowering annuals and perennials. Contains nontechnical descriptions of soil preparation, planting, maintenance, and shade. Includes color and black-and-white photographs as well as line drawings. Bibliography and appendixes provided. For home gardeners.
R: *ARBA* (1981, p. 745).

Simon and Schuster's Guide to Garden Flowers. **Guido Moggi and Luciano Giugnolini, eds.** Stanley Schuler, US ed. New York: Simon and Schuster, 1983.

A useful guide that covers details on cultivation of each type of flower, climatic suitabiltiy, food requirements, history and origin of the plants, etc. Extensive illustrations. Index is rather inadequate.
R: *ARBA* (1985, p. 504).

Rock Gardens and Water Plants in Color. Enjoy Your Garden Series. **Francis B. Stark and Conrad B. Link, eds.** Garden City, NY: Doubleday, 1973.

Each volume provides a short introduction, a glossary of terms, and an index. Profusely illustrated. Basic reference guides.
R: *ARBA* (1974, p. 661).

House Plants

The Complete Book of Terrarium Gardening. **Jack Kramer.** New York: Scribner's, 1974.

The Families of Flowering Plants Arranged to a New System Based on Their Probable Phylogeny. 3d ed. **John Hutchinson.** Reprint of 1973 edition. New York: Lubrecht and Cramer, 1979.

First edition, 1934; second edition, 1959.
A scholarly description of orders and families of numerous flowering plants. Well-organized and illustrated, the book is arranged to demonstrate parallels and relationships among the families.
R: *Journal of the Royal Horticultural Society* 85: 143 (Mar. 1969); Wal (p. 217).

The Healthy Garden Book: How to Control Plant Diseases, Insects, and Injuries. **Tom Riker.** New York: Stein & Day, 1979.
A useful guide concerning the control of plant pests and diseases. Lists major pests and their damage. Sections are alphabetically organized by the name of the plant and include illustrations of insects and damage, geographic location, and descriptive information. Covers all types of plants and examines various pesticides. Contains a detailed index. Of interest to gardeners and public libraries.
R: *ARBA* (1980, p. 708).

House Plants Indoors/Outdoors. San Francisco: Ortho Books, Chevron Chemical, 1977.
One of the best general books on house plants.

Indoor Trees. **Jack Kramer.** New York: State Mutual, 1980.
Earlier edition, 1975.
A well-organized book. Covers the latest facet of the houseplant tree.
R: *ARBA* (1976, p. 756).

The Pocket Guide to Indoor Plants. **George Seddon.** New York: Simon and Schuster, 1979.
A pocket-size reference discussing the growth and care instructions of 350 plants, including orchids and window-box plants. Contains drawings, a common name index, and lists of plants by special characteristics such as flower and ease of growth. Has a British slant. For the general public and public library collection.
R: *ARBA* (1981, p. 747).

Reader's Digest Illustrated Guide to Gardening. **Carroll C. Calkins, ed.** Pleasantville, NY: Reader's Digest Association. Distr. New York: Norton, 1978.
Profusely illustrated and comprehensive, this work contains contributions from outstanding botanists and horticulturists. Discusses many phases of gardening. Fully descriptive and comprehensively illustrated.
R: *BL* 75: 442 (Nov. 1, 1978); *LJ* 103: 1997 (Oct. 1, 1978); *ARBA* (1979, p. 760).

The Total Book of House Plants. **Russell C. Mott.** New York: Delacorte Press, 1975.
A guide to house gardening for the indoor amateur gardener. A subject index increases the practicality of the tool. Over 350 full-color paintings included.
R: *ARBA* (1976, p. 758).

Weeds and Wildflowers

The Audubon Society Field Guide to North American Wildflowers: Eastern Region. **William A. Niering and Nancy C. Olmstead.** New York: Knopf, 1979.

Compact, handy guide to over 600 species of wildflowers, vines, shrubs, and grasses that inhabit the eastern United States, with notes on 400 others. Like its companion volume, *The Audubon Society Field Guide to North American Wildflowers: Western Region,* this book presents 700 color photographs arranged by color and a text discussing the size of flower, habitat, range, history, and flowering dates for each species. Some line drawings accompany the text. Supplies index and glossary. Highly recommended for public, school, and academic libraries.
R: *Blair & Ketchum's Country Journal* 6: 119 (Nov. 1979); *BL* 77: 529 (Dec. 1, 1980); *TBRI* 45: 368 (Dec. 1979); *ARBA* (1980, p. 630).

The Audubon Society Field Guide to North American Wildflowers: Western Region. **Richard Spellenberg.** New York: Knopf, 1979.

Divided into sections, this comprehensive field guide discusses over 600 species of wildflowers and provides notes on 400 others found in western North America. Photographs are grouped by color, and the text is arranged by families. Descriptive data include habitat, range, size of flowers and leaves, months of flowering, and history. Provides glossary and index. For naturalists, students, and enthusiasts.
R: *ARBA* (1980, p. 631).

Eastern North America's Wildflowers. **Louis C. Linn.** New York: Dutton, 1978.

Contains identification keys for more than 370 wildflowers, with watercolor paintings. Grouped by color and then subgrouped by season. Pictorial information is excellent. Contains glossary and index. Highly accurate guide.
R: *BL* 76: 457 (Nov. 1, 1979).

Flowers of the Wild: Ontario and the Great Lakes Region. **Zile Zichmanis and James Hodgins, eds.** New York: Oxford University Press, 1984.

A beautiful guide of about 2,000 species of flowering plants in the region. Superb photos, glossary of terms, appendix of plant names, and indexes of both scientific and common names.
R: *ARBA* (1985, p. 521).

A Guide to Field Identification Wildflowers of North America. **Frank D. Venning, ed.** Racine, WI: Golden Press/Western Publishing, 1984.

An excellent 1-volume wildflower guide of over 1,500 species for amateurs. Excellent illustrations and index.
R: *ARBA* (1985, p. 520).

Handbook of Hawaiian Weeds. 2d ed. **E. L. Haselwood and G. G. Motter, eds.** Revised and expanded by Robert T. Hirano. Honolulu: University of Hawaii Press, 1983.

Descriptions of 226 species includes a list of characters, how the plant propagates itself, habitats of growth, history and origin of each, and more. Arranged by family in an evolutionary sequence and then alphabetically by genus in each family. The only book of its kind.
R: *ARBA* (1984, p. 652).

Jewels of the Plains: Wild Flowers of the Great Plains, Grasslands, and Hills. **Claude A. Barr.** Minneapolis: University of Minnesota Press, 1983.

The identification and cultivation of plants found in the great middle section of North America. Includes beautiful color pictures of already identified plants. For botanical and horticultural collections of all types.
R: *ARBA* (1984, p. 646).

Rocky Mountain Wildflowers. **Ronald J. Taylor.** Seattle: Mountaineers, 1982.
A colorful and simple guide to the more common flowers found in the US Rockies. Requires little knowledge of botanical terminology due to its arrangement by color and then by the general shape of the flower.
R: *ARBA* (1983, p. 627).

The Seaweed Handbook: An Illustrated Guide to Seaweeds from North Carolina to the Arctic. **Thomas F. Lee.** Boston: Mariners Press, 1977.
An easy-to-use identifying guide to the seaweeds of the US and Canadian east coast. Contains a key classification and then a description of each species. Includes pronunciation, scientific name, and plant description. Detailed and highly recommended, especially for its excellent illustrations.
R: *Choice* 15: 376 (May 1978); *ARBA* (1979, p. 673).

Weeds. 2d ed. **Walter Conrad Muenscher.** New York: Comstock Publishing Associates, 1980.
A 2-part volume on weeds. Part 1 contains general information about characteristics and control as well as numerous data tables. Part 2 deals with the identification of species of weeds. Line drawings, glossary, and literature references are provided. The appendix contains a listing of scientific and common names and a bibliography. Of interest to botanists.
R: *ARBA* (1981, p. 654).

Wild Orchids of Britain and Europe. **Paul Davies, Jenne Davies, and Anthony Huxley, eds.** Salem, NH: Chatto & Windus. Distr. Lawrence, MA: Merrimack Book Service, 1983.
A superb field guide on wild orchids in Britain and Europe. Over 300 color photographs. Comprehensive bilbiography and indexed. Recommmended for both orchidologists and laypersons.
R: *ARBA* (1985, p. 519).

Wildflower Folklore. **Laura C. Martin, ed.** Charlotte: East Woods Press, 1984.
A delightful book with a rare mixture of botany and folklore of more than 100 wild plants found in North America. Attractive full-page drawings for each plant, and index of both common and scientific names of plants.
R: *ARBA* (1985, p. 519).

Wildflowers and Weeds: A Guide in Full Color. **Booth Courtenay and James Hall Zimmerman.** New York: Van Nostrand Reinhold, 1977.
Earlier edition, 1972.
An important plant-identification book with 650 color photos.
R: *ARBA* (1974, p. 573).

Wildflowers of North America. **George S. Fichter.** New York: Random House, 1982.

A flower guide for the beginner; complete, informative, and easy to read. Limited in its use as an introduction to botany as a hobby.
R: *ARBA* (1984, p. 646).

Wild Flowers of the United States. 6 vols. **H. Rickett.** New York: McGraw-Hill, 1966–1973.

Volume 1, *Northeastern States,* 1966; volume 2, *Southeastern States,* 1967; volume 3, *Texas,* 1969; volume 4, *Southwestern States,* 1970; volume 5, *Northwestern States,* 1971; volume 6, *Central Mountains and Plains,* 1973.

An important, precise identification guide that discusses thousands of US wildflowers. Includes full-color photographs and both Latin and common English names. Highly recommended series for the layperson.
R: *American Scientist* 55: 120A, 178A (1967); *Science* 155: 65 (1967); *Choice* 7: 218 (1970); *LJ* 92: 250, 4013 (1967); *ARBA* (1970, p. 122); Jenkins (G138); Sheehy (EC32); Wal (p. 215); Win (2EC11; 3EC11).

The Wild Garden: An Illustrated Guide to Weeds. **Lys de Bray.** New York: Mayflower Books, 1978.

An illustrated guide to 160 weeds, arranged under groups by color. Indexed by common and scientific name. Well-designed identification guide.
R: *BL* 76: 523 (Nov. 15, 1979).

ZOOLOGY

Amphibians of North America: A Guide to Field Identification. **Hobart M. Smith.** Racine, WI: Golden Press, 1978.

A handbook for the identification of frogs, toads, newts, and salamanders found in the United States and Canada. Descriptions of 178 species are enhanced by range maps and color illustrations. Amphibian biology and conservation are also discussed. Supplies a short bibliography and index. For public and academic libraries.
R: *BL* 75: 861 (Feb. 1, 1979); *ARBA* (1980, p. 644).

Animal Identification: A Reference Guide. 3 vols. **R. W. Sims and D. Hollis, eds.** New York: Wiley, in cooperation with the British Museum, Natural History Department, 1980.

Volume 1, *Marine and Brackish Water Animals;* volume 2, *Land and Freshwater Animals;* volume 3, *Insects.*

In 3 volumes. Provides a comprehensive guide to scientific books and papers that can be used to identify aquatic and terrestrial animals. Arranged systematically and then geographically. Volume 1 contains identification resources for brackish water animals, volume 2 for freshwater, and volume 3 for insect groups.
R: *RSR* 8: 14 (July/Sept. 1980); *ARBA* (1981, pp. 665, 673, 676).

The Audubon Society Field Guide to North American Seashore Creatures. **Norman A. Meinkoth.** New York: Knopf, 1981.

A useful identification manual to some 850 marine invertebrates normally found in shallow water along the temperate seacoasts of the United States and Canada. Though selective in its coverage, the guide would be a fine introduction for nonspecialists.
R: *ARBA* (1983, p. 648).

A Collector's Guide to Seashells of the World. **Jerome M. Eisenberg.** New York: McGraw-Hill, 1981.

Compilation of 4,000 rare and popular specimens. Descriptions explain nomenclature, geographical distribution, and tips on preservation and cleaning for collectors. Includes photographs, glossary, bibliography, and index.
R: *BL* 77: 1178 (May 1, 1981).

Controlled Wildlife. **Carol Estes and Keith W. Sessions, comps.** Lawrence, KS: Association of Systematics Collections, 1983.

Volume 1, *Federal Permit Procedures;* volume 2, *Federally Controlled Species;* volume 3, *State Permit Procedures.*
Volume 2 is a field guide to both plant and animal wildlife species regulated by federal law. This is a major legal reference tool and should be in the collection of every large public library and scientific library.
R: *ARBA* (1985, p. 523).

A Field Guide to Dinosaurs. **The Diagram Group.** New York: Avon, 1983.

A well-written and concise guide to dinosaurs with illustrations and drawings. A good starting tool for any geology collection.
R: *LJ* 109: 434 (Mar. 1, 1984); *ARBA* (1985, p. 608).

Fieldbook of Natural History. 2d ed. **E. Laurence Palmer.** Revised by H. Seymour Fowler. New York: McGraw-Hill, 1975.

First edition, 1949.
New edition of the 25-year-old classic. Primarily devoted to the classification and description of living things. Describes some 2,000 species. The omission of technical terms makes this reference a useful item for laypersons.
R: *Science* 110: 129 (1949); *Choice* 12: 664 (July/Aug. 1975); *WLB* 50: 595 (Apr. 1975); *ARBA* (1976, p. 658); Jenkins (A84); Wal (p. 63).

Fieldbook of Pacific Northwest Sea Creatures. **Dan H. McLachlan and Jak Ayres.** Happy Camp, CA: Naturegraph, 1979.

A guide to approximately 200 common marine animals of shallow waters. Contains color plates and field keys. Includes a glossary, a short bibliography, and an index.
R: *Sea Frontiers* 25: 377 (Nov.–Dec. 1979); *TBRI* 46: 54 (Feb. 1980); *ARBA* (1980, p. 643).

Harper & Row's Complete Field Guide to North American Wildlife, Eastern Edition. **Henry Hill Collins.** New York: Harper & Row, 1981.

A revised and enlarged edition that examines 1,500 species of wildlife found in North America. Includes information on habitat, food, reproduction, and habits. Contains

extensive illustrations, some in color. A comprehensive reference volume that will be an invaluable addition to any library collection.
R: *LJ* 106: 1296 (June 15, 1981).

Harper & Row's Complete Field Guide to North American Wildlife: Western Edition. **Jay Ellis Ransom, comp.** New York: Harper & Row, 1981.
A book covering 1,800 species of birds, mammals, reptiles, amphibians, good and games fishes, and invertebrates occurring in North America. Included is a drawing and a description of each. Good for all field workers.
R: *ARBA* (1982, p. 731).

Simon and Schuster's Guide to Shells. **Bruno Sabelli.** Edited by Harold S. Feinberg. New York: Simon and Schuster, 1979.
A superb guide to shell identification and collection. Covers over 355 shells and includes 1,230 excellent illustrations. Shells are divided into 5 categories: soft surface mollusks, firm surface mollusks, coral dwellers, land and freshwater mollusks, and other marine mollusks. Descriptive information includes family, shell and body characteristics, habitat, and distribution. A classification table of text species, a glossary, and an index of entries are provided. Recommended for public or academic libraries.
R: *BL* 76: 1644 (July 15, 1980); *Choice* 18: 548 (Dec. 1980); *LJ* 105: 1648 (Aug. 1980); *ARBA* (1981, p. 675).

Wildlife of the Deserts. **Frederic H. Wagner.** New York: Abrams, 1980.
A comprehensive survey of the deserts and the wildlife they support. Text discusses general characteristics of the deserts, the 5 major desert areas, geological history, and the wildlife. Text is accompanied by over 200 excellent color photographs. Contains a glossary and an index. Suitable for public and academic libraries.
R: *BL* 77: 429 (Nov. 15, 1980); *Choice* 18: 684 (Jan. 1981).

The Wildlife Observer's Guidebook. **Charles E. Roth.** Englewood Cliffs, NJ: Prentice-Hall, 1982.
A book for observers who need exercises in observing, marking, and identifying wildlife. Also included are chapters on wildlife behavior, an index and sources for equipment. Good for wildlife novices or professionals.
R: *Choice* 20: 456 (Nov. 1982).

Wildlife of the Rivers. **William Hopkins Amos.** New York: Abrams, 1981.
A color-photographic presentation of the wildlife of the rivers. Provides scientific names and charts identifying river formations. Includes glossary and index. Useful for zoologists.
R: *BL* 77: 1283 (June 1, 1981).

Birds

Audubon's Birds of America. **George Dock, Jr. and John J. Audubon.** New York: Abrams, 1979.
Contains 30 full-page color paintings of birds and over 400 smaller plates.
R: *LJ* 104: 1266 (June 1, 1979).

The Audubon Society Field Guide to North American Birds: Eastern Region. **John Bull and John Farrand, Jr.** New York: Alfred A. Knopf. Distr. New York: Random House, 1977.

A guide to the birds of the eastern regions of the United States and Canada, arranged by habitat. Color photographs are of good quality, as are authoritative descriptions. Highly recommended.

R: *BL* 75: 403 (Oct. 15, 1978); *LJ* 102: 1746 (Sept. 1, 1977); *ARBA* (1978, p. 671).

The Audubon Society Field Guide to North American Birds: Western Region. **Miklos D. F. Udvary.** New York: Knopf, 1977.

Arranged by habitat. Geared for the inexperienced bird-watcher. Covers the western part of the United States and Canada. Provides color illustrations and photographs and detailed descriptions of various species. For all libraries.

R: *BL* 75: 403 (Oct. 15, 1978); *ARBA* (1978, p. 675).

The Audubon Society Guide to Attracting Birds. **Stephen W. Kress.** New York: Scribner's, 1985.

A well-illustrated and referenced work on attracting birds. Accurate and authoritative.

R: *LJ* 111: 86 (Feb. 1, 1986).

The Audubon Society Master Guide to Birding. 3 vols. **John Farrand, Jr., ed.** New York: Knopf, 1983.

Volume 1, *Loons to Sandpipers;* volume 2, *Gulls to Dippers;* volume 3, *Old World Warblers to Sparrows.*

A most detailed and illustrative description of North American birds. A valuable source for all levels of birders.

R: *LJ* 109: 784 (Apr. 1984).

Bird Families of the World. **Colin J. O. Harrison, ed.** New York: Abrams, 1978.

Arranged by order of birds and subdivided by family. Contains detailed articles by experts. Articles are succinct; contain physical description, feed habits, and taxonomy and are accompanied by illustrations. Table of contents lists families. A useful acquisition for various libraries.

R: *Nature* 276: 643 (Dec. 7, 1978); *BL* 76: 786 (Feb. 1, 1980); *ARBA* (1981, p. 666).

Birds of the World: A Check List. 2d ed. **James F. Clements.** New York: Two Continents, 1978.

First edition, 1974.

Revised edition includes 15 recently discovered bird species, as well as other updated information. An important reference; contains common-name index. Recommended.

R: *WLB* 53: 187 (Oct. 1978).

The Breeding Birds of Europe: A Photographic Handbook. 2 vols. **Manfred Pforr and Alfred Limbrunner, eds.** Beckenham, England: Croom Helm. Distr. Dover, NH: Tanager Books, 1983.

A superb photographic source on the subject. Each volume follows the same order-general description of each order with black-and-white drawings of each species, followed by a 2-page spread of each species (male and female).
R: *ARBA* (1985, p. 527).

The British Ornithologists' Guide to Bird Life. **Carl Fredrik Lundevall.** Poole, Dorset, England: Blandford, 1980.
R: *RSR* 8: 87 (July/Sept. 1980).

Cage Bird Identifier. **Helmut Betchel.** New York: Sterling, 1977.
Intended as an identification guide, this work is a selection of 120 bird species; with color photographs.
R: *BL* 74: 1695 (July 1, 1978); *WLB* 48: 166 (1973); *ARBA* (1974, p. 587).

Check-list of Birds of the World. 2d ed. **J. L. Peters.** Cambridge, MA: Harvard University, Museum of Comparative Zoology, 1979.
R: *RSR* 8: 14 (July/Sept. 1980).

The Complete Birds of the World. **Michael Walters.** North Pomfret, VT: David & Charles, 1980.
A listing of all world bird species. Each entry includes English and Latin names and information on food, range, nest site, habitat, and fledgling period. Indexed. Authoritative reference for all libraries.
R: *ARBA* (1981, p. 670).

Cranes of the World. **Paul Johnsgard, ed.** Bloomington, IN: Indiana University Press, 1983.
Includes detailed accounts of each of the 14 species of cranes. Illustrated and indexed. Intended for birders.
R: *ARBA* (1985, p. 526).

Crows of the World. **Derek Goodwin.** Ithaca, NY: Comstock Publishing Associates, 1976.
Describes 116 species of crows worldwide. Detailed and authoritatively written. Describes distribution and behavioral habits. Provides references. Extremely well written and produced. Highly recommended for academic natural history collections.
R: *BL* 74: 702 (Dec. 15, 1977).

A Field Guide to the Birds of Australia. **Graham Pizzey.** Princeton: Princeton University Press, 1980.
An authoritative field guide to the birds of Australia. Contains 88 plates and 725 range distribution maps. Includes descriptive information on breeding, habitat, field marks, etc. Cross-referenced. An indispensable reference for ornithologists.
R: *ARBA* (1981, p. 668).

A Field Guide to the Birds: A Completely New Guide to All the Birds of Eastern and Central North America. Peterson Field Guide Series, vol. 1. 4th ed. **Roger Tory Peterson.** Boston: Houghton Mifflin, 1980.

Earlier edition, entitled *A Field Guide to the Birds (Eastern)*, 1947.
Completely revised and rewritten edition of a major book on birds. Includes new color plates; new range maps that show winter, year-round, and breeding ranges for each species; checklists; and an index. Highly recommended to birders and all libraries.
R: *BL* 77: 492 (Dec. 1, 1980); *WLB* 55: 543 (Mar. 1981).

Field Guide to the Birds of North America. **Shirley L. Scott, ed.** Washington, DC: National Geographic Society, 1983.
A superb field guide with over 200 color plates painted by 13 artists. Invaluable tool.
R: *BL* 80: 1280 (May 15, 1984); *Choice* 21: 1442 (June 1984); *ARBA* (1985, p. 524).

A Field Guide to the Birds of South-East Asia, Covering Burma, Malaya, Thailand, Cambodia, Vietnam, Laos, and Hong Kong. **Ben F. King and Edward C. Dickinson.** Boston: Houghton Mifflin, 1975.
Some 1,200 species arranged in the *Peterson Field Guide Series* format. Over 1,000 color and black-and-white plates. The author's intention is to periodically update the guide.
R: *Choice* 12: 1290 (Dec. 1975); *ARBA* (1976, p. 683).

A Field Guide to Birds of the USSR: Including Eastern Europe and Central Asia. **V. E. Flint, ed.** Princeton: Princeton University Press, 1984.
An updated translation of a work published in Russian in 1968, this guide covers all 728 species recorded from the USSR. Illustrated and indexed by English, Latin, and Russian names. A cross-reference list with additional taxonomic and nomenclatural information.
R: *Choice* 22: 62 (Sept. 1984); *LJ* 109: 1314 (July 1984); *ARBA* (1985, p. 525).

Finding Birds around the World. **Peter Alden and John Gooders.** Boston: Houghton Mifflin, 1981.
A field guide to all aspects of birdwatching. Contains advice on locations, equipment, clothing, accommodations, and traveling routes. Provides a listing of bird species and their habits. Includes a cross-referenced taxonomic index and an alphabetical index. An invaluable aid to amateur birdwatchers.
R: *BL* 77: 906 (Mar. 1, 1981).

A Guide to North American Waterfowl. **Paul A. Johnsgard.** Bloomington, IN: Indiana University Press, 1979.
A condensed version of the identification guide, *Waterfowl of North America.* Includes shortened descriptive information to aid in the identification of geese, ducks, and swans. Species are organized into 7 tribes of waterfowl with subsections on identification and natural history. Contains 31 color photographs, as well as black-and-white sketches. Identification keys, references, and a name index are provided. A valuable addition to high school, public, and college library collections.
R: *BL* 77: 70 (Sept. 1, 1980); *LJ* 104: 2084 (Oct. 1, 1979); *ARBA* (1980, p. 640).

Gulls: A Guide to Identification. **P. J. Grant.** Vermillion, SD: Bueto Books, 1982.

A collection of some 276 black-and-white photographs of the 23 European gull species, 16 of which also are found in North America. An authoritative reference to the identification or plumage sequence recognition of the species.
R: *ARBA* (1984, p. 657).

How Birds Work: A Guide to Bird Biology. **Ron Freethy.** New York: Sterling, 1983.
This book on general biology provides an easy-to-follow description of avian physiology and anatomy. Touches upon evolution, taxonomy, flight, respiration, distribution, etc. With illustrations. Of universal interest.
R: *LJ* 108: 57 (Jan. 1983).

The Island Waterfowl. **Milton Webster Weller.** Ames, IA: Iowa State University Press, 1980.
Comprehensive coverage of island waterfowl. Contains information on distribution, colonization, adaptation, conservation, and characteristics. Includes excellent figures, black-and-white photographs, and a detailed index. Invaluable aid for students in ornithology, evolution, or zoogeography.
R: *Choice* 18: 275 (Oct. 1980).

The Larousse Guide to Birds of Britain and Europe. **Bertel Bruun.** New York: Larousse, 1978.
Contains excellent color plates, clear text, and distribution maps. A useful pocket guide to European birds.
R: *ARBA* (1979, p. 678).

Manual of Neotropical Birds. 4 vols. **Emmet R. Blake.** Chicago: University of Chicago Press, 1977–.
Volume 1, *Spheniscidae (Penguins) to Laridae (Gulls and Allies)*, 1977.
A 4-volume series of guides; volume 1 deals with neotropical birds. Considered a high-quality source of reference.
R: *BioScience* 28: 460 (July 1978); *Condor* 79: 509 (Winter 1977); *TBRI* 44: 173 (May 1978); 44: 255 (Sept. 1978).

Owls of Britain and Europe. **A. A. Wardhaugh, ed.** Dorset, England: Blandford Press. Distr. New York: Sterling, 1983.
A good summary and guide to the 13 species of owls of Britain and Europe, 7 of which also occur in North America. Well-illustrated with black-and-white photographs, and indexed.
R: *ARBA* (1985, p. 528).

Seabirds: An Identification Guide. **Peter Harrison.** Boston: Houghton Mifflin, 1982.
Concentrates on the plumage sequences and distribution of some 312 species of 1,600 birds. Provides line drawings and identification keys, color plates, distribution maps, and a bibliography, which all prove to be useful tools for the novice as well as for the expert in this field.
R: *LJ* 109: 784 (Apr. 1984).

Seabirds of the World. **Ronald M. Lockley, ed.** New York: Facts on File, 1983.

An informative photographic guide to some species of seabirds of the world. More for general reader than the birdwatcher. Most suitable for public libraries.
R: *ARBA* (1985, p. 526).

Swans of the World. **Sylvia Bruce Wilmore.** New York: Taplinger, 1979.

Earlier edition, 1974.
Provides the general reader with ample illustrative and textual material.
R: *Choice* 12: 99 (Mar. 1975); *ARBA* (1976, p. 688).

Thorburn's Birds. **James Fisher, ed.** Woodstock, NY: Overlook Press. Distr. New York: Viking Press, 1976.

A fine guide to British birds, describing distribution, etc. High-quality reference source includes illustrations. Highly recommended.
R: *ARBA* (1977, p. 668).

Wading Birds of the World. **Eric Soothill and Richard Soothill.** Dorset, England: Blandford Press. Distr. New York: Sterling, 1982.

Classifies birds by their method of hunting for food and by wading. Includes color photos, line drawings, and two indexes of Latin names and common names. Limited in scope but useful as a resource aid.
R: *ARBA* (1984, p. 660).

Waterfowl: Ducks, Geese and Swans of the World. **Frank S. Todd.** New York: Harcourt, Brace, Jovanovich, 1979.

A superb volume dealing with waterfowl of the world. Covers general characteristics, classification, and tribes of swans and geese. Text is enhanced by 788 excellent color photographs. Includes 2 appendixes: a chart describing all species and subspecies, and an examination of photography techniques. Contains a glossary. An essential reference for ornithologists, aviculturists, photographers, and sports enthusiasts.
R: *LJ* 105: 2474 (Nov. 15, 1980); *ARBA* (1981, p. 669).

Wildfowl of the World. **Eric Soothill and Peter Whitehead.** Dorset, England: Blandford Press. Distr. New York: Sterling, 1978.

Provides detailed descriptions and color photographs of 128 species of wildfowl. Information is clear and concisely presented. Contains a list of wildlife trusts and a bibliography. Well-organized and informative; a useful guide.
R: *BL* 76: 1005 (Mar. 1, 1980); *WLB* 53: 564 (May 1979).

Wild Geese of the World: Their Life History and Ecology. **Myrfyn Owen.** North Pomfret, VT: Batsford, 1980.

An excellent guide to wild geese of the world. Text examines social behavior, migration, conservation, and population dynamics. Color plates, line drawings, graphs, and 500 references complement the text. Recommended for public or academic libraries for use by upper-level undergraduate and graduate students.
R: *Choice* 18: 1292 (May 1981).

Fishes

The American Museum of Natural History Guide to Shells: Land, Freshwater, and Marine, from Nova Scotia to Florida. **William K. Emerson and Morris K. Jacobson.** New York: Knopf, 1976.

Describes and illustrates 800 North American shellfish. Well-organized by chapter. Provides common names, information on conservation, glossary, bibliography, and index. An excellent reference.

R: *Animal Kingdom* 79: 32 (Oct./Nov. 1976); *TBRI* 43: 135 (Apr. 1977); *ARBA* (1977, p. 676).

Compendium of Seashells: A Color Guide to More Than 4,200 of the World's Marine Shells. **R. Tucker Abbott and S. Peter Dance.** New York: Dutton, 1982.

Superb photographs with ecological, geographical, and natural histories of each shell. Highly recommended for the amateur collector and professional as a strong reference.

R: *ARBA* (1984, p. 667).

Dangerous Sea Creatures. A Complete Guide to Hazardous Marine Life. **Thomas Helm.** New York: Funk & Wagnalls, 1976.

A Field Guide to Pacific Coast Fishes of North America: From the Gulf of Alaska to Baja California. The Peterson Field Guide Series. **William N. Eschmeyer and Earl S. Herald.** Boston: Houghton Mifflin, 1983.

A concise account of the identification of various fish specimens, including sharks, rays, and jawless and bony fishes. Carefully prepared with illustrations and invaluable information. A must for all West Coast fishermen and beachcombers, as well as for libraries.

R: *Choice* 20: 1574 (July/Aug. 1983).

Fishes of the World. 2d ed. **Joseph S. Nelson.** New York: Wiley, 1984.

A detailed classification of an enormous variety of fishes throughout the world.

McClane's Field Guide to Saltwater Fishes of North America. **A. J. McClane, ed.** New York: Holt, Rinehart, Winston, 1978.

Describes species of saltwater fishes. Arranged alphabetically under respective family. Provides information on geographic distribution, anatomy, life span, etc. Contains color illustrations and a glossary. An authoritative pocket guide to saltwater marine life.

R: *BL* 76: 305 (Oct. 1, 1979); *WLB* 53: 278 (Nov. 1978); *ARBA* (1979, p. 682).

Poisonous and Venomous Marine Animals of the World. rev. ed. 3 vols. **Bruce W. Halstead.** Sections on chemistry by Donovan A. Courville. Princeton: Darwin Press, 1978.

Volume 1, *Invertebrates;* volumes 2 and 3, *Vertebrates.*
R: Win (2EC15a).

Popular Marine Fish for Your Aquarium. **Martyn Haywood, ed.** Blue Ridge Summit, PA: TAB Books, 1984.

A simple guide that includes an identification guide to major saltwater fishes used in home aquaria and an introduction to fish families. Color illustrations and indexed.
R: *ARBA* (1985, p. 531).

Popular Tropical Fish for Your Aquarium. **Cliff Harrison, ed.** Blue Ridge Summit, PA: TAB Books, 1984.

A simple guide to most common species of freshwater fish for home aquaria. The identification section includes information on families of tropical fish and an illustrated guide to the major freshwater fishes.
R: *ARBA* (1985, p. 531).

Simon and Schuster's Complete Guide to Freshwater and Marine Aquarium Fishes. New York: Simon and Schuster, 1977.

Contains over 300 full-color illustrations, concise key identification for pH, light, etc. Arranged by section (i.e., marine fishes, freshwater fishes, etc.) and then alphabetically within each section. Includes index of common and scientific names. A well-recommended guide.
R: *BL* 74: 1520 (May 15, 1978).

The Sharks of North American Waters. **Jose I. Castro.** College Station, TX: Texas A & M University Press, 1983.

Describes and illustrates more than 100 species of sharks reported within 500 nautical miles of the United States and Canada. For professional biologists, fishermen, scuba divers, and other marine biology enthusiasts.
R: *ARBA* (1984, p. 662).

Tropical Fish Identifier. **Braz Walker.** New York: Sterling, 1981.

Earlier edition, 1971.
An introductory identification manual covering 120 popular aquarium fish. Includes breeding data and color photos.

Insects

Aphids on the World's Crops: An Identification Guide. **R. L. Blackman and V. F. Eastop.** New York: Wiley, 1985.

A unique guide to the identification of the aphids that attack over 280 of the world's most significant crops. Systematically describes more than 400 aphid species and includes 150 halftones of mounted species.

The Audubon Society Book of Insects. **Les Line, Lorus Milne, and Margery Milne, eds.** New York: Abrams, 1983.

An exceptional guide to insects, with excellent color photographs. Intended as a general survey of the insects for the nonspecialist.
R: *ARBA* (1985, p. 533).

The Audubon Society Field Guide to North American Butterflies. **Robert Michael Pyle.** New York: Random House, 1981.

A field guide containing more than 600 species of the North American butterfly. Includes more than 1,000 color plates of butterflies in their natural habitats. Contains

glossary, index of plants eaten by caterpillars, index of butterflies, and brief listing of lepidopterist societies.
R: *BL* 78: 273 (Oct. 15, 1981).

The Audubon Society Field Guide to North American Insects and Spiders. **Lorus Johnson Milne and Margery Milne.** New York: Knopf, 1980.

Features over 700 spiders and insects of North America; notes species size, habitat, and range. Excellent color photographs of arthropods in their natural habitats enhance the text. Includes glossary and index. Silhouette thumb-tab guides allow for quick spot identifications. For all libraries.
R: *BL* 77: 657 (Jan. 15, 1981); *Choice* 18: 926 (Mar. 1981).

Butterflies: A Colour Field Guide. **M. Devarenne, ed.** London: David & Charles. Distr. New York: Hippocrene Books, 1983.

An illustrated guide of 144 butterflies of Europe and North Africa. Index covers both scientific and common names.
R: *LJ* 109: 478 (Mar. 1, 1984); *ARBA* (1985, p. 532).

The Butterflies of North America. **James A. Scott.** Stanford, CA: Stanford University Press, 1985.

The Common Insects of North America. **Lester A. Swan and Charles S. Papp.** New York: Harper & Row, 1972.

An introductory guide to the insects of the United States and Canada. Describes individual species. Does presuppose some knowledge of biology.
R: *ARBA* (1973, p. 577).

The Encyclopedia of Natural Insect and Disease Control: The Most Comprehensive Guide to Protecting Plants. **Roger B. Yepsen, Jr., ed.** Emmaus, PA: Rodale Press, 1984.

Earlier edition, *Organic Plant Protection,* 1976.
An encyclopedic guide aimed at protecting plants from bugs, diseases, and the environment at large. Over 100 color illustrations of insect pests and plant diseases.
R: *ARBA* (1985, p. 502).

A Field Guide in Color to Insects. **Jiri Zahradnik.** Translated by Olga Kuthanova. London: Octopus. Distr. New York: Mayflower Books, 1977.

A handy field guide that includes more than 800 illustrations and 120 diagrams. Insect identification is facilitated; geographic area emphasized in Europe. Provides basic discussions of insect evolution and a bibliography. A highly recommended volume.
R: *BL* 76: 62, 64 (Sept. 1, 1979).

Insect Pests of Farm, Garden, and Orchard. 7th ed. **R. H. Davidson and W. F. Lyon.** New York: Wiley, 1979.

An updated and revised introduction to insect pests of North America. Provides coverage of classification, physiology, ecology, biology, and other topics related to pest control. Also includes new pest species as well as spiders, slugs, centipedes, etc. New illustrations and conversion factors are added. Helpful to biologists and environmentalists.

The International Butterfly Book. **Paul Smart.** New York: Crowell, 1975.
An excellent guide to butterflies of the world. For the serious and amateur entomologist.
R: *ARBA* (1976, p. 690).

The Oxford Book of Insects: Pocket Edition. **John Burton et al.** New York: Oxford University Press, 1982.

Mammals

The Audubon Society Field Guide to North American Mammals. **John O. Whitaker, Jr.** New York: Knopf, 1980.
Describes over 200 mammals and provides excellent color photographs of the animals in their natural environment. Text consists of information on habitat, range, similar species, and description. Diagrams of tracks of many species are also supplied. A handy pocket volume for all libraries.
R: *BL* 77: 657 (Jan. 15, 1981); *Choice* 18: 1125 (Apr. 1981).

Bats of America. **Roger W. Barbour and Wayne H. Davis.** Reprint of 1969 edition. Lexington: University of Kentucky Press, 1979.
Guide identifies different species.
R: *LJ* 94: 3991 (Nov. 1, 1969); *ARBA* (1970, p. 124).

Cattle of the World. **Dennis Bishop and John B. Friend.** Poole, Dorset, England: Blandford Press, 1978.
Concise, well-illustrated guide to the history, development, and distribution of cattle. Covers 113 breeds in detail. Provides pertinent data. For public and academic libraries.
R: *BL* 76: 373 (Oct. 15, 1979); *WLB* 53: 564 (May 1979); *ARBA* (1980, p. 701).

The Complete Dog Book: The Photograph, History and Official Standard of Every Breed Admitted to AKC Registration, and the Selection, Training, Breeding, Care, and Feeding of Pure-Bred Dogs. 16th ed. New York: Howell Book House, 1979.
Revised edition of a standard work; contains 113 excellent photographs of different breeds. Provides information on new breeds and corrects inaccurate descriptions and standards appearing in previous edition. Details histories of 124 breeds; dog care; diseases, health and nutrition; and dog sports. Includes a glossary. A comprehensive reference for all public libraries.
R: *ARBA* (1980, p. 700).

The Complete Guide to All Cats. **Ernest H. Hart and Allan H. Hart.** New York: Scribner's, 1980.
Well-prepared, thorough reference on cats and cat care. Presents detailed information on health considerations. For public libraries.
R: *LJ* 105: 1176 (May 15, 1980); *ARBA* (1981, p. 737).

A Complete Guide to Monkeys, Apes and Other Primates. **Michael Kavanagh, ed.** New York: Viking Press, 1983.

A simply written guide for general readers. Taxonomically arranged by genus accounts. Illustrated with color photos; short bibliography. Indexed by common names.
R: *ARBA* (1985, p. 534).

A Field Guide to Dangerous Animals of North America: Including Central America. **Charles K. Levy.** Lexington, MA: Stephen Greene Press, 1983.

Deals with animals capable of causing injury to humans, those that are vectors of disease, and those directly injurious in other ways. Includes terrestrial and marine species with a description, drawing, information on characteristics, and information on medical treatment for each species. Superb, first-hand information.
R: *ARBA* (1984, p. 654).

A Field Guide to the Whales, Porpoises and Seals of the Gulf of Maine and Eastern Canada: Cape Cod to Newfoundland. **Steven K. Katona, Valerie Rough, and David T. Richardson.** New York: Scribner's, 1983.

A small and concise book filled with information on marine mammals of the Cape Cod to Newfoundland region. Useful to anyone interested in whales and seals.
R: *ARBA* (1984, p. 666).

Guide to Mammals of the Plains States. **J. Knox Jones et al.** Lincoln, NE: University of Nebraska Press, 1985.

The Larousse Guide to Horses and Ponies of the World. **Elwyn Hartley Edwards.** New York: Larousse, 1979.

Provides comprehensive coverage of different breeds of horses and ponies. Examines Arabians and thoroughbreds. Horses and ponies are arranged by country. Information covers appearance, height, characteristics, uses, origin, etc. Also gives advice on how to buy and identify good horses. Includes excellent photographs and drawings, as well as a noteworthy index. For beginners or experts on horses.
R: *LJ* 105: 394 (Feb. 1, 1980); *ARBA* (1981, p. 736).

The Mammals of North America. 2d ed. 2 vols. **Eugene Raymond Hall, ed.** New York: Wiley, 1981.

First edition, 1959.
Completely revised; sums up the most recent taxonomical studies of over 3,600 mammals from Greenland to Panama. Contains maps, illustrations, and a bibliography.
R: *RSR* 8: 14 (July/Sept. 1980); *Subscription Books Bulletin* 56: 47 (1959); Jenkins (G238); Wal (p. 237); Win (EC113).

Mammals of the Northern Great Plains. **J. Knox Jones, Jr. et al.** Lincoln: University of Nebraska Press, 1983.

Information geared to a wide range of readers interested in the free-living mammals of North Dakota, Nebraska, and South Dakota. Includes photographs and line drawings in its introduction of regional mammals; for students as well as professionals in this area.
R: *ARBA* (1984, p. 665).

Mammals of the World. 3d ed. 3 vols. (vol. 3, published in 1968 by the same press, is still in 2d ed.). **Ernest Pillsbury Walker.** Baltimore: Johns Hopkins Press, 1975.

Volumes 1 and 2 are a comprehensive account of recent genera of mammals arranged by taxonomic classification; volume 3 is a massive bibliography on all aspects of mammalogy.

R: *Nature* 208: 210 (Oct. 16, 1965); *Science* 146: 1285 (1964); *BL* 71: 528 (Dec. 1, 1975); *Choice* 12: 984 (Oct. 1975); *ARBA* (1976, p. 692); Jenkins (G244); Wal (p. 237); Win (EC114).

Mammals—Their Latin Names Explained: A Guide to Animal Classification. **A. F. Gotch.** Poole, Dorset, England: Blandford Press. Distr. New York: Sterling, 1979.

An authoritative guide to over 1,000 species of mammals. Entries consist of taxonomy of mammals and short descriptions. Taxa are given, with their derivations explained. Individual mammals are listed by their common English name under the family. Information includes scientific name, derivation, and territory. Includes general, English-names, and scientific-names indexes, as well as a bibliography on taxonomy and nomenclature. Recommended for public libraries or biologists.

R: *New Scientist* 81: 1054 (Mar. 29, 1979); *RSR* 8: 16 (July/Sept. 1980); *ARBA* (1981, p. 675).

Simon and Schuster's Guide to Cats. **Gino Pugnetti.** Edited in United States by Mordecai Siegal. New York: Simon and Schuster, 1983.

A beautifully illustrated guide to cats in 2 major sections—a 126-page general introduction, and a breeds section with detailed description of 40 long-haired and short-haired cat breeds. Over 200 color photographs; indexed.

R: *BL* 80: 778 (Feb. 1, 1984); *ARBA* (1985, p. 529).

A Standard Guide to Cat Breeds. **Grace Pond and Ivor Raleigh, eds.** New York: McGraw-Hill, 1979.

Presents standards used in judging 330 breeds and color variations of cats. Each breed is examined through descriptive, historical, and pictorial information. Discusses coat patterns, eye and coat colors, feline behavior, and genetics. Recommended for cat breeders.

R: *ARBA* (1980, p. 701).

Walker's Mammals of the World. 4th ed. **Ronald M. Nowak and John L. Paradiso.** Baltimore: Johns Hopkins University Press, 1983.

Third edition, 1975.

This expanded and revised edition includes a complete list of some 4,154 species, as well as other relevant information on living and recently extinct mammals. Essential for libraries with interests in zoology and natural history.

R: *ARBA* (1984, p. 666).

A World List of Mammalian Species. **Gordon B. Corbet and J. E. Hill.** London: British Museum of Natural History. Distr. Ithaca, NY: Comstock Publishing Associates, 1980.

A listing of the world's 4,000 mammal species. Provides information on scientific and common names, geographical distribution, habitat, and a number of species in each family. Contains indexes to genus and family and a brief bibliography. Valuable reference aid to zoologists and natural historians.
R: *Choice* 18: 774 (Feb. 1981); *ARBA* (1981, p. 674).

The World's Whales: The Complete Illustrated Guide. **Stanley M. Minasian, Kenneth C. Balcomb III, and Larry Foster, eds.** Washington, DC: Smithsonian Books. Distr. New York: Norton, 1984.

A popular guide to 76 species of whales, dolphins, and porpoises. Execellent color photographs, brief glossary, a selective bibliography, and an index of names and subjects.
R: *ARBA* (1985, p. 535).

Reptiles

The Audubon Society Field Guide to North American Reptiles and Amphibians. **John L. Behler and F. Wayne King.** New York: Knopf, 1979.

Written by 2 experts in the field of zoology. Includes good photographs of most species, brief descriptions, and small range map. Photo arrangement by body shape. Text lists animals in taxonomic order. Highly recommended for the amateur nature lover and for reference collections in public, high school, and college libraries.
R: *Choice* 17: 243 (Apr. 1980); *LJ* 105: 623 (Mar. 1, 1980); *RSR* 8: 14 (July/Sept. 1980); *ARBA* (1981, p. 677).

Easy Identification Guide to North American Snakes. **Hilda Simon.** New York: Dodd, Mead, 1979.

This basic guide for the amateur depicts different species of snakes arranged on the basis of similar physical appearance. Large illustrations and clear range maps of each species are followed by the text, which covers characteristics and habits of the snakes. A useful reference for snake identification.
R: *LJ* 105: 498 (Feb. 15, 1980); *SLJ* 26: 150 (Mar. 1980); *WLB* 54: 528 (Apr. 1980); *ARBA* (1981, p. 678).

Introduction to Canadian Amphibians and Reptiles. **Francis R. Cook, ed.** Ottawa: National Museum of Natural Sciences. Distr. Chicago: University of Chicago Press, 1984.

A definitive field guide to Canadian amphibians and reptiles. Excellent text supplemented by beautiful line drawings. Table of contents—arranged according to order, family, and species—serves as an index.
R: *ARBA* (1985, p. 536).

The New Field Book of Reptiles and Amphibians. Putnam's Nature Field Books. **Doris M. Cochran and Coleman J. Gain.** New York: Putnam, 1978.

A comprehensive identification manual that includes more than 200 photographs and diagrams. Useful to professionals.
R: *Choice* 7: 1013 (Oct. 1970); *ARBA* (1971, p. 516); Win (3EC19).

Reptiles of North America: A Guide to Field Identification. **Hobart M. Smith and Edmund D. Brodie, Jr.** Racine, WI: Western Publishing, 1982.
A guide to the identification of turtles, lizards, snakes, amphisbaenids, and crocodilians of North America north of Mexico. Uses more than 300 color illustrations, line drawings, and other methods for easy identification of species. Compact size is useful for field work.
R: *ARBA* (1984, p. 670).

EARTH SCIENCES

The Audubon Society Field Guide to North American Fossils. **Audubon Society and Ida Thompson.** New York: Random House, 1982.
A selection of 420 remains of marine and freshwater invertebrates, vertebrates, insects, and plants. Gives an introduction to geological time, a glossary, and illustrations of body parts. Good for amateur fossil hunters.
R: *ARBA* (1983, p. 619).

A Field Guide to the Atmosphere. **Vincent J. Schaefer and John A. Day.** Boston: Houghton Mifflin, 1981.
Aids in the recognition, identification, and understanding of atmospheric phenomena, including storms, rainbows, clouds, etc. Contains appendixes, glossary, bibliography, and an index. Primarily for the novice.
R: *BL* 77: 1324 (June 15, 1981); *ARBA* (1983, p. 664).

Minerals and Rocks

The Audubon Society Field Guide to North American Rocks and Minerals. **Charles W. Chesterman and Kurt E. Lowe.** New York: Knopf, 1978.
Pocket-size reference source covering some 230 mineral species and 40 types of rocks. The 800 magnificent color photographs are arranged by a color-coded scheme and are cross-referenced to the text. Includes a brief bibliography, a guide to mineral environments, a name and locality index, and a glossary.
R: *Nature* 282: 655 (Dec. 6, 1979); *BL* 76: 1151 (Apr. 1, 1980); 76: 1563 (June 15, 1980); *LJ* 104: 1550 (Aug. 1979); 105: 570 (Mar. 1, 1980); *ARBA* (1980, p. 656).

The Collector's Book of Fluorescent Minerals. **Manuel Robbins, ed.** New York: Van Nostrand Reinhold, 1983.

The Collector's Guide to Rocks and Minerals. **James R. Tindall and Roger Thornhill.** New York: Van Nostrand Reinhold, 1975.
Aims to be a general background guide to descriptive properties of rocks and minerals. Contains information on geology and rock collecting. Includes bibliography, mineral chart, and color photographs. Recommended for public libraries.
R: *ARBA* (1977, p. 696).

Colorful Mineral Identifier. **Anthony C. Tennisen.** New York: Sterling, 1981.
Earlier edition, 1972.
A pocket-sized mineral-identification guide for the amateur. Numerous color photographs.
R: *ARBA* (1973, p. 591).

The Complete Guide to Buying Gems: How to Buy Diamonds and Colored Gemstones with Confidence and Knowledge. **Antoinette Leonard Matlins and Antonio C. Bonanno, eds.** New York: Crown, 1984.
A basic consumer's guide to buying gems for personal pleasure and investment. Extensive diagrams illustrating cuts of gemstones and other relevant information. A selected list of gem-identification laboratories and a listing of national and state jewelers' associations are useful.
R: *ARBA* (1985, p. 607).

Gems: Their Sources, Descriptions and Identification. 4th ed. **Revised by B. W. Anderson.** Woburn, MA: Butterworth, 1983.
Second edition, 1970; third edition, 1975.
A useful guide for professional gemologists, serious collectors, and mineralogists. Identification tables and several appendixes, such as famous diamonds, birthstones, etc., are useful.
R: *Nature* 97: 424 (Feb. 1963); *ABL* 35: entry 404 (Aug. 1970); *ARBA* (1971, p. 521; 1976, p. 708; 1985, p. 608); Sheehy (EE18).

Fossils for Amateurs: A Guide to Collecting and Preparing Invertebrate Fossils. 2d ed. **Russell P. MacFall and Jay Wollin, eds.** New York: Van Nostrand Reinhold, 1983.
First edition, 1972.
A guide that includes information on fossil plants as well as fossil invertebrates. Includes a list of North American geological museums with good fossil displays.
R: *ARBA* (1974, p. 1594; 1985, p. 608).

An Illustrated Guide to Fossil Collecting. 3d ed. **Richard Casanova.** Happy Camp, CA: Naturegraph, 1981.
Earlier edition, 1970.
R: *Journal of Geology* (Sept. 1971).

The Larousse Guide to Minerals, Rocks and Fossils. **W. R. Hamilton, A. R. Woolley, and A. C. Bishop.** New York: Larousse, 1977.
Contains detailed information and illustrations (over 600 photographs and 300 line drawings). Index provides easy access to information. Covers 220 minerals, 90 rocks, and 300 fossils. Includes information on formula, hardness, color, streak, occurrence. Recommended for geologists.
R: *Sea Frontiers* 24: 187 (May–June 1978); *BL* 75: 408 (Oct. 15, 1978); *LJ* 102: 2149 (Oct. 15, 1977); *TBRI* 44: 246 (Sept. 1978); *WLB* 52: 503 (Feb. 1978); *ARBA* (1978, p. 694).

Minerals and Gemstones: An Identification Guide. **G. Brocardo.** New York: Hippocrene Books, 1982.

Detailed account of 151 of the nearly 2,000 mineral species. Provides clear, compact information with pictographic tables for each mineral. Emphasis on European occurrences limits its usefulness to beginners.
R: *ARBA* (1984, p. 684).

Minerals and Rocks. **Keith Lye.** New York: Arco Publishing, 1979.

An amply illustrated introductory guide to rocks and minerals. Examines 169 rocks and minerals, providing illustrations and descriptions on occurrence, identification, formation, and utilization. Suitable for any library.
R: *ARBA* (1981, p. 692).

Minerals, Rocks, and Fossils. **R. V. Dietrich and R. Wicander, eds.** New York: Wiley, 1983.

A well-illustrated and well-written guide for beginning collectors. It provides a great deal of practical advice and specific information.
R: *Journal of Geology* 92: 469 (July 1984); *BL* 79: 1309 (June 15, 1983).

Rock Hunter's Guide: How to Find and Identify Collectible rocks. **Russell P. MacFall.** New York: Crowell, 1980.

A guide to finding different rocks. Locations of rocks are arranged by name of rock within each state. For large public libraries.
R: *BL* 76: 1365 (June 1, 1980); *ARBA* (1981, p. 692).

Simon and Schuster's Guide to Rocks and Minerals. **Martin Prinz, George Harlow, and Joseph Peters, eds.** New York: Simon and Schuster, 1978.

Contains both color plates and authoritative text, comprising a field guide for the rock and mineral collector. Entries grouped by type of rock/mineral total over 370. Provides detailed data. Considered an excellent guide.
R: *American Scientist* 67: 355 (May/June 1979); *BL* 76: 738 (Jan. 15, 1980); *Choice* 16: 558 (June 1979); *LJ* 104: 201 (Jan. 15, 1979); *ARBA* (1980, p. 659); Mal (1980, p. 198).

CHAPTER 10 ATLASES AND MAPS

BIBLIOGRAPHICAL TOOLS

Dictionary Catalog of the Map Division. 10 vols. **New York Public Library, Research Libraries.** Boston: Hall, 1971.

Includes some 11,000 volumes on cartography as well as a listing of depository maps, early printed maps, and atlases.
R: Mal (1980, p. 189).

Engineering Geological Maps: A Guide to Their Preparation. **Commission on Engineering Geological Maps of the International Association of Engineering Geology.** Paris: Unesco, 1976.

Summarizes advanced techniques. Includes sections on principles, data interpretation, examples of engineering geological maps, etc.
R: *IBID* 4: 122 (June 1976).

Guide to Atlases: World, Regional, National, Thematic; An International Listing of Atlases Published Since 1950. **Gerard L. Alexander.** Metuchen, NJ: Scarecrow Press, 1971.

Supplement for 1971 through 1975 published in 1977.

Guide to Atlases: World, Regional, National, Thematic; An International Listing of Atlases Published 1971 Through 1975 with Comprehensive Indexes. **Gerard L. Alexander.** Metuchen, NJ: Scarecrow Press, 1977.

A supplement to Alexander's *Guide to Atlases.* Includes some 3,000 entries for atlases published between 1971 and 1975, and a few missed in the earlier, basic volume. Arranged in 4 sections: world atlases, regional atlases, national atlases, and thematic atlases. All entries include complete bibliographic information. Supported by indexes to publishers, authors, cartographers, and editors.
R: *BL* 74: 500 (Nov. 1, 1977); *WLB* 52: 86 (Sept. 1977).

Guide to US Government Maps. **John Andriot.** McLean, VA: Documents Index, 1976–. Irregular. Loose-leaf.

Guide to US Government Maps: Geologic and Hydrologic Maps. Prelim. ed. **Laurie Andriot and Donna Andriot.** McLean, VA: Documents Index, 1977.

Updated through August 1976.
All entries previously published in catalogs entitled *Publications of the [US] Geological Survey, 1879–1961* and *1962–1970* and supplements. Access to maps by 3 indexes: area-subject, subject-area, and coordinate index of latitude and longitude. Much information extracted from *Survey's* catalogs and therefore lacks accuracy. Until publication of final edition, this cumulation will help library clientele, who use US Geological Survey geologic and hydrologic maps.
R: *ARBA* (1978, p. 692).

International Maps and Atlases in Print. 2d ed. **Kenneth L. Winch, ed.** New York: Bowker, 1976.

First edition, 1974.

Practical guide to more than 8,000 maps and atlases of some 700 publishers. Arranged by continent and country. Almost 400 map index diagrams, a gazetteer index, and an index of universal-decimal key maps are provided.

Maps and Geological Publications of the United States: A Layman's Guide. **William R. Pampe, comp.** Falls Church, VA: American Geological Institute, 1978.

Alphabetical listing of geological publications and maps of individual states. Entries for each state cover bibliographies, general geology, mineral resources, water, etc. Handy, useful guide for nature lovers and students.

R: *ARBA* (1980, p. 656).

Map Collections in the United States and Canada: A Directory. 3d ed. **David K. Carrington and Richard W. Stephenson, eds.** New York: Special Libraries Association, 1978.

Second edition, 1970.

This completely revised and expanded directory describes 745 major map collections in the United States and Canada. Entries are arranged alphabetically by city within a state or province. Includes data about cataloging, classification, and reader services for the first time. A valuable reference tool.

R: *SL* 71: 11A (Mar. 1980).

Maps on File. New York: Facts on File, 1981.

A unique collection of some 300 maps that allow easy duplication on copying machines. Bound in loose-leaf form. Includes standard geographical maps of 150 countries, 57 specialized maps, numerous theme maps (which cover such topics as population, average income, coal reserves, and political alliance), and historical maps.

R: *BL* 77: 1332 (June 15, 1981).

Scientific Maps and Atlases: Catalogue 1976. Paris: Unesco, 1976.

R: *IBID* 4: 221 (Sept. 1976).

Union Catalog of Maps. **Robert Rountree and James A. Winkfield.** Berkeley, CA: Berkeley Documentation Center, 1974–. Bimonthly.

Lists maps, and keys to over 20 major libraries that hold them. Technical descriptions.

Also, catalogues of various map series from governmental agencies, such as the US Geological Survey Topographical Maps, US Army Topographical Maps, and so on.

World Directory of Map Collections. **Geography and Map Libraries Subsection, comp.** Edited by Walter W. Ristor. Munich: Verlag Dokumentation. Distr. New York: Unipub, 1976.

A listing of 285 collections organized alphabetically by country and city. Includes information such as name and address of collection, personnel size, collection size, geographical and subject range, reference service, classification scheme, etc. A useful reference for academic libraries.

ASTRONOMY

Astronomical Photographic Atlas. **A. Heck and J. Manfroid.** Liege, Belgium: Desoer, 1978.

An album of astronomical photographs, including the moon, sun, planets, comets, star fields, galaxies, and clusters of galaxies. Captions in English, French, German, Spanish, and Esperanto.
R: *Sky and Telescope* 57: 478 (May 1979).

Atlas of Deep-Sky Splendors. 4th ed. **Hans Vehrenberg.** New York: Cambridge University Press, 1983.

First edition, 1965; third edition, 1978.
A useful atlas consists of more than 100 charts of star fields and 400 nonstellar objects including more than 100 deep-sky objects. Appropriate for all types of libraries interested in the objects in the sky.
R: *Sky and Telescope* 57: 179 (Feb. 1979); *TBRI* 45: 132 (Apr. 1979); *ARBA* (1985, p. 602).

The Atlas of Mercury. **Charles A. Cross and Patrick Moore.** New York: Crown, 1977.

Comprehensive treatment of the planet Mercury. Based on the Mariner 10 pioneer voyage. The atlas portrays Mars' surface features in map form. Includes an additional narrative section describing atmosphere, physical conditions, and interior of planet. Profusely illustrated work features old and new artwork. Recommended to all libraries, from public to advanced astronomical research collections.
R: *Observatory* 98: 177 (Aug. 1978); *Sky and Telescope* 56: 147 (Aug. 1978); *Choice* 14: 1621 (Feb. 1978); *LJ* 102: 2268 (Nov. 1, 1977); *TBRI* 44: 243 (Sept. 1978); *ARBA* (1978, p. 634).

Atlas of the Planets. **Paul Doherty.** New York: McGraw-Hill, 1980.

A guide to the planets and the solar system for amateur astronomers. Provides detailed, readable descriptions of planetary features, observing hints, nomenclature, and other useful information. Uses data from recent spacecraft exploration. Includes numerous, instructive illustrations and diagrams of the planets as seen through a backyard telescope. Contains a bibliography, a glossary, and an index. For any public or college library.
R: *BL* 77: 292 (Oct. 15, 1980); *Choice* 18: 418 (Nov. 1980); *LJ* 105: 2093 (Oct. 1, 1980); *ARBA* (1981, p. 628).

An Atlas of Representative Stellar Spectra. **Yasumasa Yamashita, Kyoji Nariai, and Yuji Norimoto.** New York: Wiley, 1978.

Designed to supplement previously published atlases of stellar spectra. Includes numerous spectra with different dispersions from earlier atlases. Consists of 45 excellent plates depicting spectra of 197 MK standard stars by spectral type. Valuable inclusion is the numerous energy-level diagrams throughout atlas. A wise choice for special astronomy collections.
R: *Choice* 15: 1395 (Dec. 1978); *ARBA* (1979, p. 650).

The Cambridge Photographic Atlas of the Planets. **Geoffrey Briggs and Fredric Taylor.** New York: Cambridge University Press, 1982.

This text examines each planet from Mercury out to Saturn, with many photographs from spacecraft, maps, processes, and nomenclature for surface features.
R: *Sky and Telescope* 64: 559 (Dec. 1982).

Catalogue of the Universe. **Paul Murdin, David Allen, and D. Malin.** New York: Crown, 1979.

A handsome volume of photographs of galaxies, stars, and planets. Text provides background and history for each photograph. Emphasizes southern celestial objects. Includes a glossary and notes on special photographic techniques that were employed. A supplemental astronomy reference directed at both lay and professional audiences. For public and academic libraries.
R: *Nature* 283: 798 (Feb. 21, 1980); *New Scientist* 84: 196 (Oct. 18, 1979); *Physics Bulletin* 31: 175 (June 1980); *BL* 77: 1120 (Apr. 1, 1981); *LJ* 104; 2557 (Dec. 1, 1979); *WLB* 54: 463 (Mar. 1980); *ARBA* (1980, p. 612).

Deep Sky Objects; A Photographic Guide for the Amateur. **Jack Newton.** Toronto: Gall, 1977.

An atlas of star charts and Messier objects. Recommended.
R: *Astronomy* 6: 56 (May 1978); *Sky and Telescope* 55: 170 (Feb. 1978); *TBRI* 44: 209 (June 1978).

National Geographic Picture Atlas of Our Universe. **Roy Gallant.** Washington: National Geographic Society, 1980.

An atlas emphasizing the solar system but with an informative introduction to the history of astronomy and with sections on individual members of the solar system, stars and galaxies. Aimed at young adults but good for mature adults as well. Beautifully illustrated.
R: *ARBA* (1982, p. 706).

The Rand McNally New Concise Atlas of the Universe. Rev. ed. **Patrick Moore.** New York: Rand McNally, 1978.

Earlier edition, 1974.
Attractive atlas consists of 4 parts: the earth from space, the moon, the solar system, and the stars. Although informational content has been updated, the format and illustrations have undergone few revisions. Index is somewhat incomplete. Good source of general reference for school, college, public libraries, and home.
R: *Sky and Telescope* 57: 289 (Mar. 1979). *BL* 76: 370 (Oct. 15, 1979); *LJ* 104: 394 (Feb. 1, 1979); *ARBA* (1980, p. 609).

The Times Atlas of the Moon. **H. A. G. Lewis, ed.** London: Timespapers. Distr. New York: New York Times, 1969.

A big quarto volume consisting of over 60 attractive colored maps of the Near Side lunar surface. Information on techniques of lunar flight, lunar landscape, etc. A major tool on the subject, with index.
R: *Listener* 83: 256 (Feb. 19, 1970); *Science* 173: 712 (Aug. 20, 1971); *ARBA* (1971, p. 484); Mal (1980, p. 113); Wal (p. 84), Win (3EB4).

Star Atlases and Catalogs

A. P. Norton Star Atlas and Reference Handbook. 6th ed. **G. E. Satterthwaite, ed.** Edinburgh: Gall and Ingalls, 1973.
Intended for the student or amateur, this reference shows over 9,000 stars, nebulae, and clusters, with descriptive text of objects suitable for small-telescope viewing.

Apparent Places of Fundamental Stars, 1977. International Astronomical Union. Heidelberg: Astronomisches-Rechen Institut. 1940. Annual.
The 37th annual volume containing the 1,535 stars in the fourth *Fundamental Catalog.* Composed completely of tables, including information on mutations and aberrations as well as stars seen day by day.

Isophotometric Atlas of Comets. 2 pts. **W. Hogner and N. Richter.** New York: Springer-Verlag, 1979.
A selective collection of photographs needed for physical research of comets. Photographs culled from over 300 photos taken between 1902–1967. Essential atlas for the astronomy community.

Palomar Observatory Sky Atlas. 9 vols. Washington, DC: National Geographic Society, 1954–1958. Printed by the Graphic Arts Facilities of the California Institute of Technology.
Atlas of 1,618 plates. Consists of photographic reproductions of red- and blue-sensitive photographs of 879 fields, 6.6° square, and covers the entire sky month of $-27°$ declination. The most comprehensive work of its kind.

Sky Atlas 2000.0. **Wil Tirion.** New York: Cambridge University Press, 1981.
An atlas for the epoch 2000.0, including 45,300 stars to visual magnitude 8.0, and 2,500 sky objects. Contains 26 charts with stars, clusters, and nebulae, and little or no distortion of star patterns.
R: *New Scientist* 94: 442 (May 13, 1982).

The Star Atlas. 3d ed. **Jan Hevelius.** Edited by V. P. Scheheglov. Tashkent: FAN Press, 1978.
First edition, 1968; second edition, 1970.
An atlas published in 1690 by a man who observed the stars with the naked eye. The catalog of 1,564 stars is the most accurate and last one in history of pretelescopic observations. Also includes manuscripts from the seventeenth and eighteenth centuries.
R: *Observatory* 100: 9–10 (Feb. 1980).

Star Atlas. **Jacqueline Mitton and Simon Mitton.** New York: Crown, 1979.
A star guide for the neophyte astronomer. Data is applicable for all time zones. Latitudes and longitudes on a worldwide basis.
R: *Physics Teacher* 18: 74 (Jan. 1980); *Sky and Telescope* 58: 361 (Oct. 1979); *Space World* 191: 34 (Nov. 1979); *TBRI* 46: 90 (Mar. 1980).

Star Maps for Beginners. 2d ed. **I. M. Levitt and Roy K. Marshall.** New York: Simon & Schuster, 1980.

Stars and Stellar Systems: Compendium of Astronomy and Astrophysics. 9 vols. **G. P. Kuiper and B. M. Middlehurst, eds.** Chicago: University of Chicago Press, 1961–1975.

Volume 1, *Telescopes*; volume 2, *Astronomical Techniques*; volume 3, *Basic Astronomical Data*; volume 4, *Clusters and Binaries*; volume 5, *Galactic Structures*; volume 6, *Stellar Atmospheres*; volume 7, *Nebular and Interstellar Matter*; volume 8, *Stellar Structure*; volume 9, *Galaxies and the Universe*.

Subject indexes. Speculative topics have been de-emphasized to ensure the future value of this comprehensive reference.

R: Jenkins (F41); Wal (p. 84).

True Visual Magnitude Photographic Star Atlas. 3 vols. **C. Papadopoulos.** Elmsford, NY: Pergamon, 1979.

An atlas of stars shown in their apparent magnitude.

PHYSICS

An Album of Fluid Motion. **Milton Van Dyke.** Stanford: The Parabolic Press, 1982.

Arranged in 11 chapters with 279 photographs, this atlas should be essential for all graduate students in the fluid thermal engineering field.

R: *Ocean Engineering* 10: 211 (June 1983).

The Asbestos Particle Atlas. **Walter C. McCrone.** Ann Arbor: Ann Arbor Science/Butterworth, 1980.

This specialized atlas provides in-depth coverage of the identification of asbestos particle types. Includes detailed presentation on the use of the polarized light microscope. Important reference for Environmental Protection Agency officials, contractors, and architects.

R: *Choice* 18: 776 (Feb. 1981).

An Atlas of Spectral Interferences in ICP Spectroscopy. **M. L. Parsons, Alan Forster, and Donn Anderson.** New York: Plenum, 1980.

Comprises 4 tables of analytical lines that can aid an analytical spectroscopist determine trace constituents quantitatively and accurately. Concise introductions describe each table's arrangement and use. Useful to those performing multi-element analysis by eliminating identification errors.

R: *ARBA* (1981, p. 639).

The Particle Atlas: An Encyclopedia of Techniques for Small Particle Identification. 6 vols. 2d ed. **Walter C. McCrone and John G. Delly.** Ann Arbor: Ann Arbor Science, 1973–1980.

Volume 1, *Principles and Techniques*, 1973; volume 2, *Light Microscopy Atlas*, 1973; volume 3, *Electron Microscopy Atlas*, volume 4, *The Particle Analysts Handbook*, 1973; volume 5, *Light Microscopy Techniques and Atlas*; volume 6, *Electron-Optical Atlas and Techniques*, 1980.

An aid to identifying small particles by microanalysis. Aimed at all interested parties, including air and water pollution personnel, criminologists, etc. Principles of optics, mounting procedures, electron microscopy, and data compilations are included.
R: *Scientific American* 243: 44 (Oct. 1980); *Choice* 11: 740 (July/Aug. 1974); *ARBA* (1975, p. 639).

Timescale: An Atlas of the Fourth Dimension. **Nigel Calder.** New York: Viking Press, 1983.

Provides information on past and present studies of time concepts, scientific dating, and timescales. With illustrations, maps, and references to important events. For the educated layperson or the reference librarian interested in science.
R: *ARBA* (1984, p. 605).

CHEMISTRY

Analysis of Polymers, Resins, and Additives: An Atlas. 2 vols. **Dieter O. Hummel and Friedrich Scholl.** New York: Verlag Chemie, 1973–1977.

Volume 1, *Plastics, Elastomers, Fibers, and Resins*, 1977; volume 2, *Additives and Processing Aids*, 1973.
International reference tool containing spectra of industrial polymers and resins. Involves a 3-digit decimal classification scheme. Cites some 800 references. A vital tool for professionals engaged in the analysis, characterization, and structure identification of polymeric materials.

Atlas of Metal. **J. Kragten.** New York: Halsted Press, 1978.

Handy reference tool to information about the behavior of 45 common metals in the presence of some 30 ligands. Graphs for metal-ligand combinations are superposable, allowing for building of individual graphs to create new combinations. Useful in chemistry and chemical engineering collections.

Atlas of Metal-Ligand Equilibria in Aqueous Solution. **J. Kragten.** Chichester, England: Ellis Horwood. Distr. New York: Wiley, 1978.

Discusses the behavior of 45 commonly used metals in the presence of 29 ligands in aqueous solutions. Graphical format allows plots of metal-ligand systems to be superimposed on one another, creating new systems. Includes an introductory chapter, an appendix of stability constants, and an index.
R: *ARBA* (1981, p. 633).

An Atlas of Polymer Damage: Surface Examination by Scanning Electron Microscope. **L. Engel et al.** London: Wolfe Medical Publications, 1980.

Atlas of Spectral Data and Physical Constants for Organic Compounds. 2d ed. 6 vols. **J. G. Grasselli and W. Ritchey.** Cleveland: Chemical Rubber, 1975.

Organized in 3 sections: introductory, master data table, and indexes. Intended for researchers.
R: *Journal of the American Chemical Society* 95: 7189 (Oct. 17, 1973); *Nature* 245: 108 (Sept. 14, 1973); *TBRI* 39: 372 (Dec. 1973).

Atlas of Stereochemistry. 2d ed. 2 vols. **William Klyne and J. Buckingham.** London: Chapman & Hall, 1978.

First edition, 1974.

Provides stereochemical formulas for over 4,500 compounds in diagrammatic form. An essential reference for organic chemists.

R: *New Scientist* 82: 930 (June 14, 1979); *ABL* 44: entry 159 (Apr. 1979); *Choice* 12: 198 (Apr. 1975); *TBRI* 45: 247 (Sept. 1979); *ARBA* (1976, p. 649).

Atlas of Stereochemistry: Absolute Configurations of Organic Molecules. 2d ed. 2 vols. **W. Klyne and J. Buckingham.** New York: Oxford University Press, 1978.

First edition, 1974.

Major revision of previous edition; discusses compounds of pharmaceutical interest. Volume 1 covers literature through 1971; volume 2 reviews literature from 1972–1976, and includes cumulative author, title, and formulas indexes. Valuable to researchers and students.

R: *Journal of the American Chemical Society* 102: 1474 (Feb. 13, 1980); *Choice* 16: 868 (Sept. 1979); *RSR* 8: 25 (July/Sept. 1980); *ARBA* (1980, p. 614).

CRC Atlas of Spectral Data and Physical Constants for Organic Compounds. 6 vols. **Jeanette G. Grasselli and W. M. Ritchey, eds.** Cleveland: Chemical Rubber, 1976.

Presents spectral data and physical constants on nearly 21,000 compounds. Too expensive for individual purchase.

R: *Applied Spectroscopy* 31: 251 (May/June 1977); *TBRI* 43: 244 (Sept. 1977).

BIOLOGICAL SCIENCES

Atlas of Descriptive Histology. 3d ed. **Edward J. Reith and Michael H. Ross.** New York: Harper & Row, 1977.

Second edition, 1970.

Atlas of Insect Diseases. 2d ed. **Jaroslav Weiser.** Prague: Academia, 1977.

Second edition covers additional information on diagnosis, techniques, and specific insect diseases. Includes high-quality figures.

R: *Journal of Invertebrate Pathology* 31: 140 (Jan. 1978); *TBRI* 44: 110 (Mar. 1978).

Atlas of Insects Harmful to Forest Trees. Vol. 1. **Vladimir Novak et al.** New York: Elsevier Scientific, 1976.

Excellent summary of biological data on significant European forest insects. Text accompanied by high-quality color illustrations, depicting life stages and feeding damage.

R: *Journal of Forestry* 75: 600 (Sept. 1977); *TBRI* 43: 349 (Nov. 1977).

The Atlas of Insect and Plant Viruses, Including Mycoplasmaviruses and Viroids. **Karl Maramorosch, ed.** Ultrastructure in Biological Systems, vol. 8. New York: Academic Press, 1977.

Companion volume to A. J. Dalton's *Ultrastructure of Animal Viruses and Bacteriophages: An Atlas*. Consists of 4 sections: insect viruses, plant viruses, mycoviruses and viroids, mycoplasma- and spiroplasma-viruses. Material contributed by authorities in their fields. Well-illustrated with plates. Taxonomic classification conforms to recommendations of the International Committee on the Nomenclature of Viruses. Highly recommended to comparative, plant, and invertebrate virologists.
R: *RSR* 6: 20 (Oct./Dec. 1978).

Atlas of Medicinal Plants of Middle America, Bahamas to Yucatán. **Julie F. Morton.** Springfield, IL: Charles C. Thomas, 1981.

A complete reference encompassing more than 1,000 medicinal plants used by the people of northern South America, the West Indies, the Bahamas, and Central America as far north as Yucatán. Includes genera for each family and species within each genus. Informative and authoritative for any scientist involved with folk medicine, medical botany, and phytochemistry, as well as anthropology, toxicology, and ethnopharmacology.
R: *ARBA* (1983, p. 628).

Atlas of Molecular Structures in Biology. Vol. 2. **G. Fermi and M. F. Perutz.** Clarendon, NY: Oxford University, 1981.

Atlas of Plant Viruses. **R. I. B. Francki, Robert G. Milne, and T. Hatta.** Boca Raton, FL: CRC Press, 1985–.

Volume 1, 1985; volume 2, 1985.
A comprehensive collection of plant virus electron micrographs, with 192 examples. Invaluable tool for researchers in the field.

Atlas of the Netherlands Flora. Vol. 1. **J. Mennema, A. J. Quene–Boeterenbrood, and C. L. Plate, eds.** The Hague: Junk, 1980.

Volume 1, *Extinct and Very Rary Species.*
Translated from Dutch.

An Atlas of Past and Present Pollen Maps of Europe 0–13,000 Years Ago. **B. Huntley and H. J. B. Birks.** Cambridge: Cambridge University Press, 1983.

Presents a graphic representation from the latter part of the last glaciation, or over the past 13,000 years, of the changing vegetation on a continental scale. A long-awaited and organized collaboration of pollen maps from some 843 geographical locations throughout Europe. Provides useful access to data for palynologists worldwide.
R: *Nature* 308: 298 (March 1984); *Choice* 21: 1446 (June 1984); *ARBA* (1985, p. 517).

Atlas of United States Trees. Vols. 1–. **Elbert L. Little, Jr.** Washington, DC: US Forest Service. Distr. Washington, DC: US Government Printing Office, 1971–.

Volume 1, *Conifers and Important Hardwoods*, 1971; volume 2, *Alaska Trees and Common Shrubs*, 1975; volume 3, *Minor Western Hardwoods*, 1976; volume 4, *Minor Eastern Hardwoods*, 1977; volume 5, *Florida Trees*, 1978.

This series of atlases features large maps depicting the natural distribution of native tree species of the continental United States. Supported by informative introductory

material concerning explanations of maps; scientific and common names; and rare and endangered species. A must for botanical and forestry collections.
R: *ARBA* (1978, p. 666; 1980, 638).

A Geographical Atlas of World Weeds. **Leroy Holm et al.** New York: Wiley, 1979.

International atlas in chart format that provides both the countries in which researchers view the plant as a weed and the problems the weed causes. A detailed listing of 8,000 species of weeds. Uses symbols given in 10 languages to classify importance of weeds. For use by botanists and researchers. Useful to those interested in crop protection and weed science.
R: *Quarterly Bulletin of the IAALD* 25: 47 (1980); *RSR* 8: 17 (July/Sept. 1980); *ARBA* (1981, p. 653).

South American Land Birds: A Photographic Aid to Identification. **John S. Dunning with Robert S. Ridgely.** Newtown Square, PA: Harrowood Books, 1982.

A book containing descriptions and range maps for 2,500 birds. Over 1,000 were photographed close-up by catching and then releasing the birds. The second half of the book contains descriptions but no pictures of the remaining birds. Grouped by families of birds, this book is good for the amateur or experienced bird-watcher.
R: *ARBA* (1983, p. 636).

Tissues and Organs: A Text-Atlas of Scanning Electron Microscopy. **Richard G. Kessel and Randy H. Kardon.** San Francisco: W. H. Freeman, 1979.

A selective text-atlas of histology. Delves into the 3-dimensional structures of body tissues. Includes micrographs of sectioned material. Basic text for introductory biology and physiology classes; reference atlas for advanced courses in nursing, dentistry, and medicine.
R: *LJ* 104: 1577 (Aug. 1979).

World Atlas of Agriculture; Under the Aegis of the International Association of Agricultural Economists. 4 vols. plus atlas. **Committee for the World Atlas of Agriculture, ed.** Novara, Italy: Institute Geographico de Agostini. Distr. New York: Unipub, 1969– (1973).

Volume 1, *Europe, USSR, Asia Minor*; volume 2, *South and East Asia, Oceania*; volume 3, *Americas*; volume 4, *Africa*.
Atlas section contains 62 maps indicating land utilization in 20 categories.
R: *Geographical Review* 62: 136 (Jan. 1972); *IBID* 3: back cover (Mar./June 1975); Wal (p. 402); Win (3EK7).

ZOOLOGY

The Animal Kingdom. **Harold H. Hart, ed.** New York: Hart, 1977.

Pictorial representations of approximately 80 groups of animals. Includes over 2,000 pictures culled from some 46 known sources, all of which are in the public domain. Addressed primarily to artists, advertising managers, and others who require copy for works illustrated with art reproductions.
R: *BL* 74: 769 (Jan. 1, 1978).

Atlas of Animal Migration. **Cathy Jarman.** New York: Day, 1972.
Maps, illustrations, and text are combined to present the layperson with an understanding of the principal migratory groups. Includes maps of migratory bird refuges in the United States and Europe.
R: *RSR* 1: 17 (Jan./Mar. 1973); *ARBA* (1974, p. 584).

An Atlas of the Birds of the Western Palaearctic. **Colin Harrison.** Princeton: Princeton University Press, 1982.
This book contains maps that show the distribution and breeding range of 639 species of birds of the Palaearctic region. Also gives information on 167 species of allied birds. Includes details on elevation, habitats, and ranges. Highly recommended for detail and maps.
R: *ARBA* (1983, p. 637).

An Atlas of Distribution of the Freshwater Fish Families of the World. **Tim M. Berra.** Lincoln: University of Nebraska Press, 1981.
A catalog of 157 families of the world's waters with a rating of primary, secondary, or peripheral. It also contains a family account with maps, classification data, scientific and common names, and other information. Primarily for research ichthyologists and zoogeographers.
R: *Choice* 19: 888 (Mar. 1982); *ARBA* (1982, p. 738).

Atlas of an Insect Brain. **Nicholas J. Stransfeld.** New York: Springer-Verlag, 1976.
R: *Nature* 262: 332 (July 22, 1976); *TBRI* 42: 217 (June 1977).

An Atlas of Mammalian Chromosomes. Vols. 1–. New York: Springer-Verlag, 1967–.
Volume 10, 1977.

Atlas of the Orders and Families of Birds. **Charles R. Belinky and Pamela J. Sutherland.** Lyons Falls, NY: Educational Images, 1980.
A collection of color photographs of major bird groups; supplements descriptions of individual species, orders, and families. Includes a short bibliography. For school or public libraries.
R: *Choice* 18: 919 (Mar. 1981).

Atlas of the Living Resources of the Seas. **Food and Agriculture Organization of the United States.** New York: Unipub, 1981.
Third edition, 1973.
This atlas presents information on productivity, size, and distribution of fish resources, migration, and movement. In map format, the living resources are described on a worldwide basis in English, French, and Italian. Alphabetical index of fish names.
R: *Choice* 10: 1536 (Dec. 1973); *ARBA* (1974, p. 570).

An Atlas of Primate Gross Anatomy, Baboon, Chimpanzee, and Man. **Daris R. Swindler and Charles D. Wood.** Reprint ed. Melbourne, FL: Krieger, 1982.
Reprint of 1973 edition.
R: *American Scientist* 62: 600–602 (July/Aug. 1974); *Science* 183: 192 (Jan. 18, 1974).

East African Mammals: An Atlas of Evolution in Africa. Vol. 3. **Jonathan Kingdon.** New York: Academic Press, 1980.
Part A, *Carnivores*; part B, *Large Mammals.*
A tour de force of artistic and scientific merit. Lavishly illustrated with the author's own drawings. Features fine structural and anatomical details. Examines the role of each animal in an ecological, evolutionary, and functional light. A must for every biologist, naturalist, and conservationist.
R: *Scientific American* 239: 38 (Oct. 1978).

Fine Structure of Parasitic Protozoa: An Atlas of Micrographs, Drawings, and Diagrams. **E. Scholtyseck.** New York: Springer Publishing, 1979.
A compendium of electron micrographs and drawings of numerous parasitic protozoa of parasitological, veterinary, and medical value. Brief text accompanies each illustration. An innovative guide for students, teachers, and researchers.
R: *Journal of Parasitology* 65: 549 (Aug. 1979); *Transactions of the American Microscopical Society* 98: 481 (July 1979); *TBRI* 45: 368 (Dec. 1979); 46: 15 (Jan. 1980).

National Geographic Book of Mammals. **National Geographic Society.** Washington, DC: National Geographic Society, 1981.
A beautifully illustrated catalog of the more interesting mammals of the world, arranged alphabetically by common names. The introduction to the first volume discusses the major orders of the class Mammalia. Interesting and beautifully photographed.
R: *SLJ* 20 (May 1982); *ARBA* (1983, p. 646).

The World Atlas of Horses and Ponies. **Peter Churchill, ed.** New York: Crescent Books, 1980.
An atlas of a wide distribution of horse and pony breeds around the world. Stresses geographical spread through use of maps. Includes major sections devoted to the characterization of 150 different breeds, and the evolution and history of the horse. Contains large maps and color photographs. An informative, attractive reference for all libraries.
R: *ARBA* (1981, p. 736).

EARTH SCIENCES

Antarctic Map Folio Series. **American Geographical Society.** New York: American Geographical Society.
Folio 12, 1970.
R: *American Scientist* 59: 362 (May/June 1971).

Atlas of Continental Displacement: 200 Million Years to the Present. **H. G. Owen.** New York: Cambridge University Press, 1983.

The first of a 2-part publication that provides maps of the distribution of continental and oceanic crust during the last 700 million years of the earth's history. Intended for academic and large public libraries.
R: *Choice* 22: 68 (Sept. 1984); *ARBA* (1985, p. 605).

Atlas of Economic Mineral Deposits. **Colin J. Dixon.** Ithaca: Cornell University Press, 1979.

Devoted to 48 solid, noncombustible mineral deposits worldwide. Presentation in the form of maps, sections, and diagrams. Five maps show global distribution of deposits. Provides much diverse information, frequently from hard-to-get sources. Serves as an invaluable general reference tool and basic geological text.
R: *American Association of Petroleum Geologists Bulletin* 63: 1421 (Aug. 1979); *Economic Geology* 74: 1533 (Sept.–Oct. 1979); *Nature* 281: 241 (Sept. 20, 1979); *New Scientist* 83: 605 (Aug. 23, 1979); *Aslib Proceedings* 44: 430 (Oct. 1979); *LJ* 104: 1550 (Aug. 1979); *TBRI* 45: 349 (Nov. 1979); 45: 386 (Dec. 1979); 46: 32 (Jan. 1980); *ARBA* (1980, p. 657).

Atlas of Igneous Rocks and Their Textures. **W. S. MacKenzie, C. H. Donaldson, and C. Guilford.** New York: Wiley, 1982.

A colorful presentation of the common textures of igneous rocks with their definitions and photographs. Very well-presented pictorially.
R: *ARBA* (1984, p. 689).

Atlas of Infrared Spectroscopy of Clay Minerals and Their Admixtures. **H. W. van der Marel and H. Beutelspacher.** New York: Elsevier Scientific, 1976.

Comprehensive compendium of some 1,000 infrared spectra of clay and related minerals. Presents precise well-illustrated IR. Features 40-page bibliography organized by subject headings. Invaluable identification and quantitative evaluation tool for clay mineralogists and ceramists.
R: *AAPG Bulletin* 61: 120 (Jan. 1977); *Ceramic Abstracts* 55: 189 (July–Aug. 1976); *Geoderma* 17: 166 (Feb. 1977); *Marine Geology* 23: 273 (Mar. 1977); *Transactions and Journal of the British Ceramic Society* 75: ix (Nov.–Dec. 1976); *TBRI* 43: 80 (Mar. 1977); 43: 155 (Apr. 1977); 43: 235 (June 1977).

Atlas of Landforms. 3d ed. **H. A. Curran, P. S. Justis et al.** New York: Wiley, 1984.

First edition, 1966; second edition, 1974.
A fine supplement for introductory and upper-level courses that deal with landforms. Includes full-color images, topographic maps, remote sensing images, and a wide variety of ways to use and interpret landforms with current technology.
R: *Choice* 4: 28 (1967); *Librarians' Newsletter* 23: 5 (Feb. 1984); Jenkins (F78); Win (1EE11).

Atlas of Maritime History. **Christopher Lloyd.** New York: Arco, 1976.
R: *LJ* 101: 2471 (Dec. 1, 1976).

Atlas of Quartz Sand Surface Textures. **David H. Krinsley and John C. Doornkamp.** New York: Cambridge University Press, 1973.

A photographic guide for geologists in academic or industrial laboratories.
R: *ARBA* (1974, p. 605).

Atlas of Rock-Forming Minerals in Thin Section. **W. S. MacKenzie and C. Guilford.** New York: Halsted Press, 1980.

A full-color atlas of rock-forming minerals. Contains 200 photomicrographs helpful in identification. Includes clear and concise descriptions of pertinent optical data. A valuable reference for all geology and mineralogy libraries. Useful to graduate and advanced undergraduate students as well as professionals.
R: *Choice* 18: 686 (Jan. 1981); *ARBA* (1981, p. 692).

Atlas of the Textural Patterns of Basalts and Their Genetic Significance. **S. S. Augustithis.** New York: Elsevier Scientific, 1978.

A compendium of nearly 600 photographs, displaying basaltic-rock textures. Most of the illustrated matter comes from author's own collection made in Ethiopa and in the eastern Mediterranean. Photographs are very precise and accompanied by full legends. Includes full bibliography and author and subject indexes.
R: *Chemical Geology* 23: 353 (Nov. 1978); *Geological Magazine* 116: 71 (Jan. 1979); *TBRI* 45: 121 (Apr. 1979); 45: 201 (June 1979).

Coastal Landforms and Surface Features; A Photographic Atlas and Glossary. **Rodman E. Snead.** New York: Academic Press, 1981.

A photographic atlas and glossary of coastal features described during the development of the field of coastal geomorphology. A valuable aid for teachers and students of geomorphology and related fields such as geology and physiography.

A Concise World Atlas of Geology and Mineral Deposits. **Duncan Derry.** New York: Wiley, 1980.

An atlas of known mineral deposits and exploration of the geological and mineralogical processes that created these deposits. For research and large public libraries.

Ecological Atlas of Soils of the World. **Philippe Duchaufour.** Translated by C. R. DeKimpe. New York: Masson Publishing USA, 1979.

Includes Food and Agriculture Organization and US equivalents, locality, parent rock, and climatic conditions of standard soils. Also provides descriptive information and references.

Engineering Geological Maps: A Guide to Their Preparation. Paris: Unesco. Distr. New York: Unipub, 1976.

Geological World Atlas: 1:10,000,000. **Commission for the Geological Map of the World.** Paris: Unesco Press, 1977.

Complete atlas consists of 22 sheets. All maps compiled by Unesco. Explanatory matter written in English and French; describes geological features, sources of data, and references. Maps are clear, concise, and in register. Completed atlas should appeal to geologists with global interests.
R: *Nature* 272: 381, 382 (Mar. 23, 1978).

Mineral Atlas of the Pacific Northwest. **Mark Bryant et al.** Moscow, ID: University Press of Idaho, 1980.

General atlas of the economic geology of the Pacific Northwest. Describes geology, mineral resource locations, land availability and ownership, transportation, and economic impacts. Includes extensive bibliography. For the general public.
R: *Choice* 18: 1074 (Apr. 1981).

Natural Wonders of the World. **Reader's Digest, ed.** New York: Reader's Digest Association. Distr. New York: Norton, 1980.

This detailed summary of the world's geologic features is enhanced by excellent color photographs on every page. Includes the limestone cliffs, caverns, diamond mines, waterfalls. Entries present brief discussions of geologic formation and related information; concludes with a glossary and an index.
R: *BL* 77: 547 (Dec. 15, 1980).

The New York Times Atlas of the World. Rev. concise ed. **John C. Bartholomew et al., eds.** New York: New York Times Books, 1978.

First edition, 1972.
An atlas with 3 major parts: general geography, with charts of the moon and stars, general maps, and an index. In the general geography section, the handling of topics is cursory. The maps come from 2 different sources, making them look uneven.
R: *American Scientist* 67: 241 (Mar. 4, 1979).

Our Magnificent Earth: A Rand McNally Atlas of Earth Resources. New York: Rand McNally, 1979.

Emphasis on the Earth's resources: minerals, energy sources, food, forests, etc. An effective combination of text and illustrations, including photographs, maps, and drawings. Includes bibliography and detailed index. Intended for the layperson; a must for every sci-tech collection.
R: *BL* 77: 71 (Sept. 1, 1980); *LJ* 105: 571 (Mar. 1, 1980); *RQ* 19: 301 (Spring 1980); *WLB* 54: 333 (Jan. 1980).

The Rand McNally Concise Atlas of the Earth. Chicago: Rand McNally, 1976.

Consists of material culled from 2 earlier Rand McNally publications: *The Earth and Man* and *The International Atlas*. The 50 pages touch general aspects of the planet Earth and its relationship to the sea, atmosphere, solar system, mankind, etc. This textual and graphic presentation adds little to a standard reference collection, but is a worthwhile home investment.
R: *BL* 74: 63 (Sept. 1, 1977); *WLB* 51: 540 (Feb. 1977).

Regional Chemical Atlas: Orkney. London: Institute of Geological Sciences, 1979.

An atlas of regional geochemical data, based on the collection of steam sediment samples, covering Scotland. Maps illustrate the relief, geology, and mineralization. Provides in one reference tool data that was previously scattered and not widely available.
R: *Nature* 282: 538–539 (Nov. 29, 1979).

Stratigraphic Atlas of North and Central America. **T. D. Cook and A. W. Bally, eds.** Princeton: Princeton University Press, 1977.

A compendium of numerous maps and reports of geological surveys of sedimentary basins. Covers all periods and epochs during the Precambrian and Phanerozoic eras. Maps deal with hydrocarbon occurrence, underlying strata, intrusive outcrops, isopachs, etc. Legends accompany each black-and-white map. Extensive reference lists for all time periods add to reference value. A must for all collections of earth sciences.

R: *American Journal of Science* 278: 254 (Feb. 1978); *BL* 74: 502 (Nov. 1, 1977); *LJ* 102: 1482 (July 1977); *TBRI* 44: 163 (May 1978); Sheehy (EE9).

Weather Atlas of the United States. Detroit: Gale Research, 1975.

Originally published as *Climatic Atlas of the United States* by the US Government Printing Office, 1968.

Includes over 270 standardized climatic maps of the United States. Maps depict temperature ranges, precipitation, wind, sunshine, and other climatic values.

World Atlas of Geology and Mineral Deposits. **Duncan R. Derry.** New York: Halsted Press, 1980.

Provides descriptions of the geology and mineral resources of the world in map form; easy to understand. General geological background appears in the first part. Includes a small glossary and a list of sources for more information. A good addition to general collections.

R: *Choice* 18: 920 (Mar. 1981).

World Atlas of Geomorphic Features. **Rodman E. Snead.** Huntington, NY: Krieger Publishing, 1980.

A work of over 100 maps showing the distribution of 63 major landforms around the world. About 25 percent of these are in the United States. The large artificial lakes are the only man-made feature described.

R: *ARBA* (1982, p. 754).

World Map of Desertification. **United Nations Conference on Desertification, Nairobi.** New York: United Nations, 1977.

A worldwide survey of deserts and areas on the borders of deserts.

R: *IBID* 6: 400 (Dec. 1978).

ATMOSPHERIC SCIENCE

Atlas of the Air. **Claire Walter.** New York: Facts on File, 1981.

R: *BL* 77: 985 (Mar. 1, 1981).

Climatic Atlas of the Indian Ocean. 2 vols. **Stefan Hastenrath and Peter J. Lamb.** Madison: University of Wisconsin Press, 1979.

Volume 1, *Surface Climate and Atmospheric Circulation*; volume 2, *The Oceanic Heat Budget*.

A well-designed, neatly presented atlas of monthly mean oceanic and atmospheric fields. This 2-part atlas includes a useful, pithy discussion of concepts, assumptions, and equations. For university libraries serving oceanographers and meteorologists.
R: *Choice* 17: 104 (Mar. 1980); *ARBA* (1981, p. 691).

Climatic Atlas of the Tropical Atlantic and Eastern Pacific Oceans. **Stefan Hastenrath and Peter J. Lamb.** Madison: University of Wisconsin Press, 1977.

A compendium of climatological, meteorological, and oceanographic data. Presents some 97 charts with data on 12 specific subjects (sea-level pressure, resultant wind, total cloudiness, etc.). The base period for the atlas is 1911–1970. Charts are clear, concise, and uncluttered. Wired spiral-bound atlas supplemented by brief introduction, 6 tables, and a list of references.
R: *BL* 74: 499 (Nov. 1977); *ARBA* (1979, p. 701).

International Cloud Atlas. 2 vols. Geneva: World Meteorological Organization, 1975–.

Volume 1, 1975–, *Manual on the Observation of Clouds and Other Meteors*; volume 2, forthcoming.
R: *IBID* 4: 222 (Sept. 1976).

International Cloud Atlas. Abridged Atlas. 1 vol. Geneva: World Meteorological Organization. Distr. New York: Unipub, 1969.

Originally published in 1956 and reprinted in 1969.
Consists of 2 parts: descriptive and explanatory text, and plates in both black and white and color.
R: *ARBA* (1971, p. 521); Mal (1980, p. 199).

Northeast and Great Lakes Wind Atlas. **Dean DeHarpporte.** New York: Van Nostrand Reinhold, 1983.

Tables that show the relationship between wind speed and wind power, as well as maps of average wind speed, are included in this atlas. Useful for academic and medium to large public libraries.
R: *ARBA* (1985, p. 612).

Weather Atlas of the United States. **US Environmental Data Service.** Washington, DC: US Government Printing Office, 1968. Reprint. Detroit: Gale Research, 1975.

Original title: *Climatic Atlas of the United States.*
Contains 271 maps and 15 tables depicting the climate of the United States in terms of the distribution of climatic measures such as precipitation; wind; barometric pressure; relative humidity; mean, normal, or extreme values of temperature; etc.
R: *RQ* 15: 84 (Fall 1975); *ARBA* (1976, p. 707).

OCEANOGRAPHY

Atlas of Marine Use in the North Pacific Region. **Edward N. Miles et al.** Berkeley: University of California Press, 1982.
R: *LJ* (Nov. 1, 1982).

The California Water Atlas. **William L. Kahrl, ed.** North Highlands, CA: State of California, General Services, Publications Section; Los Altos, CA: William Kaufmann, 1979.

Detailed account of the development of California's modern water systems and its operation today. Includes 5 color plates of maps, graphs, and satellite photographs. This authoritative atlas is a major contribution to the literature of California and its water resources.

R: *ARBA* (1980, p. 656).

Equalant I and Equalant II: Oceanographic Atlas; Volume 1, Physical Oceanography. Paris: Unesco, 1973. Distr. New York: Unipub, 1975.

Contains 332 charts and a 400-item bibliography. Text in English, French, Russian, and Spanish. Loose-leaf format, intended for use by oceanographic and marine scientists.

R: *ARBA* (1976, p. 709).

Oceanographic Atlas of the Bering Sea Basin. **Myron A. Sayles, Knut Aagaard, and L. K. Coachman.** Seattle: University of Washington Press, 1980.

Technical reference work of the deep basin of the Bering Sea. Contains distributions of temperature and salinity, volumetric analysis, topography of the basin, and density. Detailed, thorough presentations. For specialists and oceanographers who have a large amount of prior knowledge.

R: *RQ* 19: 396 (Summer 1980); *ARBA* (1981, p. 695).

The Oceans. **Robert Barton.** New York: Facts on File, 1981.

An extensive, informative atlas of the ocean's supply of resoures and its function in protecting the ecological balance and providing mineral products, energy, and food. Includes captioned photographs, diagrams, and maps, as well as an index.

R: *BL* 77: 1067 (Apr. 1, 1981); *RSR* 9: 101 (Jan./Mar. 1981).

The Rand McNally Atlas of the Oceans. **Martyn Bramwell, ed.** New York: Rand McNally, 1977.

Thematic atlas devoted to the physical and biological aspects of oceanography. Popularity stems from the numerous, precise maps, explanatory diagrams, photographs, charts, and simplified text. Main section devoted to structure movement, marine life, and mineral resources. Most illustrations are in color. Includes 14-page subject index. Weaknesses are lack of bibliography or reference list, and few maps with scales.

R: *Oceans* 11: 70 (July 1978); *Science News* 112: 381 (Dec. 1977); *LJ* 103: 355 (Feb. 1, 1978); *RQ* 18: 213 (Winter 1978); *TBRI* 44: 330 (Nov. 1978); *WLB* 52: 585 (Mar. 1978); Sheehy (EE21).

The Times Atlas of the Oceans. **Alastair Couper, ed. and intro.** New York: Van Nostrand Reinhold, 1982.

Covers a vast amount of information concerning the sea, including physical oceanography, history, and managerial and political information. For professionals and general readers interested in the sea.

R: *LJ* 109: 785 (Apr. 1984).

World Ocean Atlas. 3 vols. **Sergei Gorshokov.** New York: Pergamon, 1976–1980.

Volume 1, 1976, *Pacific Ocean*; volume 2, 1979, *Atlantic and Indian Oceans*; volume 3, 1980, *Arctic Ocean*.

A fundamental reference to studies in meteorology, physical oceanography, marine chemistry, biology, and geology of the oceans. General information, such as navigational aids, distributions of diseases and medical establishments, etc. are also provided. Beautifully illustrated with color plates.

R: *American Scientist* 64: 561 (Sept./Oct. 1976); *IBID* 67: 610 (Sept./Oct. 1979); *Deep Sea Research* 26A: 966 (Aug. 1979); *Choice* 21: 692 (Jan. 1984); *TBRI* 46: 85 (Mar. 1980); Sheehy (EE22); *ARBA* (1985, p. 606).

ENGINEERING

Atlas of the World's Railways. **Brian Hollingsworth.** New York: Everest House, 1980.

This inclusive book encompasses all one might wish to know about railways of the world. Organized by major geographical regions and by continents, this work utilizes maps, statistics, photographs, and an index to present a broad, diverse, and exhaustive picture of the railways past and present. For businesses, travelers, and railroad enthusiasts.

R: *LJ* 105: 1746 (Sept. 1, 1980); *WLB* 55: 301 (Dec. 1980); *ARBA* (1981, p. 765).

Hydrogeologic Atlas of the People's Republic of China. **Institute of Hydrogeology and Engineering Geology, ed.** New York: Academic Press, 1982.

Written in Chinese and accompanied by an English legend.

Papermaking Fibers: A Photomicrographic Atlas. **Wilfred A. Cote.** Syracuse: State University of New York, College of Environmental Science and Forestry, Renewable Materials Institute, 1980.

Extensive revision of an original text which has larger format for detailed microphotographs of wood cells of common softwoods, hardwoods, and fiber plants used in making paper. Includes numerous figures and photographic plates. Valuable for upper-level college students.

R: *Choice* 18: 689 (Jan. 1981).

Railways Atlas of the World. Short Hills, NJ: Railways Atlas, 1982.

A collection of 86 railroad maps of the world listed in geographical order. Includes unique locations in specialized areas, though it does not include many important non-railroad details. For the library that serves those interested in international railroad information.

R: *Choice* 21: 256 (Oct. 1983).

Steel Transformation Diagrams: Atlas of Continuous Cooling Transformation Diagrams for Engineering Steels. **M. Atkins.** Sheffield, England: British Steel Corp., 1977.

Presents nearly 175 cooling transformation diagrams. Invaluable reference source for practical workers involved with heat treatment and hot processing of steels as well as researcher of steel properties and structures.
R: *Ironmaking and Steelmaking Journal* 5: 89 (1978); *TBRI* 44: 271 (Sept. 1978).

ENERGY AND ENVIRONMENTAL SCIENCES

An Atlas of Renewable Energy Resources: In the United Kingdom and North America. **J. Mustoe.** New York: Wiley-Interscience, 1984.

A description of the size and geographical distribution of the 8 principal renewable energy resources of the United Kingdom and North America. Examines the origin and methods of each energy flow and attempts to quantify the occurrence and availability of these sources compared with fossil fuels. Illustrations and maps.
R: *Librarians' Newsletter* 23: 5 (Feb. 1984).

Atlas of Selected Oil and Gas Reservoir Rocks from North America. **E. W. Biederman.** New York: Wiley, 1986.

Energy Atlas of Asia and the Far East. **United Nations. Economic Commission for Asia and the Far East.** New York: United Nations, 1970.

Serial Atlas of the Marine Environment. Folios 1–. **American Geographical Society.** New York: American Geographical Society, 1962–.
Folio 20, 1971.
R: *Geographical Journal* 128: 549 (Dec. 1962); Wal (p. 165).

The United States Energy Atlas. **David J. Cuff and William J. Young.** New York: Free Press, 1980.

This unique reference work covers renewable and nonrenewable energy resources of the United States, such as oil, gas, geothermal heat, nuclear fuel, etc. Includes many maps, graphs, tables, and photographs, as well as a glossary and a classified bibliography. For academic and research libraries.
R: *Choice* 18: 1393 (June 1981).

The World Energy Book: An A–Z Atlas and Statistical Sourcebook. **David Crable and Richard McBride, eds.** New York: Nichols, 1978.

An atlas and statistical source book relating to energy.
R: *New Scientist* 80: 955 (Dec. 21–28, 1978); *RSR* 8: 28 (Jan./Mar. 1980).

CHAPTER 11 DIRECTORIES, YEARBOOKS, BIOGRAPHICAL SOURCES

DIRECTORIES

BIBLIOGRAPHICAL SOURCES

The Directory of Directories: An Annotated Guide to Business and Industrial Directories, Professional and Scientific Rosters, and Other Lists and Guides of All Kinds. **James M. Ethridge and Cecilia Ann Marlow, eds.** Detroit: Information Enterprises. Distr. Detroit: Gale Research, 1986.

A comprehensive directory of about 5,500 directories. Volume is based on Gale Research's Directory Information Service and complements the *Encyclopedia of Associations*. Entries are grouped under 15 subject headings and include title, publisher's address, content description, frequency, indexes, price, etc. Arranged alphabetically by title under subject categories. Contains subject and title indexes. Suitable for most libraries.
R: *Choice* 18: 372 (Nov. 1980); *CRL* 41: 492 (Sept. 1980); *LJ* 105: 1723 (Sept. 1, 1980); *RQ* 20: 91 (Fall 1980); *WLB* 55: 300 (Dec. 1980); *ARBA* (1981, p. 32).

Guide to American Scientific and Technical Directories. 2d ed. **Barry T. Klein, ed.** New York: Todd, 1975.

Contains directory listings to a variety of sources pertaining to scientific and technical directories.
R: *Choice* 13: 796 (Sept. 1976); *ARBA* (1977, p. 625).

Guide to US Government Directories. **Donna Rae Larson.** Phoenix: Oryx Press, 1981.

A listing of directories published by the US government. Contains factual data on coverage of the directory, information available, types of indexes used, and ordering information. Provides a complete subject index. For any library.

International Bibliography of Directories/Internationale Bibliographie der Fachadressbuecher. 6th ed. **Helga Lengenfelder, ed.** New York: Saur, 1978.

Fifth edition, 1973.
A listing of directories previously published throughout the world in the fields of science, business, sports, the arts, libraries, health services, and many more. Written in both English and German; only geographical names in the list of directories are German. A valuable addition to large research and business library collections.
R: *RSR* 8: 2 (Apr./June 1980, p. 55).

GENERAL

Awards, Honors, and Prizes: A Source Book and Directory. 5th ed. 2 vols. **Paul Wasserman, ed.** Detroit: Gale Research, 1982.

Second edition, 1969; third edition, 1975; fourth edition, 1978.
Volume 1, *United States and Canada*; volume 2, *International and Foreign*.
R: *Choice* (Mar. 1973); *Catholic Library World* (Feb. 1970); *RQ* (Fall 1973); *ARBA* (1970, p. 123).

Business Organizations and Agencies Directory: A Guide to Trade, Business, and Commercial Organizations; Government Agencies; Stock Exchanges; Labor Unions; Chambers of Commerce; Diplomatic Representation; Trade and Convention Centers; Trade Fairs; Publishers; Data Banks and Computerized Services; Educational Institutions; Business Libraries and Information Centers; and Research Centers. **Anthony T. Kruzas and Robert C. Thomas, eds.** Detroit: Gale Research, 1980.
R: *CRL* 41: 492 (Sept. 1980).

Directory of Corporate Affiliations. Skokie, IL: National Register, 1982.

Directory of Federally Supported Information Analysis Centers. 4th ed. Washington, DC: Library of Congress, 1979.
Contains 108 entries relating mainly to science and technology. Gives information on name, address, staff, activities, services, publications, and holdings, etc. For academic libraries.
R: *BL* 76: 1497 (June 15, 1980).

Directory of Foreign Manufacturers in the United States. 2d ed. **Jeffrey S. Arpan and David A. Ricks.** Atlanta: Georgia State University, School of Business Administration, 1979.
First edition, 1975.

Directory of Manufacturers in Greater Boston. Boston: Greater Boston Chamber of Commerce, Annual.

Directory of Special Libraries and Information Centers. 3 vols. 7th ed. **Lois Lenroot-Ernt, ed.** Detroit: Gale Research, 1982.
Third edition, 1974; fifth edition, 1979.
Volume 1, *US and Canada*; volume 2, *Geographic and Personnel Index*; volume 3, *New Special Libraries (Supplements)*.
R: *Choice* 11: 1462 (Dec. 1974); *LJ* (Feb. 1969); *WLB* 49: 250 (Nov. 1974); *ARBA* (1975, pp. 77–78).

Directory of United Nations Information Systems. 2 vols. Geneva: Inter-Organization Board for Information Systems, 1980.
Summarizes the various functions of the United Nations. Provides addresses, names of centers, and more.
R: *Quarterly Bulletin of the IAALD* 23: 65 (Fall–Winter 1978).

Encyclopedia of Associations. 19th ed. 5 vols. Detroit: Gale Research, 1985.
Fourteenth edition, 1980.
Volume 1, *National Organizations of the U.S.*; volume 2, *Geographic and Executive Indexes*; volume 3, *New Associations and Projects;* volume 4, *International Organizations;* volume 5, *Research Activities and Funding Programs*.

Over 18,000 associations, such as profit, nonprofit, and citizen action groups, are included. The majority of new entries represent newly formed organizations. Provides addresses and telephone numbers and describes activities and publications.
R: *ARBA* (1985, p. 12).

Encyclopedia of Governmental Advisory Organizations: A Reference Guide to Presidential Advisory Committees, Public Advisory Committees, Interagency Committees and Other Government-Related Boards, Panels, Task Forces, Commissions, Conferences, and Other Similar Bodies of Serving in a Consultative, Coordinating, Advisory Research, or Investigative Capacity. 3d ed. **Linda E. Sullivan and Anthony T. Kruzas, eds.** Detroit: Gale Research, 1981.

Second edition, 1975.
Index to organiztion names and keywords. Supplemented by *New Governmental Advisory Organizations* between editions.
R: *BL* 77: 350 (Oct. 15, 1980); *RSR* 2: 45–46 (Oct./Dec. 1974); *ARBA* (1974, p. 149).

European Research Index. 4th ed. 2 vols. **Colin H. Williams, ed.** St. Peter Port, Guernsey, England: Hodgson. Distr. New York: International Publications, 1977.

Second edition, 1969; third edition, 1973.
Provides information on government and private establishments throughout Europe which conduct or promote scientific research. Includes organizations in Eastern Europe. Keyword cross-referenced.
R: *CRL* 28: 66 (1967); *ARBA* (1970, p. 102); Jenkins (A115); Wal (p. 45); Win (1EA23).

The Foundation Directory. 8th ed. **Marianna O. Lewis, ed.** New York: The Foundation Center. Distr. New York: Columbia University Press, 1981.

Fifth edition, 1975.
Lists over 2,500 foundations. Describes foundations possessing assets of 1 million dollars or more. Includes foundations devoted to science and technology, engineering, and medical research. Foundation, field-of-interest, donor, trustee, and administrator indexes.

Foundation Grants to Individuals. 4th ed. **Claude Barilleaux, ed.** New York: The Foundation Center, 1984.

Third edition, 1980.
A directory of all the grant programs of 975 grant-making foundations currently active in the United States. Indexed by foundation names, subject and geographic focus, travel funds, and specific educational institutions. A must for all fund seekers.
R: *ARBA* (1985, p. 11).

Guides to International Organizations. 4 vols. **Union of International Associations, ed.** New York: Saur, 1984–1985.

Volume 1: *African International Organization Directory 1984/85*, 1984; volume 2: *Arab and Islamic International Organization Directory 1984/85*, 1984; volume 3: *International Organization Abbreviations and Addresses 1984/85*, 1985; volume 4: *Intergovernmental Organization Directory 1984/85*, 1985.
A major reference set on international organizations.

The International Foundation Directory. 3d ed. **H. V. Hodson, ed.** London: Europa Publications. Distr. Detroit: Gale Research, 1983.

First edition, 1974; second edition, 1980.

An alphabetical directory of more than 700 foundations and trusts. Arranged by country and then by foundation's name. Indexed by foundation and by broad subject category. A must for every large library.

R: *Choice* 12: 370 (May 1975); *CRL* 41: 397 (July 1980); *RQ* 14: 355–356 (Summer 1975); *RSR* 8: 87 (July–Sept. 1980); *ARBA* (1985, p. 13).

International Organizations: A Dictionary and Directory. **Giuseppe Schiavone.** Chicago: St. James Press, 1983.

About 100 organizations are included in the dictionary section, accompanied by information on headquarters address, administrative head, publications, etc. Another 60 organizations are listed in the directory under 8 subject categories. Useful for academic and large public libraries.

R: *WLB* 58: 676 (May 1984); *ARBA* (1985, p. 15).

International Research Centers Directory, 1984: A World Guide to Government, University, Independent Nonprofit, and Commercial Research and Development Centers, Institutes, Laboratories, Bureaus, Test Facilities, Experiment Stations and Data Collection and Analysis Centers, as well as Foundations, Councils, and Other Organizations Which Support Research. 2d ed. **Kay Gill and Anthony T. Kruzas, eds.** Detroit: Gale Research, 1984.

First edition, 1982.

Nearly 3,000 centers in 125 countries, except the United States, are included. Useful for academic and large public libraries.

R: *ARBA* (1985, p. 13).

Libraries, Information Centers and Databases in Science and Technology: A World Guide. 1st ed. **Helga Lengenfelder, ed.** New York: Saur, 1984.

An international directory of 10,528 special libraries of more than 3,000 volumes in over 150 countires, 350 information and documentation centers, and 360 online databases from nearly 200 products. Indexed by general name and English keyword.

New Research Centers: Interedition Supplements to Research Centers Directory. 7th ed. **Robert C. Thomas.** Detroit: Gale Research, 1982.

Fifth edition, 1975.

Research Centers Directory: A Guide to University-Related and Other Non-Profit Organizations. 7th ed. **Robert C. Thomas, ed.** Detroit: Gale Research, 1982.

Fifth edition, 1975; sixth edition, 1979.

Alphabetical by institution under broad subject headings. Supplemented by *New Research Centers.*

R: *Catholic Library World* 44: 443 (Feb. 1973); *Choice* 9: 1434 (Jan. 1973); *ARBA* (1973).

Subject Directory of Special Libraries and Information Centers. 8th ed. 5 vols. **Margaret Labash Young, Harold Chester Young, and Anthony T. Kruzas, eds.** Detroit: Gale Research, 1983.

Volume 1, *Business and Law Libraries*; volume 2, *Education and Information Science Libraries*; volume 3, *Health Sciences Libraries*; volume 4, *Social Sciences and Humanities Libraries*; volume 5, *Science and Technology Libraries*.

World Guide to Abbreviations of Organizations. 6th ed. **F. A. Buttress, ed.** Detroit: Gale Research, 1980.

Fifth edition, 1975.

Lists approximately 18,000 abbreviations, 5,000 of which relate to the European economic community.

Yearbook of International Organizations 1985/86. 3 vols. **Union of International Associations, ed.** New York: Saur, 1985–86.

Volume 1: *Organization Descriptions and Index*, 22nd edition, 1985; volume 2: *International Organization Participation: Country Directory of Secretariats and Membership (Geographic Volume)*, 3d edition, 1985; volume 3: *Global Action Networks: Classified Directory by Subject and Region (Subject Volume)*, 3d edition, 1986.

An invaluable reference work that provides detailed information for over 24,000 organizations in every field of human endeavor, accompanied by a computer-generated multilingual index.

R: *Science Technology Libraries* 5:2 (Winter 1984).

GENERAL SCIENCE

American Museum Guides: Sciences. **Paul Hoffman, ed.** New York: Collier Books/Macmillan, 1983.

Alphabetically lists museums of air and space, natural history, science and technology, and special topics, with name, address, telephone number, hours, and other information for each of the 87 museums listed. A most useful guide for tourists.

R: *ARBA* (1984, p. 611).

Directory of Federal Technology. **US National Science Foundation, Federal Coordinating Council for Science, Engineering and Technology.** Washington, DC: US Government Printing Office, 1977.

A directory of federal agencies involved in technology and the dissemination of scientific and technical information. Arranged by department and agency. Includes information on contact person, functions, policies, etc.

R: *ARBA* (1978, p. 624).

Directory of Federal Technology Resources. **Center for the Utilization of Federal Technology, US Department of Commerce.** Virginia: National Technical Information Service, 1984.

Includes over 800 resources in research and engineering located in federal agencies, laboratories, and engineering centers.

R: *ARBA* (1985, p. 490).

A Directory of Natural History and Related Societies in Britain and Ireland.
A. Meenan, comp. and ed. London: British Museum of Natural History, Publications Sales, 1982.

Includes full data for 750 societies, associations, etc. involved in natural history. Details about the organizations, as well as subject and geographical indexes, are included. Good for librarians, museums, and others interested in ecology.
R: *New Scientist* 196: 50 (Oct. 1982).

European Research Centres: A Directory of Organizations in Science, Technology, Agriculture, and Medicine. 5th ed. **Trevor I. Williams, ed.** Harlow, England: Longman. Distr. Detroit: Gale Research, 1982.

Change of title; a combination of *European Research Index* and *East European Index*. Includes organizations, universities, industries, governments, and corporations in countries of Eastern and Western Europe (excluding the USSR) that are funding or performing research in the various fields of science, technology, agriculture, and medicine.
R: *ARBA* (1984, p. 610).

Federal Scientific and Technical Communication Activities. Washington, DC: National Science Foundation. Annual.

1975 report (pub. 1976) available as PB253975 from the National Technical Information Service, Springfield, Virginia.

Guide to European Sources of Technical Information. 4th ed. **Ann Pernet, ed.** Guernsey, England: Hodgson, 1976.

First edition, 1957; third edition, 1969.
This guide was designed to improve contacts in the field of scientific and technical information. The new edition covers the whole of Europe. Some 1,000 bodies are listed, and these have been indexed and keyword cross-referenced.
R: *ARBA* (1970, p. 102); Wal (p. 41).

Guide to Science and Technology in the UK. **S. E. Macreary, ed.** New York: International Publications Service, 1971.

The guide covers the subject in a series of descriptive chapters and a directory of selected British government science-based establishments and British universities.
R: Wal (p. 41).

National Science Policy and Organization of Research in. . . . **UNESCO.** New York: Unipub, 1970–. In progress.

A series of publications on different countries: *India*, 1973; *Israel*, 1970; *Philippines*, 1970; *Poland*, 1970; etc.

Science and Technology in Latin America. **Latin American Newsletters Limited.** Edited by Christopher Roper and Jorge Silva. London: Longman. Distr. Detroit: Gale Research, 1983.

Describes science and technology research country by country for those in Latin America. Very detailed.
R: *ARBA* (1984, p. 611).

Science and Technology in the Middle East: A Guide to Issues, Organizations, and Institutions. **Sardor Ziauddin, ed.** London: Longman, 1982.

Begins with a description of scientific and technological developments in the Middle East since 1975, followed by a description of 5 major regional science and technology organizations, with their origins and projects. Finally, gives an account of the organization and administration of science and technology, and government and academic research institutions with their research projects. Lists approximately 200 major establishments with addresses.

Scientific and Technical Research Centres in Australia. **Ian A. Crump.** East Melbourne, Australia: CSIRO Information. Annual.

Selected Data Resources: Physical and Science/Engineering. **John A. Feulner, comp.** Washington, DC: Library of Congress, 1980.

Provides a listing of 68 organizations that supply information on physical science and engineering. Contains names, addresses, areas of interest, information services, etc. For science and engineering libraries.
R: *LCIB* 39: 267 (Aug. 1, 1980).

Science and Engineering Career and Education

The Aero College Aviation Directory: A Comprehensive National Directory of Accredited Colleges, Universities, and Technical Schools Offering Aviation-Related Degree Training Programs. **George M. Mandins.** Fallbrook, CA: Aero Publishers, 1983.

A directory listing 377 academic institutions that give aviation courses with credit toward an associate degree or higher.
R: *ARBA* (1985, p. 616).

American Institute of Biological Sciences Directory of Bioscience Departments and Faculties in the United States and Canada. 2d ed. **Peter Gray.** New York: Wiley, 1976.

First edition, 1967.
Compiles information on approximately 2,400 bioscience departments in US and Canadian universities.
R: *SL* 59: 206 (1968); Jenkins (G29); *ARBA* (1977, p. 630); Sheehy (EC12).

Annual Guides to Graduate and Undergraduate Study. 6 bks. **Karen C. Hegener, ed.** Princeton: Peterson's Guides, 1982.

Ninth edition, 1975; fourteenth edition, 1979.
Book 1, *Accredited Institutions Offering Graduate Work: An Overview;* book 2, *Humanities and Social Sciences;* book 3, *Biological, Agricultural, and Health Sciences;* book 4, *Physical Sciences;* book 5, *Engineering and Applied Sciences;* book 6, *Peterson's Annual Guide to Undergraduate Study.*

Chemical Engineering Faculties. **D. R. Paul, ed.** New York: American Institute of Chemical Engineers, Chemical Engineering Educational Projects Committee. Annual.

Lists degree programs in the United States, Canada, Britain, Germany, and Israel, with indexes of faculty members and schools.

College Chemistry Faculties. 4th ed. **Bonnie R. Blaser and Jeanann M. Dellantonio, eds.** Washington, DC: American Chemical Society, 1977.

Directory source to American, Canadian, and Mexican colleges and universities with noteworthy departments in chemistry, biochemistry, and chemical engineering. Geographical arrangement by country and then by state or province. Supported by 2 indexes: individuals and institutions. Considered a supplementary source of data to the *National Faculty Directory*, standard college and university directories, and recent college catalogs.
R: *ARBA* (1978, p. 642).

Directory of Computer Education and Research. Int. ed. 2 vols. **T. C. Hsiao, ed.** Washington, DC: Science and Technology Press, 1978.
US edition published in 1983.

Directory of Engineering Education Institutions: Africa, Asia, Latin America. Paris: Unesco Press, 1976–.

Supplies information on some 458 institutions in 64 Third-World countries with degree-awarding status in engineering. Covers many academic aspects: students, faculty, research, specialization, etc. Useful directory in large engineering libraries and firms.
R: *RSR* 5: 23 (Oct./Dec. 1977).

Directory of Physics and Astronomy Faculties in North American Colleges and Universities. New York: American Institute of Physics, 1975–.

Geographical listing of over 2,000 institutions and faculties. Alphabetical index.

Educational Programs and Facilities in Nuclear Science and Engineering. Oak Ridge, TN: Oak Ridge Associated Universities, 1968–.

Alphabetical listing of colleges with degree programs in nuclear science and related fields.

Engineering and Technology Graduates. New York: Engineers Joint Council, Engineering Manpower Commission. Annual.

Engineering and Technology Degrees. New York: Engineers Joint Council. Annual.

Engineers Salaries—Special Industry Report. New York: Engineers Joint Council. Annual.

Graduate Assistantship Directory in the Computer Sciences. New York: Association for Computing Machinery. Annual.

Graduate Programs in Physics, Astronomy and Related Fields 1984–85. New York: American Institute of Physics, 1984. Annual.

First published in 1976.
Provides information on nearly 300 academic departments from 235 institutions in the United States, Canada, and Mexico. Appendixes of alphabetical listings of departments, graduate programs by highest-degree programs, etc. Useful to academic and special libraries.
R: *ARBA* (1985, p. 600).

Guidebook to Departments in the Mathematical Sciences in the United States and Canada. 6th ed. **Raoul Hailpern.** Mathematical Association of America, Committee on Advisement and Personnel. Washington, DC: Mathematical Association of America, 1975.
Fourth edition, 1970.

The National Faculty Directory: An Alphabetical List, with Addresses, of about 480,000 Members of Teaching Faculties at Junior Colleges, Colleges, and Universities in the United States and at Selected Canadian Institutions. 11th ed. 2 vols. Detroit: Gale Research. Annual.

National Solar Energy Education Directory. 3d ed. Washington, DC: US Government Printing Office, 1981.
Over 900 institutions providing solar-related courses are listed by state names with detailed information on the institution and its programs. Complete and easy to use.
R: *ARBA* (1983, p. 655).

Petroleum Training Directory. **John J. Connor, ed.** Boston: International Human Resources Development, 1984.
A directory of some 2,700 training resources for workers at all levels. A useful tool to industry personnel managers.
R: *ARBA* (1985, p. 509).

Professional Engineers Income and Salary Survey. **National Society of Professional Engineers.** Washington, DC: National Society of Professional Engineers. Annual.

Professional Income of Engineers. **American Association of Engineering Society.** New York: American Association of Engineering Society, 1980.

Register of Environmental Engineering Graduate Programs. 4th ed. **Association of Environmental Engineering Professors.** Compiled by P. Aarne Vesilind and Roger A. Minear. Woburn, MA: Ann Arbor Science, 1981.
A catalog of some 95 US and Canadian higher education institutions that offer environmental engineering graduate programs. Narrow in scope, but useful to prospective students, faculty, and employers of graduates in the field of environmental engineering.
R: *ARBA* (1983, p. 723).

Roster of Women and Minority Engineering Students, Graduates. **American Association of Engineering Society.** New York: American Association of Engineering Society, 1978–.

Salaries and Income of Certified Engineering Technicians. **Engineering Manpower Commission.** New York: Engineers Joint Council, 1976–.

Salaries of Engineering Technicians and Technologists. **American Association of Engineering Society.** New York: American Association of Engineering Society, 1979–.

Salaries of Engineers in Education—1980. **American Association of Engineering Society.** New York: American Association of Engineering Society, 1982.

Salaries of Engineers in Government: Special Report. **Engineering Manpower Commission.** New York: Engineers Joint Council, 1977–.

Salaries of Scientists, Engineers, and Technicians . . . A Summary of Salary Surveys. **Scientific Manpower Commission.** Washington, DC: Scientific Manpower Commission. Annual.

Science, Engineering, and Humanities Doctorates in the United States: 1981 Profile. Washington, DC: National Academy Press, 1982.

Results of a survey of those who earned science, engineering, or humanities doctorates between the years 1938 and 1980 and were residing in the United States in February 1981. Comparison of changes will be helpful to educators and/or researchers developing programs and setting goals.
R: *ARBA* (1984, p. 613).

Selected Information Resources on Science and Mathematics Education. **John Henry Hass, comp.** Washington, DC: Library of Congress, 1980.

Provides factual knowledge on 37 organizations that give information on science and mathematics education. Descriptions include name, address, publications, holdings, etc. For science libraries.
R: *LCIB* 39: 391 (Oct. 3, 1980).

Technician Education Yearbook, 1982. 10th ed. **Lawrence W. Prakken and Jerome C. Patterson, eds.** Ann Arbor: Prakken, 1982.

First edition, 1963; second edition, 1965; sixth edition, 1973; seventh edition, 1975; eighth edition, 1977.

University Curricula in the Marine Sciences and Related Fields. Washington, DC: Marine Technology Society. Annual.

A list of colleges and universities teaching marine courses, and a guide to current course offerings. Contains some scholarship information.

Who Knows? Selected Information Resources on Science Education. **John A. Feulner, comp.** Washington, DC: National Referral Center, Library of Congress, 1983.

Listings of information resources by area of interest. The volume is based on entries selected from the database of the National Referral Center of the Library of Congress. For each entry, information is given on name, address, telephone number,

NRCM code number, mission statement, description of holdings, publications, and information services. Agencies included vary greatly in size and scope.
R: *ARBA* (1985, p. 492).

World Directory of Engineering Schools. 2d ed. Easton, CT: Geographics, 1980.

This directory of 4-year undergraduate programs in engineering was derived from questionaires and visits to the schools. Other technology programs are omitted. Good for prospective students and for libraries.
R: *IEEE Proceedings* 70: 109 (Jan. 1982).

World Directory of Environmental Education Programs. **Philip W. Quigg, ed.** New York: Bowker, 1973.

Information on over 1,000 environmental education programs offered in 740 colleges and institutions in 70 countries.

World List of Forestry Schools. Rome: Food and Agriculture Organization of the United Nations, 1981.

University and nonuniversity list of forestry schools.
R: *IBID* 4: 128 (June 1976).

ASTRONOMY

See also under "Physics."

List of Radio and Radar Astronomy Observatories. Washington, DC: National Academy of Sciences, National Academy of Engineering, Committe on Radio Frequencies, 1970–. Irregular.

A directory of US and foreign observatories. Contains data about radio telescopes, including type, size, location, height, sky coverage, collecting area, etc. A useful addition to astronomy collections.

Observatories of the World. **Thornton Page.** Cambridge, MA: Smithsonian Astrophysical Observatory, 1967.

Geographical arrangement of optical and radio astronomical observatories.

US Observatories: A Directory and Travel Guide. **H. T. Kirby-Smith.** New York: Van Nostrand Reinhold, 1976.

A well-researched and thorough directory of observatories in the United States. In 2 parts. The first discusses the 15 major observatories, including location, equipment, hours of public views, etc. The second lists 300 observations, planetariums, and astronomical museums by state. Detailed, useful directory for all libraries with general astronomy collections.
R: *Scientific American* 236: 140 (Apr. 1977); *RSR* 5: 18 (Apr./June 1977); *ARBA* (1977, p. 637); Sheehy (EB10).

MATHEMATICS

Mathematical Sciences Professional Directory, 1984. Providence: American Mathematical Society, 1984.

A useful tool for anyone interested in mathematical science. A wealth of information on the administrative structure, boards and committees of the American Mathematical Society and 35 other related professional associations. Also includes addresses for individuals listed under the professional organizations; publishers and editors of mathematical journals; etc. Indexed by colleges and universities.
R: *ARBA* (1985, p. 611).

PHYSICS

Directory of Physics and Astronomy Staff Members. 24th ed. New York: American Institute of Physics, 1984. Annual.

First edition, 1959.

An annual directory that includes the names and addresses of over 30,000 physicists, astronomers, and specialists at about 2,700 academic and research institutions in the United States, Canada, Mexico, and Central America. Useful tool for academic and special libraries.
R: *ARBA* (1985, p. 600).

The Geophysical Directory. Houston: Geophysical Directory. Annual.

Comprehensive list of all companies and individuals directly connected with or engaged in geophysical exploration.
R: Win (EE63).

International Physics and Astronomy Directory. New York: Benjamin. Annual.

A hand-sized, computer-produced directory. International in scope, includes faculties of accredited universities, government research labs, societies, awards, publications, etc. Indexes include alphabetical listing by name and a geographical index. Biographical information on researchers and professors.
R: *SJ* 6: 82 (Dec. 1970); Wal (p. 80; p. 97); Win (3EG2).

Optical Industry and Systems Directory. Pittsfield, MA: Optical. Annual.

Twenty-third edition, 1976–1977.

Physics Today Buyers Guide. New York: American Institute of Physics, 1984.

This comprehensive list of North American and European suppliers offers some 2,000 products, services, instruments, components, materials, and computer hardware and software. Handy tool for physicists.

CHEMISTRY

Chem Sources. Flemington, NY: Directories, 1958–. Annual.

Comprehensive listing of chemical products of more than 700 companies.
R: Jenkins (D163); Win (ED59).

Chemical Buyer's Directory. New York: Schnell, 1975–1976. Annual.

Chemical Guide to Europe. 1st ed.–. Pearl River, NY: Noyes Data, 1961–. Annual.

Sixth edition, 1973.
A directory to the chemical firms of Western Europe. Information on ownership, plant locations, products, etc., are given.
R: *Science Reference Notes* 10: 9 (July–Oct. 1963); Jenkins (D166); Win (ED60).

Chemical Guide to the United States. 1st ed.–. Pearl River, NY: Noyes Data, 1962–. Annual.

Seventh edition, 1973.
A directory providing information on name and address, ownership, products, etc., of 400 US chemical firms.
R: *Science Reference Notes* 10: 10 (July–Oct. 1963); Jenkins (D167): Win (ED60).

Chemical Materials Catalog. New York: Reinhold. Annual.
Lists companies and availability of products.
R: *Jenkins* (D170).

Chem Product Index. 2 vols. **Friedrich W. Derz, ed.** Berlin: de Gruyter, 1976.
Lists alphabetically the names of over 300,000 chemical products. Includes molecular formulas and CAS registry number. Includes chemicals produced by some 23,000 firms. Most helpful when used in conjunction with *Chem Suppliers Directory* and *Chem Address Book.*
R: *RSR* 6: 44 (Apr./June 1978).

Chemical Week Buyer's Guide Issue. New York: Chemical Week, 1976. Annual.
A guide consists of a company directory of major producers of chemicals and packaging supplies; 2 alphabetical sections of catalog ads for chemical and raw producers, etc., and major producers and sources of supply for more than 6,000 chemical products. An index of trade names.

International Chemistry Directory, 1976/77–. New York: W. A. Benjamin, 1977–.
Directory of academic departments and faculties, laboratories, societies, grants, awards, etc. Includes journals and books-in-print sections.
R: Jenkins (D173); Win (3ED9).

OPD Chemical Buyers Directory. **Chemical Marketing Reporter.** New York: Schnell. Annual.

Sixty-fifth edition, 1977.
A working guide for the chemical engineer. Complete with product and manufacturer information on sources of supplies for chemicals and related-process materials.

BIOLOGICAL SCIENCES

A Directory of Information Resources in the United States: Biological Sciences. **US Library of Congress.** Washington, DC: US Government Printing Office, 1972. Also updates.
Alphabetical arrangement for both governmental and private agencies, including location and publication information.

Genetic Engineering and Biotechnology Firms: Worldwide Directory 1983/1984. 3d ed. **Marshall Sittig and Robert Noyes, eds.** Kingston, NJ: Sittig and Noyes, 1983.

Covers 1,200 companies of the biotechnology industry, which includes some 500 foreign firms from 29 countries. Not as detailed and consistent as it could be but still a useful source for academic or research libraries.
R: *Science Technology Libraries* 5: 2 (Winter 1984).

The International Biotechnology Directory 1984: Products, Companies, Research, and Organizations. **J. Coombs.** New York: Nature Press/Grove's Dictionaries of Music, 1983.

The directory has 3 parts—organizations and information services alphabetically arranged by type in part 1; country arrangement of government departments and societies in part 2; and lists of companies' research institutes and university departments by country and then alphabetically in part 3. Indexed.
R: *Choice* :1584 (July–Aug. 1984); *ARBA* (1985, p. 580).

International Directory of Genetic Services. 4th ed. **Daniel Bergsma and H. T. Lynch, eds.** New York: National Foundation for the March of Dimes, 1974. Also updates.

World Directory of Collections of Cultures of Microorganisms. **S. M. Martin et al.** New York: Wiley-Interscience, 1972.

Sponsored by the World Health Organization and Unesco. Microorganisms included are algae, baacteria, fungi, lichens, protozoa, tissue cultures, animal viruses, bacterial viruses, insect viruses, plant viruses, and yeasts.
R: *ARBA* (1973, p. 554).

Botany

The Complete Garden. **Arnold Leggett and Pat Falge.** Willits, CA: Oliver Press. Distr. New York: Scribner's, 1975.

Lists products of more than 400 seed and garden tool companies that sell by mail order. Similar information can be gleaned from the *Thomas Register.* Not as comprehensive as Riker's *Gardener's Catalogue, 1974.*
R: *WLB* 50: 265 (Nov. 1975); *ARBA* (1976, p. 757).

Directory of American Horticulture. Mt. Vernon, VA: American Horticultural Society. Annual.

A listing of organizations, plant societies, horticultural libraries, educational institutions, etc.

Great Botanical Gardens of the World. **Edward Hyams and William MacQuitty.** New York: Macmillan, 1969.

Detailed description of over 50 renowned botanical gardens. Includes numerous high-quality photographs and illustrative matter. Map index indicates the location of 525 worldwide botanical gardens.
R: *Geographical Journal* 136: 474 (Sept. 1970); *LJ* 95: 482 (1970); *ARBA* (1970, p. 117); Wal (p. 209); Win (3EC14).

International Directory of Botanical Gardens. 3d ed. **D. M. Henderson and H. T. Prentice.** Utrecht, Netherlands: Bohn Scheltema and Holkemema, 1977.

First edition, 1963; second edition, 1969.

Geographic arrangement of some 500 gardens, including location, staff, and publications information.

R: *Bibliography, Documentation, Terminology* 4 (May 1964); Sheehy (EC26); Wal (p. 208); Win (EC61).

The Naturalists' Directory International. **Willard H. Baetzner, comp.** South Organge, NJ: PCL Publications, 1984.

First edition, 1878, fortieth edition, 1968, forty-first edition, 1972.

Main section is a geographically arranged list of amateur and professional naturalists. Also includes directory information on natural history museums and societies. Should provide academics with a useful source for addresses.

R: *ARBA* (1973, p. 555; 1974, p. 569).

Zoology

The American Fisheries: Directory and Reference Book. **Burton T. Coffey, ed.** Camden, ME: National Fisherman. Distr. Camden, ME: International Marine Publishing, 1978.

A comprehensive directory of statistics and organizations associated with the fishing industry. The first section, arranged by geographic regions, covers various organizations. Contains directories of associations; equipment and services; and books and periodicals. The final section consists of pertinent statistics. A recommended reference for fishing industry personnel.

R: *ARBA* (1980, p. 702).

A Preliminary Directory: Aquatic Biologists of the World and Their Laboratories. 2 pts. Washington, DC: George Washington University, Biological Sciences Communication Project, 1964. Also updates.

Part 1, worldwide listing of laboratories and institutions concerned with aquatic biology and alphabetical listing of specialists; part 2, permuted index.

R: Wal (p. 196).

World Directory of Hydrobiological and Fisheries Institutions. **Robert W. Hiatt.** Washington, DC: American Institute of Biological Sciences, 1963 and updates.

Basic information for marine and freshwater laboratories and museums.

R: Wal (p. 196; 427); Win (EC111).

The World of Zoos: A Survey and Gazetter. 1st English-language ed. **R. Kirchshafer, ed.** Translated by Helda Morris. London: Batsford, 1968.

Includes introductory articles and numerous illustrative plates. Concludes with a gazetteer of zoological gardens around the world.

R: Wal (p. 225).

Zoos and Aquariums in the Americas: Including Roster of Membership, Association History, Purposes, and Objects. Wheeling, WV: American Association of Zoological Parks and Aquariums, 1930–. Biennial.

Includes directory sections on zoos in the United States and Latin America, government agencies, zoological organizations, and conservation societies, as well as various statistical information.

Agriculture

Agricultural Residues: World Directory of Institutions. 2d ed. FAO Agricultural Services Bulletins, 21, rev. 1. **Food and Agriculture Organization of the United Nations.** Rome: Food and Agriculture Organization of the United Nations, 1978.

Trilingual directory of 1,000 institutions concerned with agricultural, fishery, and forestry residues. Entries are arranged under broad subject headings with subdivision by alphabetical arrangement of titles. Listings include complete bibliographical information. Supported by subject index.
R: *IBID* 6: 261 (Sept. 1978).

Agris Forestry: World Catalogue of Information and Documentation Services. **Food and Agriculture Organization of the United Nations.** Compiled by S. Lederer Bewer, New York: Unipub, 1979.

A trilingual directory (English/French/Spanish) of 427 instituions, in more than 90 countries, that supply forestry information. Entries, organized by country, include services available, languages, and conditions of access. For agricultural research libraries.
R: *Quarterly Bulletin of the IAALD* 25: 47 (1980).

Directory of the Forest Products Industry. New York: Unipub, 1980–.

A comprehensive directory of the forest products industry. Provides a listing of more than 9,000 independent producers, major North American mills and logging operations, over 3,500 wood product wholesalers, executives, etc. Operation data is given for each mill. A valuable aid for people in the field.

Directory of Information Resources in Agriculture and Biology. **Agricultural Sciences Information Network.** Washington, DC: US Government Printing Office, 1971–.

Information on federal organizations, land grant colleges and universities, research monies, research collections at agricultural libraries, etc.

Directory of Non-Governmental Agricultural Organizations Set Up at Community Level. 6th ed. Munich: Verlag Dokumentation. Distr. New York: Saur, 1980.

This compilation of European agricultural associations includes name, address, telephone number, and director. Associations are grouped under subject headings. Written in German, French, Italian, English, Danish, and Dutch. For large university libraries.
R: *ARBA* (1979, p. 748).

Forest History Museums of the World. **Kathryn A. Fahl, comp.** Santa Cruz, CA: Forest History Society, 1983.

A directory to the location and the general content of over 300 forest history museums in 33 countries. Extensive illustrations; indexed.
R: *ARBA* (1985, p. 515).

Forest Service Organizational Directory. Washington, DC: US Department of Agriculture. Distr. Washington, DC: US Government Printing Office, 1978–.

A compilation of organization names of agencies that administer the Forest Service. Provides names, telephone numbers, functions, and projects. Contains a map designating regions.
R: *ARBA* (1979, p. 758).

International Directory of Agricultural Engineering Institutions. New York: Unipub, 1974–.

EARTH SCIENCES

Directory of Geoscience Departments: United States and Canada. 19th ed. Falls Church, VA: American Geological Institute, 1980.

All pertinent information, geographically arranged.

Directory of Meteorite Collections and Meteorite Research. New York: Unipub, 1968. Also, updates.

GMT World Register of Oceanographic Products and Services. **Robert M. Snyder, ed.** Washington, DC: Seabrook Hull, 1966.

A directory of manufacturers of oceanographic equipment and providers of services used by oceanographers.

The Geophysical Directory. 35th ed. Houston: Geophysical Directory, 1980

A listing of equipment suppliers, consultants, instruments, exploration efforts, and companies related to the geophysics field.

Geosciences and Oceanography: A Directory of Information Resources in the United States. Washington, DC: Library of Congress, 1981.

A catalog of organizations that offer geological information of some kind. Each is arranged alphabetically, with details about publications, information services, etc. Some groups are omitted entirely. Good for general reference and earth science libraries.
R: *ARBA* (1983, p. 663).

Hart's Rocky Mountain Mining Directory, 1983. **Mary Anne Dunlap, mg. ed.** Denver: Hart, 1982.

Information on mining companies, equipment suppliers, service companies, trade and professional associations, state and federal agencies, and schools involved with mining in the 11-state Rocky Mountain region. Complete and informative.
R: *ARBA* (1983, p. 728).

General Engineering

Directory of Industry Data Sources: The United States of America and Canada. **William A. Benjamin, ed.** Cambridge, MA: Ballinger, 1981.

This directory includes general reference sources, company data sources, publishers, and indexes. Reports material containing marketing data on 60 basic industries. Contains detailed bibliographic indexing. Good for undergraduate and graduate libraries.

Directory of Testing Laboratories: Commercial, Institutional. 5th ed. Philadelphia: American Society for Testing and Materials, 1975. Also late editions.

Alphabetical, geographical, and commodity arrangement of 850 US laboratories with commercial-product testing capabilities.
R: *RSR* 3: 41 (Apr.–June 1975); Wal (p. 292); Win (EA165).

Eastern Manufacturers and Industrial Directory: Buyers Guide. New York: Bell Directory. Annual.

The Engineer Buyers Guide. London: Morgan-Grampian Books, 1974–. Annual.

Information on forthcoming engineering and industrial exhibitions and events, associations, institutions, societies, manufacturers, with full names, addresses, etc. Only British societies listed.
R: Wal (p. 288).

Industrial Research in Britain. 8th ed. St. Peter Port, Guernsey, England: Hodgson. Distr. New York: International Publications, 1976 and later editions.

A standard guide that begins with directories of science-oriented government departments and public bodies, followed by a listing of science-based industrial firms and directories of trade and development associations, professional and learned societies and institutions, independent and sponsored research laboratories, semiofficial research establishments, and other science bodies.
R: *Nature* 218: 47 (Apr. 1968); Wal (p. 46; p. 288).

Industrial Research Laboratories. 19th ed. **Jaques Cattell Press, ed.** New York: Bowker, 1985.

Sixteenth edition, 1979; seventeenth edition, 1981; eighteenth edition, 1983. Provides descriptive information on 12,500 industrial organizations in the United States, including 700 companies listed for the first time. Geographic, personnel, and subject indexes available.
R: *BL* 77: 722 (Jan. 15, 1981); *ARBA* (1976, p. 638; 1978, p. 624).

Selected Sources of Information on Inventions and Product Development. **John A. Feulner, comp.** Washington, DC: Library of Congress, 1980.

Provides a listing of 40 organizations that dispense information on inventions and product development. Includes names, addresses, phone numbers, areas of interest, publications, holdings, and services. For academic libraries.
R: *LCIB* 39: 352 (Sept. 12, 1980).

AERONAUTICAL AND ASTRONAUTICAL ENGINEERING

Aircraft Directory. **Editors of Plane and Pilot Magazine.** Santa Monica, CA: Werner and Werner, 1979.

Designed mainly for North American pilots, this reference lists nearly all active private and commercial airplanes registered in the United States. Contains basic technical data and history for each plane. Includes noteworthy photographs. Although information is highly technical, the volume is suitable for both amateurs and professionals.
R: *BL* 77: 63 (Sept. 1, 1980).

Air Forces of the World: An Illustrated Directory of All the World's Military Air Powers. **Mark Hewish et al.** New York: Simon and Schuster, 1979.

An excellent comprehensive overview of 125 air forces. Contains descriptions of missions and role, defense capabilities, training, deployment of units, etc. Entries are grouped alphabetically by country under 13 geographical regions, with brief descriptions of area, population, defense expenditure, and gross national product. Includes maps, line drawings of current aircraft, and extensive color photographs. An outstanding reference for all libraries.
R: *BL* 76: 1692 (July 15, 1980); *LJ* 105: 210 (Jan. 15, 1980); *ARBA* (1981, p. 777).

Flight Directory of British Aviation. Kingston-upon-Thames, Surrey, England: Kelly's Directories, 1949–. Annual.

Formerly *"Aeroplane" Directory of British Aviation*; incorporating *Who's Who in British Aviation*.
R: Wal (p. 383).

Interavia ABC: World Directory of Aviation and Astronautics. Geneva: Interavia. Distr. New York: International Publications. Annual.

Twenty-third edition, 1975; twenty-sixth edition, 1979.
A world directory of aviation and astronautics in 97 sections. Covers administrations, operations, aircraft and power plant manufacturers, airports, and more. In English, French, German, Italian, and Spanish. Includes 40,000 entries encompassing all companies, organizations, and institutes either directly or indirectly involved in the aeronautical industry. Geographical arrangement.
R: *Engineering* 183: 632 (May 17, 1975); Wal (pp. 217, 382).

World Aviation Directory, Including World Space Directory. Vols. 1–. Washington, DC: Ziff-Davis, 1940–. Biannual.

Provides broad US, Canadian, and foreign coverage of aviation and aerospace manufacturers, distributors, officials, aviation repair stations, aerospace-oriented publications, etc.
R: *ARBA* (1977, p. 776); Jenkins (K214); Wal (p. 382); Win (EI37).

World Space Directory, Listing US and Foreign Space/Oceanology Companies, Officials, and Government Agencies. Washington, DC: American Aviation Publications, 1962–. Semiannual.

A comprehensive worldwide directory to government agencies, consultants, industrial representatives, universities, etc.
R: Jenkins (K125); 11–181; Wal (p. 392); Win (EI38).

CHEMICAL ENGINEERING

Chem Buy Direct: International Chemical Buyers Directory. **Friedrich W. Derz, ed.** Berlin: de Gruyter, 1974–.

An international reference to chemicals. The Chem Product Index contains a listing of more than 300,000 chemicals; the Chem Suppliers Directory lists more than 23,000 firms. Chem Address gives name and address of suppliers. For all chemists and chemical companies.

Chemical Industry Directory and Who's Who. London: Iliffe, 1960–. Annual.

Continues *Chemical Age Yearbook,* 1923–1958.
Ready-reference source of organizations, biographies, trade names, etc.
R: Jenkins (D168); Wal (p. 471).

Chemical Industry Directory and Who's Who. New York: Tunbridge Press, 1980–. Annual.

This useful directory contains an ABC directory of leading companies; European chemical and oil storage depots; who owns whom in the British chemical and chemical plant industries; and more.
R: *Aslib Proceedings* 36: 263 (June 1974); Wal (p. 237).

Chem Sources—Europe. Flemington, NJ: Directories Publishing, 1973–. Annual.

Lists chemical products in Europe. Arrangement is by chemical name. Indexed by product.

Corporate Diagrams and Administrative Personnel of the Chemical Industry. 14th ed. **Kenneth Kern.** Princeton: Chemical Economic Services, 1979.

Thirteenth edition, 1973.
R: *RSR* 2: 122 (Oct.–Dec. 1974); 3: 77 (Apr.–June 1975).

Cryogenics Handbook. **Beverly Law, ed.** Surrey, England: Westbury House, 1981.

A directory of companies and equipment suppliers in the cryogenics industry. Gives name, address, contact person, and description of company. It also has an index of products, and a partial list of international research projects. Good only for special libraries.
R: *ARBA* (1983, p. 610).

Davison's Textile Blue Book. Ridgewood, NJ: Davison. Annual.

The annual contains an advertiser's index; geographic list of mills, dyers, and finishers; buyer's guide showing advertisers' products, equipment, services, and supplies, etc.

Directory of Fabric Resources for 1975. New York: Geer, 1975.

Directory of Textile Plant Processes. New York: Textile World. Annual.

Directory of the Canning, Freezing, Preserving Industries. Westminster, MD: Judge, 1976. Biennial.

Lists North American packers of canned foods. Factories are listed alphabetically. Useful marketing directory.

Europlastics Yearbook. London: IPC Industrial Press. Annual.

Primarily a directory of companies and suppliers.

Food Ingredients Directory. Hasting-on-Hudson, NY: Food Ingredients Directory. Annual.

Guide to several thousand natural and synthetic ingredients in all types of food.

International Directory of the Nonwoven Fabrics Industry. New York: International Nonwovens and Disposables Association. Annual.

Japan Chemical Directory, 1985. 23rd ed. **Japan Chemical Week, ed.** Tokyo: Chemical Daily. Distr. New York: International Publications, 1985.

Eighth edition, 1970; seventeenth edition, 1979.
A directory of chemicals, plastics, fibers, fertilizers, and synthetic rubbers as well as lists of importers, domestic dealers, exporters, etc.
R: *ARBA* (1971, p. 555).

Modern Plastics Encyclopedia. New York: McGraw-Hill, 1941–. Annual.

Specific directory-type information on the industry.
R: Wal (p. 536); Win (EI64).

Plastics World. Directory of the Plastics Industry. Boston: Cahners Books. Annual.

The August issue of *Plastics World* lists suppliers in all phases of plastics manufacturing. Annual updates in the new August issues.

Post's Pulp and Paper Directory. New York: Unipub. Annual.

A comprehensive directory concerning the North American pulp and paper industry. Includes a list of more than 13,000 mill officials, as well as lists of mills, production statistics, products, machinery, research facilities, etc. An invaluable reference for those involved in the field.

Rubber Red Book: Directory of the Rubber Industry. 25th ed. New York: Palmerton, 1973. Also updates.

Lists products, and suppliers of plastics and rubber, trade names, and types of equipment. A "Who's Who" in the rubber market.

Selected Information Resources on Plastics, Polymers, and Elastomers. **John A. Feulner, comp.** Washington, DC: US Library of Congress, 1981.

A list of 47 organizations, military and federal agencies, societies, and trade associations that dispense information on polymers, plastics, and elastomers. Provides names, addresses, phone numbers, etc. For industrial and science libraries.
R: *LCIB* 40: 143 (Apr. 24, 1981).

Selected Information Resources on Pulp and Paper. **John A. Feulner, comp.** Washington, DC: US Library of Congress, 1980.

Contains a list of 27 organizations with information on pulp and paper. Includes names, addresses, holdings, services, etc. For industrial and academic libraries.
R: *LCIB* 39: 267 (Aug. 1, 1980).

Selected Information Resources on Textiles. **John A. Feulner, comp.** Washington, DC: US Library of congress, 1980.

A directory of 41 organizations that supply information on textiles. Gives facts about names, addresses, areas of interest, holdings, publications, etc. For academic libraries.
R: *LCIB* 39: 267 (Aug. 1, 1980).

Selected Sources of Information on Corrosion and Corrosion Prevention. **John A. Feulner, comp.** Washington, DC: US Library of Congress, 1980.

Contains a list of 43 organizations that provide information on corrosion and its prevention. Includes names, addresses, phone numbers, areas of interest, publications, etc. For science libraries.
R: *LCIB* 39: 267 (Aug. 1, 1980).

Worldwide Chemical Directory. London: ECN Chemical Data Services, 1977. Also updates.

Contains information about 10,000 companies that produce and market chemicals and pharmaceuticals in 95 countries. Leaves out chemical plant contractors and chemical tankers and storage companies.
R: Wal (p. 515).

Food Technology

American Breweries. **Donald Bull, Manfred Friedrich, and Robert Gottschalk.** Trumbull, CT: Bullworks, 1984.

A directory of 6,000 American breweries that have produced beer. Listed by state, then by city. Much-expanded version of *The Register of United States Breweries, 1876–1976*. Excellent tool for the subject.
R: *ARBA* (1985, p. 498).

Directory of Food and Nutrition Information Services and Resources. **Robyn C. Frank, ed.** Phoenix: Oryx Press, 1984.

A directory of information services and resources on the subject, with emphasis on nutrition education, food science management, and aspects of applied nutrition. The largest part is a listing of organizations, with additional listing of 60 databases on food and nutrition, microcomputer software programs, etc. Indexed by type of organization and geographic location. A must for nutrition library collections.
R: *ARBA* (1985, p. 498).

Quick Frozen Foods Directory of Frozen Food Processors. New York: Williams, 1946–. Annual.

Spices, Condiments, Teas, Coffees, and Other Delicacies: A Guide to the Hard to Find. **Roland Robertson.** Willits, CA: Oliver Press. Distr. New York: Scribner's, 1975. Also updates.

A directory with 2 useful indexes: the company section that describes the firm and its products, and the delicacies-by-name section.
R: *WLB* 50: 752 (June 1975); *ARBA* (1976, p. 751).

Petroleum Engineering

(See Energy)

CIVIL ENGINEERING

American Public Works Association Directory. **Rodney R. Fleming et al., eds.** Chicago: American Public Works Association. Annual.

Constructor Directory, 1976–1977. Washington, DC: Associated General Contractors of America, 1976. Annual.

Consulting Engineer Product Bulletin Directory. St. Joseph, MI: Consulting Engineer. Quarterly.

Directory of Construction Associations. 2d ed. **Joseph A. MacDonald.** Huntington, NY: Professional Publications, 1980.

A quick and easy-to-use source listing more than 2,500 contractor associations, professional and business societies, labor unions, government agencies, manufacturer and producer groups, and construction publications. Covers 87 specific categories, listing each organization alphabetically by official name.
R: *Science and Technology Libraries* 1: 109 (1980).

Energy-Efficient Products and Systems: A Comparative Catalog for Architects and Engineers. New York: Wiley-Interscience, 1983.

A listing and description of 400 product lines in conservation systems. Includes generic, brand name, and manufacturing indexes. Cross-referenced according to the Construction Specification Institute system.
R: *Librarian's Newsletter* (Oct. 1982).

Government Production Prime Contractors Directory. Washington, DC: Government Data Publications. Annual.

Handbook of Construction Resources and Support Services. **J. A. MacDonald.** New York: Wiley, 1981.

A directory of sources of information available to solve engineering and construction problems faced on construction projects. Contains listings of consultants, reference libraries, government agencies, professional societies, etc. For civil engineers.

Hydraulic Research in the United States and Canada. **US Commerce Department. National Bureau of Standards.** Washington, DC: US National Bureau of Standards, 1984–.

RIBA Directory of Manufacturers. London: RIBA Services, 1980.
A listing of approximately 5,000 building product manufacturers. Divided into 2 main sections: the first contains a list of companies, their products and services, telephone numbers, and a subject index of products. The second part consists of a list of references to publications and an alphabetical list of manufacturers, with addresses, telephone numbers, and products noted. A handy directory for reference libraries.

Selected Information Resources on Wood Products. **John A. Feulner, comp.** Washington, DC: US Library of Congress, 1981.
Lists 47 organizations, federal agencies, research centers, societies, and associations that dispense information on wood products. Gives names, addresses, phone numbers, etc. For science libraries.
R: *LCIB* 40: 143 (Apr. 24, 1981).

World Cement Directory, 1980. 2 vols. 6th ed. Malmo, Denmark: Cembureau, Cement Statistical and Technical Association, 1980 and later editions.
First edition, 1961; fourth edition, 1972; fifth edition, 1977.
Location and data on cement companies. Geographical arrangement.

World Register of Dams. New York: International Commission on Large Dams, 1973. Also later editions.
R: *Water Power* 16: 484 (Nov. 1964); Wal (p. 364).

Electrical and Electronics Engineering

China's Electronics and Electrical Products. Hong Kong: China Phone Book. Distr. Menlo Park, CA: China Phone Book, 1983.
An English–Chinese directory of 25,000 electronics and electrical products made by 4,000 manufacturers in China. An important current information tool on China.
R: *ARBA* (1985, p. 548).

Directory of Communication Organizations. Denver: Council of Communication Societies. Annual.

Electrical World Directory of Electrical Utilities. New York: McGraw-Hill. Annual.
Contains complete and current information about companies that provide electrical service and about the geographical areas they serve. One of a kind.

Electronic Engineers Master. Garden City, NY: United Technical Publications, 1976. Annual.
The first 3 volumes contain a product index and directory, a manufacturers and sales office directory, and a distributor directory. Volume 4 serves as the index to the first 3 volumes.

Electronics Buyer's Guide. New York: McGraw-Hill. Annual.

Lists over 4,000 products and services in electronics technology. Includes a directory of trade names and an advertisers' index.

IEA: The Directory of Instruments, Electronics, Automation. 13th ed. London: Morgan-Grampian Publishing, 1979.

This directory of instruments, electronics, and automation includes manufacturers' addresses, UK agents, buyer's guide for 1979, and more.
R: Wal (p. 206).

Insulation/Circuits; Directory/Encyclopedia. Libertyville, IL: Lake. Annual.

Japan's 100 Leaders in the Electric and Electronics Industries. Tokyo: Dempa. Distr. New York: International Publications. Annual.

National Electrical Safety Code, 1984. **Institute of Electrical and Electronics Engineers.** New York: Wiley, 1984.

Provides information on overhead and underground lines and supply stations, covering such topics as safe installation and maintenance of electric supply stations, and supply and communication lines.

Selected Information Resources on Electronics. **John A. Feulner, comp.** Washington, DC: US Library of Congress, 1980. Also updates.

A directory of 41 organizations, government laboratories, societies, and research centers that provide assistance and information on topics in electronics. Covers such fields as telecommunications, high fidelity, systems engineering, etc. Includes facts such as names, addresses, telephone numbers, etc. For science and engineering libraries.
R: *LCIB* 39: 482 (Dec. 12, 1980).

Telecommunications Systems and Services Directory. 2d ed. **Martin Connors, ed.** Detroit: Gale Research, 1985.

A comprehensive directory of both established and newly developed telecommunications systems, which include 150 long-distance telephone services, 125 data-communications services; over 100 electronic mail services, 75 videotex and teletext services, 150 teleconferencing services, 200 telex and telegram services, 70 local-area networks, 50 satellite services, etc. A detailed glossary and 4 detailed indexes—master index listed by names, function/service-type index, personal-name index, and geographic index. A major tool.

Computer Technology

Classroom Computer News Directory of Educational Computing Resources, 1983. **Peter Kelman, ed.** Watertown, MA: Intentional Educations, 1982.

Six sections of computer information: ideas, information, and materials; software; associations; periodicals; funding; and miscellaneous resources and other information on various sources of continuing education programs. Each provides contact people

in specific states, address information, and, if necessary, complete annotations. Up-to-date information will be useful to every educational computer user.
R: *ARBA* (1983, p. 720).

Collected Algorithms from CACM. New York: Association for Computing Machinery, 1960–. Bimonthly.
A loose-leaf service, accumulates all algorithms published in Communications of the ACM. Full program listings are usually included. The algorithm can be obtained from the originator. Useful source to both computer personnel and researchers.

Computer Dealers: 1984 Directory. Omaha: American Business Directories, 1984.
A yellow-page directory of over 22,000 dealers in the United States. Addresses, ZIP codes, phone numbers, and type of computer franchise are given.
R: *ARBA* (1985, p. 585).

Computer Directory and Buyers' Guide. Vols. 23–. Newtonville, MA: Berkeley Enterprises, 1974–. Annual.
Covers all aspects of the computer industry. Provides information on products, facilities, firms, services, and hardware leasing. Includes some technical sections. Intended for those involved in the computer industry.

Computer Graphics Marketplace. **John Cosentino, ed.** Phoenix: Oryx Press, 1981.
A directory of manufacturers, consultants and services, professional organizations, programs, and several other sources involved with the computer graphics industry and its users. Will be updated annually for more efficient use in reference collections.
R: *ARBA* (1983, p. 721).

Computer Graphics Directory '84. **Diane H. Fuller, ed.** Tulsa: PennWell Publishing, 1983.
A softbound directory that offers coverage of vendors in the first section, and also a directory of subject sections, such as turnkey systems, plotters, and hard copy and software. Indexed.
R: *ARBA* (1985, p. 588).

The Complete Guide to Computer Camps and Workshops. **Mike Benton.** Indianapolis: Bobbs-Merrill, 1984.
A guide and directory that provides a general introduction to the subject and a listing of about 500 camps arranged by state. Great majority of these camps are in the United States. No evaluation is given. A useful book for parents and public libraries.
R: *ARBA* (1985, p. 585).

Computing Marketplace: A Directory of Computing Services and Software Supplies for Word Processors, Micros, Minis, and Mainframes. 2d ed. **Bis-Pedder, ed.** Brookfield, VT: Gower Publishing, in association with Computing Services Association, 1983.

An excellent directory of micro-, mini-, and mainframe computing industry in Great Britain. It includes 10 sections, such as software vendors, hardware manufacturers, and consultants.
R: *ARBA* (1985, p. 585).

The Computer Resource Guide for Nonprofits. **Kenneth L. Gilman, ed.** San Francisco: Public Management Institute. Distr. Detroit: Gale Research, 1982.

Provides an alphabetical listing of 100 companies with a 1-page report on each; also includes a directory of funding sources for the use of nonprofit organizations. A good starting point for any nonprofit group that wants to purchase computer software.
R: *ARBA* (1984, p. 626).

Computers and Computing Information Resources Directory. **Martin Connors, ed.** Detroit: Gale Research, 1985.

A major reference directory that includes over 6,000 entries on the primary sources of information in the computer world, including associations and user groups, consultants, training organizations, special libraries, information centers, research centers, university computer facilities, computer trade shows, exhibitions, association conventions and meetings, organizations bestowing awards, online services, teleprocessing networks, newsletters, journals, and abstracting and indexing services. Indexed by names and keyword to organizations and publications, geographic locations, and personal names.

Microcomputer Market Place 1986. New York: Bowker, 1985. Annual.

An up-to-date directory source on software publishers and manufacturers, distributors, microcomputer manufacturers, supplies manufacturers, periodicals, associations, data-base services, etc.

North American Online Directory. New York: Bowker, 1985.

Formerly titled *Information Industry Marketplace.*
A useful tool for updated information on the people, companies, and services that make up the revolutionary new information industry. Includes database producers, online vendors, library networks and consortia, information brokers, information collection and analysis centers, associations, references, courses, etc.

Official Directory of Data Processing: Eastern USA Users. Gresham, OR: Official Directories and Services, 1980.

A directory of the data-processing user sites in the Eastern states, listed geographically by cities and including addresses and systems information. Systems information includes model number and programming information. Good for sales and service personnel and companies.
R: *ARBA* (1982, p. 808).

The Videodisc Book: A Guide and Directory, 1984. **R. Daynes, ed., B. Butler, assoc. ed.** New York: Wiley, 1984.

The first complete guide to all aspects of videodiscs, including creative aspects, production, manufacture, uses, programming, etc. The book also contains listings of more than 200 production and development organizations; over 100 hardware manufacturers, software developers, and integrated system dealers; and about 1,800 videodisc titles.

Computer Hardware
(Mostly Microcomputers)

Apple Computer Directory: Hardware, Software, and Peripherals. **Kelly-Grimes Corporation.** New York: Wiley, 1984.

Data Sources/Hardware. New York: Ziff-Davis. Quarterly.

A good directory source of current information on terminals and printers. Lists hundreds of printers and terminals, with major specifications arranged in tables for easy comparison.
R: *ARBA* (1985, p. 585).

Datapro Who's Who in Microcomputing, 1983. New York: Datapro/McGraw-Hill, 1983.

Information on some 2,000 computer hardware and software suppliers and companies, with relative information on each. Some previous knowledge of a company's major application area is necessary.
R: *ARBA* (1984, p. 621).

DEC Microcomputer Directory: Hardware, Software, and Peripherals. **Kelly-Grimes Corporation.** New York: Wiley, 1984.

IBM PC-Compatible Computer Directory: Hardware, Software, and Peripherals. New York: Wiley, 1984.

Latest edition, 1986.

Microcopmuter Marketplace. **Robert Driscoll, ed.** New York: Dekotek. Distr. Detroit: Gale Research, Annual.

A new directory based on information obtained from a questionnaire sent to manufacturers in the computer industry. Includes information on software publishers, distributors, suppliers, and manufacturers of microcomputer systems. Intended as an annual source.
R: *ARBA* (1984, p. 622).

1985 Microcomputer Marketplace. New York: Bowker, 1984. Semi-annually.

A useful source guide and directory to the many areas and concerns related to the world of microcomputers.

Computer Software
(Mostly Microcomputers)

Apple Software Directory: Yellow Pages to the World of Micro Computers. Overland Park, KS: PC Telemart/Vanloves. Distr. New York: Bowker, 1985. Annual.

Over 3,500 programs for Apple minicomputers are listed.

Auerbach Software Reports. Philadelphia: Auerbach Information, 1970–. Bi-monthly.

This loose-leaf service deals with software offered commercially by independent software houses and by hardware manufacturers on an unbundled (separately charged) basis. Program packages are divided into functional groups.

Business Mini/Micro Software Directory. **Information Sources, Inc., Comp.** New York: Bowker, 1984 with supplement, 1985.

About 4,500 mini and micro software packages are listed in the 1984 volume, supplemented by some 1,000 entries in the 1985 supplement.

CP/M Software Directory. New York: Bowker, 1984.

Over 3,000 programs for CP/M operating systems are listed.

A Directory of Computer Software Applications: Energy, 1977. **David W. Grooms.** Springfield, VA: US Department of Commerce, National Technical Information Service, 1977.

A specialized directory for energy researchers. Includes applications in solar energy, petroleum, electrohydrodynamic and magnetohydrodynamic resources, natural gas, nuclear fission and fusion, hydroelectric power, and geothermal energy. Well-recommended.
R: *ARBA* (1979, p. 769).

Directory of Computer Software Applications: Mathematics: 1970–April 1979. Springfield, VA: US Department of Commerce, National Technical Information Service, 1979.

This directory of reports lists computer programs and/or their documentation. Software applications are concerned with topics in mathematics such as algebra, geometry, and statistical methods. Contains complete bibliographic information for each report. Also includes corporate author and subject indexes. For mathematicians and computer programmers.
R: *ARBA* (1981, p. 625).

Directory of Information Management Software for Libraries, Information Centers, Record Centers. **Pamela Cibbarelli, Carol Tenopir, and Edward J. Kazlauskas.** Studio City, CA: Pacific Information, 1982. Also, supp. 1984.

The directory and its supplement provide descriptions of 65 currently available software packages useful to libraries and information centers.

Directory of Public Domain (and User-Supported) Software for the IBM Personal Computer. Santa Clara, CA: PC Software Interest Group, 1984.
R: *ARBA* (1985, p. 594).

The Directory of Software Publishers: How and Where to Sell Your Program. **Eric Balkan.** New York: Van Nostrand Reinhold, 1983.

A detailed description of over 150 companies, including major manufacturers and software publishers. Also a listing of 800 potential buyers.
R: *IEEE Spectrum* 20: 164 (Oct. 1983); *ARBA* (1985, p. 594).

Directory of Word Processing Systems. **Kelly-Grimes Corporation.** New York: Wiley, 1984.

The Free Software Catalog and Directory: The What, Where, Why, and How of Selecting, Locating, Acquiring, and Using Free Software. **Robert A. Froelich.** New York: Crown, 1984.

This is both a directory of free CP/M software and an encyclopedia. Each entry in the software directory, Chapter 5, has 3 parts: the header (file name, file type, disk size, etc.); keywords; and text (description). A keyword index is available.
R: *ARBA* (1985, p. 597).

IBM Software Directory. New York: Bowker, 1985.

A directory of software for IBM microcomputers.

ICP Software Directory. Carmel, IN: International Computer Programs, 1967–. Quarterly.

A directory of computer programs offered for sale, on a broad spectrum of business and technical applications. Supplemental source to *Auerbach Software Reports.* Subscription includes *ICP Software Newsletters.*

ICP Software Directory: Business Applications for Microcomputers. Bowie, MD: Brady, 1983.

A software directory with business emphasis. One-third of the software is from suppliers with addresses in Great Britain; the remaining two-thirds is from North America. Indexed.
R: *ARBA* (1985, p. 594).

International Microcomputer Software Directory. Fort Collins, CO: Imprint Software, 1981. Also updates.

A list of 5,000 programs for microcomputers. Each is accessed by machine, subject, or software house. An index of program names and a code for each software house aids in cross-referencing. Useful for users, libraries, computer stores, and software houses.
R: *ARBA* (1982, p. 806).

MacIntosh: A Concise Guide to Applications Software. **D. van Nouhuys.** New York: Wiley, 1985.

Microcomputer Directory: Applications in Educational Settings. 2d ed. **The Staff of the Monroe C. Gutman Library, comps.** Cambridge, MA: Harvard University Graduate School of Education, Gutman Library, 1982.

Includes sources of grants or descriptions of projects that have been done on the utilization of microcomputers for instructional and administrative purposes in elementary and secondary schools, colleges and universities, state education offices, museums, prisons, and computer camps in the United States. Confusion may be caused by listings that are inconsistently entered.
R: *ARBA* (1984, p. 622).

Microprogrammer's Market 1985. **Marshall Hamilton.** Blue Ridge Summit, PA: TAB Books, 1985.

A useful tool for those individuals developing or interested in developing micro-based software programs for sale to software publishers. Lists micro-software publishers in 5

major categories: business/industry, educational/tutorial, games, home use, and utilities. Indexed by publisher.
R: *Choice* 22: 64 (Sept. 1984); *ARBA* (1985, p. 592).

MicroUse Directory: Software. **Ching-chih Chen.** Newton, MA: MicroUse Information, 1984.

The first of a series of directories to be produced from the MicroUse database, which is intended to answer the question, "who uses what microcomputer, with what software, for what types of applications?" The software directory includes over 1,100 general-purpose software entries and over 350 library-specific application programs. For each entry, information is given on software name, software type, vendor (with address), language, price, hardware used, memory and peripheral requirements, and description. Five functional indexes, by type of software, vendor, hardware, operating system, and random-access memory requirement, are included.
R: *Electronic Library* 3: 117 (April 1985).

Minicomputers. New York: North Holland, 1984.

A complete international catalog of software available for minicomputers (defined as systems costing from about $10,000 to $100,000 (US). Details 5,637 programs representing over 12,000 software packages.

Office Automation and Word Processing Buyer's Guide. **Tony Webster, Richard Dougan, and Jenny Green.** New York: McGraw-Hill, 1984.

Contains an introduction to the theory of office automation systems, and office automation product reports. Includes a directory of summaries of products in various aspects of this area. Helpful in choosing the right product for home or office.
R: *ARBA* (1984, p. 618).

Online Micro-Software Guide and Directory, 1983–1984. Weston, CT: Online, 1982.

From information collected by responses to questionnaires sent to over 600 microcomputer producers in February 1982, this directory contains standardized descriptions of more than 730 software packages and other relevant information for each one. Dated in its accuracy, but useful to the micro specialist.
R: *ARBA* (1984, p. 627).

PC Clearinghouse Software Directory: Yellow Pages to the World of Microcomputers. 7th ed. **Don T. Wright, ed.** Fairfax, VA: PC Clearinghouse, 1983.

Provides information on various types of software, related literature, software and hardware vendors, operating systems, and a section on microprocessors listed by microcomputers and manufacturers. With subject, vendor, and product (title) indexes. A helpful guide to the beginning of software and hardware research.
R: *ARBA* (1984, p. 630).

PC Telemart/Vanloves IBM Software Directory: Yellow Pages to the World of Microcomputers. New York: Bowker, 1984.

The directory includes over 3,000 software packages that run on the IBM Personal Computer and IBM compatibles.
R: *ARBA* (1985, p. 595).

Pac-Finder System 34/36. Software Directory. 2d rev. ed. **Mincron SBC Corporation.** New York: North Holland, 1984. Annual.

An annual, detailed directory on third-party software available to run on the IBM System 34 and System 36, two of the most popular computers.

Programmer's Market, 1984. **Brad M. McGehee, ed.** Assisted by Robert D. Lutz. Cincinnati: Writer's Digest Books, 1983.

This guide for software writers contains necessary information on where and how they can sell their programs. Includes a directory of over 500 software publishers.
R: *ARBA* (1984, p. 627).

Science and Engineering Software. New York: North Holland, 1984.

Contains over 4,300 descriptions of software programs of interest to university and engineering company research departments. Both mini- and microcomputer programs are included. Indexed.

Software: Engineering, 1984. Minneapolis: Moore Data Management Services, 1984.

Contains information on over 1,000 programs covering all fields of engineering. Indexed by software name and vendor.
R: *ARBA* (1985, p. 538).

The Software Catalog. New York: North Holland, 1983–.
In 3 parts: *Microcomputers Including International Standard Program Numbers (ISPN)*, 1983–. Semi-annual; *Microcomputers and Minicomputers. Spring 1983 Update Including International Standard Program Numbers (ISPN)*, 1983; *Minicomputers Including International Standard Program Numbers (ISPN),* 1983–. Semi-annual.

Provides various information on several software programs, listed according to their International Standard Program Number. Extensive in details and information, but most useful for those who need a software search.
R: *ARBA* (1984, p. 628).

The Software Encyclopedia 1986. New York: Bowker, 1985. Annual.

Over 25,000 programs from 3,000 producers and vendors are listed. Indexed.

The Software Marketplace: Where to Sell What You Program. **Suzan D. Prince.** New York: McGraw-Hill, 1984.

A guide discussing the software marketplace, with a directory of publishers, magazine, and agents. Useful to beginning software programmers.
R: *ARBA* (1985, p. 598).

Software Tools Directory. Torrance, CA: Reifer Consultants, Inc. Annual.

Describes about 400 diverse automated tools that can enhance software development. Organized and easy to use.
R: *Computer* 16: 141 (Sept. 1983).

The Software Writer's Marketplace. **Dennis Joyce and John Earl Pickering.** Philadelphia: Running Press, 1984.
A directory of about 500 companies that purchase software from independent suppliers. Indexed by software interest areas, computer type, operating system, etc. Recommended for libraries serving patrons with this interest.
R: *LJ* 109: 1748 (Sept. 15, 1984); *ARBA* (1985, p. 598).

Swift's Educational Software Directory. Apple II ed. Austin: Sterling Swift, 1983–. Annual.
R: *ARBA* (1985, p. 595).

Swift's Educational Software Directory for Corvus Networks. Apple ed. Austin: Sterling Swift, 1984–. Annual.
R: *ARBA* (1985, p. 596).

Swift's Directory of Educational Software for the IBM PC. Austin: Sterling Swift, 1983–. Annual.
R: *ARBA* (1985, p. 595).

The Timex-Sinclair 1983 Directory: Where to Find Practically Everything for the Timex-Sinclair Computer. Alexandria, MN: Brown, 1983.
A useful tool for Timex-Sinclair computer owners. Illustrated.
R: *ARBA* (1985, p. 590).

Where to Find Free Programs for Your TRS-80, Apple, or IBM Microcomputer. 2 vols. **Henry Lee.** Pasadena: Pasadena Technology Press, 1983.
An excellent source for "freeware." Volume 1 is an index to about 4,000 BASIC programs included in 160 popular computing books; volume 2 indexes another 4,000 programs included in 500 computer magazines from 1979–1983. Useful tool for public libraries.
R: *ARBA* (1985, p. 595).

MARINE ENGINEERING

Biennial Directory and Information Book. New York: Society of Naval Architects and Marine Engineers.
The society is one of the largest in the field, and its directory is, correspondingly, an essential one.

Directory of Marine Sciences Libraries and Information Centers. **Carolyn P. Winn, comp.** Woods Hole, MA: International Association of Marine Science Libraries and Information Centers and Woods Hole Oceanographic Institution, 1981.
Information on marine sciences libraries and information centers throughout the United States, Canada, and the world.
R: *ARBA* (1984, p. 643)

Directory of Shipowners, Shipbuilders, and Marine Engineers. **Simon Timms and Edith Griffin, eds.** New York: International Publications. Annual.

Seventy-seventh edition, 1979.
R: Wal (p. 376).

International Shipping and Shipbuilding Directory. 2 vols. London: Benn. Distr. New York: International Publications. Annual.

Eighty-seventh edition, 1975; ninety-first edition, 1979.
Covers all aspects of shipping industry, listing shipowners, shipbuilders, repair companies, towage and salvage services, maritime organizations, subcontracting firms, etc. A standard directory for anyone in the shipping industry.
R: Wal (p. 376).

Jane's Fighting Ships. London: Low, 1898–. Annual.

Technical data on all classes of fighting ships.
R: *ARBA* (1979, p. 781); Jenkins (K182); Wal (p. 351).

Jane's Ocean Technology. New York: Watts. Annual.

Company profiles and addresses, and informative articles. Nowhere else are current commercial efforts cataloged in depth.
R: *ARBA* (1979, p. 763).

Jane's Surface Skimmers: Hovercraft and Hydrofoils. London: Jane's Yearbooks. Annual.

R: *ARBA* (1978, p. 778); Wal (p. 378).

Lloyd's Register of American Yachts. New York: Lloyd's Register of Shipping. Annual.

Lloyd's Register of Shipping-list of Shipowners. London: Lloyd's Register of Shipping. Annual.

Marine Directory. New York: Simmons-Boardman. Annual.

Merchant Vessels of the United States (Including Yachts). US Coast Guard. Washington, DC: Department of Transportation. Annual.

Oceanography Information Resources 70. Washington, DC: US National Academy of Sciences, 1970. Also, updates.

"Places" section lists pertinent industrial and commercial firms, research labs, etc. "Publications" section includes reports of the National Research Council, serials, directories, etc.

Ports of the World. **John Riethmuller.** New York: International Publications. Annual.

Twenty-ninth edition, 1975; thirty-third edition, 1980.
R: Wal (p. 234).

Sea Technology Handbook/Directory. Arlington, VA: Viking Press. Compass Books, 1975. Annual.
Earlier editions, 1974, 1973, 1972, 1970. Original title: *Undersea Technology Handbook/Directory.*
A listing of industrial firms, products, services, governmental agencies, educational institutions, oceanographic research vessels, and geophysical survey vessels.

US Directory of Marine Scientists. 3d ed. **US National Research Council, the Ocean Sciences Board, Commission on Physical Sciences, Mathematics, and Resources.** Washington, DC: National Academy Press, 1982.
First edition, 1969; second edition, 1975.
Identifies scientists in various universities and other research centers in the United States. Lists scientists alphabetically by specialty, and lists organizations alphabetically and geographically.
R: *ARBA* (1984, p. 687).

MECHANICAL ENGINEERING

Directory of Foreign Manufacturers in the United States. 3d ed. **Jeffrey S. Arpan and David A. Ricks.** Atlanta: Georgia State University, College of Business Administration, 1983.
Second edition, 1979.
Lists 3,400 firms alphabetically and indexes them by parent company. Comprises a list of foreign-owned domestic businesses. For large business reference collections.
R: *LJ* 104:1685 (Sept. 1, 1979).

The Directory of Industrial Heat Processing and Combustion Equipment: Manufacturers, 1984–1985. 4th rev. ed. Atlanta: Industrial Heating Equipment Association/Fairmont Press, 1984.
A directory of processing and combustion equipment manufactured by 37 companies, and an expanded list of manufacturers, followed by a product classifications section of types of equipment.
R: *ARBA* (1985, p. 550).

Machine and Tool Directory. Wheaton, IL: Hitchcock, 1976. Annual.
Supplies valuable information on products, dealers, and distributors.

Mechanical Engineer's Catalog and Product Directory. **American Society of Mechanical Engineers.** New York: American Society of Mechanical Engineers, 1976. Annual.
The volume is divided into 3 sections: full-line section, 4,000 leading manufacturers; product directory section, 50,000 listings of products; ASME publications section, all available ASME publications codes, standards, and periodicals with ordering instructions.

National Directory of Manufacurers' Representatives. **Herbert F. Holtje, ed.** New York: McGraw-Hill. Annual.

Formerly titled *Directory of Manufacturers' Agents Contacting Industrial Distributors.* Lists 5,000 manufacturers in over 75 market categories. Provides information on both individual representatives and geographc market. In 2 parts: nationwide alphabetical listing and state-by-state SIC market guide based on SIC system. For large reference collections.
R: *Choice* 16: 802 (Sept. 1979).

Vacuum Special Issue: World-Wide Directory of Manufacturers of Vacuum Plant, Components and Associated Equipment—1983. **Bryce Halliday, ed.** New York: Pergamon, 1983.

From *Vacuum Technology Applications and Ion Physics* (33:8). A report on 200 entries from 20 countries made available to libraries not already subscribing to the periodical.
R: *Science Technology Libraries* 5: 2 (Winter 1984).

METALLURGY

Metal Finishing: Guidebook and Directory. 40th ed. Westwood, NJ: Metals and Plastics Publications, 1972. Also later editions.

Metalworking Directory. New York: Dun & Bradstreet, 1975.

Extensive sales and marketing information on the industry, covering over 45,000 metalworking plants and metal-distributing companies.

Woldman's Engineering Alloys. 6th ed. **Robert C. Gibbons, ed.** Metals Park, OH: American Society for Metals, 1980.

A completely updated compilation of over 40,000 industrial alloys. Alphabetically arranged by trade name. A list of producers is also given, as are chemical compositions and major uses. An outstanding reference source for anyone involved with industrial alloys.
R: *RSR* 8: 25 (July/Sept. 1980).

World Mines Register: The World-Wide Directory of Mining and Mineral Processing Operations. San Francisco: World Mining. Distr. New York: Unipub. Annual.

A detailed reference to companies, products, and personnel in the mining industry worldwide. A valuable source for businesses and individuals involved with or dependent upon the mining industry.
R: *ARBA* (1983, p. 728).

NUCLEAR ENGINEERING

Directory of Nuclear Reactors. 10 vols. **International Atomic Energy Agency.** New York: Unipub, 1959–1976. Also later editions.

Volume 1, *Power Reactors*, 1959 (superseded by volume 4); volume 2, *Research, Test, and Experimental Reactors*, 1959; volume 3, *Research, Test, and Experimental Reactors*, 1960; volume 4, *Power Reactors*, 1962; volume 5, *Research, Test, and Experimental Reactors*, 1964; volume 6, *Research, Test, and Experimental Reactors*, 1966; volume 7, *Power Reactors*, 1967; volume 8, *Research, Test, and Experimental Reactors*, 1970; volume 9, *Power Reactors*, 1971; volume 10, *Power and Research Reactors*, 1976.
R: *IBID* 4: 243 (Sept. 1976); *TBRI* 43: 35 (Jan. 1977); *ARBA* (1973, p. 662); Wal (p. 310).

International Directory of Certified Radioactive Materials. **International Atomic Energy Agency, Quality Control Services, and Euratom, Central Bureau for Nuclear Measurements.** Vienna: International Atomic Energy Agency. Distr. New York: Unipub, 1976.

Lists products of 13 suppliers of radioactive materials in 10 countries. Divided into 5 groups: solutions, reagents, gaseous sources, solids, neutron sources, and miscellaneous. Provides detailed information focusing on the proper selection of calibration source. For research libraries.
R: *ARBA* (1977, p. 650).

Major Activities in the Atomic Energy Program. **US Energy Research and Department Administration.** Washington, DC: US Government Printing Office. Annual.

Annual report to Congress covers major research programs in the United States.

The Nuclear Power Issue: A Guide to Who's Doing What in the US and Abroad. **Kimberly J. Mueller, ed.** Claremont, CA: California Institute of Public Affairs, 1981.

A guide to information sources on nuclear power. Includes information on federal and state government agencies in the United States as well as various other educational, research, and professional organizations in the United States and abroad. The name, address, and purpose of the organization are given for each entry. Of interest to public and academic libraries.
R: *LJ* 106: 1297 (June 15, 1981).

Nuclear Power Stations. **Tereza Khristoforovna Marqulova.** Translated by A. Troitsky. Moscow: Mir Publishers, 1978.

Translated from The Russian. Includes index.

Nuclear Reactors Built, Being Built, or Planned in the United States as of Dec. 31, 1978. **US Energy Research and Development Adminstration.** Springfield, VA: US Department of Commerce, National Technical Information Service, 1978. Updated biannually.

Contains current information about facilities, for domestic use or export, that are capable of sustaining a nuclear chain reaction. Civilian, production, and military reactors are listed, as are reactors for export and assembly facilities. Information is given on location, owner, principal nuclear contractor, type, power rating, and start-up and shutdown dates.

Nuclear Research Index. 5th ed. St. Peter Port, Guernsey, England: Hodgson, 1976. Also later editions.

First edition, 1960; third edition, 1966; fourth edition, 1970, entitled *World Nuclear Directory: An International Reference Book.*
Geographical listing of national and international agencies connected with atomic energy. Includes addresses, telephone numbers, officers, and nuclear-activities information.
R: *Nature* 228: 1355 (Dec. 26, 1970); Jenkins (C75); Wal (p. 308); Win (3EI39).

Power Reactors in Member States. **International Atomic Energy Agency.** Vienna: International Atomic Energy Agency. Distr. New York: Unipub, 1981.

A computer-generated listing of nuclear reactors in International Atomic Energy Agency member states as of January 1981. Contains tables of reactor types, net electrical power, and projections to 1994. Also includes a list of shutdown reactors and reactor-name index.

World Survey of Major Facilities in Controlled Fusion Research. Vienna: International Atomic Energy Agency, 1984.

Earlier edition, 1973; third revised edition, 1977.
International listing of fusion reactors, both governmental and industrial.

ENERGY

Africa–Middle East Petroleum Directory. Tulsa: Pennwell. Annual.

A complete listing of petroleum companies active in Africa and the Middle East. Includes the name of the company, its key people, a personnel listing, and even some history of each company.
R: *ARBA* (1983, p. 654).

Asia–Pacific Petroleum Directory. Tulsa: Pennwell. Annual.

A complete listing of petroleum companies active in the Asia-Pacific countries. Includes the name of each company, its key people, a personnel listing, and even some history on each of the companies.
R: *ARBA* (1983, p. 654).

California Energy Directory: A Guide to Organizations and Information Resources. **Michael Paparian, ed.** Claremont, CA: California Institute of Public Affairs, 1980.

Designed as a companion to *California Environmental Directory*, this volume contains more than 600 organizations concerned with energy problems. Lists government agencies, professional and eductional organizations, and energy suppliers. Includes indexes by subject, organizations, and acronyms and initialisms. Also includes maps and full bibliographic information. For energy specialists.
R: *ARBA* (1981, p. 682).

Canadian Oil Industry Directory. Tulsa: Pennwell. Annual.

Provides a complete listing of the major oil companies in Canada and the various activities that are involved with the oil industry. Does not include a personnel index.
R: *ARBA* (1983, p. 654).

Citizens' Energy Directory. 2d rev. ed. **Jan Simpson and Ken Bossong, comps.** Washington, DC: Citizens' Energy Product, 1980. Also later editions.

First edition, 1978; second edition, 1979.
An up-to-date directory of more than 600 references to groups involved with alternative energy technologies. Entries contain complete bibliographic information and are

organized first by state, then alphabetically by name of group. Includes indexes and appendixes listing resource information. For large public and academic libraries.
R: *BL* 75: 1608 (July 15, 1979); *ARBA* (1980, p. 649).

Coal Information Sources and Data Bases. **Carolyn C. Bloch.** Park Ridge, NJ: Noyes Data Corporation, 1980.

A comprehensive guide to federal, state, and international agencies that deal with coal data in any way. Gives the address and brief description of each source. For those directly involved with the use of coal data and information.
R: *Science and Technology Libraries* 1: 73 (Summer 1981).

EIA Publications Directory: A User's Guide. Washington, DC: Energy Information Administration. Distr. Washington, DC: US Government Printing Office, 1983–. Annual.

An index by subject, report number, and title of the statistical publications of the US Energy Information Administration (EIA). A quick reference to energy statistics.
R: *ARBA* (1984, p. 675).

Energy: A Guide to Organizations and Information Resources in the United States. 2d ed. Claremont, CA: Public Affairs Clearinghouse, 1978.

First edition, 1974.
This directory is divided into 10 subjects, with sections on federal agencies, state agencies, local agencies, major producers and distributors of energy, research centers, information clearinghouses, etc. A handy source of information.
R: *BL* 75: 571 (Nov. 1978); *WLB* 53: 92 (Sept. 1978); *ARBA* (1975, p. 700; 1976, p. 686).

Energy Atlas: A Who's Who Resource to Information. Washington, DC: Fraser/Ruder and Finn, 1976.

A directory to 150 state and federal agencies in the United States, as well as 150 nongovernnmental groups. Arranged by section, information provided includes name, address, personnel, and agency jurisdiction.
R: *BL* 73: 1666 (July 1977); *ARBA* (1978, p. 687).

The Energy Directory Updates. New York: Environment Information Center, 1978–. Annual.

A comprehensive listing of governmental agencies, private industries, and public interest groups concerned with energy-related problems. Includes brief descriptions of many organizations.
R: *ARBA* (1975, p. 700); Sheehy (EJ5).

Energy for Industry and Commerce: Market Review and Directory of Energy-Saving Equipment. Cambridge: Cambridge Information and Research Services. Distr. New York: Unipub. Annual.

A 2-part directory on energy use and equipment. Part 1 deals with world trends and prospects, supplies, prices, etc. Part 2 consists of a directory of energy-saving equipment, which is indexed by type of equipment, title of equipment, and classification by company. Contains 3 appendixes: a glossary of advisory agencies; a questionnaire;

and a list of energy-conversion factors. Also includes maps and graphs. Intended mainly for British use in libraries on a technical level.
R: *ARBA* (1981, p. 682).

Energy Information Locator: A Select Guide to Information Centers, Systems, Data Bases, Abstracting Services, Directories, Newsletters, Binder Services, and Journals. **Energy Directory Update Service.** New York: Environment Information Center, 1976–. Annual.

A 192-page reference guide that describes about 1,000 publications and systems—newsletters, directories, databases, libraries, binder services, journals, etc.—and is indexed by subject, SIC code, title, and geography.
R: *SL* 67: 62 (Jan. 1976); *ARBA* (1977, p. 685).

European Petroleum Directory. Tulsa: Pennwell, 1981–. Annually.

A complete listing of petroleum companies active in the European countries. Includes the name of the company, its key people, a personnel listing, and even some history of each company.
R: *ARBA* (1983, p. 654).

Federal Energy Information Sources and Data Bases. **Carolyn C. Bloch.** Park Ridge, NJ: Noyes Data, 1979–.

A directory containing addresses and descriptions of energy information sources, such as federal agencies, departments, and offices dealing with some aspect of energy.
R: *Journal of Metals* 32: 67 (Apr. 1980).

Geothermal Resources and Technology in the United States. **National Research Council.** Washington, DC: US National Academy of Sciences, 1979–.

Geothermal World Directory. **Alan A. Tratner, ed.** Reseda, CA: Geothermal World Publications. Annual.

Lists all energy-related state agencies, US businesses involved in geothermal energy, individuals involved in research, and ongoing projects worldwide. Also contains detailed descriptions of ERDA Geothermal Program. Recommended for specialized collections.
R: *Geology* 6: 492 (Sept. 1978); *TBRI* 44: 392 (Dec. 1978); *ARBA* (1979, p. 696).

International Bio-Energy Directory. 5th ed. **Paul F. Bente, Jr., ed.** Washington, DC: Bio-Energy Council, 1984.

An international collection of over 650 reports dealing with various aspects of recent research in 60 countries and 47 states in the United States. Reports are categorized alphabetically by geographic location under 4 subject topics—biomass resources, microbial conversion, thermal conversion, and general appraisals. Contains information on title, location, objective, personnel, etc. Indexed by geographic location and organizations.
R: *ARBA* (1981, p. 681; 1985, p. 505).

International Who's Who in Energy and Nuclear Sciences. **The Longman Editorial Team, comps.** Detroit: Gale Research, 1983.

Biographical profiles of more than 3,800 individuals involved in the generation, storage, and efficient use of energy in over 70 countries. Includes chemists, physicists, engineers, economists, lawyers, and architects. International in scope, However, coverage is uneven from country to country.
R: *ARBA* (1984, p. 678).

Inventory of Power Plants in the United States. **Lucinda R. Gilliam.** Washington, DC: US Department of Energy. Distr. Washington, DC: US Government Printing Office, 1980. Annual.

An inventory of power plants in the United States. Arranged in 3 sections: existing and projected electrical generation units, jointly owned units, and projected construction units. Geographically, then alphabetically, arranged. Provides information on company name, unit identification, generator manufacturer, fuels, etc. Includes extensive maps, tables, and chart summaries. A handy ready-reference for state planners and electrical engineers.
R: *ARBA* (1979, p. 698; 1981, p. 754).

Land Drilling and Oilwell Servicing Contractors Directory. Tulsa: Pennwell. Annual.

Alphabetically lists over 800 land drilling contractors and 1,200 oil well servicing firms with name, address, telephone number, personnel, activities, etc. Cross-referenced to include all companies in the United States and beyond.
R: *ARBA* (1983, p. 655).

Latin American Petroleum Directory. Tulsa: Pennwell. Annual.

A complete listing of petroleum companies active in the Latin American countries. Includes the name of the company, its key people, a personnel listing, and even some history of each company.
R: *ARBA* (1983, p. 654).

Offshore Contractors and Equipment Directory. Tulsa: Pennwell. Annual.

Alphabetically lists 2,000 companies and 15,000 personnel of contractors and equipment in the offshore petroleum industry. Where possible, provides photographs of mobile rigs.
R: *ARBA* (1983, p. 655).

Oilfields of the World. 2d ed. **E. N. Titratsoo.** Houston: Gulf Publishing, 1976. And later edition.

First edition, 1973.
Discusses geology and geography of the world's major oil fields. Includes several maps and diagrams depicting the occurrence of hydrocarbons worldwide. Also includes crude oil production and reserve statistics. Intended for people in and out of the industry.
R: *ARBA* (1975, p. 711).

Rocky Mountain Energy Directory. Denver: Golden Bell Press.

Names, locations, job functions, and personnel listings of various companies in the fields of oil and gas, oil shale, coal, uranium, solar, base metal mining, and other

organizations. Careless editing and lack of organization make this directory of little use for those who do not already have much knowledge in the field.
R: *ARBA* (1983, p. 656).

Selected Information Resources on Solar Energy. **John A. Feulner, comp.** Washington, DC: US Library of Congress, 1980.

Consists of a list of 55 organizations that give information on solar energy. Contains names, addresses, areas of interest, services, etc. For science libraries.
R: *LCIB* 39: 391 (Oct. 3, 1980).

The Solar Energy Directory. **Sandra Oddo, Martin McPhillips, and Richard Gottlieb, eds.** New York: Wilson, 1983.

A comprehensive directory of solar energy organizations. Each chapter has an introduction, followed by the names, addresses, and descriptions of these organizations. An original work, good as a reference tool for those interested in this contemporary issue.
R: *Civil Engineering* 52: 32 (Nov. 1982); *American Libraries* 13: 688 (Dec. 1982); *ARBA* (1985, p. 511).

Synergy: A Directory of Energy Alternatives. Nos. 1–. New York: Synergy, 1974–.

Provides a variety of information concerning alternative energy sources; lists citations by energy source and then type of material (books, articles, government documents, etc.).
R: *ARBA* (1979, p. 690; 1983, p. 656).

Synfuels Project Directory. Arlington, VA: Pasha Publications, 1981–.
R: *Ulrich's Quarterly* 7: 171 (April/June 1983).

USA Oil Industry Directory. Tulsa: Pennwell. Annual.

Twentieth edition, 1981; twenty-third edition, 1984.
Provides a complete listing of the major oil companies in the United States. Includes the various activities of the company, names of key people, personnel listings, and historical data for each company.
R: *ARBA* (1983, p. 654; 1985, p. 510).

The Whole World Oil Directory. **William J. Feenberg.** Deerfield, IL: Whole World, 1985. Annual.

A listing of over 20,000 petroleum-related companies, including 1,000 new entries not found in the previous edition. Names of the companies and their key personnel are listed, as are addresses. An alphabetical index precedes each section. Good for sales personnel.
R: *Petroleum Engineer International* 55: 158 (May 1983).

World Coal Industry Report and Directory. New York: Unipub, 1979.

Thorough worldwide coverage of the coal mining industry. Lists 1,000 mines of the major coal-producing countries. Provides facts on production statistics, personnel, management techniques, equipment, exports and imports, and production. Also contains information on recent developments and maps of key mining areas. A handy directory for anyone in the coal mining industry.

World Directory of Energy Information. 3 vols. **Cambridge Information and Research Services, comp.** New York: Facts on File, 1981–1984.

Volume 1, *Western Europe*, 1981; volume 2, *Middle East, Africa and Asia/Pacific*, 1982; volume 3, *The Americas, including the Caribbean.* 1984.

A comprehensive directory set that provides, in one place, basic information on energy supply within the countries under consideration. A researcher's gold mine.

R: *BL* 77: 985 (Mar. 1, 1981); *Choice* 21: 1454 (June 1984); *ARBA* (1983, p. 659; 1985, p. 508).

World Pipelines and International Directory of Pipeline Organizations and Associations. **J. N. H. Tiratsoo, ed.** Houston: Gulf, 1983.

Over 700 organizations in 70 countries are covered. Useful to large and specialized libraries.

R: *ARBA* (1985, p. 555).

Worldwide Directory: Offshore Contractors and Equipment. Tulsa: Petroleum. Annual.

Eighth edition, 1976.

Worldwide Oilfield Service, Supply and Manufacturers Directory, 1984. 2d ed. Tulsa: Pennwell, 1984. Annual.

A complete listing of companies involved in providing oil-field services in the wholesale and/or retail sale of oil-field products, and in the design, manufacture, and construction of equipment used in the oil field.

R: *ARBA* (1985, p. 510).

Worldwide Petrochemical Directory. Tulsa: Pennwell. Annual.

Three major sections: worldwide petrochemical survey, containing various products and production per day arranged alphabetically by country, then by company; worldwide construction of plants, arranged alphabetically by country, giving the size, cost, and date of completion of each; and a personnel listing of petrochemical plants, with relevant data for each country.

R: *ARBA* (1983, p. 729).

Worldwide Petrochemical Industry. Tulsa: Pennwell. Annual.

Covers companies; capacities; processes; personnel; and plants operating, under construction, and planned.

R: *Wal* (p. 241).

Worldwide Pipeline and Contractors Directory. Tulsa: Pennwell. Annual.

Alphabetically lists over 300 pipeline operators and 1,000 engineering and construction firms with name, address, telephone number, personnel listing, activities, etc., of each firm. Includes a coal slurry pipeline section.

R: *ARBA* (1983, p. 655).

Worldwide Refining and Gas Processing Directory. Tulsa: Pennwell. Annual.

Includes statistical surveys with tables for refining production, construction, sulfur production, and lube and wax capacities arranged by country. Also includes a person-

nel listing with address, telephone, telex, major personnel, and, sometimes, a brief description of the company for each country.
R: *ARBA* (1983, p. 729).

Worldwide Synthetic Fuels and Alternate Energy Directory, 1984. 3d ed. **William R. Leek, Jr., ed.** Tulsa: Pennwell, 1983. Annual.

An updated directory of organizations involved in the development and application of synthetic fuels and alternative energy sources. Alphabetically listed by industry, government, and education with specific information on each entry. Some clarifications and specifications would be helpful.
R: *ARBA* (1984, p. 675).

ENVIRONMENTAL SCIENCES

Acid Rain Resources Directory, 1984. St. Paul, MN: Acid Rain Foundation, 1984. Annual.

A directory of public, private, and education resources currently available on acid rain. Because of the timeliness of the topic, recommended for all libraries.
R: *ARBA* (1985, p. 512).

California Environmental Directory: A Guide to Organizations and Resources. 3d ed. **Thaddeus C. Trzyna, ed.** Claremont, CA: California Institute of Public Affairs, 1980.

Contains coverage of more than 1,200 organizations concerned with California's environment. Includes maps and indexes of acronyms and initialisms, government agencies, subject, and private organizations. Subject index is cross-referenced. For environmentalists.
R: *ARBA* (1981, p. 686).

California Water Resources Directory: A Guide to Organizations and Information Resources. **Lizanna Fleming, Elizabeth G. Reifsnider, and Thaddeus C. Trzyna, eds.** Claremont, CA: California Institute of Public Affairs, 1984.

A directory to the governmental and private organizations in California concerned with water policy, development, supply, conservation, pollution, and the other aspects of water resources. Indexed by organization, program, and major project.
R: *ARBA* (1985, p. 512).

CRC Handbook of Managment of Radiation Protection Programs. **Kenneth L. Miller and William A. Weidner, eds.** Boca Raton, FL: CRC, 1986.

A comprehensive volume that details the organization and management of radiation safety programs, including both preventive and emergency response measures. Included are guidelines and checklists for managing radioactive waste-processing programs, dealing with litigation, and responding to public or news media concerns. State, federal, and international requirements for transportation of radioactive materials are also listed at the end.

Conservation Directory: A List of Organizations, Agencies, and Officials Concerned with Natural Resource Use and Management. **Jeannette Bryant, ed.** Washington, DC: National Wildlife Federation, 1982.

A listing of private and government groups concerned with environmental problems in relation to conservation studies. Includes data on agencies, national parks, committees, colleges and universities, and game refuges. Also contains a list of periodicals, audiovisual materials, and directories. Primarily for large research libraries.
R: *ARBA* (1980, p. 652).

Directory of Environmental Groups in New England. Boston: US Environmental Protection Agency, Region 1. Annual.

Directory of Government Agencies Safeguarding Consumer and Environment. **Daniel Sprecher, ed.** Alexandria, VA: Serina. Annual.

Directory of Institutions and Individuals Active in Environmentally Sound and Appropriate Technologies. Elmsford, NY: Pergamon, 1979.

A listing of some 2,000 worldwide institutions that dispense information on environmentally sound technology. Entries, organized alphabetically by country, include name, address, contact, and present activities. Useful reference for those concerned with environmentally sound technology.
R: *ARBA* (1981, p. 686).

Directory for the Environment: Organizations in Britain and Ireland, 1984–5. **Michael Barker, comp. and ed.** Boston: Routledge & Kegan Paul, 1984. Annual.

A directory of government agencies, citizens groups, and professional societies concerned with conservation, land and water use, energy, and pollution, and organizations of the "alternative movement." Alphabetical list of organizations with cross-reference list of subject headings. A subject index is available. Useful for environment collections.
R: *ARBA* (1985, p. 512).

Ecotechnics: An International Ecology Directory. New York: Fairchild Books, 1977.

A directory of firms that sell environmental control equipment. New edition includes Eastern and Western European, North American, and Japanese countries. Divided into 6 main subject areas such as air, solid waste, noise-vibration, and other areas of environmental control. Includes facts on address, telex, and names. For technical and engineering libraries.
R: *BL* 74: 1517 (May 15, 1978); *RSR* 5: 34 (Oct./Dec. 1977).

Environmental Impact Statement Directory: The National Network of EIS-Related Agencies and Organizations. **Marc Landy, ed.** New York: Plenum Press, 1981.

A comprehensive directory of organizations and agencies involved in the environmental impact statement process. Suitable for environmental designers and planners and for environmental/agricultural lawyers.

Hazardous Waste Management Directory, 1982–1983. 3d ed. Philadelphia: Pennsylvania Environmental Research Foundation, 1982. Also update edition.

An alphabetical listing, by state and then by company, of those firms within the United States involved in the design of special hazardous waste management facilities or transportation of wastes and proper disposal of them.
R: *ARBA* (1984, p. 682).

Nationwide Survey of Resource Recovery Activities. **Richard E. Hopper.** US Environmental Protection Agency. Washington, DC: US Government Printing Office, 1975. Annual.

Selected Information Resources on Solid Wastes. **John A. Feulner, comp.** Washington, DC: US Library of Congress, 1980.

A listing of 55 organizations that provides information on solid wastes. Includes facts on names, addresses, publications, holdings, etc. For industrial and academic libraries.
R: *LCIB* 39: 391 (Oct. 3, 1980).

The United States and the Global Environment: A Guide to American Organizations Concerned with International Environmental Issues. **Thaddeus C. Trzyna, ed.** Claremont, CA: California Institute of Public Affairs, 1983.

Contains information on more than 100 American organizations dealing with international environmental affairs. Indexed by name of organization, program, major project, and acronyms and initialisms.
R: *ARBA* (1985, p. 513).

World Directory of Environmental Organizations. 2d ed. **Thaddeus C. Trzyna and Eugene V. Coan, eds.** San Francisco: Sequoia Institute, 1976. Also updates.

In 4 parts: environmental concerns; intergovernmental organizations; nongovernmental organizations; and international efforts of environmental concerns. Invaluable for students, community activists, and reference librarians.
R: *ARBA* (1978, p. 688).

World Directory of Environmental Research Centers. 2d ed. **William K. Wilson, Morgan D. Dowd, and Phyllis Sholtys, eds.** Scottsdale, AZ: Oryx Press. Distr. New York: Bowker, 1974. Also updates.

Descriptions for approximately 5,000 worldwide organizations engaged in environmental research. Parts 1 and 2 arrange entries under subjects and geographical locations respectively. Arrangement and alphabetical difficulties hinder ready-reference use of this tool.
R: *BL* 70: 197 (Oct. 1, 1974); *Choice* 11: 1288 (Nov. 1974); *LJ* 99: 1796 (July 1974); *WLB* 49: 186 (Oct. 1974); *ARBA* (1975, p. 703).

World Directory of National Parks and Other Protected Areas. 2 vols. **International Union for Conservation of Nature and Natural Resources.** New York: Unipub, 1979.

Compilation of national parks and protected areas. Entries, arranged by country, include name, biotic province, location, establishment date, protections, physical and biological features, tourism, etc. Helpful reference for ecologists and park conservationists.

World Environmental Directory. 4th ed. **Beverly E. Gough, ed.** Cambridge, MA: Ballinger, 1980. Also update edition.
Second edtion, 1975; third edition, 1977.
A comprehensive listing of over 40,000 entries from all environmental fields. Entries are taken from government agencies, engineers, universities, organizations, publications, etc. Provides information on services, products, activities, and functions. An invaluable reference aid for anyone in the environmental sciences field.
R: *BL* 76: 1453 (June 15, 1980); 77: 162 (Oct. 1, 1980); *ARBA* (1976, p. 702); Sheehy (EJ21).

Pollution

APCA Directory and Resource Book. Pittsburgh: Air Pollution Control Association. Annual.
Basic information on the organization; product guide, publication list, etc.

Directory of Governmental Air Pollution Agencies. Pittsburgh: Air Pollution Control Association, 1974. Also, later eds.

Water Pollution Control: Directory Issue. Washington, DC: Water Pollution Control Federation. Annual.
Annual directory issue.

TRANSPORTATION

The Illustrated Encyclopedia of North American Locomotives: A Historical Directory of America's Greatest Locomotives from 1830 to the Present Day. **Brian Hollingsworth.** New York: Crescent Books. Distr. New York: Crown, 1984.
A historical directory of 150 notable steam, diesel, and electric locomotives, arranged chronologically from 1829 to 1983. Illustrated.
R: *BL* 81: 10 (Sept. 1, 1984); *ARBA* (1985, p. 619).

The Illustrated Encyclopedia of the World's Modern Locomotives: A Technical Directory of Major International Diesel, Electric, and Gas-Turbine Locomotives from 1879 to the Present Day. **Brian Hollingsworth and Arthur Cook.** New York: Crescent Books/Crown, 1983.
A history directory of locomotives from 1879 to 1982 in the world. Illustrated.
R: *BL* 80: 775 (Feb. 1, 1984); *ARBA* (1985, p. 619).

Innovation in Public Transportation: A Directory of Research, Development and Demonstration Projects. **Urban Mass Transportation Administration.** Washington, DC: US Department of Transportation. Annual.

Jane's World Railways. **Geoffrey Freeman Allen, ed.** Boston: Jane's Publishing, Annual.

Twenty-fifth edition, 1983.
A comprehensive directory of railways, with information arranged in sections, such as manufacturers, consultancy services, railway systems by country, and rapid transit and underground railways. Indexed.
R: *ARBA* (1985, p. 617).

New Car Dealers: Directory. Omaha: American Business Directories. Annual.
A yellow-page-type directory of about 27,000 dealership names from about 5,000 US telephone directories. For each dealer, information is given on address, Zip code, telephone number, and type of car sold.
R: *ARBA* (1985, p. 621).

Railroad Names: A Directory of Common-Carrier Railroads Operating in the United States, 1826–1982. **William D. Edson.** Potomac, MD: Edson, 1984.
Contains over 6,000 companies and common-carrier steam railroads in the United States that performed interstate service. Alphabetically arranged by corporate name. Companies operating in Alaska, the Canal Zone, Hawaii, and Puerto Rico are excluded.
R: *ARBA* (1985, p. 618).

YEARBOOKS

Britannica Yearbook of Science and the Future. Chicago: Encyclopedia Britannica, 1968–. Annual.
A popular approach to the most topical developments in science during the preceding year. Generally 16–20 state-of-the-art papers for general readers.
R: *BL* 66: 860 (Mar. 1970); *LJ* (May 1969); *ARBA* (1974, p. 546; 1971, p. 475; 1970, p. 102); Jenkins (A51); Wal (p. 16); Win (3EA13).

Facts on File Yearbook, 1941–. New York: Facts on File. Annual.
R: *BL* 77: 985 (Mar. 1, 1981).

McGraw-Hill Yearbook of Science and Technology. **Staff of McGraw-Hill Encyclopedia of Science and Technology, comp.** New York: McGraw-Hill, 1972–. Annual.
An A–Z listing of authoritative articles on the advances and developments in science and technology during the preceding year. Includes maps, charts, illustrations, etc.
R: *American Scientist* 59: 479 (July/Aug. 1971); 61: 476 (July/Aug. 1973); *BL* 71: 1020 (June 1975); *LJ* 92: 2146 (1967); *RQ* 11: 177 (Winter 1971); *ARBA* (1976, p. 635; 1973, p. 535; 1971, p. 476); Jenkins (A50); Wal (p. 17).

Nature/Science Annual. Vols. 1–. **Editors of Time-Life Books.** New York: Time-Life Books, 1970–. Annual.
An annual summary of recent trends and events in scientific discoveries. Contains excellent photographs and illustrations, which are accompanied by informative text. Also contains an excellent index. Recommended for school and public libraries.
R: *BL* 74: 1280 (Apr. 1, 1978). *ARBA* (1976, p. 637; 1973, p. 553).

Science News Yearbook. **Prepared by Science Service.** New York: Scribner's, 1969–. Annual.

A popular-language account of recent developments in science and technology. Based primarily on material published in *Science News*.
R: *Choice* 6: 1374 (Dec. 1969); *LJ* (Sept. 1969); *Saturday Review* (Dec. 6, 1969); *ARBA* (1970, p. 103).

Science Year; The World Book Science Annual. Chicago: Field Enterprises, 1965–. Annual.

An issue may be devoted to one significant scientific topic, through a series of articles written by specialists. Shorter articles summarize advances covering a wide range of topics.
R: *Science* 152: 917 (1966); *BL* 62: 725–727 (Apr. 1, 1966); *LJ* 92: 1577 (Apr. 15, 1967); *ARBA* (1971, p. 476, 1980, p. 604); Jenkins (A53); Wal (p. 17); Win (1EA12).

Technician Education Yearbook. Ann Arbor: Prakken Publications. 1963–. Biennial.

Tenth edition, 1982.
A praiseworthy reference on technician education at 4-year and 2-year schools. Lists over 2,150 institutions under 142 technical fields. Also lists state and federal officials and discusses new programs, case studies, and professional organizations. For community college, high school, and public libraries for use by counselors and administrators.
R: *Choice* 17: 206 (Apr. 1980).

Yearbook of International Organizations. **Union of International Associations, ed.** Detroit: Gale Research. Annual.

Seventeenth edition, 1978.
Includes over 8,000 descriptions of various governmental and nongovernmental organizations, associations, committees, etc. Contains 13 indexes. Useful for reference libraries.

Yearbook of Science and the Future. 1969–. Chicago: Encyclopedia Britannica. Annual.

A nontechnical yearbook that presents the past year's developments in science. Deals with all subjects, ranging from evolution of the stars to ship design. Well written and arranged with a detailed index and excellent illustrations.
R: *ARBA* (1978, p. 625).

SUBJECT SCIENCE

Annals of the International Geophysical Year. 48 vols. Oxford: Pergamon Press, 1973–1975.

Botanical Society of America Yearbook. New Brunswick, NJ: Rutgers University, Department of Botany. Annual.
R: *RSR* 2: 115 (Oct.–Dec. 1974); 3: 47 (Apr.–June 1975).

GSA Yearbook. Boulder: Geological Society of America. Annual.

International Zoo Yearbook. Vols. 1–. London: Zoological Society, 1960–. Annual.

A well-balanced reference providing information on animal populations in captivity, new developments, and breeding activities; and societies, institutions, and federations.
R: *New Scientist and Science Journal* (July 29, 1971); *Science* 162: 659 (1968); 168: 725 (May 8, 1970); *SJ* 6: 94 (July 1970); *ARBA* (1970, p. 126); Jenkins (G216); Sheehy (EC38); Wal (pp. 163, 225).

Synthetic Methods of Organic Chemistry. **W. Theilheimer, ed.** Basel: Karger, 1946–. Annual.

Volume 38, *Yearbook 1984*.
Contains papers that examine new techniques for the synthesis of organic compounds. A useful reference for organic chemists interested in the state of the art of synthetic methods.

Yearbook of Agriculture, **US Department of Agriculture.** Washington, DC: US Government Printing Office, 1895–. Annual.

Yearbook of Astronomy. **Patrick Moore, ed.** New York: Norton, 1962–.
Aimed at the amateur, this publication is both a calendar of celestial events and a series of high-interest articles on matters of general importance to astronomers.
R: *Science* 140: 288 (1963); *RSR* 6: 21 (July/Sept. 1978); *ARBA* (1971, p. 484; 1972, p. 554; 1973, p. 540; 1977, p. 638); Jenkins (E11); Wal (p. 80).

Yearbook of Fishery Statistics. **Food and Agriculture Organization of the United Nations, Fisheries Division.** New York: Unipub. Annual.

Volume 41, *Fishery Commodities,* 1975.
R: *Choice* 14: 1669 (Feb. 1978).

Yearbook of Forest Products. 32d ed. **Food and Agriculture Organization of the United Nations.** New York: Unipub, 1947–. Annual.

A classic reference to the trade and production of forest products throughout the world. Tables are organized by volume of production; volume and value of trade; and unit value in trade of selected commodities. Includes facts about pulpwood, veneer sheets, plywood, paper, wood pulp, etc. An invaluable reference for agriculture and forestry specialists and forest collections.

ENGINEERING

Engineers' Joint Council. Annual Report. New York: Engineers' Joint Council, 1962–.

Kempe's Engineer's Yearbook. London: Morgan-Grampian, 1894–. Annual.

British compilation of data, formulas, tables, and basic principles. May be more accurately described as a handbook.

Aeronautical and Astronautical Engineering

Aeronautical and Aerospace Engineering

Aerospace Yearbook. Washington, DC: Aerospace Industries Association of America, 1919–.

Formerly *The Aircraft Yearbook.*
Records annually the highlights and progress in the aerospace industry. Included data on governmental research and development programs.
R: Jenkins (K130); Wal (p. 392); Win (EI45).

Aviation Annual. Garden City, NY: Doubleday, 1943–.

Jane's All the World's Aircraft. **John W. R. Taylor, ed.** London: Jane's Publishing, 1909–. Annual.

1977–78 edition, New York: Wiley.
This useful source book of technical data on aerospace products covers home-built aircraft; sailplanes; hang gliders; air-launched missiles; satellites and spacecraft launched during 1978; and more. Includes some 1,500 photographs and silhouettes.
R: *Interavia* 25: 1397 (Nov. 1970); Wal (p. 218).

Jane's Aviation Annual. **Michael J. H. Taylor, ed.** New York: Jane's Publishing. Distr. New York: Franklin Watts, 1981–. Annual.

An update of important new developments in both civil and military aviation. A valuable resource, though it could have been indexed for efficiency purposes.
R: *ARBA* (1983, p. 733).

Jane's Avionics. **Michael Wilson, ed.** London: Jane's Publishing; Distr. Boston: Science Books International, 1982–. Annual.

Arranges information from the field of aviation electronics, first alphabetically by country under each topic, and then alphabetically by manufacturer within that country. Intended to be an annual or biannual publication.
R: *ARBA* (1983, p. 717).

Smithsonian Annals of Flight. Washington, DC: Smithsonian Institution, National Air Museum, 1964–.

Civil Engineering

American Public Works Association Yearbook. Chicago: APA, 1884–. Annual.

Concrete Industries Yearbook. Chicago: Pit and Quarry, 1938–. Annual.

The Concrete Yearbook. London: Concrete Publications, 1924–. Annual.
R: Wal (p. 497).

Engineering Foundation. Annual Report. New York: Engineering Foundation, 1929/1930–. Annual.

A report on activities and research.

The Institution of Civil Engineers Yearbook. London: Nuttall. Annual.

International Hydrographic Organization Yearbook. Monte Carlo: IHO. Annual.

Provides information on the International Hydrographic Bureau and national hydrographic offices (85 countries).
R: *IBID* 4: 123 (June 1976).

ELECTRICAL AND ELECTRONICS ENGINEERING

Computer Industry Annual. 1st ed.–. **International Computaprint Corporation, comp.** Concord, MA: Computer Design, 1969/70–.

Essays, comparative charts, and directory sections comprise this annual on all aspects of the computer industry.

Computer User's Yearbook. New York: International Publications, 1969–. Annual.

Computer Yearbook and Directory. Detroit: American Data Processing, 1966–. Annual.

Supersedes the *Data Processing Yearbook.* State-of-the-art reports, book reviews, and comprehensive directory information.
R: Jenkins (B70); Win (1EI30).

Yearbook of Consumer Electronics. **Forest H. Belt.** Indianapolis: Sams, 1973–. Annual.

While intended for consumers, much of the material is of such a technical nature as to be suitable for technicians and engineers.
R: *ARBA* (1975, p. 797).

MECHANICAL ENGINEERING

Foundry Year Book. **C. McCombe, ed.** Redhill, Surrey, England: Fuel and Metallurgical Journals, 1972–. Annual.

Lists over 1,200 foundries in the United Kingdom. Describes production facilities, machinery, etc.

The International Robotics Yearbook. Cambridge, MA: Ballinger, 1983.

The first single reference source to provide summary information on robotics. Divided into 3 parts. Part 1 is a worldwide directory of manufacturers, suppliers and services of robots and robot systems; part 2 is a directory of world research and development activities; and part 3 is a directory of grants and funding sources for robotics research and development and a directory of robot associations. Subject index to research activities is available. Excellent tool for special and research libraries in the field.
R: *Choice* 21: 1448 (June 1984); *ARBA* (1985, p. 599).

Jane's Armour and Artillery. **Christopher F. Foss, ed.** London: Jane's Publishing, 1979–. Annual.

A guide covering armor and artillery such as tanks, reconnaissance vehicles, armored personnel carriers, multiple rocket launches, ammunition, and more. Includes technical data throughout.
R: Wal (p. 208).

Mechanical World Yearbook. Manchester, England: Emott, 1887–. Annual.
Ninetieth edition, 1979.
Includes data, tables, advertisements, conversion tables, British Standard qualifications, etc.

Refrigeration and Air Conditioning Year Book. Miami: Croydon. Annual.
A reference to refrigeration and air-conditioning companies, manufacturers, suppliers, and more.
R: Wal (p. 204).

Welding Research Council Yearbook. New York: Welding Research Council. Annual.

MINING AND METALLURGY

Metal Finishing Guidebook—Directory. Vols. 1–. Westwood, NJ: Metals and Plastics Publications, 1903–. Annual.
Volume 82, 1984.

Minerals Yearbook. US Bureau of Mines. 4 vols. Washington, DC: US Government Printing Office, 1881–.
Annual report of the US mineral industry.
Volumes 1, 2, *Metals, Minerals and Fuel;* volume 3, *Area Reports: Domestic;* volume 4, *Area Reports: International.*
R: *ARBA* (1970, p. 129); Jenkins (F134; Win (EE120).

Non-Ferrous Metal Data. New York: American Bureau of Metal Statistics. Annual.
Since 1920 as the *Year Book of the American Bureau of Metal Statistics.*
Contains extensive statistical data on copper, lead, and zinc, as well as 15 other non-ferrous metals. Information is easy to read in tables and most useful to any business collection in this field.
R: *ARBA* (1984, p. 686).

Skinner's Mining International Yearbook. London: Financial Times. Annual.
Includes directories of companies and suppliers, information on professional services, and tables related to world production.

ENERGY AND ENVIRONMENTAL SCIENCES

California Environment Yearbook and Directory. Claremont, CA: Center for California Public Affairs, 1972–. Annual.

Reviews major events in California environmental affairs. Includes directory and bibliography.
R: *ARBA* (1974, p. 601).

Clean Air Yearbook. Brighton, England: National Society for Clean Air. Annual.
R: *Electrical Review* (Sept. 10, 1971); Wal (p. 369).

Environmental Quality. **US Council on Environmental Quality.** Washington, DC: US Government Printing Office. Annual.
Seventh annual report, 1976.

European Offshore Oil and Gas Yearbook. New York: International Publications, 1975–.
Contains numerous articles on offshore energy supply, drilling contractors, engineers etc.
R: *ARBA* (1977, p. 687).

International Petroleum Yearbook. Washington, DC: US Government Printing Office. Annual.
Production figures provided by the Bureau of Mines.

Offshore Oil and Gas Yearbook: UK and Continental Europe. **Martin Beudell, cons. ed.** Distr. New York: Nichols. Annual.
Fourth edition, 1982.
Combines surveys of the principal areas of oil and gas exploration and their production with a directory of supplier companies and their activities. A new addition is introductory surveys of the activity in each country discussed. An important tool for any library dealing with the petroleum industry.
R: *ARBA* (1983, p. 659).

Skinner's Oil and Gas International Yearbook. **W. Nightingale, ed.** London: Financial Times. Annual.
Production tables, lists of professional services, and an alphabetical directory of world oil and gas companies.

UK Offshore Oil and Gas Yearbook. 2d ed. London: Page, 1975.

Unesco Annual Summary of Information on Natural Disasters. Paris: Unesco. Distr. New York: Unipub, 1966–. Annual.
Arranged by type of disaster. Maps and textual material are designed for use by laypeople.
R: *Recorder* 252: 11 (May 1, 1970); *RQ* 11: 283 (Spring 1971); *ARBA* (1972, p. 595); Wal (p. 152).

Transportation

Jane's Freight Containers. **Partick Finlay, ed.** New York: Watts, 1968–. Annual.

Contains details on railway shipping facilities and manufacturers.
R: *Engineering* 209: 510 (May 1970); *ARBA* (1978, p. 777); Wal (pp. 206, 337).

Jane's Fighting Ships. **John E. Moore, ed.** London: Jane's Publishing, 1897–. Annual.

This guide to the navies of the world covers technical details of all calsses of fighting ships. Includes a ship-reference section of 142 countries, tables of naval strengths and equipment, an index of some 3,500 named ships, and more.
R: Wal (p. 209).

Jane's Military Communications. **R. J. Raggett, ed.** London: Jane's Publishing, 1979–. Annual.

A directory of acronyms and code names, equipment, and systems, with technical data throughout.
R: Wal (p. 208).

Jane's Surface Skimmers: Hovercraft and Hydrofoils. **Roy McLeavy, ed.** London: Jane's Publishing, 1967–. Annual.

Twelfth edition, 1979.
This new edition covers air cushion vehicles (ACV) manufacturers and design groups; air cushion landing systems; hydrofoils; hover plants and propulsion systems; ACV clubs and associations; ACV and hydrofoil consultants; and more.
R: Wal (p. 216).

Jane's Urban Transport Systems. **John Levett, ed.** London: Jane's Publishing. Distr. Boston: Science Books International, 1982–. Annual.

Includes information on many of the urban transport systems in the world, and is currently the only source of its kind available.
R: *ARBA* (1983, p. 735).

Jane's World Railways. **Paul J. Goldsack, ed.** London: Jane's Publishing, 1950–. Annual.

A comprehensive and detailed reference book of world railways.
R: Wal (p. 210).

Marine Engineering/Log; Annual Maritime Review and Yearbook Issue. Bristol, CT: Boardman, 1941–. Annual.

Ocean Yearbook. **Elisabeth Mann Borgese and Norton Ginsburg, eds.** Chicago: University of Chicago Press, 1978–. Every 18 months.

A series of articles on issues, transportation, marine technology, etc. There is also a section on documents and proceedings. Contains a few minor errors, but basically is a fine reference tool.
R: *Nature* 281: 162 (Sept. 13, 1979).

Railway Directory and Yearbook. **P. M. Kalla-Bishop, ed.** New York: International Publications Service. Annual.

Eighty-fifth edition, 1980.
R: Wal (pp. 210, 358).

World Motor Vehicle Data. Detroit: Motor Vehicle Manufacturers Association. Annual.

A tabular presentation of production, imports, exports, sales, and registrations of some 47 countries arranged by geographic area.
R: *ARBA* (1983, p. 740).

BIOGRAPHICAL SOURCES

REFERENCE TOOLS

American Men & Women of Science, Editions 1–14. Cumulative Index. **Jaques Cattell Press, comp.** New York: Bowker, 1983.

A one-volume index that lists 297,000 scientists alphabetically by name and editions in which it appears.

Biographical Books, 1876–1949. New York: Bowker, 1983.

A useful reference tool that includes more than 30,000 titles—biography, autobiography, collective biography, journal, diary, letter collection, biographical dictionary and directory published or distributed in the United States—classified under 12,500 personal-name headings and 3,000 LC subject headings. Indexed by vocation, author, and title.

Biographical Books, 1950–1980. New York: Bowker, 1980.

A tool that covers 45,000 biographical works published or distributed in the United States during 1950–1980. Indexed by author, title, vocation, and collective biography.

GENERAL SCIENCE

American Men and Women of Science. 15th ed. 7 vols. **Jaques Cattel Press, ed.** New York: Bowker, 1982. Also supps.

Twelfth edition, 1971–1973; thirteenth edition, 1978; fourteenth edition, 1979.
With this edition, all volumes have appeared simultaneously. The most comprehensive, up-to-date source of its kind. Alphabetically arranged profiles of some 130,000 US and Canadian scientists engaged in activities in 1,000 areas of physical, biological, and selected social sciences. The 6 biographical volumes provide pertinent information on personnel from the fields of economics, veterinary sciences, medical sciences, etc. Includes a geographic and discipline index. In keeping with the pattern of earlier editions, supplementary material may be anticipated.
R: *Nature* 301: 354 (Jan. 1983); *BMLA* 66: 263 (Apr. 1978); *Choice* 11: 1730 (Feb. 1975); *CRL* 28: 68 (1967); *LJ* 92: 2751 (1967); *RQ* (Fall 1970); *RSR* 6: 43 (Apr./June 1978); *WLB* 48: 340 (Dec. 1973); *ARBA* (1971, p. 478; 1974, p. 550); Jenkins (A87); Wal (p. 57); Win (EA183; 2EA35; 1EA34; 1EA35).

American Men and Women of Science. Microfiche. **Jacques Cattell Press, ed.** New York: Bowker, 1980.

American Men and Women of Science: Consultants, 1977. New York: Bowker, 1977. Also updates.

A specialized volume of *American Men and Women of Science*. Contains entries on 16,000 scientific consultants. Contains a discipline index. Useful for locating potential consultants.
R: *WLB* 52: 265 (Nov. 1977).

Asimov's Biographical Encyclopedia of Science and Technology: The Lives and Achievements of 1,195 Great Scientists from Ancient Times to the Present, Chronologically Arranged. 2d rev. ed. **Isaac Asimov.** Garden City, NY: Doubleday, 1982.
First edition, 1964; revised edition, 1976.
Contains 1,195 readable biographical sketches intended for students and laypersons. More a popular reference than a scholarly undertaking.
R: *New Scientist* 81: 268 (Jan. 25, 1979); *ARBA* (1973, p. 536; 1977, p. 632).

Author Biographies Master Index—Supplement to the First Edition. **Barbara McNeil and Miranda C. Herbert, eds.** Detroit: Gale Research, 1980.
R: *CRL* 42: 83 (Jan. 1981).

Biographic Encyclopedia of Science and Technology. Reprint. **Isaac Asimov.** London: David and Charles, 1978.
A repeat reprint of the 1972 edition containing biographies of 1,195 scientists. For departmental libraries.
R: *Physics Bulletin* 30: 166 (Apr. 1979).

Bio-Base: A Periodic Cumulative Master Index in Microfiche to Sketches Found in 500 Current and Historic Biographical Dictionaries. Detroit: Gale Research, 1979–1980.
This huge index to biographical sources includes 1.5 million entries. Entries include biographee's name, years of birth and death, and a code for source. Produced on microfiche, names appear exactly as they did in original sources.
R: *WLB* 54: 396 (Feb. 1980).

Biographical Dictionaries Master Index: A Guide to More than 725,00 Listings in Over Fifty Current Who's Whos and Other Works of Collective Biography. 3 vols. **Dennis La Beau and Gary C. Tarbert, eds.** Detroit: Gale Research, 1975.
An alphabetical listing of three-quarters of a million names taken from 53 current English-language biographical dictionaries. Each entry consists of biographee's name and birth date as given in index, as well as code designating the source of entry. A major tool for reference collections.
R: *BL* 73: 682 (Jan. 1, 1977).

Biographical Dictionary of American Science: The Seventeenth Through the Nineteenth Centuries. **Clark A. Elliott.** Westport, CT: Greenwood Press, 1979.
Presents essential biographical information on American scientists born between 1609 and 1867. Emphasis of inclusion is on practitioners in the physical and biological sciences. Arranged alphabetically; contains nearly 600 entries, which include full biographical information. Contains 5 appendixes; sturdily bound for frequent use. Well-indexed. A highly recommended reference.
R: *Aslib Proceedings* 45: 9 (Jan. 1980); *BL* 76: 853 (Feb. 15, 1980); 76: 1154 (Apr. 15, 1980); 76: 1157 (Apr. 15, 1980); *Choice* 16: 992 (Oct. 1979); *RQ* 19: 290 (Spring

1980); *WLB* 54: 65 (Sept. 1979); *ARBA* (1980, p. 605); Mal (1980, p. 59); Wal (p. 140).

The Biographical Dictionary of Scientists. 2 vols. **David Abbott, ed.** New York: Chambers/Facts on File, 1983.

Volume 1, *Biologists*; volume 2, *Chemists.*
Each volume contains some 200 names of scientists who have been responsible for the development and contributions of scientific concepts through their own discoveries or theories. Will eventually include 4 more volumes concerning mathematicians, astronomers, physicists, engineers and inventors. A useful guide for the younger reader interested in the development and growth of science.
R: *Nature* 308: 475 (Mar. 1984).

A Biographical Dictionary of Scientists. 3d ed. **Trevor I. Williams, ed.** New York: Wiley, 1982.

First edition, 1969; second edition, 1974.
A strictly historical compilation of eminent scientists from antiquity to the present. Biographical sketches provide pertinent data, including unique notes on family backgrounds, experiences, and contributions. Highly selective.
R: *Science* 167: 363 (Jan. 23, 1970); *Choice* 12: 516 (June 1975); *LJ* 94: 2774 (Aug. 1969); *WLB* 50: 126 (Oct. 1975); *ARBA* (1976, p. 642); Wal (p. 54); Win (3EA33).

A Biographical Encyclopedia of Scientists. 2 vols. **John Daintith, Sarah Mitchell, and Elizabeth Tootill, eds.** New York: Facts on File, 1981.

A reference of biographies for almost 2,000 scientists, of both yesterday and today. Includes a chronological listing of scientific achievement, discovery, and publications.
R: *Sky and Telescope* 63: 159 (Feb. 1982).

Biographisch-Literarisches Handwörterbuch zur Geschichte der exacten Wissenschaften. **Johoun Christine Poggendorff.** Berlin: Akademie Verlag, 1863–. Leipzig: Barth, 1863–1904. Leipzig: Verlag Chemie, 1925–1940.

An authoritative German-language directory of mathematicians, chemists, physicists, and other physical scientists. Contains full biographical information, such as name, birth date, degrees, publications, etc. Article titles are cited in the language in which they were published.

Concise Dictionary of Scientific Biography. **American Council of Learned Societies, ed.** New York: Scribner's, 1981.

Contains several articles, summarized and categorized, on over 5,000 scientists. Sufficient for smaller libraries, branches, or departmental libraries.
R: *Choice* 19: 603 (Jan. 1982); *WLB* 56: 379 (Jan. 1982).

Current Bibliographic Directory of the Arts and Sciences: An International Directory of Scientists and Scholars. 2 vols. Philadelphia: Institute for Scientific Information. Annual.

Formerly entitled *Who Is Publishing in Science.*
An expanded version that lists some 395,000 authors and includes full citation information. Mailing address and publications are given in the author listing, and the

organization section lists authors' affiliated organizations. Geographic section provides data on country or city/state. Of interest to large research libraries.
R: *SL* (May/June 1980); *ARBA* (1981, p. 50).

Dictionary of Scientific Biography. 16 vols. **Charles Coulston Gillispie, ed.-in-chief.** New York: Scribner's, 1970–1980.

Volume 12, *Ibn Rushd-Jean Servais Stas,* 1975.

With its projected completion in 14 volumes, this work will stand as the most scholarly and comprehensive reference on scientific biography for the scientist and historian of science. International in scope, it will provide well-documented biographies of deceased scientists of major importance.
R: *American Scientist* 61: 353 (May/June 1973); *ISIS* (Sept. 1971); *Physics Today* 18: 86 (1965); *Science* 170: 615 (Nov. 27, 1970); *BL* 67: 201 (1971); *CRL* 32: 38 (1971); *LJ* (July 1970); *WLB* 45: 185 (Oct. 1970); *ARBA* (1976, p. 642; 1971, p. 478); Jenkins (A88); Wal (p. 53); Win (3EA30).

Doctoral Scientists and Engineers in the United States. Profile. **US National Research Council, Commission on Human Resources.** Washington, DC: US National Academy of Science. Annual.

ISI's Who Is Publishing in Science. Annual: An International Directory of Research and Development Scientists. Philadelphia: Institute for Scientific Information. Annual.

Supplements previous *Who Is Publishing* annuals and the *International Directory*, by providing addresses and organizational affiliations by authors of articles covered by *Science Citation Index*.
R: *ARBA* (1974, p. 548).

International Directory of Research and Development Scientists. Philadelphia: Institute for Scientific Information, 1967–. Annual.

Best single source for names and addresses of scientists throughout the world.
R: *BMLA* 57: 94 (1969); *LJ* 93: 2790 (1968); Jenkins (A117); Wal (p. 40); Win (3EA22).

The International Who's Who. New York: Unipub, Annual.

Forty-fifth edition, 1981.

An outstanding compilation of facts concerning 17,000 of the world's most prominent men and women. Includes date of birth, education, nationality, career, honors and awards, publications, etc. A massive reference that is indispensable to libraries.

Landmarks in Science: Hippocrates to Carson. **Robert B. Downs.** Littleton, CO: Libraries Unlimited, 1982.

Covers over 2 millennia of the great masterworks in all areas of science, from the major trends to the development of ideas and their implementation. Arranged according to the birth dates of the authors.
R: *ARBA* (1983, p. 600).

McGraw-Hill Encyclopedia of World Biography. 12 vols. New York: McGraw-Hill, 1973.

A comprehensive 12-volume compilation that details the lives of 5,000 men and women. Written by leading scholars and educators, each article averages 800 words and contains basic facts and figures, as well as descriptions of the subject's background, character, life, and role in history. Portraits and illustrations are provided when possible. Cross-referenced and indexed. Recommended for all large public and academic libraries.

McGraw-Hill Modern Scientists and Engineers. 3 vols. **Sybil P. Parker, ed.** New York: McGraw-Hill, 1980.

A thoroughly revised and expanded edition of *McGraw-Hill Modern Men of Science*, which supplements the *McGraw-Hill Encyclopedia of Science and Technology*. Three hundred new sketches have been added, for a total of 1,140 biographies covering the leading figures in science and engineering from 1920 to 1978. Articles are either autobiographies or biographies written by subject specialists. Each sketch is accompanied by a portrait and covers date of birth, education, contributions and research, awards, a short bibliography, and cross references. A field index and an analytical index are provided. Some articles are written in highly technical language, which may be unsuitable for the general public, but the volume should be a valuable reference for academic and research libraries for use by science and engineering students, historians, and educators.

R: *Physics Today* 33: 50 (July 1980); *BL* 77: 535 (Dec. 1, 1980); *Choice* 17: 185 (Apr. 1980); *LJ* 105: 1502 (July 1980); *RQ* 20: 99 (Fall 1980); *RSR* 8: 87 (July/Sept. 1980); 8: 63 (Oct./Dec. 1980); *WLB* 55: 64 (Sept. 1980); *ARBA* (1981, p. 624).

Nobel Lectures, Including Presentation Speeches and Laureates' Biographies. 4 vols. **Nobelstiftelsen, Stockholm.** New York; American Elsevier, 1964–1967.

Includes bibliographies.

Prominent Scientists: An Index to Collective Biographies. **Paul A. Pelletier, ed.** New York: Neal-Schuman, 1981.

Index to biographies of over 10,000 scientists in various fields of science. Indexes include one alphabetized by surname and one by the scientist's associated field. Recommended to all public and academic libraries.

R: *LJ* 106: 1297 (June 15, 1981).

Scientific Elite: Nobel Laureates in the United States. **Harriet Zuckerman.** New York: Free Press, 1977.

Composed of interviews with 41 Nobel laureates from the years 1901–1972. Studies psychological and social aspects of the prize. Considered a unique and important reference for public as well as academic libraries.

R: *Physics Bulletin* 29: 476 (Oct. 1978); *Choice* 14: 1384 (Dec. 1977); *LJ* 102: 937 (Apr. 15, 1977).

Scientists and Engineers in the Federal Government. **US Civil Service Commission Library.** Washington, DC: US Government Printing Office, 1970. Also later edition.

Earlier edition, 1965.
Descriptive annotations for personnel-oriented sources received by the Civil Service Commission Library during the period 1965–1969.
R: *ARBA* (1971, p. 543).

Scientists and Inventors. **Anthony Feldman and Peter Ford.** New York: Facts on File, 1979.
Profusely illustrated reference book on over 150 inventors from Empedocles to Crick and Watson. Contains helpful subject index. Recommended for purchase by school and public libraries.
R: *BL* 76: 1151 (Apr. 1, 1980); 76: 1631 (July 1, 1980); *LJ* 104: 2362 (Nov. 1, 1979); *WLB* 54: 331 (Jan. 1980); *ARBA* (1980, p. 606).

Who Was Who in American History—Science and Technology. Chicago: Marquis Who's Who, 1976.
Consists of 10,000 brief biographical sketches of notable engineers, inventors, and scientists who are now deceased. Includes information on birth and death, education, marriage, academic/business positions and accomplishments, honors, and major publications. For science historians.

Who Is Publishing in Science. Philadelphia: Institute for Scientific Information, 1971–. Annual.
Supersedes *International Directory of Research and Development Scientists* (1967–1969). Provides address and affiliation of first authors of publications indexed by *Current Contents, Science Citation Index,* and *Social Science Abstracts.*
R: Sheehy (EA28).

Who's Who in Frontier Science and Technology. Chicago: Marquis Who's Who. Annual.
1984–1985, published in 1984.
Contains over 16,000 biographies of scientists and technologists in North America. Indexed by fields and subject specialties.
R: *Choice* 22: 255 (Oct. 1984); *LJ* 109: 1841 (Oct. 1, 1984); *WLB* 59: 149 (Oct. 1984); *ARBA* (1985, p. 489).

Who's Who in Science in Europe: A Reference Guide to European Scientists. 4 vols. 2d ed. St. Peter Port, Guernsey, England: Hodgson, 1978. Also later edition.
First edition, 1967; second edition, 1972.
Some 40,000 scientists from Eastern and Western Europe engaged in scientific research in the physical, biological, medical, and agricultural sciences.
R: *ABL* 34: entry 210 (May 1969); *CRL* 29: 328; Wal (p. 55); Win (2EA38).

World Who's Who in Science: A Biographical Dictionary of Notable Scientists from Antiquity to the Present. Chicago: Marquis Who's Who, 1968.
Some 30,000 biographical sketches, many of only historical interest.
R: *Science* 167: 363 (Jan. 23, 1970); *CRL* 30: 78 (1969); *LJ* 96: 180 (Jan. 15, 1969); Jenkins (A97); Wal (p. 54); Win (2EA39).

Who's Who of British Scientists. New York: St. Martin's Press. Annual.

Lists the top 10,000 British men and women in the biological and physical sciences. Includes information on societies, journals, research organizations, etc.
R: Wal (p. 55).

ASTRONOMY

The Whisper and the Vision: The Voyages of the Astronomers. **Donald Fernie.** Toronto: Clarke, Irwin, 1976.

A book about the lives of influential astronomers, including Herschell, Gill, etc. Recommended for astronomy and history of science collections.

This Wild Abyss: The Story of the Men Who Made Modern Astronomy. **G. E. Christianson.** New York: Free Press/Macmillan. West Drayton, UK: Collier Macmillan, 1978.

Biographies of 5 great figures in the history of astronomy are contained in this less than technical book. Histories of Copernicus, Brahe, Kepler, Galileo, and Newton, as well as Greek astronomy, are included. Christianson covers some areas very weakly and others with a stronger view.
R: *Nature* 276: 654 (Dec. 7, 1978).

MATHEMATICS

Directory of Women Mathematicians. **American Mathematical Society.** Providence: American Mathematical Society, 1973.

Makers of Mathematics. **Alfred Hooper.** New York: Random House, 1958.

A beginner's guide to the outstanding individuals in mathematics and their influence on modern mathematics.

The Men of Mathematics. **Eric Temple Bell.** New York: Simon & Schuster, 1965.

An account of 34 prominent people in the development of mathematics from its beginning, in chronological order.
R: Wal (p. 71).

Women and Mathematics, Science, and Engineering: A Partially Annotated Bibliography with Emphasis on Mathematics and with References on Related Topics. **Else Hoyrup.** Roskilde, Denmark: Roskilde University Library, 1978.

Includes index.

Women in Mathematics. **L. M. Osen.** Cambridge, MA: MIT Press, 1974.

World Directory of Mathematicians. 4th ed. Stockholm: Almquist and Wiksell, 1970.

Lists alphabetically and geographically the mathematicians of 71 countries.
R: Jenkins (B58); Wal (p. 69); Win (EF25).

Physics

Albert Einstein: His Influence on Physics, Philosophy, and Politics. **P. C. Aichelberg and R. U. Sexl, eds.** Wiesbaden, Germany: Vieweg, 1979.

A collection of essays on the influence of Einstein's physical theories. Reflects the influence of world politics. Informative volume; useful in large academic libraries.
R: *Nature* 281:324 (Sept. 27, 1979).

Albert Einstein, The Human Side: New Glimpses from His Archives. **Helen Dukas and Banesh Hoffman, eds.** Princeton: Princeton University Press, 1979.

Discusses Einstein's social values. Contains letters and other selections that shed light on the life of this great genius. Discussion of Einstein's love of music, philosophy, etc. Recommended.
R: *Nature* 282: 179 (Nov. 8, 1979).

Einstein: A Centenary Volume. **A. P. French, ed.** London: Heinemann. Cambridge, MA: Harvard University Press, 1979.

A compilation of facts on Einstein from reminiscences, scientific work, political and social ideas, and many of his letters received and written. Includes photographs and drawings. Though the book is not a complete account of Einstein's work, it does bring out some of his finer qualities.
R: *Nature* 278: 823 (Apr. 26, 1979).

Galileo at Work: His Scientific Biography. **Stillman Drake.** Chicago: University of Chicago Press, 1978.

Written by a foremost authority on Galileo, this biography fully analyzes the works of the great scientist. Provides a detailed year-by-year account. Contains a chronological list of Galileo's manuscripts. Well-recommended.
R: *Nature* 279: 457 (May 31, 1979); *Physics Today* 32: 54 (Feb. 1979); *Science* 206: 439 (Oct. 26, 1979); *Aslib Proceedings* 44: 333 (Aug. 1979).

Introduction to Isaac Newton's "Principia." **I. B. Cohen.** New York: Cambridge University Press, 1978.

A well-received history that comprises a biography of Newton's original manuscript. Well-illustrated. Scholarly, though not overly technical. A highly recommended reference.
R: *Aslib Proceedings* 44: 8 (Jan. 1979).

Isaac Newton's Papers and Letters on Natural Philosophy and Related Documents. 2d ed. **I. B. Cohen.** Cambridge, MA: Harvard University Press, 1978.

Fascimile reproductions of Newton's published works, excluding *Principia* and *Opticks*. Includes commentaries by modern historians. Comprises a good detailed source and literature guide for students of Newton's work.
R: *Aslib Proceedings* 44: 57 (Feb. 1979).

Never at Rest: A Biography of Isaac Newton. **Richard S. Westfall.** Cambridge: Cambridge University Press, 1980.

This thoroughly documented and indexed biography provides ready access to the details of the life and contributions of Isaac Newton.
R: *Optical Engineering* 22: SRO18 (Jan. 1983).

Newton and Newtoniana, 1672–1975: A Bibliography. **P. Wallis and R. Wallis.** St. Louis: Folkestone Press, 1977.

A new bibliography relating to Sir Isaac Newton, his background, and the significance of his ideas. Items are organized around such topics as optics, fluxations, and chronological and theological works.
R: *Physics Today* 31: 60 (Sept. 1978).

Pioneers of Science: Nobel Prize Winners in Physics. **Robert Lemmerman Weber.** Edited by J. M. A. Lenihan. Philadelphia: Heyden and Son, 1980.

An entertaining volume that details the lives and accomplishments of the 114 Nobel-prize-winning men and women in physics, beginning with Röntgen in 1901 and continuing through 1979. Sketches are arranged in order by date of award and include short essays and anecdotes. A portrait drawing of each laureate accompanies the description. Contains a commendable bibliography and a good index. A worthwhile purchase for all libraries, science students, and the general public.
R: *Physics Bulletin* 31: 283 (Oct. 1980); *Physics Today* 34: 84 (May 1981); *Choice* 18: 1121 (Apr. 1981).

Rutherford and Physics at the Turn of the Century. **Mario Bunge and W. R. Shea, eds.** New York: Science History, 1979.

A thorough biography of Ernest Rutherford. Discusses all his contemporary influences. Recommended for history of science collections.
R: *Nature* 279: 741 (June 21, 1979); *Physics Teacher* 17: 478 (Oct. 1979); *TBRI* 45: 358 (Dec. 1979).

Rutherford: Simple Genius. **David Wilson.** Cambridge, MA: MIT Press, 1984.

The man, the activities, and the life of Ernest Rutherford and his contributions to the field of physics are portrayed in this detailed and superb biography.
R: *Nature* 307:761 (Feb. 1984).

Physics Nobel Lectures Including Presentation Speeches and Laureates' Biographies: 1901–1921, 1922–1941, 1942–1962. 3 vols. **Nobelstiftelsen, Stockholm.** New York: American Elsevier, 1964.

A compilation of the complete Nobel lectures, including presentation speeches and laureates' biographies.
R: *NTB* 49: 269 (1964); Jenkins (A93); Win (EG43).

Selected Papers 1945–1980, with Commentary. **Chen Ning Yang.** San Francisco: W. H. Freeman, 1983.

A rather autobiographical work of the selected papers of Professor Chen Ning Yang (or Frank Yang) with his own commentary on each. For those interested in the development of fundamental physics since World War II.
R: *Nature* 306: 623 (Dec. 1983).

Women in Physics: A Roster. **American Physical Society, Committee on Women in Physics.** New York: American Institute of Physics. 1972–.

Lists women who responded to a questionnaire (1,381 out of 2,000).

CHEMISTRY

American Chemists and Chemical Engineers. **Wyndham D. Miles, ed.** Washington, DC: American Chemical Society, 1976.

Contains biographical sketches on over 500 deceased American chemists and chemical engineers. Covers a span of 300 years. Includes editors, alchemists, consultants, etc. Informative and well written.

R: *RQ* 16: 353 (Summer 1977); *ARBA* (1978, p. 645); Sheehy (ED20).

Canadian Inventors and Innovators, 1885–1950: Pioneering in Plastics. **Donald W. Emmerson.** Scarborough, Canada: Canadian Plastics Pioneers, 1979.

Covers every aspect of plastic production in Canada.

R: *Chemistry in Canada* 31: 26 (June 1979); *TBRI* 45: 275 (Sept. 1979); 46: 32 (Jan. 1980).

Great Chemists. **Eduard Farber.** New York: Wiley-Interscience, 1961.

Biographical sketches of over 100 chemists from antiquity to the twentieth century. Includes portraits.

R: *Chemical and Engineering News* 40: 501 (1962); Jenkins (D178); Wal (p. 125); Win (ED78).

Nobel Prize Winners in Chemistry, 1901–1961. Rev. ed. Eduard Farber. London: Abelard-Schuman, 1963.

For each winner there appears an overview of the prize-winning work, a portrait, and a biographical sketch. Includes 65 biographies.

R: *Science* 139: 623 (1963); Jenkins (D179); Wal (p. 125); Win (ED79).

Professionals in Chemistry. Washington, DC: American Chemical Society. Annual.

BIOLOGICAL SCIENCES

Audubon: A Biography. **John Chancellor.** London: Weidenfeld and Nicolson, 1978.

Provides a detailed history of the life of this great illustrator of birds. Vividly portrays the sometimes cantankerous artist and the events of his life.

R: *Nature* 276: 138 (Nov. 9, 1978).

Biographical Dictionary of Botanists. **Carnegie-Mellon University, Hunt Botanical Library, Hunt Institute Portrait Collection.** Boston: Hall, 1972.

Brief biographical information for botanists and horticulturists.

Biographical Notes upon Botanists: Maintained in the New York Botanical Garden Library. 3 vols. **John Hindley Barnhart, comp.** Boston: Hall, 1965.

Standard biographical information on approximately 44,000 past and present botanists, as reproduced from an annotated file.
R: Jenkins (G143); Wal (p. 208).

Dictionary of British and Irish Botanists and Horticulturists, Including Plant Collectors and Botanical Artists. 3d ed. **Ray Desmond.** London: Taylor and Francis, 1977.

This specialized reference work contains detailed information on horticulturists and botanists of the British Isles. Includes information on date of birth, education, publications, awards, herbaria, drawings, manuscripts, etc. Subject index arranged by profession, plants, or country where flora have been studied.
R: *Nature* 270: 457 (Dec. 1, 1977); *RSR* 7:7 (Jan./Mar. 1979); *TBRI* 44: 53 (Feb. 1978).

Directory of Environmental Life Scientists. 9 vols. EP1105-2-3. **Institute of Ecology.** Washington, DC: US Government Printing Office, 1974.

Supplies directory information on environmental life scientists for the United States Army Corp of Engineers. Nine volumes cover major areas of the continental United States. Arrangement by organization or personal name. Four indexes provide access to directory contents.
R: Sheehy (EJ20).

Directory of North American Entomologists and Acarologists. **William B. Hull and Gerald C. Odland, eds.** Hyattsville, MD: Entomological Society of America, 1979.

A directory of 5,000 Entomological Society of America members, divided into 3 sections: geographic, member specialty, and alphabetic index. The geographic section is arranged A–Z by state; the other sections are arranged alphabetically.
R: *RSR* 8: 17 (July/Sept. 1980).

Early Wildlife Photographers. **Charles A. W. Guggisberg.** New York: Taplinger Publishing, 1977.

A comprehensive survey of wildlife photographers. Well-illustrated with photographs. Appealing to nature and photo-historians.
R: *Choice* 14: 1384 (Dec. 1977).

The Naturalists: Pioneers of Natural History. **A. C. Jenkins.** London: Hamish Hamilton, 1978.

An introduction to the beginning of natural history, discussing all major personalities—Aristotle, John Ray, Linnaeus, Fabre, etc. Includes excellent illustrations. Recommended as an introductory biographical reference.
R: *Nature* 277: 579 (Feb. 15, 1979).

The Naturalists' Directory International. 42d ed. South Orange, NJ: PCL Publications, 1975. Also update editions.

Forty-first edition, 1972.
Contains both name and discipline indexes.
R: Sheehy (EC20).

Nobel Lectures in Molecular Biology, 1933–1975. New York: American Elsevier, 1977.
Includes bibliographical references and index.

EARTH SCIENCES

Directory of Paleontologists of the World (Excluding the Soviet Union and Continental China). Hamilton, Ontario: McMaster University, 1968.
Main section provides an alphabetical listing of paleontologists.

International Directory of Marine Scientists. Rome: Food and Agricultural Organization of the United Nations.
Lists 5,745 marine scientists from 91 countries.

US Directory of Marine Scientists. **US National Research Council, Ocean Science Committee.** Washington, DC: US Government Printing Office, 1975. Also updates.
This directory includes the names, addresses, and areas of specialization of 3,013 individuals. An author index is the key to zip codes under which main entries are organized.

Who's Who in Ocean and Freshwater Science. **A. Varley.** New York: Longman, 1978.
Contains around 40,000 entries and cross references with present positions and addresses noted.
R: Wal (p. 214).

GENERAL ENGINEERING

Americna Engineers of the Nineteenth Century: A Biographical Index. **Christine Roysdon and Linda A. Khatri.** New York: Garland STPM Press, 1978.
An index to early American engineers and technologists who died before or during 1900. Entries are arranged alphabetically and include birth and death dates, statement of field of interest, and references to biographies or obituaries. Of interest to research libraries.
R: *Choice* 16: 207 (Apr. 1979); *RSR* 8: 26 (Jan./Mar. 1980); *ARBA* (1980, p. 714).

Author Affiliation Index to the Engineering Index Annual. New York: Engineering Index. Annual.

A Biographical Index of British Engineers in the 19th Century. **S. P. Bell, comp.** New York: Garland, 1975.
Alphabetical directory of British engineers who died during or before 1900. References to obituaries and entries in other major biographical works.
R: *LJ* 100: 1534 (Sept. 1, 1975); *ARBA* (1976, p. 762).

Black Engineers in the United States: A Directory. **J. K. K. H.** Cambridge, MA: Harvard University Press, 1974.

The Consulting Engineers Who's Who and Yearbook. London: Northwood, in collaboration with the Association of Consulting Engineers, 1946–. Annual.
Information on members and firms, institutions and societies.

Directory of Engineers in Private Practice. Washington, DC: National Society of Professional Engineers. Annual.
R: *Mechanical Engineering* 90: 79 (Oct. 1968); Jenkins (K210).

Famous Names in Engineering. **James Carvill.** London: Butterworth, 1981.
A brief and informative biography of 83 famous mathematicians, physicists, and engineers who have been instrumental in forming the basis of any advanced course in engineering. Includes references, portraits, formulas, diagrams, or an illustration of each arranged both by subject and chronologically.
R: *Science and Technology Libraries* 2: 76 (Spring 1982); *ARBA* (1983, p. 144).

Great Engineers and Pioneers in Technology, From Antiquity through the Industrial Revolution. Vol. 1. **Roland Turner and Steven L. Goulden.** New York: St. Martin's, 1982.
This is the first of 3 volumes, and is divided into 5 sections, each with a brief introduction. Entries are arranged within each section in chronological order. Illustrations are reproductions and original drawings. A valuable tool for reference or reading for fun.
R: *Choice* 20: 558 (Dec. 1982).

Who Is Who: A Directory of Agricultural Engineers Available for Work in Developing Countries. **American Association of Agricultural Engineers.** St. Joseph, MI: American Association of Agricultural Engineers., 1984.

Who's Who in Technology. 2d ed. 3 vols. **Karl Strute, ed.** Zurich: Who's Who International. Distr. New York: Unipub, 1984.
First edition, 1979.
The first 2 volumes contain brief biographical sketches of individuals from 18 European countries. The third volume is an appendix, which presents both a directory of names and academic institutions. Expensive set.
R: *ARBA* (1985, p. 488).

Who's Who of British Engineers. **R. A. Baynton, ed.** New York: St. Martin's Press, 1969–. Annual.
Entries provide name, address, title, birth date, degrees, and publications.
R: *Science News* 95: 236 (1969); *Aslib Proceedings* 23: 10 (Jan. 1971); *CRL* 31: 273 (1970); *LJ* 94: 1486 (Apr. 1, 1969); *ARBA* (1970, p. 142; 1971, p. 547; 1975, p. 766.); Jenkins (K213); Wal (p. 290); Win (2EI6; 3EA32).

Who's Who in Consulting: A Reference Guide to Professional Personnel Engaged in Consultation for Business, Industry, and Government. 3 pts. 3d ed. **Paul Wasserman, ed.** Detroit: Gale Research, 1982.
Second edition, 1973.

Who's Who in Engineering. 3d ed. New York: Engineering Joint Council, 1977.

Formerly titled *Engineers of Distinction.*
A compilation of biographic data on about 4,500 engineers. Includes categories on engineering societies and related organizations, as well as society and awards indexes. For academic libraries.

Who's Who in Engineering: A Biographical Dictionary of the Engineering Profession. Vols. 1–. New York: Lewis Historical, 1922–. Irregular.

Ninth edition, 1964.
US and foreign engineers working in the United States. Includes directory information on organizations and societies.
R: Jenkins (K212); Wal (p. 29); Win (EA197).

Who's Who in Technology Today, 1982. 3d ed. **Jan W. Churchwell, mg. ed., and Louann Chaudier, assoc. ed.** Highland Park, IL: J. Dick, 1982.

A biographical directory that will prove useful in identifying experts in more than 80,000 scientific and technical areas. For scientific and technical libraries, research libraries, and information centers.
R: *ARBA* (1983, p. 600).

AERONAUTICAL ENGINEERING

America's Journeys into Space: The Astronauts of the United States. **Anthony J. Cipriano.** New York: Wanderer Books, 1979.

Portraits of 172 scientists, engineers, and astronauts who have contributed to the US space program.
R: *Science Digest* 87: 84 (Feb. 1980); *TBRI* 46: 113 (Mar. 1980).

Men of Space: Profiles of the Leaders in Space Research, Development, and Exploration. 8 vols. **Shirley Thomas.** Philadelphia: Chilton, 1960–1968.

Includes biographical material, excerpts from conversations, and information on space organizations.

Who's Who in Aviation. New York: Harwood and Charles. Distr. Elmsford, NY: Pergamon Press, 1974. Also later editions.

Comprehensive collection of individuals engaged in all aspects of the aviation industry. Includes awards and patents information, in addition to brief biographical sketches.
R: *ARBA* (1975, p. 808).

CHEMICAL ENGINEERING

Textile World's Leaders in the Textile Industry. **Laurence A. Christiansen, Jr., ed.** New York: McGraw-Hill, 1979.

Biographical listing of leading American scientists involved in the textile industry. Covers research, manufacturing, design, marketing, teaching, and public relations. Of interest to large public and academic libraries.
R: *BL* 77: 72 (Sept. 1, 1980).

Civil Engineering

A Biographical Dictionary of Railway Engineers. **John Marshall.** North Pomfret, VT: David and Charles, 1978.

This volume, designed for railroad historians, contains information on international railroad engineers, emphasizing British contributors. Covers a wide range of personalities.
R: *ARBA* (1979, p. 779).

Official Directory of Industrial and Commercial Traffic Executives. 1981. Rev ed. **Callie Possinger, ed.** Introduction by Callie Possinger. Washington, DC: Traffic Service Corporation, 1980.

Who's Who in Architecture, from 1400 to the Present. **J. M. Richards, ed.** New York: Holt, Rinehart, and Winston, 1977.

Entries on more than 600 architects from Western countries, Japan, and members of the British Commonwealth. Thoroughly cross-referenced and illustrated. Contains a wealth of biographical information.
R: *RQ* 17: 368 (Summer 1978).

Electrical and Electronics Engineering

Electrical Who's Who. London: Electrical Review, 1950–. Irregular.

Brief biographies of leading members of the professional and industrial branches of the industry.
R: *Technology and Culture* 2: 28 (1961); 2: 146 (1962); Wal (p. 110, p. 316).

McGraw-Hill's Leaders in Electronics. **Editors of Electronics Magazine, eds.** New York: McGraw-Hill, 1979.

Broad and worldwide coverage of some 5,200 leading figures involved both directly and indirectly in the electronics industry. Presented in a readable format, entries are alphabetically arranged and cover date of birth, education, professional experience, and current position; awards; publications; membership in organizations; etc. Provides an index of biographies by affiliation as well as a table of abbreviations. A recommended reference for large science and engineering libraries, as well as electronics libraries.
R: *BL* 76: 1320 (May 1, 1980); *RQ* 19: 307 (Spring 1980); *ARBA* (1980, p. 714).

Who's Who in Computer Education and Research. **T. C. Hsiao, ed.** Washington, DC: Science and Technology Press, 1975. Also updates.

Contains biographies on some 2,000 educators and researchers actively engaged in teaching. Includes US institution faculty and researchers who work in the field of computer education.
R: *ARBA* (1977, p. 761).

Who's Who In Electronics. **Jeanne Ring, ed.** Cleveland: Harris. Annual.

This massive directory lists names, addresses, and telephone numbers of more than 10,000 companies in the electronic components industry.

Who's Who and Guide to the Electrical Industry. **Kenneth Ellmore.** London: IPC Electrical-Electronic Press. Annual.
Contains brief biographies of significant members of the electronic and electrical industry.
R: *Wal* (p. 353).

MECHANICAL ENGINEERING

Mechanical Engineers in America, Born Prior to 1861: A Biographical Dictionary. **American Society of Mechanical Engineers.** New York: American Society of Mechanical Engineers, 1980.
This resource dictionary supplies essential biographical data on those American mechanical engineers active from the late eighteenth century to the early twentieth century. Includes a general listing of reference works.
R: *Science and Technology Libraries* 1: 111 (Winter 1980).

NUCLEAR ENGINEERING

Directory: Who's Who in Nuclear Energy. Hinsdale, IL: American Nuclear Society. Annual.

Who's Who in Atoms: An International Reference Book. 6th ed. 2 vols. London: Harrap. Distr. New York: International Publications, 1977. Also update edition.
Fifth edition, 1969.
Standard biographical information for more than 20,000 nuclear scientists working in 76 countries.
R: *ARBA* (1970, p. 111); Jenkins (C74); Wal (pp. 223; 309); Win (EA196; IEA37).

ENERGY

AEE Directory of Energy Professionals. **The Association of Energy Engineers.** New York: Van Nostrand Reinhold. Annual.
A directory of energy professionals with their geographic listing and government references. A helpful reference tool for academic libraries.
R: *RSR* 8: 24 (Jan./Mar. 1980).

Contact: A Guide to Energy Specialists. **World Environment Center.** New York: World Environment Center, 1982.
Each section contains a listing of individuals and organizations identified by subject area of expertise, location, and other basic information.
R: *Mechanical Engineering* 105: 95 (Apr. 1983).

International Who's Who in Energy and Nuclear Science. Detroit: Gale Research, 1983–. Irregular.

Solar Census: The Directory for the 80's. Ann Arbor: Aatec Publications, 1980.
A compilation of questionnaires completed by solar energy professionals throughout the United States. Divided into the primary activities designated by the respondents:

manufacturers; design-architecture and engineering; research and development; and education and information. Arranged alphabetically within each section.
R: *Science and Technology Libraries* 1: 74 (Summer, 1981).

ENVIRONMENTAL SCIENCE

Leaders of American Conservation. **Henry Clepper,** ed. New York: Ronald Press, 1971.
Sketches, approximately 1 page in length, of some 300 eminent conservationists.
R: *ARBA* (1973, p. 583); Mal (1980, p. 214).

MEMBERSHIP DIRECTORIES

Almost all professional societies provide annual or otherwise-regular membership directories. These are essential sources of current information on addresses, organizational affiliations, and so forth, of working scientists and engineers. See chapter on professional organizations for further information.

Biographical Memoirs. Vols. 1–. **US National Academy of Sciences,** ed. Washington, DC: US National Academy of Sciences, 1877/79–. Annual.
Contains biographies of deceased members of the National Academy of Sciences. Entries provide biographical sketch, list of publications, portrait, and chronology of events.

Biographical Memoirs of Fellows of the Royal Society: 1932–1954. 9 vols. London: The Royal Society of London, 1955.
Consists of biographies and portraits of deceased members of the Royal Society.

Sample society membership publications include the following:

Combined Membership List: American Mathematical Society, Mathematical Association of America, and Society for Industrial and Applied Mathematics, Providence: American Mathematical Society. Annual.
A complete listing of members on record as of the date of the annual volume for each of the three organizations. Includes an alphabetical listing of individuals with pertinent data, and a geographic listing by state, province, and country. The use of the *Combined Membership List* for generating mailing lists is prohibited.
R: *ARBA* (1983, p. 612).

Directory of Physics and Astronomy Staff Members. New York: American Institute of Physics, 1959–. Biennially.
Up-to-date membership information for some 30,000 physicists throughout the United States, Canada, and Mexico. With statistical information on various programs in physics and astronomy.
R: *ARBA* (1983, p. 610).

CHAPTER 12 HISTORY

GENERAL SCIENCE

Album of Science: Antiquity and the Middle Ages. **John E. Murdoch, ed.** New York: Scribner's, 1984.

An impressive single volume of the *Album of Science* series. It includes brief narrative text, several hundred illustrations, annotated bibliographies, and well-prepared indexes. A must for all libraries with collections in science.
R: *LJ* (Nov. 1, 1984, p. 2071); *ARBA* (1985, p. 494).

Album of Science: From Leonardo to Lavoisier, 1450–1800. **I. Bernard Cohen.** New York: Scribner's, 1980.

A detailed, pictorial history of science. Illustrations are taken from books and manuscripts, engravings, drawings, charts, maps, paintings, etc. A supplementary text accompanies the volume, and a bibliography is included. An invaluable reference for science historians.
R: *Scientific American* 244: 52 (Jan. 1981); *WLB* 55: 678 (Jan. 1981); *ARBA* (1981, p. 617).

Album of Science: The Nineteenth Century. **L. Pearce Williams.** New York: Scribner's, 1978.

Contains 600 pictures, maps, photographs, and diagrams from the nineteenth century illustrating the development of science and technology. Divided into sections, each with an informative introduction. Detailed indexing of illustration sources. Highly recommended for high school, public, and academic libraries.
R: *Scientific American* 239: 32 (Dec. 1978); *WLB* 52: 732 (May 1978).

Asimov's Guide to Science. Rev. ed. **Isaac Asimov.** New York: Basic Books, 1972.
R: Mal (1976, p. 281).

The Awakening Interest in Science During the First Century of Printing, 1450–1550; An Annotated Checklist of First Editions Viewed from the Angle of Their Subject Content. **Margaret Bingham Stillwell.** New York: Bibliographical Society of America, 1970.
R: *LJ* 95: 4158 (Dec. 1, 1970); *ARBA* (1971, p. 471); Wal (p. 53).

A Bibliography of Quantitative Studies on Science and Its History. **Roger Hahn.** Berkeley: University of California, 1980.

Made up of works from the International Symposium on Quantitative Methods in the History of Science, held at Berkeley in August of 1976. Covers broad topics such as methodology, bibliographic aids, manpower, input, and the like.
R: *ARBA* (1982, p. 695).

Catalogue of Scientific Papers, 1800–1900. 19 vols. **Royal Society of London.** London: Clay, 1867–1902; Cambridge, England: Cambridge University Press, 1914–1925.

With subject index.
Critical Bibliography of the History of Science and Its Cultural Influence. Nos. 1–. Beltsville, MD: History of Science Society, 1913–.

Dictionary of the History of Ideas: Studies of Selected Pivotal Ideas. 5 vols. **Philip P. Wiener, ed.-in-chief.** New York: Scribner's, 1980.
Earlier edition, 1973.

Dictionary of the History of Science. **W. F. Bynum, E. J. Browne, and Roy Porter, eds.** Princeton, NJ: Princeton University Press, 1981.
This high-quality dictionary concentrates on the development of Western post-medieval scientific thought. Also includes ancient, medieval, and non-Western theories that influenced Western science. Very helpful to students, specialists, and laypersons.
R: *WLB* 56: 542 (Mar. 1982); *ARBA* (1983, p. 598).

Early American Men of Science. **Silvio Bedini.** New York: Scribner's, 1975.
The book reflects the collector's interest in the handicraft era of instrument making—a period that ended around 1835. It provides details about little-known people. Excellent photographs and a glossary of technical terms.
R: *American Scientist* 64: 684 (Nov./Dec. 1976).

Early American Science. **Brooke Hindle, ed.** New York: Neale Watson Academic, 1976.
A collection of 22 articles on the history of science in the United States during the period ending in 1820.
R: *American Journal of Physics* 45: 112 (Jan. 1977); *Optical Society of America Journal* 67: 416 (Mar. 1977); *TBRI* 43: 85 (Mar. 1977); 43: 204 (June 1977).

A Historical Catalogue of Scientists and Scientific Books: From the Earliest Times to the Close of the Nineteenth Century. **Robert Mortimer Gascoigne, ed.** New York: Garland, 1984.
A bibliographical dictionary and a reference to scientific works. Indexed. A useful tool for all research libraries.
R: *ARBA* (1985, p. 487).

A History of Science. 2 vols. **George Sarton.** New York: Norton, 1970. Original edition, Harvard University Press, 1962.
A classic in the field.
R: *BL* 49: 120 (1952); *ARBA* (1971, p. 472); Jenkins (A28); Wal (p. 50).

History of Science. 4 vols. **Rene Taton, ed.** Translated by A. F. Pomeraus. New York: Basic Books, 1963–1966.

The History of Science and Technology in the United States: A Critical and Selective Bibliography. **Marc Rothenberg.** New York: Garland, 1982.
A bibliography of most of the important secondary works published between 1940 and 1980, including a variety of fields such as women, government, science, and religion, etc., but not medicine. A valuable tool for academic libraries.
R: *LJ* 108: 198 (Jan. 1983).

The History of Science and Technology in the United States: A Critical and Selective Bibliography. Vol. 2. **Marc Rothenberg.** New York: Garland, 1982.

Some 800 selected references dated between 1940 and 1980, with annotations, on the history of the physical sciences, social sciences, technology, and agriculture in the United States. For academic as well as public libraries.
R: *Choice* 20: 1114 (April 1983).

Information Sources in the History of Science and Medicine. **Pietro Corsi and Paul Weindling.** Stoneham, MA: Butterworth, 1983.

A 4-part coverage of the historical development of science and medicine, research methods, and a description of the major libraries and archives. Includes technical issues, along with cultural and social aspects of science and medicine.
R: *ARBA* (1985, p. 491).

International Catalogue of Scientific Literature, 1st–14th, 1901–1914. 283 vols. in 32. London: Royal Society for the International Council, 1902–1921. Reprint. New York: Johnson Reprint, 1975.

Introduction to the History of Science. 5 vols. **George Sarton.** Baltimore: Williams & Wilkins, 1927–1948. Reprint. Huntington, NY: Krieger, 1975.

ISIS Cumulative Bibliography: A Bibliography of the History of Science Formed from ISIS Critical Bibliographies, 1–90, 1913–1965. 3 vols. **Magda Whitrow, ed.;** *Personalities and Institutions*, Vol. 1. 1966–1975. **John Neu, ed.** London: Mansell/History of Science Society, 1971–1976, and 1980.

Eventually more volumes will be added in an attempt to create the most significant compilation of sources on the history of science. Exceptionally well-reviewed and well-organized tool.
R: *Nature* 237: 115–116 (May 12, 1972); *ARBA* (1973, p. 532; 1977, p. 624; 1981, p. 618); Sheehy (EA37); Wal (p. 52).

Printing Presses: History and Development from the 15th Century to Modern Times. **James Moran.** Berkeley: University of California Press, 1978.

Surveys the invention of the printing press from its beginnings in the 15th century through the 1940s.
R: *Chronicle of the Early American Industries Association* 32: 15 (Mar. 1979); *TBRI* 45: 237 (June 1979).

Science in America: A Documentary History, 1900–1939. **Nathan Reingold and Ida H. Reingold, eds.** Chicago: University of Chicago Press, 1982.

An overview of the issues, institutions, and individuals involved in science in the early twentieth century. The roles played by foundations and universities, as well as by individual scientists, are revealed through correspondence. Good for students, and useful in libraries.
R: *Choice* 19: 1423 (June 1982).

Science in America since 1820. **Nathan Reingold, ed.** New York: Science History, 1976.

This collection of insightful papers covers the revolution of scientific thought in this country.
R: *Technology and Culture* 18: 98 (Jan. 1977); *TBRI* 43: 207 (June 1977).

Science and Civilization in China. **J. Needham with Wang Ling.** Cambridge University Press, 1954.

A history of Chinese science, thought, and technology, compared with that of Asia and Europe. Contains a bibliography of Chinese and Western books and articles. Good for history of science collections and Oriental history collections.
R: Wal (p. 54).

Science in the Middle Ages. **David C. Lindberg, ed.** Chicago: University of Chicago Press, 1979.

Numerous contributions from experts bring together a vast amount of history. Includes excellent bibliographies. A detailed source of historical reference.
R: *Science* 204: 493 (May 4, 1979).

Science Since 1500: A Short History of Mathematics, Physics, Chemistry, Biology. 2d ed. **H. T. Pledge.** London: Her Majesty's Stationery Office for the Ministry of Education, Science Museum, 1966.

Scientific Books, Libraries, and Collectors: A Study of Bibliography and the Book Trade in Relation to Sciences. 3d ed., rev. **John L. Thornton and R. I. J. Tully.** London: Library Association. Distr. Detroit: Gale Research, 1971.

First edition, 1954; second edition, 1962.
A classic bibliographic history of science. Twelve updated chapters discuss the production, distribution, and storage of scientific literature from the pre-Gutenberg period to the present. Detailed index available.
R: *Science* 139: 897 (1963); *LJ* 97: 2360 (July 1972); *Library Association Record* 74: 16 (Jan. 1972); *ARBA* (1973, p. 532); Jenkins (A34); Wal (p. 10); Win (EA177).

Scientific Thought, 1900–1960: A Selective Survey. **R. Harré, ed.** New York: Oxford University Press, 1969.

A Short History of Twentieth-Century Technology c.1900–1950. **Trevor I. Williams.** Oxford: Oxford University Press, 1982.

Provides the general reader with an explanation of the necessary science and engineering involved in nuclear weapons, antibiotics, transistors, computers, and much more. Limited in its evaluation of social, economic, and political influences.
R: *Nature* 302: 90 (Mar. 1983).

The Smithsonian Book of Invention. **Staff of the Smithsonian Institution, comp.** Washington, DC: Smithsonian Institution, 1978.

Profusely illustrated volume on inventions in areas such as agriculture, industry, space, science, and medicine. Contains over 300 photographs, artworks, and drawings. Valuable addition to any history of science and technology collection.
R: *RSR* 8: 27 (Jan./Mar. 1980).

Sources for the History of Science 1660–1914. **David Knight.** Ithaca, NY: Cornell University Press, 1975.
The book describes important trends in the historical development of science.
R: *LJ* 100: 1203 (June 15, 1975); *ARBA* (1976, p. 632).

Sources in the History of Mathematics and Physical Sciences. **G. J. Toomer and M. J. Klein, eds.** New York: Springer-Verlag, 1976–.
Volume 1, *Diocles*: On *Buring Mirrors*, 1976.

Studies in the History of Mathematics and Physical Sciences. **M. J. Klein and G. J. Toomer, eds.** New York: Springer-Verlag, 1975–.
Volume 1, *A History of Ancient Mathematical Astronomy,* 1975.

Studies in the History of Modern Science. Vol. 1. **Frederick Gregory.** Boston: Reidel, 1977.
Volume 1, *Scientific Materialism in Nineteenth-Century Germany.*
Studies scientific activity in Germany during the second half of the nineteenth century. Sheds light on the proliferation of intellectual activity during this time.

Works of Science and the History Behind Them. **Isaac Asimov.** New York: New American Library, 1969.
First published in 1959.
Scientific terminology is arranged in dictionary form with a full-page discussion of the historical root and meaning of each word.
R: *ARBA* (1970, p. 100).

ASTRONOMY

Astronomy of the Ancients. **Kenneth Brechen and Michael Feirtag, eds.** Cambridge, MA: MIT Press, 1979.
A collection of articles that present information on astronomy during ancient times. Covers pictographs, first scientific instruments, Stonehenge, and naked-eye astronomy.
R: *Astronomy* 8: 57 (Feb. 1980); *TBRI* 46: 82 (Mar. 1980).

Bibliography of Astronomy, 1881–1898. Buckinghamshire, England: University Microfilms, 1970.
Prepared from standard slips of some 52,000 items recorded at the Observatoire Royal de Belgique. Available also on microfilm.

A History of Ancient Mathematical Astronomy. 3 pts. **O. Neugebauer.** New York: Springer-Verlag, 1975.
Volume 1, *Studies in the History of Mathematics and Physical Sciences.*
R: *Nature* 260: 202 (Mar. 18, 1976).

A History of Astronomy. **Antonie Pannakoek.** New York: Wiley-Interscience, 1961.
R: *Nature* 195: 316 (July 28, 1962); Jenkins (E19); Wal (p. 81).

The History of Modern Astronomy and Astrophysics: A Selected, Annotated Bibliography. **David H. DeVorkin.** New York: Garland, 1982.
R: *Choice* 20: 242 (Oct. 1982); *ARBA* (1983, p. 602).

Kepler: Four Hundred Years. New York: Oxford University Press, 1975.
Volume 18, *Vistas in Astronomy.*
R: *Nature* 259: 511 (Feb. 12, 1976).

Planets and Planetarians: A History of Theories of the Origin of Planetary Systems. **Stanley L. Jaki.** New York: Halsted Press, 1978.
Traces the history of ideas in the formation of the solar system, concentrating on basic theories. Considered an indispensable reference in history of science collections.
R: *Physics Today* 32: 84 (Jan. 1979); *TBRI* 45: 88 (Mar. 1979).

Source Book in Astronomy, 1900–1950. **Harlow Shapley, ed.** Cambridge, MA: Harvard University Press, 1960.
Useful for information on the historical development of astronomy during the first half of the twentieth century.

This Wild Abyss: The Story of the Men Who Made Modern Astronomy. **Gale E. Christianson.** New York: Free Press, 1978.
A comprehensive history of influential astronomers, containing much background information.
R: *American Scientist* 65: 638 (Sept.–Oct. 1978); *Physics Teacher* 16: 580 (Nov. 1978); *TBRI* 45: 2 (Jan. 1979).

MATHEMATICS

Bibliography and Research Manual of the History of Mathematics. **Kenneth O. May.** Buffalo, NY: University of Toronto Press, 1973.
Part 1 provides information on retrieval and storage of data. Part 2 lists 30,000 entries on the history of mathematics as it relates to information retrieval. Arranged under broad subject headings.
R: *Choice* 10: 1529 (Dec. 1973); *ARBA* (1974, p. 551).

Bibliography of Non-Euclidean Geometry. **D. M. Y. Sommerville.** London: University of St. Andrews, 1911. Reprint. New York: Chelsea, 1970.
Chronological arrangement from fourth century BC to 1911. Because of its theoretical nature and the dearth of similar works, still a useful item.
R: *ARBA* (1971, p. 481).

The History of the Calculus of Variations from the 17th Through the 19th Century. Studies in the History of Mathematics and Physical Sciences, vol. 5. **Herman H. Goldstine.** New York: Springer-Verlag, 1980.
R: *Quarterly of Applied Mathematics* 41: 124 (April 1983).

A History of Mathematics. 3d ed. **Florian Cajori.** New York: Chelsea, 1980.
Second edition, 1919.
A reprint of the second edition, which contains 200 pages on the nineteenth and early twentieth centuries. The minor errors have been corrected, and a section of notes has been added to update information or to give references. A history useful to math historians.
R: *American Scientist* 69: 106 (Jan. 2, 1981).

On the History of Statistics and Probability. **D. B. Owen, ed.** New York: Dekker, 1976.
Twenty-four papers that cover a wide range of topics on the development of statistical theory.
R: *Biometrics* 32: 701 (Sept. 1976); *TBRI* 43: 7 (Jan. 1977).

The Mathematical Papers of Isaac Newton. **Derek T. Whiteside, ed.** New York: Cambridge University Press.
Volume 1, 1664–1666: volume 2, 1667–1669; volume 3, 1670–1673; volume 4, 1667–1684 (pub. 1971); volume 5, 1683–1684 (pub. 1972); volume 6, 1684–1691 (pub. 1975); volume 7, 1691–1695 (pub. 1977).
R: *Science* 188: 826 (May 23, 1975).

Source Book in Mathematics, 1200–1800. **D. J. Struik.** Cambridge, MA: Harvard University Press, 1969.
Includes 75 excerpts from the writings of Western mathematicians from the thirteenth to the eighteenth century. Selections are limited to pure mathematics. Includes excerpts by Viete, Oresme, Newton, Leibnitz, etc.
R: *ISIS* 62: 533 (Dec. 1971); *Science* 165: 54 (July 4, 1969); *Science News* 95: 148 (Feb. 8, 1969); *Choice* 6: 796 (Sept. 1969); *NTB* 54: 116 (Apr. 1969); *ARBA* (1970, p. 104); Jenkins (B26); Wal (p. 69).

Studies in the History of Statistics and Probability. 2 vols. **M. G. Kendall and R. L. Plackett, eds.** London: Griffin, 1977.
Covers the origins of statistics from 1710 through 1973.
R: *ABL* 42: entry 370 (July 1977).

PHYSICS

Applications of Energy: Nineteenth Century. **Robert B. Lindsay, ed.** Stroudsburg, PA: Dowden, Hutchinson and Ross, 1976.
This collection of papers from the second half of the nineteenth century presents that period's concepts of energy. Well-annotated. Recommended to physicists; for all collections that deal with energy literature.
R: *Nuclear Science and Engineering* 64: 801 (Nov. 1977); *Choice* 13: 1630 (Feb. 1977); *TBRI* 43: 397 (Dec. 1977).

Bibliographical History of Electricity and Magnetism. **Paul Fleury Mottelay, comp.** London: Griffin, 1922. Reprint. New York: Arno Press, 1975.

The chronological bibliographic history spans the time from 2637 BC to Dec. 25, 1821. The book is exhaustively documented.
R: *ARBA* (1976, p. 772); Wal (p. 110); Win (EI94).

The Birth of Particle Physics. **Laurie M. Brown and Lillian Hoddeson, eds.** Cambridge: Cambridge University Press, 1983.

The results presented at the International Symposium on the History of Particle Physics in May 1980. Includes 2 panel discussions and additional papers from remarks made from the floor and physicists not present at the conference. A most unique and organized production.
R: *Nature* 308: 383 (Mar. 1984).

Early Concepts of Energy in Atomic Physics. **Robert Lindsay, ed.** New York: Academic Press, 1979.

Extracts from important studies in the history of energy and atomic physics.
R: *New Scientist* 84: 289 (Oct. 25, 1979); *TBRI* 46: 6 (Jan. 1980).

Electrical Inventions, 1745–1976. **Geoffrey W. A. Dummer.** Elmsford, NY: Pergamon, 1977.

Lists nearly 300 electrical inventions and their inventors. Considered a good research tool. For academic and research libraries.
R: *Endeavour* 2: 47 (1978); *TBRI* 44: 273 (Sept. 1978).

Electricity from Glass: The History of the Frictional Electrical Machines, 1600–1850. **W. D. Hackman.** The Hague, Netherlands: Sijthoff-Hoordhoff, 1978.

Classifies and describes the evolution of electrical machines to the beginnings of the industrial revolution. Presents the beginning of a systematic catalog of electrical inventions. Includes illustrations. Helpful in understanding the origins of electrical experimentation.
R: *Nature* 282: 763 (Dec. 13, 1979).

Electricity, Magnetism, and Animal Magnetism: A Checklist of Printed Sources, 1600–1850. **Ellen G. Gartrell, comp.** Wilmington: Scholarly Resources, 1975.

Established the holdings of 6 libraries on the topic of the history of electricity, magnetism, and animal magnetism. Divided into 2 broad areas: "Electricity and Magnetism, 1600–1850," and "Animal Magnetism to 1850," further subdivided by separate time periods: "Before 1801" and "1801–1850." Geared toward clientele interested in the subject of electricity and magnetism.
R: *ARBA* (1977, p. 762).

Electron Diffraction, 1927–1977. **P. J. Dobson et al.** Bristol, England: Institute of Physics, 1978.

A compilation of papers on advances in electron diffraction research.
R: *Acta Crystallographica* A35: 349 (Mar. 1, 1979); *TBRI* 45: 203 (June 1979).

Energy: Historical Development of the Concept. **Robert B. Lindsay, ed.** Stroudsburg, PA: Dowden, Hutchinson and Ross, 1976.

A historical reference on the history of energy. Covers papers written from antiquity through the mid-nineteenth century.
R: *Fuel* 56: 348 (July 1977); *TBRI* 43: 286 (Oct. 1977).

Essays in the History of Mechanics. **C. Truesdell.** New York: Springer-Verlag, 1968.
R: *Science* 168: 354 (Apr. 17, 1970).

Exploring the History of Nuclear Physics. Brookline, MA: Exploratory Conference on the History of Nuclear Physics, 1972.

General Relativity: An Einstein Centenary Survey. **S. W. Hawking and W. Israel.** Cambridge: Cambridge University Press, 1979.
Contains essays on the progress of the theory of relativity, marked by the hundredth anniversary of the birth of Albert Einstein. Well-arranged volume, covering progress by decade. Highly recommended.
R: *Nature* 280: 703 (Aug. 23, 1979).

General Relativity and Gravitation: One Hundred Years after the Birth of Albert Einstein. 2 vols. **A. Held, ed.** New York: Plenum Press, 1980.
Contains review and research articles on general relativity and gravitation. Contribution from international experts. A recommended reference source in the history of physics.

The Historical Development of Quantum Theory. The Discovery of Quantum Mechanics, 1925. Vol. 2. **J. Mehra and H. Rechenberg.** New York: Springer-Verlag, 1982.
R: *Physics Today* 36: 75 (July 1983).

Historical Studies in the Physical Sciences. **Russell McCormmach, ed.** *Physics Circa 1900: Personnel, Funding, and Productivity of the Academic Establishments.* Vol. 5. **Paul Forman et al., eds.** Princeton: Princeton University Press, 1975.
R: *Science* 191: 554 (Feb. 13, 1976).

History of Modern Physics 1800–1950. **The Tomash Publishers.** Los Angeles: Tomash, 1983.
Volume 1, *ALSOS* by Samuel Goudsmit; volume 2, *PROJECT Y: The Los Alamos Story*—part 1, *Toward Trinity* by D. Hawkins, part 2, *Beyond Trinity* by E. C. Truslow and R. C. Smith; volume 3, *American Physics in Transition: A History of Conceptual Change in the Late Nineteenth Century*; volume 4, *The Question of Atoms: From the Karlsruhe Congress to the First Solvay Conference.*
A major contribution in the history of modern physics. Documented, readable, and good in details.
R: *Physics Today* 36: 72 (June 1983).

History of Physics. **Spencer Weart and Melba Phillips, eds.** New York: American Institute of Physics, 1985.
A valuable library reference, a useful supplementary text for college courses, and a stimulating reading for both physicists and informed public on the history of physics.

Written by accomplished physicists including 7 Nobel Prize winners. Well documented and with over 300 photographs and illustrations.

History of the Quantum Theory. **Friedrich Hund.** Translated by Gordon Reece. New York: Barnes & Noble, 1974.

Translated from the German edition published in Mannheim, W Germany, 1967.
R: *Science* 186: 917 (Dec. 6, 1974).

History of Twentieth-Century Physics. **C. Weiner.** London: Academic Press, 1977.

A collection of papers presented at the course on the history of physics at the Italian Physical Society in 1972. The papers have a strong emphasis on atomic physics. No criteria are given for the selection of these 13 articles.
R: *Physics Bulletin* 24: 476 (Oct. 1978).

Landmark Experiments in Twentieth-Century Physics. **G. L. Trigg.** New York: Crane-Russak, 1975.

Literature on the History of Physics in the Twentieth Century. **J. L. Heilbron and Bruce R. Wheaton.** Berkeley: University of California, Office for History of Science and Technology, 1981.

The most comprehensive bibliography of twentieth-century physics literature to date. Arranged by subject, except for 3 chapters dealing with histories, biography, and institutions. Numerous cross references and an index to authors. Necessary for collections about history of science or physics.
R: *ARBA* (1983, p. 609).

Masers and Lasers: An Historical Approach. **Mario Bertolotti.** Philadelphia: Heyden, 1983.

An informative and authoritative account of the history of quantum electronics and quantum optics. For second- or third-year students of physics at the college level.
R: *Choice* 21: 735 (Jan. 1984).

The National Physical Laboratory—A History. **Edward Pyatt.** Bristol: Adam Hilger, 1983.

A complete chronological overview of the origins of the National Physics Laboratory, back to its founding. Six appendixes with biographies of principal characters, a calendar of events, and highlights of the history of standards, as well as diagrams, tables, historical records, and many photographs.
R: *Physics in Technology* 14: 253 (Sept. 1983).

The Nature of Light: An Historical Survey. **Vasco Ronchi.** Cambridge, MA: Harvard University Press, 1970.

R: *Applied Optics* (Dec. 1971); *Optical Society of America Journal* (Dec. 1971).

Niels Bohr, Collected Works. 8 vols. Amsterdam: North-Holland. 1972–. In progress.

Volume 1, *Early Work (1905–1911)*, 1972; volume 2, *Work on Atomic Physics (1912–1917)*, 1977; volume 3, *The Correspondence Principle (1918–1923)*, 1972; volume 4, *The Periodic System (1920–1923)*, 1976.

Origins in Acoustics: The Science of Sound from Antiquity to the Age of Newton. **Frederick V. Hunt.** New Haven: Yale University Press, 1978.

An in-depth history filled with extensive references from the seventeenth century forward. Comprises a source that contributes to the understanding of the origins of acoustics. Detailed, highly recommended.
R: *American Scientist* 67: 123 (Jan.–Feb. 1979); *Journal of the Acoustical Society of America* 65: 553 (Feb. 1979); *TBRI* 45: 165 (May 1979).

The Physicists: The History of a Scientific Community in Modern America. **Daniel J. Kevles.** New York: Knopf, 1978.

A primary reference source for scholars and educators in physics history.
R: *Isis* 69: 634 (Dec. 1978); *Nature* 272: 776 (Apr. 27, 1978); *TBRI* 45: 126 (Apr. 1979).

Radioactivity in America: Growth and Decay of a Science. **Lawrence Badash.** Baltimore: Johns Hopkins University Press, 1979.

Chronicles the growth of radioactivity research in the United States.
R: *Science* 206: 547 (Nov. 2, 1979); *TBRI* 46: 1 (Jan. 1980).

Radioactivity: A Science in Its Historical and Social Context. **Eric N. Jenkins.** New York: Crane, Russak, 1979.

A useful historical tool.
R: *Physics Education* 14: 389 (Sept. 1979); *TBRI* 46: 5 (Jan. 1980).

Resources for the History of Physics: 1, Guide to Books and Audiovisual Materials; 2, Guide to Original Works of Historical Importance and Their Translations into Other Languages. **Stephen G. Brush, ed.** Hanover, NH: University Press of New England, 1972.

A by-product of an international seminar held at MIT in 1970. The book contains a single author/title index to both sections. Useful particularly to a research library.
R: *ARBA* (1974, p. 560).

Revolution in Miniature: The History and Impact of Semiconductor Electronics. **Ernest Braun and Stuart MacDonald.** New York: Cambridge University Press, 1978.

Documents the beginning of the semiconductor industry in a careful, accurate manner. Deserves widespread readership.
R: *Nature* 276: 147 (Nov. 9, 1978); *Science* 201: 1217 (Sept. 29, 1978); *Scientific American* 239: 36 (Nov. 1978); *TBRI* 44: 351 (Nov. 1978); 44: 391 (Dec. 1978); 45: 32 (Jan. 1979).

Rutherford and Physics at the Turn of the Century. **Mario Bunge and William R. Shea, eds.** New York: Neale Watson Academic Publications, 1979.

Describes the state of physics at the turn of the century, focusing on such luminaries as Einstein, Rutherford, Chamberlain, Wilson, and Freud. A lively and interesting volume for history of science collections.

Thirty Years of Fusion. Fusion: Science, Politics, and the Invention of a New Energy Source. **Joan Lisa Bromberg.** Cambridge, MA: MIT Press, 1982.

The basic story of magnetic fusion as described by the 30-year-old US program to confine and control the fusion process with macroscopic electromagnetic fields.
R: *Laser Focus* 19: 126 (May 1983).

Twentieth Century Physics. **Joseph Norwood.** Englewood Cliffs: NJ: Prentice-Hall, 1975.
R: *Nature* 261: 174 (May 13, 1976).

From X-rays to Quarks: Modern Physicists and Their Discoveries. **Emilio Segre.** San Francisco: W. H. Freeman, 1980.

A clear and readable explanation of important physics research done in the last 100 years. Covers the discoveries of X-rays, relativity, radioactivity, and the formulation of modern quantum theory. Provides excellent photographs and personal anecdotes about such personalities as Marie Curie and Max Planck. Some background knowledge of chemistry and algebra is assumed. Bibliography is included. A worthwhile text for all students and professors.
R: *New Scientist* 88: 525 (Nov. 20, 1980); *Scientific American* 244: 44 (Mar. 1981); *BL* 77: 429 (Nov. 15, 1980); *Choice* 18: 547 (Dec. 1980).

CHEMISTRY

A Century of Chemistry. The Role of Chemists and the American Chemical Society. **Herman Skolnik and Kenneth M. Reese, eds.** Washington, DC: American Chemical Society, 1976.

The book traces the development of chemical science and technology in the framework of the 27 technical divisions of the Society. It includes 100-year record of ACS people and events.

Chemical, Medical, and Pharmaceutical Books Printed Before 1800, in the Collection of the University of Wisconsin Libraries. **John Neu, ed.** Madison, WI: University of Wisconsin Press, 1965.

Includes the D. I. Duveen collection in chemistry and alchemy.

Classical Scientific Papers: Chemistry. 2d series. **David M. Knight, ed.** New York: American Elsevier, 1971.

First series, 1968.
R: *Journal of Chemical Education* 98: 10 (Oct. 1971); *Journal of the American Chemical Society* 93: 15 (July 28, 1971); *Science* 162: 110 (1968); *LJ* 93: 3794 (1968); Jenkins (F52).

A History of Analytical Chemistry. **Herbert A. Laitenen and G. W. Ewing.** Washington, DC: American Chemical Society, Division of Analytical Chemistry, 1977.

A valuable source for analytic chemists and for students in all disciplines of science.
R: *Analytical Chemistry* 50: 73A (Jan. 1978); *TBRI* 44: 85 (Mar. 1978).

A History of Biochemistry. Pt 5. **Marcel Florkin, ed.** New York: American Elsevier, 1979.

Part 1, Proto-Biochemistry, 1972; part 2, From Proto-Biochemistry to Biochemistry, 1972.
R: *Science* 180: 606 (May 11, 1973).

BIOLOGICAL SCIENCES

The Art of Natural History: Animal Illustrators and Their Work. **S. Peter Dance.** New York: Viking Press, 1978.

Contains examples of zoological illustrations from paleolithic cave paintings through the nineteenth century. High-quality reproduction throughout, as well as much accurate information. Highly recommended to almost any library.
R: *LJ* 104: 200 (Jan. 15, 1979).

A Bio-Bibliography for the History of the Biochemical Sciences since 1800. **Joseph S. Fruton.** Philadelphia: American Philosophical Society, 1982.

An alphabetical listing of names with dates of birth and death, followed by a listing of biographical and bibliographical reference work. A valuable work for historians of science, although some may object to the editor's selection of names.
R: *ARBA* (1983, p. 613).

A Century of DNA: A History of the Discovery of the Structure and Function of the Genetic Substance. **Franklin H. Portugal and Jack S. Cohen.** Cambridge, MA: MIT Press, 1978.

An excellent explanatory source on the history of DNA; covers the 100-year period from 1809 forward, in the isolation of DNA and the genetic code. Discusses all major researchers involved: Franklin, Chargaff, Watson, and Crick. Also contains detailed explanations of morphological structure of the molecule as well as other facts of microbiology. A lucid source.
R: *Nature* 272: 760 (Apr. 1978); *Aslib Proceedings* 43: 322 (July 1, 1978).

The Changing Scenes in the National Sciences, 1776–1976. **Clyde E. Goulden, ed.** Philadelphia: Academy of Natural Sciences, 1977.

A collection of papers presented at a symposium in 1976, dealing with the contributions of evolution, systematics, genetics, terrestrial and aquatic ecology, and sociobiology. There is a brief introduction to each section; the approach is historical.
R: *American Scientist* 67: 122 (Jan. 2, 1979).

The Eighth Day of Creation: The Makers of the Revolution in Biology. **H. F. Judson.** New York: Simon & Schuster, 1979.

A documentary history of the discovery of DNA. In 3 sections; the first describes events leading up to the discovery of DNA; the second details the state of the art in molecular biology; and the third deals with protein crystallography. Thorough, interesting, and well-referenced. For a wide audience. A fascinating history.
R: *Nature* 281: 505 (Oct. 11, 1979).

Evolutionary History of the Primates. **Frederick S. Szalay and Eric Delson.** New York: Academic Press, 1980.

A valuable reference work on various aspects of the taxonomy and evolutionary history of primate species. Lists 1,300 fossil primate deposits in alphabetical and chronological order. Contains over 260 figures and an excellent extensive bibliography. A worthy text for graduate or upper-level undergraduate students, as well as for all primatologists.

History of American Ecology. New York: Arno, 1977.

Discusses eighteenth- and nineteenth- century developments in the 4 disciplines of ecology: limnology, oceanography, plant ecology, and animal ecology. Includes primary essays on Lamarckian and Darwinian thought. Includes illustrations and graphs.
R: *Science* 203: 429 (Feb. 2, 1979).

History of Botany, 1860–1900: Being a Continuation of Sachs' History of Botany, 1530–1860. Reprint of 1909 ed. **Joseph R. Green.** New York: Russell, 1967.

History of Entomology. **Ray F. Smith, Thomas E. Mittler, and Carroll N. Smith, eds.** Palo Alto, CA: Annual Reviews, in cooperation with the Entomological Society of America, 1973.
R: *Science* 182: 1013 (Dec. 7, 1973).

A History of Fishes. 3d ed. **J. R. Norman and P. H. Greenwood.** New York: Halsted Press, 1976.

First edition, 1931; second edition, P. H. Greenwood, ed., 1963.
A natural history of fishes, clearly written and illustrated. Analytical index.
R: Wal (p. 233).

History of Genetics; From Prehistoric Times to the Rediscovery of Mendel's Laws. **Hans Stubbe.** Translated by T. R. W. Waters. Cambridge, MA: MIT Press, 1973.

Translated from the second German edition, Jena, E Germany, 1965.
R: *Science* 181: 336 (July 27, 1973).

A History of the Life Sciences. **Lois N. Magner.** New York: Dekker, 1979.

Discusses the development of embryology, cell theory, microbiology, physiology, and evolution from the seventeenth century to the beginning of the twentieth century. Includes biographical information. For undergraduate biology libraries.
R: *Nature* 284: 83 (Mar. 6, 1980).

History of the Life Sciences: An Annotated Bibliography. **Pieter Smit.** New York: Hafner, 1974.

Originated as part of the life sciences section of Sarton's *Guide to the History of Science,* 1952. Over 4,000 entries emphasizing historical aspects of the life sciences, for scholarly research in biology and medicine. Includes works published through 1971.
R: *ARBA* (1976, p. 654); Sheehy (EC1).

Human and Mammalian Cytogenetics: An Historical Perspective. **T. C. Hsu.** New York: Springer Publishing, 1979.

Discusses the important discoveries of DNA, chromosomal structure, and other aspects of cytogenetics. Provides biographical information on key researchers. An interesting scientific history.
R: *Nature* 282: 169 (Nov. 8, 1979).

An Illustrated History of the Herbals. **Frank J. Anderson.** New York: Columbia University Press, 1977.

A scholarly history of herbal medicine. Includes reproductions of woodcuts and engravings. Contains much botanical and biographical information. Discusses a wealth of information concerning ancient herbal medicine. For college and botanical libraries.
R: *American Scientist* 66: 510 (July/Aug. 1978); *Garden* 2: 34 (Mar./Apr. 1978); *Nature* 272: 653 (Apr. 13, 1978); *Scientific American* 239: 28 (July 1978); *LJ* 102: 2438 (Dec. 1977); *TBRI* 44: 173 (May 1978); 44: 215 (June 1978).

An Introduction to the History of Virology. **A. P. Waterson and Lise Wilkinson.** London: Cambridge University Press, 1978.

An excellent history that examines the evolution of the study of viruses. Emphasizes conceptual study of viruses, including animal, bacteriophages, scrapie, and others. Important historical information. Well-referenced.
R: *Nature* 278: 482 (Mar. 29, 1979); *Aslib Proceedings* 44: 110 (Mar. 1979).

Natural Science Books in English: 1600–1900. **David M. Knight.** New York: Praeger, 1972.

A bibliographic essay on scientific development in Great Britain over 3 centuries. Does not include full bibliographic data for many entries. Should be of use to both bibliographers and bibliophiles.
R: *Choice* 10: 1574 (Feb. 1973); *ARBA* (1974, p. 564).

Notes on the History of Nutrition Research. **Clive M. McCay and F. Verzár, eds.** Bern, Switzerland: Huber. Distr. Baltimore: Williams & Wilkins, 1973.
R: *Science* 182: 377 (Oct. 26, 1973).

The Story of the Royal Horticultural Society, 1804–1968. **Harold R. Fletcher.** Distr. New York: Oxford University Press, 1969.

A comprehensive history of the society. The appendixes include a calendar of events for the society from 1804 to 1968, the original charter of the society, and a list of the past and present holders of the Victoria Medal of Honour.
R: *LJ* 94: 2772 (Aug. 1969); *ARBA* (1970, p. 117).

Studies in the History of Biology. Vols. 1–. **William Coleman and Camille Limoges, eds.** Baltimore: Johns Hopkins University Press, 1977–. Annual.

Scholarly contributions on the history of biology. A major addition to the literature. Discusses all aspects of evolution, genetics, Darwinian thought, etc.
R: *American Scientist* 67: 122 (Jan./Feb. 1979); *LJ* 102: 718 (Mar. 15, 1977).

Virus: A History of the Concept. **Sally S. Hughes.** London: Heinemann Educational, 1977.

An excellent survey that concentrates on virus research up to the turn of the century. This reference is highly recommended to microbiologists and historians.
R: *New Scientist* 76: 36 (Oct. 6, 1977); *Aslib Proceedings* 43: 235 (May 1978); *TBRI* 43: 378 (Dec. 1977).

AGRICULTURE

Agricultural Records in Britain, AD 220–1977. 2d ed. **John M. Stratton.** Hamden, CT: Shoe String Press, 1979.

A compilation of agricultural records presented chronologically. Provides an overview of the effect of weather on crops. Includes general rainfall tables and a table of agricultural prices back to 1257.

Agriculture in America, 1622–1860. **Andrea J. Tucher, comp.** New York: Garland, 1984.

An alphabetical bibliography of the printed works on agriculture issued before 1861. Highly recommended for libraries interested in the history of agriculture.
R: *Choice* 21: 543 (Dec. 1984); *ARBA* (1985, p. 496).

Bibliography of Books and Pamphlets on the History of Agriculture in the United States, 1607–1967. **John T. Schlebecker.** Santa Barbara, CA: Clio Press, 1969.

About 2,000 entries arranged alphabetically by author. Subject and author indexes.
R: *Choice* 7: 370 (1970); *LJ* 95: 1012 (1970); *RQ* 10: 166 (Winter 1970); Wal (p. 404); Win (3EK3).

Old Farm Tools and Machinery: An Illustrated History. **Percy W. Blanford.** Fort Lauderdale, FL: Gale Research, 1976.

Details the development of agricultural tools from animals to steam, to tractors. Illustrations of types of machinery include ploughs, harvesting, and feeding. An excellent addition for history of technology collections.
R: *RSR* 5: 37 (Oct./Dec. 1977).

EARTH SCIENCES

A Century of Weather Service; A History of the Birth and Growth of the National Weather Service, 1870–1970. New York: Gordon & Breach, 1970.

Besides the text and photographs, an appendix that lists chronologically the meteorological milestones of the American Weather Service, 1644–1970.
R: *ARBA* (1971, p. 521).

The Edge of an Unfamiliar World: A History of Oceanography. **Susan Schlee.** London: Hale, 1973.

Essays in History of Geology. **George Willard White.** New York: Arno, 1978.
Reprints from science journals, which include articles on early American geology.
R: *Isis* 70: 289 (June 1979); *TBRI* 45: 252 (Sept. 1979).

Geological Society of America, 1888–1930: A Chapter in Earth Science History. **Herman L. Fairchild.** Geological Society of America, 1932.

A History of American Archaeology. 2d ed. **Gordon R. Willey and Jeremy A. Sabloff.** San Francisco: W. H. Freeman, 1980.
R: *Science* 187: 425 (Feb. 7, 1975).

History of Cave Science: The Scientific Investigation of Limestone Caves, to 1900. 2 vols. **Trevor R. Shaw.** Crymych, Wales: Oldham Books, 1979.
An important work for cave scientists.
R: *NSS News* 37: 292 (Dec. 1979); *TBRI* 46: 92 (Mar. 1980).

The History of British Geology; A Bibliographical Study. **John Challinor.** New York: Barnes & Noble, 1972.
R: *Science* 177: 507–508 (Aug. 11, 1972).

History of the Earth. 2d ed. **Bernhard Hummel.** San Francisco: W. H. Freeman, 1970.
R: *Journal of Geology* (July 1971); *Journal of Geological Education* (Nov. 1971).

History of the Earth Sciences during the Scientific and Industrial Revolutions with Special Emphasis on the Physical Geosciences. **D. H. Hall.** New York: American Elsevier, 1976.

The History of the Study of Landforms or the Development of Geomorphology. London: Methuen, 1973. Distr. New York: Barnes & Noble.
R: *Science* 182: 375 (Oct. 26, 1973).

Into the Deep: The History of Man's Underwater Exploration. **Robert F. Marx.** New York: Van Nostrand Reinhold, 1978.
Traces the history of deep-sea diving from early Greek times. An interesting source of references.
R: *Oceans* 11: 68 (Nov.–Dec. 1978); *TBRI* 45: 5 (Jan. 1979).

Minerals, Lands and Geology for the Common Defense and General Welfare. Vol. 1. **Mary C. Rabbitt.** Reston, VA: US Geological Survey, 1979.
Volume 1, *Before 1879.*
First volume in the projected 4-volume history of the US Geological Survey. Excellent introduction to American geology. Includes excellent bibliography and numerous illustrations.
R: *Science* 205: 891 (Aug. 31, 1979); *LJ* 105: 571 (Mar. 1, 1980).

Scientists and the Sea, 1650–1900: A History of Marine Science. **Margaret Deacon.** New York: Academic Press, 1971.

Scripps Institute of Oceanography: Probing the Oceans, 1936 to 1976. **Elizabeth N. Shor.** San Diego: Tofua, 1978.

A clear and accurate history of oceanographic research completed by the Scripps Institute.

R: *Science* 202: 969 (Dec. 1, 1978); *TBRI* 45: 8 (Jan. 1979).

The Thermal Theory of Cyclones: A History of Meteorological Thought in the Nineteenth Century. **Gisela Kutzbach.** Boston: American Meteorological Society, 1979.

Examines thermal studies of cyclone developments in the nineteenth century. Excellent illustrations included. Mainly for university libraries, meteorologists, and science historians.

Two Hundred Years of Geology in America. **Cecil J. Schneer, ed.** Hanover, NH: University Press of New England, 1979.

A welcome addition to all geology and history of science collections. Based on a compendium of 27 papers presented at a bicentennial symposium at the University of New Hampshire.

R: *Science* 207: 49 (Jan. 4, 1980); *LJ* 105: 511 (Mar. 1, 1980).

GENERAL ENGINEERING

Bibliography of the History of Technology. **Eugene S. Ferguson.** Cambridge, MA: Society for the History of Technology and MIT Press, 1968.

Primarily an annotated list of primary and secondary sources on the history of technology. Intended as a student's guide.

R: *ABL* 34: entry 55 (Feb. 1969); *CRL* 30: 380 (1969); Wal (p. 52); Win (2EA34).

History of Control Engineering: 1800–1930. **Stuart Bennett.** London: Peregrinus, 1979.

This history of engineering covers a broad range of concepts. Detailed, well-referenced. For theorists, engineers, and students.

R: *Electronics and Power* 25: 651 (Sept. 1979); *Aslib Proceedings* 45: 25 (Jan. 1980); *TBRI* 46: 30 (Jan. 1980).

History and Philosophy of Technology. **George Bugliarello and D. B. Doner.** Urbana: University of Illinois Press, 1979.

Based on 24 papers from a symposium, this book contains a varied collection of papers on the history and philosophy of technology.

R: *New Scientist* 84: 288 (Oct. 25, 1979); *Science* 206: 1175 (Dec. 7, 1979); *TBRI* 46: 31 (Jan. 1980); 46: 71 (Feb. 1980).

History of Technology. Vol. 4. **A. Rupert Hall and Norman Smith, eds.** London: Mansell. Distr. Forest Grove, OR: International Scholarly Books, 1979. Volume 3, 1978.

Contains review papers on such diverse topics as transportation, engineering, and photography. Considered a highly authoritative reference.

R: *Marine Engineers Review* 53 (Dec. 1978); *Aslib Proceedings* 43: 284 (June 1978); 44: 19 (May 1979); *TBRI* 45: 153 (Apr. 1979).

A History of Technology. Vols. 6 and 7. **Trevor I. Williams, ed.** New York: Oxford University Press, 1979.

Volume 6, *The Twentieth Century, c.1900 to c.1950,* part 1; volume 7, *The Twentieth Century, c.1900 to c.1950,* part 2.

A multivolume treatise on the history of technology up to the twentieth century. Covers all areas of technology. A scholarly and well-researched work.

R: *Physics Bulletin* 30: 398 (Sept. 1979); *Science* 204: 747 (May 18, 1979).

A History of Technology and Invention: Progress Through the Ages. Vol. 3. **Maurice Daumas, ed.** Translated by Eileen B. Hennessy. New York: Crown, 1980.

Volume 3, *The Expansion of Mechanization, 1725–1860.*

An excellent, 3-volume set on the history of technology; each chapter is written by leading experts in their fields. Covers transportation and communication, textile industries, and military techniques. Essential source for all libraries.

R: *American Scientist* 68: 713 (Nov./Dec. 1980); *BL* 76: 1408 (June 1, 1980).

Technical Americana: A Checklist of Technical Publications Printed Before 1831. **Evald Rink.** Foreword by Eugene S. Ferguson. Millwood, NY: The Eleutherian Mills Historical Library, 1981.

AERONAUTICAL AND ASTRONAUTICAL ENGINEERING

Aeronautics and Astronautics; An American Chronology of Science and Technology in the Exploration of Space, 1915–1960. Washington, DC: US National Aeronautics and Space Administration, 1961.

A listing, by day and year, of important events in aeronautics. Subject and name index.

Airplanes of the World, 1490–1976. Rev. ed. **Douglas Rolfe.** New York: Simon & Schuster, 1978.

A 12-part reference that details aviation history from the fifteenth century to the present. For each aircraft, an introduction, drawing, and full description are provided. Manufacturer, nomenclature, and aircraft nickname indexes are included. Of interest to all library collections.

R: *ARBA* (1980, p. 725).

Astronautics and Aeronautics; Chronology on Science, Technology, and Policy. Washington, DC: US National Aeronautics and Space Administration, 1961–. Annual.

Helicopters of the World. **Michael J. H. Taylor and John W. R. Taylor.** New York: Scribner's, 1976.

Traces the history of the helicopter around the world. Arranged alphabetically; provides detailed descriptions.

R: *BL* 73: 1672 (July 1, 1977); *WLB* 51: 685 (Apr. 1977); *ARBA* (1978, p. 762).

The History of Man-Powered Flight. **David A. Reay.** Elmsford, NY: Pergamon, 1978.
Describes hundreds of man-powered flight vehicles. Discusses Daedalus, Leonardo da Vinci, etc. Includes technical details. A well-put-together reference.
R: *Canadian Aeronautics and Space Journal* 24: 386 (Nov./Dec. 1978); *Aslib Proceedings* 43: 211 (Apr. 1978); *TBRI* 45: 77 (Feb. 1979).

The Rocket: The History and Development of Rocket and Missile Technology. **David Baker.** London: New Cavendish Books, 1979.
Traces the development of rocketry from the Sung dynasty through the nineteenth century. Synthesizes data otherwise available only in many different sources.
R: *Interavia* 34: 1189 (Dec. 1979); *TBRI* 46: 111 (Mar. 1980).

Two Hundred Years of Flight in America: A Bicentennial Survey. **E. M. Emme,** ed. San Diego: Univelt, 1977.
Culls papers presented at a symposium which covered dirigibles, airplanes, and space flight. Fact-filled volume includes charts, graphs, and illustrations.
R: *Aslib Proceedings* 43: 383 (Sept. 1978).

CHEMICAL ENGINEERING

The Chemical Industry: 1900–1930: International Growth and Technological Change. **L. F. Haber.** Oxford: Oxford University Press, Clarendon Press, 1971.
R: *Endeavour* (Sept. 1971).

Frozen Foods: Biography of an Industry. **E. W. Williams.** Westport, CT: AVI, 1973.
History of the frozen-food industry from 1900 on, written in a popular style. Chapters reveal the problems faced and overcome by frozen-food manufacturers, packers, retailers, consumers, etc.

The History of the British Petroleum Company. The Developing Years, 1901–1932. Vol. 1. **R. W. Ferrier.** New York: Cambridge University Press, 1982.
Presents the detailed history of the Anglo-Iranian Oil Company, now known as British Petroleum, in its actions with the Middle East. Its thorough format will prove to be a useful source for the serious student of the Middle East.
R: *Nature* 301: 547 (Feb. 1983).

History of Chemical Engineering. **William F. Furter,** ed. Washington, DC: American Chemical Society, 1980.
R: *Chemical Engineering* 88: 119 (June 29, 1981); *Science* 212: 773 (May 15, 1981).

CIVIL ENGINEERING

History of Building Types. **Nikolaus Pevsner.** Princeton: Princeton University Press, 1976.

Surveys the history of architecture from the Middle Ages to the present. Contains a good bibliography and illustrations. A valuable reference.
R: *Technology and Culture* 18: 239 (Apr. 1977); *TBRI* 43: 277 (Sept. 1977).

History of Public Works in the United States, 1776–1976. **American Public Works Association.** Chicago: American Public Works Association, 1976.

Comprehensive discussions on the provisions of water power, waste disposal, and transportation to the general public. Emphasizes the nation's efforts to improve the human environment. A valuable collection of information. For all research and public libraries.
R: *Reclamation Era* 62: 28 (Autumn 1976); *Technology and Culture* 18: 248 (Apr. 1977); *RSR* 5: 24 (Oct./Dec. 1977); *TBRI* 43: 227 (June 1977); 43: 272 (Sept. 1977).

ELECTRICAL AND ELECTRONICS ENGINEERING

Bell Laboratories Innovation in Telecommunications: 1925–1977. **Roland Mueser, ed.** Murray Hill, NJ: Bell Laboratories, Technical Documentation Department, 1979.

Bibliography of the History of Electronics. **George Shiers.** Metuchen, NJ: Scarecrow Press, 1972.

Emphasizes the historical aspects of electronics and telecommunications since 1860. Bibliographic data on some 1,800 entries. For everyone from the electronics buff to the science historian.
R: *ARBA* (1973, p. 649).

Fifty Years of Electronic Components—1921–1971. **H. A. G. Hazeu.** Eindhoven, Netherlands: Philip's Gloeilampenfabrieken, 1971.
R: *IEEE Spectrum* (Dec. 1971).

Electrical and Electronic Technologies: A Chronology of Events and Inventors to 1900. **Henry B. O. Davis.** Metuchen, NJ: Scarecrow Press, 1981.

The volume has 6 sections, each describing a particular era, from before Christ through the nineteenth century. Each has an introduction and a chronological sequence of entries specified by the individual responsible for the innovation. Index is topically arranged.
R: *Choice* 19: 890 (March 1980); *WLB* 56: 381 (Jan. 1982).

Electronic Inventions, 1745–1976. **Geoffrey W. A. Dummer.** Elmsford, NY: Pergamon, 1977.

Covers inventions from Europe and the United States. Inventions arranged in chronological order from 1642 forward. A valuable book in history of technology collections.
R: *Electronics and Power* 23: 243 (Mar. 1977); *Aslib Proceedings* 42: 341 (June 1977); *Choice* 14: 511 (June 1977); *RSR* 5: 27 (Oct./Dec. 1977); *TBRI* 43: 231 (June 1977); *ARBA* (1978, p. 771).

A History of Engineering and Science in the Bell System: The Early Years (1875–1925). **M. D. Fagen, ed.** Murray Hill, NJ: Bell Telephone Laboratories, 1975.

The most comprehensive account of the technical evolution of a major industrial corporation. The 1,073-page volume is organized by technical categories, each covered chronologically from 1875 on.
R: *American Scientist* 64: 708 (Nov./Dec. 1976); *Physics Today* 29: 64 (Nov. 1976); *Science* 196: 49–50 (April 1, 1977).

From Spark to Satellite; A History of Radio Communication. **Stanley Leinwoll.** New York: Scribner's, 1979.

Contains biographies of scientists involved in the pioneering stages of radio communication.
R: *Science Digest* 87: 93 (Jan. 1980); *TBRI* 46: 76 (Feb. 1980).

COMPUTER ENGINEERING

An Annotated Bibliography on the History of Data Processing. **James W. Cortada, comp.** Westport, CT: Greenwood Press, 1983.

Focuses on the contributions, rather than the documentations, in hardware and computing concepts made by institutions as well as individuals. Concentrates on evolutionary eras from before 1800, 1800–1939, 1939–1955, and 1955–1982, which have been eras of major growth in the computer industry.
R: *Choice* 20: 550 (Dec. 1983); *ARBA* (1984, p. 615).

From Dits to Bits: A Personal History of Electronic Computers. **Herman Lukoff.** Portland, OR: Robotics Press, 1979.

An entertaining account of the development of the computer. For public libraries.

A History of Computing in the Twentieth Century. **N. Metropolis, J. Howlett, and Gian-Carlo Rota, eds.** New York: Academic Press, 1980.

A multiauthored book on the history of computing. Offers material by pioneers in the field. Of interest to computer scientists, mathematicians, and electrical engineers.
R: *Science* 212: 536 (May 1, 1981); *Computer* 16: 26 (Feb. 1983).

A History of Microtechnique. **Brian Bracegirdle.** London: Heinemann, 1978.

This well-known book comprises a history of the microscope and its uses in scientific research beginning in the eighteenth century. There are numerous illustrations, and a descriptive bibliography of works published before 1910. An impressive, scholarly work that contains much noteworthy information.
R: *Nature* 277: 155 (Jan. 11, 1979); *Royal Microscopical Society Proceedings* 14: 48 (Jan. 1979); *Science* 204: 748 (May 18, 1979); *TBRI* 45: 95 (Mar. 1979); 45: 134 (Apr. 1979).

History of Programming Languages. **Richard L. Wexelblat, ed.** New York: Academic Press, 1981.

Constitutes the proceedings of an ACM SIGPLAN Conference held in June 1978. A record of the early history of 13 languages that have set the tone of most of today's programming.
R: *Quarterly of Applied Mathematics* 41: 356 (Oct. 1983).

MECHANICAL ENGINEERING

American Heritage History of the Automobile in America. **Stephen W. Sears.** New York: American Heritage, 1977.
R: *TBRI* 44: 277 (Sept. 1978).

A Centennial History of The American Society of Mechanical Engineers, 1880–1980. **Bruce Sinclair.** Toronto: University of Toronto Press, 1980.

A history of The American Society of Mechanical Engineers in both the sphere of social economy and the sphere of technology. It deals with how the society serves its members and the public, and the standards it sets. Good reading for members. A fine contribution to the history of engineering.
R: *American Scientist* 69: 110 (Jan. 2, 1981).

A History of Theory of Structures in the Nineteenth Century. **T. M. Charlton.** New York: Cambridge University Press, 1982.

A uniquely comprehensive account of a century of the development of the theory of structures with a detailed and critical account of the development and application of those principles.
R: *Mechanical Engineering* 105: 88 (Jan. 1983).

History of Tribology. **D. Dowson.** London: Longman, 1978.

Discusses the chronological development of tribology, including biographical information on people who founded this new science.
R: *New Scientist* 82: 50 (Apr. 5, 1979); *TBRI* 45: 234 (June 1979).

Mechanical Engineering: A Decade of Progress: 1960–1970. **E. G. Semler, ed.** London: Institution of Mechanical Engineers, 1971.

Series of articles originally published in *The Chartered Mechanical Engineer* between 1960 and 1970.

National Historic Mechanical Engineering Landmarks. **Richard S. Hartenberg, ed.** New York: American Society of Mechanical Engineers, 1979.

Photographs and historical data on landmark discoveries in mechanical engineering. Insightful perspectives. Recommended for mechanical engineering collections.
R: *Mechanical Engineering* 101: 96 (Oct. 1979); *TBRI* 45: 387 (Dec. 1979).

NUCLEAR ENGINEERING

A World Destroyed: The Atomic Bomb and the Grand Alliance. **Martin J. Sherwin.** New York: Knopf, 1975.

A double book dealing with the development of the first atomic weapons and the role they played in American foreign policy up to 1945; and the conflict between the views of the nuclear scientists and foreign-policy makers.
R: *Science* 194: 174 (Oct. 8, 1976).

ENVIRONMENTAL SCIENCE

American Environmental History: The Exploitation and Conservation of Natural Resources. **Joseph M. Petulla.** San Francisco: Boyd & Fraser, 1977.

Provides a basic history of the US environmental resources from a conservationist and anticorporate viewpoint. Well-illustrated. Recommended for agricultural and engineering libraries.
R: *Choice* 14: 889 (Sept. 1977).

Images of the Earth: Essays in the History of the Environmental Sciences. **L. J. Jordanova and R. S. Porter, eds.** London: British Society for the History of Science, 1979.

Contains individual perspectives on the history of geology, covering a broad scope of the field.
R: *New Scientist* 84: 44 (Oct. 4, 1979); *Science* 206: 1295 (Dec. 14, 1979); *TBRI* 45: 388 (Dec. 1979); 46: 44 (Feb. 1980).

The World Environment: 1972–1982. **Martin W. Holdgate, Mohammed Kassas, and Gilbert F. White, eds.** Dublin: Tycooly International, 1983.
R: *Science* 218: 922 (Nov. 26, 1982).

CHAPTER 13 IMPORTANT SERIES AND OTHER REVIEWS OF PROGRESS

GUIDE TO SERIES PUBLICATIONS

Irregular Serials and Annuals, An International Directory. 10th ed. New York: Bowker, 1985.

Third edition, 1974–1975; sixth edition, 1980; seventh edition, 1981–1982; eighth edition, 1983.

The Serials Directory. Premier ed. Birmingham, AL: EBSCO, 1986. 3 vols.

A major reference directory that includes 113,000 titles, both annuals and irregular series along with other type serials in 1 volume. International in coverage. Features a listing of over 2,000 ceased titles by both alphabetical and subject headings. Extensive indexing.

IMPORTANT SERIES

ASTRONOMY

Advances in Astronomy and Astrophysics. Vols. 1–. New York: Academic Press, 1962–. Annual.

Volume 9, 1972.
R: *American Scientist* 51: 92A (1963), 59: 620 (Sept./Oct. 1971); *ABL* 31: entry 118 (Mar. 1966); Jenkins (E6); Wal (p. 79).

Annual Review of Astronomy and Astrophysics. Vols. 1–. Palo Alto, CA: Annual Reviews, 1963–.

Volume 24, 1986.
R: *Journal of the Franklin Institute* 292: 231 (Sept. 1971); *Observatory* 99: 22 (Feb. 1979); *Physics Today* 23: 68 (Apr. 1970); *Sky and Telescope* 55: 170 (Feb. 1980); 60: 519 (Dec. 1980); *ARBA* (1971, p. 484; 1974, p. 554; 1976, p. 645; 1981, p. 631); Jenkins (E7); Wal (p. 79).

Vistas in Astronomy. Vols. 1–. Oxford: Pergamon, 1955–. Annual.

Volume 23, 1980.
R: *American Scientist* 57: 137A (Summer 1969); *Nature* 229: 127 (Jan. 25, 1971); *Physics Today* 23: 73 (Feb. 1970); *Observatory* 98: 34 (Feb. 1978); 99: 135 (Aug. 1979); *Science* 184: 660 (May 10, 1974); *Choice* 14: 889 (Sept. 1977); Wal (p. 80).

MATHEMATICS

Advances in Mathematics. Vols. 1–. New York: Academic Press, 1961–.
R: Jenkins (B13).

Advances in Probability and Related Topics. Vols. 1–. New York: Dekker, 1971–.
Volume 6, 1980.

Applied Mathematical Sciences. Vols. 1–. New York: Springer-Verlag, 1971–.
Volume 45, 1983.

Graduate Texts in Mathematics. Nos. 1–. New York: Springer-Verlag, 1971–.
Volume 91, 1983.

International Series in Pure and Applied Mathematics. Vols. 1–. New York: Pergamon, 1956–.
Volume 108, 1980.

Lecture Notes in Biomathematics. Vols. 1–. New York: Springer-Verlag, 1974–.
Volume 47, 1983.
R: *Biometrics* 34: 332 (June 1978); *TBRI* 44: 299 (Oct. 1978).

Lecture Notes in Mathematics. Vols. 1–. New York: Springer-Verlag, 1964–.
Volume 990, 1982.

Lecture Notes in Pure and Applied Mathematics. Vols. 1–. New York: Dekker, 1971–.
Volume 81, 1982.

London Mathematical Society Lecture Note Series. Vols. 1–. New York: Cambridge University Press, 1971–.
Volume 80, 1983.

Probabilistic Analysis and Related Topics. Vols. 1–. **A. J. Bharucha-Reid, ed.** New York: Academic Press, 1978–.
Volume 3, 1983.

Progress in Mathematics. Vols. 1–. New York: Plenum, 1967–.
Volume 13, 1972.
R: *American Scientist* 59: 272 (Mar./Apr. 1971); 59: 380 (May/June 1971); *ABL* 35: entry 568 (Dec. 1970); Wal (p. 69).

Studies in Logic and the Foundations of Mathematics. Vols. 1–. New York: American Elsevier, 1954–.
Volume 102, *Set Theory: An Introduction to Independence Proofs*, 1980.

Studies in Mathematical Education. Vols. 1–. **Robert Morris, ed.** New York: Unipub, 1980–. Annual.

Symposia Mathematica. Vols. 1–. New York: Academic Press, 1969–.
Volume 26, 1982.

Physics

Advances in Atomic and Molecular Physics. Vols. 1–. New York: Academic Press, 1965–.
Volume 20, 1984.
R: *American Scientist* 58: 431 (July/Aug. 1970); 59: 746 (Nov./Dec. 1971); *Physics Bulletin* 30: 27 (June 1979); 30: 534 (Dec. 1979); *Physics Today* 22: 75 (Nov. 1969); 24: 45 (Feb. 1971).

Advances in Chemical Physics. Vols. 1–. New York: Wiley-Interscience, 1958–. Irregular.
Volume 61, 1985.
R: *American Scientist* 60: 782 (Nov./Dec. 1972); *Chemistry in Britain* 16: 667 (Dec. 1980); *Journal of the American Chemical Society* 102: 7824 (Dec. 17, 1980); *Physics Today* 22: 111 (Jan. 1969); Jenkins (C18); Wal (p. 126).

Advances in Geophysics. Vols. 1–. New York: Academic Press, 1952–.
Volume 26, 1984.
R: *American Scientist* 58: 678 (Nov./Dec. 1970); Wal (p. 151).

Advances in Liquid Crystals. Vols. 1–. New York: Academic Press, 1975–.
Volume 6, 1983.
R: *Physics Bulletin* 30: 309 (June 1979); *Physics Today* 29: 51 (July 1976); *TBRI* 43: 121 (Apr. 1977).

Advances in Microwaves. Vols. 1–. New York: Academic Press, 1966–.
Volume 8, 1974.
R: *IEEE Spectrum* (July 1971); *ARBA* (1971, p. 562).

Advances in Optical and Electron Microscopy. Vols. 1–. New York: Academic Press, 1966–.
Volume 9, 1983.
R: *American Scientist* 58: 675 (Nov./Dec. 1970).

Advances in Plasma Physics. Vols. 1–. New York: Wiley-Interscience, 1968–.
Volume 6, 1976.
R: *American Scientist* 58: 328 (May/June 1970); *Physics Today* 22: 77 (Nov. 1969); 24: 56 (May 1971); *Science* 163: 803 (1969); *Choice* 6: 250 (1969); Jenkins (C21); Wal (p. 102).

Advances in Quantum Electronics. Vols. 1–. New York: Academic Press, 1970–.
Volume 3, 1975.
R: *American Scientist* 59: 617 (Sept./Oct. 1971).

Advances in Quantum Physics. New York: Academic Press, 1973–.
Volume 3, 1975.

Advances in Structural Research by Diffraction Methods. Vols. 1–. Oxford: Pergamon Press, 1964.
Volume 6, 1975.

Benchmark Papers in Acoustics. Vols. 1–. **R. B. Lindsay, ed.** New York: Academic Press, 1975–.
Volume 16, *Acoustical Measurements,* 1982.
R: *Journal of Sound and Vibration* 58: 605 (June 22, 1978); *TBRI* 44: 317 (Oct. 1978).

Benchmark Papers in Optics. Vols. 1–. **S. S. Ballard, ed.** New York: Academic Press, 1975–.
Volume 3, *Light in the Sea,* 1977.

Benchmark Papers in Physical Chemistry and Chemical Physics. Vols. 1–. **Joyce J. Kaufman and Walter S. Koski, eds.** New York: Academic Press, 1978–.
Volume 2, *X-ray Photoelectron Spectroscopy,* 1978; volume 3, *Ion-Molecule Reactions,* part 1, *Kinetics and Dynamics,* part 2, *Elevated Pressures and Long Reaction Times,* 1979; volume 5, 1982.
These collections of research papers are culled from the published literature and discuss current and relevant theoretical and experimental concepts in a chosen field.
R: *Chemistry in Britain* 15: 152 (Mar. 1979); 15: 260 (May 1979); 16: 390 (July 1980); *Physics Bulletin* 30: 161 (Apr. 1979).

Comments on Modern Physics. Vols. 1–. New York: Gordon & Breach, 1967–. Bimonthly.
Issued in 6 parts: *Comments on Nuclear and Particle Physics; Comments on Solid State Physics; Comments on Astrophysics and Space Physics; Comments on Atomic and Molecular Physics; Comments on Plasma Physics and Controlled Fusion; Comments and Communications in Statistical Physics.*

Infrared and Millimeter Waves. Vols. 1–. New York: Academic Press, 1979–.
Volume 12, 1984.

Electron Spectroscopy: Theory, Techniques, and Applications. Vols. 1–. **C. R. Brundle and A. D. Baker, eds.** New York: Academic Press, 1977–.
Volume 2, 1978; volume 3, 1979; volume 4, 1981.

Lecture Notes in Physics. Vols. 1–. New York: Springer-Verlag, 1969.
Volume 175, 1983.

Methods in Computational Physics. Vols. 1–. New York: Academic Press, 1963–.
Volume 18, 1982.

Methods of Experimental Physics. Vols. 1–. New York: Academic Press, 1959–.
Volume 21, 1983.
R: *American Scientist* 63: 218 (Mar./Apr. 1975); 63: 570 (Sept./Oct. 1975); *Journal of the American Chemical Society* 101: 2508 (Apr. 25, 1979); *Observatory* 99: 100 (June 1979); *Physics Bulletin* 32: 216 (July 1981); *Physics Today* 22: 91 (July 1969).

Methods of Surface Analysis. Vols. 1–. New York: American Elsevier, 1975–.
R: *American Scientist* 64: 442 (July/Aug. 1976).

Physical Acoustics. Vols. 1–. **W. P. Mason, ed.** New York: Academic Press, 1964–.
Volume 16, 1982.

Physical Sciences Data. Vols. 1–. New York: Elsevier North-Holland, 1978–.
Volume 12, 1982.

Physics and Chemistry in Space. Vols. 1–. New York: Springer-Verlag, 1970–.
Volume 10, *Nonlinear Phenomena in the Ionosphere*, 1978.
R: *Science* 203: 429 (Feb. 2, 1979).

Physics of Quantum Electronics. Vols. 1–. Reading, MA: Addison–Wesley, 1974–.
Volume 7, 1980.

Progress in Heat and Mass Transfer. Vols. 1–. Oxford: Pergamon Press, 1969–.
Volume 19, no. 10, 1977.
R: *American Scientist* 59: 476 (July/Aug. 1971).

Progress in Low Temperature Physics. Vols. 1–. New York: American Elsevier, 1955–. Irregular.
Volume 3, 1961; volume 6, 1970; volumes 7a and 7b, 1979; volume 8, 1983; volume 9, 1984.
R: *American Scientist* 59: 752 (Nov./Dec. 1971); *Physics Bulletin* 31: 99 (Mar. 1980); *Physics Today* 22: 107 (Jan. 1969); Wal (p. 109).

Progress in Nuclear Magnetics Resource Spectroscopy. Vols. 1–. Oxford: Pergamon Press, 1966–.
Volume 15, 1984.
R: *Chemistry in Britain* 7: 530 (Dec. 1971).

Progress in Optics. Vols. 1–. New York: North-Holland, 1961.
Volume 20, 1983.
R: *Physics Bulletin* 31: 20 (Jan. 1980); 32: 110 (Apr. 1981); *TBRI* 43: (July 1977).

Progress in Quantum Electronics. Vols. 1–. Oxford and New York: Pergamon Press, 1969–.
Volume 6, 1981.
R: *American Scientist* 66: 508 (July/Aug. 1978).

Progress in Surface Science. Vols. 1–. Oxford: Pergamon Press, 1971–.
Volume 12, 1983.

Reviews of Plasma Physics. Vols. 1–. New York: Consultants Bureau, 1965–.
Volume 8, 1980.
R: *Physics Bulletin* 30: 535 (Dec. 1979).

Solid-State Physics: Advances in Research and Applications. Vols. 1–. **Frederick Seitz, David Turnbull, and Henry Ehrenreich, eds.** New York: Academic Press, 1955–.
Volume 37, 1982.
R: *Physics Bulletin* 29: 531 (Nov. 1978); 30: 226 (May 1979).

Springer Series in Chemical Physics. Vols. 1– **I. I. Sobelman.** New York: Springer-Verlag, 1979-.
Volume 24, 1983.

Springer Series in Computational Physics. Vols. 1–. **W. Beiglbock and H. Cabannes, eds.** New York: Springer–Verlag, 1977–.
Volume 7, 1983.

Springer Series in Electrophysics. Vols. 1–. **W. Engl et al., eds.** New York: Springer-Verlag, 1977–.
Volume 10, 1982.

Springer Series in Solid-State Sciences. Vols. 1–. New York: Springer-Verlag, 1978–.
Volume 44, 1983.

Springer Tracts in Modern Physics. Vols. 1–. New York: Springer-Verlag, 1922–

Volume 100, 1983.
R: *Physics Bulletin* 31: 321 (Nov. 1980).

Techniques of Physics. Nos. 1–. **N. H. March and H. N. Daglish, eds.** New York: Academic Press, 1973–.
Number 5, *Green's Function and Condensed Matter,* 1981.

Topics in Applied Physics. Vols. 1–. New York: Springer-Verlag, 1973–.
Volume 45, 1985.
R: *Journal of the American Chemical Society 100:* 357 (Jan. 4, 1978); *Physics Bulletin* 30: 534 (Dec. 1979); 31: 205, 207 (July 1980): *Physics Today* 29: 76 (May 1976).

Biophysics

Advances in Biological and Medical Physics. Vols. 1–. New York: Academic Press, 1948–.
Volume 17, 1980.

Advances in Biophysics. Vols. 1–. Baltimore: University Parch Press, 1970–.
Volume 14, 1981.
R: *American Scientist* 60: 780 (Nov./Dec. 1972).

Annual Review of Biophysics and Bioengineering. Vols. 1–. **Donald M. Engelman, ed.** Palo Alto, CA: Annual Reviews, 1972–.
Volume 15, 1986.

Current Topics in Bioenergetics. Vols. 1–. **D. R. Sanadi, ed.** New York: Academic Press, 1966–.
Volume 13, 1984.

Progress in Biophysics and Molecular Biology. Vols. 1–. Oxford: Pergamon Press, 1950–.
Volume 42, 1984.
R: *Pharmaceutical Journal* 194: 159 (Feb. 13, 1965); Jenkins (G77); Wal (p. 203).

Nuclear Physics

Advances in Mass Spectrometry. Vols. 1–. Barking, Essex, England: Applied Science, 1959–.
Volume 6, 1974; volume 8, 1980, New York: John Wiley.

Advances in Nuclear Physics. Vols. 1–. New York: Plenum, 1968–.
Volume 14, 1984.
R: *American Scientist* 59: 369 (Jan./Feb. 1971); *Physics Bulletin* 29: 183 (Apr. 1978); 30: 309 (June 1979); *Physics Today* 22: 107 (Apr. 1969); 23: 74 (June 1970); *Science* 160: 1331 (1968); *ABL* 36: entry 473 (Nov. 1971); 44: entry 466 (Nov. 1979); Jenkins (C20); Wal (p. 308).

Advances in Particle Physics. Vols. 1–. New York: Wiley-Interscience, 1968–.
Irregular.

Annual Review of Nuclear and Particle Science. Vols. 1–. Palo Alto, CA: Annual Reviews, 1950–. Annual.
Volume 36, 1986.
Original title: *Annual Review of Nuclear Science.*

Progress in Nuclear Physics. Vols. 1–. Oxford: Pergamon Press, 1950–. Irregular.
Volume 13, 1977.
R: *American Scientist* 58: 677 (Nov./Dec. 1970); Wal (p. 308).

Chemistry

Advances in Analytical Chemistry and Instrumentation. Vols. 1–. New York: Wiley-Interscience, 1960–.
Volume 11, 1974.
R: *Nature* 190: 944 (June 10, 1961); Wal (p. 130).

Advances in Catalysis. Vols. 1–. New York: Academic Press, 1948–.
Volume 27, 1978; volume 28, 1979; volume 30, 1981.
Original title: *Advances in Catalysis and Related Subjects.*
R: *American Scientist* 67: 492 (July/Aug. 1979), 68: 463 (July/Aug. 1980), *Chemistry in Britain* 14: 90 (Jan. 1978); *Journal of the American Chemical Society* 101: 2255 (Apr. 11, 1979); 102: 7826 (Dec. 17, 1980).

Advances in Chemistry. Vols. 1–. Washington, DC: American Chemical Society, 1950–.
Volume 178, 1979.
R: *Chemistry in Britain* 7: 438 (Oct. 1971); *Science* 192: 1121 (June 11, 1976); *TBRI* 44: 11 (Jan. 1978).

Advances in Chromatography. Vols. 1–. New York: Dekker, 1966–.
Volume 24, 1984.
R: *American Scientist* 59: 751 (Nov./Dec. 1971); *Chemistry in Britain* 14: 620 (Dec. 1978); *TBRI* 44: 161 (May 1978); *ARBA* (1971, p. 487); Wal (p. 131).

Advances in Clinical Chemistry. Vols. 1–. **Harry Sobotka and C. P. Stewart, eds.** New York: Academic Press, 1958–.
Volume 23, 1983.

Advances in Electrochemistry and Electrochemical Engineering. Vols. 1–. New York: Wiley, 1961–.
Volume 12, 1981.

Advances in Photochemistry. Vols. 1–. **James N. Pitts, Jr. et al., eds.** New York: Wiley, 1963–.
Volume 13, 1986.
R: *Journal of the American Chemical Society 100:* 4923 (July 19, 1978); 102: 5131 (July 16, 1980); *Journal of the Electrochemical Society* 124: 433C (Dec. 1977); *TBRI* 44: 41 (Feb. 1978).

Advances in Quantum Chemistry. Vols. 1–. New York: Academic Press, 1964–.
Volume 16, 1982.
R: *Journal of the American Chemical Society* 100: 7788 (Nov. 22, 1978); Jenkins (D40); Wal (p. 125).

Advances in Radiation Chemistry. Vols. 1–. New York: Wiley-Interscience, 1969–.
Volume 5, 1976.
R: *American Scientist* 58: 677 (Nov./Dec. 1970); *Chemistry in Britain* (Oct. 1971); *Journal of the American Chemical Society* 100: 1017 (Feb. 1, 1978); *TBRI* 43: 241 (Sept. 1977); Wal (p. 127).

Advances in X-ray Analysis. Vols. 1–. New York: Plenum, 1965–.
Volume 25, 1982.
R: *Chemistry in Britain* 7: 436 (Oct. 1971).

The Alkaloids: Chemistry and Physiology. Vols. 1–. **R. H. F. Manske and H. L. Holmes, eds.** New York: Academic Press, 1950–.
Volume 19, 1981.

Annual Reports on Analytical Atomic Spectroscopy. Vols. 1–. Washington, DC: American Chemical Society, 1971–. Annual.
R: *Chemistry in Britain* 14: 250 (May 1978); 17: 344 (July 1981); *Laboratory Practice* 28: 145 (Feb. 1979).

Annual Reports in Medicinal Chemistry. Vols. 1–. **C. K. Cain, ed.** New York: Academic Press, 1966–.
Volume 17, 1982.

Annual Reports on NMR Spectroscopy. Vols. 1–. New York: Academic Press, 1968–.
Volume 15, 1984.
R: *Journal of the American Chemical Society* (July 14, 1971).

Annual Reports on the Progress of Chemistry. Vols. 1–. London: Chemical Society, 1904–.
Volume 75, 1979.
R: *Chemistry in Britain* 16: 158 (Mar. 1980); *Journal of the American Chemical Society* 100: 7787 (Nov. 22, 1978); 101: 3423 (June 6, 1979); 102: 1477 (Feb. 13, 1980); 102: 5434 (July 30, 1980); 102: 7825 (Dec. 17, 1980).

Annual Review of Physical Chemistry. Vols. 1–. Palo Alto, CA: Annual Reviews, 1950–.
Volume 37, 1986.
R: *Chemistry and Industry* p. 1047 (Aug. 8, 1970); p. 847 (July 24, 1971); *Chemistry in Britain* 14: 622 (Dec. 1978); *Journal of the American Chemical Society* 102: 7822 (Dec. 17, 1980); *Physics Today* 22: 81 (Dec. 1969); *ARBA* (1970, p. 109; 1971, p. 486; 1972, p. 556; 1976, p. 647); Jenkins (D43); Wal (p. 126).

Catalysis Reviews. Vols. 1–. New York: Dekker, 1967–.
Volume 3, 1970.
R: *Chemistry in Britain* 7: 347 (Aug. 1971); *ARBA* (1971, p. 487).

Essays in Chemistry. Vols. 1–. New York: Academic Press, 1970–.
Volume 8, 1982.
R: *Chemistry in Britain* 14: 465 (Sept. 1978); *Chemistry and Industry* 31: 882 (July 31, 1971).

Electrochemistry. Vols. 1–. Washington, DC: American Chemical Society, 1970–. Annual.
Volume 7, 1980.

Journal of Chromatography Library. Vols. 1–. Amsterdam: Elsevier Scientific, 1973–.
Volume 21, 1982.

Macromolecular Chemistry. Vols. 1–. **A. D. Jenkins and J. F. Kennedy.** London: Royal Society of Chemistry, 1980–. Annual.
Volume 1, *A Review of the Literature Published during 1977 and 1978.*

Photochemistry. Vols. 1–. Washington, DC: American Chemical Society, 1970–.
Volume 10, 1979.

Progress in Analytical Chemistry. Vols. 1–. New York: Plenum, 1968–.
Volume 8, 1976.
R: *Journal of the American Chemical Society* 100: 1331 (Feb. 15, 1978); *TBRI* 43: 370 (Dec. 1977).

Progress in Solid-State Chemistry. Vols. 1–. Oxford: Pergamon Press, 1964–.
Volume 14, 1983.
R: *TBRI* 43: 47 (Feb. 1977); Jenkins (D47); Wal (p. 125).

Progress in Thin-layer Chromatography and Related Methods. Vols. 1–. Ann Arbor, MI: Ann Arbor–Humphrey, 1970-.
R: *ARBA* (1971, p. 488).

Reactivity and Structure. Vols. 1–. New York: Springer-Verlag, 1975–.
Volume 15, 1982.

Recent Advances in Phytochemistry. Vols. 1–. New York: Appleton-Century-Crofts, 1968–.
Volume 13, 1979.
R: *American Scientist* 58: 108 (Jan./Feb. 1970); *Chemistry in Britain* 14: 516 (Oct. 1978); *Journal of the American Chemical Society* 101: 1359 (Feb. 28, 1979).

Reports on the Progress of Applied Chemistry. Vols. 1–. London: Society of Chemical Industry, 1916–. Annual.
Volume 56, 1971.
R: Wal (p. 471).

Surfactant Science Series. Vols. 1–. New York: Dekker, 1967–.
Volume 10, 1982.
R: *Journal of the Society of Dyers and Colourists* 94: 221 (May 1978); *TBRI* 44: 272 (Sept. 1978).

Survey of Progress in Chemistry. Vols. 1–. New York: Academic Press, 1963–.
Volume 10, 1983.
R: *Journal of the American Chemical Society* 103: 2145 (Apr. 22, 1981); Jenkins (D48); Wal (p. 122).

Techniques of Chemistry. Vols. 1–. New York: Wiley, 1971–.
Volume 17, *Applications of Lasers to Chemical Problems*, 1982.

Theoretical Chemistry: Advances and Perspectives. Vols. 1–. **Douglas Henderson and Henry Eyring, eds.** New York: Academic Press, 1975–.
Volume 5, 1980.

Topics in Carbon-13 NMR Spectroscopy. Vols. 1–. New York: Wiley-Interscience, 1974–.
Volume 3, 1979.
R: *Science* 188: 141 (Apr. 11, 1975).

Topics in Inorganic and General Chemistry. Vols. 1–. New York: American Elsevier, 1964–.
Volume 18, 1980.

Topics in Stereochemistry. Vols. 1–. New York: Wiley, 1967–.
Volume 13, 1982.
R: *American Scientist* (Sept./Oct. 1971); *Chemistry in Britain* (Aug. 1971).

Topics in Stereochemistry. Vols. 1–. New York: Wiley, 1969–.
Volume 15, 1984.
R: *Chemistry in Britain* 15: 467 (Sept. 1979); 16: 506 (Sept. 1980); *Journal of the American Chemical Society* 100: 1330 (Feb. 15, 1978).

Biochemistry

Advances in Carbohydrate Chemistry and Biochemistry. Vols. 1–. New York: Academic Press, 1945–. Annual.
Volume 42, 1984.
R: *Chemistry in Britain* 15: 38 (Jan. 1979); Wal (p. 138).

Annual Review of Biochemistry. Vols. 1–. Palo Alto, CA: Annual Reviews,1932–. Annual.
Volume 45, 1976.
R: *American Scientist* 59: 635 (Sept./Oct. 1971); *ARBA* (1970, p. 109; 1974, p. 566; 1976, p. 647; 1981, p. 638); Jenkins (D42); Wal (p. 202).

Chemical and Biochemical Applications of Lasers. Vols. 1–. **C. Bradley Moore, ed.** New York: Academic Press, 1974–.
Volume 2, 1977; volume 3, 1977; volume 4, 1979; volume 5, 1980.
R: *Physics Today* 31: 50 (July 1978).

Comprehensive Biochemistry. Vols. 1–. **Marcel Florkin and Elmer H. Stotz, eds.** New York: American Elsevier, 1962–.
Volume 34A, 1984; volume 34B, 1984.

Computers in Chemical and Biochemical Research. Vols. 1–. **Charles E. Klopfenstein and Charles L. Wilkins.** New York: Academic Press, 1972–.
Volume 2, 1974.

Contemporary Topics in Immunochemistry. Vols. 1–. New York: Plenum, 1972-.
Volume 9, 1980.
R: *Science* 179: 888 (March 2, 1973).

Essays in Biochemistry. Vols. 1–. **Paul N. Campbell and G. D. Greville.** New York: Academic Press, 1965–.
Volume 19, 1983.

Horizons in Biochemistry and Biophysics. Vols. 1–. **E. Quagliariello, F. Palmieri, and Thomas P. Singer, eds.** Reading, MA: Addison-Wesley, 1974–.
Volume 5, 1978.
R: *American Scientist* 66: 501 (July/Aug. 1978).

International Review of Biochemistry. Vols. 1–. Baltimore: University Park Press, 1974–.
Volume 21, 1979.
R: *Journal of the American Chemical Society* 102: 5134 (July 16, 1980).

Methods of Biochemical Analysis. Vols. 1–. New York: Wiley-Interscience, 1954–

Volume 30, 1984.
R: *Choice* 7: 1220 (Nov. 1970); *ARBA* (1971, p. 487); Jenkins (D102).

Progress in Bioorganic Chemistry. Vols. 1–. New York: Wiley-Interscience, 1971–.
Volume 4, 1976.

Progress in the Chemistry of Fats and Other Lipids. Vols. 1–. Oxford: Pergamon Press, 1952–.
Volume 16, 1978.
R: Wal (p. 138).

Progress in Pesticide Biochemistry. Vol. 1–. **D. H. Hutson and T. R. Roberts.** New York: Wiley, 1981–.
A topical review of pesticide biochemistry. Covers such topics as new modes of action, biochemical toxicology in mammals, and environmental chemistry.
Volume 4, *Progress in Pesticide Biochemistry and Toxicology*, 1984.

Recent Advances in Biochemistry. 4th ed. London: Churchill, 1960–. Irregular.
R: *Nature* 190: 569 (May 13, 1961); Wal (p. 202).

Reviews in Biochemical Toxicology. Vols. 1–. New York: American Elsevier, 1979–.
Volume 7, 1985.

Topics in Lipid Chemistry. New York: Wiley-Interscience, 1970–.
R: *American Scientist* 59: 618 (Sept./Oct. 1971); 61: 80 (Jan./Feb. 1973); *Chemistry and Industry* 43: 1231 (Oct. 23, 1971).

Vitamins and Hormones: Advances in Research and Applications. Vols. 1–. **Robert S. Harris and Kenneth V. Thimann, eds.** New York: Academic Press, 1943–.
Volume 41, 1983.

Inorganic Chemistry

Advances in Inorganic Chemistry and Radiochemistry. Vols. 1–. New York: Academic Press, 1959–
Volume 28, 1984.
R: *Chemistry in Britain* 14: 46 (Jan. 1978); 17: 80 (Feb. 1981); *Journal of the American Chemical Society* 101: 2790 (May 9, 1979); 102: 5136 (July 16, 1980); 102: 7828 (Dec. 17, 1980); Jenkins (D35).

Advances in Molten Salt Chemistry. Vols. 1–. New York: Plenum, 1971–.
Volume 5, 1983.

Annual Reports in Inorganic and General Syntheses. Vols. 1–. **Kurt Niedenzu and Hans Zimmer, eds.** New York: Academic Press, 1973–.
Volume 5, 1978.
R: *Chemistry in Britain* 15: 142 (March 1979).

Benchmark Papers in Inorganic Chemistry. Vols. 1–. **H. H. Sisler, ed.** New York: Academic Press, 1972–.
Volume 6, 1977.

Inorganic Chemistry Concepts. Vols. 1–. New York: Springer-Verlag, 1977–.
Volume 8, 1983.

Inorganic Chemistry of the Main Group Elements. Vols. 1–. **C. C. Addison, ed.** Washington, DC: American Chemical Society, 1973–.
Volume 2, 1974; volume 3, 1976; volume 4, 1977; volume 5, 1978.

Inorganic Synthesis. Vols. 1–. New York: Wiley, 1939–. Annual.
Volumes 1–17, published by McGraw-Hill; volume 20, 1980.

Progress in Inorganic Chemistry. Vols. 1–. New York: Wiley, 1959–.
Volume 32, 1984.
R: *American Scientist* 58: 683 (Nov./Dec. 1970); *Chemistry in Britain* 7: 301 (July 1971); *Journal of the American Chemical Society* 100: 7129 (Nov. 25, 1978); 101: 7140 (Nov. 7, 1979); 102: 7406 (Nov. 19, 1980); 102: 3616 (June 17, 1981); *TBRI* 43: 328 (Nov. 1977).

Transition Metal Chemistry. Vols. 1–. New York: Dekker, 1965–. Volume 9, 1984.
R: *ARBA* (1971, p. 486).

Organic Chemistry

Advances in Free-Radical Chemistry. vols. 1–. New York: Academic Press, 1965–.
Volume 5, 1975; volume 6, 1980.

Advances in Heterocyclic Chemistry. Vols. 1–. New York: Academic Press, 1963–.
Volume 37, 1985. R: *Chemistry in Britain* 15: 204 (Apr. 1979); 17: 140 (Mar. 1981); *Journal of the American Chemical Society* 100: 1327 (Feb. 15, 1978); 101: 7136 (Nov. 7, 1979); 101: 3418 (June 6, 1979); 102: 4284 (June 4, 1980); 102: 6906 (Oct. 22, 1980); 103: 2145 (Apr. 22, 1981); Wal (p. 138).

Advances in Macromolecular Chemistry. Vols. 1–. New York: Academic Press, 1968–.
Volume 2, 1971.
R: *Journal of the American Chemical Society* 93: 3311 (July 14, 1971); *Science* 165: 167 (July 11, 1969).

Advances in Organometallic Chemistry. Vols. 1–. New York: Academic Press, 1964–.
Volume 23, 1984.
R: *Chemistry in Britain* 15: 33 (Jan. 1979); *Journal of the American Chemical Society* 100: 7442 (Nov. 8, 1978); 102: 7620 (Dec. 3, 1980); 102: 895 (Jan. 16, 1980); Wal (p. 137).

Advances in Physical Organic Chemistry. Vols. 1–. New York: Academic Press, 1963.
Volume 20, 1984.
R: *Chemistry in Britain* 15: 540 (Oct. 1979); *journal of the American Chemical Society* 93: 7350 (Dec. 29, 1971); 100: 6799 (Oct. 11, 1978); 102: 434 (Jan. 2, 1980).

Advances in Protein Chemistry. Vols. 1–. New York: Academic Press, 1944–.
Volume 36, 1983.
R: Wal (p. 138).

The Alkaloids: Chemistry and Physiology. Vols. 1–. New York: Academic Press, 1950–.
Volume 19, 1981.

Annual Reports in Organic Synthesis. Vols. 1–. New York: Academic Press, 1970–.
Volume 13, 1983.
R: *Journal of the American Chemical Society* 101: 3419 (June 6, 1979); *NTB* 60: 102 (Mar. 1975).

Annual Review of Photochemistry. Vols. 1–. New York: Wiley-Interscience, 1969–.
Volume 1, 1969, *Survey of 1967 Literature.*
R: *Choice* 5: 1191 (Nov. 1969); *ARBA* (1970, p. 159); Wal (p. 127).

Benchmark Papers in Organic Chemistry. Vols. 1–. **C. A. VanderWerf, ed.** New York: Academic Press, 1974–.
Volume 6, 1977.

Carbohydrate Chemistry. Vols. 1–. Washington, DC: American Chemical Society, 1968–. Annual.
Volume 11, 1979.

Chemical Analysis of Organo-Metallic Compounds. Vols. 1–. **T. R. Crompton.** New York: Academic Press, 1973–.
Volume 5, 1977.

Organic Chemistry: A Series of Monographs. Vols. 1–. New York: Academic Press, 1964–. Irregular.
Volume 43, 1980.
R: *Chemistry in Britain* 7: 344 (Aug. 1971).

Organic Photochemistry. Vols. 1–. **O. L. Chapman, ed.** New York: Dekker, 1967–.
Volume 2, 1969; volume 3, 1973; volume 4, 1979; volume 5, 1981; volume 6, 1983.

Organic Reaction Mechanisms. Vols. 1–. New York: Wiley-Interscience, 1965–.
Comprehensive series of annual surveys on the subject. Author and subject indexes.
R: *Choice* 7: 530 (June 1970); *ARBA* (1971, p. 488).

Organic Reactions. Vols. 1–. New York: Wiley, 1942–. Annual.
Volume 33, 1984.
R: *Chemistry in Britain* 7: 528 (Dec. 1971); Jenkins (D126); Wal (p. 136).

Organic Synthesis. Vols. 1–. New York: Wiley, 1921–. Annual.
Volume 61, 1983.
R: *Chemistry and Industry* 30: 849 (July 24, 1971); *Chemistry in Britain* 7: 480 (Nov. 1971); *Laboratory Practice* 27: 1067 (Dec. 1978); Jenkins (D127); Wal (p. 136).

Organometallic Chemistry. Vols. 1–. Washington, DC: American Chemical Society, 1971–. Annual.

Volume 8, 1980.
R: *Journal of the American Chemical Society* 101: 278 (Jan. 3, 1979); 102: 4854 (July 2, 1980); 103: 4654 (July 29, 1981).

Organophosphorus Chemistry. Vols. 1–. **Stuart Trippett.** London: The Chemical Society, 1970–. Annual.
Volume 11, 1978/1979.

Progress in Electrochemistry of Organic Compounds. Vols. 1–. **A. N. Frumkin and A. B. Ershler, eds.** New York: Plenum, 1971–.

Progress in Organic Chemistry. Vols. 1–. London: Butterworth, 1952–. Irregular.
Volume 8, 1973.
R: Wal (p. 135).

Progress in Physical Organic Chemistry. Vols. 1–. New York: Wiley-Interscience, 1970–.
Volume 15, 1985.
R: Jenkins (D46); Wal (p. 135).

Progress in Total Synthesis. Vols. 1–. **Sarah E. Danishefsky and Samuel Danishefsky.** New York: Plenum, 1971–.

Reviews in Macromolecular Chemistry. Vols. 1–. New York: Dekker, 1967–.
Volume 4, 1970.
R: *ARBA* (1971, p. 486).

Studies in Organic Chemistry. Vols. 1–. New York: Dekker, 1973–.
Volume 9, 1980.
R: *Chemistry in Britain* 16: 45 (Jan. 1980).

Synthetic Methods of Organic Chemistry. Vols. 1–. Basel: S. Karger, 1948–. Approx. Annual.
Volume 33, *Yearbook 1979,* 1979.
R: *Journal of the American Chemical Society* 100: 1329 (Feb. 15, 1978); 101: 3419 (June 6, 1979); 102: 2136 (Mar. 12, 1980).

Polymer Chemistry

Advances in Polymer Science. Vols. 1–. Berlin and New York: Springer-Verlag, 1958–.
Volume 40, 1981.
R: *American Scientist* 63: 719 (Nov./Dec. 1975); *Journal of the American Chemical Society* 102: 5133 (July 16, 1980); 102: 6188 (Sept. 10, 1980; Jenkins (D39).

Progress in Polymer Science. Oxford: Pergamon Press, 1967–. Irregular.
Volume 2, 1970; volume 3, 1971; volume 6, 1980.
R: *Chemistry in Britain* 7: 528 (Dec. 1971).

Progress in Polymer Science, Japan. New York: Halsted Press, 1972–.
Volume 8, 1975.
R: *American Scientist* 63: 719 (Nov./Dec. 1975); 63: 241 (Mar./Apr. 1975).

BIOLOGICAL SCIENCES

General

Advances in Cancer Research. New York: Academic Press, 1953–.
Volume 42, 1985.
R: *American Scientist* 63: 355 (May/June 1975).

Advances in Cyclic Nucleotide Research. New York: Raven Press, 1972–.
Volume 15, 1983.
R: *American Scientist* 62: 494 (July/Aug. 1974); *Science* 179: 889 (Mar. 2, 1973); *TBRI* 44: 172 (May 1978).

Advances in Experimental Medicine and Biology. New York: Plenum, 1967–. Irregular.
Volume 125, 1980.
R: *American Scientist* 62: 734 (Nov./Dec. 1974); *New Scientist* (Dec. 9, 1971).

Advances in Lipid Research. Vols. 1–. New York: Academic Press, 1963–.
Volume 20, 1983.
R: *American Scientist* 59: 269–270 (Mar./Apr. 1971); 61: 594–595 (Sept./Oct. 1973); *Journal of the American Chemical Society* 68: 699 (Nov./Dec. 1980); 102: 4554 (June 18, 1980); 103: 2146 (Apr. 22, 1981); *TBRI* 43: 173 (May 1977).

Advances in Marine Biology. Vols. 1–. New York: Academic Press, 1963–.
Volume 21, 1984.
R: *Nature* 222: 60 (Apr. 5, 1969); *TBRI* 44: 52 (Feb. 1978); Wal (p. 195).

Advances in Radiation Biology. Vols. 1–. New York: Academic Press, 1964–.
Volume 10, 1983.

Advances in Substance Abuse: Behavioral and Biological Research: A Research Annual. Vols. 1–. **Nancy K. Mello, ed.** Greenwich, CT: Jai Press, 1980–.

Advances in Teratology. Vols. 1–. New York: Academic Press, 1966–.
Volume 5, 1972.
R: *American Scientist* 61: 95 (Jan./Feb. 1973).

Advances in the Biosciences. Vols. 1–. Oxford: Pergamon Press, 1969–.
Volume 29, 1984.

Advances in the Study of Behavior. Vols. 1–. New York: Academic Press, 1965–.
Volume 14, 1984.
R: *American Scientist* 58: 111 (Jan./Feb. 1970).

Annual Review of Pharmacology and Toxicology. Vols. 1–. Palo Alto, CA: Annual Reviews, 1961–.
Volume 24, 1984.

The Antigens. Vols. 1–. **Michael Sela, ed.** New York: Academic Press, 1973–.
Volume 2, 1974; volume 3, 1975: volume 4, 1977; volume 5, 1979; volume 6, 1982.

Applied Biology. Vols. 1–. **T. H. Coaker, ed.** New York: Academic Press, 1976–.
Volume 2, 1977; volume 3, 1978; volume 4, 1979; volume 5, 1980.

Basic Life Sciences. Vols. 1–. **Alexander Hollaender, ed.** New York: Plenum Press, 1973–.
Volume 10, 1978.

Benchmark Papers in Biological Concepts. Vols. 1–. **P. Gray, ed.** New York: Academic Press, 1973–.
Volume 3, *Carotenoproteins in Animal Coloration,* 1977.

Current Topics in Developmental Biology. Vols. 1-. New York: Academic Press, 1967–.
Volume 8, 1980.

Current Topics in Experimental Endocrinology. Vols. 1–. New York: Academic Press, 1971–.
Volume 1, 1972: volume 2, 1974; volume 3, 1978; volume 4, 1983; volume 5, 1983.
R: *American Scientist* 61: 488 (July/Aug. 1973); 63: 352 (May/June 1975); *Science* 176: 1228 (June 16, 1972).

Current Topics in Membranes and Transport. Vols. 1–. New York: Academic Press, 1970–.
Volume 20, 1984.
R: *American Scientist* (July/Aug. 1971).

International Review of Experimental Pathology. Vols. 1–. **G. W. Richter and M. A. Epstein.** New York: Academic Press, 1962–.
Volume 26, 1984.

International Review of Neurobiology. Vols. 1–. New York: Academic Press, 1959–.
Volume 25, 1984.
R: *American Scientist* 63: 585 (Sept./Oct. 1975).

Methods in Membrane Biology. Vols. 1–. New York: Plenum, 1974–.
Volume 10, 1979.
R: *American Scientist* 63: 469 (July/Aug. 1975).

Pictured-key Nature Series. Dubuque, IA: Brown, 1946–.

Progress in Neurobiology. Vols. 1–. Oxford: Pergamon Press, 1973–.
Volume 20, 1984.

Progress in Nucleic Acid Research and Molecular Biology. Vols. 1–. New York: Academic Press, 1963–.
Volume 30, 1983.
R: *American Scientist* 59: 267 (Mar./Apr. 1971); 59: 756 (Nov./Dec. 1971).

Progress in Surface and Membrane Science. Vols. 1–. New York: Academic Press, 1964–.
Volume 14, 1981.
R: *Journal of the American Chemical Society* 101: 1912 (Mar. 28, 1978).

Progress in Theoretical Biology. Vols. 1–. New York: Academic Press, 1967–. Irregular.
Volume 6, 1981.

Progress in Toxicology. Vols. 1–. Berlin: Springer-Verlag, 1973–.
Volume 2, 1975.
R: *American Scientist* 62: 608 (Sept./Oct. 1974).

Recent Progress in Hormone Research. Vols. 1–. New York: Academic Press, 1947–.
Volume 38, 1982.
R: *American Scientist* 58: 558 (Sept./Oct. 1970); 60: 630 (Sept./Oct. 1972); *Science* 168: 1335 (June 12, 1970).

The Proteins. 3d ed. Vols. 1–. **Hans Neurath and Robert L. Hill, eds.** New York: Academic Press, 1975–.
Volume 4, 1979.

Residue Reviews. Vols. 1–. **F. A. Gunther, ed.** New York: Springer-Verlag, 1962–.
Volume 83, 1982.

Survey of Biological Progress. Vols. 1–. New York: Academic Press, 1949–. Irregular.
Volume 4, 1962.

Topics in Bioelectrochemistry and Bioenergetics. Vols. 1–. **G. Milazzo, ed.** New York: Wiley, 1976–.
Volume 5, 1983.

Agriculture and Food Science

Advances in Agronomy. Vols. 1–. New York: Academic Press, 1949–.
Volume 34, 1981.
R: *American Scientist* 58: 560 (Sept./Oct. 1970); 61: 482 (July/Aug. 1973); *TBRI* 43: 67 (Feb. 1977).

Advances in Food Research. Vols. 1–. New York: Academic Press, 1948–.
Volume 29, 1983.
R: Jenkins (J107); Wal (p. 485).

Analytical Methods for Pesticides and Plant Growth Regulators and Food Additives. Vols. 1–. New York: Academic Press, 1963–.
Volume 13, 1984.

Annual Review of Nutrition. Vols. 1–. Palo Alto, CA: Annual Reviews, 1981–.
Volume 6, 1986.

Review of the World Wheat Situation. **International Wheat Council.** London: International Wheat Council. Annual.
Seventeenth annual, 1974/1975, 1976.
R; *IBID* 4: 106 (June 1976).

The State of Food and Agriculture, 1976. World Review; Review by Regions; Population, Food Supply, and Agricultural Development. Rome: Food and Agriculture Organization of the United Nations, 1978. Annual.
Annual survey of developments in the world food and agricultural situation and report on the outlook for the future.
R: *IBID* 3: 135 (Sept. 1975).

Botany

Advances in Botanical Research. Vols. 1–. New York: Academic Press, 1963–.
Volume 10, 1984.
R: Wal (p. 207).

Annual Review of Phytopathology. Palo Alto, CA: Annual Reviews, 1963–.
Volume 24, 1986.
R: *American Scientist* 60: 255 (Mar./Apr. 1972); *ARBA* (1976, p. 664; 1981, p. 649); Wal (p. 209).

Annual Review of Plant Physiology. Palo Alto, CA: Annual Reviews, 1950–.
Volume 37, 1986.
R: *ARBA* (1976, p. 664; 1981, p. 649); Wal (p. 209).

Experimental Botany: An International Series of Monographs. Vols. 1–. **J. F. Sutcliffe, ed.** New York: Academic Press, 1965–.
Volume 16, 1982.

Horticultural Reviews. Vols. 1–. **Jules Janick, ed.** Westport, CT: AVI Publishing, 1979–. Annual.
Volume 6, 1984.
R: *ARBA* (1980, p. 708).

Cell and Molecular Biology

Advances in Cell and Molecular Biology. Vols. 1–. New York: Academic Press, 1971–.
Volume 3, 1974.
R: *Nature* 234: 283 (Dec. 3, 1971); *Science* 175: 510 (Feb. 4, 1972).

Advances in Cell Biology. Vols. 1–. New York: Appleton-Century-Crofts, 1970–.
Volume 21, part A, 1980.
R: *American Scientist* 60: 790 (Nov./Dec. 1972); *ARBA* (1971, p. 493).

Amino Acids, Peptides, and Proteins. Vols. 1–. Washington, DC: American Chemical Society, 1970–. Annual.
Volume 9, 1976.

Annual Review of Cell Biology. Vols. 1–. Palo Alto, CA: Annual Reviews, 1985–.
Volume 2, 1986.

Current Topics in Cellular Regulation. Vols. 1–. New York: Academic Press, 1969–. Annual.
Volume 24, 1984.
R: *American Scientist* (Nov./Dec. 1971).

International Review of Connective Tissue Research. Vols. 1–. **David A. Hall, ed.** New York: Academic Press, 1963–.
Volume 10, 1983.

International Review of Cytology. Vols. 1–. **G. H. Bourne and J. F. Danielli, eds.** New York: Academic Press, 1952–.
Volume 90, 1985.
R: Wal (p. 198).

Laboratory Techniques in Biochemistry and Molecular Biology. Vols. 1–. Amsterdam: North-Holland, 1969–.
Volume 11, 1983.

Methods in Cell Biology. Vols. 1–. New York: Academic Press, 1964–.
Volume 26, 1982. Continues, *Methods in Cell Physiology.*
R: *American Scientist* 64: 445, 452 (July/Aug. 1976); 67: 613 (Sept./Oct. 1979).

Progress in Molecular and Subcellular Biology. Vols. 1–. Berlin and New York: Springer-Verlag, 1969–.
Volume 8, 1983.
R: *Science* 170: 1072 (Dec. 4, 1970).

Results and Problems in Cell Differentiation. Vols. 1–. New York: Springer-Verlag, 1968–.
Volume 11, *Differentiation and Neoplasia*, 1980.
R: *Science* 205: 1370 (Sept. 28, 1979).

Enzymology

Advances in Enzyme Regulation. Vols. 1–. Oxford: Pergamon Press, 1963–.
Volume 20, 1982.

Advances in Enzymology and Related Areas of Molecular Biology. Vols. 1–. **A. Meister.** New York: Wiley, 1941–.
Earlier title: *Advances in Enzymology.*
Volume 58, 1985.
R: *American Scientist* 63: 584 (Sept./Oct. 1975); Wal (p. 203).

Methods in Enzymology. Vols. 1–. New York: Academic Press, 1955–.
Volume 99, 1983.
R: *American Scientist* 63: 710 (Nov./Dec. 1975).

Genetics

Advances in Genetics. Vols. 1–. New York: Academic Press, 1947–.
Volume 22, 1984.
R: *American Scientist* 60: 86 (Jan./Feb. 1972); Wal (p. 197).

Advances in Human Genetics. New York: Plenum, 1970–.
Volume 13, 1983.
R: *American Scientist* 61: 593 (Sept./Oct. 1973); 64: 219 (Mar./Apr. 1976); *Science* 170: 1296 (Dec. 18, 1970); 192: 1346 (June 15, 1976).

Annual Review of Genetics. Vols. 1–. Palo Alto, CA: Annual Reviews, 1967–.
Volume 20, 1986.
R: *American Scientist* 60: 637 (Sept./Oct. 1972); *Science* 159: 1091 (1948); *ARBA* (1977, p. 651); Jenkins (G163).

Benchmark Papers in Genetics. Vols. 1–. **D. L. Jameson, ed.** New York: Academic Press, 1974–.
Volume 11, *Hybridization: An Evolutionary Perspective*, 1979.

Evolutionary Biology. Vols. 1–. New York: Plenum, 1967–.
Volume 16, 1983.
R: *ARBA* (1971, p. 492).

Microbiology and Immunology

Advances in Applied Microbiology. Vols. 1–. New York: Academic Press, 1959–. Annual.
Volume 30, 1984.
R: *ARBA* (1971, p. 493); Jenkins (G170); Wal (p. 199).

Advances in Aquatic Microbiology. Vols. 1–. **M. R. Droop and H. W. Jannasch, eds.** New York: Academic Press, 1977–.
Volume 2, 1980.

Advances in Immunology. Vols. 1–. New York: Academic Press, 1961–.
Volume 36, 1984.
R: *American Scientist* 61: 590 (Sept./Oct. 1973).

Advances in Microbial Ecology. Vols. 1–. **M. Alexander, ed.** New York: Plenum Press, 1977–.
Volume 7, 1983.

Advances in Microbial Physiology. New York: Academic Press, 1967–.
Volume 25, 1984.
R: *American Scientist* 60: 790 (Nov./Dec. 1972).

Advances in Virus Research. Vols. 1–. New York: Academic Press, 1953–.
Volume 29, 1984.

Annual Reports on Fermentation Processes. Vols. 1–. **D. Perlman, ed.** New York: Academic Press, 1977–. Annual.
Volume 5, 1982.
R: *Chemistry in Britain* 15: 146 (March 1979).

Annual Review of Immunology. Vols. 1–. Palo Alto, CA: Annual Reviews, 1983–.
Volume 4, 1986.

Annual Review of Microbiology. Vols. 1–. Palo Alto, CA: Annual Reviews, 1947–.
Volume 40, 1986.
R: *Nature* 190: 662 (May 20, 1961); *ARBA* (1976, p. 656; 1981, p. 646); Jenkins (G171); Wal (p. 199).

Benchmark Papers in Microbiology. Vols. 1–. **W. W. Umbreit, ed.** New York: Academic Press, 1973–.
Volume 19, 1983.

Contemporary Topics in Immuno-biology. Vols. 1–. New York: Plenum, 1972–.
Volume 9, 1980.
R: *American Scientist* 62: 491 (July/Aug. 1974); 63: 469 (July/Aug. 1975).

Contemporary Topics in Molecular Immunology. Vols. 1–. New York: Plenum, 1972–.
Volume 7, 1978; volume 8, 1981; volume 9, 1983.
R: *American Scientist* 62: 491 (July/Aug. 1974).

Current Topics in Microbiology and Immunology. Berlin and New York: Springer-Verlag, n.d.
Volume 99, 1982.

Methods in Microbiology. Vols. 1–. New York: Academic Press, 1969–.
Volume 16, 1983.
R: *American Scientist* 59: 766 (Nov./Dec. 1971); *Chemistry and Industry* 42: 1196 (Oct. 16, 1971).

Microbiology. Vols. 1–. Washington, DC: American Society for Microbiology, 1974–. Annual.
R: *Journal of the American Chemical Society* 100: 7443 (Nov. 8, 1978); *ARBA* (1980, p. 623).

Progress in Indusrial Microbiology. Vols. 1–. Edinburgh: Churchill Livingstone. Distr. New York: American Elsevier, 1959–.
Volume 17, 1983.
R: *Chemistry and Industry* 40: 1128 (Oct. 2, 1971); *Journal of the American Chemical Society* 102: 5134 (July 16, 1980); *ABL* 30: entry 163 (Mar. 1965); *ARBA* (1976, p. 656); Wal (p. 481).

Physiology

Advances in Comparative Physiology and Biochemistry. Vols. 1–. New York: Academic Press, 1962–.
Volume 7, 1978.

Advances in Reproductive Physiology. Vols. 1–. London: Science, 1966–.
Volume 6, 1973.

Annual Review of Physiology. Vols. 1–. Palo Alto, CA: Annual Reviews, 1939–.
Volume 48, 1986.
R: *ARBA* (1976, p. 656).

Benchmark Papers in Human Physiology. Vols. 1–. **L. L. Langley, ed.** New York: Academic Press, 1973–.
Volume 17, 1983.

Reviews of Physiology, Biochemistry, and Pharmacology. Berlin and New York: Springer-Verlag.
Volume 99, 1983.

Zoology

Advances in Insect Physiology. Vols. 1–. New York: Academic Press, 1963–.
Volume 17, 1984.
R: *American Scientist* 58: 559 (Sept./Oct. 1970), 61: 88 (Jan./Feb. 1973).

Advances in Parasitology. Vols. 1–. New York: Academic Press, 1963–.
Volume 23, 1984.
R: *American Scientist* 58: 215 (Mar./Apr. 1970); 59: 111 (Jan./Feb. 1971).

Annual Review of Entomology. Palo Alto, CA: Annual Reviews, 1956–.
Volume 31, 1986.
R: *American Scientist* 59: 374 (May/June 1971); 60: 792); *Nature* 987: 542 (Aug. 13, 1960); 190: 945 (June 10, 1961); *ARBA* (1970, p. 125; 1976, p. 676; 1981, p. 673); Jenkins (G155); Wal (p. 231).

Benchmark Papers in Behavior. Vols. 1–. **M. W. Schein and S. W. Porges, eds.** New York: Academic Press, 1974–.
Formerly entitled: *Benchmark Papers in Animal Behavior.*
Volume 16, 1982.

International Review of General and Experimental Zoology. Vols. 1–. New York: Academic Press, 1964–. Irregular.
Volume 4, 1970.

Research in Protozoology. New York and Oxford: Pergamon Press, 1967–. Irregular.
Volume 4, 1972.

Earth Sciences

Advances in Geology. Vols. 1–. New York: Academic Press, 1965–.
R: Jenkins (F25).

Advances in Geophysics. Vols. 1–. New York: Academic Press, 1952–.
Volume 26, 1984.

Advances in Microbiology of the Sea. Vols. 1–. New York: Academic Press, 1968–.
R: *American Scientist* 57: 130A (Summer 1969).

Annual Review of Earth and Planetary Sciences. Palo Alto, CA: Annual Reviews, 1973–. Annual.
Volume 14, 1986.
R: *Science* 182: 706 (Nov. 16, 1973); *ARBA* (1976, p. 703; 1981, p. 690).

Benchmark Papers in Geology. Vols. 1–. **R. W. Fairbridge, ed.** New York: Academic Press, 1972–.
Volume 78, 1983.

International Geophysics Series. Vols. 1–. **J. Van Mieghem, ed.** New York: Academic Press, 1960–.
Volume 29, 1982.
R: *EOS; Transactions, American Geophysical Union* 59: 204 (Apr. 1978); *TBRI* 44: 250 (Sept. 1978).

Oceanography and Marine Biology: An Annual Review. New York: Hafner, 1963–. Annual.
Volume 21, 1984.
Highly technical review articles. Reviews often have addenda that discuss papers written since the completion of the reviews.
R: *ARBA* (1972); Wal (p. 164).

Progress in Oceanography. Vols. 1–. New York: Pergamon Press, 1963–.
Volume 12, 1984.
R: *Science* 144: 987 (1964); Jenkins (F146); Wal (p. 125).

AERONAUTICAL AND ASTRONAUTICAL ENGINEERING

Advances in Space Science and Technology. Vols. 1–. New York: Academic Press, 1959–. Annual.
Volume 11, 1972.
R: *American Scientist* 59: 481 (July/Aug. 1971); *ABL* 35: entry 555 (Nov. 1970); Jenkins (K45); Wal (p. 392).

Advances in the Astronautical Sciences. Vols. 1–. Tarzana, CA: American Astronautical Society, 1959–.
Volume 13, 1963.

Progress in Aerospace Sciences. Vols. 1–. Oxford: Pergamon Press, 1961–.
Volume 19, 1982.
R: *American Scientist* 57: 382A (Winter 1969); *Engineering* 210: 584 (Nov. 27, 1970).

BIOENGINEERING

Advances in Biochemical Engineering. Vols. 1–. Berlin and New York: Springer-Verlag, 1970–.
Volume 27, 1983.
R: *Chemistry and Industry* 13: 472 (July 1, 1978); 19: 776 (Oct. 4, 1980); 23: 849 (Dec. 1, 1979); *Journal of the American Chemical Society* 102: 5134 (July 16, 1980); 102: 6392 (Sept. 24, 1980); 103: 1000 (Feb. 25, 1981).

Advances in Biomedical Engineering. Vols. 1–. New York: Academic Press, 1971–.
Volume 7, 1979.
R: *American Scientist* 62: 747 (Nov./Dec. 1974); 63: 473 (July/Aug. 1975); *TBRI* 43: 350 (Nov. 1977).

Annual Review of Biophysics and Bioengineering. Vols. 1–. Palo Alto, CA: Annual Reviews, June 1972–.
Volume 13, 1984.
R: *Journal of the American Chemical Society* 100: 7443 (Nov. 8, 1978); 102: 6905 (Oct. 22, 1980); *ARBA* (1976, p. 651; 1981, p. 641).

Biotechnology International. **A. H. Sheppard.** London: Imsworld Publications,1982–. Annual.
Three parts: Europe; United States and Canada; Japan, Asia, and Australia.
R: *Ulrich's Quarterly* 7: 8 (Jan. 1983).

Chemical Engineering

Advances in Chemical Engineering. Vols. 1–. New York: Academic Press, 1956–.Volume 12, 1983.

Advances in Corrosion Science and Technology. Vols. 1–. New York: Plenum.
Volume 7, 1980.

Advances in Cryogenic Engineering. Vols. 1–. New York: Plenum, 1960–.
Volume 30, 1984.

Annual Reviews of Industrial and Engineering Chemistry. Vols. 1–. Washington, DC: American Chemical Society, 1970–.
Volume 2, 1974.

Chemical Processing and Engineering. Vols. 1–. New York: Dekker, 1975–.
Volume 11, 1977.

Reviews in Polymer Technology. Vols. 1–. New York: Dekker, 1972–.

Review of Textile Progress. Vols. 1–. New York: Plenum, 1949–.
Volume 18, 1966/1967.
R; Jenkins (J137); Wal (p. 524).

Civil Engineering

Advances in Hydroscience. Vols. 1–. New York: Academic Press, 1964–.
Volume 13, 1982.
R: *American Scientist* 64: 463 (July/Aug. 1976).

Highway Research in Progress. **US Highway Research Board.** Washington, DC: National Research Council, 1965–. Annual.
Number 7, 1974.
R: Wal (p. 361).

Progress in Construction Science & Technology. Vols. 1–. New York: Barnes & Noble, 1971–.
Volume 2, 1973.
R: *American Scientist* 61: 241 (Mar./Apr. 1973).

ELECTRICAL AND ELECTRONICS ENGINEERING

Advances in Communication Systems; Theory and Applications. Vols. 1–. New York: Academic Press, 1965–.
Volume 4, 1975.

Advances in Computers. Vols. 1–. New York: Academic Press, 1960–.
Volume 11, 1971, contains cumulative author and subject indexes to all previous volumes. Volume 23, 1984.
R: *Scientific Information Notes* 1: 290 (1969); Jenkins (B59); Wal (p. 545).

Advances in Computing Research. Greenwich, CT: Jai Press, 1983–. Annual.
Volume 1, 1983.

Advances in Control Systems. Vols. 1–. New York: Academic Press, 1964–.
Volume 8, 1971.
R: *American Scientist* 57: 72A (Spring 1969).

Advances in Electronics and Electron Physics. Vols. 1–. New York: Academic Press, 1948–. Annual.
Volumes 1–5 entitled *Advances in Electronics*; volume 65, 1985.
R: *American Scientist* 58: 326 (May/June 1970); 58: 684 (Nov./Dec. 1970); 60: 246 (Mar./Apr. 1972); *Physics Bulletin* 30: 440 (Oct. 1979); 30: 534 (Dec. 1979); *Sky and Telescope* 55: 251 (Mar. 1978); *NTB* 60: 209 (June 1975); Jenkins (C19); Wal (p. 322).

Advances in Image Pickup and Display. Vols. 1–. New York: Academic Press, 1974–.
Volume 6, 1983.

Annual Review in Automatic Programming. Vols. 1–. **R. E. Goodman, ed.** Oxford: Pergamon Press, 1960–.
Volume 90, 1981.
R: Wal (p. 545).

Annual Review of Computer Science. Vols. 1–. Palo Alto, CA: Annual Reviews, 1986–.
Volume 1, 1986.

Benchmark Papers in Electrical Engineering and Computer Science. Vols. 1–. **J. B. Thomas, ed.** New York: Academic Press, 1973–.
Volume 27, 1983.
R: *Choice* 13: 1458 (Jan. 1977).

Lecture Notes in Computer Science. Vols. 1–. New York: Springer-Verlag, 1973–.
Volume 1, 1973.

Progress in Control Engineering. New York: Academic Press, 1962–. Annual.

Progress in Cybernetics and Systems Research. Vols. 1–. Washington, DC: Hemisphere Publishing, 1975–.
Volume 11, 1982.

Progress in Dielectrics. Vols. 1–. London: Newnes, 1959–.
Volume 4, 1962.

MATERIALS SCIENCE

Advances in Corrosion Science and Technology. Vols. 1–. New York: Plenum Press, 1970–.
Volume 7, 1979.

Advances in Materials Research. Vols. 1–. New York: Wiley-Interscience, 1967–.
R: Wal (p. 295).

Annual Review of Materials Science. Vols. 1–. Palo Alto, CA: Annual Reviews, 1971–.
Volume 16, 1986.
R: *Journal of Metals* 30: 24 (Feb. 1978); *Journal of the American Chemical Society* 100: 2592 (Apr. 12, 1978); 102: 3308 (Apr. 23, 1980); 102: 2986 (May 21, 1980); *Science* 175: 1452 (Mar. 31, 1972); Wal (p. 291).

Materials Science Research. Vols. 1–. New York: Plenum Press, 1963–.
Volume 13, *Sintering Processes*, 1980.

Progress in Materials Science. Vols. 1–. Oxford: Pergamon Press, 1961–.
Volumes 1–8 appeared as *Progress in Metal Physics;* volume 13, 1969, volume 27, 1983.
R: *Physics Today* 25: 69 (Jan. 1972); Wal (pp. 287, 291, 504).

Semiconductors and Semimetals. Vols. 1–. **Robert K. Willardson.** New York: Academic Press, 1966–.
Volume 20, 1985.

MECHANICAL ENGINEERING

Advances in Applied Mechanics. Vols. 1–. New York: Academic Press, 1948–. Irregular.
Volume 24, 1984.
R: *American Scientist* 61: 496 (July/Aug. 1973); 62: 616 (Sept./Oct. 1974); 63: 719 (Nov./Dec. 1975); 64: 228 (Mar./Apr. 1976); *Nature* 188: 352 (Oct. 29, 1960); Jenkins (K42); Wal (p. 100).

Advances in Heat Transfer. Vols. 1–. **Thomas F. Irvine, Jr. and James P. Hartnett, eds.** New York: Academic Press, 1964–.
Volume 16, 1984.
R: *American Scientist* 67: 621 (Sept./Oct. 1979); *Science* 218: 151 (Oct. 8, 1982).

Advances in Machine Tool Design and Research. Vols. 1–. New York: Pergamon Press, 1963–.
Volume 9, 1971.

Annual Review of Fluid Mechanics. Vols. 1–. Palo Alto, CA: Annual Reviews, 1969–.
Volume 18, 1986.
R: *American Scientist* 59: 483 (July/Aug. 1971); 63: 478 (July/Aug. 1975); *Journal of Fluid Mechanics* (49: 415) (Sept. 29, 1971); *Physics Bulletin* 31: 57 (Feb. 1980); *Science* 166: 489 (Oct. 24, 1969); *ARBA* (1977, p. 766).

Control and Dynamic Systems: Advances in Theory and Application. Vols. 1–. **C. T. Leondes, ed.** New York: Academic Press, 1964–. Annual.
Volumes 1–8, *Advances in Control Systems*; volume 12, 1984.

NUCLEAR ENGINEERING

Advances in Nuclear Science and Technology. Vols. 1–. New York: Academic Press, 1962–. Biennial.
Volume 16, 1983.
R: *American Scientist* 58: 680 (Nov./Dec. 1970); Wal (p. 308).

Annual Review of Nuclear and Particle Science. Vols. 1–. CA: Annual Reviews, n.d.
Volume 35, 1985.

Annual Review of Nuclear Science. Vols. 1–. Palo Alto, CA: Annual Reviews, 1952–. Annual.
Volume 27, 1977.
R: *American Scientist* 59: 756 (Nov./Dec. 1971); 60: 628 (Sept./Oct. 1972); 61: 359 (May/June 1973); *ARBA* (1976, p. 651); Jenkins (C23); Wal (p. 308).

Progress in Nuclear Energy. Vols. 1–. Oxford: Pergamon Press, 1956–. Irregular.
Volume 12, 1984.
R: *American Scientist* 58: 678, 690 (Nov./Dec. 1970); Jenkins (K46).

ENERGY

Advances in Energy Systems and Technology. Vols. 1–. **Peter Auer, ed.** New York: Academic Press, 1979–.

Volume 4, 1983.
R: *Mechanical Engineering* 101: 106 (July 1979); *New Scientist* 84: 544 (Nov. 15, 1979); *TBRI* 45: 310 (Oct. 1979); 46: 68 (Feb. 1980).

Advances in Solar Energy. **Karl W. Boer and John A. Duffie.** 2 vols. Boulder, CO: American Solar Energy Society, 1982–.
Volume 1, *An Annual Review of Research and Development in 1981*, 1982; volume 2, *Advances in Solar Energy*, 1985.

Advances in Transport Processes. Vols. 1–. **A. S. Mujumdar, ed.** New York: Wiley, 1983–.
Volume 4, 1985.

Annual Review of Energy. Vols. 1–. Palo Alto, CA: Annual Reviews, 1976–. Annual.
Volume 11, 1986.
R: *Journal of the American Chemical Society* 101: 1910 (Mar. 28, 1979); *LJ* 101: 1512 (July 1976); *TBRI* 43: 350 (Nov. 1977); *ARBA* (1977, p. 686).

Benchmark Papers on Energy. Vols. 1–. **R. B. Lindsay and M. E. Hawley, eds.** New York: Academic Press, 1975–.
Volume 10, 1983.

Progress in Biomass Conversion. Vols. 1–. **Kyosti V. Sarkanen and David A. Tillman, eds.** New York: Academic Press, 1979–. Annual.
Volume 4, 1983.

ENVIRONMENTAL SCIENCES

Advances in Ecological Research. Vols. 1–. New York: Academic Press, 1962–.
Volume 14, 1984.
R: *British Book News* p. 42 (Jan. 1972); Jenkins (G148); Wal (p. 203).

Advances in Environmental Science and Engineering. Vols. 1–. **James R. Pfafflin and Edward N. Ziegler, eds.** New York: Gordon & Breach, 1979–.
Volume 4, 1981.
R: *Choice* 17: 246 (Apr. 1980); *ARBA* (1981, p. 687).

Advances in Environmental Science and Technology. Vols. 1–. New York: Wiley-Interscience, 1969–.
Volume 10, 1980.
R: *Journal of the American Chemical Society* 102: 7627 (Dec. 3, 1980); Wal (pp. 64, 367); *Science* 169: 463 (July 31, 1970).

Annual Review of Ecology and Systematics. Vols. 1–. Palo Alto, CA: Annual Reviews, 1970–.
Volume 17, 1986.
R: *Science* 173: 713 (Aug. 20, 1971); *ARBA* (1974, p. 566; 1976, p. 655); Wal (p. 204).

Benchmark Papers in Ecology. Vols. 1–. **F. B. Golley, ed.** New York: Academic Press, 1974–.
Volume 11, 1982.

Current Topics in Environmental and Toxicological Chemistry. Vols. 1–. New York: Gordon and Breach, 1975–.
Volume 2, *Recent Advances in Environmental Analysis*, 1979.
R: *ARBA* (1981, p. 687).

Developments in Toxicology and Environmental Sciences. Vols. 1–. New York: American Elsevier, 1977–.
Volume 8, 1980.

Ecological Studies. Vols. 1–. New York: Springer-Verlag, 1973–.
Volume 44, 1983.

Environment and Man. Vols. 1–. **John Lenihan and William W. Fletcher, eds.** New York: Academic Press, 1976–.
Volume 10, *Economics of the Environment*, 1979.

Pollution Technology Review. Nos. 1–. Park Ridge, NJ: Noyes Data, 1973–.
Number 95, 1982.

Studies in Environmental Science. Vols. 1–. **Michel M. Benarie, ed.** New York: Elsevier Scientific, 1978–.
Volume 2, 1982.
R: *TBRI* 45: 111 (Mar. 1979).

Zoophysiology. Vols. 1–. New York: Springer-Verlag, 1971–.
Formerly entitled: *Zoophysiology and Ecology*.
Volume 13, 1983.
R: *Science* 196: 157 (Apr. 8, 1977); *TBRI* 43: 212 (June 1977).

OTHER REVIEWS OF PROGRESS

Besides the above selected series, there are numerous journal and nonjournal publications that provide authoritative reviews of progress in a subject field. The following is a sample list of such publications, a few of which have already been listed in the chapter on periodicals.

AAAS Reviews of Science. Washington, DC: American Association for the Advancement of Science, 1976–.

The following volumes are available: *Energy: Use, Conservation, and Supply; Population: Dynamics, Ethics, and Policy; Food: Politics, Economics, Nutrition, and Research, Materials: Renewable and Nonrenewable Resources*. A series of compendium volumes that includes authoritative articles originally published in *Science*. Publications deal with today's critical issues.

Biological Reviews. Cambridge, England: Cambridge Philosophical Society, 1925–. Quarterly.

Chemical Reviews. Vols. 1–. Washington, DC: American Chemical Society, 1924–. Bimonthly.

Modern Chemical Engineering. New York: Reinhold, 1963–. Irregular.

Monthly Weather Review. Vols. 1–. Washington, DC: US Government Printing Office, 1872–.
A periodical devoted to recently observed weather phenomena.
R: Mal (1980, p. 189).

Physical Reviews. Vols. 1–. New York: American Institute of Physics, 1893–. Semimonthly.

Quarterly Review of Biology. Vols. 1–. Stony Brook, NY: Stony Brook Foundation, 1962–.
Each volume provides critical review articles on topics of current interest in the biological sciences. Addressed to both researchers and students.

Reports of Progress in Physics. Vols. 1–. London: Physical Society, 1934–. Annual.

Review of Scientific Instruments. 1930–. Monthly.

Reviews of Geophysics and Space Physics. Washington, DC: American Geophysical Union, 1963–. Quarterly.

Reviews of Modern Physics. Vols. 1–. New York: American Institute of Physics, 1929–. Quarterly.
Important reviewing journal for advanced students and working research physicists.
R: Jenkins (C27).

Viewpoints in Biology. London: Butterworth, 1962–. Irregular.

CHAPTER 14 TREATISES

ASTRONOMY

Astronomy: Fundamentals and Frontiers. 3d ed. **Robert Jastrow and Malcolm H. Thompson.** New York: Wiley, 1977.
Second edition, 1974.

The Solar System. 4 vols. **G. P. Kuiper and B. M. Middlehurst, eds.** Chicago: University of Chicago Press, 1953–1963.
Volume 1, *The Sun*; volume 2, *The Earth as a Planet*; volume 3, *Planets and Satellites*; volume 4, *The Moon, Meteorites, and Comets*.
R: *Science* 119: 548 (1954); Jenkins (E40); Wal (p. 83).

MATHEMATICS

Fundamentals of Mathematics. 3 vols. **H. Behnke et al., eds.** Cambridge, MA: MIT Press, 1974.
A comprehensive reference source translated from the second German edition of *Grundzüge der Mathematik*, Göttingen, Vandenhoeck and Ruprecht, 1960.
Volume 1, *Foundations of Mathematics; The Real Number System and Algebra*; volume 2, *Geometry*; volume 3, *Analysis*. R: *Nature* 256: 152 (July 10, 1975).

London Mathematical Society Monographs. Nos. 1–. **D. A. Edwards and P. M. Cohn, eds.** New York: Academic Press, 1970–.
Number 16, *Convexity Theory and Its Applications in Functional Analysis*, 1981.

Mathematics in Science and Engineering: A Series of Monographs and Textbooks. Vols. 1–. **Richard E. Bellman, ed.** New York: Academic Press, 1965–.
Volume 157, *Introduction to Algebraic System Theory*, 1981.

North-Holland Mathematics Studies. Vols. 1–. New York: American Elsevier, 1970–.
Volume 42, *Cohomology of Completions*, 1980.

North-Holland Series in Applied Mathematics and Mechanics. Vols. 1–. New York: American Elsevier, 1967–.
Volume 24, *Theoretical Kinematics*, 1979.

Perspectives in Mathematical Logic. 3 vols. **R. O. Gandy et al., eds.** New York: Springer-Verlag, 1975–1978.
Volume 1, *Admissible Sets and Structures*, 1975; volume 2, *Recursion-Theoretic Hierarchies*, 1978; volume 3, *Basic Set Theory*, 1978.

Pure and Applied Mathematics: A Series of Monographs and Textbooks. Vols. 1–. **Paul A. Smith and Samuel Eilenberg, eds.** New York: Academic Press, 1949–.
Volume 96, *Bounded Analytical Functions*, 1981.

PHYSICS

Applied Optics and Optical Engineering: A Comprehensive Treatise. 5 vols. **R. Kingslake, ed.** New York: Academic Press, 1965–1980.

Volume 1, *Light: Its Generation and Modification*, 1965; volume 2, *The Detection of Light and Infrared Radiation*, 1965; volume 3, *Optical Components*, 1965; volume 4, *Optical Instruments*, part 1, 1967; volume 5; *Optical Instruments*, part 2, 1969; volume 7, 1979; volume 8, 1980.

R: *Journal of the Optical Society of America*, p. 1423 (Oct. 1970); *Sky and Telescope* 36: 38 (1968); Jenkins (K121).

Essays in Physics. 6 vols. **G. K. Conn and G. N. Fowler, eds.** New York: Academic Press, 1970–1976.

Methods of Experimental Physics. Vols. 1–. **Robert Celotta and Judah Levine, eds.** New York: Academic Press, 1970–.

Volume 22, *Solid-State Physics: Surfaces*. Edited by Robert L. Park and Max G. Lagally, 1985.

An ongoing reference source on all aspects of experimental physics. A must for all academic and research libraries.

Methods of Modern Mathematical Physics. Rev. ed. Vol. 1. **Michael Reed and Barry Simon.** New York: Academic Press, 1980–.

Volume 1, *Functional Analysis*, first edition, 1972; volume 2, *Fourier Analysis Self-Adjointness*, 1975; volume 3, *Scattering Theory*, 1979; volume 4, 1978.

Photographic Techniques in Scientific Research. Vols. 1–. **A. Newman, ed.** New York: Academic Press, 1973–.

Volume 1, edited by J. Cruise and A. Newman, 1973; volume 2, 1976; volume 3, 1978.

Progress in High-Temperature Physics and Chemistry. 5 vols. Oxford: Pergamon Press, 1967–1973.

R: *American Scientist* 59: 367 (May/June 1971); *ABL* 35: entry 67 (Feb. 1970); Wal (p. 109).

Pure and Applied Physics: A Series of Monographs and Textbooks. Vols. 1–. **H. S. W. Massey and Keith A. Brueckner, eds.** New York: Academic Press, 1970–.

Volume 40, *Nuclear Spectroscopy and Reactions*, 1974.

Semiconductors and Semimetals. Vols. 1–. **R. K. Willardson and Albert C. Beer, eds.** New York: Academic Press, 1967–.

Volume 14, *Lasers, Junctions, and Structure*, 1979; Volume 23, *Pulsed Laser Processing of Semiconductors*, edited by R. F. Wood, C. W. White, and R. T. Young, 1985.

A valuable set of treatises that provide state-of-the-art information on the physics of semiconductors and semimetals. A must for all institutions working on semiconductor compounds. Extensive references.

CHEMISTRY

The Chemistry of the Nitro and Nitroso Groups. 2 vols. **Henry Feuer, ed.** Melbourne, FL: Krieger, 1981.

Earlier edition, 1969.
There has not been a book on this subject before. The chapters cover not only classical descriptive chemistry, but also photochemistry, spectroscopy, etc.
R: *Journal of the American Chemical Society* 91: 7789 (Dec. 31, 1969); *TBRI* 36: 35 (Feb. 1970).

The Chemical Thermodynamics of Actinide Elements and Compounds. 11 vols. **F. L. Oetting et al., eds.** Vienna: International Atomic Energy Agency, 1977. Avail. New York: Unipub.

Volume 1, *Actinide Elements;* volume 2, *Actinide Aqueous Ions.*
Series divided into 11 parts, each containing a main text and an appendix. The text gives a critical evaluation of published data found in the literature up to early 1975. Each appendix contains the thermodynamic tables for the actinide materials pertaining to that part.
R: *IBID* 4: 141 (June 1976).

Chemistry of Carbon Compounds: A Modern Comprehensive Treatise. 2d rev. ed. Vols. 1–. New York: American Elsevier, 1964–.

The most comprehensive English-language treatise on the subject.
R: *Chemistry and Industry* (Feb. 20, 1971); *NTB* 51: 125 (1966); Jenkins (D119); Wal (p. 132).

Compendium of Organic Synthetic Methods. Vols. 1–. **Louis S. Hegedus and Leroy G. Wade, Jr., eds.** New York: Wiley, 1973–.

Volume 3, 1977; volume 4, 1980.
Part 4 of a treatise on organic transformations. Up-to-date with a clear format. Lists reactions, reagents, yield, and references. Valuable for upper-level undergraduate and graduate chemistry students, as well as for professional synthetic chemists.
R: *Choice* 18: 1292 (May 1981).

Comprehensive Inorganic Chemistry. **J. C. Bailar et al., eds.** Elmsford, NY: Pergamon Press, 1973.

Comprehensive survey of elements and associated compounds from the time of discovery through current uses. Intended for a wider reading audience than professional chemists' guides such as Mellor's *Comprehensive Treatise* and Gmelins' *Handbuch der Anorganischen Chemie.*
R: *ARBA* (1974, p. 556).

Comprehensive Organic Chemistry: The Synthesis and Reactions to Organic Compounds. 6 vols. **Derek Barton et al., eds.** Elmsford, NY: Pergamon, 1979.

An outstanding 6-volume set on all aspects of organic chemistry. Covers the reactions and properties of major classes of compounds. Provides references. Indexes include molecular formula, author, subject, reaction, and reagent access. An indispensable reference for chemistry collections.

R: *Chemistry in Britain* 16: 222 (Apr. 1980); *Journal of the American Chemical Society* 102: 893 (Jan. 16, 1980); *Nature* 285: 421 (June 5, 1980); *New Scientist* 85: 259 (Jan. 24, 1980); *RSR* 8: 26 (July/Sept. 1980).

Comprehensive Organometallic Chemistry. 6 vols. **Geoffrey Wilkinson, ed.** Elmsford, NY: Pergamon, 1982.

Written by more than 100 leading researchers in the field.

Comprehensive Treatise of Electrochemistry. Vols. 1–. **J. O'M. Bockris et al., eds.** New York: Plenum Press, 1980–.

Volume 3, *Electrochemical Energy Conversion and Storage*, 1981; volume 4, *Electrochemical Materials Science*, 1981.

A series of volumes dealing with various aspects of electrochemistry. Volume 3 covers modern electrochemistry, including state-of-the-art information on energy conversion and storage. Volume 4 examines corrosion processes. An outstanding treatise that is of use to electrochemists, chemists, and materials scientists.

A Comprehensive Treatise on Inorganic and Theoretical Chemistry. 16 vols. **J. W. Mellor.** New York: Halsted Press, 1922–1937. Also, supps., 1956–.

Arrangement of volumes is by elements based on the periodic table. Volume 16 is a general index. An authoritative treatise for a research library.

R: *Chemistry and Industry* 7: 648 (Apr. 7, 1962); Jenkins (D108); Wal (p. 131); Win (ED62).

Critical Reports on Applied Chemistry. Vols. 1–. **A. S. Teja.** New York: Halsted Press, 1979–.

Volume 3: *Chemical Engineering and the Environment*, 1981.

Crystal Structures. 6 vols. 2d ed. **Ralph W. G. Wyckoff.** New York: Wiley-Interscience, 1963–1971.

Originally a loose-leaf publication. This multivolume hard-cover set provides comprehensive descriptions of each crystalline compound, with diagrams, formulas, etc.

R: Wal (p. 139); Win (EE71).

Experimental Chemical Thermodynamics. Vols 1–. **Stig Sunner and Margret Mansson, eds.** Elmsford, NY: Pergamon, 1979–.

Volume 1, *Combustion Calorimetry*.

The first volume of a new series concerned with experimental chemical thermodynamics; this comprehensive reference deals with aspects of combustion calorimetry. Examines such subjects as units and physical constants, assignment of uncertainties, reduction of data, history, and various metallic and organometallic compounds. Includes bibliographies, tables, and line drawings. Recommended for all chemistry libraries and undergraduate physical chemistry students.

R: *Choice* 17: 103 (Mar. 1980).

Growth of Crystals. 12 vols. **A. A. Chernov, ed.** New York: Plenum Press, 1960–.

Volumes 2–10, edited by A. V. Shubnikov and N. N. Sheftal; volume 11, 1979; volume 12, in preparation.
English translation of original Russian series dealing with crystal growth. Covers nucleation at surfaces, layered growth, impurity trapping, and kinetic aspects.
R: *Aslib Proceedings* 44: 470 (Nov. 1979).

Inorganic Chemistry Concepts. 3 vols. New York: Springer-Verlag, 1977–1978.
Volume 1, *Lasers and Excited States of Rare Earths*, 1977; volume 2, *Magnetic Properties of Transition Metal Compounds*, 1977; volume 3, *Mossbauer Spectroscopy and Transition Metal Chemistry*, 1978.
A 3-volume treatise.

Perspectives in Structural Chemistry. 4 vols. **D. Dunitz and J. A. Ibers, eds.** New York: Wiley, 1967–1971.
R: *American Scientist* (Sept. 1971).

Physical Chemistry: An Advanced Treatise in Eleven Volumes. 11 vols. **H. Eyring, W. Jost, and D. Henderson, eds.** New York: Academic Press, 1971–1975.
R: *Journal of the American Chemical Society* 100: 1977 (Mar. 15, 1978).

Physical Chemistry: A Series of Monographs. Vols. 1–. **Ernest M. Loebl, ed.** New York: Academic Press, 1960–.
Volume 37, *Theoretical Foundations of Electron Spin Resonance*, 1978. Volumes 1–12 were edited by Eric Hutchinson and P. Van Rysselberghe.

Recent Developments in Separation Science. 7 vols. **Norman N. Li, ed.** Cleveland: Chemical Rubber, 1972–1981.
Volume 1, 1971; volume 2, 1972; volume 3, 1976; volume 4, 1978; volume 5, 1979; volumes 6 and 7, 1981.

Rodd's Chemistry of Carbon Compounds: A Modern Comprehensive Treatise. 2d ed. Vols. 1–. **S. Coffey, ed.** New York: American Elsevier, 1977–.
Volume 4, 1979.
This volume deals with the chemistry of heterocyclic compounds. Format is like that of earlier editions. Valuable for graduate academic libraries and organic chemists.
R: *Journal of the American Chemical Society* 100: 4639 (May 24, 1978); 100: 6547 (Sept. 27, 1978); 100: 7788 (Nov. 22, 1978); 101: 3418 (June 6, 1979).

Standard Methods of Chemical Analysis. 6th ed. 3 vols. Melbourne, FL: Krieger, 1975.
Volume 1, *Elements*; volume 2, *Industrial and Natural Products and Noninstrumental Methods*; volume 3, *Instrumental Methods*.
Most widely accepted and readily applied methods of analysis.

Techniques of Chemistry. **Arnold Weissberger and Bryant W. Rossiter, eds.** New York: Wiley–Interscience, 1971–. Also, supp. to vol. 1, 1976.
Volume 17, *Applications of Lasers to Chemical Problems*, 1982.
The 5-part comprehensive treatment of the physical methods of chemistry.

Techniques of Organic Chemistry. 3d ed. 14 vols. (some in 2d ed.). **Arnold Weissberger et al., eds.** New York: Wiley-Interscience, 1949–.
A comprehensive presentation with chapter-end references. Author and subject indexes.
R: Jenkins (D122); Wal (p. 134).

Topics in Nucleic Acid Structure. **S. Neidle.** New York: Halsted Press, 1981.
A series reviewing current knowledge of the structure of nucleic acids. Each section is described in terms of x-ray crystallographic and solution nuclear magnetic resonance data.

Treatise on Analytical Chemistry. 2d ed. Pts. 1–. Vols. 1–. **I. M. Kolthoff and P. J. Elving, eds.** New York: Wiley, 1978–.
First edition, part 1, 12 volumes, 1959–1976; part 2, *Analytical Chemistry of Inorganic and Organic Compounds*, volume 10, 1978; part 3, *Analytical Chemistry in Industry*, 4 volumes, 1967–1977; second edition, volume 7, part 1, *Theory and Practice of Analytical Chemistry*, 1981; second edition, volume 14, part 1, *Theory and Practice*, 1986.
Discusses in detail thermodynamics and quantum theory. Well-organized, authoritative reference. Belongs in every university or polytechnic library.
R: *Chemistry in Britain* 15: 34 (Jan. 1979): 15: 258 (May 1979); *Journal of the American Chemical Society* 100: 7787 (Nov. 22, 1978); *Laboratory Practice* 27: 575 (July 1978).

Treatise on Solid State Chemistry. **N. Bruce Hannay, ed.** New York: Plenum, 1973–.
Volume 1, *The Chemical Structure of Solids*; volume 2, *Defects in Solids*; volume 3, *Crystalline and Non-crystalline Solids*; volume 4, *Reactivity of Solids*; volume 5, *Changes of State*; volume 6A, *Surfaces I*; volume 6B, *Surfaces II*.
R: *Journal of Metals* 30: 73 (Jan. 1978); *Science* 185: 689 (Aug. 23, 1974).

BIOCHEMISTRY

Applied Biochemistry and Bioengineering. Vols. 1–. **Lemual B. Wingard, Jr., Ephraim Katchalski-Katzir, and Leon Goldstein, eds.** New York: Academic Press, 1976–.
Volume 2, *Enzyme Technology*, 1979.
A reference text examining industrial applications of enzymes. Mainly for special and research libraries.

Bioorganic Chemistry. Vols. 1–. **E. E. van Tamelen, ed.** New York: Academic Press, 1977–.
Volume 1, *Enzyme Action*; volume 2, *Substrate Behavior*, 1978; volume 3, *Macro- and Multi-molecular Systems*, 1977; volume 4, *Electron Transfer and Energy Conversion; Cofactors; Probes*, 1978.

Chemical and Biochemical Applications of Lasers. Vols. 1–. **C. Bradley Moore, ed.** New York: Academic Press, 1974–.
Volume 5, 1980.

Comprehensive Biochemistry. **M. Florkin and E. H. Stotz, eds.** New York: American Elsevier, 1962–. In progress.
Volume 34, 1982.
Authoritative treatise on the subject with extensive bibliographic references.
R: *Nature* 258: 27 (Nov. 6, 1975); *Science* 140: 1201 (1963); *Scientific American* 220: 126 (Feb. 1969); Jenkins (D101).

BIOLOGICAL SCIENCES

The Bacteria: A Treatise on Structure and Function. Vols. 1–. **I. C. Gunsalus and R. Y. Stanier, eds.** New York: Academic Press, 1960–.
Volume 1, *Structure*; volume 2, *Metabolism*, 1961; volume 3, *Biosynthesis*, 1962; volume 4, *Physiology of Growth*, 1962; volume 5, *Heredity*, 1964; volume 6, *Bacterial Diversity*, 1978; volume 7, *Mechanisms of Adaptation*, 1979.

The Biochemistry of Plants: A Comprehensive Treatise, Vols. 1–. **P. K. Stumpf and E. E. Conn, eds.** New York: Academic Press, 1980–.
Volume 1, *The Plant Cell*; volume 2, *Metabolism and Respiration*, 1980; volume 3, *Carbohydrates, Structure and Function*, 1980; volume 4, *Lipids: Structure and Function*, 1980; volume 5, *Amino Acids and Derivatives*, 1980; volume 6, *Proteins and Nucleic Acids*, 1981.
Contains up-to-date information on all aspects of plant biochemistry. Volumes are interlinked by a cumulative index. Authoritative for all scientists concerned with plant biochemistry.

Biology of the Reptilia. Vols. 1–. **Carl Gans, ed.** New York: Academic Press, 1969–.
Volume 9, *Neurology A*, 1979; volume 10, *Neurology B*, 1979.

Biotechnology. A Comprehensive Treatise. 8 vols. **H. J. Rehm and G. Reed, eds.** Deerfield Beach, FL: Verlag Chemie International, 1982–1984.
A series that will become the first standard reference tool in the discipline.

Cell Biology: A Comprehensive Treatise. Vols. 1–. **Lester Goldstein and David M. Prescott, eds.** New York: Academic Press, 1977–.
Volume 1, *Genetic Mechanisms of Cells*; volume 2, *The Structure and Replication of Genetic Material*, 1979; volume 3, *Gene Expression: The Production of RNA's*, 1980; volume 4, *Gene Expression: Translation and the Behavior of Proteins*, 1980.
A multivolume treatise that comprises an in-depth series on the state of the art in cell biology. Among subjects covered are genetic expression, cell differentiation, enzymology, and DNA.
R: *Nature* 291: 174 (May 14, 1981); *RSR* 6: 20 (Oct./Dec. 1978).

The Cell Nucleus. 7 vols. **Harris Busch, ed.** New York: Academic Press, 1974–1979.
Volume 1, 1974; volume 2, 1974; volume 3, 1974; volume 4, *Chromatin*, part A, 1978; volume 5, *Chromatin*, part B, 1978; volume 6, *Chromatin*, part C, 1978; volume 7, *Chromatin*, part D, 1979.
Seven volumes on the cell nucleus. Multi-authored volumes; highly sophisticated.

Chemical Zoology. Vols. 1–. **Marcel Florkin and Bradley J. Scheer, eds.** New York: Academic Press, 1967–.

Volume 1, *Protozoa*, 1967; volume 2, *Porifera, Coelenterata, and Platyhelminthes*, 1968; volume 3, *Echinodermata, Nematoda, and Acanthocephala*, 1969; volume 4, *Annelida, Echiura, and Sipuncula*, 1969; volume 5, *Arthropoda*, 1970; volume 6, *Arthropoda*, 1971; volume 7, *Mollusca*, 1972; volume 8, *Deuterostimians, Cyclostomes, and Fishes*, 1974; volume 9, *Amphibia and Reptilia*, 1975; volume 10, *Aves*, 1978; volume 11, *Mammalia*, 1979.
Thus far, an 11-volume treatise on various aspects of chemical zoology. Covers majors classes.

The Enzymes. 3d ed. 14 vols. **Paul Boyer, ed.** New York: Academic Press, 1970–1981.

Volume 13, *Oxidation-Reduction Part C Dehydrogenases (II) Oxidases (II) Hydrogen Peroxide Cleavage*, 1976; volume 14, *Nucleic Acids*, part A, 1981.
A comprehensive treatise on oxidation-reduction enzymes; considered a valuable reference.

Evolution and the Genetics of Populations. 4 vols. **Sewall Wright.** Chicago: University of Chicago Press, 1968–1978.

Volume 1, *Genetic and Biometric Foundations*, 1968; volume 2, *The Theory of Gene Frequencies*; volume 3, *Experimental Results and Evolutionary Deductions*, 1977; volume 4, *Variability within and among Natural Populations*, 1978.
A treatise that contains classic studies in genetics, such as the works of Haldane, Mayr, and Fisher. A highly recommended, comprehensive reference.
R: *Nature* 272: 561 (Apr. 1978); *Nature* 275: 569 (Oct. 12, 1978); *Science* 207: 173 (Jan. 11, 1980).

Fish Physiology. **W. S. Hoar and D. J. Randall, eds.** New York: Academic Press, 1969–.
Volume 8, *Bioenergetics and Growth*, 1979.

Foundations of Mathematical Biology. 3 vols. **Robert Rosen, ed.** New York: Academic Press, 1972–1973.

Volume 1, *Subcellular Systems*; volume 2, *Cellular Systems*; volume 3, *Supercellular Systems*.
R: *American Scientist* 62: 489 (July/Aug. 1974); 63: 360 (May/June 1975).

The Fungi: An Advanced Treatise. 4 vols. in 5. **G. C. Ainsworth and Alfred Sussman.** New York: Academic Press, 1965–1973.

An in-depth treatment of the subject. Bibliographies after each chapter. Author and subject indexes.
R: Sheehy (EC34).

Hormonal Proteins and Peptides. Vols. 1–. **Choh H. Li, ed.** New York: Academic Press, 1973–.

Volume 2, 1975; volume 3, 1975; volume 4, 1977; volume 5, *Lipotropin and Related Peptides*, 1978; volume 6, *Thyroid Hormones*, 1978; volume 7, *Hypothalmic Hormones*, 1979; volume 8, *Prolactin*, 1980; volume 9, *Techniques in Protein Chemistry*, 1980.

Hyman Series in Invertebrate Biology: Zoological Sciences. 6 vols. **Libbie Henrietta Hyman.** New York: McGraw-Hill, 1940–1967.

Advanced treatise on invertebrate zoology.

R: *Science* 96: 219 (1940); Jenkins (G211); Wal (p. 227); Win (2EC16).

An Introduction to Physiology. Vol. 4. **H. Davson and M. B. Segal.** New York: Grune & Stratton, 1978.

A multivolume treatise on physiology. Discussions are clear and succinct. Copiously illustrated, well-recommended.

R: *Nature* 281: 411 (Oct. 4, 1979).

Marine Ecology: A Comprehensive, Integrated Treatise on Life in Oceans and Coastal Waters. 5 vols. **Otto Kinne, ed.** New York: Wiley, 1970–.

Volume 1, *Environmental Factors*, 1970; volume 2, *Physiological Mechanisms*, 1975; volume 3, *Cultivation*, 1976–1978; volume 4, *Dynamics*, 1978; volume 5, in preparation. An exhaustive treatise on marine biology. Contains contributions from experts; emphasizes laboratory testing methods. Contains bibliographic references from worldwide studies. Indexed by subject, author, and taxonomic category. Highly recommended.

R: *Nature* 272: 380, 381 (Mar. 23, 1978); *Choice* 13: 1622 (Feb. 1977).

Microbial Toxins: A Comprehensive Treatise. Vols. 1–. **Samuel J. Ajl, Solomon Kadis, and Thomas C. Montie, eds.** New York: Academic Press, 1970–.

Volume 1, *Bacterial Protein Toxins*, 1970; volume 2A, *Bacterial Protein Toxins*, 1971; volume 3, *Bacterial Protein Toxins*, 1970; volume 4, *Bacterial Endotoxins*, 1971; volume 5, *Bacterial Endotoxins*, 1971; volume 6, *Fungal Toxins*, 1971; volume 7, *Algal and Fungal Toxins*, 1971; volume 8, *Fungal Toxins*, 1972.

The Mycoplasmas. 3 vols. **M. F. Barile et al., eds.** New York: Academic Press, 1979.

Volume 1, *Cell Biology*; volume 2, *Human and Animal Mycoplasmas*; volume 3, *Plant and Insect Mycoplasmas*.

Physical Techniques in Biological Research. 8 vols. **William Nastuk.** New York: Academic Press, 1955–1973.

Critical, 8-volume survey of the techniques used in dealing with organic substances, specifically written for the biologist. Sections on optical techniques, cells and tissues, special methods, and electrophysical methods.

Plant Disease: An Advanced Treatise. 5 vols. **James G. Horsfall and Ellis B. Cowling, eds.** New York: Academic Press, 1977–1980.

Volume 1, *How Disease Is Managed*, 1977; volume 4, *How Pathogens Induce Disease*, 1979; volume 5, *How Plants Defend Themselves*, 1980.

R: *Nature* 290: 806 (Apr. 30, 1981); *Science* 199: 289 (Jan. 20, 1978); *TBRI* 44: 110 (Mar. 1978).

Plant Pathology: An Advanced Treatise. 3 vols. **James Gordon Horsfall.** New York: Academic Press, 1959–1960.
Emphasizes disease concepts for plants from around the world.
R: *Science* 131: 1368 (1960); 132: 30 (1960); Jenkins (G104).

Plant Physiology: A Treatise. 11 vols. **F. C. Steward, ed.** New York: Academic Press, 1959–1972.
Volume 1A, *Cellular Organization and Respiration*, 1960; volume 1B, *Photosynthesis and Chemosynthesis*, 1960; volume 2, *Plants in Relation to Water and Solutes*, 1959; volume 3, *Inorganic Nutrition of Plants*, 1963; volume 4A, *Metabolism: Organic Nutrition and Nitrogen Metabolism*, 1965; volume 4B, *Metabolism: Intermediary Metabolism and Pathology*, 1966; volume 5A, *Analysis of Growth: Behavior of Plants and Their Organs*, 1969; volume 5B, *Analysis of Growth: The Responses of Cells and Tissues in Culture*, 1969; volume 6A, *Physiology of Development: Plants and Their Reproduction*, 1972; volume 6B, *Physiology of Development: The Hormones*, 1972; volume 6C, *Physiology of Development from Seeds to Sexuality*, 1972.

Reproduction of Marine Invertebrates. Vols. 1–. **Arthur C. Giese and John S. Pearse, eds.** New York: Academic Press, 1974–.
Volume 1, *Acoelomate and Pseudocollomate Metazoans*; volume 2, *Entoprocts: Lesser Callomates*, 1975; volume 5, *Molluscs: Pepecypods and Lesser Classes*, 1979.

Sisson and Grossman's—The Anatomy of the Domestic Animals. 5th ed. 2 vols. **Robert Getty.** Philadelphia: Saunders, 1975.

Traite de Zoologie: Anatomie, Systematique, Biologie. 17 vols. **Pierre Paul Grasse.** Paris: Masson, 1948–.
Authoritative, comprehensive work on zoology. Includes bibliographies.

The Yeasts. **A. H. Rose and J. S. Harrison, eds.** London and New York: Academic Press, 1969–1971.
Volume 1, *Biology of Yeasts*; volume 2, *Physiology and Biochemistry of Yeasts*; volume 3, *Yeast Technology*.
Author and analytical subject-indexes.

EARTH SCIENCES

Compendium of Meteorology: For Use by Class 1 and Class 2 Meteorological Personnel. 2 vols. **Aksel Wiin-Nielson, ed.** New York: Unipub, 1973–.
Volume 1, part 1, *Dynamic Meteorology*; part 2, *Physical Meteorology*; part 3, *Synoptic Meteorology*; volume 2, part 1, *General Hydrology*; part 2, *Aeronautical Meteorology*; part 3, *Marine Meteorology*.
A series of lectures covering all areas of meteorology. For meteorology students and instructors.
R: *IBID* 8: 13 (Spring 1980).

The Sea: Ideas and Observations on Progress in the Study of the Seas. Vols. 1–. **M. N. Hill and Edward D. Goldberg, eds.** New York: Wiley-Interscience, 1962–.

Volume 1, *Physical Oceanography*, 1962; volume 2, *The Composition of Sea Water: Comparative and Descriptive Oceanography*, 1963; volume 3, *The Earth beneath the Sea: History*, 1963; volume 4, *New Concepts of Sea Floor Evolution*, 1971; volume 5, *Marine Chemistry*, 1974; volume 6, *Marine Modeling*, 1977.
Authoritative, critical, and detailed survey.
R: Wal (p. 163).

Treatise on Invertebrate Paleontology. **Raymond C. Moore et al., eds.** Prepared for the Joint Committee on Invertebrate Paleontology. New York: Geological Society of America and University of Kansas Press, 1953–. In progress.
Parts A–W, 1953–1981.
R: Win (EE135; 1EE22; 2EE23; 3EE14).

Water—A Comprehensive Treatise. 6 vols. **Felix Franks, ed.** New York: Plenum Press, 1979.
Volume 6, *Recent Advances*.
Examines recently developed methods of testing the chemistry and physics of aqueous systems, solutions, and hydrates.
R: *Nature* 261: 728 (June 24, 1976).

World Survey of Climatology. 15 vols. **H. E. Landsberg, ed.** Amsterdam: Elsevier, 1969–.
R: Sheehy (EE15); Win (3EE7).

CHEMICAL ENGINEERING

Chemical Engineering Monographs. 8 vols. **S. W. Churchill, ed.** New York: American Elsevier, 1975–.
Volume 1, *Polymer Engineering*, 1975; volume 2, *Filtration Post-Treatment Processes*, 1975; volume 3, *Multicomponent Diffusion*, 1976; volume 4, *Transport in Porous Catalysts*, 1977; volume 7, *Twin Screw Extrusion*, 1977; volume 8, *Fault Detection and Diagnosis in Chemical and Petrochemical Processes*, 1978.
Concise, authoritative text on chemical engineering. Among subjects covered are polymer engineering, filtration processes, and fault detection. Each subject treated is clear and focused.

Chemical Engineering Practice. 12 vols. **Herbert W. Cremer and T. Davies.** New York: Academic Press, 1956–1965.
Authoritative encyclopedic survey of chemical engineering.
R: *Nature* 203: 681 (Aug. 15, 1964); Wal (p. 473); Sheehy (EJ87).

Treatise on Adhesion and Adhesives. 5 vols. New York: Dekker, 1967–1981.
Volume 1, *Theory*, 1967; volume 2, *Materials*, 1969; volume 5, 1981.

MATERIALS SCIENCE

Advances in High-Pressure Research. 4 vols. **R. S. Bradley, ed.** New York: Academic Press, 1966–1974.
R: *American Scientist* 58: 432 (July/Aug. 1970); Wal (pp. 112, 301, 475).

Composite Materials. 8 vols. **L. J. Broutman et al.** New York: Academic Press, 1974–1975.
Volume 1, *Interfaces in Metal Matrix*; volume 2, *Mechanics of Composite Materials*; volume 3, *Engineering Applications*; volume 4, *Metallic Matrix Composites*; volume 5, *Fracture and Fatigue*; volume 6, *Interfaces in Polymer Matrix Composites*; volume 7, *Structural Design and Analysis*; volume 8, *Structural Design and Analysis*, 1975.

Glass: Science and Technology. Vols. 1–. **D. R. Uhlmann and N. J. Kreidl, eds.** New York: Academic Press, 1983–.
Volume 1, *Glass-Forming Systems*, 1983; volume 2, *Processing I*, 1984; volume 5, *Elasticity and Strength in Glasses*, 1984.
A multivolume treatise that covers a full range of topics on glass, from fundamentals of structure and properties to highly applied areas. An invaluable source.

Phase Diagrams: Materials Science and Technology. Vols. 1–. **A. M. Alper.** New York: Academic Press, 1970–.
Volume 1, *Theory, Principles, and Techniques of Phase Diagrams*; volume 5, *Crystal Chemistry, Stoichiometry, Spinodal Decomposition, Properties of Inorganic Phases*, 1978.

Strength of Materials. Vol. 1–. **J. M. Alexander.** New York: Halsted Press, 1981–.
Volume 1, *Fundamentals*.

Techniques of Metals Research. Vols. 1–. Melbourne, FL: Krieger, 1968–.
Volume 7, part 2, 1976.
A compilation of review articles on the various techniques of metals research. Extensive bibliographic references; author and subject indexes for each volume.
R: *Choice* 6: 37 (1969); Win (2E142).

Treatise on Materials Science and Technology. Vols. 1–. **Herbert Herman, ed.** New York: Academic Press, 1972–.
Volume 10, *Properties of Solid Polymeric Materials*, part A, 1977; volume 16, *Erosion*, 1979; volume 17, *Glass II*, 1979; volume 18, *Ion Implantation*, 1980; volume 19, *Experimental Methods*, part A, 1980; volume 20, *Ultrarapid Quenching of Liquid Alloys*, 1981; volume 21, *Electronic Structure and Properties*, 1981; volume 22, *Glass III*, 1982; volume 26, *Glass IV*, 1985.
A comprehensive reference on the fundamentals of research applied to materials, including metals and alloys. Details procedures of analysis. For metallurgists, ceramists, physicists, and crystallographers.
R: *American Scientist* 69: 239 (Mar./Apr. 1981); *Journal of Metals* 32: 64 (Apr. 1980); *Journal of the American Chemical Society* 101: 5108 (Aug. 15, 1979); 102: 3664 (May 7, 1980).

MECHANICAL ENGINEERING

Machine Tools. **Manfred Weck.** 4 vols. New York: Wiley, 1983–1984.
Volume 1, *Types of Machines, Forms of Construction, and Applications*, 1983; volume 4, *Metrological Analysis and Performance Tests*, 1984.

NUCLEAR SCIENCE ENGINEERING

Nuclear Science and Technology: A Series of Monographs and Textbooks. Vols. 1–. **V. L. Parsegian, ed.** New York: Academic Press, 1961–.
Volume 12, *Nuclear Reactor Safety*, 1977.
R: *Physics in Technology* 9: 169 (July 1978); *TBRI* 44: 313 (Oct. 1978).

ENERGY

Energy. Reading, MA: Addison-Wesley, 1974–1976.
Three-volume set resulted from lectures given to the undergraduates at the University of California at San Diego. Volume 1, *Demands, Resources, Impact, Technology, and Policy*, edited by S. S. Penner and L. Icerman, 1974; volume 2, *Non-nuclear Technologies*, edited by S. S. Penner and L. Icerman, 1975; and volume 3, *Nuclear Energy and Energy Policies*, edited by S. S. Penner, 1976.
R: *American Scientist* 65: 94 (Jan.–Feb. 1977); *Physics Today* 30: 66–67 (Mar. 1977).

Solar Energy Research and Development in the European Community. Netherlands: Reidel, 1981–.
In 8 parts: *Series A: Solar Energy Applications to Dwellings*, 1982; *Series B: Thermomechanical Solar Power Plants*, 1983; *Series C: Photovoltaic Power Generation*, 1982; *Series D: Photochemical, Photoelectrochemical and Photobiological Processes*, 1982; *Series E: Energy from Biomass*, 1981; *Series F: Solar Radiation Data*, 1982; *Series G: Wind Energy*, 1983; *Series H: Solar Energy in Agriculture and Industry.*
R: *Ulrich's Quarterly* 7: 171 (April/June 1983).

ENVIRONMENTAL SCIENCE

Air Pollution. 3d ed. 5 vols. **Arthur Stern.** New York: Academic Press, 1976–77.
Second edition, 1968.
Greatly revised and expanded; contains contributions by experts in the field.
R: *Choice* 13: 1627 (Feb. 1977); Sheehy (EJ24).

Environmental Quality and Safety: Chemistry, Toxicology, and Technology. 5 vols. **Frederick Coulston and Friedhelm Korte, eds.** New York: Academic Press, 1972–1976.
Volume 1, *Global Aspects of Chemistry, Toxicology and Technology as Applied to the Environment*, 1972; volume 2, 1973; volume 3, 1974; volume 4, 1975; volume 5, 1976.

A treatise on 5 volumes; deals with various aspects of technology and toxicology. Discusses many different environments, applications, and situations.

Environmental Science: An Interdisciplinary Monograph Series. Nos. 1–. **Douglas H. K. Lee, E. Wendell Hewson, and Daniel Okun, eds.** New York: Academic Press, 1970–.

Number 17, *Asbestos and Disease*, 1978.

A multivolume treatise on various aspects of ecology, pollution, and the environment.

Marine Science Instrumentation. Vols. 1–5. Pittsburgh: ISA, 1973.

These volumes contain the proceedings of the Instrument Society of America's Marine Sciences Instrumentation Symposia.

Physiological Ecology: A Series of Monographs, Texts and Treatises. Nos. 1–. **T. T. Kozlowski, ed.** New York: Academic Press, 1971–.

Number 18, *Functional Adaptations of Marine Organisms*, 1981.

Water Pollution: A Series of Monographs. Vols. 1–. **K. S. Spiegler and J. I. Bregman, eds.** New York: Academic Press, 1974–.

Volume 5, *Water Quality Management under Conditions of Scarcity: Israel as a Case Study*, 1980.

CHAPTER 15   ABSTRACTS AND INDEXES, AND CURRENT-AWARENESS SERVICES

ABSTRACTS AND INDEXES

GUIDES TO ABSTRACTS AND INDEXES

The following are examples of sources that provide both a guide to available abstracts and indexes in science and technology and an insight into the role that such publications play throughout the disciplines.

Abstracting Scientific and Technical Literature. Repr. ed. **Robert E. Maizell, Julian F. Smith, and T. E. R. Singer.** Melbourne, FL: Krieger, 1979.
First edition, 1971.
An introductory guide and text for scientists, abstractors, and indexers.

Abstracts and Indexes in Science and Technology: A Descriptive Guide. **Dolores B. Owen and Marguerite M. Hanchey.** Metuchen, NJ: Scarecrow Press, 1974.
Provides a comprehensive listing of indexing tools in scientific and technical subjects. Entries, described in outline form, are arranged under general headings.
R: *Choice* 11: 1459 (Dec. 1974); *ARBA* (1975, p. 639); Sheehy (EA10).

Biological Indicators of Environmental Quality: A Bibliography of Abstracts. **William A. Thomas, William H. Wilcox, and Gerald Goldstein.** Ann Arbor: Ann Arbor Science, 1973.
A unique collection of abstracts on environmental monitoring. Aimed at aquatic biologists, plant physiologists, foresters, etc.
R: *ARBA* (1975, p. 692).

A Guide to Chemical Abstracts. **Patricia J. Delks.** Boca Raton, FL: Science Media, 1978.
A slide-tape presentation on the use of *Chemical Abstracts*. Includes 43 slides (35mm), audio cassette, and 8-page script, which lasts approximately 20 minutes. A high-quality presentation for undergraduate science students.

A Guide to the World's Abstracting and Indexing Services in Science and Technology: National Federation of Indexing and Abstracting Services Report Number 102. Washington, DC: Science and Technology Division, Library of Congress, 1963. Reprint. Boston: Gregg Press, 1972.
Some 1,800 useful but dated services arranged by title.
R: *ARBA* (1974, p. 549); Win (EA62).

Guide to Special Issues and Indexes of Periodicals. 3d ed. **Miriam Whlan, ed.** New York: Special Libraries Association, 1982.

Second edition, 1976.
A guide to the contents of special issues of American and Canadian trade, technical, and consumer periodicals. Entries are organized alphabetically by journal title with a listing of special features, issues, sections appearing on a recurring basis, and editorial and advertiser indexes. Includes a detailed subject index.

A World Bibliography of Bibliographies and of Bibliographic Catalogues, Calendars, Abstracts, Digests, Indexes, and the Like. 4th and final ed., rev. and greatly enl. 5 vols. Lausanne: Societas Bibliographica, 1965–1966.

The most comprehensive work of its kind.
R: Jenkins (A11); Wal (p. 11).

GENERAL SCIENCE

Bulletin Signalétique. Paris: Centre National de la Recherche Scientifique, 1956–. Monthly for most sections.

Multiple sections, mostly monthly, carry about 400,000 abstracts per year from about 8,000 periodicals. Each issue of each part has an author index. Annual subject and author indexes.
R: Wal (p. 5).

Composite Index for CRC Handbooks. 2d ed. **Robert Weast, ed.** Cleveland: Chemical Rubber, 1977.

First edition, 1971.
Provides access to one of the largest sources of scientific and technical literature, covering all 49 handbooks in the physical and life sciences.
R: *Aslib Proceedings* 24: 273 (May 1972); *RSR* 5: 22 (Oct./Dec. 1977); 7: 15 (Apr./June 1979); *ARBA* (1979, p. 643); Wal (p. 49).

Cumulative Index to Science Education: 1916–1976. Vols. 1–60. **Audrey B. Champagne and Leopold B. Klopfer, eds.** New York: Wiley, 1978.

A valuable index for science educators and researchers. Provides quick access to information on research topics, teaching objectives, curricula, equipment, etc.
R: *ARBA* (1980, p. 604).

General Science Index. Vols. 1–. New York: Wilson, 1978–.

Indexes some 90 English-language general science periodicals. Covers such fields as biology, earth sciences, environment, physics, food science, etc. Includes subject entries written in simple language and an author listing of citations to book reviews. Contains numerous cross references. A valuable reference geared for the nonspecialist. For all public, high school, and undergraduate libraries. Also available online through WILSONLINE.
R: *Choice* 16: 202 (Apr. 1979); *LJ* 106: 1297 (June 15, 1981); *NLW* 79: 115 (June 1978); *WLB* 53: 409 (Jan. 1979); *ARBA* (1980, p. 605); Sheehy (EA11).

Government Reports Announcements and Index. Springfield, VA: US National Technical Information Service, 1938–. Semimonthly.

Formerly *US Government Research and Development Reports Index.*
Announcements and *Index* sections combined into one publication in 1975. Abstracts of scientific and technical-report literature emanating from over 225 government organizations. Over 50,000 abstracts are produced annually. Entries arranged under 22 major subject categories, with numerous subdivisions. Index section includes subject, personal and corporate-author, contract-number, and accession/report indexes.
R: Katz (p. 14).

A Guide to "Referativnyi Zhurnal." 2d ed., rev. **E. J. Copley.** London: National Reference Library of Science and Invention, 1972.
R: Owen (p. 13).

Index to Book Reviews in the Sciences. Nos. 1–. Philadelphia: Institute for Scientific Information, 1980–.
Includes more than 35,000 book reviews appearing in over 3,000 journals. Covers all major areas of science, such as engineering, physical and chemical sciences, agriculture, biology, and environmental sciences. Entries are arranged alphabetically by editor and author of book, and full bibliographic information is provided. Also included is a *Permuterm Subject Index to Book Titles.* The index is published monthly with semiannual cumulations. An essential reference tool for all academic and research libraries.
R: *Choice* 18: 216 (Oct. 1980); *ARBA* (1981, p. 622).

Index to Scientific Reviews. Vols. 1–. Philadelphia: Institute for Scientific Information, 1975–. Semiannual.
An eclectic index to scientific review articles. Contains author listings and permuted title-subject indexes. Indexes some 3,000 journals.
R: *ARBA* (1976, p. 639).

Index to Scientific and Technical Proceedings. Vols. 1–. Philadelphia: Institute for Scientific Information, 1978–. Monthly.
An index to about 90,000 conference papers and articles. Indexes include author, subject, conference sponsor, meeting location, author's organization, and title words. An invaluable reference aid for all scientific libraries.

Index to US Government Periodicals. **Ivan A. Walters, Jr., ed.** Chicago: Infordata International, 1984. Quarterly, with 4th issue as cumulative.
Fourteenth annual, 1984.
A significant contribution to accessing journal articles published by the US government. Indexing dates back to 1970, and 176 journals are indexed, only 50 of which are included by standard indexing sources. Author and subject indexes available. A must for all libraries interested in government publications.
R: *ARBA* (1975, p. 10; 1976, p. 5; 1985, p. 16).

Information Science Abstracts. New York: Plenum Publishing, 1966–. Quarterly.
Continues *Documentation Abstracts.* Covers "information" in the broad sense, including its generation, publication, collection, documentation, etc. Classified arrangement, with author and annual subject indexes.
R: Katz (p. 18).

Pandex Current Index to Scientific and Technical Literature. New York: Pandex, to 1969; CCM Information, 1967–. Biweekly.
A multidisciplinary index covering technical reports, books and approximately 2,000 scientific journals. Arranged in 2 sections: subjects; and authors and permuted titles.
R: *American Documentation* 19: 357 (1968); *CRL* 29: 72 (1968); *Sci-Tech News* 23: 19 (1969); *SL* 58: 728 (1967); Jenkins (A21); Wal (p. 8).

Referativnyi Zhurnal. Moscow: Akademiya Nauk, 1953–.
The most comprehensive abstracting service. In 1966, 21,000 periodicals as well as 6,000 monographs, etc. from 110 countries were covered. Over 1 million abstracts per year. There are 61 series.
R: Wal (p. 8).

Science Abstracts. Vols. 1–. London: Institution of Electrical Engineers, 1898–.
Originally issued in 2 sections: series A, *Physics Abstracts;* and series B, *Electrical and Electronics Abstracts.* Since 1966, series C, *Computer and Control Abstracts,* has been added.
The 3 abstract journals contain journal, patent, report, book, and conference information. Subject and author indexes. All abstracts are now part of the INSPEC (Information Service in Physics, Electrotechnical, and Control) system. For more information on each abstract, see under separate entry.
R: Katz (p. 24); Sheehy (EG1); (EJ36); (EJ37); Wal (pp. 9, 35); Win (EG12).

Science Citation Index. Vols. 1–. Philadelphia: Institute for Scientific Information, 1961–. 6/yr.
The components of *Science Citation Index* are: *Citation Index, Source Index,* and *Permuterm Subject Index.*
A comprehensive computer-produced index that provides access to related articles by listing both cited and citing (source) authors and works. The vast majority of citations are from over 3,200 journals, though patents, reports, meetings, etc. are included.
R: *Chemical and Engineering News* 42: 55–56 (Aug. 31, 1964); *Chemistry and Industry* 10: 416 (Mar. 6, 1965); *Nature* 211: 556–557 (Aug. 6, 1966); 277: 1173 (Sept. 12, 1970); *Science* 145: 142 (1964); *Journal of Documentation* 21: 139–141 (1965); *LJ* 89: 2735–2737 (1964); *LRTS* 9: 478 (1965); 12: 415 (1968); *LT* 16: 374 (1968); *ARBA* (1976, p. 640); Jenkins (A20); Katz (p. 25); Wal (p. 10); Win (2EA15).

Science Fair Project Index 1973–1980. **Akron-Summit County Public Library, Science and Technology Division.** ed. Metuchen, NJ: Scarecrow Press, 1983.
Lists books and articles dealing with science fair projects from grade 5 through high school. A good source for project ideas.
R: *ARBA* (1984, p. 613).

Science Research Abstracts Journal. Vols. 1–. Riverdale, MD: Cambridge Scientific Abstracts, 1973–. 10/yr for each pt.
Part A, "Superconductivity; Magnetohydrodynamics, Plasmas, Theoretical Physics and Superconductivity Research"; part B, "Laser and Electro-optic Reviews; Quantum Electronics."

Unconventional energy sources. Approximately 32,000 abstracts annually. Subject and author indexes.

Scientific American Cumulative Index, 1948–1978: Index to the 362 Issues from May 1948 Through June 1978. New York: Scientific-American Illustrated Library, 1979.

A detailed, subject-access index to *Scientific American*; also includes all tables of contents. Useful for school and academic libraries.

R: *Choice* 16: 804 (Sept. 1979); *LJ* 104: 1241 (June 1, 1979); *RSR* 8: 23 (Jan./Mar. 1980); *WLB* 54: 65 (Sept. 1979); *ARBA* (1980, p. 605).

Astronomy

Astronomy and Astrophysics Abstracts. Vol. 26. **S. Bohme et al., eds.** New York: Springer-Verlag, 1980.

Volume 18, 1977; volume 19, 1977, volumes 20, 21, 22, 1978; volumes 23, 24, 25, 1979; volume 26, 1980.

A semiannually published index to literature in astronomy and astrophysics. Arranged by broad subject categories that include space research, applied mathematics and physics, the planetary system, etc. First section deals with bibliographic tools, conference proceedings, and monographs. Includes author and subject index. An essential reference for all astronomy collections.

R: *Observatory* 98: 35 (Feb. 1978); 99: 23 (Feb. 1979); *Sky and Telescope* 55: 255 (Mar. 1978); 55: 430 (May 1978); 56: 563 (Dec. 1978); 57: 478 (May 1979); 58: 458 (Nov. 1979); 59: 328 (Apr. 1980); 59: 416 (May 1980); *Choice* 12: 97 (Mar. 1975); *ARBA* (1980, p. 609).

Astrophysical Abstracts. New York: Gordon & Breach, 1969– .

Volume 1 consists of over 400 abstracts under broad categories. Author and subject indexes.

R: Wal (p. 82).

Mathematics

American Mathematical Monthly. Index Volume (1894–1973). **Kenneth O. May, comp.** Washington, DC: Mathematical Association of America, 1977.

Lists articles within each volume of the journal by chronological order. Well-arranged for easy access to information. A significant contribution for mathematical researchers.

R: *RSR* 6: 18 (July/Sept. 1978).

An Author and Permuted Title Index to Selected Statistical Journals. **Brian L. Joiner et al.** Washington, DC: US Government Printing Office, 1970.

R: *Journal of the American Statistical Association* (Dec. 1971).

Author Index of Mathematical Reviews 1965–1972. 4 vols. Providence: American Mathematical Society, 1974.

Continues *The Twenty Year Author Index, 1940–1959* and the *Author Index, 1960–1964*, published in 1961 and 1964 respectively. Index includes complete citations and references to volumes of *Mathematical Reviews* that contain corresponding abstracts.
R: *ARBA* (1976, p. 643).

CompuMath Citation Index: An International, Interdisciplinary Index to the Literature of Applied Mathematics, Computer Science, Statistics, Operations Research, and Related Disciplines. Philadelphia: Institute for Scientific Information. Annual.

Eighth edition, 1984.

A spinoff from the large database that also generates *Science Citation Index* and other products. More than 270,000 source items containing more than 3,256,000 references are listed. Covers 300 journals completely and another 6,000 selectively. The source index lists each article included; citation index is arranged by the references, and subject access is provided by a Permuterm index.
R: *ARBA* (1985, p. 493).

Index of Mathematical Papers. Vols. 1–. Providence: American Mathematical Society, July–Dec. 1971–. Semiannually, 1971–1972; annually since 1973.

Volumes 9 and 10, 1979.

Originally, the index was published twice annually and contained bibliographic information on articles processed by the Mathematical Title Service during the preceding 6 months. Articles were arranged by author and by subject. Since volume 4, 1973, the index has become an annual author and subject index to *Mathematical Reviews*. It supplements the cumulative author indexes to *Mathematical Reviews*, which cover the periods 1940–1959, 1960–1964, and 1965–1972.
R: *ARBA* (1976, p. 643); *ARBA* (1985, p. 611).

Index to Statistics and Probability. **Ross and Turkey,** Los Altos, CA: R and D Press, 1972–1979.

Volume 2, *Citation Index*, 1973; volumes 3 and 4, *Permuted Titles*, 1975; volume 5, *Location and Author*, 1973; volume 6, *Index to Minimum Abbreviations*, 1979.

Mathematical Reviews. Providence, RI: American Mathematical Society, 1940–. Monthly.

International coverage of mathematical publications, including books, articles, translations, research papers, etc. Signed abstracts are arranged by subject and numbered consecutively. Primarily English-language material, but does contain French and German abstracts. Author and brief subject indexes. Cumulative indexes available. Cumulative author indexes available for volumes 1–20, 21–28.
R: *ARBA* (1976, p. 643); Jenkins (B10); Owen (p. 16); Sheehy (EF3); Win (EF11).

Mathematical Reviews Annual Index. 4 vols. Providence, RI: American Mathematical Society. Annual.

Statistical Theory and Methods Abstracts. Vols. 1–. 1959–. Edinburgh: Oliver & Boyd, for the International Statistical Institute 1959–1970; London: Longmans, 1971–. Quarterly.

Formerly *International Journal of Abstracts: Statistical Theory and Methods,* 1954–1963. Classified arrangement for such subjects as probability, distribution, experiment design, etc. Author index.
R: Wal (p. 35); Win (EF12).

Zentralblatt für Mathematik und Ihre Grenzgebiete. Berlin: Deutsche Akademie für Wissenschaften zu Berlin, 1931–. 26/yr.

Covers all fields of pure and applied mathematics. Abstracts in German, English, French, or Italian. Author and cumulative subject indexes.
R: Owen (p. 17).

PHYSICS

Abstracts of Photographic Science and Engineering Literature. **Society of Photographic Scientists and Engineers.** New York: Engineering Index, 1962–[n.d.].

Has ceased publication.
Included aerial photography and photogrammetry.
R: Jenkins (K16); Katz (p. 2).

Acoustics Abstracts. Vols. 1–. Brentwood, Essex, England: Multi-Science, 1967–. Monthly.
R: Wal (p. 103).

Applied Health Physics Abstracts and Notes. Ashford, England: Nuclear Technology, 1975–. Quarterly.

Cumulative Index to Contemporary Acoustical Literature. **Acoustical Society of America.** New York: American Institute of Physics, 1971.
Covers 1964–1968.

Current Physics Index. Vols. 1–. New York: American Institute of Physics, 1975–. Quarterly.

Index to 44 physics journals covering current physics research. Entries, arranged by subject, include title, author, and bibliographic information. A worthwhile addition to science libraries.

Index to the Literature of Magnetism. Vols. 1–. New York: American Institute of Physics, 1961–.

Laser Abstracts. Vols. 1–. Evanston, IL: Lowry-Cocroft Abstracts, 1963–. Weekly.

Comprehensive coverage of articles, patents, reports, and papers of proceedings on the subject. Intended primarily for specialized collections.
R: Jenkins (K34); Wal (p. 318).

Physical Review Abstracts. 1976–. Semimonthly.

Contains abstracts of papers accepted for publication in the *Physical Review* and *Physical Review Letters,* thus provides advance information to the physics community about work to be published.

Physics Abstracts. Vols 1–. London: Institution of Electrical Engineers, 1898–. Semimonthly.

Section A of *Science Abstracts.*
The major abstracting journal for physics. Over 85,000 entries are included annually, as obtained from 120 core journals as well as books, conference proceedings, patents, and selected technical reports. Author and subject indexes. Supplemented by *Current Papers in Physics.*
R: Jenkins (C8); Katz (p. 24); Wal (pp. 93, 305).

Retrospective Index to Theses of Great Britain and Ireland 1716–1950. Vol. 4. **Rolla Turner, comp.** Edited by Roger R. Bilboul. Santa Barbara, CA: ABC-Clio Press, 1976.

Volume 4, *Physical Sciences.*
Part of a series of indexes that provide comprehensive coverage of thesis work done in Great Britain. Contains over 5,000 entries, each with complete information. Cross-referenced to broader and narrower fields; offers bibliographic control of an extensive body of dissertations.
R: *ARBA* (1978, p. 626).

CHEMISTRY

Access. Columbus, OH: Chemical Abstracts Service, American Chemical Society, 1969–. Annual, with quarterly supps.

Beginnings in 1971, new title is *Chemical Abstracts Service Source Index.* Supersedes the Society's *List of Periodicals Abstracted by Chemical Abstracts.*
R: *Chemical and Engineering News* 47: 46 (Mar. 31, 1969); Jenkins (D15a); Wal (p. 122).

Analytical Abstracts. Vols. 1–. Cambridge, England: Society for Analytical Chemistry, 1954–. Monthly.

Continues Section C of the lapsed *British Abstracts.* Covers all aspects of analytical chemistry.
R: Wal (p. 129).

Biochemistry Abstracts: Nucleic Acids. Pt 2. Bethesda, MD: Cambridge Scientific Abstracts, 1971–. Monthly.

Formerly: *Nucleic Acids Abstracts.*
Covers the primary literature of the science of nucleic acids. Arranged by broad subjects and subdivided into narrower categories. Author and annual subject indexes.
R: Owen (p. 31); Wal (pp. 138, 201).

Chemical Abstracts: Key to the World's Chemical Literature. Vols. 1–. Easton, PA: American Chemical Society, 1907–. Weekly.

One of the most ambitious and comprehensive scientific abstracting services available. Descriptive annotations arranged in 80 sections under broad headings. Surveys over 12,000 journals, international patents, conference proceedings, etc., over half of which are in English. Seven types of indexes offer a variety of points of entry into the literature: author, subject; chemical substance; formula; ring systems; patent concordance; numerical patent. Five-year collective and decennial indexes are also available.

Associated publications include *CA Subject Index Alert; Chemical-Biological Activities* and *Polymer Science and Technology.* For computerized data base, see CA CONDENSATES.
R: Katz (p. 9); Sheehy (ED2).

Chemical Abstracts Service Source Index 1907–1979 (CASSI). Columbus, OH: American Chemical Society, 1980.

Provides access to 50,000 scientific publications internationally. Refers user to original source, library holdings, publishers directory, etc. Supplies hard-to-find bibliographic data, updating supplements.
R: *RSR* 8: 22 (July/Sept. 1980).

Chemo-Reception Abstracts. London: Information Retrieval, 1973–. Quarterly.

International coverage of areas such as psychophysics, neuroanatomy, peripheral and central sensory mechanisms, etc. Includes book notices and notifications of proceedings. Author indexes and cumulative subject indexes.

Current Abstracts of Chemistry and Index Chemicus. Vols. 1–. Philadelphia: Institute for Scientific Information, 1960–. Weekly.

Original title: *Index Chemicus, 1960–1969.*
Computer-produced abstracting service containing information on the synthesis and identification of new compounds and new chemical reactions or syntheses. Molecular formulas, author, subject, and journal indexes.
R: Owen (p. 29); Wal (p. 115).

Index of Reviews in Organic Chemistry. **D. A. Lewis, comp.** New York: State Mutual, 1980.

A volume containing 2,800 items that have been selected from the period of January 1979 through November 1980.
R: *Chemistry in Britain* 18: 581 (Aug. 1982).

Index to Reviews, Symposia Volumes, and Monographs in Organic Chemistry. New York: Pergamon Press, 1964–. Biennial.
R: *Choice* 4: 804 (1967); Jenkins (D27); Win (ED71; 2ED15).

Source Index Quarterly. Vols. 1–. Columbus, OH: Chemical Abstracts, 1970–. Quarterly.

Provides complete identification of over 10,000 source publications included in *Chemical Abstracts,* that are generally cited by abbreviated titles. Identifies libraries at which original documents are available. Various notes are provided with complete bibliographical information. Supersedes *Access.*

Thin-Layer Chromatography Abstracts, 1971–1973. **Ronald M. Scott.** Ann Arbor: Ann Arbor Science, 1973.

First edition, *Thin-Layer Chromatography: An Annotated Bibliograhy,* 1965.
A continuation of *Thin-Layer Chromatography Abstracts, 1968–1971.* Present volume contains over 1,000 abstracts divided into groups of compounds. Annotations emphasize practical aspects.
R: *ARBA* (1975, p. 653; 1973, p. 542).

Biological Sciences

Abridged Index Medicus. **US National Library of Medicine.** Washington, DC: US Government Printing Office, 1970–. Monthly.

Rapid access to 100 English-language journals of immediate interest to the practicing physician.
R: Katz (p. 1).

Behavioural Biology Abstracts. London: Information Retrieval, 1973–. Quarterly.

References and abstracts arranged under subject categories and subdivided by taxonomic groups. All abstracts in English, though foreign titles are included. International in scope.
R: Owen (p. 69).

Biological Abstracts. Vols. 1–. Philadelphia: Biosciences Information, 1926–. Semimonthly.

The major abstracting service available on the biological sciences. Approximately 140,000 abstracts of articles from some 8,000 serials, as well as books, reports, etc., are listed annually under 85 subject categories. Various indexes on colored paper: author; biosystemic or taxonomic; CROSS-referring abstract numbers to major categories and subheadings; and BASIC, or subjects in context based on titles. The computerized base is BA PREVIEW. From the BIOSIS database several smaller abstracts such as *Abstracts of Entomology, Abstracts of Mycology,* etc., are generated. *BA* is supplemented by *Bioresearch Index.*
R: Katz (p. 7); Jenkins (G7); Wal (pp. 191, 194, 205, 222); Win (EC7).

Biochemistry Abstracts. Biological Membranes. Pt 1. Bethesda, MD: Cambridge Scientific Abstracts, 1973–. Monthly.

Formerly *Biological Membrane Abstracts.*
International coverage of chemical components, physical properties, model systems, and theoretical aspects of membrane functions, etc.
R: Owen (p. 79).

Biochemistry Abstracts. Amino Acids, Peptides, and Proteins. Pt 3. Bethesda, MD: Cambridge Scientific Abstracts, 1972–. Monthly.

Formerly *Amino Acids Peptide and Protein Abstracts.*
Surveys literature covering the major aspects of the subject. Arranged under broad subject areas. Includes author, peptide, and protein indexes and an annual subject index.
R: Owen (p. 32); Wal (p. 138).

Biological Abstracts/RRM (Reports, Reviews, Meetings). Vols. 17–. Philadelphia: Biosciences Information Service, 1980–. Monthly.

Original title, *Bioresearch Index,* 1965–1980.
Supplements *Biological Abstracts* by providing complete citations to over 125,000 symposia, reports, articles, conference proceedings, letters, etc., not included in *Biological Abstracts.* Gives author, biosystematic, concept, generic, and subject indexes. Citations

are arranged under broad subject categories. Recommended to undergraduate students and those involved in the biological sciences.
R: *Choice* 18: 61 (Sept. 1980); *RSR* 8: 13 (July/Sept. 1980); Wal (p. 191); Win (1EC1; 2EC3).

Biology: Current Titles in the Biological Sciences. **Stephen R. Edwards.** Lawrence, KS: Allen Press, 1983–. Quarterly.

Current Index to Statistics: Applications, Methods, and Theory. Vols. 1–. **Brian L. Joiner, ed.** Washington, DC: American Statistical Association and Institute of Mathematical Statistics, 1976–. Annual.

Indexes articles from 60 core journals that relate to statistical applications in the biological sciences.
R: *RSR* 6: 18 (July/Sept. 1978).

Genetics Abstracts. Vols. 1–. New York: CCM Information, 1967.

Index Medicus. **US National Library of Medicine.** Washington, DC: US Government Printing Office, 1960–. Monthly.

Produced by the National Library of Medicine's MEDLARS (Medical Literature Analysis and Retrieval System).
Nearly 3,000 journals with about 200,000 citations annually. Prior to 1976, coverage was limited to periodical literature. From 1976 on, conference publications, monographs, etc., are included.
R: Katz (p. 17).

Index of Human Ecology. **Owen J. Jones and Elizabeth A. Jones.** New York: International Publications Service, 1974.

A valuable index, given the cross-disciplinary nature of the field. Primarily a subject index to the abstracting journals that cover these subjects.
R: *RSR* 3: 25 (Apr./June 1975); *ARBA* (1975, p. 708).

International Abstracts of Biological Sciences. Vols. 1–. London: Pergamon Press, 1954–. Monthly.

Original title: *British Abstracts of Medical Sciences.*
Arranged by broad subject categories, with emphasis on physiological and biochemical aspects of anatomy, animal behavior, biochemistry, and experimental medicine.
R: Jenkins (G12);

Microbiology Abstracts. Vols. 1–. Bethesda, MD: Cambridge Scientific Abstracts, 1965–. Monthly.

Section A: industrial and applied microbiology; section B: bacteriology; section C: algology, mycology, and protozoology; published since 1972. Approximately 15,000 abstracts yearly.
R: Jenkins (G175); Owens (p. 77); Wal (p. 198).

Nutrition Abstracts and Reviews. **Commonwealth Bureau of Animal Nutrition, prep.** Farnham Royal, UK: Commonwealth Agricultural Bureau, 1931–.

Quarterly to 1972; monthly from 1973; since 1977, separated into 2 parts: series A: Human and Experimental; series B: Livestock Feeds and Feeding.
R: Owen (p. 125).

Virology Abstracts. Vols. 1–. London: Information Retrieval, 1967–. Monthly.

Abstracts that focus on all topics of virology research covering culture, immunology, isolation and identification methods, viral infections. Also includes information about proceedings and new books. Virus name, author, patentee indexes are included.

Agriculture

Agrindex. Vols. 1–. Rome: AGRIS. Distr. New York: Unipub, 1975–. Monthly.

Aids in identifying scientific and technical literature in major agricultural libraries. Information is now synthesized into a computer database. Includes books and reports available from both libraries and organizations. Provides bibliographic information. Recommended for all Food and Agriculture Organization depository and large research libraries.
R: *ARBA* (1977, p. 728).

Bibliography of Agriculture. New York: CCM Information, 1942–. Monthly.

An index to current literature, domestic and foreign, received by the National Agricultural Library. Includes sections arranged by main entry, subject and author indexes, and checklists of new government publications. More extensive than the Wilson *Biological and Agricultural Index.*
R: *LT* 15: 882 (Apr. 1967); *RSR* 7: 57 (Apr./June 1979); *SL* 53: 531–536 (Nov. 1962); Katz (p. 6); Wal (p. 395).

Bibliography of Agriculture. Vols. 1–. Phoenix: Oryx Press, 1942–. Monthly.

Latest volume, 1981.

A monthly compilation of the journal literature in agriculture and allied sciences. Each issue includes 35,000 subject index references to government documents, journal articles, proceedings, and pamphlets; and more than 15,000 main entry citations. Data provided by the National Agricultural Library, United States Department of Agriculture. An annual cumulation combines the past 12 subject, author, and corporate author indexes. For science libraries.

Biological and Agricultural Index: A Cumulative Subject Index to Periodicals in the Fields of Biology, Agriculture, and Related Sciences. Vols. 1–. New York: Wilson, 1916/1918–. Monthly.

Originally entitled *Agricultural Index,* through 1964.

Primarily a student's and layperson's guide to literature in the areas of agriculture, botany and forestry, conservation, etc. Comprehensive analytic indexing, including book reviews.
R: *ARBA* (1976, p. 660); Katz (p. 7); Wal (pp. 191, 395); Win (EC6; EK24).

Farm and Garden Index. Vols. 1–. Mankato, MN: Minnesota Scholarly Press, 1978–. Quarterly.

An index to approximately 119 horticultural and agricultural periodicals. Provides thorough coverage and extensive cross references. For agricultural libraries.
R: *Choice* 16: 502 (June 1979); *LJ* 104: 89 (Jan. 1, 1979); *RQ* 19: 95 (Fall 1979); *ARBA* (1980, p. 698).

Soils and Fertilizers. **Commonwealth Bureau of Soils.** Farnham Royal, UK: Commonwealth Agricultural Bureau, 1938–. 6/yr., 1938–1972; 12/yr. since 1973.

A review article precedes abstracts in each issue. Annual author and subject indexes.
R: Jenkins (J30); Owen (p. 103); Wal (pp. 409, 411).

World Agricultural Economics and Rural Sociology Abstracts. Farnham Royal, UK: Commonwealth Agricultural Bureau, 1959–. Quarterly, 1959–1972; monthly since 1973.

Each volume has cumulative author, subject, and geographical indexes.
R: Jenkins (J20); Wal (p. 397); Win (EK27).

Botany

Abstracts of Mycology. Vols. 1–. Philadelphia: Biosciences Information, 1967–. Monthly.

A collection of references and abstracts relating to fungi and lichens reported in *Biological Abstracts* and *Bioresearch Index.* Various indexes give access to approximately 10,000 items annually.
R: Jenkins (G91); Wal (p. 218).

Botanical Abstracts. 15 vols. Baltimore: Williams & Wilkins, 1918–1926.

Though superseded by *Biological Abstracts,* for retrospective purposes this is the best source.
R: Jenkins (G81); Wal (p. 205); Win (EC7).

Complete Index for the Six Volumes of Wild Flowers of the United States. **Harold W. Rickett.** New York: New York Botanical Garden and McGraw-Hill, 1975.

An index volume to a 6-part series that provides identification information for wild flowers. Includes over 1,300 color plates. Useful as a guide for the layperson. Well recommended.
R: *ARBA* (1977, p. 660).

Consolidated Index to Flora Europaea. **G. Halliday and M. Beadle.** New York: Cambridge University Press, 1983.

A combined index to all 5 volumes of *Flora Europaea,* covering families, genera, and species. A useful tool for any botanical library.
R: *ARBA* (1985, p. 518).

Excerpta Botanica. Stuttgart: Fischer, 1959–. Sec. A, 10/yr; sec. B, quarterly.

Section A includes abstracts both of articles on systematic botany and of periodical literature on herbana, gardens, museums, etc. Section B is an unannotated listing of monographic world literature on plant geography and ecology.
R: *RSR* 7 : 8 (Jan./Feb. 1979).

Forestry Abstracts. **Commonwealth Forestry Bureau.** Farnham Royal, UK: Commonwealth Agricultural Bureau, 1939–. Quarterly 1939–1972; monthly since 1973.

Covers all aspects of forestry, including forest products and their utilization. Scans books, maps, and pamphlets as well as journal literature.
R: Jenkins (J80); Owen (p. 103); Wal (pp. 414, 515); Win (EK44).

Gray Herbarium Index. 10 vols. **Harvard University.** Boston: Hall, 1968.

Photoduplication of over 250,000 cards of the Gray Herbarium catalog. Primary emphasis on name and literature citations of established Western Hemispheric vascular plants. Some duplication of *Index Kewenis*.
R: *ARBA* (1970, p. 116); Jenkins (G120).

Herbage Abstracts. **Commonwealth Bureau of Pastures and Field Crops.** Farnham Royal, UK: Commonwealth Agricultural Bureau, 1931–. Quarterly, 1931–1972; monthly since 1973.

International coverage of grasslands, rangelands, fodder crops, agrometeorology, environmental contamination, etc. Author indexes and annual subject index.
R: Jenkins (J25); Owen (p. 98).

Horticultural Abstracts. **Commonwealth Bureau of Horticulture and Plantation Crops.** Farnham Royal, UK: Commonwealth Agricultural Bureau, 1931–. Quarterly, 1931–1972; monthly since 1973.

Topically arranged abstracts on ornamental plants, plantation crops, fruits and vegetables, etc. Quarterly author and annual author and subject indexes.
R: Jenkins (J26); Owen (p. 98) Wal (p. 418).

Index to American Botanical Literature, 1886–1966. 4 vols. **Torrey Botanical Club, comp.** Boston: Hall, 1969.

First supplement, 1967–1976.
An author catalog of over 100,000 books and papers published since 1886. Annuals from 1959 to 1972 may be purchased as files of cards. Book-form supplements are expected every 10 years.
R: *ARBA* (1970, p. 116); Jenkins (G87); Sheehy (EC24); Wal (pp. 206, 215); Win (EC44; 3EC7).

Index of Fungi. **Commonwealth Mycological Institute.** Kew, England: Commonwealth Mycological Institute, 1940–. Semiannual.

Primarily a register of new genera and species of fungi as gleaned from the world literature. Was originally issued as a supplement to the *Review of Applied Mycology*, 1922–.

Index to Grass Species. 3 vols. **Agnes Chase and Cornelia D. Niles, comps.** Boston: Hall, 1963.

Contains over 62,000 entries dealing with the species, varieties, and genera of the world's grasses. Genera are listed alphabetically, and full bibliographic information is provided for each grass. Indexes grass literature from 1763 to 1962.

Index Herbarium: A Guide to the Location and Contents of the World's Public Herbaria. 5th edition. Vols. 1–. **Joseph Lanjouw and F. A. Stafleu.** Utrecht, Netherlands: International Bureau for Plant Taxonomy and Nomenclature of the International Association for Plant Taxonomy, 1964–.

Part 1, "Herbaria," gives detailed directory information about the herbaria arranged alphabetically by city. Part 2, "Collectors," is an alphabetical list of collectors, together with scope of collections.

R: Jenkins (G145); Sheehy (EC27); Wal (p. 209); Win (EC62; 1EC7).

Index Kewenis. Plantarum Phanerogamarum Nomina et Synonyma Omnium Generum et Specierum a Linnaeo Usque ad Annum MDCCCLXXXV Complectens. 2 vols. London: Oxford University Press, 1893–1895.

Index Kewenis. Supplementum. Vols. 1–. for 1866–1895. Oxford: Oxford University Press, Clarendon Press.

Volume 16 for 1971–1976, published in 1980.

Supplements have appeared every 5 years since 1895 for this repository of names for flowering plants, generic and binomial, and the sources of their original publication. Not so comprehensive as the *International Plant Index,* 1962–.

R: Jenkins (G121); Wal (p. 216); Win (EC52; 3EC9).

Index to Illustrations of Living Things Outside North America: Where to Find Pictures of Flora and Fauna. **Lucile Thompson Munz and Nedra G. Slauson.** Hamden, CT: Shoe String Press, 1981.

A listing of where to find illustrations of more than 9,000 plants and animals in 206 books. A useful reference tool for all libraries, especially school libraries.

R: *Library Professional Publications* (Spring 1983, p. 28); *ARBA* (1983, p. 620).

Index Muscorum. Vols. 1–. **R. van der Wijk.** Utrecht, Netherlands: International Bureau for Plant Taxology, 1959–. In progress.

R: Wal (p. 220).

Index to Plant Distribution Maps in North American Periodicals Through 1972. **W. Louis Phillips and Ronald L. Stuckey, comps.** Boston: Hall, 1976.

Over 28,000 entries, which cover the entire plant kingdom; alphabetically arranged by taxa. Entries provide journal citation, geographic location, type of map. A useful research tool for botanists and taxonomists. Highly recommended.

R: *RSR* 5: 6 (Apr./June 1977).

Weed Abstracts. **A. R. C. Weed Research Organization.** Farnham Royal, UK: Commonwealth Agricultural Bureau, 1954–. Bimonthly, 1962–1972; monthly since 1973.

Approximately 400 journals and 100 annual reports are scanned.

R: Jenkins (J31); Owen (p. 100); Wal (p. 412).

Zoology

Abstracts of Entomology. Vols. 1–. Philadelphia: Biosciences Information, 1970–. Monthly.

Coverage of pure and applied studies of insects and other arachnids. British equivalent is *Entomology Abstracts*, vols. 1–, 1969–.
R: Wal (p. 229).

Aquatic Sciences and Fisheries Abstracts. Biological Sciences and Living Resources. Pt 1. Ocean Technology, Policy and Non-Living Resources. Pt 2. Vols. 1–.
Bethesda, MD: Cambridge Scientific Abstracts, 1969–. Monthly.
Originally titled *Aquatic Biology Abstracts*, 1969–1971.
Some 3,000 journals are culled, resulting in some 15,000 abstracts annually. Emphasis on books and articles in physical and chemical oceanography, limnology, and pollution and its effects on fishes.
R: Katz (p. 5); Owen (p. 68); Wal (p. 425).

Fishing Industry Index International. London: Haymarket Press, 1970–. Annual.
Original title: *Fisheries Yearbook and Directory,* 1952–1970.
Primarily a directory of suppliers, producers, distributors, buyers' guides, etc. Includes advertisers index.
R: Wal (p. 426).

International Index of Laboratory Animals. 4th ed. **Michael F. W. Festing, ed.** Carshalton, Surrey, England: Medical Research Council Laboratory Animals Centre, 1980.

Laboratory Animal Science: A Review of the Literature. Argonne, IL: Argonne National Laboratory, Biological and Medical Research Division, 1966–. Quarterly.
Approximately 155 international journals are culled for brief abstracts on laboratory-animal technology and medicine. Species index.
R: Mal (1980, p. 174).

Marine Fisheries Abstracts. **US Department of Commerce, National Oceanic and Atmospheric Administration.** Washington, DC: US Government Printing Office, 1948–. Monthly.
Formerly *Commercial Fisheries Abstracts.*

Wildlife Abstracts: A Bibliography and Index of the Publications Abstracted in Wildlife Review. **US Fish and Wildlife Service.** Washington, DC: US Government Printing Office, 1935–.
Titles arranged by coded subject headings. Includes issue and page numbers from *Wildlife Review.* Author and subject indexes.
R: Win (EK45).

World Fisheries Abstracts. Rome: Food and Agriculture Organization of the United Nations, 1948–1973.

This now-ceased publication was a quarterly review of technical literature on fisheries and related industries. Included author and subject indexes.
R: Owen (p. 83); Wal (p. 426).

Zoological Record . . . Being the Record of Zoological Literature Relating to the Year. . . . Vols. 1–, 1864–. London: Zoological Society of London: 1865–. Annual.

The primary retrospective bibliography in the field. Time lag of approximately 3 years. Section 20 lists new genera and subgenera reported. Has 22 annual sections, each published separately with indexes.
R: *LT* 15: 831 (Apr. 1967); Jenkins (G193); Wal (p. 223); Win (EC99).

EARTH SCIENCES

General

Abstracts of North American Geology. Washington, DC: US Geological Survey, 1966–1971.

Contains abstracts of technical papers and other materials, including citations to maps on the geology of North America. It superseded *Geological Abstracts,* 1953–1958, and *Geo-Science Abstracts,* 1959–1966, and has been continued by *Bibliography and Index of Geology.*
R: Jenkins (F12); Owen (p. 47); Wal (p. 150); Win (1EE7).

Abstract Journal in Earthquake Engineering. Vols. 1–. Berkeley: University of California, Earthquake Engineering Research Center, 1971–.

Bibliography and Index of Geology. Vols. 33–. Boulder, CO: Geological Society of America in cooperation with the American Geological Institute, 1969–. Monthly.

Volumes 1–32 published as *Bibliography and Index of Geology Exclusive of North America.* The major source for an academic geology library, providing access to geological literature, including books, maps, monographs, etc. Arranged under 21 broad categories, and over 36,000 citations are produced annually. Information in this index is coordinated with the AGI's magnetic tape file and the Geological Reference File (GEO-REF). Author and subject indexes.
R: Katz (p. 6); Wal (p. 143); Win (EE25; 2EE9; 3EE1).

British Geological Literature. Vols. 1–. Borune End, England: Coredon Press, 1964–. Quarterly.

About 1,200 references a year. Author arrangement with subject indexes. Limited to geological articles published in Great Britain.
R: Wal (p. 143); Win (EE10).

Earthquake Engineering Reference Index. **International Association for Earthquake Engineering.** London: British National Section, 1963.

Dated but still valuable to civil and geo-engineers.
R: Wal (p. 356).

Geo Abstracts. 7 Sects. Norwich, England: Geo Abstracts, 1966–. Bimonthly.
Formerly: *Geographical Abstracts,* 1966–1971.
The basic abstract journal for geography. Arranged by subject, with each section available separately. Complete citation and brief abstracts. Comprehensive annual indexes.
R: Jenkins (F76); Katz (pp. 5, 14); Owen (p. 44).

Geodex System/Structural Information Service. Sonoma, CA: Geodex International, 1956–. 3/year.
Formerly: *Geodex Structural Information Service.*
The service is based on punched cards, manually manipulated rather than read by machine. The holes punched in the card correspond to the numbers of the abstracts in the accompanying abstracts' volumes to which the keyword applies. By combining the punch cards, one locates abstracts of articles that deal with one's specific problem.

Geological Abstracts. 6 vols. New York: Geological Society of America, 1953–1958.
Retrospective guide to abstracts of US and translated Russian journals. Superseded by *Geoscience Abstracts.*
R: Win (EE31).

Geophysical Abstracts: Abstracts of Current Literature Pertaining to the Physics of the Solid Earth and to Geological Exploration. **US Geological Survey.** Washington, DC: US Geological Survey, 1929–. Monthly.
Worldwide bibliographic coverage of literature pertaining to the physics of the solid earth and to geophysical exploration. Alphabetic arrangement by 24 subject categories. All abstracts are in English.
R: Jenkins (F80); Owen (p. 48); Wal (p. 151); Win (EE36; EE148).

Geophysics Abstracts. Norwich, England: GeoAbstracts, 1977–.
Indexed annually.

Geoscience Abstracts. 8 vols. Washington, DC: American Geological Institute, 1959–1966.
Offers coverage of geological literature published in North America for the period. Replaced *Geological Abstracts,* 1953–1958. Continued by *Abstracts of North American Geology* (1966–1971).
R: Jenkins (F16); Wal (p. 143); Win (EE34).

Geotechnical Abstracts. Vols 1–. Essen, W Germany: Deutsche Gesellschaft dur Erd-und Grundbau, for International Society for Soil Mechanics and Foundation Engineering, 1970–. Monthly.
English edition of *Dokumentation Bodernmechanik und Grundbau.*
Continuation of *Geodex Soil Mechanics Information Service,* 1969–1972. About 500 subject journals scanned. All abstracts in English despite origin of article.
R: Wal (p. 355).

An Index of State Geological Survey Publications Issued in Series Supplement, 1963–1980. **John B. Corbin, comp.** Metuchen, NJ: Scarecrow Press, 1982.

Prior to 1963, published in 1965.
An index of monographs issued in numbered series, arranged alphabetically by state and by series within the state. Updates the original by including South Carolina, Hawaii, and Alaska, but still omits New York State.
R: *ARBA* (1983, p. 664).

Meteorological and Geoastrophysical Abstracts. Vols. 1–. Washington, DC: American Meteorological Society, 1950–. Monthly.

Volumes 1–10 entitled *Meteorological Abstracts and Bibliography.*
Provdies current abstracts on primary sources. Citations include author, abstracts, decimal classification, subject headings, etc. The major abstracting service of its kind. Author, subject, and geographic indexes.
R: Jenkins (F95); Katz (p. 20); Wal (p. 151: 168); Win (EE80; EE82).

World Index of Strategic Minerals: Production, Exploitation, and Risk. **David Hargreaves and S. Fromson.** New York: Facts on File, 1983.

A political and economic overview of strategic minerals based on the critical nature of their end use. Detailed and informative.
R: *ARBA* (1984, p. 685).

Oceanography

Deep-Sea Research with Oceanographic Literature Review. New York: Pergamon Press, 1953–. 24/year.

Contains research papers on various aspects of the field, including morphology, geology, and instrumentation.

Desalination Abstracts. Tel Aviv: National Center of Scientific and Technological Information, 1967–. Quarterly.

Approximately 800 abstracts annually on the scientific and economic aspects of water desalination. Covers wide range of primary literature. Name, subject, and patent number indexes.
R: Katz (p. 4); Wal (p. 367).

Ocean Research Index: A Guide to Ocean and Freshwater Research, Including Fisheries Research. 2d ed. New York: International Publications Service, 1976.

First edition, 1970.
Primarily a directory of research organizations that are concerned with marine and freshwater science. Addresses, scope of interest, etc.
R: Wal (p. 165); Win (3EE11).

Oceanic Abstracts. Vols. 1–. La Jolla, CA: Oceanic Library and Information Center, 1964–. Bimonthly.

Formerly *Oceanic Index and Oceanic Citation Journal.*
Summarizes, organizes, and indexes the world's technical literature on oceans. In cludes information on meetings, exhibits, and conferences. Author, subject, and institution indexes.
R: *ARBA* (1985, p. 610); Jenkins (F143); Katz (p. 24).

Oceanographic Index. 16 vols. **Marine Biological Laboratory and Woods Hole Oceanographic Institution.** Boston: Hall, 1972–1976.

Author Cumulation 1946–1970, 3 volumes; *1971–1974*, 1 volume; *Regional Cumulation 1946–1970*, 1 volume; *1971–1974*, 1 volume; *Subject Cumulation 1946–1970*, 4 volumes; *1971–1974*, 2 volumes; *Journals Catalog*, 1 volume; *Organismal Cumulation 1946–1973*, 3 volumes.

A library catalog emphasizing biological and physical oceanography.

R: *ARBA* (1973, p. 593); Wal (p. 163).

Offshore Abstracts. London: Taylor, 1974–. Bimonthly.

Coverage of primary literature related to the offshore oil and gas industries. Emphasizes British sources. In many instances, overlaps with *Oceanic Abstracts*. Annual subject indexes.

R: Katz (p. 7).

General Engineering

Applied Science and Technology Index. New York: Wilson, 1913–. Monthly.

Formerly *Industrial Arts Index*, 1913–1957.

A cumulative subject index to over 200 English-language periodicals. Emphasizes both theoretical aspects of sciences and their engineering applications. Aimed at layperson and high school and college students. Quarterly and annual cumulations.

R: Jenkins (K11); Katz (p. 4); Wal (p. 5); Win (EA63); *ARBA* (1985, p. 493);

Bioengineering Abstracts. Vols 1–. New York: Engineering Index, 1974–. Monthly.

Service provides access to abstracts that appear in *Engineering Index* and are relevant to engineering in the life sciences. Weighted heavily in favor of reseach rather than clinical work. Conference proceedings, symposia, monographs, standards, and selected book reviews are included. Subject guide is given in preliminary pages.

Current Technology Index. London: Library Association, 1962–. Monthly.

Formerly *British Technology Index*.

An alphabetical subject index to articles in British technical journals. Covers all phases of engineering, chemical technology, and manufacturing processes. Author indexes.

R: *ARBA* (1983, p. 716).

Engineering Index Monthly and Author. Vols. 1–. New York: Engineering Index, 1884–. Monthly.

Formerly: *Engineering Index*.

The most comprehensive English-language abstracting service in engineering and all its subdisciplines. Scans and abstracts reports, patents, books, symposium papers, and articles from over 3,500 journals. Author indexes. Annual cumulation. Related publications and services (including online) include *Engineering Index Monthly*, 1962–; *Engineering Index Annual*, 1884–; *COMPENDEX* (online); *CARD-A-LERT*.

R: *LJ* 94: 706 (1969); Jenkins (K13); Katz (p. 13); Wal (p. 284); Win (2EI1).

PIE: Publications Indexed for Engineering. New York: Engineering Index, 1972. Also later editions.

Lists almost 3,000 current serial and nonserial publications abstracted and indexed by this service. Includes list of abbreviations for scientific and engineering terms.
R: *ARBA* (1973, p. 632).

SHE: Subject Headings for Engineering. New York: Engineering Index, 1972. Also later editions.

Alphabetical list of terms used by technical editors at Engineering Index. Primarily a tool for indexers.
R: *ARBA* (1974, p. 666).

Technical Abstracts. **US Goddard Space Flight Center, Greenbelt, MD.** Washington, DC: National Aeronautics and Space Administration, 1962–.

Technical Literature Abstracts. Vols. 1–. New York: Society of Automotive Engineers, 1975–.

AERONAUTICAL AND ASTRONAUTICAL ENGINEERING

Index Aeronauticus: Journal of Aeronautical and Astronautical Abstracts. London: Ministry of Aviation, 1945–1968. Monthly.

Abstracts the contents of approximately 500 journals and arranges them by universal decimal classification. Also included is a list of translations. Annual subject and author indexes.
R: Wal (p. 379).

Industrial Aerodynamics Abstracts. Vols. 1–. Cranfield, Bedford, England: British Hydromechanics Research Association, 1970–.
R: Wal (p. 103).

International Aerospace Abstracts. Vols. 1–. New York: American Institute of Aeronautics and Astronautics, 1961–. Semimonthly.

Continues *Aeronautical Engineering Index*, 1947–1957; *Aerospace Engineering Index*, 1958; and *Aero/Space Reviews*, 1959–1960.

A worldwide abstracting service issued alternately with *STAR*. Covers periodicals, books, meeting papers, translations, and proceedings in the fields of aeronautics and space science. Author, subject, report-number, and accession-number indexes are issued quarterly, semiannually, and annually.
R: Jenkins (K33); Katz (p. 18); Wal (p. 387); Win (EI19).

STAR (Scientific and Technical Aerospace Reports): A Semimonthly Abstract Journal with Indexes. Washington, DC: US National Aeronautics and Space Administration, 1963–. Semimonthly, with quarterly indexes.

Comprehensive coverage of over 1 million entries. Comprehensive abstracting and indexing journal covering worldwide technical report literature in space science and aeronautics. Emphasizes National Aeronautics and Space Administration reports but also includes myriad publications of government agencies, research organizations, and universities. Annual cumulated index. Companion volume: *Guide to the Subject Indexes for Scientific and Technical Aerospace Reports,* 1964.
R: Jenkins (A168); Katz (p. 25).

Pacific Aerospace Index. Vols. 1–. **Pacific Aerospace Library.** Los Angeles, CA: Pacific Aerospace Library, 1944–. Irregular.

Titled *Uniterm Index to Periodicals*, 1955–1964, annual.
Library accessions list of aeronautical and related periodicals. Arranged by subject. Includes journals providing cover-to-cover translations from Russian.
R: Wal (p. 379).

CHEMICAL ENGINEERING

Electroanalytical Abstracts. Vols. 1–. Basel: Birkhauser Verlag, 1963–. Bimonthly.

A continuation of the abstract section of the *Journal of Electroanalytical Chemistry*. Covers all aspects of fundamental, physicochemical, and analytical electrochemistry. Abstracts are not generally of a critical nature.
R: Wal (p. 130); Win (ED20).

Food Science and Technology Abstracts. Bucks, England: Commonwealth Agricultural Bureau, 1969–. Monthly.

Abstracts over 1,000 journals.
R: *Aslib Proceedings* 21: 505 (Dec. 1969); 23: 330 (July 1971); Wal (p. 484).

RAPRA Abstracts. Shrewsbury, Shropshire, England: Rubber and Plastics Research Association, 1923–. Biweekly.

Extensive coverage of latest journal literature and patent information.
R: Wal (p. 529).

Rheology Abstracts: A Survey of World Literature. Vols. 1–. New York: Pergamon Press, for the British Society of Rheology, 1958–. Quarterly.

Covers more than 100 journals. Annual author and subject indexes.
R: Jenkins (C13); Wal (p. 102).

Textile Technology Digest. Charlottesville, VA: Institute of Textile Technology, 1944–. Monthly.

International coverage of all aspects of textile technology. Author index in individual issues. In 1972 cumulative author, patent-concordance, and subject indexes published.
R: Owen (p. 106).

Theoretical Chemical Engineering Abstracts. Vols. 1–. London: Technical Information, 1964–. Bimonthly.

Classified arrangement. Emphasizes fluid dynamics, heat transfer, and mass transfer.
R: Wal (p. 472); Win (EI55).

World Textile Abstracts. Manchester, England: Shirley Institute, 1969–. Semimonthly.

Continues *Textile Abstracts*, 1967–1969. Abstracts 600 textile and related journals. Includes British and US patents.
R: *Chemical and Engineering News* 47: 52 (Mar. 17, 1969); Jenkins (J138); Owens (p. 107); Wal (p. 522).

Civil Engineering

Building Science Abstracts. **Great Britain. Ministry of Technology.** London: Her Majesty's Stationery Office, 1928–.
R: Jenkins (K22); Wal (p. 552).

Civil Engineering Hydraulics Abstracts. Bedford, England: British Hydromechanics Research Association, 1976–. Monthly.

More oriented to the applications of fluid mechanics to civil and structural engineering than either *Engineering Index* or *Oceanic Abstracts*. Worldwide coverage of hydraulics as it pertains to coastal and ocean waters, instrumentation, etc.

Concrete Abstracts. Vols. 1–. Detroit: American Concrete Institution, 1972–. Bimonthly.
Volume 3, 1974.
Covers primary literature. Keyword index.

Documentation Cards (Hydraulic Research Abstracts). Delft, Netherlands: Delft Hydraulics Laboratory, 1956–.
Weekly abstracting card service.

Fluid Power Abstracts: Worldwide Coverage of Hydraulics and Pneumatics Literature. Cranfield, Bedford, England: British Hydromechanics Research Association, 1970–. Monthly.
Coverage of more than 400 technical journals.
R: Wal (p. 362).

Highway Research Abstracts. Vols. 1–, nos. 1–. Washington, DC: US Highway Research Board, National Research Council, National Academy of Sciences, National Academy of Engineering, 1931–. Monthly.
A guide to current literature for the highway administrator and engineer.
R: Jenkins (K32); Katz (p. 14); Wal (p. 360); Win (EI84).

HRIS Abstracts. Washington, DC: Highway Research Information Service, Highway Research Board, Division of Engineering, National Research Council, National Academy of Sciences, National Academy of Engineering, 1968–. Quarterly.

Classified arrangement of approximately 1,000 abstracts in each issue. Subject, source, and author indexes. Related publications include the less inclusive *Highway Research Abstracts,* 1931–, and its *Index to Publications,* 1921–1949, updated by irregular supplements.
R: *Engineering,* p. 893 (Dec. 1, 1967); Wal (p. 360).

International Civil Engineering Abstracts. London: Telford, 1974–. Monthly.

Formerly: until 1982, *ICE Abstracts;* until 1975, *European Civil Engineering Abstracts.* Published for the Institution of Civil Engineers in Great Britain, with the cooperation of the American Society of Civil Engineers. Provides informative abstracts of articles appearing in journals received in the International Civil Engineering library. Keyword index.

Transportation Research Abstracts. Vols. 1–. Washington, DC: Transportation Research Board, National Research Council, 1931–. Monthly.

Formerly, until 1974, *Highway Research Abstracts.*
Abstracts from books, reports, and articles focusing on various topics of transportation.

Urban Mass Transportation Abstracts. Washington, DC: UMTA, 1972–. Bimonthly. Avail. National Technical Information Service.

ELECTRICAL AND ELECTRONICS ENGINEERING

General

ERA Abstracts. Vols 1–. Leatherhead, Surrey, England: Electrical Research Association, 1946–. Weekly.

Abstracts literature on such subjects as electric fields, dielectrics, power systems, instruments, etc. No indexes.
R: Wal (p. 312).

Electrical and Electronics Abstracts. Vols. 1–. London: Institution of Electrical Engineers, 1898–. Monthly.

Section B of *Science Abstracts.*
Over 40,000 items per year are abstracted from 89 core journals from 13 countries as well as secondary sources. Author and subject indexes. Supplemented by *Current Papers in Electrical and Electronics Engineering.*
R: *LT* 15: 872 (Apr. 1967); Jenkins (K27); Katz (p. 24); Wal (p. 312); Win (EI95).

Electronics and Communications Abstracts Journal. Vols. 1–. Riverdale, MD: Cambridge Scientific Abstracts, 1967–. Bimonthly.

Formerly: *Electronics Abstracts Journal.*
Indexed periodicals, government reports, conference proceedings, books, dissertations, and patents.
R: Jenkins (K28); Wal (p. 320).

Electronics Engineering Master Index: A Subject Index to Electronic Engineering Periodicals. New York: Master Index, 1925–.

Index to IEEE Publications. New York: Institute of Electrical and Electronics Engineers, 1971–. Annual.

Extensive coverage of abstracts, book reviews, conference papers, articles, and communications originally published in the *IEEE Transactions, IEEE Journals, Proceedings of the IEEE, IEEE Spectrum,* and other publications. Complete bibliographic information and various user's aids.
R: *ARBA* (1975, p. 775).

Key Abstracts. Vols. 1–. London: Institution of Electrical Engineers; New York: Institute of Electrical and Electronics Engineers, 1975–. Monthly.

There are 8 monthly parts: *Solid-State Devices; Electronic Circuits; Systems Theory; Communication Technology; Power Transmission and Distribution; Industrial Power and Control Systems; Electrical Measurements and Instrumentation; Physical Measurements and Instrumentation.*
Designed for individual use of research and development specialists and for smaller, specialized companies and libraries. Each *Key Abstract* is drawn from the INSPEC database and covers only the most important worldwide developments within its specific subject discipline—no more than 250 items in each issue. Issues are restricted to 24 pages and contain complete bibliographic information.

Technical News Bulletin. Manchester, England: Associated Electrical Industries, 1926–. Weekly.

Some 3,000 abstracts annually on electrical engineering, electronics, and related fields.

Computer Technology

British Commercial Computer Digest. 17th ed. **Computer Consultants Ltd.** New York: Pergamon Press, 1976. Annual.

Tenth edition, 1969.

Collected Algorithms for CACM. New York: Association for Computing Machinery, 1960–. Bimonthly.

Contains algorithms originally published in the association's *Communications.* Issued bimonthly and serially numbered to facilitate use in loose-leaf binders. Subject indexes.

Computer Abstracts. St. Helier, Jersey, British Channel Islands: Technical Information, 1957–. Monthly.

Relatively small number of abstracts (4,000/yr.) covering the broad range of computer literature. Includes reports, conference proceedings, and patents as well as journal literature. Not as comprehensive as INSPEC's *Computer and Control Abstracts.*
R: Jenkins (B60); Katz (p. 11); Wal (p. 541).

Computer and Control Abstracts. Vols. 1–. London: Institution of Electrical Engineers; New York: Institute of Electrical and Electronics Engineers, 1966–. Monthly.

Section C of *Science Abstracts.*
Over 24,000 items are abstracted from 89 core journals as well as secondary sources such as books, conferences, patents, etc. Author and subject indexes. Supplemented by *Current Papers in Computers and Control.*
R: Jenkins (K25); Katz (p. 24); Wal (p. 301, 541); Win (1EI25).

Computer and Information Systems Abstract Journal. Vols. 1–. Bethesda, MD: Cambridge Scientific Abstracts, 1962–. Monthly.

Formerly, *Computer and Information Systems; Information Processing Journal.*
Worldwide abstracting journal for material on computer software, applications, and

electronics. Covers all types of primary sources. Includes acronym, author, and subject indexes.
R: *Information* 3: 321 (Nov./Dec. 1971); Katz (p. 11); Wal (p. 542).

Computer Industry Abstracts. Vols. 1–. Las Mesa, CA: Data Analysis Group, 1984–. Quarterly.

Available both in hard-copy version and Lotus 1-2-3 formatted data disk for IBM PC-compatibles. A source for market information on computers, related equipment, and software.

Computer Literature Index. Annual Cumulation. Phoenix: Applied Computer Research, 1984.

Subject-arranged index to periodicals, conference reports, special reports, and books of the computer profession, with an author and publisher index. Not easy to locate information.
R: *ARBA* (1985, p. 587).

Computer Simulation 1951–1976: An Index to the Literature. **Per Holst.** London: Mansell, 1979.

Index includes 6,000 entries taken from various journals, abstracts, symposia, reports, etc., from 1951 to 1976. Composed of a permuted title index and an author index. A helful reference for those interested in computer simulation.
R: *ARBA* (1980, p. 719).

Computing Reviews. Vols. 1–. New York: Association for Computing Machinery, 1960–. Monthly.

The journal produces critical reviews and abstracts in all branches of the computer field. Books, articles, conference proceedings, selected theses, etc. are included. Author indexes and a cumulative "KWIC-Subject" and "Permuted Index" are available.
R: Katz (p. 11).

Computers and Information Processing World Index. **Suzan Deighton, John Gurnsey, and Janet Tomlinson, eds.** Phoenix: Oryx Press, 1984.

A compilation of various national and international information sources, agencies, and publications dealing with computers. Part 1 is a directory of organizations, part 2 consists of reference works, part 3 is a listing of texts and journals, and part 4 lists journals alphabetically under 39 countries. Information is not as up-to-date as it should be.
R: *ARBA* (1985, p. 587).

New Literature on Automation: International Classified Abstract Journal on ADP. Vols. 1–. Amsterdam: Netherlands Automatic Information Processing Research Center, 1961–. Monthly.

In universal decimal classification order, each issue has over 100 abstracts from over 30 journals. Author and subject indexes. Cumulative index available for 1961–1965.
R: Wal (p. 542).

Marine Engineering

Marine Engineering/Shipbuilding Abstracts. Nos. 1–. **Institute of Marine Engineers.** London: Institute of Marine Engineer, 1938–. Monthly.
Published as part of the Transactions of the Institute.
R: Wal (p. 375).

Ship Abstracts. Vols. 1–. Oslo: Norwegian Center for Informatics, 1973–. 10/yr.
Formerly *Scandinavian Ship Abstract Journal.*

Materials Science

Ceramic Abstracts: Journal of the American Ceramic Society. Vols. 1–. Columbus: American Ceramic Society, 1922–. Bimonthly.
Subject and author indexes.
R: Jenkins (K23); Wal (p. 495).

Master Index to Materials and Properties. **Yeram Sarkis Touloukian and C. Y. Ho.** New York: Plenum Press, 1979.
An index to the 13-volume *TPRC Data Series.* Presented in tabular form, the material covers substance name and physical properties, as well as volume number and page. The index is invaluable to any scientific libraries that own the series.
R: *ARBA* (1980, p. 620).

Solid-State Abstracts Journal: An Abstract Journal Involving the Physics, Metallurgy, Crystallography, Chemistry, and Device Technology of Solids. Bethesda, MD: Cambridge Scientific Abstracts, 1957–. Bimonthly.
Formerly: *Solid State Abstracts,* 1960–1965.
R: Jenkins (C14); Wal (p. 112).

Mechanical Engineering

Applied Mechanics Review, Assessment of World Literature in Engineering Science. Vols. 1–. Easton, PA: American Society of Mechanical Engineers, 1948–. Monthly.
Critical coverage of primary literature, including some 400 periodicals, in applied mechanics and related sciences. Other features include book reviews and informative articles. Arranged by subject.
R: Jenkins (K19); Katz (p. 4); Wal (p. 295); Win (EA70).

Automotive Literature Index, 1947–1976: A Thirty-Year Guide to "Car and Driver," "Motor Trend," and "Road & Track." **A. Wallace, comp.** Distr. Osceola, WI: Motorbooks International, 1981.
A most specific indexing of articles listed by auto make, model, and year, with subject heading lists. Mainly for historical research from 1947–1976. However, a supplement is following for the years 1977–1981, which would be of great service for those with automotive interests.
R: *ARBA* (1983, p. 738).

Corrosion Abstracts. Vols. 1–. Houston: National Association of Corrosion Engineers, 1962–. Bimonthly.
Originally appeared as a section of NACE's journal *Corrosion,* 1945–1961. Presently a classified arrangement of abstracts by subheadings. Subject indexes. Monthly cumulations of NACE's abstracts appear in *Bibliographic Survey of Corrosion,* Houston, 1945–.
R: Wal (p. 293).

Corrosion Control Abstracts. Vols. 1–. London: Scientific Information Consultants, 1966–. Monthly.
English translation of monthly *Referativnyi zhurnal: Korroziya i zashchita ot korrozi.*
R: Wal (p. 294).

Fluid Flow Measurements Abstracts. Vols. 1–. Cranfield, England: BHRA Fluid Engineering, 1974–. Bimonthly.
Information guide devoted to abstracting world literature on the measurement of flow, and other parameters and properties, of fluids.

Fluidics Feedbacks; Abstracts, Reviews, Industrial News, Products, Patents: A Monthly Review of the World of Fluidics. Vols. 1–. Cranfield, Bedford, England: British Hydromechanics Research Association, 1967–. Monthly.
Surveys on a worldwide basis the primary literature of hydraulic engineering. Includes annual, subject, personal, and corporate-author indexes.
R: Wal (p. 362).

METALLURGY

Alloys Index. London, England, and Metals Park, OH: The Metals Society and the American Society for Metals, 1974–. Monthly.
Can be used in conjunction with *Metals Abstract.* Provides names of original documents. Based on a detailed classification scheme, arranged alphabetically by alloy name.
R: *RSR* 5: 10 (Apr./June 1977).

International Copper Information. Nos. 1–. Potters Bar, Herts, England: Copper Development Association, 1971–. Quarterly.
Formerly: *Copper Abstracts.*
R: Wal (p. 510).

IMM Abstracts: A Survey of World Literature on Economic Geology, Mining, Mineral Dressing, Extraction Metallurgy, and Allied Subjects. Vols. 1–. London: Institution of Mining and Metallurgy, 1950–. Bimonthly.
Abstracts presented in a universal decimal classification order. Coal and ferrous extraction metallurgy are not covered.
R: Wal (pp. 174, 341).

Index of Current Literature on Coal Mining and Applied Subjects. Vols. 1–. London: Institution of Mining Engineers, 1963–.

Emphasizes English-language material, accumulating approximately 600 references yearly. Arranged by subject and universal decimal classification order.
R: Wal (p. 343).

Metals Abstracts. Vols. 1–. London: Institute of Metals and the American Society for Metals, 1968–. Monthly.

A hybrid resulting from the merger of *Review of Metal Literature,* 1944–1967 and *Metallurgical Abstracts,* 1909–1967. Currently provides international coverage of all aspects of metallurgy and related fields. Scans both primary and secondary literature. Abstracts are primarily descriptive and number some 25,000 annually.
R: *Scientific Information Notes* 10: 24 (Apr./May 1968); Jenkins (K37); Katz (p. 20); Wal (p. 500); Win (2EI33).

Mineralogical Abstracts; A Quarterly Journal of Abstracts in English Covering the World Literature of Mineralogy and Related Fields. Vols. 1–. London: Mineralogical Society of Great Britain, 1920–.

International coverage of primary literature in mineralogy. Arranged by subject with author, subject, and annual topographical index.
R: Jenkins (F115); Wal (pp. 140, 174); Win (EE102).

World Aluminum Abstracts. Metals Park, OH: American Society for Metals, 1968–. Monthly.

Formerly *Aluminum Abstracts,* 1963–1969.
Coverage of over 1,600 journals. Arranged under broad subject headings.
R: Wal (p. 511).

MILITARY SCIENCE

Library Index to Military Periodicals. Maxwell Air Force Base, AL: Air University, 1949–.

A subject index to significant articles and news items from some 65 military and aeronautical periodicals. Intended for the sociologist and historian as well as for military personnel.
R: Katz (p. 3); Win (EI22).

Military Science Index: A Monthly List of Papers of Military Interest from Current Periodicals Received in the Library of the Royal Military College of Science. Vols. 1–. Shrivenham, Wiltshire, England: Royal Military College of Science, 1962–. Monthly.

Subject arrangement of articles emphasizing the engineering aspects of military science. Annual cumulations. Limited circulation.
R: Wal (p. 345).

NUCLEAR ENGINEERING

INIS ATOMINDEX. Vols. 1–. Vienna: International Nuclear Information System, 1970–. Semimonthly.

Volumes 1–10 published under title: *List of Bibliographies on Nuclear Energy.*
With the discontinuation of *Nuclear Science Abstracts* on June 30, 1976, *ATOMINDEX* and *ERDA Research Abstracts* became the major sources of information on the subject, as gleaned from research reports, books, journal literature, etc. The work is international in scope, the result of a cooperative effort between the US Energy Research and Development Administration and the International Nuclear Information System. Indexed similarly to *NSA*, including personal and corporate-author, subject, and report-number indexes.
R: *IBID* 4: 141 (June 1976); Sheehy (EJ64).

Nuclear Science Abstracts. Vols. 1–33 (no. 12). Washington, DC: US Energy Research and Development Administration, 1948–June 30, 1976. Semimonthly.

Prior to February 1975, issued by US Atomic Energy Commission.
Abstracts the technical report literature sponsored by the US Atomic Energy Commission and other government agencies, universities, and foreign research establishments. It also provides international coverage of patents, books, periodical articles, translations, etc., in nuclear science and technology. Complete bibliographic information as well as corporate, personal-author, subject, and report-number indexes. Ordering procedures also available. Has been replaced by *INIS ATOMINDEX.*
R: Jenkins (A167); Katz (p. 20); Sheehy (EJ65); Wal (p. 304); Win (EI202.

Energy

Energy Abstracts. New York: Engineering Index, 1974–. Monthly.

Includes 5 sections: *Energy Sources Abstracts; Energy Production, Transmission, and Distribution Abstracts; Energy Utilization Abstracts; Energy Conversion Abstracts; Energy Conservation Abstracts.*
These cover reprints of energy-related citations from *Engineering Index.*

Energy Abstracts for Policy Analysis. Washington, DC: US Government Printing Office, Nov. 1975–. Monthly.

Formerly: *NSF-RANN Energy Abstracts,* Jan. 1973–Oct. 1974.
Publishes abstracts of nontechnical information pertaining to broad areas of social and economic aspects of energy resources; production, consumption, conservation, and environmental considerations. Joint project of the US Energy Research and Development Administration, the Federal Energy Administration, and the National Science Foundation.

Energy Bibliography and Index. 4 vols. Houston: Gulf Publishing, 1978–1981.

Provides quick reference to bibliographic data on some 20,000 books, maps, and technical documents, each with abstract included. Includes government documents, US Library of Congress call numbers, author indexes, subject index, keyword index. Comprises a complete source for the energy researcher. International in scope, updated quarterly. Well-recommended.

The Energy Index. New York: Environment Information Center, 1973–. Annual.

Volume 1 is a selected guide to energy information since 1970 (but published in 1973). It includes worldwide coverage to literature on all aspects of energy. Approximately 3,000 items are abstracted, 8,000 multiple-entry cross references are provided, and more than 7,500 authors are indexed. Includes various features (journal articles, conference proceedings, research reports, and films), and a subsection on energy-related patents. *Energy Index 74: A Select Guide to Energy Documents, Laws and Statistics,* 1974, is volume 2 of this work, which is slated for further supplements.
R: *ARBA* (1975, p. 706); Katz (p. 4).

Energy Information Abstracts. New York: Environment Information Center, 1976–. Monthly.

Under 21 energy categories, the journal emphasizes nontraditional sources (reports, surveys, documents, etc.) from international sources. Also covers selections from 1,000 scientific and trade journals. Indexed by subject, SIC code, and author.
R: *RSR* 5: 23 (Oct./Dec. 1977); *ARBA* (1977, p. 680).

Energy Research Abstracts. Vols. 1–. US Department of Energy. Oak Ridge, TN: Technical Information Center, 1976–. Semimonthly.

Continues *ERDA Research Abstracts.*
Provides abstract coverage of all scientific and technical reports issued by the US Department of Energy, including monographs, theses, patents, etc.
R: Sheehy (EJ54).

Energy Review. Santa Barbara: CA: Energy Research, 1974–. Bimonthly.

A review-abstracting service on all aspects of energy. Emphasizes research and technical information.
R: Katz (supp., p. 4).

Fuel and Energy Abstracts; A Summary of World Literature on all Technical and Scientific Aspects of Fuel and Power. London: Institute of Fuel, 1960–. Monthly.

Formerly: *Fuel Abstracts and Current Titles.*

Gas Abstracts. Chicago: Institute of Gas Technology, 1956–. Monthly.

International Petroleum Abstracts. Applied Science, 1969–. Quarterly.

Formerly *Institute of Petroleum Abstracts,* 1969–1972.

Petroleum Abstracts. Tulsa, OK: University of Tulsa, 1961–. Weekly.

Solar Energy Index: The Arizona State University Solar Energy Collection. **George Machovec.** Elmsford, NY: Pergamon, 1980.

ENVIRONMENTAL SCIENCES

General

EPA Index: A Key to US Environmental Protection Agency Reports and Superintendent of Documents and NTIS Numbers. **Cynthia E. Bower and Mary L. Rhoads, eds.** Phoenix: Oryx Press, 1983.

A much needed reference tool that lists over 21,000 US Environmental Protection Agency-numbered documents that have been distributed to federal depository libraries and/or cited in the US Government Printing Office *Monthly Catalog* prior to 1982. Indexed by number and title.
R: *Choice* 21: 1442 (June 1984); *RQ* 23: 472 (Summer 1984); *ARBA* (1985, p. 514).

Environment Abstracts. New York: Environment Information Center, 1974–. Monthly.
Volume 8, number 10, 1978.
Brief abstracts of articles pertaining to all aspects of the environment. Thoroughly indexed, easy-to-use arrangement. Highly recommended.
R: *ARBA* (1975, p. 706; 1979, p. 693).

Environmental Effects on Materials and Equipment. Vols. 1–. Washington, DC: US National Academy of Sciences, National Research Council, Prevention of Deterioration Center, 1961–. Monthly.
Section A is a continuation of *Prevention of Deterioration Abstracts,* volumes 1–19, 1946–1962; section B continues *Environmental Effects on Materials and Equipment,* volume 1, 1961. Subject and author indexes.
R: Wal (p. 293).

Envrionmental Quality Abstracts. Vols. 1–. Louisville: Data Courier, 1975–. Quarterly.
Abstracts from some 200 journals that pertain to the environmental sciences. Arranged under the following topics: public policy, population and health, endangered species, resources, and recycling. Geared toward high school and undergraduate use. Indexed by keyword and author.
R: *ARBA* (1977, p. 680).

Hydro-Abstracts. Vols. 1–. Urbana, IL: American Water Resources Association, 1968–.
Formerly (until 1980); *Water Resources Abstracts.*
Published in convenient loose-leaf format.
R: Katz (p. 27); Wal (p. 166).

Index to Ecology. Los Angeles: National Information Center for Educational Media. Annual.
Multimedia.

Land Use Planning Abstracts: A Select Guide to Land and Water Resources Information. Vols. 1–. New York: Environment Information Center, 1974–. 2/yr.
Includes reviews, abstracts, and index sections. Most abstracts are from technical reports and journals, though many are from obscure miscellaneous publications. Subject, keyword, and author indexes.
R: *BL* 71: 593 (Dec. 15, 1975); *Choice* 13: 1418 (Jan. 1977); *ARBA* (1976, p. 699; 1979, p. 696).

Nuclear Waste Management Abstracts. **Richard A. Heckman and Camille Minichino.** New York: IFI/Plenum, 1982.

Defines the problem of nuclear waste management, and then presents abstracts from recent literature dealing with this issue. Abstracts are indexed by keywords, title, and author. A useful guide for those interested in this area.
R: *ARBA* (1984, p. 679).

Sea Grant Newsletter Index. Rockville, MD: Environmental Sciences Information Center, 1968–.

Latest volume, 1973.

Index to all issues of newsletters that have been produced with Sea Grant support.

Selected Water Resources Abstracts. Vols. 1–. Washington, DC: Water Resources Scientific Information Center, 1968–. Avail. National Technical Information Center. Monthly.

Coverage of primary literature on such topics as water cycle, properties of water, conservation, resources, legislation, etc. Includes subject-author, organization, and accession-number indexes.
R: Katz (p. 26); Wal (p. 365); Win (2EE13).

Pollution

Abstracts on Health Effects of Environmental Pollutants. Vols. 1–. Philadelphia: Biosciences Information Science, 1972–. Monthly.

Composed of selected material from two databases: BIOSIS and MEDLARS. Contains bibliographic information for approximately 1,000 research articles each issue. Author, subject, cross indexes.

Air Pollution Abstracts. Vols. 1–. Research Triangle Park, NC: Air Pollution Technical Information Center, 1970–. Subscriptions from US government Printing Office. Monthly.

While similar material may be gained by consulting *Environment Index,* the above tool covers a wider range of literature on air pollution. In essence, a current-awareness service for literature accessioned by APTIC.
R: Katz (p. 3); Wal (p. 368).

Pollution Abstracts. Vols. 1–. Bethesda, MD: Cambridge Scientific Abstracts, 1970–. Bimonthly.

Abstracts a variety of literature, including papers, patents, articles, and reports. Comprehensive coverage of all known pollutants. Supplementary material includes calendars of meetings and legal and legislative developments.
R: Katz (p. 21); Wal (p. 368); Win (3EI2).

Water Quality Abstracts: A Special Compendium from Pollution Abstracts, Inc. La Jolla, CA: Pollution Abstracts, 1973–.

A second compendium, 1974.

Beginning in 1973, all abstracts on water quality, dating back to 1969, that appeared in *Pollution Abstracts* have been culled and presented here in a more accessible format. Abstracts are approximately 200 words in length, with full bibliographic data. There are no indexes.
R: *ARBA* (1975, p. 708).

CURRENT-AWARENESS SERVICES

GENERAL

Card-a-Lert. New York: Engineering Index.
A weekly current-awareness service available on 3-by-5-inch cards.

Current Contents. Philadelphia: Institute for Scientific Information, 1957–. Weekly.

A current-awareness table-of-contents service. Sections relevant to science and technology are the following: *Physical, Chemical, and Earth Sciences,* 1961–; *Life Sciences,* 1958–; *Engineering, Technology, and Applied Sciences,* 1970–; *Agricultural, Biological, and Environmental Sciences,* 1970–; *Clinical Practice,* 1973–; *Social and Behavior Sciences,* 1974–. Each section covers upwards of about 1,000 journals, or approximately 200 journals per week.
R: Jenkins (G11); Katz (p. 12; supp., p. 3); Wal (pp. 6, 92, 116, 192, 284, 396); Win (3EA12).

Weekly Abstracts Newsletter. Springfield, VA: National Technical Information Service. Weekly.

Formerly: *Weekly Government Abstracts.*
Weekly newsletters describe most unclassified, federally funded research as it is completed. The abstracts are issued under the following sections: *Administration and Management; Agriculture and Food; Behavior and Society; Building Industry Technology; Biomedical Technology and Human Factors Engineering; Business and Economics; Chemistry; Civil Engineering; Communications; Computer, Control, and Information Theory; Energy; Environmental Pollution and Control; Government Inventions for Licensing; Health Planning and Health Services Research; Industrial and Mechanical Engineering; Library and Information Sciences; Material Sciences; Medicine and Biology; NASA Earth Resources Survey Program; Natural Resources and Earth Sciences; Ocean Technology and Engineering; Physics; Problem-Solving Information for State and Local Governments; Transportation; Urban and Regional Technology and Development.*

SUBJECT

Asher's Guide to Botanical Periodicals. Amsterdam: Asher, 1973–. 15/yr.

Changed to *Guide to Botanical Periodicals* from volume 3, number 1, January 1976 on. A table-of-contents service covering selected botanical journals, published approximately every 3 weeks. Some 5,000 international periodicals are searched, resulting in some 60,000 titles each year. Includes author-subject index and an annual issue devoted to the publications of conference proceedings and symposia.

Biochemical Title Index. Vols. 1–. Philadelphia: Biosciences Information, 1962–. Monthly.

Over 20,000 reference annually culled from 500 journals. For abstracts one must consult *Biological Abstracts* and *Biochemical Sections of Chemical Abstracts,* 1962–.
R: Wal (p. 201); Win (ED15).

Chemical-Biological Activities. Vol. 1, nos. 1–. Columbus: American Chemical Society, 1965–. Biweekly.

Computer-based abstracting service covering literature on biological aspects of organic compounds. Indexed by authors, molecular formulas, and KWIC.
R: *Chemical and Engineering News* 42: 64–65 (Nov. 16, 1974); *Journal of Chemical Documentation* 3: 81–86 (1963); Wal (p. 201); Win (IEC12).

Chemical Titles. Vols. 1–. Easton, PA: American Chemical Society, 1960–. Biweekly.

Current-awareness service. A computer-generated KWIC index to contents of over 700 international chemical periodicals. Titles are generally listed in the same month in which they appear in the original journals. There are 3 points of access to the publication: the keyword indexes, a bibliography section, and an author index.
R: Jenkins (D16); Katz (p. 10); Wal (p. 115); Win (ED18).

Computer Contents. **Catherine H. Fay.** Northbrook, IL: Management Contents, 1983–. Biweekly.

Current Chemical Engineering Papers. Frankfurt: DECHEMA, 1965–. Monthly.
International coverage with the exception of the USSR.

Current Chemical Papers. Vols. 1–. London: Chemical Society, 1954–. Monthly.
Continues section AI-II of discontinued *British Abstracts.* Current-awareness service.
R: Jenkins (D18); Wal (p. 115).

Current Mathematical Publications. Providence: American Mathematical Society, 1969–. Biweekly.
Formerly: *Contents of Contemporary Mathematical Journals and New Publications,* 1969–1975.
Current-awareness service issued in conjunction with *Mathematical Reviews.*

Current Papers in Computers and Control. Vols. 1–. London: Institution of Electrical Engineers, 1966–. Monthly.

Current Papers in Electrical and Electronics Engineering. Vols. 1–. London: Institution of Electrical Engineers, 1964–. Monthly.

Current Papers in Physics. Vols. 1–. London: Institution of Electrical Engineers, 1966–. Semimonthly.

Contains about 65,000 titles of research articles from more than 900 of the world's physics journals. The low-cost companion current-awareness journals to the 3 sections of *Science Abstracts,* with its same coverage and arrangement.

Current Physics Advance Abstracts. New York: American Institute of Physics, 1972–. Monthly.

Provides abstracts of articles that are to appear in leading physics journals 2–4 months in the future. Intended for physicists who require advance knowledge of articles accepted for publication.

Current Physics Titles. New York: American Institute of Physics, 1972–. Monthly.

Table-of-contents service. Divided into 3 areas: nuclei and particles; solid state; atoms and waves. Indicates the nature (theoretical or experimental) of articles.

General Physics Advance Abstracts. Vols. 1–. New York: American Institute of Physics, 1985–. Semimonthly.

Provides the scientific community with prepublication abstracts of articles to appear in some 33 American Institute of Physics and member society journals. Abstracts are grouped by journal, providing quick access to specialized areas of interest.

Geotitles Weekly. London: Geosystems Publications, 1969–.

Current-awareness publication for geoscience. International in scope, with an author and source index.
R: Wal (p. 144).

Hydata. Vols. 1–. Urbana, IL: American Water Resources Association, 1965–. Monthly.

Table-of-contents service in the field of water resources and hydrology.
R: *Science* 148: 1449 (1965); Jenkins (F90); Wal (p. 166); Win (1EE12).

ISMEC Bulletin ("Information Service in Mechanical Engineering"). London: Institution of Mechanical Engineering and Institution of Electrical Engineering, 1973–. Monthly.

Provides notification of currently published papers worldwide in all aspects of engineering management, mechanical, and production engineering.
R: Katz (p. 6).

Marine Science Contents Tables. Vols. 1–. Rome: Fishery Resources Division, Food and Agriculture Organization of the United Nations, 1966–. Monthly.

A table-of-contents service to 100 periodicals in many languages. Useful guide to future conferences on marine science. Free upon request.
R: Katz (supp. p. 7).

Monthly Review of Technical Literature. Vols. 1–. London: British Railways, Research Department, 1930–. Monthly.

Guide to current literature in railroad engineering.
R: Wal (p. 357).

PC News Watch. Vols. 1–. Nos. 1–. Andover, MA: Hyatt Research, 1983–. Monthly.

A very organized newsletter containing abstracts of articles related to personal computers from 20 professional trade magazines and journals.
R: *ARBA* (1984, p. 624).

Solid Waste Information Retrieval System Accession Bulletin. Vols. 1–. Washington, DC: US Government Printing Office, 1970–.

CHAPTER 16 PERIODICALS

REFERENCE SOURCES

Periodicals are, unquestionably, the most important primary-information sources for scientists. In this chapter, only a limited number of major-subject journals are listed. Readers should consult the comprehensive serial bibliographic source, *Ulrich's International Periodical Directory*, for more detailed information on the journals listed and for further pertinent journal titles.

The following are a few sample tools that can provide general and specific reference information on scientific and technical periodicals:

ABI/SELECTS: The Annotated Bibliography of Computer Periodicals. Louisville: Data Courier, 1983.

A bibliography of 533 serials on the subject of computer and computer-related areas. Indexed by title and by publisher.
R: *LJ* 109: 575 (Mar. 15, 1984); *ARBA* (1985, p. 580).

The Abstract Journal, 1790–1920: Origin, Development, and Diffusion. **Bruce M. Manzer.** Metuchen, NJ: Scarecrow Press, 1977.

Covers the emergence and growth of the abstract journal. Describes, categorizes, and analyzes abstract journals. Discusses various sponsors in both Europe and America. Useful appendixes include a chronological list of journals, bibliographic notes, tables and graphs. Recommended to all those interested in the historical development of the journal.
R: *Journal of the American Society for Information Science* 29: 213 (July 1978).

Bibliographie der Zeitschriftenliteratur zum Stand der Technkik: Bibliography of Periodical Literature on the State-of-the-Art for the Areas of Technology. Vol. 1, pt. 1. **Utz-Friedebert Taube, ed.** New York: Saur, 1978.

The first of a 4-part bibliographic series covering various aspects of the state-of-the-art of science and technology. Articles are taken from journals published in 1976 and are intended to give citations and patent documentation to technical and nonpatent literature. Entries are grouped by International Patent Classification number and include title, author, and bibliographic citation. For science libraries.
R: *ARBA* (1980, p. 600).

Computer Publishers and Publications 1985–86: An International Directory and Yearbook. 2d ed. **Efrem Siegel and Frederica Evan, eds.** Communications Trends. Distr. Detroit: Gale Research, 1985. Annual.

Provides updated information on over 275 publishers of books on computers and on more than 600 periodicals concerned with computers. With several indexes.
R: *Choice* 21: 1438 (June 1984); *RQ* 23: 468 (Summer 1984); *Science & Technology Libraries* 5: 2 (Winter 1984); *WLB* 59: 290–291 (Dec. 1984); *ARBA* (1985, p. 585).

Directory of Canadian Scientific and Technical Periodicals. **Canada, National Science Library.** Ottawa: National Research Council of Canada, 1969.
R: *Aslib Proceedings* 22: 185 (May 1970); Wal (p. 39).

Directory of Japanese Scientific Periodicals. 3d ed. New York: International Publishers Service, 1976.
Earlier editions: 1964, 1967, 1974.
Some 5,000 periodicals in a classified arrangement.
R: Sheehy (EA6); Win (2EA12).

Directory of Publishing Sources; The Researcher's Guide to Journals in Engineering and Technology. **Sarojini Balachandran.** New York: Wiley-Interscience, 1982.
This directory of almost 300 technical journals has a detailed subject keyword index to help locate specific subject areas. Good for engineers and scientists who wish to publish their articles.
R: *IEEE Spectrum* 20: 13 (Jan. 1983); *Choice* 19: 1023 (Apr. 1982); *ARBA* (1983, p. 714).

A History of Scientific and Technical Periodicals: The Origins and Development of the Scientific and Technical Press, 1665–1790. 2d ed. **David A. Kronick.** Metuchen, NJ: Scarecrow Press, 1976.

List of Serial Publications in the British Museum of Natural History. 3 vols. 3d ed. Charlottesville, VA: University Press of Virginia, 1980.

Microcomputing Periodicals: An Annotated Bibliography. 8th ed. **George Shirinian, comp.** Toronto, Canada: 1983.
A unique directory to periodical information sources on microcomputers. Includes a list of periodical indexes, a list of periodicals currently in print, and a list of periodicals that have ceased publication or changed names. For librarians and users.
R: *Serials Librarian* 9: 99 (Fall 1984).

Scientific and Technical Journals. **Jill Lambert.** Hamden, CT: Shoe String, 1985.
Introduces the uses and importance, as well as the problems and possible drawbacks, of journals in scientific and technical development. Includes surveys and studies on this issue.

The Scientific Journal. **A. J. Meadows, ed.** London: Aslib, 1979.
A compilation of 20 papers dealing with all aspects of the scientific journal. Covers history, refereeing, economics of publishing, patterns of communications, networks of citations, and the future. For university libraries.
R: *CRL* 41: 540 (Nov. 1980); *Journal of Documentation* 36: 271 (Sept. 1980).

The Scientific Journal: Editorial Policies and Practices; Guidelines for Editors, Reviewers, and Authors. **Lois DeBakey et al.** St. Louis: Mosby, 1976.

Scientific Journals in the United States: Their Production, Use, and Economics. **Donald W. King et al.** New York: Van Nostrand Reinhold, 1981.

Scientific Periodicals: Their Historical Development, Characteristics and Control. **Bernard Houghton.** London: Bingley; Hamden, CT: Linnet Books, 1975.

The book covers the history of scientific periodicals, provides a bibliography of bibliographies, and discusses the problem of identifying the core journals in the various branches of science.
R: *ARBA* (1976, p. 631).

The Serials Directory. Premier ed. Birmingham, AL: EBSCO, 1986. 3 vols.

A major reference directory that includes 113,000 titles, both annuals and irregular series along with other type serials in 1 volume. International in coverage. Features a listing of over 2,000 ceased titles by both alphabetical and subject headings. Extensive indexing.

US Government Scientific and Technical Periodicals. **Philip A. Yannerella and Rao Aluri, comps.** Metuchen, NJ: Scarecrow Press, 1976.

A listing of over 260 US government scientific and technical periodicals. Covers latest title, date of issuance, frequency, issuing agency, subscription price, etc.

Ulrich's International Periodicals. 2 vols. 23d ed. New York: Bowker, 1984. Annual.

A must for every library interested in finding information on periodicals.

World List of Aquatic Sciences and Fisheries Serial Titles. Rome: Food and Agriculture Organization of the United Nations, 1981.

Bibliographic information on more than 1,000 core serials.
R: *IBID* 4: 125 (June 1976).

World List of Scientific Periodicals Published in the Years 1900–1960. 4th ed. 3 vols. London: Butterworth, 1963–1965.

A major tool for locating scientific and technical periodicals. It is continued in *British Union-Catalogue of Periodicals, Incorporating World List of Scientific Periodicals, New Periodicals Titles,* 1964—quarterly with annual cumulations.

ABBREVIATIONS

Periodicals frequently are indicated by codes of journal titles for machine entry, or cited by abbreviated titles in references. The following few sample tools demonstrate how information on standard periodical abbreviations can be obtained.

CODEN for Periodical Titles: An Aid to the Storage and Retrieval of Information and to Communication Involving Journal References. 2 vols. **American Society for Testing and Materials.** Philadelphia: ASTM, 1967. Also, supps. 1968–.

Supersedes the 1963 edition.
An A–Z list of 4-letter codes (CODEN) for titles of scientific periodicals designed for information retrieval. Covers over 40,000 periodical titles.
R: *American Documentation* 4: 54 (1953); Wal (p. 33).

INSPEC List of Journals. London: Institution of Electrical Engineers.

Used by INSPEC for processing-units indexing and abstracting services. In addition to providing abbreviated and full titles of publications, countries of origin, and publishers' names and addresses, a feature of the latest edition is the inclusion of CODEN.

List of Journals Indexed by the National Agricultural Library, 1974–76. 2 vols. US National Agricultural Library. Beltsville, MD: US National Agricultural Library, 1978, and updates.

Includes bibliographic information on more than 6,000 serial publications indexed in the *Bibliography of Agriculture*. Volume 1 consists of an alphabetic listing of titles, and volume 2 contains a geographic index. Provides worldwide coverage of journals and is a valuable reference source.
R: *RSR* 7: 13 (July/Sept. 1979).

List of Serials with Coden, Title Abbreviations, New, Changed, and Ceased Titles. Philadelphia: Biosciences Information Service. Annual.

Includes significant serial publications covered by BIOSIS. International in scope. Contains 8,000 serial titles published in 107 countries.

Periodical Title Abbreviations. 2 vols. 4th ed. **Leland G. Alkire, Jr., ed.** Detroit: Gale Research, 1983.

Second edition, 1977. Volume 1, abbreviations by abbreviation; volume 2, abbreviations by title.
Translates about 20,000 commonly used abbreviations for periodicals appearing in books and journals. Includes titles from medicine, science, and technology. For academic and large public libraries.
R: *BL* 74: 708 (Dec. 15, 1977).

PIE: Publications Indexed for Engineering. New York: Engineering Index, 1972. Also, updates.

List of 2,285 titles indexed for *Engineering Index*.

Source Index Quarterly. Vols. 1–. Columbus OH: Chemical Abstracts Service, 1970–. Quarterly.

Provides complete identification of more than 10,000 source publications included in *Chemical Abstracts* that are generally cited by abbreviated titles. Identifies libraries at which original documents are available. Various notes are provided with complete bibliographical information. Supersedes *Access*.
R: Mal (7–12).

CUMULATIVE INDEXES

To some popular- or major-subject journals, cumulative indexes from commercial sources, in addition to journals' own cumulative indexes, are available, as seen in the following examples:

American Scientists. Cumulative Index to Vols. 34–61 (1946–1973). New Haven: American Scientists, 1975.

A continuation of the index to volumes 1–33 (pub. 1945). Nearly 1,100 articles and research reports indexed. Consists of 2 author and title keyword sections.

Chemical Abstracts Service Source Index. Columbus, OH: Chemical Abstracts Service, 1969–. Quarterly.

Cumulative Index to the Mossbauer Effect Data Indexes. **John G. Stevens, Virginia E. Stevens, and William L. Gettys, eds.** New York: Plenum Press, 1980.

Provides quick access to the 10,000 references contained in first 9 volumes of the *Mossbauer Effect Data Index (MEDI)*. Supplies 3 key indexes by author, subject, and isotope.
R: *ARBA* (1981, p. 641).

Japanese Physics: Combined Cumulative Index of Three Major Journals, 1946–1974, by Subject and Author. 2 vols. New York: Nichigai Associates, 1975.

Combined cumulative author and subject indexes of the following journals: *Journal of the Physical Society of Japan* (1946–1974); *Japanese Journal of Applied Physics* (1962–1974); *Progress in Theoretical Physics and Supplement* (1946–1974).
R: *SL* 66: 5A (Aug. 1975).

Journal of Applied Physics and Applied Physics Letters. Combined Cumulative Index—Subject and Author, 1962–1973. New York: Nichigai Associates, 1974.

Physical Review and Physical Review Letters. Combined Cumulative Subject Index, 1951–1973. New York: Nichigai Associates, 1974.

Scientific American, Cumulative Index, 1948–1978. New York: Scientific American, 1979.

Consists of various sections including authors of articles, title index, "longer" book-review index, and others.
R: *ARBA* (1974, p. 549).

SELECTIVE TITLES

GENERAL SCIENCE

American Scientist. 1913–. Bimonthly.

Bulletin of Science, Technology and Society. 1981–. Bimonthly.

Franklin Institute Journal. 1826–. Monthly.

ISIS. 1913–. Quarterly.

Nature. 1869–. Weekly.
In 3 weekly editions: *Nature; Nature Physical Sciences; Nature New Biology.*

New Scientist. 1956–. Weekly.

Royal Society. Proceedings. 1854–. Irregular.

Science 1880–. Weekly.

Scientific American. 1845–. Monthly.

ASTRONOMY

American Astronomical Society. Bulletin. 1969–. 4/yr.

Astronomical Journal. 1849–. Monthly (10/yr.).

Astronomical Society of the Pacific. Publications. 1889–. Bimonthly.

Astronomy and Astrophysics. 1969–. Monthly.

Astrophysical Journal. 1895–. Bimonthly.

Astrophysics and Space Science. 1968–. Monthly.

British Astronomical Association. Journal. 1890–. 6/year.

Observatory. 1877–. Bimonthly.

Royal Astronomical Society. Monthly Notices. 1827–. 14/yr.

Royal Astronomical Society. Quarterly Journal. 1960–. Quarterly.

Sky and Telescope. 1941–. Monthly.

Soviet Astronomy. 1957–. Bimonthly.

MATHEMATICS

Advances in Applied Mathematics. 1980–. Quarterly.

Advances in Mathematics. 1967–. Monthly.

American Journal of Mathematical and Management Sciences. 1981–. Quarterly.

American Mathematical Monthly. 1894–. 10/yr.

American Mathematical Society. Bulletin. 1894–. Bimonthly.

American Mathematical Society. Notices. 1953–. 8/yr.

American Mathematical Society. Proceedings. 1950–. Monthly.

American Mathematical Society. Transactions. 1900–. Monthly.

Annals of Mathematics. 1884–. Bimonthly.

International Journal of Computer Mathematics. 1964–. 8/yr.

JASA. Journal of the American Statistical Association. 1888–. Quarterly.

Mathematical Methods in the Applied Sciences. 1979–. Quarterly.

Mathematics of Computation. 1943–. Quarterly.

Mathematics Magazine. 1926–. Bimonthly.

Quarterly Journal of Mathematics. 1930–. Quarterly.

SIAM Review. 1959–. Quarterly.

PHYSICS

Academy of Sciences of the USSR. Bulletin. 1954–. Monthly.

Acoustical Society of America. Journal. 1929–. Monthly plus 3 supplements.

Advances in Physics. 1952–. Bimonthly.

American Journal of Physics. 1933–. Monthly.

American Physical Society. Bulletin. 10 issues/yr.

Annals of Physics. 1957–. 14/yr.

Applied Optics. 1962–. Semimonthly.

European Journal of Physics. 1980–. Quarterly.

Fusion Power Report: Complete Monthly Coverage of Worldwide Fusion Developments. 1980–. Monthly.

Hadronic Journal. 1978–. Bimonthly.

International Journal of Thermophysics. 1980–. Quarterly.

Journal of Applied Physics. 1931–. Monthly.

Journal of Chemical Physics. 1931–. Semimonthly.

Journal of Mathematical Physics. 1969–. Monthly.

Journal of Physics. 1874–.
In 7 sections: A, *Mathematical and General*, 1874–, monthly; B, *Atomic and Molecular Physics*, 1874–, monthly; C, *Solid-State Physics*, 1874–, 24/yr.; D, *Applied Physics*, 1950–, monthly; E, *Scientific Instruments*, 1922–, monthly; F, *Metal Physics*, 1971–, monthly; G, *Nuclear Physics*, 1975–, monthly.

Laser and Particle Beams. 1983–. Quarterly.

Nuclear Physics. 2 sects. 1956–. Weekly.

Optical Society of America. Journal. 1917–. Monthly.

Physical Review. 1893–. 72/yr.
In 4 sections: A, *General Physics*, monthly; B, *Condensed Matter*, semimonthly; C, *Nuclear Physics*, monthly; D, *Particles and Fields*, semimonthly.

Physical Review Letters. 1958–. Weekly.

Physics Today. 1948–. Monthly.

Reports on Progress in Physics. 1937–. Monthly.

Reviews of Modern Physics. 1929–. Quarterly.

Review of Scientific Instruments. 1930–. Monthly.

Soviet Physics—JETP. 1955–. Monthly.

Surveys in High-Energy Physics; An International Journal. 1980–. Quarterly.

CHEMISTRY

Accounts of Chemical Research (ACS). 1968–. Monthly.

American Chemical Society. Journal. 1879–. Semimonthly.

Analyst. 1876–. Monthly.

Analytical and Industrial Molecular Spectroscopy. 1984–. Quarterly.

Analytical Chemistry. 1929–. Monthly.

CRC Critical Reviews in Analytical Chemistry. 1970–. Quarterly.

Chemical Engineering News. 1923–. Weekly.

Chemical Reviews. 1924–. Bimonthly.

Electrochemical Society. Journal. 1948–. Monthly.

Inorganic Chemistry. 1962–. Semimonthly.

Journal of Chemical and Engineering Data. 1959–. Quarterly.

Journal of Chemical Education. 1924–. Monthly.

Journal of Chemical Information and Computer Sciences. 1960–. Quarterly.

Journal of Computational Chemistry. 1980–. Quarterly.

Journal of Organic Chemistry. 1936–. Semimonthly.

Journal of Physical Chemistry. 1896–. Semimonthly.

Royal Society of Chemistry. Journal.
In 5 sections: *Faraday Transaction 1*, 1972–, monthly; *Faraday Transaction 2*, 1972–, monthly; *Perkin Transaction 1*, 1972–, semimonthly; *Perkin Transaction 2*, 1972–, monthly; *Dalton Transaction*, 1972–, semimonthly.

Royal Society of Chemistry. Reviews. 1947–. Quarterly.

BIOLOGICAL SCIENCES

General

Biological Agriculture and Horticulture: An International Journal. 1982–. Quarterly.

Biological Bulletin. 1898–. Bimonthly.

Biologicke Listg/Biological Review. 1912–. Quarterly.

BioScience. 1951–. Monthly.

Biotechnology Advances. 1983–. 2/year.

Biotechnology News. 1981–.

Current Genetics. 1980–. Bimonthly.

Developmental Biology. 1959–. Monthly.

Ecology. 1920–. Bimonthly.

Human Biology. 1929–. Quarterly.

Journal of Experimental Biology. 1923–. Bimonthly.

Journal of Microbiological Methods. 1983–. Bimonthly.

Journal of Theoretical Biology. 1961–. Bimonthly.

Agriculture

Agricultural Engineering. 1920–. Monthly.

Agricultural Meteorology. 1964–. Monthly.

Agricultural Science Review. 1963–. Quarterly.

Agronomy Journal. 1907–. Bimonthly.

Farm Journal. 1877–. Monthly.

Journal of Agricultural and Food Chemistry. 1953–. Bimonthly.

Journal of Cereal Science. 1983–. Quarterly.

Journal of Dairy Science. 1917–. Monthly.

Rice Abstracts. New York: Unipub, 1978–. Monthly.

Seed Abstracts. New York: Unipub, 1978–. Monthly.

Soil Science. 1916–. Monthly.

World Crops. 1949–. Bimonthly.

Biochemistry and Biophysics

Biochemical Journal. 1911–. Semimonthly.
Part 1: *Cellular Aspects*; part 2: *Molecular Aspects.*

Biochemica et Biophysica Acta. 1946–. 120/yr.

Biochemistry. 1964–. Semimonthly.

Biophysical Journal. 1960–. Monthly.

Journal of Biological Chemistry. 1905–. Semimonthly.

Journal of Molecular Biology. 1959–. 36/yr.

Quarterly Reviews of Biophysics. 1968–. Quarterly.

Botany

American Journal of Botany. 1914–. 10/yr.

American Naturalist. 1867–. Monthly.

Annals of Botany. 1887–. Monthly.

Botanical Review. 1935–. Quarterly.

Journal of Experimental Botany. 1960–. Monthly.

Plant Physiology. 1926–. Monthly.

Genetics

Biochemical Genetics. 1967–. Monthly.

Genetical Research. 1960–. Bimonthly.

Genetics. 1916–. Monthly.

Heredity. 1947–. Bimonthly.

Journal of Heredity. 1910–. Bimonthly.

Microbiology

Applied and Environmental Microbiology. 1953–. Monthly.

CRC Critical Reviews in Microbiology. 1971–. Quarterly.

Journal of Bacteriology. 1916–. Monthly.

Journal of General Microbiology. 1947–. Monthly.

Nutrition

British Journal of Nutrition. 1947–. Bimonthly.

Food and Nutrition. 1971–. Bimonthly.

Food and Nutrition News. 1929–. 5/yr.

Journal of Applied Nutrition. 1947–. Semiannually.

Journal of Nutrition. 1928–. Monthly.

Journal of Nutrition Education. 1969–. Quarterly.

Nutrition News. 1937–. Quarterly.

Nutrition Reviews. 1942–. Monthly.

Physiology

American Journal of Physiology. 1898–. Monthly.

Journal of Applied Physiology. 1948–. Monthly.

Journal of General Physiology. 1918–. Monthly.

Journal of Physiology. 1878–. Monthly.

Physiological Reviews. 1921–. Quarterly.

Zoology

American Journal of Anatomy. 1901–. Monthly.

American Zoologist. 1961–. Quarterly.

Journal of Experimental Zoology. 1904–. 18/yr.

Journal of Morphology. 1887–. Monthly.

Physiological Zoology. 1928–. Quarterly.

Systematic Zoology. 1952–. Quarterly.

EARTH SCIENCES

American Association of Petroleum Geologists. Bulletin. 1917–. Monthly.

American Mineralogist. 1916–. Bimonthly.

Biological Oceanography Journal. 1981–. Quarterly.

Earth Science Reviews. 1966–. Quarterly.

Geological Magazine. 1864–. Bimonthly.

Geological Society. Journal. 1845–. Bimonthly.

Geological Society of America. Bulletin. 1888–. Monthly.

Journal of Geology. 1893–. Bimonthly.

Journal of Geophysical Research. 1896–.

Journal of Petrology. 1960–. Quarterly.

Journal of Structural Geology. 1979–. Bimonthly.

Mineralogical Record. 1970–. Bimonthly.

Reviews of Geophysics and Space Physics. 1963–. Quarterly.

Atmospheric Sciences

American Meteorological Society. Bulletin. 1920–. Monthly.

Journal of Atmospheric Sciences. 1944–. Monthly.

Journal of Climate and Applied Meteorology. 1962–. Monthly.

Royal Meteorological Society. Quarterly Journal. 1871–. Quarterly.

WMO Bulletin. 1952–. Quarterly.

Weather. 1946–. Monthly.

Weatherwise. 1948–. Bimonthly.

Oceanography

Deep-sea Research with Oceanographic Literature Review. 1953–. Semimonthly.

Journal of Marine Research. 1937–. Quarterly.

Journal of Physical Oceanography. 1971–. Monthly.

Limnology and Oceanography. 1956–. Bimonthly.

Marine Biology. 1967–. 18/yr.

Marine Chemistry. 1972–. Bimonthly.

Marine Geology. 1964–. Semimonthly.

Marine Technology Society. Journal. 1966–. Quarterly.

Ocean Engineering. 1968–. Bimonthly.

Ocean Industry. 1966–. Monthly.

GENERAL ENGINEERING

Engineer. 1856–. Weekly.

Engineering. 1866–. 1866–1971, Weekly; 1971–, Monthly.

Engineering Journal. 1918–. Bimonthly.

Instruments and Control Systems. 1928–. Monthly.

International Journal of Engineering Science. 1963–. Monthly.

Journal of Engineering and Applied Sciences. 1981–. Quarterly.

Materials Engineering. 1929–. Monthly.

Professional Engineer. 1934–. Quarterly.

Recent Awards in Engineering. 1982–. Quarterly.

Technology and Culture. 1960–. Quarterly.

AERONAUTICAL AND ASTRONAUTICAL ENGINEERING

AIAA Journal. 1963–. Monthly.

Aeronautical Journal. 1897–. Monthly.

Aeronautical Quarterly. 1949–. Quarterly.

Astronautics and Aeronautics. 1957–. Monthly.

Aviation, Space, and Environmental Medicine. 1930–. Monthly.

Aviation Week and Space Technology. 1916–. Weekly.

Flight International. 1909–. Weekly.

Flying. 1927–. Monthly.

IEEE Transactions on Aerospace and Electronic Systems. 1965–. Bimonthly.

Journal of Aircraft. 1964–. Monthly.

Journal of Astronautical Sciences. 1954–. Quarterly.

Journal of Spacecraft and Rockets. 1964–. Bimonthly.

Space Science Reviews. 1962–. Monthly.

CHEMICAL ENGINEERING

AIChE Journal. 1955–. Bimonthly.

Chemical and Engineering News. 1923–. Weekly.

Chemical Engineering. 1902–. Biweekly.

Chemical Engineering Progress. 1947–. Monthly.

Chemtech. 1970–. Monthly.

Industrial and Engineering Chemistry. Quarterlies. 1962–. Quarterly.
Published with 3 ACS quarterlies: *Industrial and Engineering Chemistry: Fundamentals*; *Industrial and Engineering Chemistry: Process Design and Development*; *Product Research and Development.*

Journal of Chemical Technology and Biotechnology. 1951–. Monthly.

Journal of Chemical and Engineering Data. 1959–. Quarterly.

CIVIL ENGINEERING

American Society of Civil Engineers. Division Journals.

American Society of Civil Engineers. Proceedings. 1873–. Monthly.

Civil Engineering. 1930–. Monthly.

Civil Engineering Systems. 1983–. Quarterly.

Engineering Geology. 1965–. 8/yr.

Engineering Issues. 1971–. Quarterly.

Engineering News Record. 1874–. Weekly.

International Journal for Numerical and Analytical Methods in Geomechanics. 1977–. Quarterly.

Magazine of Concrete Research. 1949–. Quarterly.

New Civil Engineer. 1972–. Weekly.

Structural Engineer. 2 pts. 1908–. Monthly.

ELECTRICAL AND ELECTRONICS ENGINEERING

Electrical Engineering. 1887–. Monthly.

Electrical Review. 1872–. Weekly.

Electronic Engineering. 1928–. Monthly.

Electronics. 1930–. Biweekly.

Electronics and Power. 1955–. Monthly.

Electronics Letters. 1965–. Biweekly.

IEEE Proceedings. With 9 pts. 1871–. Bimonthly.

IEEE Spectrum. 1964–. Monthly.

IEEE Transactions. 1952–. Thirty-six separate transactions.

Journal of Electronic Materials. 1972–. Bimonthly.

Microelectronic Engineering. 1983–. Quarterly.

Radio and Electronic Engineer. 1939–. Monthly.

RCA Review. 1936–. Quarterly.

Computer Technology

ACM Computing Surveys. 1969–. Quarterly.

Annals of the History of Computing. 1979–. Quarterly.

Artificial Intelligence Report. 1983–. 10/yr.

Association for Computing Machinery. Journal. 1954–. Quarterly.

Communications of the ACM. 1958–. Monthly.

Computer and Structures. 1971–. Bimonthly.

Computer Journal. 1958–. Quarterly.

Computers and People. 1951–. Monthly.

Computers and the Humanities. 1966–. Quarterly.

Database Update. 1983–. Monthly.

IBM Systems Journal. 1962–. Quarterly.

InfoWorld. 1979–. Weekly.

International Journal of Mini and Microcomputers. 3/yr.

International Journal of Robotics Research. 1982–. Quarterly.

Journal of Robotic Systems. 1984–. Quarterly.

Journal of Systems and Software. 1979–. Quarterly.

Journal of Telecommunication Networks. 1982–. Quarterly.

Journal of VLSI and Computer Systems. 1983–. Quarterly.

Microcomputer Applications. 1980–. 3/yr.

Microcomputer Review. 1983–. 3/yr.

Micro Software Today. 1984–. Monthly.

New Generation Computing; An International Journal Devoted to Research on Fifth-Generation Computer. 1983–. Quarterly.

Personal Computing. 1977–. Monthly.

SIAM Journal on Scientific and Statistical Computing. 1980–. Quarterly.

Simulation; Journal of the Society for Computer Simulation. 1963–. Monthly.

Software: Practice and Experience. 1971–. Monthly.

INDUSTRIAL ENGINEERING

Ergonomics. 1957–. Monthly.

Industrial Engineering. 1969–. Monthly.

Industrial Management. 1952–. Monthly.

International Journal of Production Research. 1961/1962–. Bimonthly.

Operations Research. 1952–. Bimonthly.

Plant Engineering. 1947–. Biweekly.

Plant Management and Engineering. 1941–. Monthly.

Systems Research. 1983–. 3/year.

MECHANICAL ENGINEERING

American Society of Mechanical Engineers. Transactions. 1880–. Quarterly.

The current 6 sections: *Journal of Engineering for Power; Journal of Engineering for Industry; Journal of Heat Transfer; Journal of Basic Engineering; Journal of Applied Mechanics; Journal of Lubrication Technology.*

Chartered Mechanical Engineer. 1954–. Monthly.

Experiment in Fluids; Experimental Methods and Their Application to Fluid Flow. 1983–. Quarterly.

Heat Transfer Engineering. 1979–. Quarterly.

International Journal of Mechanical Sciences. 1960–. Monthly.

Journal of Applied Mathematics and Mechanics. 1958–. Bimonthly.

Journal of Applied Mechanics. 1935–. Quarterly.

Journal of Engineering for Industry. 1970–. Quarterly.

Journal of Fluid Mechanics. 1956–. Monthly.

Journal of Heat Transfer. 1970–. Quarterly.

Journal of Lubrication Technology. 1967–. Quarterly.

Journal of Structural Mechanics. 1972–. Quarterly.

Mechanical Engineering. 1906–. Monthly.

Mechanics of Materials: An International Journal. 1982–. Quarterly.

Wear: International Journal of the Science and Technology of Friction Lubrication and Wear. 1958–. Monthly.

Nuclear Engineering

American Nuclear Society Transaction. 1958–. 4 vols/yr.

Annals of Nuclear Energy. 1954–. Monthly.

Atomic Energy Review. 1963–. Quarterly.

Bulletin of the Atomic Scientists. 1945–. Monthly.

Journal of Nuclear Materials. 1959–. Semimonthly.

Journal of Nuclear Science and Technology. 1964–. Monthly.

Nuclear Energy. 1962–. Bimonthly.

Nuclear Engineering and Design. 1964–. 21/yr.

Nuclear Engineering International. 1956–. Monthly.

Nuclear News. 1959–. Monthly.

Nuclear Safety. 1959–. Bimonthly.

Nuclear Science and Engineering. 1956–. Monthly.

Nuclear Technology. 1965–. Monthly.

Nuclear Technology/Fusion. 1981–. Quarterly.

Soviet Atomic Energy. 1956–. Monthly.
English translation of *Atomnaya Energiya*.

Energy

See also "Nuclear Engineering."

ASCE Journal of Power Division. 1958–. Irregular.

Biomass. 1981–. Quarterly.

Electrical World. 1896–. Monthly.

Energy. 1976–. Monthly.

Energy and the Environment. 1980–. Monthly.

Energy Conversion and Management. 1961–. Quarterly.

Energy Policy. 1973–. Quarterly.

Energy Sources. 1973–. Quarterly.

Energy Systems and Policy. 1973–. Quarterly.

Energy Today. 1974–. Biweekly.

IEEE Transactions on Power Apparatus and Systems. 1965–. Monthly.

International Journal of Solar Energy. 1982–. Bimonthly.

JPT. Journal of Petroleum Technology. 1949–. Monthly.

Journal of Engineering for Power. (Transactions of the American Society of Mechanical Engineers, series A.) 1970–. Quarterly.

Nuclear Fusion. 1960–. Monthly.

Power. 1882–. Monthly.

Power Engineering. 1896–. Monthly.

Power Engineering. USSR Academy of Sciences. 1974–. Bimonthly.

Society of Petroleum Engineers Journal. 1961–. Bimonthly.

Solar Energy. 1957–. Monthly.

Thermal Engineering. 1964–. Monthly.

World Power Data. 1949–. Monthly.

ENVIRONMENTAL SCIENCES

Air Pollution Control Association. Journal. 1951–. Monthly.

American Water Works Association Journal. 1914–. Monthly.

Atmospheric Environment. 1967–. Monthly.

CRC Critical Reviews in Environmental Control. 1970–. Quarterly.

Environmental Pollution: An International Journal. Series A, B. 1970–. Monthly.

Environmental Science and Technology. 1967–. Monthly.

Environment International; A Journal of Science Technology, Health, Monitoring, and Policy. 1978–. Bimonthly.

International Journal of Environmental Studies. 1970–. Quarterly.

Journal of Environmental Sciences. 1958–. Bimonthly.

Journal of Environment Science and Health. 1975–. 14/yr.
Part A: *Environmental Science and Engineering.* 1968–. 14/yr. Part B: *Pesticides, Food Contaminants, and Agricultural Wastes.* 1975–. 14/yr.

Water, Air, and Soil Pollution. 1971–. 8/yr.

Water Engineering and Management. 1981–. Monthly.

Water Pollution Control. 1901–. 5/yr.

Water Pollution Control Federation. Journal. 1928–. Monthly.

Water Research. 1967–. Monthly.

Water Resources Bulletin. 1965–. Bimonthly.

Water Resources Research. 1965–. Bimonthly.

CHAPTER 17 TECHNICAL REPORTS AND GOVERNMENT DOCUMENTS

TECHNICAL REPORTS

Since World War II, technical-report literature has become an indispensable aspect of scientific and technical communication. The literature itself is generally the result of research supported by governmental grants and contracts and may be issued in various formats, including individual-author preprints, corporate-proposal reports, progress reports, state-of-the-art surveys, and the final report of a technical contract.

The majority of the literature emanating from certain agencies is issued in technical-report format. Among them are the National Aeronautics and Space Administration, the Energy Research and Development Administration, and the Defense Documentation Center. The major abstracting tools of the agencies (which are discussed in more detail elsewhere), such as *Scientific and Technical Aerospace Reports (STAR)* and *Nuclear Science Abstracts (NSA)*, now *Atomindex*, are indispensable aids in accessing report literature.

It is the responsibility of the National Technical Information Service, Springfield, Virginia, to act as a clearinghouse for all unclassified reports of federally sponsored research and development. Librarians should consider that, because NTIS must remain fiscally self-sufficient; there are frequent price increases for NTIS publications.

From the international scene, international organizations such as the International Atomic Energy Agency, Federation of Agriculture Organizations, and United Nations Educational, Scientific, and Cultural Organization are all major producers of technical report literature in the fields of science and technology. Lists of their publications, including technical reports, can be obtained directly from those organizations.

The following list provides current and retrospective sources for accessing technical reports, as well as a source designed to promote a better understanding of this type of literature. Chapter 23, Databases, contains online information from technical reports and document sources. They are essential sources and should not be ignored.

GENERAL REFERENCE TOOLS

Dictionary of Report Series Codes. 2d ed. **Lois E. Godfrey and Helen F. Redman, eds.** New York: Special Libraries Association, 1973 and later edition. First edition, 1962.
Comprehensive guide to the alphanumeric codes used in identifying technical report literature. The work is divided into 3 color-coded sections: explanations of series designations or assigners' methods in expanding on a series designation; list of report series code by letter, with corresponding agency; and corporate entries with corre-

sponding report series codes. Also includes a bibliography of articles on technical codes. Indispensable for anyone needing control over the vast array of report literature.
R: *ARBA* (1974, p. 547).

The Federal Data Base Finder: A Directory of Free and Fee-Based Data Bases and Files Available from the Federal Government. **Sharon Zarozny and Monica Horner.** Potomac, MD: Information USA, 1984.

A directory of 3,000 databases and files available from the federal government. Most of the entries included in each database are either government technical reports or government documents.
R: *ARBA* (1985, p. 18).

Federal Technology Catalog 1983: Summaries of Practical Technology. US Department of Commerce, National Technical Information Service, Center for the Utilization of Federal Technology. Springfield, VA: National Technical Information Service, 1984.

A compilation of summaries of and an index to over 1,000 *Tech Notes* announced in 1983. Annual updates. A subject index follows the main entry section.
R: *ARBA* (1985, p. 493).

Government Reports Announcement and Index. Vol. 75, nos. 7–. Springfield, VA: National Technical Information Service, Apr. 1975–. Biweekly.

Announcement and *Index* sections published separately, 1971 to 1975.
Lists and indexes available reports from National Technical Information Service under 22 subject fields. Subject, personal-author, corporate-author, report-number, and accession-number indexes.

Government-Wide Index to Federal Research and Development Reports. 71 vols. Washington, DC: US Clearinghouse for Federal Scientific and Technical Information, 1965–1971.

A subject, author, report-number, and accession-number index to the following publications: *US Government Research and Development Reports; Technical Abstract Bulletin; Scientific and Technical Aerospace Reports; Nuclear Science Abstracts.* This information is currently available in *Government Reports Announcement and Index.*

A Guide to US Government Scientific and Technical Resources. **Rao Aluri and Judith Schiek Robinson.** Littleton, CO: Libraries Unlimited, 1983.

A useful guide to publications of US government publications in science and technology. Some entries do not have complete bibliographic information.

Guidelines for Format and Production of Scientifc and Technical Reports. **American National Standards Institute.** New York: American National Standards Institute.

Index of Administrative Publications: Regulations, Circulars, Pamphlets, Posters, Joint Chiefs of Staff Publications, Department of Defense, and Miscellaneous Publications. **US Department of the Army.** Washington, DC: US Government Printing Office, 1978.

An army publication that lists administrative publications, including circulars, regulations, pamphlets, and posters. Conveniently arranged in 4 sections to allow quick retrieval of information.
R: *ARBA* (1979, p. 785).

Index to US Government Periodicals. **Ivan A. Watters, Jr., ed.** Chicago: Infordata International, 1970–. Quarterly; 4th issue is annual cumulative.

Fourteenth edition, 1984.
A major contribution to accessing articles in US government journals. 176 titles are indexed, only 50 of which are included in other standard indexing sources. Most of the 176 journals are available in depository libraries.
R: *ARBA* (1975; 1976; 1985, p. 16).

Technical Abstract Bulletin. **US Defense Documentation Center, Defense Supply Agency.** Springfield, VA: Clearinghouse for Federal Scientific and Technical Information, 1946–. Semimonthly.

Lists classified scientific research. Restrictions on availability and use.

Technical Report Standards. **Lawrence Harvill and T. L. Kraft.** La Mesa, CA: Banner Books International, 1979.

Standards for technical writers. Contains logical guidelines; explains report writing, etc.
R: *Microwaves* 17: 83 (Feb. 1978); *TBRI* 44: 154 (Apr. 1978).

Unclassified Publications of Lincoln Laboratory. Vols. 1–. Lexington, MA: MIT Lincoln Laboratory, 1967–.

US Government Research and Development Reports. 71 vols. Washington, DC: US Clearinghouse for Federal Scientific and Technical Information, 1946–1971.

Superseded by *Government Reports Announcement and Index.* Picked up reports not included in *Nuclear Science Abstracts* and *STAR.*

Use of Reports in Literature. **Charles P. Auger, ed.** Woburn, MA: Butterworth, 1975.

A systematic guide that deals with the nature and development of reports; the acquisition, bibliographical control, organization, etc., of reports. Two appendixes provide keys to reports, series code, and trade literature. Invaluable reference for librarians and researchers.
R: *ARBA* (1976, p. 631).

Washington Information Workbook: The Encyclopedia of Sources. 7th ed. Washington, DC: Washington Researchers, 1983.

This convenient volume highlights the 13 cabinet departments of the executive branch, the legislative and judicial branches, and the Library of Congress. Publications of these departments and branches are also included.
R: *ARBA* (1985, p. 17).

Weekly Abstract Newsletter. Springfield, VA: National Technical Information Service. Weekly.

Formerly, *Weekly Government Abstracts.*
See chapter on abstracts and indexes.

SUBJECT REFERENCE TOOLS

The following is only a sample list of both private and governmental sources that bear on the report literature of a specific subject. As far as private sources are concerned, professional societies are essential ones. Readers are well advised to contact each professional society in a specialized field for a list of this type of publication.

Air Pollution Technical Publications of the US Environmental Protection Agency. **Air Pollution Technical Information Center.** Research Triangle Park, NC: APTI, 1973–. Semiannually.

Lists EPA's air pollution reports with report numbers, prices, and order numbers.

Biological Abstracts/RRM (Reports, Reviews, Meetings). Vols. 17–. Philadelphia: Biosciences Information Service, 1980–. Monthly.

Original Title: *Bioresearch Index,* 1965–1980.
Supplements *Biological Abstracts* by providing complete citations to over 125,000 symposia, reports, articles, conference proceedings, letters, etc. not included in *Biological Abstracts.* Gives author, bio-systematic, concept, generic, and subject indexes. Citations are arranged under broad subject categories. Recommended to undergraduate students and those involved in the biological sciences.
R: *Choice* 18: 61 (Sept. 1980); *RSR* 8: 13 (July/Sept. 1980); Wal (p. 191); Win (1EC1; 2EC3).

Building Technology Publications 1965–1975. US National Bureau of Standards, Center for Building Technology. Washington, DC: US Government Printing Office, 1976.

Presents reports of the Building Technology Publications; deals with all publications including handbooks, notes, and consumer information. Includes a bibliography.
R: *ARBA* (1978, p. 766).

Directory of Engineering Document Sources. 2d ed. **D. P. Simonton.** Newport Beach, CA: Global Engineering Documentation, 1974.

Similar in coverage to the *Dictionary of Report Series Code,* being a consolidated cross index of document initialisms in the engineering field.
R: *ARBA* (1973, p. 632; Mal (1976, p. 250).

Index to ASTM Technical Papers and Reports. Philadelphia: American Society for Testing and Materials.

Five-year indexes, 1960–1965; 1966–1970; 1971–1975; 1976–1980; 1980–1985.

Indexed Bibliography of Office of Research and Development Reports, Updated to January 1975, EPA-600/9-74-001. **US Environmental Protection Agency.** Washington, DC: US Government Printing Office.

List of reports to the EPA arranged by report number under major series. Similar information may be gleaned from *Government Research Reports*, but not so quickly.
R: *ARBA* (1976, p. 697).

Intergovernmental Oceanographic Commission Technical Series. Paris: Unesco. Irregular.

Report Availability Notice: Ocean Science and Technology. **US Office of Naval Research.** Washington, DC: US Government Printing Office. Continuous.

A collection of abstracts of recent reports and reprints produced through ONR Ocean Science support in the areas of marine geophysics, ocean technology, chemical oceanography, etc.

Science and Government Report International Almanac. **D. S. Greenberg and A. D. Norman, eds.** Washington, DC: Science and Government Report. Annual.

A valuable almanac of international reports of applied science and technological research and developments. Covers all major industrialized nations. An essential source for scientific administrators.
R: *Nature* 282: 164 (Nov. 8, 1979).

Solar Energy, TID-3351-RIPI. 2 vols. **US ERDA Technical Information Center.** Oak Ridge, TN: US Energy Research and Development Administration, 1975.

A massive compilation of references to nearly 10,000 reports and documents on solar energy. Citations arranged under broad subject categories. Corporate and personal, subject, and report-number indexes available.

Title List of Documents Made Publicly Available. Vols. 1–. **US Nuclear Regulatory Commission.** Washington, DC: US Nuclear Regulatory Commission, 1979–.

US Energy Information Administration. Quarterly Coal Report. Washington, DC: US Energy Information Administration, 1983.
R: *Ulrich's Quarterly* 7: 171 (Apr./June, 1983).

GOVERNMENT DOCUMENTS

Both international and national governments and their contractors write and print numerous publications intended for use by scientists, engineers, professionals, and the general public. For example, the International Atomic Energy Agency, the US Department of Agriculture, and their respective parallel organizations are all major producers of scientific and technical literature. Throughout this book, we find government publications in every category of the literature. In fact, many of the significant indexing and abstracting tools in the fields of science and technology are products of governmental agencies and/or professional societies.

Although access to such information is generally considered a complex matter, there are numerous bibliographic tools that provide control over and access to unclassified and general governmental publications; some are listed under the category "General Bibliographic Sources."

Technical reports may formally be considered government documents, but their nature and importance is such that they were discussed separately in the earlier section of this chapter.

The following sources, published by both government and commercial publishers, are the most significant titles for both determining the availability of documents and providing a sound understanding of the scope and use of government publications in general. A small selected subject guide to scientific and technical government publications follows the general listing and illustrates the types of tools available.

GENERAL BIBLIOGRAPHIC SOURCES

Bibliographic Guide to Governmental Publications—US: 1978. **The Research Libraries of the New York Public Library and the US Library of Congress.** Boston: Hall, 1979.

British Official Publications. 2d ed. **John E. Pemberton.** New York: Pergamon Press, 1974.

Updates and revises the 1971 edition, including a new chapter on non-HMSO publications.
R: *SL* 65: 252 (May/June 1974); Chen (p. 373).

Catalog of the Public Documents of the Congress and of all Departments of the Government of the United States for the Period from March 4, 1893–Dec. 31, 1940 (Document Catalog). Washington, DC: US Government Printing Office, 1896–1945.

This major retrospective guide is an analytic dictionary catalog of significant congressional and departmental publications. Of historical interest to scientists.

Checklist of Major US Government Series. **John Androit.** McLean, VA: Documents Index, 1973–.

Volume 1, *Department of Agriculture.*
Projected 30 volumes covering major series published by government departments and independent agencies.

A Consumers', Researchers', and Students' Guide to Government Publications. New York: Wilson, 1983.

A practical guide with 3 sections. Section 1 gives a general introduction to government publications; section 2 introduces the major indexes and explains their use and scope; and section 3 consists of 7 subject chapters. Indexed by title and subject.
R: *ARBA* (1985, p. 16).

Cumulative Subject Index to the Monthly Catalog of US Government Publications, 1900–1971. **William W. Buchanan and Edna M. Kanely.** Washington, DC: Carrollton Press, 1973–.

Projected completion in 15 volumes.

Developing Collections of US Government Publications. **Peter Hernon and Gary R. Purcell.** Greenwich, CT: JAI Press, 1982.

Government Publications: Key Papers. **Bernard M. Fry and Peter Hernon.** Oxford: Pergamon Press, 1981.

Government Publications Reviews, Including Acquisition Guide. Vols. 1–. Elmsford, NY: Pergamon, 1974–. Bimonthly.

Formerly, *Government Publications Reviews.*
Covers the field of documents distribution, library handling, and use of documents produced by all levels of government—federal, state, and municipal—and by the United Nations, international agencies, and other countries.

Government Reference Books: A Biennial Guide to US Government Publications. **LeRoy C. Schwarzkopf, comp.** Littleton, CO: Libraries Unlimited, 1968–.

Eighth edition, 1984.
A comprehensive annotated bibliography of works published by the US government. Personal-author, title, and subject indexes.
R: *RQ* 19: 95 (Fall 1979); *ARBA* (1985, p. 17).

A Guide to Popular Government Publications: For Libraries and Home Reference. **Linda C. Pohle.** Littleton, CO: Libraries Unlimited, 1972. Also newer edition.

The new guide describes some 2,000 government publications, covering more than 100 topics of popular interest, that include consumer education, environment, etc. Indexed by subject.
R: *Choice* 9: 1434 (Jan. 1973); *LJ* (Feb. 1, 1973); *RQ* 12: 198 (Winter 1972); *WLB* 47: 195 (Oct. 1972).

Guide to US Government Publications. **John Andriot.** McLean, VA: Documents Index. Annual.

Fundamental information for over 2,000 agencies that issue series or periodicals. Annotated entries arranged by SUDOCS classification number.

Guide to US Government Statistics. 4th ed. **John Andriot.** McLean, VA: Documents Index, 1973.

Third edition, 1961.
An annotated guide to more than 1,700 recurring government publications and 3,000 titles in statistical numbered series.

Introduction to United States Public Documents. 2d ed. **Joe Morehead.** Littleton, CO: Libraries Unlimited, 1978.

First edition, 1975.
The usefulness of this text is enhanced by many illustrations and a detailed index. The first textbook designed for use in library-school government-documents courses.
R: *CRL* (July 1975); *RQ* 14: 363 (Summer 1975).

Locating United States Government Information: A Guide to Sources. **Edward Herman.** Buffalo, NY: William S. Hein, 1983.
A useful guide with over 120 illustrations in addition to tables. Indexed by title and subject.
R: *Journal of Academic Librarianship* (Mar. 1984, p. 34); *ARBA* (1985, p. 15).

Monthly Catalog of US Government Publications. **US Superintendent of Documents.** Washington, DC: US Government Printing Office, 1895–.
Most comprehensive bibliography of US government publications. Does not include much of the material issued as technical reports and contracted for by such agencies as NASA, ERDA, and DOD. Does include significant publications from agencies such as the Bureau of Mines, National Bureau of Standards, etc. Monthly indexes with annual cumulations.

Monthly Checklist of State Publications. Washington, DC: US Government Printing Office, 1910–. Monthly.
State documents do not figure to be of overwhelming importance to scientists, but this source is the most complete bibliographic tool available.

Municipal Government Reference Sources: Publications and Collections. **Peter Hernon et al.** New York: Bowker, 1978.

New Guide to Popular Government Publications for Libraries and Home Reference. **Walter L. Newsome.** Littleton, CO: Libraries Unlimited, 1978.
A listing of more than 2,800 publications of the US government. Topically arranged; includes information on government audiovisual resources. Helpful to both the layperson and the specialist. A highly recommended reference tool.
R: *LJ* 104: 394 (Feb. 1, 1979).

Public Access to Government Information: Issues, Trends, and Strategies. **Peter Hernon and Charles R. McClure.** Norwood, NJ: Ablex, 1984.

Publications of the National Bureau of Standards, Catalog; A Compilation of Abstracts and Key Word and Author Indexes. 1966–67; 1968–69; 1970–. US National Bureau of Standards. Washington, DC: US Government Printing Office, 1969–. Annual.
R: Sheehy (EA49).

State Government Reference Publications: An Annotated Bibliography. 2d ed. **David Parish.** Littleton, CO: Libraries Unlimited, 1981.
First edition, 1974.
Covers only selected publications.

UNDOC: Current Index: United Nations Documents. New York: United Nations, 1950–. 10/yr.

Formerly, *UNDEX*.
Covers only United Nations documents and publications. Checklist and subject index cumulate annually.

Unesco List of Documents and Publications, 1972–1976. 2 vols. New York: Unesco. Distr. New York: Unipub, 1979. Also annual list.

Lists all publications and documents, as well as articles appearing in periodicals, published by Unesco in 1972–1976. Divided categories include lists of acronyms, annotated list, personal-name index, subject index, and conference index.

US Government Manual. Washington, DC: US Government Printing Office, 1935–. Annual.

Standard reference and guide to government agencies and internal structure.

US Government Scientific and Technical Periodicals. **Philip A. Yannarella and Rao Aluri, comps.** Metuchen, NJ: Scarecrow Press, 1976.

A comprehensive listing of 266 government scientific and technical periodicals. Includes complete bibliographic information sources of microform edition, location, SUDOCS number, etc. A well-recommended reference.
R: *BL* 73: 205 (Sept. 15, 1976); *Choice* 13: 1122 (Nov. 1976); *LJ* 101: 1001 (Apr. 15, 1976); *WLB* 50: 812 (June 1976); *ARBA* (1977, p. 631).

SUBJECT BIBLIOGRAPHIC SOURCES

Catalogue of ICAO Publications. **International Civil Aviation Organization.** Montreal, Canada: ICAO. Annual. Also, monthly supps.
R: *IBID* 4: 148 (June 1976); *IBID* 7: 167 (Summer 1979).

Catalogue of Publications of the World Meteorological Organization, 1980 and Supplement. Geneva, Switzerland: World Meteorological Organization, 1982.

A master list of the World Meteorological Organization's and Global Atmospheric Research Programme's publications, with prices and an order form. Also available in English, French, Russian, and Spanish.
R: *Journal of Aerosol Science* 14: 695 (1983).

Historic Amerian Engineering Record Catalog, 1976. **US National Park Service.** Compiled by Donald E. Sackheim. Washington, DC: US Government Printing Office, 1977.

Lists historic documents, organized alphabetically by state, city, town. Provides exact geographic location of significant engineering and industrial sites. Illustrated with photographs, line and engineering drawings.
R: *ARBA* (1978, p. 758).

International Atomic Energy Agency Publications: Catalogue 1976/77. New York: Unipub, 1976.

Lists all publications of the International Atomic Energy Agency issued from 1958 up to the end of March 1976 and still available. Includes ordering information.
R: *IBID* 4: 241 (Sept. 1976); Wal (p. 304).

Publications of the World Meteorological Organization, 1951–1977. World Meteorological Organization. New York: Unipub, 1977. Also later edition.
R: *IBID* 6: 167 (June 1978).

Publications on Toxic Substances: A Descriptive Listing. Washington, DC: Interagency Regulatory Liaison Group. Distr. Washington, DC: US Government Printing Office, 1979.

A listing of publications about common toxic substances from 4 government agencies: Food and Drug Administration, Occupational Safety and Health Administration, Environmental Protection Agency, and Consumer Product Safety Commission. Entries are arranged by where the toxic substance is found. Includes ordering information, a list of agencies, and a subject index. Recommended for academic and public libraries.
R: *ARBA* (1981, p. 688).

SAMPLE GOVERNMENT DOCUMENTS

Government documents in science and technology cover a broad range from popular works to reports of highly technical research. The following publications, which are listed under issuing agencies with SUDOCS numbers, are intended to serve as an example of the types of literature available from the government:

Bureau of Mines

List of Bureau of Mines Publications and Articles. **Rita Sylvester.** Washington, DC: US Department of the Interior, Bureau of Mines. Annual.

Reports from the bureau arranged by report number, including abstracts for each citation. Author and subject indexes. Supplements to the bureau's 50-year list, 1910–1960 and 5-year lists, 1960–1964, 1965–1969, 1970–1974, and 1975–1979.
R: Sheehy (EJ60; EJ276; EJ277).

Mineral Facts and Problems, I28.3: 630. 1968.

Encyclopedic in scope. Covers history, prices, uses, and research.

Minerals Yearbook, I28.37: Item 639. 4 vols. Annual.

Department of Agriculture

Bibliography of Agriculture. New York: Macmillan, 1908–. Monthly.
Formerly *Agricultural Library Notes*, 1926–1942.
Compilation of literature received in the National Agricultural Library.
R: *RSR* 7: 57 (Apr./June 1979).

Soil Survey Reports, A57.38: item 102.

Series of soil maps superimposed on aerial photographs.

Wood Handbook: Wood as an Engineering Material, A1.76.

Environmental Protection Agency

Report to Congress, Disposal of Hazardous Wastes, EP1.17; 115.

Selected Publications on the Environment, EP1.21: EN 8/4.

Federal Aviation Administration

The FAA Statistical Handbook of Civil Aviation. TD4.20; item 431-C-14. Annual.
Statistics on aeronautical production, federal highways, general aviation, etc.

United States Civil Aircraft Register, TD 4.18/2; item 431-F-5. Annual.
Combines *Statistical Study of US Civil Aircraft* and *US Active Civil Aircraft by State and County.*

Geological Survey

Glaciers: A Water Resource I19.2: G45.

The Journal of Research. 1973–. 6/yr.
Papers written by members of the US Geological Survey.

The National Atlas of the United States, I19.2: N21a; item 621.
First national atlas in the country.

Publications of the Geological Survey. **US Geological Survey.** Washington, DC: US Government Printing Office, 1965–.
1879–1961 (pub. 1965); 1962–1970 (pub. 1971).
A permanent catalog of myriad types of material. A supplement covering the years 1962–1970 offers similar access to bulletins, maps, reports, etc. *New Publications* is a monthly listing updating the above, from 1971.
R: Jenkins (p. 80); Mal (1980, p. 191); Win (EE27).

National Highway Safety Bureau

National Highway Safety Bureau Corporate Author Authority List: A Supplement to Highway Safety Literature. Washington, DC: NHSB.

National Highway Safety Bureau Subject Category List: A Supplement to Highway Safety Literature. Washington, DC: NHSB.

CHAPTER 18 CONFERENCE PROCEEDINGS, TRANSLATIONS, DISSERTATIONS, AND RESEARCH IN PROGRESS, PREPRINTS, AND REPRINTS

CONFERENCE PROCEEDINGS

As a primary source of information, conferences and their proceedings can be of considerable value to the scientist and engineer. The conferences, congresses, and symposia themselves may range from small meetings of professional societies to elaborate international scientific conventions. The papers that are called for and presented at the events very often provide access to original research months before its appearance in periodical literature. The comments and rebuttals generated by such presentations also play important roles in scientific and technical communication. Adequate bibliographic control of such information is contingent on both prior notification of meetings and details of the publications arising from them. The following are the major tools in the area:

CALENDARS AND FORTHCOMING MEETINGS

General

Forthcoming International Scientific and Technical Conferences. London: Aslib. Quarterly.

Lists conferences in chronological order with date, title, location, and address. Subject, location, and sponsoring-organization indexes.
R: Wal (p. 44).

International Congress Calendar. Munich, New York: Saur. Distr. Detroit: Gale Research, 1961–. Annual; 4 issues/yr.

Twenty-sixth edition, 1986.
A useful guide to over 7,000 international events scheduled for the next 12 to 15 months. Organized in three sections—geographical, chronological, and subject/organizations index.
R: *ARBA* (1985, p. 12).

Scientific Meetings. Vols. 1–. New York: Special Libraries Association, 1956–. Quarterly.

Alphabetical list of scientific and technical organizations and universities that are sponsoring regional, national, and international meetings and institutes. Subject index.

World Meetings Outside USA and Canada. Newton Centre, MA: Technical Meetings Information, 1968–. Quarterly.

A 2-year registry, revised and updated each quarter, of future medical, scientific, and technical meetings. Indexed by sponsor, date, keyword, location, and deadline for papers.

World Meetings: United States and Canada. Newton Centre, MA: Technical Meetings Information, 1963–. Quarterly.

Subject

Details of scientific meetings in specific subject fields can be found in series issued by professional societies or agencies such as the following, although such series are not the only effective sources.

Meetings on Atomic Energy. Vienna: International Atomic Energy Agency, 1969–. Quarterly.

Calendar of upcoming conferences and training courses on atomic energy.
R: Wal (p. 309).

World Calendar of Forthcoming Meetings: Metallurgical and Related Fields. London: Iron and Steel Institute, 1965–. Quarterly.

Entries are chronological, with access by title, location, and organizing body.
R: Wal (p. 503).

Some of the best such sources would probably be the "meeting" sections of various official news journals in the fields, such as:

Chemical Engineering Progress. 1947–. Monthly.

Chemistry and Industry. 1881–. Semimonthly.

Mechanical Engineering. 1906–. Monthly.

Nature. 1869–. Weekly.

Nuclear News. 1959–. Monthly.

Physics Today. 1948–. Monthly.

Science. 1880–. Weekly.

Published Proceedings

Major Bibliographical Tools

BLL Conference Index, 1964–1973. **British Library, Lending Division.** Boston Spa, England: British Library, Lending Division, 1974.

A cumulated publication of all conferences received by the National Lending Library through 1973. Includes 46,500 entries from all fields of science and technology. Well arranged and indexed.
R: Sheehy (EA31).

Bibliographic Guide to Conference Publications: 1975–. Boston: Hall, 1976–. Annual.

Latest volume, *Bibliographic Guide to Conference Publications: 1979,* 2 vols., 1980.
An annual listing of recent purchases of the New York Public Library and the Library of Congress. Composed of conference proceedings, meetings, and symposia held mainly in the United States or printed in English. Arranged in alphabetical order, conferences are grouped into broad categories: engineering, physical sciences, earth sciences, etc. Helpful to acquisition and cataloging departments and to research libraries.
R: *Choice* 18: 769 (Feb. 1981); Sheehy (EA30).

Computext Book Guides: Conference Publications. Vols. 1–. Boston: Hall, 1974–. Monthly.
R: *ARBA* (1976).

Current Index to Conference Papers in Engineering. Vol. 1, nos. 1–. New York: Crowell Collier, 1969–.

Indexed by subject, author, and meeting.
R: Wal (p. 288).

Current Index to Conference Papers: Science and Technology. Vols. 2–. New York: CCM Information, 1971–. Monthly.

Combines and supersedes the following publications: *Current Index to Conference Papers in Chemistry,* 1969; *Current Index to Conference Papers in Engineering,* 1969; *Current Index to Conference Papers in Life Sciences.* Lists conferences and papers presented. Subject and author indexes.
R: Win (3EA25).

Current Programs. Vols. 1–. Chestnut Hill, MA: World Meetings Information Center, 1973–.

A monthly current-awareness service that provides the titles of some 120,000 scientific and technical papers presented at about 1,200 worldwide meetings annually. Covers the life sciences, chemistry, physical sciences, geosciences, and engineering. Subject, author, and meetings indexes.
R: Chen (p. 380).

Directory of Published Proceedings, Series SEMT—Science/Engineering/Medicine/ Technology. Vols. 1–. White Plains, NY: Inter Dok, 1965–. Annual cumulation.

A monthly chronological listing of proceedings, meetings, symposia, congresses, etc.
R: *ARBA* (1971, p. 477); Jenkins (A136); Wal (p. 45); Win (1EA29).

Index of Conference Proceedings Received. Boston Spa, England: British Lending Library, 1973–. Monthly.

Formerly *Index of Conference Proceedings Received by the NLL,* 1964–1973. An accession list of conference proceedings with brief annotations and a keyword subject index.

Index to Scientific and Technical Proceedings. Vols. 1–. Philadelphia: Institute for Scientific Information, 1978–. Monthly.

An index to about 90,000 conference papers and articles. Lists proceedings in all formats, including all addresses of first-named authors. Organized by meeting topic, geographic location, and sponsor. Helpful in achieving bibliographic control by conference literature. Includes permuted index, information on where to locate reprints, etc. Considered an indispensable tool for large science libraries.
R: *BMLA* 66: 497 (Oct. 1978); *Catholic Library World* 50: 126 (Oct. 1978); *NLW* 79: 257 (Jan. 1978); *ARBA* (1979, p. 644); Sheehy (EA35).

International Meeting Reports. Brussels: Union of International Associations. Annual.

NATO Advanced Study Institutes Series. New York: Plenum, 1975–.

NATO Conference Series. New York: Plenum, 1962–.

Nobel Symposia. New York: Raven Press, 1966–.

Vols. 1–22 published by Halsted Press; from vol. 24 publisher will vary.
Number 37, *Substance P,* Ivon Euler and B. Pernow, eds., 1977.

Proceedings in Print. Vols. 1–. Arlington, MA: Proceedings In Print. 1964–. Bimonthly.

Volume 1, numbers 1,2-volume 3, number 2, published by the Aerospace Division, Special Libraries Association. Originally meant as an index to conference proceedings related to aerospace technology, but beginning with volume 3, number 3, an index to all published conference proceedings regardless of subject. A subject and agency index.
R: Win (1EA30).

Yearbook of International Congress Proceedings: Bibliography of Reports Arising Out of Meeting Held by International Organizations During the Year, 1960–67. Union of International Associations, Pub. no. 211, etc. **Eyvind S. Tews, ed.** Brussels, Belgium: Union of International Associations, 1969–.

Conference reports are listed in English, German, Spanish, and French and are arranged chronologically. Includes an index of organizations plus an author-editor index.

Sample Subject Biblographical Tools

Availability of Nuclear Science Conference Literature. Oak Ridge, TN: US Atomic Energy Commission, 1963–. Irregular.

Brief Subject and Author Index of Papers Published in the "Proceedings," 1847–1950. London: Institution of Mechanical Engineers, 1951.

Superseded by same indexes covering 1951–1964, 1960–1969.
R: Wal (p. 296).

Conference Proceedings in the IAEA Library. Vienna: International Atomic Energy Agency, 1972. Annual.
The 1972 volume is a computer-produced listing of proceedings received from 1957–1971.

Index to Conferences Assigned CONF-Numbers. Oak Ridge, TN: US Atomic Energy Commission, 1962–1974.
KWIC index and conference-number index of nuclear-related meetings covering the period 1962–1973.

Index to Conferences Relating to Nuclear Science. Oak Ridge, TN: US Atomic Energy Commission, 1971–.
A frequently revised KWIC index of published and unpublished conference proceedings. Updated by *Availability of Nuclear Science Conference Literature.*

It is necessary to state the importance of abstracting and indexing journals as location tools for conference publications. Most of the major sources listed in the chapter on abstracts and indexes cover conference publications extensively. The following are samples of these sources:

Atomindex

Bibliography and Index of Geology

Biological Abstracts/RRM (Reports, Reviews, Meetings). Formerly, *Bioresearch Index*

Chemical Abstracts

Energy Index

Engineering Index Monthly and Author Index. Formerly, *Engineering Index*

Environmental Index

International Aerospace Abstracts

Mathematical Reviews

Metals Abstracts

INIS Atomindex

Oceanic Abstracts

Science Abstracts (3 sections)

Science Citation Index

Scientific and Technical Aerospace Reports

CONFERENCE PUBLICATIONS

Besides the above general bibliographic tools for locating information on conference publications, information on the actual contents of specific-subject conferences can be obtained from various sources, such as professional organizations' conference proceedings series, symposium series issued either by commercial publishers or conference sponsoring organizations, and periodicals devoted to providing either abstracts or full contents of papers presented at various meetings. The following are some of these types of sources, arranged by subject:

Astronomy

International Astronomical Union: Symposium. **G. O. Abell and P. J. E. Peebles, eds.** Boston: D. Reidel,
Number 92, *Objects of High Redshift*, 1982.

Mathematics

American Mathematical Society. Proceedings. Providence: AMS, 1950–. Monthly.

American Mathematical Society. Transactions. Vols. 1–. Providence: American Mathematical Society, 1900–. Monthly.

Regional Conference Series in Mathematics. Nos. 1–. Providence: American Mathematical Society, 1970–.
Number 44, 1980.

Symposium in Pure Mathematics: Proceedings. Vols. 1–. Providence: American Mathematical Society, 1947–.
Volume 36, 1980.

Physics

AIP Conference Proceedings Series. **American Institute of Physics.** New York: American Institute of Physics, 1971–.
Number 1, *Feedback and Dynamic Control of Plasmas*, 1971; . . . number 132, *Hadron Spectroscopy*, 1985; *Hadronic Probes and Nuclear Interactions*, 1985; *The State of High Energy Physics*, 1985; *Energy Sources: Conservation and Renewables*, 1985.
Essential reference tool for all scientific libraries, providing up-to-date coverage of the latest research in physics and related topics. Topics include astrophysics, atoms and molecules, condensed matter, general physics, nuclear physics, optics, and particle fields.

American Physical Society Bulletin. Vols. 1–. New York: American Institute of Physics, 1956–. Monthly.
Abstracts of speeches delivered at society meetings. Includes other news relating to the society and its membership.

Annals of the Israel Physical Society. New York: American Institute of Physics, 1977–.
Annals 1, *Atomic Physics in Nuclear Experimentation,* 1977; annals 2, *Statistical Physics—"Statphs 13,"* 1978; annals 3, *Group Theoretical Methods in Physics,* 1980; annals 4, *Molecular Ions, Molecular Structure, and Interaction with Matter,* 1981; annals 5, *Percolation Structure and Processes,* 1984; annals 6, *Vacuum Ultraviolet Radiation Physics VUV VII,* 1983; annals 7, *Electromagnetic Properties of High Spin Nuclear Levels,* 1984.
A series of hardbound conference proceedings that promotes interaction within the international scientific community.

International Conference on Atomic Physics: Atomic Physics Proceedings. Nos. 1–. New York: Plenum Press, 1968–.
Number 7, 1981.

Proceedings of the Annual Conference on Application of X-ray Analysis. Vols. 1–. New York: Plenum Press, 1960–.
Volume 22, *Advances in X-ray Analysis,* 1979.
Discusses all major important developments, includings energy production, materials optimization, mineral characterization.

Proceedings of the International School of Physics, "Enrico Fermi." New York: North-Holland.
Number 64, 1977.

Chemistry

Advances in Chemistry Series: "Collected Papers from the Symposia on . . ." Vols. 1–. Washington, DC: American Chemical Society, 1950–.
Volume 175, 1979.
Full text of brief papers.

American Chemical Society Abstracts of Meeting Papers. Washington, DC: American Chemical Society, 1947–. Annual.
One-hundred and seventieth, 1975.
Inclusion of papers to be presented at forthcoming meetings of the American Chemical Society. Includes author and KWIC title indexes.
R: *ARBA* (1976, p. 647).

American Chemical Society: ACS Symposium Series. Nos. 1–. Washington, DC: American Chemical Society, 1974–.
Number 146, 1981.
R: *TBRI* 43: 249 (July 1977).

Conference on Analytical Chemistry in Energy Technology. No. 23. **W. S. Lyon, ed.** Ann Arbor: Ann Arbor Science Publishers, 1980.
Number 21, 1977.

International Congress of Pure and Applied Chemistry. Conference Proceedings. London: Butterworth, 1961–.

Polymer Symposia. Nos. 1–. New York: Wiley, 1962–.
Number 1, *Journal of Polymer Science, Part C, Polymer Symposia*; number 66, 1979.

Biological Sciences

CIBA Foundation Symposia. Nos. 1–. New York: Elsevier, 1973–. Number 61, 1979.
The CIBA Foundation sponsors numerous symposia. Sample titles are *Outcome of Severe Damage to the Central Nervous System*, no. 34; *Health and Disease in Tribal Societies*, no. 49; *The Control of Cerebral Vascular Smooth Muscles*, no. 56.

Cold Spring Harbor Symposia on Quantitative Biology. Vols. 1–. 1933–. Annual.
Volume 44, 1980; volume 47, 1983.
A publication of the proceedings of the symposia, covering the texts of all papers delivered. Arranged by subject.

Excerpta Medica. International Congress Series. Nos. 1–. Amsterdam: Excerpta Medica Foundation, 1952–.
Number 471, 1978.
R: *Brain* 98: 186 (Mar. 1975); *Journal of Neurosurgery* 43: 380 (Sept. 1975).

Federation of European Biochemical Societies: FEBS Meeting. Vols. 1–. Elmsford, NY: Pergamon, 1964–.
Volume 63, 1980.

Federation Proceedings. **Federation of American Societies for Experimental Biology.** 1942–. Monthly.

Society for Experimental Biology: Symposia. Vols. 1–. New York: Cambridge University Press, 1947–.
Volume 33, 1979.

Society for General Microbiology: Symposium. Nos. 1–. New York: Cambridge University Press, 1949–.
Number 30, 1980.

Symposia of the Society for Experimental Biology. Vols. 1–. New York: Cambridge University Press, 1947–. Annual.
This series represents a publication of the papers read at the Society's annual meeting.

Earth Sciences

Geological Society of America. Proceedings. New York: The Society, 1933–.

IEEE International Conference on Engineering in the Ocean Environment. New York: IEEE, 1971–. Annual.

International Geological Congress Proceedings. Every 4 yrs.

Marine Technology Society Annual Meeting. Proceedings, 1965–. Washington, DC: MTS, 1965–. Annual.

Rare Earth Research Conference. **Gregory J. McCarthy, James J. Rhyne, and Herbert B. Silber, eds.** New York: Plenum Press.
Number 14, *The Rare Earths in Modern Science and Technology.*
Includes bibliographical references and indexes.

Engineering

International Conference on Recent Advances in Biomedical Engineering. No. 10. London: Biological Engineering Society, 1980.

Aeronautical Engineering

AGARD Conference Proceedings. Nos. 1–. New York: Scholium International, 1968–.
Number 11, 1970.
A series of AGARD conference proceedings on various topics in rocketry and spaceflight instrumentation. Highly technical; for special libraries.

AIAA Guidance, Control, and Flight Mechanics Conference. New York: American Institute of Aeronautics and Astronautics. Annual.

Industrial Astronautical Congress. (Editors and publishers vary), 1950–. Annual.
Technical papers in English, French, and German.

Space Research: Proceedings of the Plenary Meetings of COSPAR. Vols. 1–. **Committee on Space Research.** (Editors and publishers vary), 1960–. Annual.
Includes technical papers on all aspects of space research.

Chemical Engineering

American Institute of Chemical Engineers: AIChE Symposium Series. Nos. 1–. New York: American Institute of Chemical Engineers, 1951–.
Number 201, 1980.

Applied Polymer Symposia. Nos. 1–. New York: Wiley, 1960–.
Number 34, 1978.

Chemical Engineering Progress Symposium Series. New York: American Institute of Chemical Engineers, 1951–.

International Union on Pure and Applied Chemistry Conferences. (Publishers vary.)

Proceedings of the Offshore Technology Conference. Nos. 1–. Dallas: Offshore Technology Conference, 1969–.

Numbers 1–6, entitled: *Offshore Technology Conference Preprints;* number 7, 1975.

Record of Conference Papers of the Petroleum and Chemical Industry Conference. Nos. 13–. New York: Institute of Electrical and Electronics Engineers, 1966–.

Civil Engineering

American Society of Civil Engineers Proceedings.
Eleven separate journals of the various divisions of the society.

Institution of Civil Engineers Proceedings. 2 pts. 1952–. Quarterly.

Research on Transport Economics. Paris: European Conferences of Ministers of Transport, 1968–. Annual.

Electrical Engineering

AFIPS Conference Proceedings. Vols. 1–. Montvale, NJ: American Federation of Information Processing Societies (AFIPS) Press, 1951–.
Volume 47, 1978.

Conference Proceedings of the Symposium on Computer Architecture. Nos. 1–. New York: Institute of Electrical and Electronics Engineers, 1973–. Annual.
Number 8, 1981.

IEEE Proceedings. 1913–. Monthly.
Refers to proceedings of meetings sponsored or cosponsored by IEEE.

Industrial and Commercial Power System Technical Conference Record. Nos. 1–. New York: Institute of Electrical and Electronics Engineers, 1964–.
Number 15, 1979.

International Council of the Aeronautical Sciences. Proceedings. Washington, DC: Spartan Books, 1958–.

Proceedings of the International Conference on Data Bases. Nos. 1–. Philadelphia: Heyden & Son, 1980–.

Marine Engineering

Transactions of the Annual Marine Technology Society Conference. 11 vols. Washington, DC: MTS, 1975.

Transactions of the Society of Naval Architects and Marine Engineers. Vols. 1–. New York: Society of Naval Architects and Marine Engineers, 1893–. Annual.
Volume 85, 1977.

Materials Engineering

Army Materials Technology Conference. Nos. 1–. Chestnut Hill, MA: Brook Hill Publishing, 1972–.
Number 6, 1979.

Materials Research Symposium. Nos. 1–. Washington, DC: US Government Printing Office, 1966–.
Number 10, 1978.

Mechanical Engineering

Institution of Mechanical Engineers Proceedings. 1847–. Irregular.

Naval Engineering

Symposium on Naval Hydrodynamics. Nos. 1–. Washington, DC: US Government Printing Office, 1956–.
Number 13, 1980.

Nuclear Engineering

International Conference on Peaceful Uses of Atomic Energy. Vienna: International Atomic Energy Agency, 1955–. Irregular.
Technical papers from one of the largest conferences relating to nuclear science.

Proceedings of the IEEE Annual Conference on Nuclear and Space Radiation Effects. Nos. 1–. Piscataway, NJ: Institute of Electrical and Electronics Engineers, 1964–.

Energy

Proceedings of the National Conference on Technology for Energy Conservation. Nos. 1–. Silver Spring, MD: Information Transfer, 1977–.
Number 4, 1979.

Workshop on Wind Energy Conversion Systems: Workshop Proceedings. Nos. 1–. Washington, DC: US Government Printing Office, 1974–.
Number 3, 1978.

TRANSLATIONS

Although much of the world's scientific literature is in English and thus presents no great problems to the American scientist, the years since World

War II have seen a considerable increase in significant scientific contributions in languages not generally known by American and British scientists and engineers. To facilitate scientific investigation, numerous translation services have been devised to keep English-speaking scientists informed about the work of their foreign counterparts. On the other hand, non-English-speaking scientists' needs for translations from foreign languages to their native languages are much more obvious. The following tools are essential to people in search of both information on translations and individual translations themselves:

GENERAL SOURCES

British Reports, Translations and Theses. London: Her Majesty's Stationery Office, 1971–. Monthly.
Formerly: *BLLD Announcement Bulletin.*

Bulletin des traductions—CNRS. Paris: Centre National de la Recherche Scientifique.
Leading source for French translators.

Cumulative Index to English Translations, 1948–1968. 2 vols. Boston: Hall, 1973.

Index Translationum. Vols. 1–. New York: Unipub, 1949–.
Volume 29, 1976; volume 31, 1978.
A classic reference aid to more than 50,400 translated works published in over 60 countries. Information is arranged by country and includes author, original and translated title, subject, translator, and place of publication. Classified by the Universal Decimal Code. An invaluable tool for anyone in any discipline.

Monthly Catalog. **US Superintendent of Documents.** Washington, DC: US Government Printing Office, 1895–.
Lists members of Joint Publication Research Service series that are primarily translations of Russian and Chinese scientific literature.

Nachweise von Vebersetzunger. Hanover, W Germany: Technische Informationsbibliothek der Technischen Universitat.
Leading tool for German translations.

Transdex. Bibliography and Index to the US Joint Publications Research Service Translations. New York: Macmillan, 1970–.
Translations generated by the Joint Publications Research Service.

Translations Register-Index. New York: National Translations Center, 1967–. Monthly.
Succeeds *Technical Translations,* Washington, DC: Clearinghouse for Federal Scientific and Technical Information, 1959–1967.

Lists accessions of the translations center and thus provides the scientist with unpublished translations into English from the world literature in the natural, physical, medical, and social sciences.

Translation and Translators: An International Directory and Guide. **Stefan Congrat-Butlar, ed.** New York: Bowker, 1979.
A directory of translation organizations and translators involved in the scientific and technical, industrial, and literary fields. Contains a listing of translators by language as well as translation associations and training programs and guidelines. Also provided are 60 translation periodicals and an international bibliography of over 230 works. Useful reference for all science libraries.
R: *CRL* 41: 93 (Jan. 1980).

World Transindex. Delft, Netherlands: European Translations Centre, 1967–. Monthly.
Formerly, *World Index of Scientific Translations.*
Translations acquired by the European Translation Centre, listed by subject and citation index.

World Transindex: Announcing Translations in All Fields of Science and Technology. Vols. 1. Delft: International Translation Centre, 1978–. Monthly.
An index of approximately 32,000 translations from Asiatic and East European terms into Western terms, and then from Western terms into French terms.
R: Wal (p. 56).

ABSTRACTING AND INDEXING SOURCES

Like conference publications, abstracting and indexing tools are essential for obtaining access to translations of research reports, conference papers, and other sources of both primary and secondary information. All titles listed in the conference proceedings section also covers translation information. Government agencies have heavy translation activities, thus sources such as *Government Reports Announcements and Index, Nuclear Science Abstracts* (now *Atomindex*), and *Scientific and Technical Aerospace Abstracts* are major tools covering the government's translation report series, such as AEC-tran- and NASA-TT-.

SUBJECT SOURCES

Various tools exist that provide translations of materials in specific subject areas. The following is an indicative list:

American Mathematical Society: Translations. Series 2. Vols. 1–. Providence: American Mathematical Society, 1955–.
Series 1, 11 volumes, 1962; series 2, volume 115, 1980.

Index to Translations Selected by the American Mathematical Society. Vols. 1, 2. Providence: American Mathematical Society, 1966–1973.

Proceedings (Trudy) of the P. N. Lebedev Physics Institute. **D. V. Skobel'tsyn, ed.** New York: Consultants Bureau, 1964–.
Volume 67, 1975.
Translations of works published either by or for the Academy of Sciences of the USSR. Consists of technical research papers. Each volume covers 1 subject area.

SAO Russian Translation Series. Cambridge, MA: Smithsonian Astrophysical Observatory, 1962–. Irregular.

Translation Bulletin. Amsterdam: Elsevier, 1960–. Monthly.
Collects and indexes translations in the nuclear-energy field.

Translations of Mathematical Monographs. Vols. 1–. Providence: American Mathematical Society, 1962–.
Volume 51, *Equations of Mixed Type,* 1978.

Cover-to-Cover Translations

The "cover-to-cover" translation is a phenomenon dating back only to the late 1940s. Publications chosen for such treatment are generally primary-research journals, though some secondary sources of information are also available. As the labor involved in undertaking such a translation is often not justified by the worth of many of the articles themselves, the process has in recent years been modified to include the translations of only selected articles from foreign journals and selected articles that are composites of several originals. The following are examples of useful bibliographic sources for the identification of both cover-to-cover and selected translations of scientific and technical journals:

A Guide to Scientific and Technical Journals in Translation. 2d ed. **Carl J. Himmelsbach and Grace E. Brociner, comps.** New York: Special Libraries Association, 1972.
First edition, 1968.
Lists some 278 cover-to-cover translations and 53 miscellaneous journals. Primary emphasis is on Russian journals. Titles are transliterated and arranged alphabetically.
R: *CRL* 30: 83 (1969); *LJ* 93: 2790 (1968); *ARBA* (1974, p. 548); Jenkins (A155); Wal (p. 39); Win (2EA8).

NASA Technical Translations. Washington, DC: US National Aeronautics and Space Administration, 1959–.
Cover-to-cover translations of books, articles, and theses. Translations issued as technical reports series, NASA-TT-. Indexed in *GRA* and *STAR.*

Translations Journals: List of Journals Translated Cover to Cover, Abstracted Publications, and Publications Containing Selected Articles. Delft, Netherlands: European Translation Center, 1970.

For book translation, standard bibliographical tools such as *Books in Print,* and abstracting and indexing services such as *Mathematical Reviews* should also be fruitful sources of information. It is worth noting that there are a very small number of secondary abstracting journals that are translated cover to cover such as the following:

Cybernetics Abstracts. 1964–. Monthly.

Soviet Abstracts: Mechanics. 1970–. Quarterly.

Both are translated cover to cover from the corresponding sections of *Referativnyi Zhurnal.*

DISSERTATIONS AND RESEARCH IN PROGRESS

Dissertations

As primary sources of information, theses and dissertations can play a vital role in medical and scientific communication. Although they are rarely used with such frequency as is the journal article, patent, or technical report, satisfactory bibliographic control over such items is nevertheless essential. For practical purposes, the following list will include major tools that attempt to provide information on completed dissertations as well as theses and research in progress:

General Tools

American Doctoral Dissertations. Ann Arbor: University Microfilms. Annual.

Complete listing of all doctoral dissertations accepted by American and Canadian universities.

British Reports, Translations, and Theses. London: Her Majesty's Stationery Office, 1971–. Monthly.

Formerly, *BLLD Announcement Bulletin.*

Comprehensive Dissertation Index, 1861–1972. 37 vols; *1973–1977.* 19 vols. Ann Arbor: Xerox University Microfilms, 1973–. Supp. annually.

Succeeds *Dissertation Abstracts International Retrospective Index,* 1970. Some 417,000 entries from US and foreign universities. Eliminates the need for retrospective compilations such as H. W. Wilson's *Doctoral Dissertations Accepted by American Universities, 1933/34–1954/55.*

DATRIX (Direct Access to Reference Information, a Xerox Service).

A university microfilms service that provides title-keywords searches back to 1938. Covers some 275,000 dissertations.
See chapter on databases.

Dissertation Abstracts International. Ann Arbor: University Microfilms, 1969–. Monthly.

Volumes 1–11, 1938–1951, as *Microfilm Abstracts;* volumes 12–29, 1952–1969, as *Dissertation Abstracts.*

Primary emphasis on those emanating from US academic institutions. Section B is entitled *Physical Sciences & Engineering*. For earlier French theses, consult *Catalogue des theses de doctorat en sciences naturelles soutenues à Paris de 1891 à 1954.* Paris: Person, 1956.
For German theses, consult *Jahresverzeichnes der Deutschen Hochschulschriften.*
R: Jenkins (A2); Wal (p. 31).

The Doctoral Dissertation as an Information Source: A Study of Scientific Information Flow. **Calvin James Boyer.** Metuchen, NJ: Scarecrow Press, 1973.

Guide to Theses and Dissertations: An Annotated, International Bibliography of Bibliographies. **Michael M. Reynolds, ed.** Detroit: Gale Research, 1974.
Identifies and annotates more than 2,000 bibliographies of theses and dissertations in the United States and throughout the world. Indexed by institution, name and title, and subject. A key source for academic librarians and researchers.
R: *LJ* (Nov. 1, 1975).

Index to Theses Accepted for Higher Degrees in the Universities of Great Britain and Ireland. London: Aslib, 1953–. Annual.
Emphasizes theses relating to science and engineering. Arranged by school and subarranged by subject and author.

Master Abstracts. Ann Arbor: University Microfilms International. 1962–. Quarterly.
Covers only selected universities. Cumulative author and subject index.

Masters Theses in the Pure and Applied Sciences Accepted by Colleges and Universities of the United States and Canada. Vols. 1–. New York: Plenum Press, 1957–. Annual.
Volume 25, 1981.
Covers over 10,000 masters theses from nearly 250 US and Canadian universities. Theses are listed alphabetically by author within 44 study disciplines, excluding mathematics and the life sciences. For the academic research library.
R: *Journal of the American Chemical Society* 103: 2912 (May 20, 1981); *ARBA* (1981, p. 618).

MIT Abstracts of Theses Accepted in Partial Fulfillment of the Requirement for the Doctor's Degree with a Listing of the Titles of Theses Accepted for the Engineer's Degree and the Master's Degree. Cambridge, MA: MIT Press, 1953–.
Part I, *Abstracts of Doctor's Theses;* part II, *Titles of Engineer's and Master's Theses.* From 1953–1967, general volumes are divided by subject area. From 1968, separate volumes for each subject.

Theses and Dissertations as Information Sources. **Donald Davinson.** London: Bingley & Hamden, Linnett, 1977.
Discusses the bibliographic control of theses and dissertations. For librarians involved in technical services.
R: *Sci-Tech News* 32: 50 (Apr. 1978).

University of London Theses and Dissertations Accepted for Higher Degrees. London: University of London. Annual.
R: Chen (p. 389).

Subject Guides

Lists of subject theses are often available on a regular basis from the institutions granting the degrees.

Abstracts of Theses: Department of Civil Engineering. Cambridge, MA: Massachusetts Institute of Technology, 1963–. Annual.
Prior to 1970, the abstracts were entitled *Abstracts of Theses Submitted for Doctors', Engineers' and Masters' Degrees in Civil Engineering Awarded.* Similar abstracts in other subjects are also available from MIT.

Aerospace Thesis Topics List. New York: American Institute of Aeronautics and Astronautics, 1965.

A Bibliography of Doctoral Research on Ecology and the Environment, 1938–1970. Ann Arbor: University Microfilms, 1971. Also later edition.

Bibliography of Theses in Geology, 1967–1970. **Dederick C. Ward, ed.** Boulder, CO: Geological Society of America, 1973. Also later edition.
Series dates back to 1958. Includes both master's and doctoral theses.
R: *ARBA* (1974, p. 606; 1971, p. 520); Win (1EE6, 3EE3).

Dissertations in Physics: An Indexed Bibliography of All Doctoral Theses Accepted by American Universities, 1861–1959. **M. L. Marckworth et al.** Stanford, CA: Stanford University Press, 1961.

Energy: A Key-Phrase Dissertation Index. Ann Arbor: University Microfilms, 1976. Annual supps.
Over 5,000 citations are selected from the University Microfilm's over .5-million-dissertation database. Arranged by a key-phrase system with author index.
R: *RSR* 5: 23 (Oct./Dec. 1977); *ARBA* (1978, p. 689).

Forestry Theses Accepted by Colleges and Universities in the United States: January 1956–June 1966. **Michael P. Kinch, comp.** Corvallis, OR: Oregon State University Press, 1978.

Forestry Theses Accepted by Colleges and Universities in the United States: July 1966–June 1973. **Michale P. Kinch, comp.** Corvallis, OR: Oregon State University Press, 1979.
A listing of master's theses and Ph.D. dissertations accepted by colleges and universities from 1956–1973. Theses are grouped by state and then by college or university. Under this category is an alphabetical listing of the authors. Subject index is included. For academic libraries with strong holdings in forestry.
R: *ARBA* (1980, p. 706).

Readers should keep in mind that subject abstracting and indexing journals, such as those listed in the sections on conference proceedings, are also important sources for subject thesis information.

RESEARCH IN PROGRESS

General Science

Annual Register of Grant Support, 1969–. Los Angeles: Academic Media, 1969–. Annual.

Supersedes *Grant Data Quarterly.* One of 4 major sections is devoted to science and as such describes grant-support programs of government agencies, foundations, business, and professional organizations.

Current Research on Scientific and Technical Information Transfer. National Science Foundation, Division of Science Information. New York: Norton, 1977.

Contains state-of-the-art papers on research progress in information science. Full text of papers is included on 7 microfiche inside the book cover. All papers included are the results of government (National Science Foundation) studies. A useful reference for information managers and for public and academic libraries.
R: *JASIS* 30: 373 (Nov. 1979).

Directory of Researchers by Discipline. 7 vols. **Japan Society for the Promotion of Science, ed.** Tokyo: Japan Society for the Promotion of Science, 1979.

First edition, entitled *Directory of University Researchers by Subject,* 1956; second edition, 1961; third edition, entitled *Directory of University Researchers by Specialized Subject,* 1971.

An enlarged and updated 7-volume edition of a comprehensive directory of researchers in Japan. Some 110,000 entries span all disciplines of science: human, natural, and social. Entries are arranged under specialized disciplines by name. Includes short biographies, research interests, and publications. Indexes cover researchers and institutions. A recommended reference for academic libraries.
R: *LCIB* 39: 235 (July 4, 1980).

Directory of Research Grants. **William K. Wilson and Betty L. Wilson, eds.** Phoenix: Oryx Press. Annual.

Latest edition, 1984.

Organized by academic fields, this directory contains a list of research grants awarded to academic departments. Information covered includes description of grant, amount, sponsor, and application date. Indexes contain grant names, sponsoring organizations, and sponsoring organizations by type.
R: *CRL* 41: 492 (Sept. 1980).

Directory of Scientific Research Institutes in the People's Republic of China. 3 vols. **Susan Swannack-Nunn.** Washington, DC: US National Council for US–China Trade, 1977–1978.

Includes information on agriculture, chemicals, electronics, and machine industries in the People's Republic of China. Arranged by subject area, includes full information.
R: Sheehy (EA27).

Directory of Soviet Research Organizations. US Central Intelligence Agency, National Assessment Center. Seattle: University Press of the Pacific, 1979.
R: *CRL* 41: 93 (Jan. 1980).

Federal Funds for Research Development and Other Scientific Activities. **US National Science Foundation.** Washington, DC: US Government Printing Office. Annual.

Government Research Centers Directory. 2d ed. **Anthony T. Kruzas and Kay Gill, eds.** Detroit: Gale Research, 1984.

First edition, 1980.
Describes some 1,600 research and development centers operated by the government. Covers all disciplines, including science and technology, life sciences, agriculture, conservation, etc. A recommended reference for all libraries, businesses, government agencies, and media professionals.
R: *AL* 13 (Apr. 1982); *Choice* 18: 636 (Jan. 1981); *RQ* 20: 211 (Winter 1980).

Grants and Awards. US National Science Foundation. Washington, DC: US Government Printing Office. Annual.

Grant Data Quarterly. Vols. 1–. Los Angeles, CA: Academic Media, 1967–. Quarterly.

Each issue devoted to a specific type of grant support (e.g., government support programs, foundation support programs). Entries include the following data: type, purpose, eligibility, financial data, duration, application information, deadlines, etc. Includes organization and subject indexes.
R: *MRW* S1, p. 30.

Industrial Research Laboratories. 19th ed. **Jaques Cattell Press, ed.** New York: Bowker, 1985.

Sixteenth edition, 1979; seventeenth edition, 1981; eighteenth edition, 1983.
Provides descriptive information on 12,500 industrial organizations in the United States, including 700 companies listed for the first time. Geographic, personnel, and subject indexes available.
R: *BL* 77: 722 (Jan. 15, 1981); *ARBA* (1976, p. 638; 1978, p. 624).

Index to Scientific Reviews. Vols. 1–. Philadelphia: Institute for Scientific Information, 1975–. Semiannual.

An eclectic index to scientific review articles. Contains author listings and permuted title-subject indexes. Indexes approximately 3,000 journals.
R: *ARBA* (1976, p. 639).

National Patterns of R&D Resources: Funds and Manpower in the United States. US National Science Foundatin. Washington, DC: US Government Printing Office, Irregular.
1953–1972, 1972.

Research and Development in the Federal Budget. American Association for the Advancement of Science. Washington, DC: American Association for the Advancement of Science. Annual.

Research Centers Directory: A Guide to University-Related and Other Nonprofit Research Organizations Established on a Permanent Basis and Carrying on Continuing Research Programs in Agriculture, Business, Conservation, Education, Engineering and Technology, Government, Law, Life Sciences, Mathematics, Area Studies, Physical and Earth Sciences, Social Sciences, and Humanities. 8th ed. **Archie M. Palmer, ed.** Detroit: Gale Research, 1983.

First edition, entitled *Directory of University Research Bureaus and Institutes,* 1960.
A directory of over 6,000 nonprofit organizations covering all academic fields. Entries are organized alphabetically by parent institution or organizational title into 15 general subject fields. Each citation includes address, phone, sources of finances, staff, research areas, facilities, activities, etc. Contains 3 indexes: titles of research agencies, subject, and names of parent institutions. A valuable addition to reference collections.
R: *BL* 76: 1440 (June 1, 1980); *RQ* 19: 96 (Fall 1979).

Research Grant Index, DHEW Pub. no. 76–200. 2 vols. **US National Institutes of Health.** Washington, DC: US Government Printing Office. Annual.

Fifteenth, 1976.
The first volume contains approximately 9,000 subject headings under which appear the identification numbers of pertinent projects. The second volume contains project identification date, etc.

Research in British Universities, Polytechnics and Colleges. 3 vols. **RBUPC Office, the British Library, ed.** Boston Spa, Wetherby, West Yorkshire, England: British Library, 1979–.

Volume 1, *Physical Sciences,* 1979; volume 2, *Biological Sciences,* 1980; volume 3, *Social Sciences,* in preparation.
Three volumes that cover the following areas of research: physical sciences, biological sciences, and social sciences. Will benefit a wide variety of students.
R: *Physics Bulletin* 30: 441 (Oct. 1979); *NLW* 80: 115 (June 1979); Wal (p. 149, p. 166).

SSIE Science Newsletter. Vols. 1–. Washington, DC: Smithsonian Science Information Exchange, 1971–.

Volume 8, no. 9, August 1979.
Outlines information on some 200,000 ongoing and recently completed projects in all fields of basic and applied research active during the present and past 2 years. Project information gleaned from over 1,300 organizations that support research: federal, state, and local government agencies; nonprofit associations and foundations; colleges and universities; and foreign research organizations. Contains extensive section on medical sciences research. Also available online.

Science and Technology Research in Progress 1972–1973. 7 vols. Orange, NJ: Academic Media, in cooperation with Smithsonian Science Information Exchange, 1973. Also, updates annually.

Volume 1, *Engineering Sciences;* volume 2, *Chemistry and Chemical Engineering;* volume 3, *Earth and Space Sciences;* volume 4, *Electronics and Electrical Engineering;* volume 5, *Materials;* volume 6, *Mathematics;* volume 7, *Physics.*
Provides current and accurate information on research in progress. Contains investigator index, research-organization index, funding-organization index, and subject index.

Science Research in Progress. **US National Science Foundation.** Washington, DC: Smithsonian Institution, 1949–. Annual.
Printout of unpublished but planned research and research in progress.

Scientific and Technical Research Centers in Australia. New York: State Mutual, 1982.
A directory of all the governmental, institutional, and private research organizations in Australia. Covers research, contact personnel, and scope of work. For science reference libraries.

Scientific Research in British Universities and Colleges. **Great Britain Department of Education and Science.** London: Her Majesty's Stationery Office, 1951–. Annual.
Most recent volume, 1984/85.
To be superseded by *Research in British Universities, Polytechnics, and Colleges.* New series also in 3 sections: physical sciences; biological sciences; social sciences.
R: Sheehy (EA26; EA182).

A Summary of Research in Science Education—1983. **William G. Golliday, Barry L. S. McGuire, Stanley L. Helgeson, and Patricia E. Blosser.** New York: Wiley, 1985.
Published as volume 69, number 3 of *Science Education.* It reviews the results of 422 research projects on such issues as student and teacher characteristics, teacher education, and instructional materials and strategies.

World Directory of Research Projects, Studies, and Courses in Science and Technology Policy. Paris: Unesco, 1981.
Contains the results of a survey of institutions and projects, which the main entry section details. The index section lists parent organizations and their courses and projects, which welcome visiting researchers, and a list of publications. Good for policy-makers.
R: *Unesco Journal of Information Science, Librarianship and Archives Administration* 4: 135 (Apr. 1982); *ARBA* (1983, p. 599).

Mathematics

AMS Grants and Proposals. **American Mathematical Society.** Providence: American Mathematical Society. Annual.

Physics

Health Physics Research Abstracts. **International Atomic Energy Agency.** New York: Unipub. Irregular.

Number 7, 1977.
Contains current research abstracts from international health physics experts, including some 250 reports. Also contains statements of ongoing investigations.
R: *IBID* 6: 71 (Mar. 1978).

Research Fields in Physics at United Kingdom Universities and Polytechnics. 7th ed. London: Institute of Physics, 1984.
Fifth edition, 1978; sixth edition, 1981.
A detailed guide to work currently being done in the United Kingdom.
R: *NTB* 60: 18 (Jan. 1975); *RSR* 3: 73 (Apr./June 1975).

Chemistry

Accounts of Chemical Research. Easton, PA: American Chemical Society. Monthly.
Concise, critical reviews of research areas under active investigation.

Directory of Graduate Research. Vols. 1–. **American Chemical Society, Committee on Professional Training.** Washington, DC: American Chemical Society, 1953–. Biennial.
A directory of the faculties, publications, and research theses in academic chemistry, chemical engineering, biochemistry, and pharmaceutical departments at American and Canadian universities.
R: Win (ED11).

Directory of UK Polymer Research. **Plastics and Rubber Institute.** London: Plastics and Rubber Institute, 1979–.
For industrial researchers; discusses a broad scope of research activities on polymeric materials.
R: *Rubber World* 181: 66 (Nov. 1979); *TBRI* 46: 76 (Feb. 1980).

Biological Sciences

Human Adaptability: A History and Compendium of Research in the International Biological Programme. **K. J. Collins and J. S. Weiner.** London: Taylor & Francis, 1977.
In 2 sections. The first provides a history of the International Biological Programme; the second contains a synthesis of research on topics such as human growth, genetics, nutritional status, etc.
R: *Aslib Proceedings* 42: 331 (June 1977).

National Biomedical Research Directory. Betheda, MD: National Health Directory, Annual.
Provides information on biomedical research organizations, research institutions, publications, research/medical libraries, and key NIH personnel. Useful for every person engaged in biomedical research of administration.

Research Contracts in the Life Sciences. **US Energy and Research Development Agency (ERDA).** Springfield, VA: National Technical Information Service. Annual.

Listing of ERDA-funded research projects.

Research in Biological and Medical Sciences; Annual Progress Report. Vols. 1–. **Walter Reed Army Institute of Research.** Washington, DC: US Government Printing Office, 1975/1976–.

Includes research relating to biochemistry, communicable diseases and immunology, internal medicine, physiology, psychology, surgery, and veterinary medicine.

Agricultural Sciences

Agricultural Research Index: A Guide to Agricultural Research, Including Dairy Farming, Fisheries Food, Forestry, Horticulture, and Veterinary Science. 6th ed. 2 vols. **J. Burkett, ed.** Harlow, England: Francis Hodgson, 1978. Also newer edition.

A listing of organizations that conduct research in agricultural fields. Alphabetically organized by country. Includes a general title index. For agricultural collections.
R: *Quarterly Bulletin of the IAALD* 24: 110 (Fall/Winter 1979).

Research Highlights. **Centro Internacional de Agricultura Tropical.** New York: Unipub, 1977. Also supps.

Reviews the research activities of the Centro Internacional de Agricultura Tropical. Covers such topics as beans, rice, cassava, genetic resources, swine flu, and library and information services.
R: *IBID* 6: 382 (Dec. 1978).

Earth Sciences

Marine Research, 1973. **Federal Council for Science and Technology; Interagency Committee on Marine Science and Engineering.** Washington, DC: US Government Printing Office. Annual.

Essential for access to ocean engineering work in progress. Work is indexed by subject, investigator, contractor, and supporting agency.

Ocean Research Index: Including Freshwater Science. 2d ed. St. Peter Port, Guernsey, England: Hodgson, 1976.

First edition, 1970.
A reference directory of world establishments engaged in ocean or freshwater research. Associated fields such as fishing are included. More than 100 countries are covered, and there is an international section. There is also a selected listing of marine museums of the world and oceanographic data centers or designated national agencies. Cross-referenced and indexed.

General Engineering

Directory of Current Research. Cambridge, MA: MIT Industrial Liason Office. Annual.

Engineering College Research and Graduate Study. Washington, DC: American Society for Engineering Education. Annual.

Industrial Research in Britain. 8th ed. New York: International Publishers Service, 1976. Also later edition.

Aeronautical and Astronautical Engineering

Aerospace Research Index. 4th ed. St. Peter Port, Guernsey, England: Hodgson, 1969.
Worldwide guide to institutions conducting or promoting research in aerospace and related fields.
R: Wal (p. 392).

Roster of US Government Research and Development Contracts in Aerospace and Defense. **Frost & Sullivan, Inc., New York.** Washington, DC: Bowker Associates, 1965.
Lists 7,500 contracts awarded by some 300 government agencies involved in space and defense research. The dated source is still useful for obtaining information on potential funding agencies.

Space Research in the UK. Royal Society of London, Science Research Council, 1967–.

Space Research in United Kingdom Universities. London: Royal Society of London. Annual.

Civil Engineering

Water Resources Research Catalog. Vols. 1–. **US Office of Water Resources Research.** Washington, DC: US Government Printing Office, 1965–.
Describes current research on water resources problems. Abstracts included.
R: *Geotimes* 9: 24 (May/June 1965); Jenkins (F93); McDonald (p. 5).

Electrical and Electronics Engineering

Annual Summary of Research in Electronics. New York: Polytechnic Institute. Annual.

Directory of Computer Education and Research: International Edition. 2 vols. **T. C. Hsiao, ed.** Schenectady, NY: Science and Technology Press, 1978.

Nuclear Engineering

Nuclear Research Report. Vols. 1–. Columbus, OH: Nuscience Publications. 1972–.
Annual compilation of ongoing research projects in nuclear research. Includes titles of projects, author affiliations, and subject and author indexes.

US Energy Research and Development Administration. Materials Sciences Program.
US ERDA. Springfield, VA: National Technical Information Service. Annual.

Water Reactor Safety Research Program: A Description of Current and Planned Research. **US Nuclear Regulatory Commission.** Washington, DC: US Nuclear Regulatory Commission, 1979–.

World Nuclear Directory: A Guide to Organizations and Research Activities in Atomic Energy. 6th ed. **C. W. J. Wilson, ed.** St. Peter Port, Guernsey, England: Hodgson, 1981.

A list of names, addresses, and telexes of those organizations doing research in atomic energy. Entries are numbered and arranged alphabetically by country. Some headings in the subject index do not correspond to entries. Poor binding and type. Good for libraries that need the data.
R: *Choice* 19: 1222 (May 1982); *ARBA* (1983, p. 657).

Energy

Energy Research Programs. 1st ed.– **Jaques Cattell Press, ed.** New York: Bowker, 1980–.

Alphabetically arranged compilation of energy activities throughout the United States, Canada, and Mexico. Includes all kinds of research facilities, full address information, geographical index, personnel index, and subject index by industrial classification. Directory covers production, experimentation, consultation, application, and transmission. Useful for libraries seeking energy information.
R: *BL* 77: 1310 (June 1, 1981); *Choice* 18: 1234 (May 1981); *CRL* 41: 567 (Nov. 1980); *RQ* 20: 209 (Winter 1980); *ARBA* (1981, p. 682).

Information on International Research and Development Activities in the Field of Energy. US National Science Foundation. **David F. Hersey, comp.** Washington, DC: US Government Printing Office, 1976–.

Contains information on research projects in Canada, Italy, Germany, France, Netherlands, and the United Kingdom. Contains a subject, investigator, organization, and sponsor index. Provides numerous cross references.
R: *RSR* 5: 33 (Oct./Dec. 1977).

Solar Energy and Research Directory. **Ann Arbor Science Special Task Group, prep.** Ann Arbor: Ann Arbor Science. Annual.

A useful listing of companies associated with solar energy. Includes names, addresses, etc. Numerous cross references, geographical index, classification and subclassification systems are also included.
R: *LJ* 103: 722 (Apr. 1, 1978); *ARBA* (1979, p. 700).

World Energy Directory: A Guide to Organizations and Research Activities in Non-Atomic Energy. New York: Longman, 1981.

The first directory of worldwide non-atomic energy research. Includes more than 1,200 companies and organizations in 80 countries.
R: *American Libraries* 13 (July/Aug. 1982), *ARBA* (1983, p. 657).

Environmental Sciences

Environmental and Civil Programs of the Federal Government. Greenwich, CT: DMS, 1975–.

Market intelligence reports cover over 75 programs in 11 market areas of federal spending: air pollution, antipoverty, education, housing and urban development, law enforcement, health, transportation, environment, water resources, post office automation, and solid waste disposal.

A Review of the US EPA Environmental Research Outlook. **US Congress, Office of Technology Assessment.** Washington, DC: US Government Printing Office, 1976–. Annual.

Summaries of Solid Wastes Research and Training Grants, US PHS Pub.-1596. **US Environmental Control Administration.** Washington, DC: US Government Printing Office, 1966–. Annual.

R: McDonald (p. 5).

Pollution

Directory of National and International Pollution Monitoring Programs. Smithsonian Institution, Center for Short-lived Phenomena, 1974–.

R: *IBID* 4: 31 (Mar. 1976).

Environmental Pollution: A Guide to Current Research. Produced from data gathered by Science Information Exchange, Smithsonian Institution, Washington, DC. New York: CCM Information, 1971–.

Materials on projects are classified by subject. For each project, information on name, organization and address, brief description of objectives, progress, and supporting organization are given.

Pollution Research Index: A New Reference Guide. St. Peter Port, Guernsey, England; Hodgson, 1975–.

This new world guide covers the whole field of research into the causes and prevention of water, air, land, and noise pollution at government, industry, and university levels. The guide is fully indexed and cross-referenced.

Projects in the Industrial Pollution Control Program. **US Environmental Protection Agency. Office of Research & Monitoring.** Washington, DC: US Government Printing Office. Annual.

PREPRINTS

Current use of the term *preprints* implies a broad range of documents from informal communications to finished manuscripts awaiting publication in a journal. While bibliographic control is often troublesome, those preprints that more closely resemble advance copies of journal articles or transaction papers are often adequately controlled by the issuance of preprint series by professional organizations.

Among those societies that announce preprints are the American Society of Mechanical Engineers and the Society of Automotive Engieners. Preprint announcements for these societies are made in *Mechanical Engineering and Automotive Engineering,* respectively. Other organizations committed to the dissemination of information available in preprints include the American Institute of Chemical Engineers and the American Institute of Aeronautics and Astronautics. For a more complete listing of preprint series consult:

Directory of Engineering, Scientific and Management Document Sources. **D. Simonton.** Newport Beach, CA: Global Engineering Documentation Services, 1974.

The following are some specific sources:

American Chemical Society—Preprints. Vols. 1–. Washington, DC: American Chemical Society, Division of Fuel Chemistry, 1957–.
Volume 20, 1975.
A cumulative index to volumes 1–19, 1957–1974, is also available.

ASLE Annual Meeting Preprint. American Society of Lubrication Engineers. Annual.

OTC Preprint Index 1969–1973. Dallas: OTC, 1973.

Polymer Preprints. **American Chemical Society, Division of Polymer Chemistry.** Washington, DC: American Chemical Society, 1975.

REPRINTS

REFERENCE TOOLS

Out-of-print or discontinued publications are commonly acquired through reprint services and microreprography. Reprinted material is available from various institutes and organizations as well as from more familiar reprint publishers.

The useful tools for locating these reprints are catalogs of reprint publishers, such as Johnson's, and the following general sources:

Encyclopedia Reprints Series. 5 vols. **N. M. Bikales, ed.** New York: Wiley-Interscience, 1971.

Guide to Reprints. **Carol Wade, ed.** Washington, DC: NCR Microcard Editions, 1967–. Annual.
Lists books, journals, and various materials available as reprints from US and foreign publishers. Includes a directory of publishers. Supplemented by *Announced Reprints,* Washington, DC: NCR Microcard Editions, 1969–. Quarterly.

Subject Tools

On the other hand, it should be kept in mind that scientists are generally concerned with "reprints" of current publications (most popular journal articles) obtained from publishers at the same time of publication for ready dissemination. These play an important role in the scientific and technical communication process. In addition, research centers and laboratories, professional organizations, and governmental agencies also frequently issue reprint series of papers produced by their staff scientists. The following are a few selected sample series:

Astrophysics Today. **A. G. W. Cameron, ed.** New York: American Institute of Physics, 1984.
Reprinting news items and articles from the past decade from *Physics Today* on the frontiers of research on the solar system, stars, galactic physics, and cosmological physics. Valuable as supplementary reading materials.

Atlantic Oceanographic and Meteorological Laboratories. Collected Reprints. Washington, DC: US Government Printing Office. Annual.

Collected Reprints. Woods Hole, MA: Woods Hole Oceanographic Institution. Annual.

Current Physics Reprints (CPR). New York: American Institute of Physics.
AIP provides a photocopy service for quick access to articles in American Institute of Physics and member society journals. Currently, the cost is $10.00 per article copy (postage included) for articles up to 20 pages.

Ocean Research Institute. Collected Reprints. Tokyo: University of Tokyo. Annual.

Selected Reprint Series. New York: American Institute of Aeronautics and Astronautics. Irregular.

CHAPTER 19 PATENTS AND STANDARDS

PATENTS

Patents are a significant part of technical literature for health scientists and engineers engaged in research and development. Accurate knowledge of existing and pending patents saves the researcher from the possibility of duplicating the efforts of other scientists and, more importantly, reduces the possibility of legal complications resulting from infringement upon a patent belonging to another party. In certain fields, such as chemical research, no literature search is complete without an examination of existing patents. *Chemical Abstracts* provides a patent index and concordance for such purposes.

The patent itself is a document that represents a contract between a government and inventor or patentee. As patents are honored only in the country in which they are issued, necessary tools have been developed to provide international coverage of patent literature. These tools will be listed with those that provide coverage of the US patent system. The Patent Office Search Center in Arlington, Virginia, maintains files of issued patents and their supporting records. Interested parties may order copies of original patents or consult them at regional patent copy depository libraries.

As patents may present considerable complications for science librarians and scientists or engineers who have no access to patent attorneys or agents, the following list of sources is intended to provide a fundamental understanding of patent classification, procedure, and practice. See chapter 23, "Data Bases," for online patent information.

Mainly on Patents: The Use of Literature of Industrial Property and Its Literature. **Felix Liebesny, ed.** Hamden, CT: Archon Books, 1973.

Emphasis on British publications.
R: *ARBA* (1974, p. 545).

Manual of Classification of Patents. **US Patent Office.** Washington, DC: US Government Printing Office, 1974–. Irregular.

There are some 3,000 major classes and approximately 60,000 subclasses under which a patent may be placed. This tool explains the system and provides class and subclass numbers that facilitate the use of the *Official Gazette*.

Manual of Patent Examining Procedures. 3d ed. **US Patent Office.** Washington, DC: US Government Printing Office, 1949–.

First edition, 1949; third edition, 1969.
Material pertaining to all practices and procedures used by examiners at the Patent Office.
R: *ARBA* (1970, p. 146).

Technical Information Sources: A Guide to Patent Specifications, Standards, and Technical Reports Literature. 2d ed. **Bernard Houghton.** Hamden, CT: Shoestring Press, 1972.
Primarily a guide to British patent information for librarians.

INDEXES TO US AND INTERNATIONAL PATENTS

Abstracting and indexing journals can be valuable sources of patent information. Unquestionably *Chemical Abstracts* is the most important one. *Science Citation Index, Engineering Index,* and *Computer Abstracts* are a few additional sample tools that make available such information.

Journals often also include abstracts or notes on new patents. Among the numerous titles offering such features are *Journal of Applied Chemistry, Electrical Review, Metal Finishing,* and *Textile Manufacturer.*

The most important guides to the patent literature are various indexes issued either by government patent offices or commercial firms. The following list is indicative of the range of these tools. Some of the databases of these indexes are machine searchable, and readers should consult the chapter on databases for further information.

SOURCES ABOUT PATENTS

About Patents; Patents as a Source of Technical Information. **Great Britain: The Patent Office, Department of Trade and Industry.** London: Her Majesty's Stationery Office, n.d.
R: *Engineering* (Oct. 14, 1971).

Development and Use of Patent Classification Systems. **US Patent Office.** Washington, DC: US Government Printing Office, 1966.

The Encyclopedia of Patent Practice and Invention Management. **Robert Peyton Calvert, ed.** New York: Van Nostrand Reinhold, 1974.
A 2-volume encyclopedia that presents comprehensive statements of principles and procedures in many phases of patent practice and invention management.
R: *Choice* 2: 567 (1966); Win (1EA40).

Encyclopedia of UK and European Patent Law. **T. A. Blanco-White et al, eds.** London: Sweet and Maxwell, 1977.
A loose-leaf guide covering several UK patent acts, conventions, and treatises along with those conventions of Paris and Strasbourg.
R: Wal (p. 261).

European Patents Handbook. **Chartered Institute of Patent Agents.** London: Oyez, 1978.
Contains workings of the EPC, PCT, and CPC as well as other sources and materials.
R: Wal (p. 259).

Foreign Patents: A Guide to Official Patent Literature. **Francis J. Kase.** Dobbs Ferry, NY: Oceana, 1973.

Foreign patent material arranged alphabetically by country, including information on foreign patent offices, English translations of printed specifications, patent journals, data on trademarks, etc. Primarily for use by searchers in the United States.
R: *Choice* 10: 946 (Sept. 1973); *LJ* 98: 2325 (Aug. 1973); *ARBA* (1974, p. 548).

General Information Concerning Patents: A Brief Introduction to Patent Matters. Washington, DC: US Government Printing Office, 1973.

Practical information on the general subject of patents. Intended for company and private-inventor use.

Handbooks of Patent Technology. New York: Plenum, 1982–.

How to Find Out About Patents. **Frank Newby.** New York: Pergamon Press, 1967.

Practical information on typical patents.
R: *NTB* 52: 267 (1967); Jenkins (A182); Wal (p. 240).

How to Patent Without a Lawyer. **Albert Thumann.** Atlanta: Fairmont, 1978.

A useful guide to patenting for the engineer or individual inventor.
R: *Invention Management* 3: 5 (Nov. 1978); *TBRI* 45: 120 (Mar. 1979).

International Classification of Patents. **Council of Europe.** West Wickham, England: Morgan-Grampian, 1968.

The Inventor's Patent Handbook. Rev. ed. **Stacy V. Jones.** New York: Dial Press, 1969.

Earlier edition, 1969.
Practical information on obtaining a patent, selling patents, contract agreements, etc.
R: *Choice* 4: 400 (1967); *RQ* 10: 82 (Fall 1970); *ARBA* (1971, p. 545); Jenkins (A181); Win (2EA41).

Isocyanate Polymers. Vol. 1. US Patents 1979–1981. **Patricia A. Dorler, ed.,** 1982.

Lasers. Vol. 1. US Patents 1978–1981. **Rick W. Myrick, ed.,** 1983.

These handbooks deal with isocyanate polymers and lasers as search tools designed to aid scientists, inventors, and information specialists. Contains abstracts of several patent documents. This first of a series of books was designed with the user in mind.

Patent It Yourself! How to Protect, Patent and Market Your Inventions. **David Pressman.** New York: McGraw-Hill, 1979.

This guide for inventors outlines legal and commercial steps to take to safeguard and market ideas. Chapters include record keeping, patent searching, deciding whether ideas will sell, and more.
R: *Chemical Engineering* 87: 16 (Jan. 14, 1980).

Searching for Foreign Patents. **Bert W. Whitehurst.** Boston: Galleon Publications, 1978.

For beginners and others, a step-by-step guide to searching foreign patents. Includes a diagrammatic foreign patent searching strategy, and covers mechanical, chemical, electrical, plant, and design patents. Useful for engineering libraries.
R: *RSR* 8: 28 (Jan./Mar. 1980).

What Every Engineer Should Know About Patents. **William G. Konold et al.** New York: Dekker, 1979.

A book on patents and patent law. Covers what can be patented, patent searches, patent applications, interferences, rights and their enforcement, licensing, and more. Briefly covers trade secrets, copyrights, and trademarks.
R: *Chemical Engineering* 87: 16 (Jan. 14, 1980); *Laboratory Practice* 29: 35 (Jan. 1980).

US PATENTS

Index to Classification. **US Patent Office.** Washington, DC: US Government Printing Office, 1947–.

Index to classification by subject descriptor.

Index to Patents Issued from the United States Patent and Trademark Office. **US Patent Office.** Washington, DC: US Government Printing Office, 1920–. Annual.

Two main sections: part 1, an alphabetical list of patentees, persons, or companies with new or reissued patents; part 2, index of subjects of inventions arranged according to Patent Office classification.

Index to the US Patent Classification. **US Patent and Trademark Office.** Washington, DC: US Patent and Trademark Office, 1977.

An alphabetical listing of subject headings of the patent classification system.

Index of Trademarks Issued from the United States Patent Office. **US Patent Office.** Washington, DC: US Government Printing Office, 1927–. Annual.

Official Gazette of the United States Patent and Trademark Office. **US Patent and Trademark Office.** Washington, DC: US Government Printing Office, 1872–. Weekly.

Lists and briefly describes patents on a weekly basis. Includes sections on notices, suits, and reissues of former patents. Numerical listing under 4 groups: general and mechanical; chemical; electrical; and design. Includes numbers, title, inventor, and assignees. Patentee, classified, and geographical indexes. Most comprehensive US source.

US Patent Previews. New York: Bowker, 1965–1970.

Lists 60,000 patent applications.

INTERNATIONAL PATENTS

Only major international sources of patent information have been included here. For information on the patent offices of the major industrial countries and their official journals, consult the front section of the patent index in *Chemical Abstracts*.

IINPADOC (International Patent Documentation Center). Arlington, VA: IFI/ Plenum.

A microfiche service offered in collaboration with World Intellectual Property Organization, covering patent family, classification, and applicant services.

The International Index of Patents. Williamsport, PA: Bro-Dart Books, 1964–.

Not limited to general science.

Derwent Publications, Ltd. Documentation Services. London.

This firm is the publisher of the most important current-awareness services for worldwide patent literature. The database contains patent specifications from 24 countries and is added to at the rate of some 12,000 weekly. For machine-readable databases see the chapter on databases.

The following are Derwent's most significant publications in the field:

Central Patents Index—CPI. 1970–. Weekly.

Two "Alerting Bulletins," Arranged by country and by systematic classification. Includes patent number, CPI class, patentee, and basic number or patent concordance indexes.

General Patents Index—GPI. 1970–. Weekly.

Contains 3 printed abstracts journals: *P, General, Q, Mechanical, R, Electrical*. Also issued on cards and microfilm.

World Patents Index. 1974–. Weekly.

General, electrical, mechanical, and chemical sections. Easy access through numerous indexes.

SUBJECT GUIDE TO PATENTS (SELECTIVE)

Science

Chemical Abstracts Patent Index and Concordance. Columbus, OH: American Chemical Society.

One of the number of indexes issued annually in *Chemical Abstracts*. A major tool to chemical patent literature.

Commercial Food Patents, US, 1979. **Hallie B. North.** Westport, CT: AVI Publishing, Annual.

Presents indexed and subject-classified abstracts of recent US food patents issued in 1979. Stresses composition, processing, and packaging of food items in commercial

plants. Classification headings include cereal grains and flours, confections, bakery produces, and alcoholic beverages. Abstracts include diagrams and charts. Indexed by inventor, patent number, and patent assignee. For those in the food-processing industry, government agencies, and university researchers.
R: *ARBA* (1981, p. 762).

International Patents Digest of Foamed Plastics: Guide to the Published Patent Literature for 1971–73. **Technomic Research Staff.** Westport, CT: Technomic, 1975.

Liquid Fertilizers. **M. S. Casper.** Park Ridge, NJ: Noyes Data, 1973.

Information on this source is based on US patents since 1960, relating to the manufacture and application of liquid fertilizers. Includes company, inventor, and US patent indexes.
R: *ARBA* (1974, p. 687).

Oceanic Patents. **Evelyn Sinka.** La Jolla, CA: Ocean Engineering Information Center, 1969.

Contains full information on patents on equipment, techniques, and applications in oceanography and related fields.

Patenting in the Biological Sciences: A Practical Guide for Research Scientists in Biotechnology and the Pharmaceutical and Agrochemical Industries. **R. S. Crespi.** New York: Wiley, 1982.

An introductory guide, presented in a simple and comprehensive manner, for biological scientists in interpreting and using the basic patent laws with their research.
R: *Nature* 298: 587 (Aug. 5, 1982).

Patents for Chemical Inventions. Advances in Chemistry Series, 46. **American Chemical Society.** Edited by Elmer J. Lawson and Edmund A. Godula. Washington, DC: American Chemical Society, 1964.

A collection of symposium papers on chemical patent law for research chemists and managers.

Science Abstracts Part A: Physics Abstracts. London: Institution of Electrical Engineers.

In every semiannual cumulative author index, numerous patents can be located in the section entitled "Small Index."

Understanding Chemical Patents: A Guide for the Inventor. **John T. Maynard.** Washington, DC: American Chemical Society, 1978.

A highly readable guide to the basic philosophy of patent systems. Includes useful glossary and up-to-date bibliography. Highly recommended for the research chemist.
R: *The Chemist* 55: 21 (July 1978); *Journal of Coatings Technology* 50: 126 (Dec. 1978); *Rubber Chemistry and Technology* 51: G66 (July–Aug. 1978); *TBRI* 44: 287 (Oct. 1978); 44: 368 (Dec. 1978); 45: 128 (Apr. 1979).

Uniterm Index to US Chemical and Chemically Related Patents. Vols. 1–. New York: IFI/Plenum, 1950–. Bimonthly.

Separate patentee and assignee indexes.

Engineering

Commercial Food Patents, US. **Hallie B. Nort.** Westport, CT: AVI. Annual.

Dyeing of Synthetic Fibers: Recent Devvelopments. **Keith Johnson.** Park Ridge, NJ: Noyes Data, 1974.

Information on synthetic fibers based on US patents.

Genetic Engineering/Biotechnology Patents: 1980–1981. **Loren Hickman, ed.-in-chief and McGraw-Hill Special Publications Center, eds.** New York: McGraw-Hill, 1982.

An easy-to-use sourcebook containing a listing of genetic engineering and biotechnological patents with the full texts and detailed descriptions of some 40 key patents.

A Guide to Literature and Patents Concerning PVC Technology. 2d ed. **L. R. Whittington, ed.** Stamford, CT: Society of the Plastics Industry, 1964.

Includes 18,000 informative abstracts of chemical literature and patents, 1838–1962. Lists libraries carrying card files of US patents.

R: *British Plastics* 37: 628 (Nov. 1964); Wal (p. 537).

Mineral Exploration, Mining, and Processing Patents, 1975. **Oliver S. North.** Arlington, VA: North, 1976.

NASA Patents Abstracts Bibliography. Washington, DC: US Government Printing Office, 1969–. Semiannual.

A semiannual publication listing abstracts of NASA-owned inventions covered by US patents or applications. Published as a service to firms and individuals seeking new and licensable products for the commercial market. Index is cumulative, and abstracts section covers 6-month periods.

Nonwoven Fabric Technology. **M. McDonald.** Park Ridge, NJ: Noyes Data, 1971.

An industrially oriented summary of some important US patents concerning the state of the art in nonwoven technology.

Patent Abstract Series No. 5: Electrical and Electronic Apparatus: Government Owned Inventions Available for License. **US Department of Commerce, Office of Technical Services.** Washington, DC: US Government Printing Office, (n.d.).

Patent Abstracts: Solid Waste Management, 1945–1969. Part 1: United States; Part 2: International. **US Environmental Protection Agency.** Washington, DC: US Government Printing Office 1973.

Diagrams, charts, and drawings accompany the short abstracts of the patents. Includes patent number, inventor, and assignee index.

Pollution Control in the Petroleum Industry. Pollution Technology Review, no. 4. **H. R. Jones.** Park Ridge, NJ: Noyes Data, 1973.

Information on patented processes as well as data drawn from various sources of government reports and patents.

It is worth keeping in mind that there are several publishers of subject sources based on US patent information. The Noyes Data Corporation of Park Ridge, New Jersey is one. The following is a sample list of these kinds of patent publications from Noyes Data Corporation:

Fresh Meat Processing. **E. Karmas.** 1970.

Fruit Processing. **Milton Gutterson.** 1971.

Poultry Processing. **G. H. Weiss.** 1971.

Sausage Processing. **E. Karmas.** 1972.

Vegetable Processing. **Milton Gutterson.** 1971.

STANDARDS

Standardization, for our purposes, applies to the rules, techniques, and various conditions which must be adhered to in industrial practices, trading conditions, units of measurement, engineering design, terminology, and so on. Standards may be voluntary or mandatory, a condition that varies from country to country, and within each country, depending on what sector of the national economy is involved. Interestingly enough, the American Society for Testing and Materials defines standardization as "a democratic procedure for evolving accepted rules of behavior." Hence the rules for ensuring the performance of activities in an orderly way are contingent on the cooperation of all concerned, be they producers, consumers, or general-interest groups.

While many organizations are involved in providing standards and recommended practices throughout American government, trade, and industry, three may be considered to be in the vanguard of such activities.

American National Standards Institute (ANSI)
1430 Broadway, New York, NY 10018

Selected publications: *The Annual Catalog; The Magazine of Standards,* a quarterly journal; *The ANSI Reporter,* a biweekly newsletter; and *Standards Action,* a bimonthly journal.

American Society for Testing and Materials (ASTM)
1916 Race St., Philadelphia, PA 19103

Selected publications: *Book of ASTM Standards; ASTM Standardization News,* monthly; *Journal of Testing and Evaluation,* 6/yr.; *Special Technical Publications; ASTM Proceedings,* annually.

National Bureau of Standards (NBS)
Washington, DC 20245

Selected publications: *Circulars; Dimension/NBS,* monthly; *Journal of Research:* Section A, *Physics and Chemistry,* bimonthly; section B, *Mathematics and Mathematical Physics,* quarterly; section C, *Engineering and Instrumentation,* quarterly; *Technical News Bulletin.*

On the international and foreign scene, the International Organization for Standardization (ISO) is one of the most important organizations, and every country usually has at least one national standards organization that is responsible for the nation's standardization activities. For example, Deutsches Institut für Normung e V (DIN), or the German Institute for Standardization, is a nonprofit, independent agency that coordinates all standardization work in the Federal Republic of Germany. DIN is analogous to the American National Standards Institute (ANSI) and coordinates about 19,500 DIN existing standards and 4,800 draft ones. All DIN standardization literature is published by Beuth Verlag in Berlin; several of its publications have been selected for inclusion in the following sections for illustration purpose.

GUIDES TO STANDARDS

Directory of United States Standardization Activities. NBS Special Publication, no. 417. **US National Bureau of Standards.** Washington, DC: US Government Printing Office, 1975.
R: Sheehy (EA47).

Guide to Specifications and Standards of the Federal Government. **US General Services Administration.** Washington, DC: US Government Printing Office, 1963–.
A general guide to the many specifications and standards set by the federal government.

Index and Directory of US Industry Standards. **Information Handling Services.** Englewood, CO: Information Handling Services, 1985–. Annual.

Standards and Specifications Information Sources, MIG No. 6. **Erasmus J. Struglia.** Detroit: Gale Research, 1965–.
Annotated bibliography of literature of standardization, catalogs and indexes of standards and specifications, and guides to periodical indexes that include material on standards. Author-title and subject indexes.
R: SL 57: 260 (1966); Jenkins (K206); Win (2EA42).

Standards for Engineering Qualifications: A Comparative Study in Eighteen European Countries. New York: Unipub, 1975–.
Primarily aimed at providing basic qualification standards for different categories of technological careers and establishing criteria for evaluating education programs.
R: *IBID* 3: 155 (Sept. 1975).

Standards in Nuclear Science and Technology: A Bibliography. TID-3336. **US Atomic Energy Commission.** Springfield, VA: National Technical Information Service, 1973.

Technical Information Sources: A Guide to Patent Specifications, Standards and Technical Reports Literature. 2d ed. **Bernard Houghton.** Hamden, CT: Shoe String Press, 1972.

First edition, 1967.
Five chapters on patents, three on standards, and two on technical-report literature for the library student and professional. Emphasizes British publications but presents suitable coverage of US sources. Useful bibliographies.
R: *AL* 3: 926 (Sept. 1972); *ARBA* (1973, p. 626); Jenkins (pp. 2, 23); Wal (p. 240); Win (2EA40).

Technical Standards: An Introduction for Librarians. **Walt Crawford.** White Plains, NY: Knowledge Industry, 1985.

A guide for librarians, writers, and researchers on the use of technical standards. Includes glossary and bibliography.

MAJOR BIBLIOGRAPHIC TOOLS—GENERAL

ANSI Catalog. New York: American National Standards Institute, 1923–. Annual.

Latest edition, 1976. Present name since 1969.
Supplemented by the *Listing of New and Revised American National Standards,* alternate months. More than 6,000 American and 4,000 international standards approved by ANSI. The 1977 catalog began a new format by issuing 2 catalogs: one listing American National Standards and a second, international standards and recommendations. Both catalogs list standards by subject and by the sponsor's designation.

ASTM Standards in Building Codes. Philadelphia: American Society for Testing and Materials, 1962–.

Over half of the standards in this edition have been approved by the US American Standards Institute.

Book of ASTM Standards. Philadelphia: American Society for Testing Materials, 1939–. Annual.

While the adoption of ASTM standards is purely voluntary, they do represent a common ground between producers and consumers. The 48 parts contain current ASTM standards and tentative specifications, test methods, recommended practices, definitions, proposed methods, etc. Each annual supersedes the previous edition. Essential material for the engineer.
R: *ARBA* (1976, p. 761; 1971, p. 545); Jenkins (K201); Wal (pp. 534, 537); Win (EA204).

British Standards Yearbook. London: British Standards Institution, 1937–. Annual.

Complete annotated list of British standards, plus handbooks and codes of practice. Kept up to date by *B&I News.*

Canadian Standards Association. Catalogue of Standards. Rexdale, Ontario: Canadian Standards Association. Annual.

Catalogue, English Translations of German Standards. Berlin: Beuth Verlag. Distr. Philadelphia: Heyden. Annual.

Twentieth edition, 1984.

A catalog of all currently available Deutsches Institut für Normung e V (DIN) standards in English translation, arranged by subject groups based on the universal decimal classification. Indexed by DIN number and by keywords in the title.

R: *ARBA* (1985, p. 538).

DIN Catalogue of Technical Rules. 2 vols. Berlin: Beuth Verlag. Distr. Philadelphia: Heyden. Annual.

The latest edition, 1984.

This annual 2-volume catalog contains lists of approximately 36,000 standards, codes of practice, instruction sheets, and statutory regulations issued by the Federal Republic of Germany and taken from 68 collections of German technical rules in engineering and technology. Indexed by German subject, numerical number, English subject, etc.

R: *ARBA* (1985, p. 539).

Federal Information Processing Standards Index, National Bureau of Standards FIPS-Pub-12-1. **US National Bureau of Standards.** Washington, DC: US Government Printing Office, 1972.

A bibliography, directory, and handbook in one package.

An Index of International Standards. **Sophie Cnumas, ed.** Prepared for the US National Bureau of Standards. Washington, DC: US Government Printing Office, 1974.

Computer-produced index containing almost 3,000 standard titles of the ISO, IEC, CEE, OIML, and CISPR.

R: *ARBA* (1975, p. 770); Sheehy (EA48).

Index of Specifications and Standards. **US Department of Defense.** Washington, DC: US Government Printing Office, 1951–. Monthly.

Intended for suppliers of the military establishment.

An Index to US Voluntary Engineering Standards, NBS Special Publication 329. **W. J. Slattery, ed.** Washington, DC: National Bureau of Standards, 1971.

Covers standards, specifications, and text methods issued by national standardization organizations in the United States.

NBS Standard Reference Materials Catalog. Washington, DC: US National Bureau of Standards. Distr. Washington, DC: US Government Printing Office.

A listing of more than 1,000 certified chemical composition standards, physical property standards, and engineering-type standards. Materials are grouped under 3 cate-

gories: standard reference materials, research materials, and special reference materials. Includes a price list. For academic research libraries.
R: *ARBA* (1981, p. 756).

Publication of the NBS. **US National Bureau of Standards.** Washington, DC: US Government Printing Office.
Publications for 1901–1947, 1947–1957, 1957–1960, etc. Available also with annual supplements.

US Federal Supply Service: Index of Federal Specifications and Standards. **US General Services Administration.** Washington, DC: US Government Printing Office, 1952–. Monthly.
A guide to the standards and specifications to which suppliers of goods to the US Government must conform. Arranged by subject. Does not include military suppliers.

Subject Guides to Standards

Professional societies and organizations at all levels are major sponsoring bodies of numerous standards in subject areas. While a small percentage of them have been approved by ANSI and thus are incorporated and included in the *ANSI Catalog*, a great majority of them can be readily identified only through the use of the sponsoring organizations' standards catalogs, such as the *IEEE Standards*, New York, Institute of Electrical and Electronics Engineers, and so on.

On the other hand, journals, particularly official news publications of those professional societies that are involved in standardization activities, are essential information sources of current standards. For example, *Nuclear News*, the monthly official news journal of the American Nuclear Society, has a "Standards" column in every issue.

Besides the above-mentioned subject guides, the following works are indicative of titles that are available on particular applications of standardizations. ASTM has advanced the concept that these applications fit most frequently into 4 major categories: (1) units of measurement; (2) terminology and symbolic representation; (3) products and processes; and (4) safety of persons and goods.

As a majority of publications seem to fall within the "products and processes" category, a subject breakdown has been included for this section.

Units of Measurement

See also SI Units in tables section.

Electric Units and Standards. **P. Vigoreux.** London: Her Majesty's Stationery Office, 1970.

Metric Standards for Engineering. London: British Standards Institution, 1966.
Based on existing British standards and specifications used by traditional metric-issuing countries.

Model State Packaging and Labelling Regulations, as Adopted by the National Conference on Weights and Measures. Washington, DC: US National Bureau of Standards, 1972.

Metrication Handbook: The Modernized Metric System Explained. Neenah, WI: Keller, 1975.

Provides history and basic information regarding the metric system and its use in the United States. Helpful for the undergraduate and the layperson.
R: *ARBA* (1977, p. 635).

Quantities, Units, and Symbols. 2d ed. **Royal Society of London, Symbols Committee.** London: Royal Society of London, 1975.

Emphasizes standard symbols in physics, though it includes some chemical and mathematical symbols as well. Contains a short list of physical constants and a bibliography.
R: *RSR* 5: 17 (Apr./June 1977).

SI Metric Handbook. **John L. Feirer and The Metric Company.** New York: Scribner's, 1977.

A reference book in 2 parts: the metric system, and international standards. Contains conversion tables, charts, applied metrics. For high school and college students.
R: *LJ* 102: 937 (Apr. 15, 1977).

SI Units in Engineering and Technology. **S. H. Qasim.** Oxford: Pergamon, 1977.

An alphabetical listing of conversion units. Helpful for quick retrieval.
R: *Aslib Proceedings* 42: 446 (Aug. 1977).

Units of Weight and Measure: International (Metric) and US Customary, National Bureau of Standards Miscellaneous Publication 286. **L. J. Chisholm.** Washington, DC: National Bureau of Standards. Reprint. Detroit: Gale Research, 1975.

Provides conversions from metric to US and US to metric.
R: *ARBA* (1976, p. 634).

VNR Metric Handbook of Architectural Standards. **Patricia Tutt and David Adler.** New York: Van Nostrand Reinhold, 1979.

A summary of design standards for specific types of buildings. Discusses other building features including heating, lighting, ventilation, and landscaping. Intended for British architects; standards are in metric units.
R: *ARBA* (1981, p. 759).

World Metric Standards for Engineering. **Knut O. Kverneland.** New York: Industrial Press, 1978.

Includes standard data on metric conversion as reflected by the International Organization for Standardization. Contains helpful index and tables. One of the few single-volume sources of internationally accepted standards.
R: *Journal of Metals* 31: 52 (Nov. 1979); *Mechanical Engineering* 100: 96 (Dec. 1978); *TBRI* 45: 74 (Feb. 1979); *ARBA* (1979, p. 764).

Terminology and Symbolic Representation

Abbreviations for Use on Drawings and in Text. **American National Standards Institute.** New York: American Society of Mechanical Engineers, 1972.

Attempts to establish standard abbreviations for words and terms in engineering and scientific writing. Much of this material is a revision of American National Standard's *Abbreviations for Use on Drawings,* 1950, and *Abbreviations for Scientific and Engineering Terms,* 1941.
R: *ARBA* (1974, p. 664).

Abbreviations for Use on Drawings and in Text. **American National Standards Institute.** New York: American Society of Mechanical Engineers, 1972.

Attempts to establish standard abbreviations for words and terms in engineering and scientific writing.

American National Standards Vocabulary for Information Processing. New York: American National Standard Institute, 1970.

Encompasses more than 1,200 terms and definitions.

Architectural Graphic Standards. 6th ed. **C. G. Ramsey and H. R. Sleeper.** New York: Wiley, 1981.

Sixth edition, 1970.
Standards provided for draftsmen and engineers.

Companies Holding Certificates of Authorization for Use of Code Symbol Stamps. **Boiler and Pressure Vessel Committee, American Society of Mechanical Engineers.** New York: ASME, 1972.

IEEE Standard Dictionary of Electrical and Electronics Terms. New York: Wiley-Interscience, 1972.

About 13,000 technical terms used by electrical and electronics engineers. Terms originated in *IEEE Standards, American National Standards,* and *IEC Recommendations.*

International Stratigraphic Guide: A Guide to Stratigraphic Classification, Terminology, and Procedure. **International Union of Geological Sciences, International Subcommission on Stratigraphic Classification.** New York: Wiley, 1976.

An international standard that replaces national and regional codes of stratigraphic use. Considered an essential reference; contains bibliography.
R: Sheehy (EE6).

Standard Graphical Symbols: A Comprehensive Guide for Use in Industry, Engineering and Science. **A. Arnell.** New York: McGraw-Hill, 1963.

A compendium of symbols approved as standards by eminent scientific and engineering institutions. Some 9,000 illustrations and symbols.

USA Standard Glossary of Terms in Nuclear Science and Technology. **Atomic International Forum.** New York: American Standards Institute, 1967.

Emphasis on highly specific terms and definitions.

Safety of Persons and Goods

Code of Safety for Fishermen and Fishing Vessels. 2 vols. London: Inter-Governmental Maritime Consultive Organization, 1975.

R: *IBID* 4: 145 (June 1976).

Designer's Guide to OSHA: A Design Manual for Architects, Engineers, and Builders to the Occupational Safety and Health Act. **Peter S. Hopf.** New York: McGraw-Hill, 1975.

Over 500 of the most frequently confronted federal standards are interpreted and illustrated. Intended primarily for independent businessmen and designers.

R: *ARBA* (1976, p. 761).

Evaluation of the Microbiology Standards for Drinking Water. **US Environmental Protection Agency, Office of Drinking Water.** Washington, DC: US Environmental Protection Agency, 1978.

Food Chemicals Codex. 3d ed. **US National Academy of Sciences, National Research Council, Committee on Food Protection.** Washington, DC: US Government Printing Office, 1981.

Second edition, 1972.

Includes 639 monographs. Deals not only with food additives, but also with substances that are not added directly to food but come into contact with foods, such as food-processing aids. Presents standards, descriptions, and identification methods.

R: *Journal of Pharmaceutical Sciences,* 62: 347 (Feb. 1973).

IAEA's Safety Standards and Measures. **International Atomic Energy Agency.** New York: Unipub, 1976.

Coverage includes definitions, general information, application of safety standards and measures to assisted operations, and changes in safety standards and measures.

R: *IBID* 4: 241 (Sept. 1976).

An Index to the Hazardous Materials Regulations. **US Department of Transportation.** Washington, DC: US Government Printing Office, 1972.

R: *ARBA* (1973, p. 658).

International Maritime Dangerous Goods Code. 4 vols. **Intergovernmental Maritime Consultative Organization.** New York: Unipub, 1977.

Contains packing regulations; includes illustrations and a glossary.

R: *IBID* 6: 315 (Sept. 1978).

National Fire Codes: A Compilation of NFPA Codes, Standards, Recommended Practices, and Manuals. 18 vols. Boston: National Fire Protection Association. Annual.

A main source of regulations for hazardous materials.

Regulations for the Safe Transport of Radioactive Materials. Safety series, 6. Rev. ed. **International Atomic Energy Agency and the World Health Organization, spons.** New York: Unipub, 1980.

First edition, 1974.
R: *IBID* 4: 149 (June 1976).

Specifications for Identity of Food Colours, Flavouring Agents and Other Food Additives. **Food and Agriculture Organization of the United Nations.** Rome: Food and Agriculture Organization of the United Nations, 1979.

A listing of specifications and tests for identification of 68 substances. Examines analytical methods, conditions for gas-liquid chromatographic assays, determination of aromatic hydrocarbons, etc. Useful for chemists and food scientists.
R: *IBID* 8: 16 (Spring 1980).

United States Food Laws, Regulations, and Standards. **Y. H. Hui.** New York: Wiley, 1979.

Contains up-to-date standards for food laws and regulations. Discusses agencies and their structure, food inspection, publications, governmental responsibilities, and controversial microbiological quality standards. Also acts as a guide for further study. Of interest to exporters, practitioners, college students, and anyone wishing to know more about food laws.
R: *Chemical Engineering* 87: 14 (Jan. 14, 1980).

Products and Processes

Biology

Biological Substances: International Standards, Reference Preparations, and Reference Reagents, 1979. Geneva: World Health Organization, 1979.

Consists of revised listings of international biological standards, international biological reference reagents, and international biological reference preparations. Substances are alphabetically arranged within specific groups (antibiotics, antigens, blood products, etc.). For science libraries.
R: *IBID* 7: 363 (Winter 1979).

Lists of International Biological Standards, International Biological Reference Preparations, and International Biological Reference Reagents, 1982. Albany, NY: World Health Organization Publication Centre USA, 1982.
R: *Science* 218: 402 (Oct. 1982).

Chemistry

Reagent Chemicals: American Chemical Society Specifications. 5th ed. **American Chemical Society, Committee on Analytical Reagents.** Washington, DC: American Chemical Society, 1974.
First edition, 1950; third edition, 1960; fourth edition, 1968.
Updates information in previous volumes; serves as a precise analytical work. Also includes much new data.
R: *Chemical Engineering News* 47: 76 (Oct. 14, 1968); *Choice* 5: 1423 (Jan. 1969); *RSR* 5: 11 (Apr./June 1977); Jenkins (D162); Win (2ED7).

Aeronautical Engineering

SAE Aerospace Index of Aerospace Standards, Aerospace Recommended Practice, Aerospace Information Report. **Society of Automotive Engineers.** New York: Society of Automotive Engineers, 1975.
A numerical index of all currently available SAE aerospace standards, recommended practices, and information reports. A subject index and an index of cancelled items.

NASA Specifications and Standards. NASA-SP-9000. **US National Aeronautics and Space Administration.** Springfield, VA: National Technical Information Service, 1967–.
Three sections: alphanumerical list of standards and specifications; subject index; originator index.

Chemical Engineering

ASTM Standards Relating to Miscellaneous Petroleum Products. Philadelphia: American Society for Testing and Materials, 1975–.

Product Standards Index. **V. L. Roberts.** New York: Pergamon Press, 1977.
An index with information on 144 organizations concerned with consumer protection here and abroad. Standards are listed by subject classification with numerous cross references. Also includes a topical index and an index listing organizations. Simplifies identifying standards organizations tremendously. Good for engineering libraries.
R: *RSR* 8: 26 (Jan./Mar. 1980).

World Index of Plastics Standards. **Leslie H. Breden, ed.** Washington, DC: National Bureau of Standards. Avail. US Government Printing Office, 1971.
A computer-produced index including 9,000 national and international standards for plastics and related materials.

Civil Engineering

BS Handbook 3: Summaries of British Standards for Building. 4 vols. **British Standards Institution.** London: British Standards Institution, 1977.

A summary of 1,349 standards, such as product, codes of practice, glossaries, methods of tests, and lists of symbols to be used in the construction industry. Not intended as a guide for manufacturing or testing.
R: British Standards Institute *Yearbook* (1978, p. 702); Wal (p. 625).

Energy Conservation Standards for Building Design, Construction, and Operation. **Fred S. Dubin and Chalmers G. Long, Jr.** New York: McGraw-Hill, 1978.
An outgrowth of a set of manuals written for the Federal Energy Administration. Offers a detailed checklist for the design of energy-management programs. Intended for use by architects, engineers, etc. Considered a comprehensive source of reference. Presents energy standards helpful in conservation; includes metric conversion, glossary, and bibliography.
R: *Plant Engineering* 33: 156 (June 28, 1979); *TBRI* 45: 275 (Sept. 1979).

Standards in Building Codes. 15th ed. 2 vols. Philadelphia: American Society for Testing and Materials, 1980. Also newer edition.
A review of all building standards with indications of those approved by the USA Standards and the USA Standards Institute.

Structural Design Guide to the ACI Building Code. 2d ed. **P. F. Rice and Edward S. Hoffman.** New York: Van Nostrand Reinhold, 1979.
First edition, 1972.

Uniform Building Code and 1975 Accumulative Supplement. 7 vols. Whittier, CA: International Conference of Building Officials, 1972–1975.

The following standards are published by the American Association of State Highway Officials, Washington, DC:

Guide Specification for Highway Constructions.

Standard Specifications for Highway Bridges.

Standard Specifications for Movable Highway Bridges.

Standard Specifications for Transportation Materials and Methods of Sampling and Testing.

Standard Specifications for Welding of Structural Steel Highway Bridges.

Electrical Engineering

Changes in the National Electrical Code. **George W. Flach.** Englewood Cliffs, NJ: Prentice-Hall. Annual.
The book includes the latest changes in the National Electrical Code. Includes tables and examples. For special libraries.

EIA Standards. New York: Electronics Industries Association.
Emphasis on the areas of solid-state devices, electron tubes, component parts, consumer and government products.

Guide to the National Electrical Code. **Thomas Harman and Charles Allen.** Englewood Cliffs, NJ: Prentice-Hall. Annual.

IEEE Standard Atlas Test Language. 2d ed. **Institute of Electrical and Electronics Engineers.** New York: Institute of Electrical and Electronics Engineers, 1979.

Aids in understanding test specifications and procedures in commercial, industrial, and military applications.

IEEE Standard Techniques for High-Voltage Testing. 6th ed. New York: Institute of Electrical and Electronics Engineers, 1978.

A revised reference for standard IEC high-voltage testing. Includes wet test procedure and application guide for measuring devices.

National Electrical Code. Boston: National Fire Protection Agency. Annual.

Actual code as followed by practicing electricians.

National Electrical Code and Blueprint Reading. 10th ed. **Kenneth Gebert.** Chicago: American Technical Society, 1983.

Sixth edition, 1972; eighth edition, 1977.

National Electrical Code Handbook. **W. I. Summers.** Boston: National Fire Protection Agency. Annual.

National Electrical Code Reference Book. 4th ed. **J. D. Garland.** Englewood Cliffs, NJ: Prentice-Hall, 1984.

Second edition, 1978; third edition, 1981.
Covers the 4,000 rules of the National Electrical Code. Includes comprehensive discussion, fundamentals of electricity, diagrams, and illustrations. An essential reference.

National Electrical Safety Code. **Institute of Electrical and Electronics Engineers.** New York: Wiley, 1984.

Periodically revised.
A basic reference source for those requiring information on safe installation of electric supply lines. A standard reference for engineering and science libraries.
R: *Choice* 15: 106 (Mar. 1978).

Computer Technology

IEEE Standards for Local Area Networks: Carrier Sense Multiple Access with Collision Detection (CSMA/CD) Access Method and Physical Layer Specifications. **Institute of Electrical and Electronics Engineers.** New York: Wiley, 1985.

IEEE Standards for Local Area Networks: Token-Passing Bus Access Method and Physical Layer Specifications. **Institute of Electrical** and Electronics Engineers. New York: Wiley, 1985.

IEEE Standard Logical Link Control–Local Area Networks Standard 802.2. **Institute of Electrical and Electronics Engineers.** New York: Wiley, 1985.

McGraw-Hill's Compilation of Data Communications Standards. **Harold C. Folts and Harry R. Karp, eds.** New York: McGraw-Hill, 1978.

A compilation of data communications standards from 5 groups: US government; International Telegraph and Telephone Consultative Committee; International Organization for Standardization; American National Standards Institute; and the Electronics Industries Association. Contains detailed cross references. Recommended for network users, manufacturers, and telecommunication carriers.
R: *Choice* 16: 1054 (Oct. 1979); *ARBA* (1980, p. 718);

Software Engineering Standards. **Institute of Electrical and Electronics Engineers.** New York: Wiley, 1985.

This is a complete software standards set forth by IEEE. Includes the Standard Glossary of Software Engineering Terminology, the Standard for Software Quality Assurance Plans, the Standard for Software Configuration Management Plans, etc.

Standardized Development of Computer Software. Standards. Pt 2. **Robert C. Tausworthe.** Englewood Cliffs, NJ: Prentice-Hall, 1979.

Provides detailed standards for computer program design. Discusses quality assurance, level of documentation, and production systems. For computer science collections.

A Technical Guide to Computer-Communications Interface Standards. **US National Bureau of Standards.** Washington, DC: US Government Printing Office, 1974–.

A technical summary for federal standards on data communications pertaining to computer networking. Includes some international standards and industry practices.
R: *ARBA* (1975, p. 793).

Marine Engineering

International Safety Standard Guideline for the Operation of Undersea Vehicles. **Marine Technology Society.** Washington, DC: Marine Technology Society, 1979.

Rules and Regulations for the Construction and Classification of Steel Ships. London: Lloyd's Register of Shipping. Annual.

Rules for Building and Classing Steel Barges for Offshore Service, 1976. New York: American Bureau of Shipping. Annual.

Rules for Building and Classing Steel Vessels. New York: American Bureau of Shipping. Annual.

Rules for the Construction and Classification of Mobile Offshore Units. London: Lloyd's Register of Shipping. Annual.

Mechanical Engineering

ASHRAE Standards. New York: American Society of Heating, Refrigeration, and Air-Conditioning Engineers, 1958–. Loose-leaf.

Periodically updated collection of standards, methods of rating, and testing and recommended practices.

ASME Standards. New York: American Society of Mechanical Engineers. Irregular.

Publication containing specific recommended practices.

DIN Standarization Handbooks. Berlin: Beuth Verlag. Distr. Philadelphia: Heyden.

The following 6 are presently available in English translations, and more are promised from 1984–. No. 1, *Mechanical Standards;* no. 4, *Iron and Steel Quality Standards;* no. 8, *Welding 1;* no. 15, *Steel Pipelines;* no. 28, *Iron and Steel Dimensional Standards;* and no. 155, *Iron and Steel Quality Standards, 2.*

Engineering Specifications and Statistical Issue. **Automotive Industries.** Radnor, PA: Chilton. Annual.

Fifty-seventh edition, 1975.

Handbook of Comparative World Steel Standards. **International Technical Information Institute.** Tokyo: International Technical Information Institute, 1976.

Contains data on international steel standards. Highly useful for engineering and business libraries.
R: *RSR* 5: 36 (Oct./Dec. 1977).

International Guide to Screw Threads: Symbols, Profiles, and Designations of Threads in Standards of Various Countries. **Hans-Peter Grode and Manfred Kaufmann, eds.** Berlin: Beuth Verlag. Distr. Philadelphia: Heyden, 1983.

A useful specialized reference tool for engineering libraries. Indexed by subject and by keyword.
R: *ARBA* (1985, p. 539).

Mechanical Standards: Basic Standards. Berlin: Beuth Verlag. Distr. Philadelphia: Heyden, 1984.

Contains basic mechanical standards of Germany. Indexed.
R: *ARBA* (1985, p. 540).

Standard for Classification and Application of Welded Joints for Machinery and Equipment. New ed. **American Welding Society Committee on Machinery and Equipment.** Miami: AWS, 1977–.

Standards and Practices of Instrumentation: A Compilation of Complete ISA Standards and Recommended Practices, Abstracted Standards for Other Organizations,

Key-Word Index. 5th ed. **Mark. T. Yothers, ed.** Pittsburgh: Instrument Society of America, 1977.

First edition, 1963; third edition, 1970.

A compendium of standards relevant to instrumentation. In 3 sections: index to all available information standards; subject index; and complete listing of 45 standards published by the ISA. Comprehensive; considered an essential reference work for all engineering and technical libraries.

R: *ARBA* (1978, p. 775).

Steel Pipelines 1: Standards on Dimensions and Technical Delivery Conditions. Berlin: Beuth Verlag. Distr. Philadelphia: Heyden, 1984.

The first English edition based on the German original. Indexed.

R: *ARBA* (1985, p. 540).

Structural Welding Code. Miami: American Welding Society. Annual.

Ninth edition, 1975.

Welding 1: Standards Dealing with Filler Metals, Manufacture, Quality, and Testing. Berlin: Beuth Verlag. Distr. Philadelphia: Heyden. Annual.

Contains standards dealing with filler metals, manufacture, quality, and testing. Indexed by DIN number and keyword.

R: *ARBA* (1985, p. 540).

Nuclear Engineering

Environmental Radiation Measurement. **National Council on Radiation Protection and Measurements.** Washington, DC: National Council on Radiation Protection and Measurements, 1976.

Presents a systematic set of standards for environmental radiation measurements, emphasizing weak radiation fields and specific radionuclides.

R: *International Journal of Radiation Biology* 32: 307 (Sept. 1977); *TBRI* 43: 370 (Dec. 1977).

Environmental Standard Review Plans for the Environmental Review of Construction Permit Applications for Nuclear Power Plants. **US Office of Nuclear Reactor Regulation.** Washington, DC: US Nuclear Regulatory Commission, 1979.

Includes bibliographical references.

Implementation of Long-Term Environmental Radiation Standards: The Issue of Verification. **US National Academy of Sciences, Committee on Radioactive Waste Management.** Washington, DC: US National Academy of Sciences, 1979.

Index of US Nuclear Standards. **William J. Slattery.** Washington, DC: US National Bureau of Standards. Distr. Washington, DC: US Government Printing Office, 1977.

Contains permuted titles of more than 1,200 nuclear-related standards, specifications, test methods, etc. published by 34 US government agencies, technical societies and

organizations. Entries include title, date of issuance, price, date of approval, acronym, standard number, etc. Also contains alphabetical list of organizations from which standards can be ordered.
R: *ARBA* (1979, p. 697).

National and International Standardization of Radiation Dosimetry. 2 vols. Proceedings of International Symposium held by the International Atomic Energy Agency, Dec. 1977. Vienna: International Atomic Energy Agency, 1978.

In English and French. Discusses laboratory activities, international aspects of dosimetry standardization and measurement of radioprotection. Includes rules and regulations for use of ionized radiation.
R: *IBID* 6: 430 (Dec. 1978); *IBID* 7: 51 (Spring 1979).

Nuclear Safeguards Technology, 1978. 2 vols. New York: Unipub, 1979.

Discusses all facets of safeguards for nuclear materials. Considered a timely evaluation of current research.

Nuclear Standards News. Hinsdale, IL: American Nuclear Society, 1970–. Monthly.

Intended to provide the latest developments in nuclear standards. Distributed to selected libraries and members of ANS's Information Center on Nuclear Standards.

Nuclear IEEE Standards. 2 vols. **Institute of Electrical and Electronics Engineers.** New York: Wiley, 1981.

Volume 1, *Criteria for Protection Systems for Nuclear Power Generating Stations;* volume 2, *IEEE Recommended Practice for Installation Design and Installation of Large Lead Storage Batteries for Generating Stations and Substations.*
In 2 volumes, the complete set of nuclear Institute of Electrical and Electronics Engineers standards. Contains all essential criteria, tests, methods, procedures, safety protection systems, power supplies, instrumentation, etc. For collections in nuclear and particle physics.

CHAPTER 20 TRADE LITERATURE

Trade literature, while often a useful source of health science information, is often ignored by many science librarians. As the quality of such publications increases, however, librarians would do well to appreciate such literature for its information value rather than denigrate it for its commercial overtones. Often it is highly specialized, as in the case of the literature of pharmacy and pharmacology, where trade publications are likely to provide detailed information on the biochemical properties and chemical structure of drugs.

While organizations are naturally amenable to providing details of their products to potential customers, an individual undertaking a deliberate collection of trade literature would start by consulting buyers' guides such as *Thomas' Register of American Manufacturers.* Such sources are available in various formats, including comprehensive listings of manufacturers' names and addresses, specific subject area directories, and directories that include examples of trade literature for each concern.

Also common is the compendium of individual brochures and pamphlets that are reissued as manufacturers' catalogs and distributed to scientists and technicians. As an ample listing of directories is included elsewhere in this book, the reader is advised to consult these works for appropriate companies and manufacturers.

The literature itself takes various forms, including that of journals, bulletins, monographs, audiovisual presentations, handbooks, preprints, and even scholarly publications such as the *IBM Journal of Research and Development.* Much of it is free for the asking, with the notable exception of many publications in the electronics industry. Frequently, companies will also issue annotated guides to their publications with ordering information. The International Nickel Company's irregularly appearing *Review of Available Literature* is one such publication. The following sections include examples of significant trade literature as well as a guide to sources of other free and inexpensive materials.

GUIDE TO TRADE LITERATURE

Free and Inexpensive Materials. 2d ed. **Robert Monahan.** Belmont, CA: Pitman Learning for Preschool and Early Childhood, 1977.

First edition, 1973.

Gebbie House Magazine Directory. Sioux City, IA: House Magazine, 1946–.

Alphabetical, title, geographic, and printers listings. Also arranged by standard industrial classification numbers.

Guide to Special Issues and Indexes of Periodicals. 3d ed. **Miriam Whlan, ed.** New York: Special Libraries Association, 1982.

Second edition, 1976.
A guide to the contents of special issues of American and Canadian trade, technical, and consumer periodicals. Entries are organized alphabetically by journal title with a listing of special features, issues, sections appearing on a recurring basis, and editorial and advertiser indexes. Includes a detailed subject index.

Magazine Industry Market Place. New York: Bowker, Annual.

Latest edition, 1986.
A major directory of the companies and individuals in the magazine business. Over 2,000 consumer, trade, professional, literary, and scholarly periodicals are included. Each entry provides publisher's name, circulation, publication frequency, advertising rates, trim size, cost per issue and per subscription, and a brief description.

TRADE REFERENCE TOOLS

Publications on manufacturers and industrial companies concerning their products, devices, and equipment include all types of information sources, such as handbooks, manuals, catalogs, buyers' guides, directories, etc. Many of these reference sources can be found in this book in the chapters on handbooks, manuals, guides, directories, and so on, and thus will not be repeated again. To illustrate the various types of reference publications available from trade industries, the following selective publications of the RCA Electric Company in Harrison, New Jersey, alone should be adequate for readers to imagine the similar types of publications from companies such as IBM, General Electric, Westinghouse, General Motors, etc.

National Trade and Professional Associations of the United States and Canada and Labor Unions. **Craig Colgate, Jr., ed.** Washington, DC: Columbia Books. Annual.

Vertical File Index: A Subject and Title Index to Selected Pamphlet Material. New York: Wilson, 1932–. Monthly.

World Guide to Trade Associations. 3d ed. **Barbara Verrell, ed.** New York: Saur, 1985.

First edition, 1973; second edition, 1980.
A valuable directory of more than 31,000 national and international trade associations arranged in almost 400 trade categories. For the first time, in this edition, publications of associations are included and over 5,500 periodicals are listed.

RCA Cos-mos Integrated Circuits Manual

RCA Electro-Optics Handbook

RCA Linear Integrated Circuits

RCA Power Transistor Manual

RCA Power Transistors

RCA RF Power Transistor Manual

RCA Receiving Tube Manual

RCA Silicon Controlled Rectifier: Experimenters' Manual

RCA Solid State Hobby Circuits Manual

RCA Solid State Servicing

RCA Transistor, Thyristor, and Diode Manual

SELECTED TRADE JOURNALS

The trade periodical publications from industrial companies and manufacturers are very rich information sources. They range from technical and research publications to more trade-oriented ones. The following are samples of those with more technical contents:

Bell Laboratories Record. 1925–. Monthly.

Bell System Technical Journal. 1922–. 10/year.

Ciba-Geigy Journal. 1957–. Quarterly.

Commodore Microcomputers. Bimonthly.

Computing Reporting in Science and Engineering. 1965–. Quarterly.

Digital Review: The Independent Guide to DEC Computing. Monthly.

Exxon Aviation News Digest. 1947–. Weekly.

IBM Journal of Research and Development. 1957–. Bimonthly.

Lotus: Computing for Managers and Professionals. 1985–. Bimonthly.

Marconi Instruments Contact. 1967–. Quarterly.

Microwave Journal. 1958–. Monthly.

Package Engineering. 1956–. Monthly.

Philips Technical Review. 1936–. Monthly.

PLASDOC—Plastics and Polymers Patents Documentation. 1963–. Weekly.

RCA Review. 1936–. Quarterly.

Shell Aviation News. 1931–. Bimonthly.

Western Electric Engineer. 1957–. Quarterly.

COMMERCIAL PRODUCT INFORMATION

To facilitate the use of information available from catalogs of suppliers to certain industries, it is possible to obtain vendor catalog services from commercially available services. For example, the USA Information Handling Services, Inc. provides a service called VSMF, which supplies cassettes of microfilmed catalogs with index books. The UK Technical Indexes, Ltd. offers similar services.

CHAPTER 21 NONPRINT MATERIALS

As new technology becomes more and more commonly used, there is a clear trend toward the production of more types of information sources in various nonprint formats. These sources range from the more traditional microforms to the newest digital compact disks and videodisks. Nonprint materials of all kinds currently enjoy an important role in the development and dissemination of scientific and technical information and will continue to be even more prominent. In another area of development, electronic publishing, many scientific journals and publications are available in electronic form in addition to the traditional print format. The following sources, both general and specific, are intended as a selective guide to the nonprint information sources.

GENERAL REFERENCE SOURCES

GUIDES AND INDEXES

Audio Video Market Place 1985. New York: Bowker, 1985. Annual.

A 1-volume reference directory to thousands of AV products and services with 5,500 profiles of AV companies. Entries can be found under either geographical or subject categories.

Basic US Government Micrographic Standards and Specifications. **National Microfilm Association.** Silver Springs, MD: NMA, 1976. Also, later edition.

This book contains 464 pages providing copies of 23 standards and specifications, in effect all federal government standards and specifications.

Books on Cassette. New York: Bowker, 1985.

An annotated listing of more than 7,000 titles in 84 subject areas in which all subject areas include science, technology, and computers. Six indexes—author, title, subject, producers, reader/performers, and producers' works—are available.

Dictionary of Audio-Visual Terms. **British Kinomatograph Sound and Television Society.** London: Focal Press, 1983.

About 2,000 terms used in the audiovisual field, including the preparation of pictures and sound by film and video as well as by tape-slide, film-strip, and multivision. Uses some illustrations. For libraries specializing in the field.
R: *Science and Technology Libraries* 4: 122 (Winter 1983).

Educational Film Locator. 3d ed. **The Consortium of University of Film Centers.** New York: Bowker, 1983.

First edition, 1978; second edition, 1980.
A large volume that provides complete access to educational films, including over 40,000 titles, and provides bibliographic information, translated titles, major subject

groupings, brief annotations, rental and buying information, and holdings of film libraries. Indexed by author, title, audience level, and foreign title.

Educational Media Catalogs on Microfiche. 3d ed. **Walter J. Carroll, ed.** New York: Olympic Media, 1984. Updated twice/yr.

First edition, 1978; second edition, 1982.
Collection of supplier catalogs, films, filmstrips, and audiocassettes.

Educational Media Yearbook. **James W. Brown and Shirley N. Brown, eds.** Denver: Libraries Unlimited. Annual.

Tenth edition, 1984.

Educators' Guide to Free Films. **Mary Foley Horkheimer and John C. Diffor, eds.** Randolph, WI: Educators Progress Service. Annual.

Thirty-ninth edition, 1979; forty-fourth edition, 1984.
R: *CRL* 41: 93 (Jan. 1980).

Feature Films on 8mm, 16mm, and Videotape: A Directory of Feature Films Available for Rental, Sale, and Lease in the United States and Canada. 7th ed. **James L. Limbacher.** New York: Bowker, 1982.

A comprehensive guide on where to buy and how to rent 8mm and 16mm films and videotapes. Provides information on 23,000 films in the United States and Canada. Includes a new section on foreign-language films. For public, school, and college libraries.

Guide to Microforms in Print. Washington, DC: NCR Microcard Editions, 1961–. Annual.

Books, journals, etc., available on microforms from US publishers. Does not cover theses and dissertations. For subject approach use *Subject Guide to Microforms in Print*, Washington, DC: NCR Microcard Editions, 1962–1963–, biennial.

Index to Educational Records. **National Information Center for Educational Media.** Los Angeles: University of Southern California, 1971–.

First edition, 1971; fifth edition, 1980.

Index to Educational Slides. **National Information Center for Educational Media.** Los Angeles: University of Southern California, 1971–.

First edition, 1971; second edition, 1974; fourth edition, 1980.

Index to Educational Videotapes. **National Information Center for Educational Media.** Los Angeles: University of Southern California, 1971–.

First edition, 1971; fifth edition, 1980.

Index to 8mm Motion Cartridges. **National Information Center for Educational Media.** Los Angeles: University of Southern California, 1971–.

First edition, 1969, published by R. R. Bowker; second edition, 1971; sixth edition, 1980.

Index to Educational Overhead Transparencies. **National Information Center for Educational Media.** Los Angeles: University of Southern California, 1969–.
Second edition, 1971; sixth edition, 1980.

Index to 16mm Educational Films. **National Information Center for Educational Media.** Los Angeles: University of Southern California, 1971–.
First edition, 1969, published by R. R. Bowker; fifth edition, 1975; seventh edition, 1980.

Index to 35mm Educational Filmstrips. **National Information Center for Educational Media.** Los Angeles: University of Southern California, 1971–.
First edition, 1969, published by Bowker; third edition, 1971; seventh edition, 1980.

International Index to Film Periodicals. **Frances Thorpe, ed.** New York: St. Martin's Press, 1979–.

International Index to Multimedia Information. **Audio-Visual Association.** Monterey Park, CA: AVA. 1970–. Quarterly.
Formerly *Film Review Index.*

International Micrographics Source Book. New York: Microfilm Publishing, 1980–. Annual.
A single-volume directory of international sources for micrographic products, service bureaus, dealers, trade names, publishers, storage centers, etc. Includes bibliography and keyword index. A major reference tool for librarians.

Library of Congress Catalog: Motion Pictures and Filmstrips. **US Library of Congress.** Washington, DC: US Government Printing Office. Annual.

Media Review Digest. Ann Arbor: Pierian Press. Annual.
Volume 10, 1980.
Formerly *Multi Media Reviews Index.*

Microfilm Source Book. 1973–1974 ed. New Rochelle, NY: Microfilm Publishing, 1973.
A single-source reference guide to the microfilm industry, including products and services. Gives the sources of supply for every important microfilm service and piece of equipment. Also included are an industrywide name and address section; trademark/trade name reference guide; guide to consultants; list of microfilm publishers and their products; directory of associations and officers; listings of service companies and their services; storage centers; employment services; bibliographies; etc.

Microform Review. Weston, CT: Microform Review, 1972–. Quarterly.
Critical analysis of many new publications, including an abundance of scientific material.

Microforms: The Librarians' View. 2d ed. **Alice H. Bahr.** White Plains, NY: Knowledge Industry, 1978.

First edition, 1976.
A useful update to trends in microforms. Explains the proposed American National Standards Institute (ANSI) microforms standards and their implications for librarians. Covers the 5 different methods for handling micromaterials. Profusely illustrated.
R: *Journal of Academic Librarianship* (Mar. 1979).

Non-Book Materials in Libraries: A Practical Guide. **Richard Fothergill and Ian Butchart.** Hamden, CT: Shoe String Press, 1978.

A useful guide to nonbook materials and to the abstracting and indexing of these materials. Includes information on video and video disc. Well-recommended.
R: *Library Association Record* 81: 28 (Jan. 1979).

Practical Video: The Manager's Guide to Applications. **John A. Bunyan, James C. Crimmins, and N. K. Watson.** White Plains, NY: Knowledge Industry, 1978.

A practical guide on the day-to-day application of video. Case histories and actual users' experiences are included.

Previews: Audiovisual Software Reviews. Vols. 1–. New York: Bowker, 1972–. Monthly.

A monthly publication that reviews audiovisual materials. Multidisciplinary, the reviews are separated by subject, such as consumer education, holidays, etc. Also reviews films and slide tapes related to health and medicine. Useful for medical and undergraduate libraries; provides pertinent annotations indicating audience level.

Princeton Guide to Microforms: Serials. Princeton, NJ: Microfilm Corporation. Annual.

Princeton Telephone Guide to Microforms. Princeton, NJ: Princeton Microfilm Corporation. Annual.

Selecting Instructional Media: A Guide to Audiovisual and Other Instructional Media Lists. 2d ed. **Mary Robinson Sive.** Littleton, CO: Libraries Unlimited, 1978.
R: *RQ* 19: 96 (Fall 1979).

Television and Management: The Manager's Guide to Video. **John Bunyan and James Crimmins.** White Plains, NY: Knowledge Industry, 1977.

The User's Guide to Standard Microfiche Formats. **National Microfilm Association.** Silver Spring, MD: NMA, 1975.

A 16-page illustrated booklet providing accurate descriptions of all standard microfiche formats. Includes the latest NMA, ANSI, DOD, and ISO standards in one publication. A glossary is also provided.

Video in Libraries: A Status Report. **Alice H. Bahr.** White Plains, NY: Knowledge Industry, 1977–.

Second edition, 1979.
Highlights the current status of library involvement with video and the rationale for offering video services. A helpful guide for any library using or planning to use video.
R: *Computers and the Humanities* (Sept./Oct. 1977); *Video Systems* (Sept. 1977).

The Video Guide. **Charles Bensinger.** Santa Barbara, CA: Video-Infor Publications, 1977–.

Second edition, 1979; third edition, 1982.
A comprehensive guide to all aspects of video technology. Includes a glossary, manufacturers' addresses, video program sources, bibliography, and index. Valuable reference tool for the experienced user.
R: *JAL* 5: 107 (May 1979); *ARBA* (1980, p. 716).

The Video Programs Index. 4th ed. **Ken Winslow, ed.** Syosset, NY: The National Video Clearinghouse, 1979.

Covers subjects, formats, and rental/purchase information for about 400 major video program distributors.

Video Register. White Plains, NY: Knowledge Industry, 1979–.

First edition, 1979; second edition, 1982; third edition, 1983; fourth edition, 1984.
Lists major users in business, government, health, and education, with names and addresses; manufacturers of nonbroadcast video gear; dealers, production/post-production houses, and other service companies; and video producers/publishers/distributors that have programs for sale or for rent.

Video Source Book. Syosset, NY: The National Video Clearinghouse, 1979–. Annual.

Sixth edition, 1984; seventh professional edition, 1986.
Lists approximately 40,000 videotapes and discs in every field and their descriptions, producers, casts, ratings, and awards. Includes directory information on distributors. Indexed by subject and title.
R: *LJ* 105: 601 (Mar. 1, 1980).

Video-based Information Systems: A Guide to Educational, Business, Library, and Home Use. **William Saffady.** Chicago: American Library Association, 1985.

A general introduction to the function and uses of information systems that utilize video technologies. Topics included are video recording, digital television, video disk systems for entertainment and educational use, cable TV, videotext, microfacsimile systems, optical disks, and others.

Video User's Handbook. **Peter Utz.** Englewood Cliffs, NJ: Prentice-Hall, 1980–.

First edition, 1980; second edition, 1982.
A guide to the use and maintenance of television equipment, such as receivers, cameras, videotape recorders, and other devices. Contains pictures, diagrams, and other visual aids. Aimed at school media directors, technicians, and industrial users. For public and community college libraries.

The Videodisc Book: A Guide and Directory. **R. Daynes and B. Butler, eds.** New York: Wiley, 1984.

The first complete guide to virtually all facets of videodiscs, including creative aspects, production, manufacture, uses, programming, and existing titles. Contains listings of more than 200 production and development organizations; over 100 hardware manufacturers, software developers, and integrated systems dealers; and about 1,800 videodisc titles.

Audiovisual Equipment

Audiovisual Equipment and Materials: A Basic Repair and Maintenance Manual. **Don Schroeder and Gary Lare.** Metuchen, NJ: Scarecrow Press, 1979.

Discusses equipment maintenance and repair of all types of projectors, slide viewers, screens, cassette recorders, cameras, and public address systems. Appendixes provide helpful information on tools, lubricants, cleaners, basic electricity and magnetism, cables and connectors. A well-illustrated, useful book for media specialists, school administrators, and librarians.
R: *LJ* 104: 1269 (June 1, 1979); *ARBA* (1980, p. 716).

The Audio-Visual Equipment Directory. Fairfax, VA: National Audio-Visual Association. Annual.

Twenty-eighth edition, 1982–83.
The only comprehensive, up-to-date audiovisual equipment guide. Contains information on over 2,000 currently available audiovisual equipment items. Includes various photographs.

Audiovisual Market Place: A Multimedia Guide. New York: Bowker. Annual.

Sixth edition, 1976; eleventh edition, 1981; latest edition, 1985–86.
Now an annual, quick-reference guide. Consists of 3 major areas: audiovisual software, hardware, and reference sources. Listings provide company names, addresses, and product lines. Classified indexes.

Buyer's Guide to Microfilm Equipment, Products, and Services. Silver Spring, MD: National Microfilm Association. Annual.

Buyer's Guide to Micrographic Equipment, Products, and Services. Silver Spring, MD: National Microfilm Association. Annual.

A listing of sustaining members of NMA by product and service. Designed as a concise introduction and continuing reference for present and potential users of micrographic equipment, products, and services.

The Complete Handbook of Videocassette Recorders. 2d ed. **Harry Kybett.** Blue Ridge Summit, PA: TAB Books, 1983.

First edition, 1977.
An effective reference on the use of video cassettes. Discusses studio work, editing, and sound dubbing. For technical libraries.
R: *Choice* 14: 1530 (Jan. 1978).

Guide to Micrographic Equipment. 8th ed. 3 vols. **National Micrographics Association.** Silver Spring, MD: National Micrographics Association, 1983.

Sixth edition, 1975; seventh edition, 1979.
Volume 1, *Production Equipment* (cameras, processors, duplicators, and inspection apparatus); volume 2, *User Equipment* (readers, reader-printers, enlargers, and automatic retrieval units); volume 3, *COM Recorders* (computer-output micrographics).
Illustrated directories provide specifications, prices, and pictures of micrographic equipment.

Index to Producers and Distributors. 5th ed. **National Information Center for Educational Media.** Los Angeles: National Information Center for Educational Media, 1980.

International File of Micrographics Equipment and Accessories. **William Saffady, ed.** Westport, CT: Meckler Publishing, 1977–.

First edition, 1977; second edition, 1979; third edition, 1983.
A vendor catalog of micrographic equipment providing information on micrographic equipment, from automated storager and retrieval systems to fiche readers. Contains complete information, an index for easy access to information. A detailed source, listing all significant micrographic equipment worldwide.

SUBJECT SOURCES

GENERAL

AAAS Audiotape Cassette Album Series. Washington, DC: American Association for the Advancement of Science.

AAAS Science Book and Films. (Variant title: *Science Book and Films*). Washington, DC: American Association for the Advancement of Science. 1965–. 5/yr.

Reviews approximately 1,000 new science books as well as over 250 new science films produced by commercial firms, universities, and government agencies each year. A quick and reliable source for the evaluation of the latest science books and films.
R: *ARBA* (1979, p. 639).

AAAS Science Film Catalog. **Ann Seltz-Petrash and Kathryn Wolff, eds.** New York: Bowker. Annual.

Contains ordering information for 5,600 science films for elementary through adult grades. These films can be bought, borrowed, or rented from US producers/distributors. Topics include philosophy and related subjects, social sciences, pure sciences, and technology.

Educators' Guide to Free Science Materials: A Multimedia Guide. **Mary H. Saterstrom, ed.** Randolph, WI: Educators Progress Service. Annual.

Seventeenth edition, 1976; twenty-fifth edition, 1984.
Lists nearly 2,000 selected free science materials, including films, filmstrips, slides, charts, etc. New edition includes many additions. Up-to-date and comprehensive, highly recommended.
R:*BL* 73: 1601 (June 15, 1977).

Films in the Sciences—Reviews and Recommendations. **Michele M. Newman and Madelyn A. McRae, comp., eds.** Washington, DC: American Association for the Advancement of Science, 1980.

A reference of 1,000 films from all areas of science. Includes detailed summaries, evaluations, Dewey decimal categorizing, subject-title and film distributor indexes. Essential for librarians, science teachers, and anyone with an interest in this area.
R: *Science* 215: 317 (Jan. 15, 1982).

Science Books & Films. **American Association for the Advancement of Science.** Washington, DC: AAAS. Quarterly.

Quarterly review magazine, which each year gives reviews of 1,000 new science trade/text books and 250 new 16mm science films. The reviews are both descriptive and critical. Ordering information, 9 explicit level designations (kindergarten through professional), and 4 ratings (highly recommended to not recommended) are also given.

SUBJECT SPECIFIC

Archival Microfilm Packages. New York: American Institute of Physics. 16mm, 35mm, reel, and cartridge.

Offers microfilm edition of the 46 primary journals and the conference proceedings published by the American Institute of Physics and its member societies. Packages 1 and 2 include journals published by AIP and its affiliated member societies, and those translated by AIP; packages 3 and 4 contain the journals of the Institute of Physics (UK) and several other journals no longer publicly available.

Audiovisual Resources in Food and Nutrition. 2d ed. Phoenix: Oryx Press, 1984.
First edition, 1979.
A listing of audiovisual materials in food and nutrition. Covers 3 subject areas with 1,200 annotations. Each entry contains complete bibliographic information and an abstract. For libraries with strong collections in food, cooking, and diet.
R: *ARBA* (1981, p. 738).

Catalog of Museum Publications and Media: A Directory and Index of Publications and Audiovisuals Available from United States and Canadian Institutions. 2d ed. **Paul Wasserman, ed.** Detroit: Gale Research, 1980.
R: *CRL* 41: 188 (Mar. 1980).

Current Physics Microform (CPM). New York: American Institute of Physics. Available in 16mm, 35mm, reel, and cartridge.

Section 1 consists of 26 English-language journals and the American Institute of Physics Conference Proceedings Series ($6,951); section 2, of 20 Soviet and Chinese translation journals ($9,295).
A time, space and money saver for corporate and institutional libraries, CPM offers 2 subscription packages of the microfilm edition of the 46 primary journals and the conference proceedings published by the American Institute of Physics and its member societies.

Energy: A Multimedia Guide for Children and Young Adults. **Judith H. Higgins.** Santa Barbara, CA: ABC-Clio Press, 1979.

Offers over 400 annotated entries to energy information sources. A 4-part guide: Energy sources; curriculum guides and courses; selected tools; and alternate information sources in the energy field. Intended for school and public libraries.
R: *WLB* 53: 653 (May 1979).

Energy: Sources of Print and Nonprint Materials. **Maureen Crowley, ed.** New York: Neal-Schuman Publishers, 1980.

A listing of over 770 organizations that provide information on energy. Consists mainly of US organizations such as businesses, government agencies, and civic and educational groups. Entries are grouped by type of organization and contain name, address, telephone number, key activities and programs, and a list of publications. Includes an appendix of grass-roots groups as well as information on free and inexpensive sources. Title, subject, and sources indexes are provided. A highly recommended reference for academic and public libraries, as well as for laypersons and undergraduates.
R: *BL* 77: 721 (Jan. 15, 1981); *CRL* 41: 396 (July 1980); *LJ* 105: 1375 (June 15, 1980); *RQ* 20: 94 (Fall 1980); *WLB* 55: 141 (Oct. 1980); *ARBA* (1981, p. 681).

The Environment Film Review: A Critical Guide to Ecology Films. New York: Environment Information Center, 1972–.

A comprehensive guide to environmental films; basic information on each film, such as length, color or B/W, cost, where to obtain, intended audience, etc., are given. Materials are organized by EIC's 21 major environmental subject classifications. Highly recommended.

Films in the Mathematics Classroom. **Barbara J. Bestgen and Robert E. Reys.** Reston, VA: National Council of Teachers of Mathematics, 1982.

An extensive catalog of over 200 films related to mathematics for the classroom. Includes information on the uses of films for this purpose, reviews of each film, ratings, and other pertinent data to help teachers of mathematics, in the primary grades through college, in their lesson plans.
R: *ARBA* (1983, p. 611).

NMAC Catalog. US National Medical Audiovisual Center. Atlanta: National Medical Audiovisual Center. Annual.

100 Short Films about the Human Environment. **William R. Ewald, ed.** Santa Barbara, CA: ABC-Clio Press, 1982.

Describes 100 films selected for the ACCESS Collection, with title, date of release, running time, producer, description, and awards received. An essential tool for academic, public, and school film collections.
R: *ARBA* (1984, p. 1377).

Selected Instrumentation Films. Pittsburgh: Instrument Society of America. Annual.

It is also important to keep in mind that professional organizations and government agencies are major producers of nonprint materials. Users are advised to consult the audiovisual catalogs for individual sources.

Subtidal Marine Invertebrates of North America. 2 pts. **George S. Zumwalt and Diane M. Zumwalt.** Lyons Falls, NY: Educational Images, 1979.

Part 1, *Porifera, Cnidaria, Annelida, Mollusca*; part 2, *Arthropods, Brachiopods, Bryozoa, Echinoids, Urochordates*.

Comprises a series of slides and guides that provide descriptions and explanations, general information, and references. Useful for marine biology, geology, and evolution courses in high schools and universities.

R: *Choice* 18: 1125 (Apr. 1981).

CHAPTER 22 PROFESSIONAL SOCIETIES AND THEIR PUBLICATIONS

Professional organizations are generally one of the most valuable sources of information for scientific and technical subjects. Whether as producers of computerized databases, sponsors of conferences, or publishers of abstracting and indexing journals, periodicals, monographs, reports or symposia, standards, and so on, professional organizations remain a vital link in scientific communication. Useful sources of information on societies and their publications, sample catalogs of society publications, and the names of some major societies themselves are included here for the reader's convenience. As these lists do not attempt to be comprehensive, the current-year catalogs of appropriate societies should be requested for complete and up-to-date information.

DIRECTORIES

(See also chapter 11.)

Australian Scientific Societies and Professional Associations. 2d ed. **Ian A. Crump, ed.** Victoria, Australia: Commonwealth Scientific and Industrial Research Organization. Distr. Forest Grove, OR: International Scholarly Book Service, 1978. Also update edition.

First edition, 1971.

This expanded edition consists of 399 alphabetically arranged entries. Covers such information as name of society, address, phone number, affiliated societies, history, purpose, activities, fields of interest, publications, and membership. Contains list of initials, names, publications, awards, and subject indexes.

R: *RQ* 19: 192 (Winter 1979); *ARBA* (1980, p. 603).

Directory of British Associations and Associations in Ireland: Interests, Activities and Publications of Trade Associations; Scientific and Technical Societies; Professional Institutes; Learned Societies; Research Organizations; Chambers of Trade and Commerce; Agricultural Societies; Trade Unions; Cultural, Sports and Welfare Organizations in the United Kingdom and in the Republic of Ireland. 7th ed. **G. P. Henderson and S. P. A. Henderson, eds.** Beckenham, Kent: C. B. D. Research. Distr. Detroit: Gale Research, 1984.

Fourth edition, 1974; sixth edition, 1980.

R: *CRL* 41: 492 (Sept. 1980); *RSR* 9: 94 (Jan./Mar. 1981).

Directory of Engineering Societies and Related Organizations. 11th ed. **Rudolph J. Yacyshyn, ed.** New York: American Association of Engineering Societies, 1984.

Sixth edition, 1970; seventh edition, 1974; ninth edition, 1976.
Presents basic information on 485 national, regional, Canadian, and international organizations related to engineering. Indexed by official initials of the societies, geographical areas, and keyword. A useful tool for engineering libraries.
R: *ARBA* (1985, p. 537).

Directory of European Associations. 2 vols. 3d ed. **R. W. Adams, ed.** Detroit: Gale Research, 1983–1984.
Second edition, 1976–1979. Volume 1, *National Industrial, Trade, and Professional Associations*, 1983; volume 2, *National Learned, Scientific, and Technical Societies*, 1984.
Covers learned, scientific, and technical societies in all countries except Great Britain and Ireland, which are included in *Directory of British Associations*.
R: *ARBA* (1985, p. 11).

Encyclopedia of Associations. 18th ed. 3 vols. **Denise Akey, ed.** Detroit: Gale Research, 1983–1984.
Thirteenth edition, 1979; fourteenth edition, 1980; fifteenth edition, 1981.
Volume 1, *National Organizations of the United States*; volume 2, *Geographic and Executive Indexes*; volume 3, *New Associations and Projects*.
Fully describes activities and publications. Provides addresses and telephone numbers.
R: *BL* 77: 350 (Oct. 15, 1980); 77: 473 (Nov. 15, 1980); *RQ* 12: 314 (Spring 1973); 19: 95 (Fall 1979); *SL* 66: 350 (July 1975); *ARBA* (1973, p. 101).

Guide to World Science Series. 25 vols. St. Peter Port, Guernsey, England: Hodgson, 1974.
Good source to associations and services on international level.

National Trade and Professional Associations of the United States, 1984. 19th ed. **Craig Colgate, Jr. and Stephen J. Freedman, eds.** Washington, DC: Columbia Books, 1984. Annual.
First published in 1966.
An annual directory of about 6,000 national trade associations, labor unions, professional and scientific societies, and other national organizations. A must for every library.
R: *ARBA* (1985, p. 14).

North American Horticulture: A Reference Guide. **Barbara W. Ellis, ed.** New York: Scribner's, 1982.
Directory of national, state, and Canadian horticultural societies, associations, etc., and requirements for admission. Also includes the US Department of Agriculture programs and regulations. Gives a bibliography of reference works and covers subjects of interest to the gardener.
R: *ARBA* (1983, p. 712).

World Guide to Scientific Associations and Learned Societies. 2d ed. New York: Bowker, 1978.
First edition, 1974.
Revised edition includes over 1,000 new entries. Fully informative. For libraries with science collections.

World Guide to Scientific Associations and Learned Societies. 4th ed. **Barbara Verrel and Helmut Opitz.** New York: Saur, 1984.

First edition, 1974, and second edition, 1977, published by Verlag Dokumentation in Munich and distributed by Bowker in New York.

A thoroughly revised and expanded edition that includes about 22,000 international, national, and regional associations from 150 countries in every area of science, culture, and technology. Over 12,000 periodicals and bulletins of these associations are also included with a special index of official association abbreviations. Subject index of activities, organized by country, is available.

R: *Choice* 11: 1459 (Dec. 1974); *LJ* 99: 3124 (Dec. 1, 1974); *ARBA* (1975, p. 641).

World Guide to Trade Associations. 3d ed. **Barbara Verrell, ed.** New York: Saur, 1985.

First edition, 1973; second edition, 1980.

A valuable directory of more than 31,000 national and international trade associations arranged in almost 400 trade categories. For the first time, in this edition, publications of associations are included and over 5,500 periodicals are listed.

SELECTIVE LIST OF PROFESSIONAL ORGANIZATIONS

American Association for the Advancement of Science (AAAS)
1515 Massachusetts Avenue, NW
Washington, DC 20005

American Chemical Society (ACS)
115 16th Street, NW
Washington, DC 20036

American Geological Institute (AGI)
5205 Leesburg Pike
Falls Church, VA 22041

American Geophysical Union (AGU)
2000 Florida Avenue, NW
Washington, DC 20009

American Institute of Aeronautics and Astronautics (AIAA)
1290 Avenue of the Americas
New York, NY 10019

American Institute of Biological Sciences (AIBS)
104 Wilson Boulevard
Arlington, VA 22209

American Institute of Chemical Engineers (AICHE)
345 East 47th Street
New York, NY 10017

American Institute of Physics (AIP)
335 East 45th Street
New York, NY 10017

AIP member societies
  American Physical Society
  Optical Society of America
  Acoustical Society of America
  Society of Rheology
  American Association of Physics Teachers
  American Crystallographic Association
  American Astronomical Society
  American Association of Physicists in Medicine
  American Vacuum Society

American Mathematical Society (AMS)
PO Box 6248
Providence, RI 02940

American Meteorological Society (AMS)
45 Beacon Street
Boston, MA 02108

American Nuclear Society (ANS)
555 N Kensington Avenue
LaGrange Park, IL 60525

American Society for Engineering Education (ASEE)
1 Dupont Circle
Suite 400
Washington, DC 20036

American Society for Metals (ASM)
Metals Park, OH 44073

American Society for Testing and Materials (ASTM)
1916 Race Street
Philadelphia, PA 19103

American Society of Civil Engineers (ASCE)
345 East 47th Street
New York, NY 10017

American Society of Mechanical Engineers (ASME)
345 East 47th Street
New York, NY 10017

American Statistical Association (ASA)
806 15th Street, NW
Washington, DC 20005

Association for Computing Machinery (ACM)
1133 Avenue of the Americas
New York, NY 10036

Chemical Society (CS)
Burlington House
London W1V OBN
England

Engineers Joint Council (EJC)
345 East 47th Street
New York, NY 10017

Federation of American Societies for Experimental Biology (FASEB)
9650 Rockville Pike
Bethesda, MD 20014

Geological Society of America (GSA)
3300 Penrose Place
Boulder, CO 80301

Institute of Electrical and Electronics Engineers (IEEE)
345 East 47th Street
New York, NY 10017

Institute of Environmental Sciences (IES)
940 East Northwest Highway
Mt. Prospect, IL 60056

Institution of Chemical Engineers (ICE)
165-171 Railway Terrace
Rugby CV21 3HQ
England

Institution of Civil Engineers (ICE)
1–7 Great George Street
London SWIP 3AA
England

Institution of Electrical Engineers (IEE)
LTD Department, 445 Hoes Lane
Piscataway, NJ 08854

Institution of Mechanical Engineers (IME)
Box 23, Bury Street
Edmunds, Suffolk 1P32 6BN
England

International Astronomical Union (IAU)
IAU-UAI Secretariat
61, Avenue de l'observatoire
75014 Paris
France

International Union of Biological Sciences (IABS)
51 Boulevard de Montmorecy
F-75016 Paris
France

National Society of Professional Engineers (NSPE)
2029 "K" Street, NW
Washington, DC 20006

Royal Society (RS)
6 Carlton House Terrace
London SW1Y 5AG
England

Society for Industrial and Applied Mathematics (SIAM)
Suite 1405 Architects Building
117 South 17th Street
Philadelphia, PA 19103

Society of Automotive Engineers (SAE)
400 Commonwealth Drive
Warrendale, PA 15096

Society of Chemical Industry (SCI)
50 East 41st Street
Suite 92
New York, NY 10017

SOCIETY PUBLICATIONS

Guide

Scientific, Engineering and Medical Societies Publications in Print, 1980–1981. 4th ed. **James M. Matarazzo and James M. Kyed, eds.** New York: Bowker, 1976.
First edition, 1974; second edition, 1976; third edition, 1979–1980.
A one-stop reference intended to provide coverage of published materials of 350 US scientific and engineering societies. Provides ordering information.

R: *Choice* 11: 1458 (Dec. 1974); *LJ* 99: 1927 (Aug. 1974); *ARBA* (1975, p. 638). Sheehy (EA3).

Sample Society Catalogs and Bibliographic Tools

Catalogs

ACI Publications. Detroit: American Concrete Institute. Annual.

American Institute of Physics Catalog of Journals, Books, Microforms, Bibliographic Databases. New York: American Institute of Physics, 1985. Annual.

American Society for Testing and Materials Publications. Philadelphia: American Society for Testing and Materials. Annual.

National Fire Protection Association, Catalog, Publications, and Visual Aids. Boston: NFPA. Annual.

Publications Catalog. **Institution of Electrical Engineers.** Somerset, NJ: Institution of Electrical Engineers. Annual.

Includes indexes, order forms, as well as a list of publications.

Indexes

ASCE Combined Index. New York: American Society of Civil Engineers, 1962–. Annual.

Subject arrangement of ASCE-generated material. Includes author index.

ASCE Publications Information. New York: American Society of Civil Engineers, 1966–. Bimonthly.

Formerly (until 1982): *ASCE Publication Abstracts.*
Gives comprehensive coverage of the society's publications when used in conjunction with the *ASCE Combined Index*, 1962–.

American Society of Mechanical Engineers. Seventy-seven Year Index: Technical Papers, 1880–1956. Easton, PA: American Society of Mechanical Engineers, 1957.

Indexes author and subject to selected articles published in the various journals of the society. Continued by the society's *Society Records: Indexes to Publications*, 1956–, annual.

Brief Subject and Author Index of Papers Published by the Institution 1951–1964. London: Institution of Mechanical Engineers, 1966.

Supplemented by annual subject and author indexes in the Institution's *Journal of Strain Analysis* and *Proceedings.*

Cumulative Index: SAE Papers 1965–1973. Warrendale, PA: Society of Automotive Engineers, 1974.

Over 8,000 entries, a large portion of which are from *SAE Transactions*. Coverage of space, air, and watercraft as well as the automotive materials. Includes subject and author indexes and chronological listings. Primarily of interest to research collectors.
R: *ARBA* (1976, p. 763).

Cumulative Index to ASCE Publications, 1975–1979. New York: American Society of Civil Engineers, 1980.

Three volumes to date, offering retrospective coverage of American Society of Civil Engineers publications for the periods 1950–1959, 1960–1969, and 1970–1974. Especially useful for bibliographic control over *ASCE Proceedings*, 1950–1959; *Transactions*, 1935–
1959; and *Civil Engineering*, 1930–1959.
R: Wal (p. 352).

Index-Abstracts to SAE Transactions and Literature. **Society of Automotive Engineers.** New York: SAE. Annual.

Index of SNAME Publications, 1961–1973. New York: Society of Naval Architects and Marine Engineers, 1974.

Update of index covering the years 1961–1969. Arranged by subject and author, in reverse chronological order. Complete ordering information available.
R: *ARBA* (1976, p. 780).

CHAPTER 23   DATABASES

There is no need to elaborate on the importance of databases for those who are concerned with scientific and technical information. The following are good examples of sources of information about databases:

REFERENCE SOURCES

Answers Online: Your Guide to Information Databases. **B. Newlin.** New York: McGraw-Hill, 1985.

A guide for those microcomputer users who wish to use online information services via their microcomputers.

The Computer Data and Database Source Book. **Matthew Lesko.** New York: Avon, 1984.

A massive directory of 172 government and over 1,000 commercial and public access databases and sources of information for use with computers.
R. *LJ* 109: 2057 (Nov. 1, 1984; *ARBA* (1985, p. 13).

The Computer Phone Book. **Mike Kane.** New York: New American Library, 1983.

A directory of listings and description information of over 400 systems and services that can be connected via computer modems. These systems include *BRS After Dark, CompuServe, Delphi, Dow Jones, Knowledge Index (DIALOG),* and *The Source.* Essential tool for all those who are interested in database access and electronic mail.
R: *ARBA* (1985, p. 591).

Computer-Readable Data Bases: A Directory and Data Sourcebook. 2 vols. **Martha E. Williams, ed.** Chicago: American Library Association, 1985–.

Earlier editions published by Knowledge Industry.
Volume 1, *Science, Technology, Medicine*; volume 2, *Business, Law, Humanities, Social Sciences.*
A comprehensive directory listing over 1,000 databases worldwide. Each entry contains the name and producer of the database, its coverage, year of origin, number of items in the base, availability in batch or online mode, pricing, and other information. Indexed by subject, producer, processor, and database name. Useful tool for users of databases, producers of databases, and information suppliers, students, and teachers.
R: *Information Technology and Libraries* 4: 369 (Dec. 1985).

Data Dictionary/Directory Systems: Administration, Implementation and Usage. **Belkis W. Leong-Hong and Bernard K. Plagman.** New York: Wiley, 1982.

A reference to database application and data dictionary/directory system nomenclature with a thorough overview of the responsibilities of a data administrator.
R: *ARBA* (1984, p. 617).

Database: A Bibliography. Vols. 1–. **Yahiko Kambayashi.** Washington, DC: Computer Science, 1981–.

A printout of a computerized bibliography of 3,912 articles covering database research from the 1970s. For the graduate and professional areas.
R: *Choice* 19: 606 (Jan. 1982).

DataBase Directory. White Plains, NY: Knowledge Industry Publications, 1985. Semiannual updates.

A directory that covers more than 2,000 databases and database systems in all types of subjects. For each entry, information is given on the subject, subject access, producer/producer services, time coverage and file data, vendor and price, etc. Indexed by subject, producer/vendor, and alternate name of database.

Directory of Computerized Data Files and Related Software. Washington, DC: National Technical Information Service, 1974–. Annual.

Directory information on more than 500 machine-readable data files available from 60 agencies.
R: *ARBA* (1976, p. 770; 1978, p. 767).

Directory of Online Databases. Santa Monica, CA: Cuadra Associations. New York: Elsevier Science Publishing, 1985–.

A directory of 2,764 online databases. International in coverage. Quarterly updates. Data on type of database, subject, producer, content, coverage, online service, and updating. Lists producers and addresses, and includes subject, service, producer, and database name indexes.
R: *CRL* 41: 188 (Mar. 1980).

Encyclopedia of Information Systems and Services. 2 vols. 1985–1986. 6th ed. **John Schmittroth, Jr., ed.** Detroit: Gale Research, 1984–1985.
First edition, 1971; second edition, 1974; third edition, 1978; fourth edition, 1980; fifth edition, 1982.

Volume 1, *International*, 1984; volume 2, *United States*, 1985.
Provides detailed descriptions of more than 3,500 organizations (2,350 of which are in the United States) that produce and/or provide access to computerized information in all subject areas. Coverage includes database producers and publishers, online vendors, computer time-sharing companies, videotex/teletext services, fee-based information services, document delivery sources, consultants, etc. Has 27 indexes.
R: *BL* 76: 1438 (June 1, 1980); 77: 350 (Oct. 15, 1980); *RQ* 14: 170 (1974); *RSR* 3: 13 (June 1975); *ARBA* (1975, p. 78); Sheehy (EJ48).

Eusidic Database Guide, 1983. Medford, NJ: Learned Information, 1983.

Gives 982 entries of organizations that are located in the United States, Europe, Africa, and Asia and are associated with either bibliographic databases or databanks of information available in Europe via online or batch processing. A useful source for finding organizations throughout the world.
R: *ARBA* (1984, p. 620).

The Federal Data Base Finder: A Directory of Free and Fee-based Data Bases and Files Available from the Federal Government. **Sharon Zarony and Monica Horner, eds.** Chevy Chase, MD: Information USA, 1984. Annual.

A complete guide to 3,000 databases and files available from the US Federal government. Useful appendix on the data-collection patterns of the Bureau of the Census, the National Technical Information Service, and the National Archives and Records Service.
R: *ARBA* (1985, p. 18).

Hi-Tech Data Base Buyers' Guide. 1985–1986. Chevy Chase, MD: Information USA, 1985.

Information on 1,000 online and off-line databases and files from over 100 vendors, of which, 517 are on hi-tech industries, 193 on energy and the environment, 197 on computers and technology, 107 on chemistry and engineering, 46 on aerospace and defense, 23 on patents, and 23 on hi-tech information clearinghouses. Indexed by industry and vendor. Available in book form or in dBASE or Lotus 1-2-3 formats on diskettes.

Libraries, Information Centers and Databases in Science and Technology: A World Guide. 1st ed. **Helga Lengenfelder, ed.** New York: Saur, 1984.

An international directory of libraries, online databases and documentation centers in all areas of pure and applied sciences. Lists institutions by country. Includes 360 on-line databases from nearly 200 producers are included. Indexed by general name and English keyword.

Nonbibliographic Data Banks in Science & Technology. **Stephan Schwartz et al. eds.** New York: Unipub, 1985.

North American Online Directory. New York: Bowker, 1985.
Formerly *Information Industry Market Place.*

A useful tool for updated information on the people, companies, and services that make up the revolutionary new information industry. Includes database producers, online vendors, library networks and consortia, information brokers, information collection and analysis centers, associations, references, courses, etc.

Numeric Databases. **Ching-chih Chen and Peter Hernon, eds.** Norwood, NJ: Ablex, 1984.

A general overview on numeric databases in all fields.
R: *Choice* 22: 456 (Nov. 1984); *Government Information Quarterly* 2: 223 (1985); *Government Publications Review* 12: 261 (1985); *Library Hi-Tech News* no. 8: 92 (1985); *New Library Science Annual* 1: 67 (1985); *Online* 9: 100 (Jan. 1985); *Online Review* 9: 44 (Feb. 1985).

Omni Online Database Directory. **Mike Edelhart and Owen Davies.** New York: Collier Books/Macmillan, 1983.

An introductory reference work for those interested in online searching on personal computers. There are 3 parts—introductory text, directory section of more than

1,000 information files under some 40 broad subject areas, and appendixes including online vendors. Two sections on *The Source* and *CompuServe*.
R: *Choice* 21: 1274 (May 1984); *ARBA* (1985, p. 11).

Online Bibliographic Databases. **James L. Hall and Majorie J. Brown, eds.** London: Aslib. Distr. Detroit: Gale Research, 1983.

A directory of 200 online bibliographic databases accessible through 40 online service suppliers throughout the world. Each entry has information on name and acronym of database, supplier's name and address, subject field, printed versions (if any), file details, charge, and more. Indexed.

Online Database Search Services Directory: A Reference and Referral Guide to Libraries, Information Firms, and Other Sources Providing Computerized Information Retrieval and Associated Services Using Publicly Available Online Databases. 2 vols. **John Schmittroth, Jr. and Doris Morris Maxfield, eds.** Detroit: Gale Research, 1984.

Information on more than 600 organizations in the United States and Canada that provide online search capabilities and activities.
R: *Information Technology and Libraries* 4: 370 (Dec. 1985). *Science Technology Libraries* 5: 2 (Winter 1984).

Online International Command Chart. Weston, CT: Online, 1985.

A useful tool for information retrieval. It compares—command-by-command—the retrieval commands of 16 major US, Canadian, and European systems, from DIALOG II to VU/TEXT to CAN/OLE, etc. Instant reference to 41 retrieval, document delivery, and other vital commands.

Online Searching: A Dictionary and Bibliographic Guide. **Greg Byerly.** Littleton, CO: Libraries Unlimited, 1983.

Contains all of the online commands and everyday terms used in online searching. Defines each database in a helpful and descriptive way. For easy and efficient reference.
R: *Choice* 21: 399 (Nov. 1983).

Telecommunications Systems and Services Directory. 2d ed. **Martin Connors, ed.** Detroit: Gale Research, 1985.

A directory on today's high-technology communications systems and services—voice and data communications, from long-distance telephone services to teleconferencing and electronic mail. Master index and indexes of personal name and geographic locations. Includes telecommunications glossary.

DATABASE SEARCH GUIDES AND REFERENCE MANUALS

In order to search each database effectively, every database vendor publishes a search guide and/or reference manual for its database(s). Their newsnotes and technical memos are also very useful. The following are only a few sample publications of this type:

BIOSIS Previews Search Guide. Philadelphia: BIOSIS, 1985.

Bluesheets. Palo Alto, CA: DIALOG, 1985–.

BRS Brief System Guide Sign-on Procedures. Latham, NY: Bibliographical Retrieval Service, 1984.

BRS Directory and Database Catalog. Latham, NY: Bibliographical Retrieval Service. Annual.

BRS System Reference Manual. 2 vols. Latham, NY: Bibliographical Retrieval Service, 1979–.

Chronolog. Palo Alto, CA: DIALOG. Monthly.

DIALOG Database. Palo Alto, CA: DIALOG. Annual

Guide to DIALOG Searching. Palo Alto, CA: DIALOG, 1979–.

Orbit: New User Training. Santa Monica, CA: System Development Corporation.

Orbit User Manual. Santa Monica, CA: System Development Corporation, 1982. Also updates.

SDC Search Service Quick Reference Guide. Santa Monica, CA: System Development Corporation.

Technical Memo: Enhancements to the DIALOG Search System. Palo Alto, CA: DIALOG, 1984–.

WILSONLINE Guide and Documentation. Bronx, NY: Wilson, 1984.

WILSONLINE Quick Reference Guide. Bronx, NY: Wilson, 1984.

SELECTIVE DATABASES

The directories included in the beginning of this chapter list thousands of databases; it is thus impossible to list all the databases related to science and technology here. There are numerous commercial vendors of databases, and the largest is DIALOG, which provides access to more than 200 databases. The System Development Corporation (SDC) and the Bibliographic Retrieval Service (BRS), though providing access to smaller numbers of databases, are also major vendors. The following is a quick guide to some of the databases in science and technology accessible through DIALOG, SDC, or BRS.

Since the rates for these databases vary greatly from one to another and change substantially over time, readers are advised to consult the newest edition of database directories or to write to the database vendors directly for detailed and up-to-date information.

Database Name	Subject and Source	Entry Date	Items (Approx.)	Vendor
AEROSPACE DATABASE	Provides references, abstracts, and controlled vocabulary indexing of key scientific and technical documents covering aerospace research and development in over 40 countries.	1984–present	21,000	DIALOG
AGRICOLA (formerly CAIN)	International journal and monographic literature in agriculture and related fields. Prepared by US National Agricultural Library.	1970–present	2,826,000	BRS DIALOG SDC
AIM/ARM	Abstracts of instructional and research materials in vocational and technical education. From the Center for Vocation Education.	Sept. 1967–1976 (records from 1977–present are included in ERIC)	17,000	DIALOG
AMERICAN CHEMICAL SOCIETY PRIMARY JOURNAL DATABASE (CJFTX)	Online availability provides immediate access to the most current information in chemistry research.	1980–present		BRS
AMERICAN MEN AND WOMEN OF SCIENCE	Database is a biographical registry of eminent active American and Canadian scientists. Included are all areas of the physical and biological sciences. From Bowker.		127,250	BRS DIALOG

APIPAT	Refining patents, both US and non-US. Prepared by the Central Abstracting and Indexing Service of the American Petroleum Institute.	1964–present	105,000	SDC
APILIT	Worldwide refining literature. Prepared by the Central Abstracting and Indexing Service of the American Petroleum Institute.	1964–present	420,000	SDC
APTIC	Covers all aspects of air pollution, including the social, political, legal, and administrative aspects. From the Manpower and Technical Information Branch, US Environmental Protection Agency.	1966–Sept. 1976 (with some additions through 1978)	87,000	DIALOG
AQUACULTURE	Provides access to information on the growing of marine, brackish, and freshwater organisms. Topics include disease, economics, engineering, food and nutrition, growth requirements, and legal aspects of water organisms. From the National Oceanic and Atmospheric Administration.	1970–present	10,400	DIALOG

cont'd.

Database Name	Subject and Source	Entry Date	Items (Approx.)	Vendor
AQUALINE	Provides access to information on all aspects of water, wastewater, and the aquatic environment. Topics include water resources development and management, drinking-water quality, water treatment, sewage systems, sludge disposal, groundwater pollution, river management, tidal waters, quality monitoring, and environmental protection. From the Water Research Centre.	1960–present	80,000	DIALOG
AQUATIC SCIENCES AND FISHERIES ABSTRACTS	Database on life sciences of the seas and inland waters and related legal, political, and social topics. Covers aquatic biology, oceanography, fisheries, and water pollution. From the NOAA/Cambridge Scientific Abstracts.	1978–present	180,000	DIALOG
ARTHUR D. LITTLE/ONLINE	An index to the nonexclusive information sources of Arthur D. Little, Inc., its divisions, and subsidiaries. Covers industries, technologies, and management. From Arthur D. Little, Decision Resources.	1977–present	680	DIALOG

cont'd.

ASI	Statistical publications of the US Government: periodicals, annuals, biennials, surveys, reports, etc. Prepared by Congressional Information Service.	Jan. 1973	102,500	SDC DIALOG
A–V ONLINE (formerly NICEM)	Offers coverage of nonprint educational material, from preschool to professional and graduate school levels. Provides references to all educational media—16mm films, 35mm filmstrips, overhead transparencies, audiotapes, videotapes, phonograph records, motion picture cartridges, and slides. From the National Information Center for Educational Media, Access Innovations, Inc.		304,000	DIALOG
BIOSIS PREVIEWS	Contents of *Biological Abstracts* (BA) and *BioResearch Index* (BioI). Approximately 9,000 serial publications, as well as books. Prepared by Biological Sciences Information Service.	1969–present	4,566,000	BRS DIALOG SDC
BIOTECHNOLOGY	Coverage includes all technical aspects of biotechnology. Sources include over 1,000 scientific and technological journals and conference proceedings.	July 1982– present	45,000	SDC

cont'd.

Database Name	Subject and Source	Entry Date	Items (Approx.)	Vendor
BLS PRODUCER PRICE INDEX	Contains time series of consumer price indexes calculated by the US Bureau of Labor Statistics.	Dates of coverage vary	11,000	DIALOG
CA SEARCH	Database containing bibliographic data, keyword phrases, and index entries for all documents covered by Chemical Abstracts Service. From the Chemical Abstracts Service.	1967–present	6,840,000	BRS DIALOG
CAB ABSTRACTS	A file of agricultural and biological information containing all records in the main abstract journals published by Commonwealth Agricultural Bureaux. from the Commonwealth Agricultural Bureaux.	1972–present	1,760,000	DIALOG
CAS82/CAS77/CAS72/ CAS67	Coverage includes chemical science literature from over 12,000 journals; patents from 26 countries; new books; conference proceedings; and government research reports.	1982–present (1982) 1977–1981 (1977) 1972–1976 (1972) 1967–1971 (1967)	1,365,000 2,220,000 1,800,000 1,300,000	SDC

cont'd.

CASSI	Chemical Abstract Source Index compiles bibliographic and library holdings information for scientific and technical primary literature relevant to the chemical sciences.	current	50,000	SDC
CDI	Dissertations accepted for academic doctoral degrees granted by US and some non-US universities. Materials stem from *Dissertation Abstracts International*. Prepared by University Microfilms International.	1861	530,000	DIALOG SDC
CEH80/CEH132/CE-HINDEX	Provides annual supply and price data for many of the 1,300 major chemical groups, chemical-related industries, and US economic indicators.	current	2,000	SDC
CHEMCON (CA CONDENSATES)	Corresponds to *Chemical Abstracts*. Prepared by Chemical Abstracts Service.	Jan. 1972	1,641,000	BRS DIALOG SDC
CHEMDEX/ CHEMDEX2/ CHEMDEX3	Chemical dictionary files that cite literature from 1972 to date.	1972–present	2,600,000	SDC

cont'd.

Database Name	Subject and Source	Entry Date	Items (Approx.)	Vendor
CHEMICAL EXPOSURE	Database of chemicals that have been identified in both human tissues and body fluids and in feral and food animals. Spans the range of body-burden information—information related to human and animal exposure to food, air, and water contaminants in addition to pharmaceuticals. From the Chemical Effects Information Center.	1974–present	11,300	DIALOG
CHEMICAL INDUSTRY NOTES (CIN)	Covers business-oriented periodicals and those dealing with recent events in the chemical industry. From Predicasts and Chemical Abstracts Service.	1974–present	557,000	DIALOG
CIN	Citations to the business literature in the chemical industry. Covers over 80 periodicals. Prepared by Chemical Abstracts Service.	1974–present	520,000	SDC
CHEMICAL REGULATIONS AND GUIDELINES SYSTEM (CRGS)	Index to US federal regulatory material relating to the control of chemical substances, covering federal statutes; promulgated regulations; and available federal guidelines, standards, and support documents. From the US Interagency Regulatory Liaison Group, CRC Systems, Inc.	1982–present	4,600	DIALOG

cont'd.

CHEMNAME	Provides substance searching and identification on the basis of nomenclature, trade names, synonyms, and other substructure searching techniques. From DIALOG Information Services, Inc. and Chemical Abstracts Service.	1967–present	1,572,000	DIALOG
CHEMSEARCH	Dictionary listing of the most recently cited substances in CA SEARCH and is a companion file to CHEMNAME. From DIALOG Information Services, Inc. and Chemical Abstracts Service.	Current	103,000	DIALOG
CHEMSIS	Dictionary, nonbibliographic file containing those chemical substances cited once during a Collective Index Period of Chemical Abstracts. From DIALOG Information Services, Inc. and Chemical Abstracts Service.	1982–present	1,037,000	DIALOG
CHEMZERO	Dictionary, nonbibliographic file containing those chemical substances for which there are no citations in DIALOG's CA SEARCH files. From DIALOG Information Services, Inc. and Chemical Abstracts Service.	1965–present	1,421,000	Dialog

cont'd.

Database Name	Subject and Source	Entry Date	Items (Approx.)	Vendor
CIS INDEX	Covers US Congress publications. Prepared by Congressional Information Service.	1970–present	188,000	DIALOG
CLAIMS/CHEM	US chemical and chemically related patents plus some foreign equivalents. From IFI/Plenum Data Company.	1947–present	3,276,000	Dialog
CLAIMS/CITATION	Files reference over 5 million patent numbers cited in US. Each record includes a US patent number plus patent numbers cited to that patent by other US patents. From Search Check, Inc. and IFI/Plenum Data Company.	1947–present	3,276,000	Dialog
CLAIMS/CLASS	Classification code and title Dictionary for all classes and all subclasses of the US Patent Classification System. Corresponds to the US Patent and Trademark Office's *Manual of Classification*. From IFI/Plenum Data Company.	Current	110,500	Dialog
CLAIMS COMPOUND REGISTRY	Dictionary of specific chemical compounds. Each record contains the IFI compound term number, main compound name, synonyms, molecular formula, and fragment codes and terms. From IFI/Plenum Data Company.	Current	13,900	Dialog

cont'd

Name	Description	Coverage	Records	Vendor
CLAIMS/GEM	US general, electrical, and mechanical patents. From IFI/Plenum Data Company.	Jan. 1975	60,700	Dialog
CLAIMS/US PATENT ABSTRACTS	Files contain patents listed in the general, chemical, electrical, and mechanical sections of the *Official Gazette* of the US Patent Office. From IFI/Plenum Data Company.	1982–present	220,500	Dialog
CLAIMS/US PATENT ABSTRACTS WEEKLY	Companion to CLAIMS/US PATENT ABSTRACTS. Includes the most current weekly update and records from the current month. From IFI/Plenum Data Company.	Current	1,300–7,800	Dialog
CLAIMS/UNITERM	Gives access to chemical and chemically related patents. Contains subject indexing for each chemical patent from a controlled vocabulary designed to facilitate retrieval of chemical structures and polymers. From IFI/Plenum Data Company.	1950–present	1,574,000	Dialog
COLD	Covers all aspects of Antarctica, the Antarctic Ocean, and subantarctic islands.	1962–present	74,000	SDC
COMPENDEX	Corresponds to *Engineering Index Monthly*. Prepared by Engineering Index.	1970–present	1,415,000	BRS Dialog SDC

cont'd.

Database Name	Subject and Source	Entry Date	Items (Approx.)	Vendor
COMPUTER DATABASE	Contains abstracts from 530 journals, newsletters, tabloids, proceedings, and meeting transactions. Also business books and self-study courses, covering all aspects of computers, telecommunications, and electronics. From Management Contents, Inc.	1983–present	117,000	Dialog
CORROSION	Provides data on the effects of over 600 agents on the most widely used metals, plastics, nonmetallics, and rubbers over a temperature of 40° to 560° F.	current	2,400	SDC
CRDS	Contains up-to-date information on new developments in the field of synthetic organic chemistry.	1944–present	90,000	SDC
CRECORD	Covers highly current coverage of the activities on the floor of Congress. Prepared by Capitol Services.	Jan. 1976	40,000	SDC
CRIS/USDA	Database for agriculturally related research projects. Covers current research in agriculture and related sciences. From the US Department of Agriculture.	1983–present	35,700	Dialog

cont'd.

DATA PROCESSING AND INFORMATION SCIENCE CONTENTS	Informative table-of-contents and bibliographic database that provides subject access to leading microcomputer journals.	1982–present	BRS
DISSERTION ABSTRACTS ONLINE	Informative guide including author, subject, and title for every American dissertion accepted at an accredited institution since 1861.	1861–present	872,500	Dialog
DOE ENERGY	Database of the US Department of Energy. Contains coverage of journal articles, report literature, conference papers, books, patents, dissertations, and translations. Topics include nuclear, wind, fossil, geothermal, tidal, and solar energy. From the US Department of Energy.	1974–present	1,470,000	Dialog
EBIB	Contains worldwide literature on energy. From the Texas A&M Library collection.	1919–present	14,000	SDC
EI ENGINEERING MEETINGS	An index to significant proceedings of engineering and technical conferences, symposia, meetings, and colloquia.	1979–present	286,900	Dialog

cont'd.

Database Name	Subject and Source	Entry Date	Items (Approx.)	Vendor
EIMET	Covers significant papers from published proceedings of engineering and technical conferences, symposia, meetings, and colloquia.	Mid-1982–present	400,000	SDC
ELECTRIC POWER DATABASE	Provides references to research and development projects of interest to the electric power industry and corresponds to the printed work *Digest of Research in the Electric Utility Industry*. From the Electric Power Research Institute.	1972–present	15,200	Dialog
ELECTRONIC YELLOW PAGES-CONSTRUCTION DIRECTORY	Provides online yellow page information for contractors and construction agencies.	Current	668,500	Dialog
EMBASE (formerly EXCERPTA MEDICA)	Consists of abstracts and citations of articles from over 4,000 biomedical journals published throughout the world. Covers the field of human medicine and related disciplines. From the Environmental Studies Institute. From *Excerpta Medica*.	June 1974–present	2,999,500	Dialog
ENCYCLOPEDIA OF ASSOCIATIONS	Contains detailed information on several thousand trade associations, professional societies, and other voluntary organizations.	current	18,200	Dialog

cont'd.

ENERGYLINE	Covers over 200 core journals and other selected journals, and reports, monographs, and newspaper articles. Prepared by the Environment Information Center.	1971–present	54,800	Dialog SDC
ENERGYNET	Database provides up-to-date, directory-type information on almost 3,000 organizations and 8,000 people in energy-related fields.	current	2,750	Dialog
ENVIRONLINE	Covers environmental literature book/film reviews and extracts from *Daily Federal Register.* From the Environmental Information Center.	1971–present	115,500	Dialog SDC
ENVIRONMENTAL BIBLIOGRAPHY	Covers the fields of general human ecology, atmospheric studies, energy, land resources, water resources, and nutrition and health.	1973–present	275,000	Dialog
EPIA	Contains access to literature on electric power plants and related facilities.	1975–present	20,000	SDC
ERIC	Covers report and periodical literature in many education and education-related areas. Prepared under funding by the US National Institute of Education, from the Educational Resources Information Center.	1966–present	561,500	BRS Dialog SDC

cont'd.

Database Name	Subject and Source	Entry Date	Items (Approx.)	Vendor
FLUIDEX	Provides indexing and abstracting of fluid engineering, including theoretical research as well as the latest technology and applications. From BHRA, the Fluid Engineering Centre.	1973–present	158,000	Dialog
FOOD SCIENCE AND TECHNOLOGY ABSTRACTS	Provides access to research and new development literature in areas related to food science and technology. Includes allied disciplines such as agriculture, chemistry, biochemistry, and physics. From the International Food Information Service.	1969–present	281,000	Dialog
FOODS ADLIBRA	Provides up-to-date information on the latest developments in food technology and packaging. From Foods Adlibra Publications.	1974–present	94,500	Dialog
FEDERAL RESEARCH IN PROGRESS	Provides access to information about ongoing federally funded research projects in the fields of physical sciences, engineering, and life sciences.	current	66,200	Dialog
FEDREG	Includes rules, proposed rules, public law notices, meetings, hearings, and presidential proclamations on different subjects.	1977–present	162,000	SDC

cont'd.

DATABASES

FOREST	Provides worldwide literature pertinent to the entire wood products industry.	1947–present	16,058	SDC
FOUNDATION DIRECTORY	Descriptions of over 3,500 foundations from the Foundation Center.	current	4,000	Dialog
FOUNDATION GRANTS INDEX	Cumulation of grants records of more than 400 US philanthropic foundations from the Foundation Center.	1973–present	209,000	Dialog
FSTA	Provides information covering the entire food science and technology.	1969–present	310,000	SDC
GEOARCHIVE	A geoscience database indexing more than 100,000 references each year. From Geosystems.	1969–present	538,500	Dialog
GEOREF	Geological reference file. Covers geosciences literature from journals, plus conferences, symposia, and monographs. Prepared by the American Geological Institute.	1967–present	1,005,000	Dialog SDC
GPO MONTHLY CATALOG	Includes records of reports, studies, fact sheets, maps, handbooks, and conference proceedings issued by all US federal government agencies.	July 1976–present	227,300	Dialog

cont'd.

Database Name	Subject and Source	Entry Date	Items (Approx.)	Vendor
GRANT INFORMATION SYSTEM	Covers grants offered by all levels of governments, commercial organizations, associations, and private foundations in over 88 disciplines. Prepared by Oryx Press. (Began November 1976.)	current	3,600	Dialog
HEILBRON	70,000 entries covering more than 150,000 chemical substances.	current	70,000	Dialog
ABI/INFORM	Covers business management periodical literature from over 550 journals. Prepared by ABI, a division of Data Courier.	Aug. 1971–present	276,500	BRS Dialog SDC
INFORMATION SCIENCE ABSTRACTS	Covers literature on information science and related areas. Includes indexing and abstracts for journal articles, patents, proceedings, monographs, government documents and reports, series, and other publications. From IFI/Plenum Data Company.	1966–present	83,400	Dialog

cont'd.

INSPEC	Corresponds to the printed *Physics Abstracts, Electrical and Electronics Abstracts, Computer and Control Abstracts,* and *IT Focus*. Subject areas covered are atomic and molecular physics; computer programming and applications; computer systems and equipment; and elementary particle physics. From the Institution of Electrical Engineers (IEE).	1970–1976; 1977–present 1969–present	2,494,000	BRS Dialog SDC
ISMEC	Information service in mechanical engineering from Data Courier.	Jan. 1973	161,000	BRS Dialog
INSPEC-PHYSICS	Corresponds to *Physics Abstracts* from the Institution of Electrical Engineers.	1969–present	463,900	BRS Dialog
INSPEC-ELEC/COMP	Corresponds to the *Electrical and Electronics Abstracts* and *Computer and Control Abstracts*.	1969–present	348,000	BRS Dialog
INTERNATIONAL PHARMACEUTICAL ABSTRACTS	Provides information on all phases of the development and use of drugs and on professional pharmaceutical practice. From American Society of Hospital Pharmacists.	1970–present	106,000	Dialog

cont'd.

Database Name	Subject and Source	Entry Date	Items (Approx.)	Vendor
IRIS	Abstracts of educational and instructional materials on water quality and water resources. Includes journal articles, books, brochures and pamphlets, films, filmstrips, videotapes, slides, and overhead transparencies. From the US Environmental Protection Agency Information Project.	1979–present	6,400	Dialog
LIBCON/E LIBCON/F LIBCON/S	Covers all subject areas in monographic literature and audiovisual materials, and includes MARC records from the Library of Congress and LC-cataloged items. LIBCON/E (English materials); LIBCON/F (non-English materials); and LIBCON/S (current materials).	Jan. 1965	691,700 707,700 289,500	SDC
LIFE SCIENCES COLLECTION	Contains abstracts of information in the fields of animal behavior, biochemistry, ecology, entomology, genetics, immunology, microbiology, toxicology, and virology. From Cambridge Scientific Abstracts.	1978–present	730,000	Dialog

cont'd.

LISA (LIBRARY AND INFORMATION SCIENCE ABSTRACTS)	Contains international materials in the field of library and information science. Covers librarianship and library services, online information retrieval, videotex, electronic publishing, word processing, teleconferencing, information storage and retrieval, and abstracting and indexing services. From Library Association Publishing.	1969–present	66,700	Dialog
MATHFILE	Provides coverage of the mathematical research literature of the world. From the American Mathematical Society.	1973–present	420,000	BRS Dialog
MDF/I	This file provides current designation and specification numbers for ferrous and nonferrous metals and alloys.	current	15,000	SDC
MEDLINE	Corresponds to *Index Medicus*, *Index to Dental Literature*, and *International Nursing Index*. Covers every subject in the field of biomedicine. Produced by the US National Library of Medicine.	1966–present	4,687,000	BRS Dialog

cont'd.

Database Name	Subject and Source	Entry Date	Items (Approx.)	Vendor
MENTAL HEALTH ABSTRACTS	Database citing worldwide information relating to the general topic of mental health. From the National Clearinghouse for Mental Health Information, through 1982; IFI/Plenum Data Company, from 1983 to present.	1969–present	465,500	Dialog
MENU—THE INTERNATIONAL SOFTWARE DATABASE	Includes listings of commercially available software for any type of mini- or microcomputer. From the International Software Database Corporation.	current	21,400	Dialog
METADEX	*Metals Abstracts Index* and *Alloys Index*. From the American Society for Metals.	1966–present	596,500	BRS Dialog SCD
METEOROLOGICAL AND GEOASTRO-PHYSICAL	Meteorological and geoastrophysical literature from the American Meteorological Society and the National Oceanic and Atmospheric Administration	1972–present	110,000	BRS Dialog

cont'd.

MICROCOMPUTER INDEX	A subject and abstract guide to magazine articles. Includes general articles about the microcomputer world, book reviews, software reviews, discussions of applications in various milieus, and descriptions of new microcomputer products. From Microcomputer Information Services.	1981–present	31,600	Dialog
MICROSEARCH	Contains coverage of reviews and instructional articles from microcomputer-related literature.	1982–present	17,000	SDC
NATIONAL FOUNDATIONS	Provides records of all 22,100 US foundations that award grants, regardless of the assets of the foundation or of the total amount of grants awarded annually.	current	22,000	Dialog
NICSEM/MIMIS	Provides descriptions of media and devices for use with handicapped children.	1978 edition	36,000	Dialog
NONFERROUS METAL ABSTRACTS	Includes all aspects of nonferrous metallurgy and technology.	1961–Dec. 1983	121,000	Dialog
NTIS	Corresponds to the *Weekly Government Abstracts* and the semimonthly *Government Reports Announcements*. Prepared by National Technical Information Service (NTIS).	1964–present	1,122,000	BRS Dialog SDC

cont'd.

Database Name	Subject and Source	Entry Date	Items (Approx.)	Vendor
NUC/CODES	Contains the names, complete addresses, and National Union Catalog codes for all libraries cited in the CASSI database.	current	400	SDC
NURSING AND ALLIED HEALTH (CINAHL)	Provides access to nursing journals, publications of the American Nurses' Association and the National League for Nursing, and primary journals in allied health disciplines. Prepared by the Cumulative Index to Nursing & Allied Health Literature.	1983–present	30,900	Dialog
OCCUPATIONAL SAFETY AND HEALTH (NIOSH)	Covers over 150 journals in health and safety and related fields.	1972–present	109,500	Dialog
OCEANIC ABSTRACTS	Organizes and indexes technical literature on oceanography, marine biology, marine pollution, ships and shipping, geology and geophysics, meteorology, and governmental and legal aspects of marine resources. From the Cambridge Scientific Abstracts.	1964–present	156,500	Dialog

cont'd.

ONLINE CHRONICLE	Provides information on major online industry events, new databases, computer equipment, search aids, and people in the online world. Prepared by JOB-LINE.	1981–present	2,300	Dialog
ONTAP COMPENDEX	Contains references and abstracts of the engineering and technological literature to be used for training new searchers, and practice in searching technical literature.	Jan.–Apr. 1980	34,000	Dialog
KIRK-OTHMER ENCYCLOPEDIA OF CHEMICAL TECHNOLOGY	Features precise retrieval and display of extensive tabular data, index terms, plus cited and general references.	3d edition (1978)		BRS
PACKAGING SCIENCE AND TECHNOLOGY ABSTRACTS	Provides access to research and development literature in all aspects of packaging science.	1981–present	7,500	Dialog
PAPERCHEM	Over 1,000 periodicals and patents related to paper industries. Prepared by Institute of Paper Chemistry.	July 1967–present	211,500	Dialog SDC
PATLAW	Covers the sources of intellectual property decisions of the US Supreme Court.	1967–present	87,600	Dialog

cont'd.

Database Name	Subject and Source	Entry Date	Items (Approx.)	Vendor
P/E NEWS	Covers 5 major publications in the petroleum and energy fields. Prepared by Central Abstracting Indexing Service of the American Petroleum Institute.	1975–present	359,000	SDC Dialog
PESTDOC/PESTDOC-II/PESTDOC UDB	Covers worldwide literature on pesticides, herbicides, and plant protection. Designed specifically for the information requirements of agricultural chemical manufacturers. Exclusive of fertilizers.	1968–1984 1975–1984 1985–present	144,500 total	SDC
PHARMACEUTICAL NEWS INDEX	Provides current news about pharmaceuticals, cosmetics, medical devices, and related health fields. From Data Courier, Inc.	Dec. 1975–present	140,000	Dialog
PIE	This database contains biological, ecological, physical, and socioeconomic information on the Pacific Islands.	1927–present	22,000	SDC
POLLUTION	Corresponds to *Pollution Abstracts*. Prepared by Pollution Abstracts, Data Courier.	Jan. 1970	102,000	Dialog SDC
POWER	Contains catalog records for books, monographs, proceedings, and other material in the book collection of the Energy Library.	1950–present	38,500	SDC

cont'd.

PSYCALERT	Companion file to PSYCINFO. Provides bibliographic information and indexing for all material included in PSYCINFO. From the American Psychological Association.			Dialog SDC
PSYCINFO	Covers the world's literature in psychology and related disciplines in the behavioral sciences.	1967–present	493,500	Dialog
PTS DOMESTIC STATISTICS	Time series and forecasts on US economics, production, etc. From Predicasts.	Jul. 1971–present	140,000	Dialog
PTS EIS PLANTS	Data on 110,000 industrial plants in the continental US. From Predicasts.	current	2,543,000	Dialog
PTS F&S INDEXES (FUNK & SCOTT)	Concise information on articles relevant to all aspects of business. From Predicasts.	1972–present	2,588,000	Dialog
PTS INTERNATIONAL STATISTICS	Time series and forecasts on foreign economics, production, etc. From Predicasts.	Jan. 1972	140,000	Dialog
PTS MARKET ABSTRACTS (CMA & EMA)	Covers *Chemical Market Abstracts* and *Equipment Market Abstracts*. From Predicasts.	Feb. 1972	100,000	Dialog
PTS MARKET DEFENSE AND TECHNOLOGY	Weekly current awareness for *PTS Market Abstracts* and *F&S Indexes*. From Predicasts.	1982–present	70,900	Dialog

cont'd.

Database Name	Subject and Source	Entry Date	Items (Approx.)	Vendor
PTS US FORECASTS	Consists of abstracts of published forecasts for the United States from trade journals, business and financial publications, key newspapers, government reports, and special studies.	July 1971–present	350,000	Dialog
PTS US TIME SERIES	Includes about 500 time series on the United States, giving historical data and projected consensus of published forecasts through 1990.	Years included vary.	46,800	Dialog
RINGDOC (includes UDB, Ring 6475)	Covers over 400 scientific journals of the pharmaceutical literature. Prepared by Derwent Publications.	1964–present	800,000	SDC
ROBOTICS	Significant resource providing access to current literature in all aspects of robotics.	1980–present		BRS
SAE	Provides access to papers of technical meetings and conferences.	1965–present	16,000	SDC
SCISEARCH	Corresponds to *Science Citation Index*. From the Institute for Scientific Information.	1974–present	6,189,000	Dialog
SDF	A listing of 7,500 known drugs and other commonly occurring compounds.	current	7,500	SDC

cont'd.

SOVIET SCIENCE AND TECHNOLOGY	Provides access to scientific and technical information from Soviet bloc countries.	1975–present	86,400	Dialog
SPIN	Provides the most current indexing and abstracting of a selected set of the world's most significant physics journals.	1975–present	243,000	Dialog
SSIE	Covers ongoing and recently completed research in the life, physical, and social sciences. Prepared by the Smithsonian Science Information Exchange.	1974	262,400	SDC
SSIE CURRENT RESEARCH	Contains reports of governmental and privately funded scientific research projects, either in progress or started and finished during 1978–1982. Includes all fields of basic and applied research in the life, physical, social, and engineering sciences. Prepared by the National Technical Information Service, US Dept. of Commerce.	1978–Feb. 1982	439,000	Dialog
SOCIAL SCISEARCH	Multidisciplinary database index covering every area of the social and behavioral sciences. It corresponds to the printed *Social Science Citation Index*. From the Institute for Scientific Information.	1972–present	1,483,000	Dialog

cont'd.

Database Name	Subject and Source	Entry Date	Items (Approx.)	Vendor
SOCIOLOGICAL ABSTRACTS	Includes the world's literature in sociology and related disciplines in the social and behavioral sciences. Prepared by Sociological Abstracts, Inc.	1963–present	153,500	Dialog
STANDARDS	Covers a diverse range of standards and specifications developed by the SAE and ASTM.			SDC
STANDARDS & SPECIFICATIONS	Provides access to all government and industry standards.	1950–present	109,000	Dialog
SUPERINDEX	Consists of back-of-the-book indexes from almost 2,000 professional-level reference books in science, engineering, and medicine.	recently published		BRS
TELEGEN	Includes information related to the fields of genetic engineering and biotechnology. It corresponds to the printed *Telegen Reporter*. From the Environment Information Center, Inc.	1973–present	14,500	Dialog
TEXTILE TECHNOLOGY DIGEST	Contains international coverage of the literature of textiles and related subjects.	1978–present	116,100	Dialog

cont'd.

TRINET COMPANY DATABASE	Contains current address, financial, and marketing information on US single- and multi-establishment companies.	current	290,000	Dialog
TRINET ESTABLISHMENT DATABASE	Contains current address, financial, and marketing information on US establishments with 20 or more employees.	current	438,000	Dialog
TRIS	Provides transportation research information on air, highway, rail, and maritime transport; mass transit; and other transportation modes.	1968–present	218,000	Dialog
TROPAG	Covers worldwide literature on tropical and subtropical agriculture.	1975–present	60,000	SDC
TSCA INITIAL INVENTORY	A nonbibliographic dictionary listing chemical substances in commercial use in the United States.	May 1983 edition	56,400	Dialog
TULSA	Covers oil and gas exploration, development, and production. Prepared by the Information Services Department of the University of Tulsa.	1965–present	360,000	SDC

cont'd.

Database Name	Subject and Source	Entry Date	Items (Approx.)	Vendor
USCLASS	Contains all US classifications, cross-reference classification, and unofficial classifications for all patents issued from 1790 to date.	1790–present		SDC
USPA/USP77/USP70	Contains all US patents, continuations, divisionals, and defensive documents from 1971 to the present.	1971–present		SDC
VETDOC UDB/VETDOC	Covers worldwide journal literature on veterinary applications of drugs, hormones, vaccines, and growth promotants.	1983–present 1968–1982	68,000 total	SDC
WATER RESOURCES ABSTRACTS	Covers water resource topics including water resource economics ground and surface water hydrology, metropolitan water resources planning and management, and water-related aspects of nuclear radiation and safety. Prepared by the US Department of the Interior.	1968–present	176,000	Dialog
WATERNET	Provides an index of the publications of the American Water Works Association and the WWA Research Foundation. From the American Water Works Association.	1971–present	15,200	Dialog

cont'd.

WELDASEARCH	Provides primary coverage of the international literature on all aspects of the joining of metals and plastics and related areas.	1967–present	76,500	Dialog
WORLD PATENTS INDEX	These files contain data from 3 million inventions represented in more than 6 million patent documents from 30 patent-issuing authorities around the world.	1963–present	3,000,000	Dialog
WORLD ALUMINUM ABSTRACTS	Provides coverage on the world's technical literature on aluminum.	1968–present	108,500	Dialog
WORLD TEXTILES	Contains indexes on the world literature of the science and technology of textile and related materials.	1970–present	130,000	Dialog
WPI/WPIL	Authoritative file of data relating to patent specifications issued by the patent offices of 24 major industrial countries.	1977–present	more than 1,000,000	SDC
ZOOLOGICAL RECORD	Provides worldwide coverage of zoological literature with special interest on systematic/taxonomic information. Prepared by the BioSciences Information Service and the Zoological Society of London.	1978–1981	286,500	Dialog

NLM Databases (selected)

CHEMLINE. Chemical Dictionary online. About 270,000 chemical substance names representing 76,355 unique substances.

MEDLARS. Medical Literature Analysis and Retrieval System Online. Its database produces *Index Medicus*. About 3,000 biomedical journals are included. Jan. 1964–. Accessible also through BRS (uniquely among existing commercial vendors).

MEDLINE. Medlars-on-line. Supplemented by BACK 66 (BACK 72 for literature from 1966 to 1972). Current year's citations plus 2 previous years.

TOXLINE. Includes 375,000 references. The component subfiles are *Chemical-Biological Activities* (CBAC), 1965–; *Abstracts on Health Effects of Environmental Pollutants*, 1972–; *International Pharmaceutical Abstracts*, 1970–; *Toxicity Bibliography*, 1968–; *Health Aspects of Pesticides Abstracts Bulletin*, 1966–; *Hayes File on Pesticides*, 1940–1966.

ELECTRONIC PUBLISHING AND VIDEODISC AND CD-ROM PRODUCTS

Since 1984, micro-based optical videodisc technology has exploded in a variety of information environments. Recent developments in videodisc technology and in linking this technology with computerized databases have created great potential for those applications in large-scale data storage and retrieval in subject and applications areas. Chen has offered a general overview ["Micro-Based Optical Videodisc Applications," *Microcomputers for Information Management* 2 (4): 217–240 (Dec. 1985)], and this is only the beginning of a flux of information sources on CD-ROM. The following are sample listings of databases in science and technology that are available in CD-ROM discs and can be subscribed by end-users for use on their microcomputers with appropriate CD-ROM drives, via various electronic publishing services or vendors:

- Digital Equipment Corporation—10 CD-ROM serials are available now, with more to come:

    COMPENDEX: Electrical and Computer Engineering
    COMPENDIX: Chemical Engineering
    COMPENDIX: Aerospace Engineering
    NTIS: Environmental Health and Safety
    NTIS: Computers, Communication and Electronics
    NTIS: Medicine, Health Care, and Biology
    NTIS: Aeronautics, Aerospace, and Astronomy
    CHEMICAL ABSTRACTS: Health and Safety in Chemistry

RSC: Current Biotechnology Abstracts

Fraser Williams FINE Chemicals Director

- Information Access Company InfoTrac Project started in March 1985 with videodiscs containing over a half-million citations from major business and technical publications, such as those included in *Business Index*, *Magazine Index*, *National Newspaper Index*, and *Trade and Industry Index*.
- BRS offers a complete package of services with CD-ROM technology. It includes database design; data conversion; data inversion; premastering; mastering; retrieval software BRS/SEARCH, driver software for Philips, Hitachi, and others; and training. This enables one to consider electronic publishing of private scientific and technical files.
- Cambridge Scientific Abstracts has captured its entire *Aquatic Sciences and Fisheries Abstracts* online database on a single CD-ROM disc.
- Elsevier Science Publishers, Amsterdam, The Netherlands, has captured its Excerpta Medica database, the EMBASE, on CD-ROM.
- Silver Platter Information Services, Wellesley Hills, Massachusetts, offers access to large reference databases including ERIC, PsycLIT, EMBASE, NICEM, and others on CD-ROM.

CD-ROM products have exploded in the marketplace. It is said that in mid-1986, there were already over 40 CD-ROM products on the information market, and it is fair to expect that there will be very fast development in this area related to the use of scientific and technical databases in the years to come.

NAME INDEX

A. R. C. Weed Research Organization, 550
ACTION/ Peace Corps, Information Collection and Exchange, 107
Aagaard, Knut, 390
Abbott, David, 450
Abbott, R. Tucker, 363
Abel, Dominick, 315
Abell, G. O., 607
Abercrombie, M., 125
Abercrombie, Stanley A., 145
Abikoff, W., 303
Abraham, Guy E., 172
Acoustical Society of America, 542
Ad Hoc Committee of Transportation Librarians, 28
Adams, Catherine F., 83, 254
Adams, Charles K., 207
Admas, J. T., 191
Adams, Leon D. 316
Adams, R. W., 668
Addison, C. C., 501
Adler, David, 642
Aero Publishers Aeronautical Staff, 276
Agajanian, A. H., 49, 50
Agricultural Sciences Information Network, 408
Agris, Paul F., 37
Ahmad, Shamin, 230
Aichelberg, P. C., 455
Ainsworth, G. C., 529
Air Pollution Technical Information Center, 594
Aires, Sidney John, 144
Ajl, Samuel J., 530
Akey, Denise, 668
Akron-Summit County Public Library, 539
Albright, Arlene, 33
Alden, Peter, 360
Alexander, Gerard L., 373
Alexander, J. M., 511, 533
Ali, Salim, 179
Aliev, A. I., 249
Alkire, Leland G. 575
Allaby, Michael, 148, 149
Allan, E., 192

Al-Layla, Muhammad A., 230
Allen, B. L., 285
Allen, Charles, 648
Allen, Charles E., 324
Allen, David, 376
Allen, G. R., 73
Allen, Geoffrey F., 439
Alper, A. M., 533
Alter, Dinsmore, 343
Altman, Norman H., 177
Altman, Philip L., 170, 243, 253
Aluri, Rao, 15, 592, 599
Amboy, Perth, 325
American Association for the Advancement of Science, 307, 621, 664
American Association of Agricultural Engineers, 460
American Association of Engineering Society, 401, 402
American Chemical Society, 6, 623, 635, 646
American Chemical Society of Polymer Chemistry, 628
American Concrete Institute, 278
American Consulting Engineers Council, 276
American Council of Learned Societies, 450
American Cyanamid Company, 222
American Fabrics Magazine, 86
American Federation of Information Processing Societies, 611
American Geographical Society, 384, 392
American Geographical Society, New York, 9
American Institute of Architects, 322
American Institute of Physics, 607
American Institute of Timber Construction, 280
American Machinist Editors, 289
American Mathematical Society, 454, 614, 615, 622
American Museum of Natural History Library, 4
American National Standards Committee, 138
American National Standards Institute, 134, 592, 639, 643

American Philosophical Society, 5
American Physical Society, 457
American Public Works Association, 485
American Radio Relay League, 202
American Society for Metals, 97
American Society for Testing and
  Materials, 250, 574, 639
American Society of Civil Engineers, 324
American Society of Mechanical
  Engineers, 214, 463, 427, 643,
American Society of Metals, 290
American Welding Society Committee on
  Machinery and Equip., 650
Amos, S. W., 137, 203
Amos, William H., 357
Andersen, H. H., 47
Anderson, B. W., 371
Anderson, Donn, 378
Anderson, Edwin P., 334
Anderson, Frank J. 479
Anderson, S., 254
Andrews, Henry N., 318
Andriot, Donna, 373
Andriot, John, 264, 373, 596, 597
Andriot, Laurie, 373
Angel, Heather, 319
Angell, Ian O., 330
Angell, Madeline, 347
Angelo, Joseph A., 133
Angier, Bradford, 347
Anglemyer, Mary, 59
Anglin, D., 146
Anglin, Donald L., 208
Antelman, Marvin S., 82
Antony, Arthur, 18
Arbuckle, J. Gordon, 228
Arctic Institute of North America, 43
Ardern, Richard, 21
Arecchi, F. T., 160
Arem, Joel E., 77
Arlt, Reiner, 58
Arlt, W., 251
Arnell, A., 643
Arnould, Michael, 101
Arpan, Jeffrey S., 394, 427
Asfahl, C. R., 214
Ash, Irene, 82, 85, 257
Ash, Lee, 17
Ash, Michael, 82, 85, 257
Asimov, Isaac, 449, 465, 469

Askland, Carl L., 89
Association of Energy Engineers, 463
Association of Environmental
  Engineering Professors, 401
Astle, Melvin J., 165
Atkins, Harold, 279
Atkins, M., 391
Atkinson, Bernard, 170
Atomic Energy Commission, 639
Atomic International Forum, 644
Audio-Visual Association, 659
Audubon, John J., 357
Audubon Society, 370
Auer, Peter, 518
Auger, C. P., 47, 52
Auger, Charles P., 593
Augustithis, S. S., 386
Aurand, L., 298
Austin, D. G., 257, 321
Automotive Industries, 650
Avallone, Eugene, 215
Axford, W. Ian, 153
Ayensu, Edward S., 313, 347
Ayres, Jak, 356
Azad, Hardam S., 231
Azzi, A., 296

Back, Philippa, 314
Bacus, J., 173
Badash, Lawrence, 475
Baer, Eleanora A., 5
Baetzner, Willard H., 407
Bahr, Alice H., 659, 660
Bailar, J. C., 524
Bailey, A. E., 238
Bailey, Ethel Zoe, 127
Bailey, Liberty Hyde, 127
Bair, Frank E., 256
Baird, H. W., 294
Baker, A. D., 492
Baker, David, 342, 484
Baker, Joseph T., 181
Baker, Robert F., 193
Balachandran, Sarojini, 25, 573
Balbach, Margaret, 296
Balcomb, Kenneth C., 369
Balje, O. E., 335
Balkan, Eric, 421
Ballard, S. S., 492
Ballentyne, Denis W. G., 114, 115,

INDEX

Bally, A. W., 388
Bance, S., 167
Banov, Abel, 210
Bapton, Douglas, 161
Baratsabal, Pierre, 299
Barbour, Roger W., 366
Bard, Allen J., 67
Barile, M. F., 530
Barilleaux, Claude, 395
Barin, Ihsan O. K., 252
Barker, John A., 137
Barker, Michael, 437
Barner, Herbert E., 168
Barnes, Ervin H., 273, 274
Barnett, J. A., 310
Barnett, V., 238
Barnhart, John H., 457
Baron, Robert E., 244
Baron, Stephen P. E., 279
Barr, Avron, 205
Barr, Claude A., 353
Barrett, Paul H., 98
Bartholomew, John C., 387
Barton, Derek, 524
Barton, Robert, 390
Bashkin, Stanley, 248
Basselman, J. A., 130
Bassett, Frank E., 255
Batellier, Jean-Pierre, 299
Bates, Robert L., 130
Battey, M. H., 131
Battle, G. C., 34
Baughman, Gary L., 227
Bauman, Laurence, 163
Baumeister, Theodore, 215
Baumel, J. J., 129
Baynton, R. A., 460
Beach, Edward L., 147
Beadle, M., 548
Beadle, W. E., 281
Beale, Helen Purdy, 39
Bean, W. J., 314
Beaver, Bonnie, 316
Bebee, Charles N., 38
Becher, Paul, 66, 82
Beck, Rasmus, 241
Becker, Ernest I., 272
Beckett, Gillian, 70
Beckett, Kenneth, 70
Bedini, Silvio, 466

Bednar, Henry H., 218
Behler, John L., 369
Behnke, H., 522
Beiglbock, W., 494
Beilstein, Friedrich, 163
Belding, William G., 155
Belinky, Charles R., 383
Bell, E. A., 72
Bell, Eric Temple, 454
Bell, F. G., 255
Bell, S. P., 459
Bellance, Nicolo, 186
Bellman, Richard E., 522
Belloni, Lanfrance, 34
Belove, C., 195, 199
Belt, Forest H., 444
Ben-Yomi, M., 105
Benarie, Michel M., 520
Bender, Arnold E., 135
Benjamin, William A., 410
Bennett, Gary F., 231
Bennett, Stuart, 482
Bensinger, Charles, 661
Bente, Paul F., 432
Benton, Mike, 329, 418
Beran, J. A., 294
Bereny, J. A., 339
Bergmeyer, H. U., 296
Bergsma, Daniel, 406
Bernstein, I. M., 221
Bernstein, Richard B., 308
Berra, Tim M., 383
Berrios, Jose, 107
Berry, Richard W., 133
Bertolotti, Mario, 474
Besancon, Robert M., 66
Bestgen, Barbara J., 665
Betchel, Helmut, 359
Betz Laboratories, 184
Beudell, Martin, 446
Beutelspacher, H., 385
Bevan, Stanley C., 121
Bever, Michael B., 90
Bewer, Susan L., 7
Bewick, Micheal W. M., 184
Beyer, William H., 236, 238
Bharucha-Reid, A. J., 490
Biboul, Roger R., 543
Bickel, John O., 194
Biddle, Wayne, 115

Bieber, Doris M., 55
Biederman, E. W., 392
Bier, Robert A., 45
Bierman, Howard, 161
Biery, Terry L., 180
Bikales, N. M., 628
Bilan, Dennis, 293
Billett, Michael, 217
Billings, Clayton H., 279
Birks, H. J. B., 381
Bishop, A. C., 371
Bishop, Dennis, 366
Bishop, Roy L., 154
Bisset, Ronald, 58
Black, George W. Jr., 29
Blackburn, Graham, 93
Blackman, R. L., 364
Blackwelder, Richard E., 42
Blackwell, Richard J., 29
Blake, A., 214, 216
Blake, Emmet R., 361
Blake, L. S., 189
Blake, Leslie S., 257
Blanchard, J. Richard, 310
Blanchard, Robert O., 348
Blanco-White, T. A., 631
Bland, Roger C., 300
Blandford, Percy W., 480
Blaser, Bonnie R., 400
Bliss, Lawrence C., 296
Bloch, Carolyn C., 25, 431, 432
Bloome, S., 302
Blosser, Patricia E., 622
Bock, Rudolph, 165
Bockris, J. O'M., 525
Bodson, Dennis, 144
Boer, Karl W., 519
Bohme, S., 540
Bolz, Ray E., 244
Bolz, Roger W., 143
Bonanno, Antonio C., 371
Bond, P. S., 147
Bond, Richard G., 227, 229
Bonnington, S. T., 52
Bonny, J. B., 192
Booser, E. R., 215
Borgaonkar, Digamber S., 37
Borgerson, Mark J., 329
Borgese, Elisabeth M., 447
Bosch, Robert, 146

Bossong, Ken, 430
Botez, Dan, 144
Bott, J. F., 158
Bottle, R. T., 19, 20
Boulos, Loufty, 176
Bourgeois, Joanne, 78
Bourne, G. H., 509
Boustead, I., 226
Bovay, H., 198
Bower, Cynthia E., 566
Bownam, Hamilton B., 217
Boxshall, G. A., 148
Boyer, Calvin J., 617
Boyer, Paul, 529
Bracegirdle, Brian, 486
Braden, James F., 333
Bradfield, Valerie J., 23
Bradley, Elihu F., 290
Bradley, R. S., 533
Brady, George S., 214
Bragdon, Clifford R., 27
Bramwell, Martyn, 390
Brand, John R., 198
Brandes, Eric A., 333
Brandup, J., 188
Brater, Ernest F., 193
Braun, Ernest, 475
Brechen, Kenneth, 469
Brechner, Irv, 141
Breden, Leslie H., 646
Bregman, J. I., 535
Bregman, Jacob I., 233
Brenchley, David L., 231
Bretherick, L., 167
Brewer, Richard, 284
Bridgewater, A. V., 233
Briggs, D., 160
Briggs, Geoffrey, 376
Brimmer, Frances M., 124
British Kinematograph Sound and Television Society, 657
British Library, 603
British Standards Institution, 639, 646
Brobeck, U., 296
Brocardo, G., 371
Brociner, Grace, E., 615
Brodie, Edmund D., 370
Bromberg, Joan Lisa, 476
Brookes, John, 323
Brookhart, J. M., 172

Brooks, Hugh, 137
Broutman, L. J., 533
Brown, Dennis A., 313
Brown, J., 303
Brown, James W., 658
Brown, Lauren, 348
Brown, Laurie M., 472
Brown, Lester O., 92
Brown, Marjorie J., 678
Brown, P. R., 304
Brown, Sam, 342
Brown, Shirley N., 658
Brown, Tom, 314
Brown, Vinson, 152
Browne, E. J., 466
Bruce, Robert, 303
Brueckner, Keith A., 523
Brundle, C. R., 492
Brush, Stephen G., 34, 475
Bruun, Bertel, 361
Bruzek, A., 121
Bryant, Jeannette, 436
Bryant, Mark, 387
Buchanan, Robert E., 273, 310
Buchanan, William W., 597
Buchsbaum, Walter H., 87
Buchwald, Vagn F., 181
Buckingham, J. B., 122, 380, 379
Buckland, W. R., 118
Bucksch, H., 101
Budnick, Jim, 340
Bugliarello, George, 482
Bull, Donald, 414
Bull, John, 358
Bunge, Mario, 456, 475
Bunyan, John A., 660
Burington, R., 156
Burington, Richard S., 236
Burkart, A., 242
Burke, John Gordon, 27
Burkett, J., 624
Burnham, Robert, 153
Burroughs, Stephen M., 293
Burton, John, 366
Burton, Maurice, 75
Burton, Philip E., 141
Burton, Robert E., 40, 75
Busch, Harris, 528
Buschsbaum, Walter H., 258
Butchart, Ian, 660

Butkovskiy, A. G., 155
Butler, B., 419, 662
Butler, Marian, 2
Buttress, F. A., 95, 397
Byerly, Greg, 678
Bynum, W. F., 466

Cabannes, H., 494
Cain, C. K., 497
Cajori, Florian, 471
Cakir, A., 282
Calder, Nigel, 379
Calkins, Carroll C., 352
Callaham, Ludmilla I., 105
Calvert, Robert P., 631
Calvert, Seymour, 232
Cambridge Information & Research Services, 435
Cameron, A. G. W., 629
Cameron, Derek, 235
Camougls, G., 341
Campbell, Bruce, 127
Campbell, Paul N., 500
Canada, National Science Library, 572
Canuto, V., 154
Cap, Ferdinand F., 160
Carlander, Kenneth D., 178
Carley, Larry W., 335
Carlson, G. E., 136
Carlson, Harold, 219
Carmicheal, Robert S., 212
Carnegie-Mellon University, 457
Caroli, S., 241
Carpenter, M. Stahr, 31
Carr, Anna, 179
Carr, David, 346
Carr, Joseph J., 201, 202
Carriere, Gerardus, 109
Carrington, David K., 374
Carroll, Walter J., 658
Carson, J. W., 282
Carter, David E., 321
Carter, E. F., 115
Carter, F. T. C., 103
Carvill, James, 460
Casanova, Richard, 371
Case, Gerard R., 319
Casper, M. S., 635
Castle, F., 236
Castro, Jose I., 364

Cator, Lynn E., 48
Catsky, J., 40
Caulfield, H. J., 159
Cavinato, Joseph L., 151
Celotta, Robert, 523
Centre National de la Recherche Scientifique, 613
Centro International de Agricultura Tropical, 624
Chaballe, L. Y., 111
Chabot, Leon, 160
Chadwick, Alena F., 40
Chaiwel, Alain, 277
Challinor, John, 130, 481
Chalmers, R. A., 294
Champagne, Audrey B., 537
Champion, Richard, 328
Chan, Harvey T., 187
Chancellor, John, 457
Chandler, Harry E., 153
Chandor, Anthony, 142
Chaney, J. F., 250, 258
Chang, S., 195
Chang, Sheldon S. L., 204
Change, Raymond, 54
Chant, Christopher, 93
Chant, S. R., 71
Chao, Jing, 245
Chapey, Nicholas P., 184
Chapman, D. R., 116
Chapman, O. L., 503
Charlton, T. M., 487
Charlwood, B. V., 72
Chartered Institute of Patent Agents, 631
Chartrand, Mark R., 343
Chase, Agnes, 549
Chase, Chevy, 677
Chaudier, Louann, 461
Chauliaguet, Charles, 299
Chemical Abstracts Service, 169
Chemical Marketing Reporter, 405
Chen, Ching-chih, 1, 16, 327, 422, 677
Chen, E. C. S., 295
Chen, J. C. P., 186
Cheremisinoff, Paul N., 90, 158, 184, 212, 217, 227, 228, 231, 232, 284, 262
Chering, W. L., 249
Cherman, Carl M., 154
Chernov, A. A., 525

Chernukhin, A. E., 105
Chesterman, Charles W., 370
Chisholm, L. J., 235, 642
Chiu, Hong-yee, 101
Chiu, Yishu, 114
Christensen, Fritjof E., 294
Christensen, James J., 167, 251
Christian, Jeffrey M., 335
Christiansen, Donald, 197
Christiansen, Laurence A., 461
Christianson, G. E., 454
Christianson, Gale E., 470
Chrysler Corporation, 102
Chryssostomidis, Marjorie, 22, 56
Chu, Alfred E., 206
Chubin, Daryl E., 30
Churacek, J., 169
Church, Helen N., 253
Church, Horace K., 191
Churchill, Peter, 384
Churchill, S. W., 532
Churchwell, Jan W., 461
Cibbarelli, Pamela, 421
Cioffari, Bernard, 293
Cipriano, Anthony J., 461
Clark, Brian D., 58
Clark, Hannah, R., 54
Clark, John, 255
Clark, Neal, 178
Clark, P. F., 148
Clark, T., 165
Clarke, Donald, 62, 301
Clason, W. E., 110, 111
Clauser, Henry R., 212, 214, 89
Clay, H. M., 227
Clay, Katherine, 207
Clayton, F. E., 210
Clayton, G. D., 210
Cleevely, R. J., 45
Clements, James F., 358
Clepper, Henry, 464
Clifford, Martin, 239
Close, H. A., 323
Clyde, James E., 322
Cnumas, Sophie, 640
Coachman, L. K., 390
Coaker, T. H., 506
Coakley, W. A., 164
Coan, Eugene V., 438
Coblans, Herbert, 18

Cochran, Doris M., 369
Coffer, Zvi, 286
Coffey, Burton T., 407
Coffey, David J., 73
Coffey, S., 526
Cofield, R. E., 282
Cogswell, F. N., 321
Cohen, I. Bernard, 44, 455, 465
Cohen, Jack S., 477
Cohen, Morris, L., 30
Cohn, P. M., 522
Cole, Andrew R. H., 241
Cole, Richard J., 172
Coleman, William, 479
Colgate, Craig, 654, 668
Colin, Henri, 18
Collazo, Javier L., 107
Collings, E. W., 288
Collins, Henry Hill, 356
Collins, K. J., 623
Collins, Mike, 6
Collins, Ramond A., 197
Collison, Robert L., 100
Commission for the Geological Map of the World, 386
Commission of the European Communities, 97, 100
Committee for the World Atlas of Agriculture, 382
Committee on Space Research, 610
Commonwealth Bureau of Animal Nutrition, 546
Commonwealth Bureau of Horticulture and Plantation Crops, 549
Commonwealth Bureau of Pastures and Field Crops, 549
Commonwealth Bureau of Soils, 548
Commonwealth Forrestry Bureau, 549
Commonwealth Mycological Institute, 549
Compassero, B. P., 46
Computer Consultants Ltd., 560
Conference of Biological Editors Style Manual Committee, 274
Congrat-Butlar, Stefan, 614
Conlan, J., 303
Conn, E. E., 528
Conn, G. K., 523
Conniffe, Patricia, 139
Connolly, T. F., 34, 35

Connor, John J., 401
Connors, Martin, 678, 417, 419
Conseil International de la Langue Francaise, 150
Considine, Douglas M., 63, 66, 67, 84, 225
Considine, Glen D., 84
Consortium of University of Film Centers, 657
Constance, John D., 335
Constantine, Albert, 323
Conway, H. McKinley, 181
Conway, Richard A., 229
Cook, Arthur, 439
Cook, David, 317
Cook, Francis R., 369
Cook, J. Gordon, 189
Cook, Steve, 302
Cook, T. D., 388
Cooke, Edward I., 123
Cooke, Richard W. I., 123
Coombs, Clyde F., 201
Coombs, J., 406
Coon, Nelson, 126
Cooper, J. R., 235
Copley, E. J., 538
Corbet, Gordon B., 368
Corbin, John B., 553
Corkhill, T., 134
Corkhill, Thomas, 125
Corliss, William R., 154, 287, 318
Cornell, G., 303
Cornell, James, 255
Cornog, Martha, 26
Corsi, Pietro, 467
Cortada, James W., 486
Cosentino, John, 418
Cote, Rosalie J., 97
Cote, Wilfred A., 391
Coulston, Frederick, 534
Council of Europe, 632
Counihan, Martin, 147
Couper, Alastair, 390
Court, Doreen, 314
Courtenay, Booth, 354
Cousteau, Jacques Yves, 262
Cousteau Society Staff, 262
Cowan, Henry J., 297
Cowan, R. S., 20
Cowan, Samuel T., 129

Cowling, Ellis B., 530
Cox, George W., 296
Cox, Peter, 347
Cox, Richard H., 172
Coyle, J., 286
Coyle, Patrick L., 85
Crabbe, David, 91, 271, 392
Crabtree, Koby T., 295
Crafts-Lighty, A., 22
Craigie, J. S., 175
Craker, Lyle E., 40
Cramp, Stanley, 179
Crane, John, 298
Crawford, S. R., 285
Crede, C. E., 162
Creedy, John, 293
Cremer, Herbert W., 532
Crespi, R. S., 635
Cresser, M. S., 294
Criddle, W. J., 170
Crimmins, James C., 660
Croall, Stephen, 223
Crockford, H. D., 294
Croft, L. R., 172
Croft, Terrell, 196
Crompton, T. R., 503
Cronquist, Arthur, 176, 300
Crook, Leo, 132
Crosby, Marshall R., 126
Cross, Charles A., 375
Crouse, W., 146
Crouse, William H., 208
Crowley, Ellen T., 95
Crowley, Maureen, 56, 665
Crowson, Phillip, 181
Crump, Ian A., 399, 667
Csbaki, F., 305
Cuff, David J., 392
Culberson, Chicita F., 312
Cullen, J., 351
Culp, Russell L., 228
Cunningham, John E., 201, 217
Cunningham, John J., 97
Cunningham, William A., 82
Curran H. A., 385
Czapowskyj, Miroslaw M., 57

D'Aleo, Richard J., 264
Daglish, H. N., 494
Daintith, John, 120, 123, 450

Dalal-Clayton, D. B., 130
Dales, R. P., 296
Dance, S. Peter, 363, 477
Danheiser, R. L., 251
Danielli, J. F., 509
Danishefsky, Samuel, 504
Danishefsky, Sarah E., 504
Darbre, A., 170
Darmstadter, Joel, 268
Darton, Nelson H., 44
Dasbach, Joseph M., 30
Dasenbrock, David H., 143
Daumas, Maurice, 483
Davidson, A., 215
Davidson, R. H., 179, 365
Davidson, Robert L., 188
Davies, Evan John, 221
Davies, Jenne, 354
Davies, Owen, 677
Davies, Paul, 354
Davies, T., 532
Davinson, Donald, 617
Davis, Bob J., 61
Davis, C., 189
Davis, Elisabeth B., 20
Davis, Henry B. O., 485
Davis, M. S., 100
Davis, P. H., 351
Davis, Richard C., 19, 69
Davis, Wayne H., 366
Davson, H., 530
Dawson, E. Yale, 313
Day, David, 305
Day, John A., 370
Day, Robert A., 299
Daynes, R., 419, 662
De Bray, Lys, 355
De Coster, Jean, 108, 146
De Kerchove, Rene, 112
De Padirac, B., 99
DeBakey, Lois, 573
DeBlase, Anthony F., 275
DeChiara, Joseph, 192
DeFilipps, Robert A., 347
DeHarpporte, Dean, 389
DeLuca, H. F., 166
DeMaw, M. F. "Doug", 204
DeSola, Ralph, 95
DeVorkin, David H., 470
DeVries, Louis, 102, 103, 122

Deacon, Margaret, 481
Dean, Amy E., 332
Dean, Bashford, 41
Dean, John A., 169
Deasington, R. J., 330
Deason, Hilary J., 3
Debelius, JoAnne R., 11
Deeson, Eric, 120
Degremont Company Editors, 194
Deighton, S., 51
Deighton, Suzan, 561
Dekker, Marcel, 515
Delks, Patricia J., 536
Dellantonio, Jeanann M., 400
Delly, John G., 378
Delpar, Helen, 77
Delson, Eric, 477
Demaison, Henri, 107
Dempsey, Michael, 62
Dempsey, Michael W., 307
Denney, R. C. A., 120, 122
Denver Public Library, 12
Derivation and Tabulation Associates, 258
Derry, Duncan R., 386, 388
Derz, Friedrich W., 405, 412
Deskins, Barbara B., 315
Desmond, Ray, 458
Desprairies, P., 261
Devarenne, M., 365
Dexter, S. C., 213
Dhangra, Onkar D., 312
Diagram Group, 356
Dickinson, Colin, 69, 127
Dickinson, Edward C., 360
Dickinson, Terence, 342
Dickinson, William C., 227
Dietrich, Richard V., 244, 245, 372
Diffor, John C., 658
Dinkel, John, 137
Dittmer, Dorothy S., 253
Dixon, Colin J., 385
Dixon, John, 297
Dixon, Robert S., 235
Dobrzynski, L., 159
Dobson, P. J., 472
Dock, George, 357
Doherty, Paul, 375
Doig, Alison G., 32, 264
Dolphin, David, 241

Donaldson, C. H., 385
Doner, D. B., 482
Doornkamp, John C., 386
Dorf, Richard, 261
Dorian, A. F., 101, 103
Dorler, Patricia A., 632
Dorling, Alison R., 17
Dougan, Richard, 423
Douglas, James S., 343
Douglas-Young, John, 87, 161
Dow Chemical Company, 242
Dowd, Morgan D., 438
Downing, Douglas, 89
Downs, Robert B., 451
Dowson, D., 487
Doyle, Rodger P., 186
Drake, Stillman, 455
Drazil, J. V., 114
Driscoll, Frederick F., 281
Driscoll, Robert, 420
Driscoll, Walter G., 159
Droop, M. R., 511
Drouillard, Thomas F., 32
Drury, George H., 335
DuBois, Harry, 187
Dubin, Fred S., 302, 647
Dubuigne, Gerard, 136
Duchaufour, Philippe, 386
Dudas, Janice L., 291
Dudley, Darle W., 216
Duffie, John A., 519
Dugdale, Chester B., 349
Dukas, Helen, 455
Dummer, Geoffrey W., 472, 485
Dunitz, D., 526
Dunlap, Mary Anne, 409
Dunning, John S., 382
Dunthorne, Gordon, 8
Dupayrat, J., 96
Durney, Lawrence J., 208
Durrant, C. J., 121
Dyer, Jon C., 230
Dyer, K. R., 180
Dykeman, Peter A., 347

Eadie, Donald, 330
Earley, Bert, 161
Earney, Fillmore C. F., 53
Eastman, C. R., 41
Eastop, V. F., 364

Eberhard, A. von G., 66
Ebison, Maurice, 246
Edelhart, Mike, 677
Edlin, Herbert L., 70, 314
Edson, William D., 440
Edwards, D. A., 522
Edwards, Elwyn H., 367
Edwards, Ernest P., 295
Edwards, Stephen R., 546
Efron, Vera, 134
Eggers-Lura, A., 340
Ehrenreich, Henry, 494
Eilenberg, Samuel, 522
Eiselt, H. A., 157
Eisen, Sydney, 31
Eisenberg, Jerome M., 356
Eisenreich, Gunther, 106
Electronics Magazine, editors, 462
Electronics Today International, editorial staff, 202
Elerick, Maria Luz, 29
Elias, Thomas S., 346, 347
Elizanowski, Z., 245
Elliot, Clark A., 449
Ellis, Barbara W., 668
Ellis, G. P., 170
Ellmore, Kenneth, 463
Elmaghraby, Salah E., 157
Elonka, Stephen M., 282, 290
Elving, P. J., 527
Elwell, D., 308
Embrey, Peter G., 283
Emerson, William K., 363
Emiliani, Cesare, 116
Emme, E. M., 484
Emmerson, Donald W., 457
Energy Directory Update Service, 432
Energy Information Administration, 261
Energy Resources Center, 339
Engel, L., 379
Engelman, Donald M., 495
Engineering Manpower Commission, 402
Engineering Societies Library, New York, 11
Engineering and Mining Journal, 220
Englisch, W., 241
Englund, Harold M., 232
Ensminger, A. H., 84
Ensminger, M. E., 73
Environmental Protection Agency, 266

Epstein, M. A. 506
Epstein, Samuel S., 92
Erdtmann, Gerhard, 239
Ernst, Richard, 102
Ershler, A. B., 504
Eschmeyer, William N., 363
Eshbach, Ovid Wallace, 181
Estes, Carol, 356
Ethridge, James M., 393
Eubanks, Gordon, 303
European Brewery Convention, 110
European Communities Commission, 98
European Heating and Ventilation Assn., representatives, 146
Eustace, Harry F., 200
Evan, Frederica, 12, 572
Evans, D. S., 282
Evans, Frank L., 183
Evans, G., 162
Evans, L. S., 332
Evensen, Harold A., 341
Everest, F. Alton, 161
Everett, Thomas H., 71
Everling, W., 297
Evinger, William R., 264
Ewald, William R., 665
Ewing, G. W., 477
Exton, Harold, 156
Eyring, H., 526
Eyring, Henry, 499
Eyring, LeRoy, 168

Fabian, M. E., 162
Fagen, M. D., 486
Fahl, Kathryn A., 409
Fairbridge, Rhodes W., 77, 78, 513
Fairchild, E. J., 227
Fairchild, Herman L., 481
Fairchild, Wilma B., 27
Falge, Pat, 406
Fantel, Hans, 305
Farago, Francis T., 209
Farber, Eduard, 457
Farmer, Penny, 338
Farrall, Arthur W., 130
Farrand, John, 72, 358
Farrell, Lois, 310
Fasman, Gerald D., 171
Faust, Joan Lee, 349, 351
Fay, Catherine H., 570

Federal Aviation Administration, 320
Federal Council For Science and
  Technology, 624
Federation Internationale de la
  Precontrainte, 113
Federation of American Societies for
  Expermiental Biology, 609
Federation of Societies for Coatings
  Technology, 135
Feenberg, William J., 434
Feigenbaum, Edward, 205
Feiler, Howard D., 230
Feirer, John L., 642
Feirtag, Michael, 469
Feldman, Anthony, 453
Feldman, Edwin B., 225
Feldzamen, A. N., 300
Ferguson, Eugene S., 482
Fermi, G., 381
Fernelius, W. C., 98
Fernie, Donald, 454
Ferrier, R. W., 484
Festing, Michael F. W., 551
Feuer, Henry, 524
Feulner, John A., 399, 402, 410, 413,
  414, 416, 417, 434, 438
Feys, R., 118
Fichter, George S., 354
Field, Barry C., 26
Field, Richard L., 301
Fieser, Louis F., 252
Fieser, M., 251
Fieser, Mary, 252
Fink, Donald G., 194, 195, 197
Finkel, L., 303
Finkle, C. W., 78
Finlay, Patrick, 446
Fisher, James, 76, 362
Fisher, Ronald A., 239
Fitch, F. B., 118
Flach, George W., 647
Flanagan, W. M., 199
Flatau, Courtney L., 182
Fleischer, Michael, 131
Fleming, James Evans, 51
Fleming, Lizanna, 436
Fleming, Rodney R., 415
Fletcher, Alan, 236
Fletcher, Harold R., 479
Fletcher, Leonard, 136

Fletcher, W. W., 176
Fletcher, William W., 520
Flint, V. E., 360
Flood, Walter Edgar, 116
Flores, Ivan, 205
Florkin, Marcel, 477, 499, 528, 529
Flugge, S., 66
Folts, Harold C., 649
Food and Agriculture Organization of
  the United Nations, 8, 645
Fookes, P. G., 181
Ford, Peter, 453
Ford, Richard A., 251
Forman, Paul, 473
Forster, Alan, 378
Forsythe, William E., 240
Foss, Christopher, 444
Foster, J. S., 270
Foster, Leslie T., 60
Foster, Michael S., 313
Foster, Timolthy R. V., 288
Foster, W., 230
Fothergill, Richard, 660
Fowler, G. N., 523
Fowler, W. H., 282
Fraga, Serafin, 248, 249
Francki, R. I. B., 381
Frank, Robyn C., 414
Franks, Felix, 532
Frederick, Richard G., 124
Frederiksen, Thomas M., 298
Freedman, Alan, 140
Freedman, Deborah A., 299
Freedman, George, 299
Freedman, Stephen J., 668
Freeman, Henry G., 104
Freeman, R. B., 29
Freeman, R. L., 282
Freeman, Robert L., 41
Freeman, Roger L., 203
Freethy, Ron, 361
Frein, Joseph B., 192
Freiser, H., 97
French, A. P., 455
Freudenberger, Robert, 335
Frick, William, 150
Friedrich, Manfred, 414
Friend, John B., 366
Friends of the Earth, 180
Fripiat, J. J., 212, 259

Frishman, B. L., 280
Froehlich, Robert A., 422
Fromherz, Hans, 102
Fromson, S., 554
Frost, J. M., 203
Frost & Sullivan, Inc., 625
Frumkin, A. N., 504
Fruton, Joseph, 477
Fry, Bernard M., 597
Frye, Keith, 77
Fryer, J. D., 177
Fuhr, J. R., 33
Fuller, Diane H., 418
Fuller, John P., 283
Furia, Thomas E., 185, 186
Furter, William F., 484
Fusonie, Alan E., 7

GB Department of Education and Science, 622
GB Deparment of Trade and Industry, 631
GB Institute of Geological Sciences, 270
GB Ministry of Technology, 558
Gadney, Alan, 327
Gaffney, Matthew P., 31
Gain, Coleman J., 369
Galland, Frank J., 140
Gallant, Roy, 376
Gandy, R. O., 522
Gans, Carl, 528
Gantt, Elizabeth, 175
Garber, Max B., 147
Gardiner, Crispin W., 157
Gardner, John W., 36
Gardner, William, 98
Garfield, Eugene, 106
Garland, J. D., 648
Garland, Sarah, 345
Garner, Robert J., 174
Garoogian, Andrew, 51
Garrison, Paul, 81
Gartrell, Ellen G., 472
Gascoigne, Robert M., 466
Gasparic, Jiri, 169
Gatland, Kenneth, 64
Gatley, William S., 341
Gaydon, A. G., 251, 294
Gaylord, Charles N., 194
Gaylord, Edwin H., 194

Gebert, Kenneth, 648
Gehm, Harry W., 233
Gellert, W., 65
Genders, Roy, 345, 350
General Analytics, Inc., 305
Geoffrey, Philip West, 127
Geography and Map Libraries Subsection, 374
George, Jennifer J., 76
George, John David, 76
Gere, James M., 280
Gerking, Shelby D., 295
Gerrish, Howard H., 115
Getty, Robert, 531
Gettys, William L., 576
Getzen, F. W., 294
Ghosh, Pradip K., 290, 315, 338
Giacoletto, L. J., 195, 196
Gibbons, Norman E., 273
Gibbons, Robert C., 428
Gibbs, G. Ian, 118
Gibson, Carol, 119
Gibson, Robert W., 15
Gieck, Kurt, 181, 256
Giefer, Gerald J., 21
Gierloff-Emden, H. G., 285
Giese, Arthur C., 531
Gieseking, Audrey, 322
Gilbert, Pamela, 20
Gill, Kay, 396, 620
Gilliam, Lucinda R., 433
Gillispie, Charles C., 451
Gilman, Kenneth L., 330, 419
Gilpin, Alan, 148, 149
Gingerich, Owen, 287
Ginsberg, G. L., 325
Ginsburg, Norton, 447
Giugnolini, Luciano, 351
Glasby, J. S., 86
Glasby, John Stephen, 67
Glass, Robert L., 328
Glasstone, Samuel, 224
Glick, Phyllis G., 349
Glover, T., 273
Gmelin-Institut, 164
Gmiehling, J., 251
Godel, Jules B., 23
Godfrey, Lois E., 591
Godfrey, Robert S., 268
Godfrey, Tony, 286

Godman, Arthur, 115
Goedecke, D. Werner, 106
Goedecke, Dipl-Ing W., 145
Gold, Seymour M., 27
Goldberg, Edward D., 532
Golden, Jack, 262
Goldsack, Paul J., 447
Goldstein, Gerald, 536
Goldstein, Leon, 527
Goldstein, Lester, 528
Goldstine, Herman H., 470
Golez, Alfred R., 193
Golley, F. B., 520
Golliday, William G., 622
Gomersall, A., 61
Gomes, Kenneth J., 333
Good, Phillip I., 327, 328
Gooders, John, 75, 360
Goodman, R. E., 516
Goodman, Robert L., 304
Goodwin, Brian L., 166
Goodwin, Derek, 359
Goodyear, P., 303
Gordon, Arnold J., 251
Gordon, Charlotte W., 154
Gordon, Michael, 96
Gorshokov, Sergei, 390
Gotch, A. F., 368
Gottleber, Timothy T., 98
Gottlieb, Richard, 434
Gottschalk, Robert, 414
Gough, Beverly E., 438
Gould, Frank W., 348
Gould, S. H., 105
Goulden, Clyde E., 477
Goulden, Steven L., 460
Gradshteyn, I. S., 237
Graf, Alfred Byrd, 274
Graf, Rudolf F., 68, 125, 138, 297
Graham, John, 144
Graham, Roger, 303
Grandilli, Peter A., 188
Grant, P. J., 360
Grant, Sheila, 350
Graselli, J. G., 379
Grasse, Pierre P., 531
Grasselli, Jeanette G., 380
Gray, Dwight, 157
Gray, Peter, 68, 310, 399, 506
Grayson, Donald K., 44

Grayson, Martin, 82, 86, 87, 90
Greeenstein, Carol Horn, 141
Green, Don W., 183
Green, Jenny, 423
Green, Joseph R., 478
Green, Orville, 182
Green, W. R., 330
Green, William, 183
Greenberg, Arnold, 285
Greenberg, D. S., 595
Greenberg, Emanuel, 315
Greenberg, Madeline, 315
Greenfield, J. D., 207
Greenwood, N. N., 251
Greenwood, P. H., 478
Gregory, Frederick, 469
Gregory, Tony, 71
Greville, G. D., 500
Griffin, Edith, 426
Grimes, Dennis J., 327
Grode, Hans-Peter, 650
Grogan, Denis, 16
Grooms, David W., 421
Gross, R. W., 158
Grossman, Fred, 245
Groves, Donald G., 80
Grzimek, Bernhard, 74, 75, 92
Gschneidner, Karl A., 168
Guedes, Pedro, 86, 87
Guggisberg, Charles A. W., 458
Guilford, C., 385, 386
Gunsalus, I. C., 528
Gunston, Bill, 94, 133
Gunther, F. A., 507
Gupta, R., 158
Gurnsey, John, 561
Gurs, K., 241
Gusev, B. V., 104
Gutman, Frederick T., 335
Gutteridge, Anne C., 124
Gutterson, Milton, 637

Haas, Leonard E., 134
Haber, L. F., 484
Habercom, Guy E., 53
Haberkamp de Anton, Gisela, 108
Hackman, W. D., 472
Haensch, Gunther, 108
Hahn, Roger, 465
Hailpern, Raoul, 401

Hala, E., 51
Halamka, John, 327
Hale, Leslie J., 253
Halevy, Abraham H., 174
Hall, A. J., 189
Hall, A. Rupert, 482
Hall, Carl W., 145, 147
Hall, D. H., 481
Hall, David A., 509
Hall, David C., 296
Hall, Elizabeth C., 8
Hall, Eugene R., 367
Hall, G. K., 613
Hall, H. W., 5
Hall, James L., 678
Halliday, Bryce, 428
Halliday, G., 548
Hallmark, Clayton L., 258, 306
Hallowell, Elliot R., 277
Halpin, Anne M., 71
Halstead, Bruce W., 363
Hamblin, W. K., 296
Hamilton, Chris J., 20
Hamilton, Marshall, 422
Hamilton, W. R., 371
Hammond, Kenneth A., 27
Hammond, Ogden H., 244
Hampel, Clifford A., 123
Hanchey, Marguerite, 536
Hancock, G. F., 226
Handley, William, 210
Hanlon, Joseph F., 209
Hannah, Harold W., 311
Hannay, N. Bruce, 527
Hansen, Eldon R., 237
Hansen, R. Gaurth, 254
Hanzak, Jan, 73
Hardin, Lloyd, 324
Hare, G., 337
Hargreaves, David, 554
Harkins, C., 307
Harlow, George, 372
Harman, Thomas, 648
Harman, Thomas L., 324
Harnly, Caroline D., 45
Harper, Charles A., 197, 205
Harre, R., 468
Harris, Cyril M., 162, 230
Harris, Robert S., 501
Harris, Susan E., 153

Harrison, Cliff, 364
Harrison, Colin, 383
Harrison, Colin J. O., 358
Harrison, J. S., 531
Harrison, Peter, 361
Harrison, T. S., 220
Hart, Allan H., 366
Hart, D. J., 282
Hart, Ernest H., 366
Hart, Harold H., 382
Hartenberg, Richard S., 487
Harter, H. Leon, 29
Harter, Harman L., 238
Hartley, H. O., 238
Hartman, Tom, 263
Hartmann, William K., 117
Hartnett, J. P., 235
Hartnett, James P., 517
Harvard University, 549
Harvey, Anthony P., 14
Harvey, Arthur F., 33
Harvey, Philip D., 260
Harvill, Lawrence, 593
Harwood, G. E., 273
Haselwood, E. L., 353
Hass, John H., 402
Hastenrath, Stefan, 388, 389
Hatch, John E., 332
Hatch, Stephan L., 313
Hatheway, Jean P., 293
Hatta, T., 381
Hawking, S. W., 473
Hawkins, D., 473
Hawkins, Donald T., 33, 36, 48
Hawkins, S. E., 296
Hawksworth, D. L., 38
Hawkweed Group, 323
Hawley, Gessner G., 122, 123
Hawley, M. E., 519
Hay, Roy, 125, 348
Hayward, A. T. J., 331
Hayward, P. H. C., 147
Haywood, Martyn, 363
Haywood, T. W., 272
Hazen, Margaret H., 42,
Hazen, Robert M., 42, 44
Hazeu, H. A. G., 485
Head, K. H., 279
Heath, Henry B., 289
Heath, R., 329

INDEX

Heck, A., 375
Heckman, Richard A., 567
Heftmann, Erich, 163
Hegedus, Louis S., 524
Hegener, Karen C., 399
Heilbron, J. L., 35, 474
Heiserman, David L., 205
Heisler, S. I., 319
Helbing, W., 242
Held, A., 473
Held, G., 304
Helgeson, Stanley L., 622
Hellebust, Johan A., 175
Helm, Thomas, 363
Helms, Harry, 207
Helms, Harry L., 329
Helms, Howard D., 49
Hemingway, T. K., 196
Hemphill, John, 348
Hemphill, Rosemary, 348
Henderson, D., 526
Henderson, D. M., 407
Henderson, Douglas, 499
Henderson, G. P., 667
Henderson, S. P. A., 667
Hendrickson, Robert, 255
Hepler, Opal, 274
Herald, Earl S., 363
Herbert, Miranda C., 449
Herbert, W. J., 128
Herde, Faye, 300
Heriteau, Jacqueline, 174
Herman, Edward, 598
Herman, Herbert, 533
Herner, Saul, 14
Hernon, Peter, 677, 597, 598
Herrick, Clyde N., 304
Herrington, Donald, 245
Hersey, David, 626
Hetsroni, G., 158
Hevelius, Jan, 377
Hewish, Mark, 411
Hewlett-Packard, 281
Hewson, E. Wendell, 535
Heywood, V. H., 71, 254
Hiatt, Robeert W., 407
Hibbert, D. B., 134
Hickman, C. J., 125
Hickman, Loren, 636
Hicks, Tyler G., 182

Higgins, Alex, 290
Higgins, David, 329
Higgins, Judith H., 665
Higgins, Lindley R., 192, 210
Hill, J. E., 368
Hill, L. R., 129
Hill, M. N., 532
Hill, R., 286
Hill, Robert L., 507
Hillier, Harold, 126
Hillson, Charles J., 314
Hindle, Brooke, 466
Hinman, W., 147
Hinsdill, Ronald D., 295
Hirano, Robert T., 353
Hirshfeld, Alan, 247
Hites, Ronald A., 227
Hiza, M. J., 35
Hnatek, Eugene R., 208, 330
Ho, C. Y., 250, 562
Ho, J. K. K., 459
Hoar, W. S., 529
Hoch, Irving, 269
Hochman, S., 102
Hoddeson, Lillian, 472
Hodgins, James, 353
Hodson, H. V., 396
Hoffman, Banesh, 455
Hoffman, Edward S., 647
Hoffman, Paul, 397
Hofmeister, Richard A., 143
Hogg, Ian V., 94
Hogner, W., 377
Holdgate, Martin W., 488
Hollaender, Alexander, 506
Hollingsworth, Brian, 390, 439
Hollis, D., 355
Holm, LeRoy G., 311, 382
Holmes, H. L., 497
Holmes, Richard T., 127
Holmes, Sandra, 124, 350
Holst, P., 50
Holst, Per, 561
Holtje, Herbert F., 427
Holum, John R., 92
Holzbecher, Zavis, 167
Home, R. W., 34
Honacki, James H., 179
Honda, Shojo, 32
Honkala, Barbara H., 41

Hood, W., 121
Hooper, Alfred, 454
Hoover, Norman K., 311
Hope, C. E. G., 74
Hopf, Peter S., 323, 644
Hopkins, Jeanne, 118
Hopper, Richard, 438
Hora, Bayard, 71
Hord, Michael, 182
Hordeski, Michael, 142
Horkheimer, Mary Foley, 658
Horn, P., 203
Hornbostel, Caleb, 212
Horner, Monica, 677, 592
Horsfall, James G., 530, 531
Horton, Holbrook L., 218
Hoshizaki, Barbara J., 274
Houghton, Bernard, 573, 631, 639
Houghton, David D., 180
Howard, J. D., 296
Howatson, A. M., 244
Howe, Robin, 135
Howell, Yvonne, 339
Hower, Rolland O., 295
Howlett, J., 486
Howson, A. G., 119
Hoyrup, Else, 454
Hsiao, T. C., 400, 462
Hsieh, Kitty, 55
Hsu, T. C., 479
Hubbard, Freeman, 89
Hubbard, Stuart W., 140
Hucek, H., 220
Hucek, H. J., 221
Hudak, Joseph, 350
Hufault, J. R., 303
Huffman, Jim, 303
Hughes, Frederick, 324, 325
Hughes, Sally S., 480
Hui, Y. H., 645
Hull, William B., 458
Humm, Harold J., 310
Hummel, Bernhard, 481
Hummel, Dieter O., 379
Hummelsbach, Carl J., 615
Hund, Friedrich, 474
Hunt, Daniel V., 148, 218, 340
Hunt, Frederick V., 475
Hunt, James W., 330
Hunt, Lee M., 80

Hunt, Peter, 70
Hunt, Roy E., 278
Hunting, Athony, 86
Huntington, W. C., 322
Huntley, B., 381
Hurd, B., 46
Hurlbut, C., 276
Hurlbut, C. S., 79, 283
Huskins, D. J., 169
Hutchinson, John, 351
Hutson, D. H., 500
Hutton, G., 56
Huxley, Anthony, 354
Hyams, Edward, 406
Hyman, Charles J., 120
Hyman, Libbie H., 530

IWT Verlag GmbH, 104
Ibers, J. A., 526
Idlin, Ralph, 120
Igoe, Robert S., 135
Illingworth, Valerie, 117, 140
Immergut, E. H., 188
Information Handling Services, 638
Information Sources, Inc., 421
Innes, Clive, 346
Institute for Scientific Information, 620
Institute of Ecology, 458
Institute of Electrical and Electronics Engineers, 417, 611, 612, 648, 649, 652
Institute of Marine Engineers, 131, 562
Institute of Physics, 623
Institution of Civil Engineers, 275
Institution of Electrical Engineers, 673
Intergovernmental Maritime Consultative Organization, 284, 644
International Association for Earthquake Engineering, 552
International Association of Engineering Geology, 373
International Atomic Energy Agency, 99, 260, 284, 298, 336, 337, 341, 428, 429, 430, 612, 622, 644, 645
International Civil Aviation Organization, 277, 599
International Computaprint Corporation, 444
International Council of Scientific Unions, 4, 248

Internatonal Dairy Federation, 108
International Institute of Refrigeration, 113
International Plastics Selector, 259
International Road Federation, 271
International Techinical Information Institute, 650
International Union for Conservation of Nature and Natural Resources, 438
International Union of Biochemistry, 98
International Union of Geological Sciences, 643
International Union of Pure and Applied Chemistry, 97, 99, 272
International Wheat Council, 508
Iqbal, M., 334
Irvine, Thomas F., 235, 517
Irving, H. M. N. H., 97
Isaacs, Alan, 113, 116
Ishikawa, Kaoru, 331
Israel, W., 473
Isshiki, Koichiro R., 327
Iyanaga, Shokichi, 119
Iyengar, K. T. S. R., 244
Izatt, Reed M., 167

J. N. Lythgoen Society For Testing & Materials, 273
Jablonski, David, 78
Jackson, Albert, 305
Jackson, G. N., 74
Jackson, Joseph H., 343
Jackson, Julia A., 130
Jacobsen, Niels, 312
Jacobson, I. D., 146
Jacobson, Morris K., 363
Jacques Cattell Press, 448
Jaeger, R. C., 223
Jain, M. K., 166
Jaki, Stanley L., 470
Jaklovsky, Jozef, 53
James, A. M., 120, 134
James, Glen, 119
James, R. C., 119
Jameson, D. L., 510
Janick, Jules, 508
Jannasch, H. W., 511
Japan Chemical Week, 413
Japan Lunar & Planetary Observers Network, 343

Japan Society for the Promotion of Science, 619
Jaques, Harry E., 300, 301
Jaques Cattell Press, 448, 626
Jarman, Cathy, 383
Jarmul, S., 337
Jaschek, Carlos, 247
Jastro, Robert, 522
Jay, Frank, 138
Jeffrey, Charles, 97
Jenkins, A. C., 458
Jenkins, A. D., 498
Jenkins, Eric N., 475
Jerrerd, H. G., 114
Jessop, G. R., 282
Jewkes, J., 287
Johnsgard, Paul A., 359, 360
Johnson, Arnold H., 84
Johnson, C. R., 286
Johnson, Carl R., 99
Johnson, David E., 197
Johnson, Hugh, 84
Johnson, Keith, 636
Johnson, Leon J., 297
Johnson, M. L., 125
Johnson, Norman L., 65
Johnstone, William D., 249
Joiner, Brian L., 540, 546
Jones, C. Eugene, 126
Jones, Elizabeth A., 546
Jones, Franklin D., 218
Jones, H. R., 637
Jones, J. K., 254
Jones, J. Knox, 367
Jones, Kenneth Glyn, 155
Jones, Lincoln D., 302
Jones, Owen J., 546
Jones, Stacy V., 632
Jones, Thomas H., 196
Jongerius, A., 112
Jordan, S., 276
Jordanova, L. J., 488
Jorgensen, Finn, 201
Jorgensen, S. E., 229
Joseph, Marjory, 322
Jost, W., 526
Jousset, Jack, 102
Joyce, Dennis, 425
Judd, B. Ira, 175
Judson, H. F., 477

Junk, W. Ray, 187
Justis, P. S., 385

K. Mardia and Zemroch, 238
Kadis, Solomon, 530
Kahrl, William L., 390
Kaigai, Gijutsu S., 285
Kaiser, James F., 49
Kaisler, Stephen, 330
Kalber, Frederick, 181
Kalbus, Barbara, 296
Kalla-Bishop, P. M., 447
Kallard, Thomas, 293
Kambayashi, Yahiko, 676
Kamen-Kaye, Maurice, 43
Kaminskii, Alexander A., 35
Kane, Mike, 675
Kanely, Edna M., 597
Karassik, Igor J., 218
Kardon, Randy H., 382
Karmas, E., 637
Karp, H., 325
Karp, Harry R., 649
Karrow, Paul F., 189
Kartesz, John T., 350
Kase, Francis J., 632
Kassas, Mohammed, 488
Katchalski-Katzir, Ephraim, 527
Katona, Steven K., 367
Katz, Dorothy D., 170, 243
Katz, Harry S., 187
Katz, Samuel, 65
Kaufman, Joyce J., 492
Kaufman, Milton, 197, 198
Kaufmann, Manfred, 650
Kavanagh, Michael, 366
Kavass, Igor I., 55
Kawada, Yukiyosi, 119
Kaye, G. W. C., 235, 238
Kaye, Joseph, 245
Kazlauskas, Edward J., 421
Keenan, Joseph H., 245
Keesee, Anne M., 34
Kegyesy, E., 218
Keller, J. J., 273
Keller, Mark, 134
Keller, R. A., 241
Kelly, Brian W., 327
Kelly-Grimes Corporation, 420, 421
Kelman, Peter, 417

Kemmer, N., 307
Kemp, Peter, 92, 93
Kemper, Alfred M., 190
Kendall, M. G., 118, 471
Kendall, Maurice G., 32, 264
Kennard, Olga, 38
Kennedy, J. F., 498
Kenny, Peter, 152
Kern, Kenneth, 412
Kern, R. F., 333
Kerr, Alistair, 165
Kerrich, G. J., 38
Kessell, Richard G., 382
Kestin, Joseph, 292
Kettridge, J. O., 102
Kevles, Daniel J., 475
Keyes, John H. 338
Keys, John D. 345
Kezdi, Arpad, 194
Khandpur, Raghbir S., 167
Khatri, Linda A., 459
Kidnay, A. J., 35
Kinch, Michael P., 618
King, A. L., 52
King, Alexander, 102
King, Ben F., 360
King, Donald W., 573
King, F. Wayne, 369
King, Horace W., 193
King, Linda S., 19
King, Michael M., 19
King, Ralph W., 209
King, Robert C., 171
King-Hale, D. G., 240, 245
Kingdon, Jonathan, 384
Kingslake, R., 523
Kingzett, Charles T., 67
Kinman, Kenneth E., 179
Kinne, Otto, 530
Kirby-Smith, H. T., 403
Kirchshafer, R., 407
Kirsch, Debbie, 37
Kirschenbaum, Donald M., 35
Kirschman, J. D., 315
Kjar, Kathie J., 313
Klapper, M., 257
Klein, Barry T., 393
Klein, C., 276, 283
Klein, M. J., 469
Klein, Stanley, 74

Klema, Ernest D., 54
Klement, Alfred W., 59, 227
Klerer, Melvin, 245
Klopfenstein, Charles E., 500
Klopfer, Leopold B., 537
Klos, Joel, 141
Kloss, Albert, 324
Klyne, William, 380
Knaack, Marcelle S., 94
Knacke, O., 252
Knapp, D. R., 164
Knight, Allen W., 28
Knight, David M., 469, 476, 479
Koch, J. S., 256
Koenig, E., 260
Koeppl, James W., 179
Kohan, Anthony L., 334
Kolb, Helga, 122
Kolb, John, 210
Kolthoff, I. M., 527
Komar, Paul D., 227
Koncz, T., 279
Kondo, Kohji, 109
Kong, F. K., 192
Konold, William G., 633
Koolhaas, J., 250
Koolish, Ruth K., 290
Koral, Richard L., 216
Koren, Herman, 229
Korte, Friedhelm, 534
Korwowski, Jacek, 248
Koshkin, N. I., 158
Koski, Walter S., 492
Kothandaraman, C. P., 249
Kotrly, S., 165
Kovalenko, E. G., 105
Kozlowski, T. T., 535
Kraft, T. L., 593
Kragton, J., 379
Kramer, Jack, 351, 352
Krawinkler, Helmut, 280
Kreider, Jan F., 226
Kreidl, N. J., 533
Kreiger, M., 301
Kreith, Frank, 226
Krell, Erich, 166
Kremer, S., 260
Kress, Stephen W., 177, 358
Kreuterch, Marie-Luise, 274
Krieger, R., 325

Krinsley, David H., 386
Krogmann, David, 296
Kronick, David A., 19, 573
Kruss, Lenelis, 58
Kruzas, Anthony T., 394, 395, 396, 397, 620
Kryt, Dobromila, 108
Kubaschewski, O., 252
Kues, Barry S., 44
Kuesel, T. R., 194
Kuiper, G. P., 378, 522
Kulsis, A., 166
Kunkel, Barbara K., 15
Kunz, W., 240
Kurstak, E., 175
Kurtz, Edwin B., 200
Kurtz, Max, 323
Kut, D., 337
Kut, David, 147
Kuthanova, Olga, 365
Kutz, M. P., 215
Kutzbach, Gisela, 482
Kuznetsov, B. V., 106
Kverneland, Knut O., 642
Kybett, Harry, 662
Kyed, James M., 3, 672
Kyle, James, 305

L'Association Quebecoise des Industries de la Peinture, 102
La Beau, Dennis, 449
Laby, T. H., 235, 238
Lafayette College, 284
Laitenen, Herbert A., 477
Lajtha, Abel, 167
Lama, R. D., 213
Lamb, Brian, 70, 71,
Lamb, Edgar, 70, 71
Lamb, George H., 291
Lamb, Peter J., 388, 389
Lambeck, Raymond P., 334
Lambert, Jill, 573
Lambert, Mark, 79
Landa, Henry C., 226
Landsberg, H. E., 532
Landy, Marc, 150, 437
Lang, Jovian P., 2
Lang, K. R., 248
Lang, Kenneth R., 287
Lange, H., 242

Langenkamp, R. D., 184
Langenkamp, Robert D., 131
Langley, L. L., 512
Lanjouw, Joseph, 550
Lansford, Edwin, 67
Lapedes, Daniel N., 62, 79, 84, 90, 115, 121, 125
Lapinski, Michael, 193
Lare, Gary, 662
Larson, Donna Rae, 393
Laskin, Allen I., 172
Latin American Newsletters Limited, 398
Lavenberg, Stephen S., 204
Law, Beverly, 412
Law, Donald, 274
Lawson, Simon, 271
Lazenby, David W., 191
Leathart, Scott, 350
Leatherwood, Stephen, 179
Lechevalier, Hubert A., 172
Lederer, Charles M., 240
Ledermann, Walter, 156
Ledin, George, 143
Lee, Douglas H. K., 535
Lee, Henry, 425
Lee, Kaiman, 47, 86, 91
Lee, Reginald, 136
Lee, Thomas F., 354
Leecraft, Jodie, 131
Leek, William R., 436
Lefevre, Marc J., 272
Lefkovits, Henry C., 142
Lefkovits, Sandra L., 142
Leftwich, A. W. A., 128
Legget, Robert F., 189
Leggett, Arnold, 406
Lehner, Philip N., 171
Lehr, Jay H., 193
Lehrman, H. P., 235
Leinwoll, Stanley, 486
Lemp, Helena B., 272
Lengenfelder, Helga, 393, 396
Lenihan, J. M. A., 161, 456
Lenihan, John, 520
Lenk, John D., 197, 198, 199, 202, 205, 206, 305, 324
Lennette, Edwin H., 274
Lenroot-Ernt, Lois, 394
Lenton, Roberto, 316
Leondes, C. T., 518

Leong-Hong, Belkis W., 675
Leopold, M., 128
Lerner, Rita G., 65, 66
Lerro, Joseph P., 236
Leschonski, K., 103
Lesko, Matthew, 675
Leung, Albert Y., 83
Levett, John, 447
Levi, Leo, 307
Levine, Judith, 523
Levine, Raphael B., 294
Levine, Richard J., 12
Levitan, Eli L., 195
Levitt, I. M., 377
Levy, Charles K., 348, 367
Levy, Sidney, 187
Lewis, D. A., 544
Lewis, H. A. G., 376
Lewis, Marianna O., 395
Lewis, Richard S., 64
Lewis, T., 238
Lewis, Walter H., 150
Li, Choh H., 529
Li, Norman N., 526
Liebe, Janice T., 164
Liebesny, Felix, 630
Lien, David A., 203
Lightman, Bernard, 31
Lilley, G. P., 19
Lilly, L. C. R., 334
Lima, Robert F., 106
Limbacher, James L., 658
Limbrunner, Alfred, 358
Limoges, Camille, 479
Lin, Wen C., 204
Lincoln, Brayton, 333
Lincoln, R. J., 148
Lindberg, David C., 468
Lindeburg, Michael R., 276
Linden, David, 197
Lindkvist, R. G. T., 213
Lindsay, R. B., 492, 519
Lindsay, Robert B., 471, 472
Line, Les, 316, 364
Linek, J., 51
Link, Conrad B., 351
Link, R. F., 256
Linn, Louis C., 353
Lippert, Herbert, 235
Lipscomb, David M., 231

Liston, Linda, 181
Little, Elbert L., 41, 344, 345, 381
Little, R. John, 126
Little, Richard, 334
Little, Sherry, 334
Livesey, W. A., 218
Lloyd, Christopher, 385
Lloyd's Register of Shipping, 269
Lockley, Ronald M., 362
Loebl, Ernest M., 526
Loewenfeld, Claire, 314
Loewenstein, Werner R., 172
Loewer, Peter, 301
Loftness, Robert L., 224
Logie, Gordon, 112
Logsdon, Gene, 310
Lohwater, A. J., 105
Long, C. R., 40
Long, Chalmers G., 647
Long, Larry, 280
Long, Leslie "Doc", 278
Longley, D., 141, 303
Longley, Dennis, 207
Longman Editorial Team, 432
Looney, John J., 47
Loosjes, T. P., 14
Lord, Harold W., 341
Lorenz, O. A., 176
Loshak, L., 280
Louden, Louise, 61
Lovett, D. R., 114, 115
Lowe, Joseph D., 101
Lowe, Kurt E., 370
Loyd, S., 56
Lubin, George, 213
Lucas, John, 69, 127
Lucas, Ted, 302
Luchsinger, Arlene E., 20
Ludwig, Raymond H., 258
Lukoff, Herman, 486
Lund, Herbert F., 232
Lund, P. G., 244
Lundevall, Carl F., 359
Lutz, Robert D., 424
Lye, Keith, 372
Lyman, Warren J., 165
Lynch, H. T., 406
Lynch, Wilfred, 185
Lyon, W. F., 365
Lyon, W. L., 179

Lyon, W. S., 609
Lyons, Jerry L., 89
Lytle, Robert J., 173, 190, 191

Mabie, H. H., 259
MacDonald, Joseph A., 415
MacDonald, Stuart, 475
MacFall, Russell P., 371, 372
MacKenzie, W. S., 385, 386
MacQuitty, William, 406
Macdonald, David, 74
Machovec, George, 566
Machu, Willibald, 205
Macinko, George, 27
Mackay, Alan L., 246
Mackenroth, Donald R., 87
Mackie, Raymond K., 309
Macreary, S. E., 398
Macura, P., 109, 110
Magee, Dennis W., 313
Magel, Charles R., 42
Magid, J., 209
Magill, Robert E., 126
Magner, Lois N., 478
Maiorana, Charlie, 15
Maizell, Robert E., 18, 536
Major, David, 316
Makepeace, R. J., 177
Makower, Joel, 143
Malin, D., 376
Malinowsky, H. Robert, 16
Mallow, Alex, 160
Malloy, John F., 219
Mandava, N. Bhushan, 173
Mandis, George M., 399
Mandle, Matthew, 280, 282
Manfroid, J., 375
Mann, J. Y., 51
Mano, M. Morris, 204
Mansfield, Jerry W., 25
Manske, R. H. F., 497
Mansson, Margaret, 525
Manthy, Robert S., 269
Manufacturing Chemists Association, 309
Manzer, Bruce M., 572
Maramorosch, Karl, 380
March, N. H., 494
Marchworth, M. L., 618
Margulis, Lynn, 309
Marien, Michael, 27

Marine Biological Laboratory and Woods Hole Oceanograph. Institute, 555
Marine Technology Society, 332, 649
Marion, Jerry, 293
Markus, John, 137, 281, 325
Marlow, Cecilia Ann, 393
Marmion, Daniel M., 185
Marqulova, Tereza K., 429
Marr, Iain L., 165
Marre, Louis A., 334
Marsden, Brian G., 342
Marsh, Paul, 137
Marshall, Frederick C., 297
Marshall, John, 462
Marshall, Roy K., 377
Martin, E. A., 124
Martin, Franklin W., 186
Martin, G. A., 33
Martin, Hubert, 173
Martin, Laura C., 354
Martin, M. D., 214
Martin, Richard Mark, 73
Martin, Robert E., 275
Martin, S. M., 406
Martin, Sarah S., 31
Martin, Therese A., 121
Martin, W. Keble, 176
Martna, Maret, 43
Martz, Christopher, 292
Marx, Robert F., 481
Masloff, Jacqueline, 91
Maslow, Philip, 259
Mason, C. W., 165
Mason, W. P., 493
Massey, H. S. W., 523
Masters, Deborah C., 2
Masters, K., 188
Masuy, L., 111
Matarazzo, James M., 3, 672
Mathematical Association of America, 6
Mathematical Society of Japan, 119
Mathison, Ruby L., 57
Matisoff, Bernard S., 198
Matlins, Antoinette L., 371
Mavituna, Ferda, 170
Maxfield, Doris M., 678
May, Daryl N., 230
May, Kenneth O., 470, 540
Mayall, R. Newton, 342
Maykuth, D. J., 221

Maynard, D. N., 176
Maynard, Jeff, 141
Maynard, John T., 635
Mazria, Edward, 292, 339
McAleese, Frank G., 160
McAllister, D., 196
McAninch, Sandra, 57
McBride, Richard 91, 392
McCaig, M., 308
McCann, Margaret T., 284
McCarthy, Gregory J., 610
McCay, Clive M., 479
McClane, A. J., 363
McClure, Charles R., 598
McCombe, C., 444
McCormack, Mairi, 134
McCormmach, Russell, 473
McCrone, Walter C., 378
McCullagh, James C., 340
McDonald, Elvin, 174
McDonald, M., 636
McDonald, Rita, 23
McEntee, Howard G., 202
McGehee, Brad M., 424
McGill, Marion, 315
McGill University, 9
McGraw-Hill Encyclopedia of Science and Technology, staff, 440
McGuire, Barry L. S., 622
McGuire, G. E., 272
McKerrow, W. S., 318
McKetta, John J., 82
McLachlan, Dan H., 356
McLeavy, Roy, 447
McLeish, C. W., 162
McMullan, J. T., 338
McNeil, Barbara, 449
McNeill, D. B., 114
McNeill, Joseph G., 323
McNiff, Martin, 302, 303
McPartland, J. F., 199, 200
McPhillips, Martin, 261, 340, 434
McRae, Alexander, 291, 292
McRae, Madelyn A., 664
McRoy, C. Peter, 175
McVickar, Malcolm H., 312
McWilliams, M., 297
McWilliams, Peter A., 326
Meacham, Stanley, 245
Meacock, H. M., 162

Meade, George P., 186
Meadows, A. J., 573
Meek, B. L., 329
Meeks, Chester L., 88
Meenan, A., 398
Mehra, J., 473
Meiksin, Z. H., 281
Meinck, R., 149
Meinkoth, Norman, A., 355
Meister, A., 510
Melby, Edward C., 177
Mellan, I., 212
Mellan, Ibert, 168
Mello, Nancy K., 505
Mellor, J. W., 525
Melnick, Daniel, 300
Melton, L. R. A., 18
Melzer, J. E., 240
Menditto, Joseph, 37
Mennema, J., 381
Menzel, Donald H., 342
Merritt, Frederick S., 189, 190
Meshenberg, Michael J., 26, 59
Messier, Charles, 342
Metric Company, 642
Metropolis, N., 486
Metz, Karen S., 26
Meyers, Robert A., 225, 226
Miasek, M. A., 40
Michaelson, M., 235
Mickadeit, R. E., 322
Mickel, John T., 300
Middlebrooks, E. Joe, 230, 291
Middleditch, B. S., 262
Middlehurst, B. M., 378, 522
Middleton, A. H., 335
Middleton, H. A., 325
Middleton, Robert G., 157
Mieghem, J. Van, 514
Mieners, H. F., 293
Mikes, Otakar, 168
Milazzo, G., 507
Milazzo, Guilio, 241
Miles, Edward N., 389
Miles, Wyndham D., 457
Milewski, J. V., 187
Miller, B. J., 33
Miller, Ethel B., 277
Miller, Gary M., 198
Miller, Howard A., 301

Miller, Kenneth L., 436
Miller, Merl, 330
Miller, Paul R., 113
Miller, R. C., 35
Miller, R. W., 215
Miller, Richard K., 158
Miller, Ryle, 277
Miller, Samuel D., 47
Millington, T. Alaric, 118
Millington, William, 118
Mills, Dick, 68
Millspaugh, Charles F., 344
Milne, Lorus J., 364, 365
Milne, Margery, 364, 365
Milne, Robert G., 381
Miloren, K. W., 22
Mims, Forrest M., 196
Minasian, Stanley M., 369
Mincron SBC Corporation, 424
Mindell, H. L., 302
Minear, Roger A., 401
Minichino, Camille, 567
Misckhe, C. R., 219
Mitchell, E. N., 272
Mitchell, Jerome F., 182
Mitchell, Richard S., 132
Mitchell, Sarah, 450
Mittler, Thomas E., 478
Mitton, Jacqueline, 118, 377
Mitton, Simon, 63, 377
Moder, Joseph J., 157
Moerman, Daniel E., 125
Moffat, Donald W., 245
Moggi, Guido, 351
Mohle, H., 149
Monahan, Robert, 653
Mondey, David, 80, 81, 263
Money, S. A., 259
Monkhouse, F. J., 149
Monroe C. Gutman Library Staff, Harvard University, 422
Monselise, Shaul P., 174
Montagne, Prosper, 85
Montgomery, R., 338
Montgomery, Richard H., 340
Montie, Thomas C., 530
Montone, W. V., 158
Moody, Alton B., 285
Moody, Richard, 318
Moore, Alick, 253

Moore, Bradley, 499
Moore, C. Bradley, 527
Moore, Harry P., 197
Moore, John E., 447
Moore, Patrick, 117, 155, 247, 342, 343, 375, 376, 442
Moore, Peter D., 313
Moore, Raymond C., 532
Moore, Richard H., 253
Moore, T. C., 275
Moore, W. G., 130
Moran, James, 467
Morehead, Joe, 597
Morehouse, L. G., 176
Morgan, R., 338
Morgan, S. W. K., 222
Morisawa, Marie, 297
Morres, Angelo C., 284
Morresi, A. C., 228
Morressi, A. C., 262
Morris, Francesca, 174
Morris, Robert, 490
Morrison, James W., 223
Morrow, L. C., 210
Morton, Chloe, 110
Morton, Ian D., 110
Morton, Julia F., 381
Moschytz, G. S., 203
Moselle, Gary, 269
Moses, Alfred J., 152
Mosher, Doug, 326
Moskowitz, Lester R., 214
Moss, Stephen J., 165
Moss, T. S., 159
Motor Vehicle Manufacturers Association, 448
Mott, Russell C., 352
Mottelay, Paul Fleury, 471
Motter, G. G., 353
Motz, Lloyd, 117
Mount, Ellis, 21
Mueller, Kimberly J., 429
Muenscher, Walter C., 354
Mueser, Roland, 485
Muhlberg, Helmut, 346
Muir, G. D., 168
Muirden, James, 153, 247
Mujumdar, A. S., 519
Muller, Dieter, 108

Multhauf, Robert P., 48
Mumford, C. J., 233
Munz, Lucile T., 550
Murdin, Paul, 376
Murdoch, John E., 465
Muroga, Saburo, 299
Murphy, Vreni, 181
Murray, J. W., 318
Murray, R. B., 338
Murray, Spence, 306
Mustoe, J., 392
Mycological Society of America, 349
Myers, Arnold, 24
Myers, Darlene, 24
Myers, Patricia A., 121
Myrick, Rick W., 632

NALCO Chemical Company, 231
Nania, Georges, 107
Nariai, Kyoji, 375
Nass, Leonard I., 85
Nastuk, William, 530
National Audiovisual Association, 662
National Council on Radiation Protection and Measurements, 651
National Engineering Laboratory, 34
National Geographical Society, 384
National Information Center for Educational Media, 658, 659, 663
National Microfilm Association, 657, 660
National Micrographics Association, 663
National Oceanic and Atmospheric Administration, 266
National Research Council, 432
National Science Foundation, 619
National Society of Professional Engineers, 401
Naus, Joseph I., 238
Nayler, J. L., 145
Naylor, G. H. F., 145
Neal, Kenneth G., 296
Neave, H. R., 239
Needham, J., 468
Needles, Howard L., 189
Neff, Norman D., 238
Neidle, S., 527
Nelson, John A., 192
Nelson, Joseph S., 363
Nemeth, Bela A., 242

Nero, Anthony V., 288
Nettles, J. E., 188
Neu, John, 467, 476
Neufeld, Lynne M., 26
Neufert, Ernst, 190
Neugebauer, O., 469
Neumann, Charles J., 270
Neurath, Hans, 507
New York Botanical Garden Library, 8
New York Public Library, 373
Newby, Frank, 632
Newlin, B., 675
Newman, A., 523
Newman, Donald G., 302
Newman, Michele M., 664
Newrock, Melody, 326
Newsome, Walter L., 598
Newton, Jack, 376
Nichols, Herbert L., 298
Nichols, Joseph C., 303
Nicholson, Iain, 117
Nicita, Micheal, 24
Nicolson, Iain, 117
Niedenzu, Kurt, 501
Niering, William A., 352
Nightingale, W., 446
Nijdam, J., 110
Niles, Cornelia D., 549
Nobelstiftelsen, Stockholm, 452, 456
Noel, Joel V., 147
Noel, John V., 151
Nof, S. Y., 217
Nof, Simon Y., 216
Noiseux, Ronald A., 328
Nonis, U., 349
Norback, Craig, 224
Norback, Peter, 224
Norimoto, Yuji, 375
Norman, A. D., 595
Norman, J. R., 478
North, Hallie B., 634, 636
North, Oliver S., 636
Northrop, Stuart A., 44
Norton, Harry A., 211
Norton, Thomas W., 299
Norwood, Joseph, 476
Novak, Vladimir, 380
Nowak, Ronald M., 368
Nowell, J. W., 294

Noyes, Robert, 406
Nuclear Regulatory Commission, 625
Null, Gary, 286
Nyvall, Robert F., 173

O'Bannon, Loran S., 144
O'Brien, Jacqueline, 264
O'Donoghue, Michael, 77, 133
O'Hanlon, John F., 308
O'Neill, P. J., 235
Oberg, Erik, 218
Ockerman, Herbert W., 287
Ocran, Emanuel B., 55,132
Ocvirk, F. W., 259
Oddo, Sandra, 434
Odland, Gerald C., 458
Oetting, F. L., 524
Offerd, Robin, 309
Oklahoma Cooperative Wildlife Research Unit, 36
Okun, Daniel, 535
Olcott, William T., 342
Olmstead, Nancy C., 352
Olphen, H. Van, 259
Olsen, George H., 196
Onken, U., 251
Opitz, Helmut, 669
Oppermann, Alfred, 102, 103, 104
Oram, Richard, 61
Orchin, Milton, 123
Organic Gardening Magazine Staff, 69
Orr, W. I., 202
Osborne, Adam, 303
Osen, L. M., 454
Osman, Tony, 63
Otten, L., 213
Ottensmann, J. R., 304
Owen, D. B., 471
Owen, Dolores B., 536
Owen, Donald B., 238
Owen, H. G., 384
Owen, Myrfyn, 362
Owings, Loren C., 59

Pacholczyk, A. G., 154
Pacific Aerospace Library, 557
Packard, Ruby, 221
Page, Chester H., 235
Page, Thorton, 403

Pagel, Annabel, 272
Palen, J. W., 288
Palmer, Archie M., 621
Palmer, E. Laurence, 356
Palmer, Joseph, 147
Palmer, Ralph S., 178
Palmieri, F., 500
Pampe, William R., 374
Pancoast, Harry M., 187
Pannakoek, Antonie, 469
Pansini, Anthony J., 324
Papadopoulos, C., 378
Paparian, Michael, 430
Papp, Charles S., 365
Paradiso, John L., 368
Parikh, G., 206
Parish, David, 598
Parker, Albert, 233
Parker, Sybil P., 62, 64, 66, 67, 79, 80, 89, 91, 92, 116, 123, 127, 132, 133, 142, 452
Parmley, R. O., 278
Parmley, Robert O., 219
Parris, N. A., 294
Parrish, A., 283
Parsegian, V. L., 534
Parsons, M. L., 378
Parsons, Michael L., 158
Paruit, Bernard H., 112
Pasto, D. J., 286
Patterson, George J., 49
Patterson, Jerome, 402
Paul, D. R., 399
Paul, M. A., 130
Paxton, Albert S., 269
Payne, E. M., 115
Payne, Gordon A., 225
Payne, R. W., 310
Pearman, William, 292
Pearse, John S., 531
Pearse, Reginald W., 251, 294
Pearson, Carl E., 156
Pearson, Egon S., 238
Peck, Theodore P., 22
Peckner, Donald, 221
Pedde, Lawrence D., 273
Peebles, P. J. E., 607
Pelczar, Michael J., 295
Pelletier, Paul A., 452
Pemberton, John E., 596

Pennells, E., 278
Pennington, Jean A. T., 253
Peray, Kurt E., 208
Perbal, Bernard, 311
Pereira, Norman C., 229
Pergolizzi, Robert G., 286
Perlman, D., 511
Permutit Company, 263
Pernet, Ann, 14, 398
Perry, Frances, 348
Perry, Robert H., 183, 276
Perry, Roger, 232
Person, Russell V., 307
Perutz, M. F., 381
Pestemer, M., 242
Peters, Barbara J., 292
Peters, George A., 292
Peters, J. L., 359
Peters, Joseph, 372
Peterson, Lee, 347
Peterson, Martin S., 84
Peterson, Roger T., 359
Petrusha, Ronald, 24
Petulla, Joseph M., 488
Pevsner, Nikolaus, 484
Pfafflin, James R., 91, 519
Pforr, Manfred, 358
Phelps, Frederick M., 239
Philips, Melba, 473
Phillips, A. L., 222
Phillips, Gary, 88
Phillips, John P., 252
Phillips, Paul, 191
Phillips, Roger, 350
Phillips, Ronald C., 175
Phillips, W. Louis, 550
Pickering, James S., 286
Pickering, John Earl, 425
Pinkepank, Jerry A., 334
Pirone, Pascal P., 346
Pirson, A., 70
Pitt, E., 209
Pitt, M. J., 209
Pitt, Valerie H., 121
Pitts, J., 279
Pitts, James N., 496
Pizzey, Graham, 359
Plackett, R. L., 471
Placzek, Adolf K., 87
Plagman, Bernard K., 675

Plane and Pilot Magazine, editors, 411
Plastics & Rubber Institute, 623
Plate, C. L., 381
Pledge, H. T., 468
Plummer, R. D., 281
Plung, D. L., 307
Pluscauskas, Martha, 2
Poggendorff, Johoun C., 450
Pohle, Linda C., 597
Poinar, George O., 295
Pointon, A. J., 308
Pollard, Hazel L., 113
Pollard, J. H., 157
Pond, Grace, 368
Poniatowski, Michael, 10
Poole, Lon, 302
Pope, Carl, 92
Popenoe, Cris, 12
Porges, S. W., 513
Porter, Alan L., 319
Porter, James H., 244
Porter, R. S., 488
Porter, Roy, 44, 466
Portugal, Franklin H., 477
Pospisilova, J., 41
Possinger, Callie, 462
Potter, James, 182
Potteron, David, 346
Pouchert, Charles J., 250
Powals, Richard, 230
Powell, F. C., 235
Powell, P. C., 321
Powell, Russell H., 30
Powers, J. P., 322
Powers, Thomas K., 206
Prakken, Lawrence W., 402
Premier ed., 574
Prenis, John, 139
Prentice, H. T., 407
Prepared by Science Service, 441
Prescott, David M., 528
Prescott, George W., 300
President's Commission on Coal, 261
Pressman, David, 632
Preston, D. W., 293
Preston, Richard J., 177
Priestley, B., 305
Primack, Alice Lefler, 14
Primack, Richard B., 348
Primrose, S. B., 287

Prince, Alan, 52
Prince, David J., 143
Prince, Suzan D., 424
Prinz, Martin, 372
Pronen, Monica, 28
Pugh, Eric, 96
Pugnetti, Gino, 368
Purcell, Gary R., 597
Putman, R. E., 136
Putnam, T. M., 250, 258
Pyatt, Edward, 474
Pye, Orrea, 315
Pyle, Robert M., 364

Quagliariello, E., 500
Quasim, S. H., 642
Quene-Boeterenbrood, A. J., 381
Quigg, Philip W., 403
Quimme, Peter, 316

RBUPC Office, the British Library, 621
Rabbitt, Mary C., 481
Rabiner, Lawrence R., 49
Raggett, R. J., 447
Rahn, Frank J., 336
Rainbird, George, 315
Raleigh, Ivor, 368
Ralston, Anthoy, 88
Ralston, Valerie, 46
Ramalingom, T., 145
Ramdas, V., 250
Ramsey, C. G., 643
Ramu, S. A., 244
Randall, B. A. O., 131
Randall, D. J., 529
Ransom, Jay Ellis, 357
Rappoport, Zvi, 242
Rau, John G., 228
Ray, Johnny, 197
Rayed, M. E., 213
Raznjevic, Kuzman, 168
Read, P. G., 145
Reader's Digest, 387
Reader's Digest Association, 71
Reay, David A., 226, 484
Rechcigl, Miloslav, 186
Rechenberg, H., 473
Reddig, Jill Swanson, 27
Redding, James L., 186
Redman, Helen F., 591

Reebie Associates, 28
Reed, G., 528
Reed, Gerald, 310
Reed, Michael, 523
Reedman, J. H., 283
Reehl, William F., 165
Reese, Kenneth M., 476
Reeves, Randall R., 179
Rehm, H. J., 528
Reichelt, Jon, 286
Reid, W. Malcolm, 20
Reifsnider, Elizabeth G., 436
Reiger, R., 124
Reilly, Edgar M., 177
Reilly, Edwin D., 88
Reiner, L., 322
Reingold, Ida H., 467
Reingold, Nathan, 467
Reith, Edward J., 380
Remer, Daniel, 328
Reproduction Research Information Service, 42
Reschke, Robert C., 191
Research Libraries of the New York Public Library, 29, 596
Reynolds, Charles E., 193
Reynolds, Michael M., 617
Reys, Robert E., 665
Rhoads, Mary L., 566
Rhyne, James J., 610
Ricciuti, Edward, 316
Rice, James O., 211
Rice, P. F., 647
Richards, J. M., 462
Richards, W. G., 36
Richardson, David T., 367
Richardson, Jeanne M., 16
Richter, H. P., 302
Richter, Herbert P., 200
Richter, N., 377
Rickards, Clarence, 96
Rickards, T., 282
Rickett, H., 354
Rickett, Harold W., 548
Ricks, David A., 394, 427
Rider, Robin E., 31
Ridge, John D., 43
Ridgely, Robert S., 382
Ridpath, Ian, 64
Riethmuller, John, 426

Riker, Tom, 352
Riley, Sharon J., 98
Rinaldi, Augusto, 345
Ring, Jeanne, 462
Ringold, Paul L., 255
Rink, Evald, 483
Ripley, S. Dillon, 179
Ristor, Walter W., 374
Ritcher, G. W., 506
Ritchey, W. M., 379, 380
Ritchie, James D., 292
Robards, Terry, 316
Robbins, Manuel, 370
Robbins, Peter, 333
Roberts, D., 286
Roberts, M. W., 171
Roberts, T. R., 500
Roberts, V. L., 646
Robertson, Forbes S., 297
Robertson, Roland, 415
Robinson, Anthony, 81, 93
Robinson, J. Hedley, 247
Robinson, J. W., 159
Robinson, Joseph F., 290
Robinson, Judith, 15
Robinson, Judith S., 592
Robinson, Julian Perry, 53
Robinson, W., 232
Rodwell, Peter, 326
Roe, Keith E., 124
Rogers, Dilwyn J., 57
Rohlf, F. James, 239
Rolfe, Douglas, 483
Ronan, Colin, 64
Ronchi, Vasco, 474
Ronen, Naomi, 30
Ronen, Yigal, 223
Roose, Robert W., 225
Root, Waverley, 136
Roper, Christopher, 398
Rosalen, Robert C., 211
Rose, A. H., 531
Rose, J. W., 235
Rosen, Robert, 529
Rosen, Sherman J., 284
Rosen, W. J., 322
Rosenberg, Jerry M., 140
Rosenblatt, David H., 165
Rosler, H. J., 242
Ross, E. J. F., 251

Ross, J. F., 202
Ross, Michael H., 380
Ross, Richard D., 229
Ross, Robert B., 221
Ross, Steve, 28, 228
Ross, Steven S., 210
Ross and Turkey, 541
Rossiter, Bryant W., 526
Rostron, M., 56
Rota, Gian-Carlo, 486
Roth, Charles E., 357
Rothenberg, Marc, 466, 467
Rothman, Harry, 35
Rough, Valerie, 367
Rountree, Robert, 374
Roush, W., 251
Rowbotham, George E., 209
Rowley, Gordon, 70
Royal Society of London, 237, 465, 642
Roycroft, Roy, 85
Roysdon, Christine, 459
Rozenberg, Mikhail B., 105
Rubenstein, H. M., 341
Rucera, A., 115
Rudd, Robert L., 27
Ruffner, James A., 243, 255, 256
Rugh, Archie, 2
Rule, Wilfred, 297
Rush, Richard D., 190
Rushby, N. J., 329
Rutherford, G.K., 112
Rycroft, Robert W., 55
Ryzhik, Iosif M., 237

Sabelli, Bruno, 357
Sable, Martin H., 37
Sabloff, Jeremy, 481
Sabnis, Gajanan M., 192
Sachs, Margaret, 64
Saffady, William, 661, 663
Safford, Edward Jr., 281
Safford, Edward L., 216
Sahai, Hardeo, 107
Sainsbury, Diana, 129
Salvato, Joseph A., 291
Salvendy, Gavriel, 209
Sams Engineering Staff, 201
Samsonov, G. V., 169
Samsonov, Grigorii V., 213
Sanadi, D. R., 495

Sanders, Paul A., 184
Sands, Leo G., 87
Sanks, Robert L., 331
Sargent, Frederick, II, 27
Sarjeant, William A. S., 45
Sarkanen, Kyosti, 519
Sarton, George, 466, 467
Satas, Donatas, 217
Saterstrom, Mary H., 663
Satterthwaite, G. E., 377
Sattler, Helen Roney, 128
Saveskie, Peter N., 203
Sax, N. Irving, 332
Sax, Newton I., 257
Saxena, K. M. S., 248
Say, M. G., 280
Sayigh, A. A. M., 340
Sayles, Myron A., 390
Scalisi, Philip, 317
Scanlon, L., 280
Scarpa, Ioannis S., 289
Schaaf, William L, 32
Schaefer, Barbara K., 17
Schaefer, Vincent J., 370
Schaff, W. L., 31
Scheer, Bradley J., 529
Scheheglov, V. P., 377
Schein, M. W., 513
Scheuerman, Ricard V., 168
Schiavone, Giuseppe, 396
Schintlmeister, J., 240
Schlebecker, John T., 480
Schlee, Susan, 480
Schmidt, Gerald D., 178
Schmidt, Richard A., 90
Schmittroch, John, 676, 678
Schneer, Cecil J., 482
Scholl, Friedrich, 379
Scholle, Peter A., 318
Scholtyseck, E., 384
Schroder, Don, 662
Schroder, Klaus, 211
Schuler, Stanley, 302
Schultz, Claire K., 100
Schultz, Vincent, 57, 59
Schulz, Joachim, 107
Schulz-Dubois, E. D., 160
Schumacher, M., 222
Schure, Alexander, 305
Schwann, W. Creighton, 200, 302

Schwartz, Diane, 40
Schwartz, Karlene V., 309
Schwartz, Lazar M., 295
Schwartz, M. M., 212
Schwartz, Maurice L., 79
Schwartz, Mel M., 283
Schwartz, Melvin M., 289
Schwartz, Mortimer D., 58
Schwartz, Paula, 295
Schwarzkopf, Leroy C., 597
Schweitzer, Phillip, 185
Schwimmer, Sigmund, 289
Scientific Manpower Commission, 402
Scorer, Richard S., 76
Scott, James A., 365
Scott, John S., 136, 137, 149
Scott, Ronald M., 544
Scott, Shirley L., 360
Scott, Thomas G., 312
Scripps Institution of Oceanography, 11
Seagreaves, Eleanor R., 59
Seal, Robert A., 17, 31
Sears, Stephen W., 487
Seddon, George, 352
Segal, M. B., 530
Segre, Emilio, 476
Seidman, Arthur H., 197, 198, 205
Seitz, Frederick, 494
Sela, Michael, 506
Selby, Samuel M., 236
Seldman, A. H., 206
Seller, J. P., 239
Seltx-Petrash, Ann, 663
Semler, E. G., 487
Sessions, Keith W., 356
Sessions, Ken, 305
Sessions, Kendall W., 289
Sestak, Z., 40
Sexl, R. U., 455
Shah, Vishu, 188
Shain, M., 141, 303
Shain, Micheal, 207
Shakman, Robert A., 263
Shands, William E., 39
Shapiro, Max S., 65
Shapley, Harlow, 470
Sharma, V. K., 241
Sharpe, R. S., 211
Shaw, Ralph R., 15
Shaw, Robert B., 348

Shaw, Trevor R., 481
Shea, W. R., 456
Shea, William R., 475
Sheppard, A. H., 515
Sherr, S., 325
Sherr, Sol, 324
Sherwin, Martin J., 487
Shields, J., 187
Shiers, George, 485
Shigley, J. E., 219
Shipp, James F., 104, 105
Shirinian, George, 573
Shirkevich, M. G., 158
Shirley, Virginia, 240
Shoemaker, Thomas M., 200
Sholtys, Phyllis, 438
Shor, Elizabeth N., 482
Shortall, J. W., 223
Shoup, Terry E., 330
Shriner, Ralph L., 294
Shugar, G., 183
Shugar, Gerston J., 163
Shugar, Ronald A., 163
Shvarts, Vladimir V., 104
Sibley, Edgar H., 142
Sidgwick, J. B., 153
Siegel, Efrem, 12, 572
Signeur, Austin V., 18, 36, 48
Silber, Herbert B., 610
Silberhorn, Gene M., 345
Silva, Jorge, 398
Silvester, Peter P., 329
Simmons, Mary Ann, 28
Simon, Andre L., 135
Simon, Barry, 523
Simon, Hilda, 369
Simon, James E., 40
Simon, Noel, 76
Simonds, John O., 284
Simons, William W., 161
Simonton, D., 628
Simonton, D. P., 594
Simpson, Jan, 430
Simpson, Peter, 221
Sims, Phillip L., 16
Sims, R. W., 38, 355
Sinclair, Bruce, 487
Sinclair, James B., 312
Singer, T. E. R., 536
Singer, Thomas P., 500

Singleton, Alan, 96
Singleton, Paul, 129
Sington, Adrian, 79
Sinha, Evelyn, 46, 635
Sinkankas, John, 180
Sinnott, Roger W., 247
Sippl, C. J., 144
Sippl, Charles J., 139, 142
Sippl, Roger, 139
Sisler, H. H., 501
Sittig, Marshall, 83, 168, 341, 406
Sittler, Carolyn J., 45
Sive, Mary R., 660
Skeist, Irving, 213
Skerman, V. B. D., 310
Skobel'tsyn, D. V., 615
Skolnik, H., 19
Skolnik, Herman, 476
Skrokov, M. Robert, 207
Skvarcius, R., 304
Slattery, W. J., 640
Slattery, William J., 651
Slauson, Nedra G., 550
Sleeper, H. R., 643
Slesser, Malcolm, 148
Sliney, David, 162
Small, J., 149
Smart, Paul, 366
Smeaton, Robert W., 201
Smit, Pieter, 478
Smith, Carroll N., 478
Smith, Craig B., 224, 338
Smith, David G., 76
Smith, David M., 309
Smith, F. G. Walton, 181
Smith, Hobart M., 355, 370
Smith, Julian F., 536
Smith, M. J., 191
Smith, Norman, 482
Smith, Paul A., 522
Smith, Paul G., 149
Smith, R. C., 473
Smith, Ray F., 478
Smith, Richard A., 255
Smith, Roger C., 20
Smithe, Frank B., 311
Smithells, Colin J., 260
Smithsonian Astrophysical Observatory, 615
Smithsonian Institution Staff, 468

Snead, Rodman E., 386, 388
Sneddon, Ian N., 119
Snyder, Robert M., 409
Sobel, Lester A., 261
Sobelman, I. I., 494
Sobotka, Harry, 496
Society of American Bacteriologists, 273
Society of Automotive Engineers, 151, 219, 646, 674
Society of Manufacturing Engineers, 219
Society of Naval Architects and Marine Engineers, 612
Society of Photographic Scientists and Engineers, 542
Society of the Plastics Industry, 188
Sokal, Robert R., 239
Solarova, J., 41
Somerville, S. H., 130
Sommerville, D. M. Y., 470
Sonneborn, George, 235
Soothill, Eric, 362
Soothill, Richard, 362
Sordillo, Donald A., 281
Sorenson, Ann W., 254
Sorenson, K., 189
Sorum, C. H., 294
Souders, Mott, 181
Soule, J., 126
Sournia, A., 275
Sparrow, Christopher, 60
Spellenberg, Richard, 353
Spencer, Donald D., 139, 140, 142
Spencer, E. Y., 311
Spiegel, Zane, 60
Spiegler, K. S., 535
Splittstoesser, Walter E., 174
Sprecher, Daniel, 437
Spring, Harry M., 334
Stack, Barbara, 216
Stafleu, F. A., 550
Stafleu, Frans A., 20
Stamp, L. D., 131
Stamper, Eugene, 216
Stangl, Martin, 69
Stanier, R. Y., 528
Stanley, Claude M., 208
Stark, Francis B., 351
Starr, Philip, 292
Stecher, Paul G., 68
Steele, Guy L., 142

Steele, James H., 171
Steen, Lynn Arthur, 31
Steffek, Edwin F., 70
Stein, J. Stewart, 136
Stein, Janet R., 175
Stepan, Jan, 30
Stephen, David, 73
Stephenson, F. W., 201
Stephenson, Richard W., 374
Sterba, Gunther, 69
Stern, Arthur, 534
Stevens, J. G., 249
Stevens, John G., 576
Stevens, V. E., 249
Stevens, Virginia E., 576
Steward, F. C., 531
Stewart, C. P., 496
Stewart, D. C., 262
Stewart, T. F. M., 282
Stiegler, Stella E., 130
Stiles, W. S., 248
Stillwell, Margaret B., 465
Stokes, Adrian V., 88
Stokes, Donald W., 316
Stokes, Lillian Q., 316
Stoliarov, D. E., 106
Stoner, John O., 248
Storey, Jill, 4
Stotz, E. H., 528
Stotz, Elmer H., 499
Stout, David F., 205, 207
Stransfeld, Nicholas J., 383
Stratton, Charles, 128
Stratton, John M., 480
Straub, Conrad P., 227, 229
Strelka, C. S., 280
Strong, Myron Reuben, 41
Struglia, Erasmus J., 638
Struik, D. J., 471
Strute, Karl, 460
Stuart, Malcolm, 69
Stubbe, Hans, 478
Stubbendieck, J., 313
Stuckey, Ronald L., 550
Stumpf, P. K., 528
Sturge, John M., 161
Sube, Ralf, 106
Subramanyam, Krishna, 5
Subramanyan, S. N., 240, 249
Suess, M. E., 333

Suess, Michael J., 232, 285
Sullivan, Linda E., 395
Sullivan, Thomas F. P., 225
Summers, Kelly W., 60
Summers, W. I., 648
Summers, W. K., 45
Summers, Wilford L., 200
Suncha, L., 165
Sunner, Stig, 525
Sussman, Alfred, 529
Sutcliffe, J. F., 508
Sutherland, Pamela J., 383
Swaminathan, M., 187
Swan, Lester A., 365
Swanborough, Gordon, 319, 320
Swannack-Nunn, Susan, 619
Swanson, Ellen, 273
Swanson, Gerald, 47
Swanson, Roger W., 131
Swindler, Daris A., 384
Sydenham, P. H., 156
Sydenham, Peter H., 214
Sykes, Peter, 309
Sylvester, Rita, 600
Synge, Patrick M., 125
Szalay, Frederick S., 477
Szokolay, S. V., 191

Tackaberry, John A., 136
Tampion, John, 253, 346
Tanaka, Tyozaburo, 72
Tanner, James T., 316
Tarbert, Gary C., 449
Tasch, P., 319
Taton, Rene, 466
Tattar, Terry A., 348
Taube, U., 572
Tausworthe, Robert C., 649
Taylor, A. C., 213
Taylor, Fredric, 376
Taylor, Gordon E., 154
Taylor, John W. R., 81, 263, 319, 320, 443, 483
Taylor, Michael J. H., 81, 263, 443, 483
Taylor, Ronald J., 354
Tchobanoglous, George, 60
Technische Informationslibliothek der Technischen Universit., 613
Technomic Research Staff, 635
Tedeschi, Frank P., 304

Teja, A. S., 525
Tennisen, Anthony C., 371
Tenopir, Carol, 421
Terrell, Edward E., 312
Terres, John K., 72
Tews, Eyvind S., 605
Texas A&M University Libraries, 55
Texas Instruments, 286
Thackray, Philip C., 281
Tharp, Gerald D., 295
Theilheimer, W., 442
Thewlis, J., 119, 120
Thimann, Kenneth V., 501
Thomas, G. M., 295
Thomas, J. B., 516
Thomas, Richard, 222
Thomas, Robert C., 396
Thomas, Robert G., 394
Thomas, Shirley, 461
Thomas, William A., 536
Thomason, Micheal, 206
Thompson, Ida, 370
Thompson, Malcolm H., 522
Thomson, J. B., 123
Thomson, W. A. R., 175
Thomson, William A. R., 349
Thornhill, Roger, 370
Thornton, John L., 5, 468
Thorpe, Frances, 659
Thumann, Albert, 339, 632
Thuronyi, Geza T., 43
Tigg, George L., 293
Tillman, David A., 519
Time-Life Books, editors, 440
Timms, Simon, 426
Tindall, James R., 370
Tiratsoo, E. N., 339, 443
Tiratsoo, J. N. H., 435
Tirion, Wil, 377
Todd, Arthur H. J., 150
Todd, Frank S., 362
Todd, J. D., 244
Tolstoi, D. M., 105
Tomash Publishers, 473
Tomkeieff, S. I., 131
Tomlinson, Janet, 561
Toomer, G. J., 469
Tootill, Elizabeth, 124, 126, 450
Torrey Botanical Club, 549
Touloukian, Y. S., 250

Touloukian, Yeram S., 562
Tower, Keith, 332
Towers, T. D., 208
Traister, John E., 137, 199
Traister, Robert, 324
Traister, Robert J., 137, 153
Tratner, Alan A., 432
Treibel, Walter A., 206
Tremaine, Maria, 43
Tremlett, H. F., 283
Trigg, G. L., 65, 474
Trigg, George L., 66
Trippett, Stuart, 504
Troitsky, A., 429
Trost, Stanley R., 327
Truesdell, C., 473
Truslow, E. C., 473
Tryon, Alice F., 174
Tryon, Rolla M., 174
Trzyna, Thaddeus C., 436, 438
Tsai, J. C. C., 281
Tucher, Andrea J., 480
Tully, R. I. J., 5, 468
Tuma, Jan J., 155, 159
Turnbull, David, 494
Turner, Ben, 85
Turner, Errett, 35
Turner, L. W., 195
Turner, Roland, 460
Turner, Rolla, 543
Turner, Rufus P., 138, 300
Turner, Wayne C., 224
Turner, William C., 219
Tutt, Patricia, 642
Tuve, George L., 244
Tver, David F., 117, 132, 133, 143, 148
Tyckoson, David A., 45
Tyndalo, Vassili, 345
Tyrkiel, Eugeniusz, 109

UK National Physical Laboratory, 237
UN, 113, 262, 268
UN Conference on Desertification, 388
UN Economic Commission for Asia and Far East, 392
UN Economic Commission for Europe, 268
UN Food and Agriculture Organization, 38, 100, 271, 315, 383, 408, 442
UNCTAD Secretariat, 194

US Atomic Energy Commission, 223
US Atomic Energy Commission, Division of Technical Info., 25
US Bureau of Land Management, 247
US Bureau of the Census, 267
US Central Intelligence Agency, 267
US Civil Aeronautics Board, 266
US Civil Service Commission Library, 452
US Coast Guard, 426
US Commerce Department, 416
US Congress, 255
US Congress Office of Technology Assessment, 627
US Congress, House of Representatives, 116
US Congressional Information Service, 264
US Council on Environmental Quality, 446
US Defense Documentation Center, 593
US Department of Agriculture, 7, 265, 267, 442
US Department of Air Force, 147
US Department of Commerce, 265, 397, 551, 636
US Department of Defense, 640
US Department of Energy, 339
US Department of Transportation, 644
US Department of the Army, 592
US Department of the Interior, 10
US ERDA, 626
US ERDA Technical Information Center, 595
US Energy Research and Development Administration, 429
US Energy and Research Development Agency, 624
US Environmental Data Service, 389
US Environmental Protection Agency, 595, 636, 644
US Environmental Science Services Administration, 10
US Federal Highway Administration, 266
US Federal Power Commission, 267
US Federal Works Agency, 237
US Fish and Wildlife Service, 551
US General Services Administration, 638, 641
US Geological Survey, 10, 553, 601

US Goddard Space Flight Center, 556
US Government Printing Office, 636
US Highway Research Board, 515
US House of Representatives, 244
US Institute of Hydrogeology and Engineering Geology, 391
US Library of Congress, 15, 16, 261, 405, 596, 659
US Library of Congress, Science and Technology Division, 43, 60
US Motor Vehicle Manufacturers Association, 268, 269
US National Air and Space Museum Library, 183
US National Academy of Sciences, 464, 644, 651
US National Aeronautics and Space Administration, 646
US National Agricultural Library, 7, 8
US National Bureau of Standards, 236, 242, 638, 640, 641, 649
US National Highway Traffic Safety Administration, 61
US National Institute for Occupational Safety and Health, 341
US National Institutes of Health, 621
US National Library of Medicine, 60, 545, 546
US National Medical Audiovisual Center, 665
US National Oceanic and Atmospheric Administration, 243, 255, 265
US National Park Service, 599
US National Research Council, 234, 427, 451, 459
US National Research Council, Committee on Digest of Liter., 49
US National Science Foundation, 246, 397, 620, 622,
US National Solar Energy Education Campaign, 56
US National Technical Information Service, 148
US Nautical Almanac Office, 240, 246, 247, 258
US Naval Observatory Library, 6
US Naval Oceanographic Office, 243
US Nuclear Regulatory Commission, 595
US Office of Naval Research, 595

US Office of Nuclear Reactor Regulation, 651
US Office of Surface Mining, 52
US Office of Transportation Planning Analysis, 266
US Office of Water Resources Research, 625
US Patent Office, 630, 631, 633
US Patent and Trademark Office, 633
US Research and Special Programs Administration, 266
US Superintendent of Documents, 613, 598
US Weather Bureau, 262
USDI, 278
USSR Academy of Sciences, 589
Udvardy, Miklos D. F., 358
Uhlmann, D. R., 533
Ullman, J. E., 182
Umbreit, W. W., 511
Unesco, 398
Unesco, Secretariat, 99
Unger, J. H. W., 50
Union of International Associations, 397, 441, 395
Union of Leather Technologists' and Chemists' Societies, 135
United Nations, 19, 22, 24, 26, 46
University of California, Berkeley, Water Resource Center, 21
University of California, Water Resources Center Archives, 10
Unterweiser, Paul M., 333
Urban Mass Transportation Administration, 439
Urbanski, Jerzy, 187
Urdang, Lawrence, 63
Utz, Peter, 661
Uvarov, E. B., 116

Vallentine, John F., 16
Van Balen, John, 9
Van Cleemput, W. M., 50
Van Derveer, Paul D., 134
Van Dyke, Milton, 378
Van Nouhuys, D., 422
Van Olphen, H., 212
Van Oss, J. F., 81
Van Tamelen, E. E., 527
Van Zyll de Jong, C. G., 178

Van der Leeden, Frits, 270
Van der Marel, H. W., 385
Van der Wijk, R., 550
Vandenberghe, J. P., 111
VanderWerf, C. A., 503
Vankat, J. L., 349
Vargaftik, N. B., 241
Varley, A., 459
Vasaru, Gheorge, 53
Vaughan, William, 159
Vaughn, P. R., 181
Vehrenberg, Hans, 375
Venning, Frank D., 353
Verma, V. N., 170
Verna, Rajni K., 239
Verney, Peter, 180
Vernick, Arnold S., 230
Verrell, Barbara, 669
Verschueren, Karel, 166, 229
Verzar, F., 479
Veselovsky, Zdenek, 73
Vesilind, Aarne, 401
Vestling, Martha M., 164
Veziroglu, T. Nejat, 90
Vigoreux, P., 641
Vigoureux, P., 235
Villate, Jose T., 149
Vince, John, 140
Vincent, Jack, 76
Vinitskii, I. M., 213
Vollnhals, O., 111
Von Frajer, H., 157
Von Randow, R., 32
Vulkan, G., 61
Vutukuri, V. S., 213

W. Engl et al, 494
Waddington, Janet, 243
Wade, Carol, 628
Wade, Leroy G., 524
Wagner, Frederic H., 357
Wahl, M., 220
Wahll, M. J., 221
Walford, A. J., 15
Walker, Braz, 364
Walkeer, Ernest P., 368
Walker, G., 331
Walker, William, 312
Walker, Winifred, 344
Walker-Smith, D., 323

Wall, Alexander C., 139
Wall, Clifford N., 294
Wall, Elizabeth S., 139
Wall, F. J., 157
Wall, James D., 228
Wallace, A., 562
Wallace, Richard E., 37
Wallis, P., 456
Walls, Jerry G., 74
Walter, Claire, 388
Walter Reed Army Institute of Research, 624
Walters, Ivan A., 538
Walters, LeRoy, 36
Walters, Michael, 359
Walton, Anne, 308
Walton, Charles F., 218
Walton, E. K., 131
Wamsley, Stuart H., 183
Wang, Lawrence K., 229
Wang, Ling, 468
Ward, Dederick C., 21, 45, 618
Ward, Jack W., 23
Wardhaugh, A. A., 361
Ware, George W., 83
Ware, George W., 252
Waring, R. H., 216
Warlaw, A. C., 287
Warren, Betty, 58
Warren, Kenneth S., 19
Warring, R. H., 217
Wasser, Clinton H., 312
Wasserman, Paul, 264, 393, 460, 664
Wasson, J. T., 318
Waterfield, M. D., 170
Waterford, Van, 326
Waters, Farl J., 304
Waters, T. R. W., 478
Waterson, A. P., 479
Wathern, Peter, 58
Watson, Allan, 128
Watson, N. K., 660
Watt, John H., 302
Watters, Ivan A., 593
Watznauer, Adolf, 103
Weart, Spenser, 473
Weast, Robert, 537
Weast, Robert C., 152, 165
Webb, J. A., 313

Webb, Maurice J., 218
Weber, Marvin J., 157
Weber, R. David, 25
Weber, Robert L., 456
Weber, Samuel, 258
Webster, Tony, 326, 328, 423
Weck, Manfred, 534
Wedepohl, K. H., 166
Weeks, John, 94
Weidner, William A., 436
Weigart, Alfred, 63
Weik, Martin H., 144
Weindling, Paul, 467
Weiner, C., 474
Weiner, J. S., 623
Weiner, Jack, 48
Weinshank, Donald J., 98
Weir, Donald M., 171
Weiser, Jaroslav, 380
Weismantel, G., 210
Weiss, G. H., 637
Weissbach, Oskar, 163
Weissberger, Arnold, 526, 527
Weller, Milton W., 361
Wellman, Frederick L., 253
Wenger, Karl F., 174
Wennrich, Peter, 95, 96
Werger, Joanne, 40
Werner, Mario, 163
West, Geoffrey P., 73
West, Robert L., 54
West, T. S., 97
Westcott, Cynthia, 177
Westfall, Richard S., 455
Westinghouse Electric Corporation, 200
Westlund, D. A., 280
Wexelblat, Richard L., 486
Wexler, Philip, 22
Whalen, George J., 68, 125, 297
Whalley, Paul E. S., 128
Wheaton, Bruce R., 35, 474
Wheeler, Alwyne, 128
Wheeler, Marjorie W., 21, 45
Whelchel, Leslie, 292
Whiffen, David Hardy, 272
Whitaker, John O., 366
White, George W., 481
White, Gilbert F., 488
White, Jerry, 88

Whitehead, Harry, 129
Whitehead, Peter, 362
Whitehurst, Bert W., 633
Whiteside, Derek T., 471
Whitfield, Philip, 75
Whitrow, Magda, 30
Whitten, D. G. A., 132
Whittington, L. R., 636
Whittington, Lloyd R., 135
Whlan, Miriam, 536, 653
Wicander, R., 372
Wichterle, I., 51
Wick, Alexander E., 241
Wick, Susanne R., 310
Wiener, Philip P., 466
Wieser, Peter F., 222
Wiin-Nielsen, Aksel, 531
Wilcox, William H., 536
Wilkins, Charles L., 500
Wilkins, G. A., 247
Wilkinson, Geoffrey, 525
Wilkinson, Lise, 479
Wilkinson, P. C., 128
Willardson, R. K., 523
Willardson, Robert K., 517
Willey, Gordon R., 481
Williams, Alan, 148
Williams, Andrew E., 351
Williams, Arthur B., 196, 204
Williams, Conlin H., 395
Williams, E. W., 484
Williams, Gareth, 126
Williams, John G., 351
Williams, L. Pearce, 465
Williams, Martha, 675
Williams, Roger John, 67
Williams, Trevor I., 398, 450, 468, 482
Williams, W. John, 164
Williamson, D. H., 296
Williamson, Edward H., 259
Willis, Cleve E., 26
Willis, Jerry, 330
Wills, B. A., 283
Wilmore, Sylvia B., 362
Wilson, Betty L., 619
Wilson, C. W. J., 625
Wilson, David, 456
Wilson, David G., 230
Wilson, Michael, 443

Wilson, William K., 438, 619
Winch, Kenneth L., 374
Windholz, Martha, 241
Windsor, Philip, 339
Wingard, Lemaul B., 527
Wingate, Isabel B., 134
Winkfield, James A., 374
Winn, Carolyn P., 425
Winneberger, J. H. T., 284
Winner, Walter P., 277
Winslow, Ken, 661
Winter, Ruth, 133, 135
Wiren, Harold N., 24
Wise, David B., 93
Wisniak, J., 249
Witrow, Magda, 467
Wittman, Alfred, 141
Wolbarsht, Myron, 162
Wolf, Helmut F., 158
Wolfe, William L., 160
Wolff, Kathryn, 4, 663
Wolfheim, Jaclyn H., 42
Wollin, Jay, 371
Wolseley, Pat, 319
Wong, H. Y., 184
Wood, Charles D., 384
Wood, D. N., 21
Wood, Gerald L., 178
Wood, Phyllis, 311
Woods, Alvin E., 298
Woolen, Anthony, 277
Wooley, A. R., 371
Woolley, Alan, 79
Woolrich, Willis R., 277
Wooten, David C., 228
World Energy Conference, 111, 114
World Environment Center, 463
World Health Organization, 645
World Meteorological Organization, 233, 266, 276
Worthing, C. R., 173
Worthing, Charles R., 278
Worthley, Jean Reese, 309
Wrathall, Claude P., 96
Wright, Don T., 423
Wright, John I., 313
Wright, Sewall, 529
Wright, Wilfred, 308
Wyatt, H. V., 20

Wyckoff, Ralph W. G., 525
Wyllie, Thomas D., 176
Wyman, Donald, 72
Wyse, Bonita W., 254
Wyszecki, Gunter, 248

Yacyshyn, Rudolph, 667
Yamada, Hiroshi, 113
Yamaguchi, Mas, 275
Yamashita, Yasumasa, 375
Yanarella, Ann-Marie, 55
Yanarella, Ernest J., 55
Yang, Chen Ning, 456
Yannarella, Philip A., 599, 574
Yarrow, D., 310
Yarwood, T. M., 237
Yates, Frank, 239
Yaws, Carl L., 252
Yepsen, Roger B., 365
Yescombe, E. R., 22
Yothers, Mark T., 651
Young, Carol, 138
Young, George, 297
Young, Harold C., 397
Young, Margaret L., 397
Young, R. A., 232
Young, Robert J., 232
Young, Roland S., 170
Young, William J., 392
Yudkin, Michael, 309

Zahler, P., 296
Zahradnik, Jiri, 365
Zalubas, Romuald, 33
Zaremba, Joseph, 17
Zarony, Sharon, 677, 592
Zbar, Paul, 297
Zhong Wai Publishing Dictionary Editing Group, 101
Ziauddin, Sardor, 399
Zichmanis, Zile, 353
Ziegler, Edward N., 91, 519
Zimmer, Hans, 501
Zimmerman, Helmut, 63
Zimmerman, James Hall, 354
Zimmerman, M., 106
Zimmerman, M. H., 70
Zissis, George J., 160
Zombeck, M. V., 154
Zubibi, Fabio, 101

Zuckerman, Harriet, 452
Zumwalt, Diane M., 666
Zumwalt, George S., 666
Zurick, Timothy, 145

TITLE INDEX

Publications ending with a period are short titles.

100 Short Films about the Human Environment, 665
1001 Questions Answered About Astronomy, 286
1979 National Survey of Compensation Paid Scientists and., 266
1980 Handbook of Agricultural Charts, 254
1984–1985 International Micrographics Source Book, 289
1985 Microcomputer Marketplace, 420
5,000 Questions Answered Maintaining, Repairing, and Improv., 302

A–V ONLINE (formerly NICEM), 683
A–Z of Astronomy, 117
A–Z of Clinical Chemistry, 121
A–Z of Offshore Oil and Gas, 129
A. P. Norton Star Atlas and Reference Handbook, 377
AAAS Audiotape Cassette Album Series, 663
AAAS Reviews of Science, 520
AAAS Science Book and Films, 663
AAAS Science Book List, 3
AAAS Science Book List Supplement, 4
AAAS Science Film Catalog, 663
ABC's of Calculus, 300
ABC's of Electronics, 304
ABC's of Hi-Fi and Stereo, 305
ABI/INFORM, 696
ABI/SELECTS: Annotated Bibliography of Computer Periodicals, 572
ACI Manual of Concrete Inspection, 278
ACI Publications, 673
ACM Computing Surveys, 586
ACM Guide to Computing Literature, 23
ACS Laboratory Guide, 308
AEE Directory of Energy Professionals, 463
AEROSPACE DATABASE, 680
AFIPS Conference Proceedings, 611
AGARD Conference Proceedings, 610
AGLINET Union List of Serials, 4
AGRICOLA, 680

AGRIS Forestry, 408
AGRIS Forestry World Catalog of Information and Doc., 7
AIA Energy Notebook, 223
AIA Metric Building and Construction Guide, 322
AIAA Guidance, Control, and Flight Mechanics Conference, 610
AIAA Journal, 584
AIChE Journal, 585
AIM 65 Laboratory Manual and Study Guide, 280
AIM/ARM, 680
AIP Conference Proceedings, 607
AMERICAN CHEMICAL SOCIETY PRIMARY JOURNAL DATABASE (CFTX), 680
AMERICAN MEN AND WOMEN OF SCIENCE, 680
AMS Grants and Proposals, 622
ANSI Catalog, 639
APCA Directory and Resource Book, 439
APILIT, 681
APIPAT, 681
APTIC, 681
AQUACULTURE, 681
AQUALINE, 682
AQUATIC SCIENCES AND FISHERIES ABSTRACTS, 682
ARCO Motor Vehicle Dictionary, 106
ARTHUR D. LITTLE/ONLINE, 682
ASCE Combined Index, 673
ASCE Journal of Power Division, 588
ASCE Publications Information, 673
ASHRAE Standards, 650
ASI, 683
ASLE Annual Meeting Preprint, 628
ASM Handbook of Engineering Mathematics, 155
ASM Metals Reference Book, 211
ASM Thesaurus of Metallurgical Terms, 97
ASME Guide for Gas Transmission and Distribution Piping Sys., 333
ASME Handbook, 214

ASME Standards, 650
ASTM Standards Relating to Miscellaneous Petroleum Products, 646
ASTM Standards in Building Codes, 639
Abbreviations Dictionary, 95
Abbreviations for Use on Drawings and in Text, 643
About Patents, 631
Abridged Index Medicus, 545
Abstract Journal in Earthquake Engineering, 552
Abstract Journal, 1790–1920, 572
Abstracting Scientific and Technical Literature, 536
Abstracts and Indexes in Science and Technology, 536
Abstracts of Entomology, 550
Abstracts of Mycology, 548
Abstracts of North American Geology, 552
Abstracts of Photographic Science and Engineering Literature, 542
Abstracts of Theses, 618
Abstracts on Health Effects of Environmental Pollutants, 568
Academy of Sciences of the USSR. Bulletin, 578
Access, 543
Accounts of Chemical Research, 579, 623
Acid Rain, 60
Acid Rain Resources Directory, 436
Acoustic Emission, 32
Acoustical Society of America. Journal, 578
Acoustics Abstracts, 542
Acronyms, Initialisms, and Abbreviations Dictionary, 95
Active Filter Design, 203
Adhesive Handbook, 187
Adhesives, 320
Advanced Oscilloscope Handbook for Technicians and Engineers, 157
Advanced Stereo System Equipment, 324
Advances in Agronomy, 508
Advances in Analytical Chemistry and Instrumentation, 495
Advances in Applied Mathematics, 577
Advances in Applied Mechanics, 517
Advances in Applied Microbiology, 511

Advances in Astronomy and Astrophysics, 489
Advances in Atomic and Molecular Physics, 491
Advances in Biochemical Engineering, 514
Advances in Bilogical And Medical Physics, 494
Advances in Biomedical Engineering, 514
Advances in Biophysics, 494
Advances in Botanical Research, 508
Advances in Cancer Research, 505
Advances in Carbohydrate Chemistry and Biochemistry, 499
Advances in Catalysis, 496
Advances in Cell Biology, 509
Advances in Cell and Molecular Biology, 509
Advances in Chemical Engineering, 515
Advances in Chemical Physics, 491
Advances in Chemistry, 496
Advances in Chemistry Series, 608
Advances in Chromatography, 496
Advances in Clinical Chemistry, 496
Advances in Communication Systems, 516
Advances in Comparative Physiology and Biochemistry, 512
Advances in Computers, 516
Advances in Computing Research, 516
Advances in Control Systems, 516
Advances in Corrosion Science and Technology, 515, 517
Advances in Cryogenic Engineering, 515
Advances in Cyclic Nucleotide Research, 505
Advances in Ecological Research, 519
Advances in Electrochemistry and Electrochemical Engineering, 496
Advances in Electronics and Electron Physics, 516
Advances in Energy Systems and Technology, 518
Advances in Environmental Science and Engineering, 519
Advances in Enzyme Regulation, 510
Advances in Enzymology and Related Areas of Molecular Bio., 510

Advances in Experimental Medicine and
   Biology, 505
Advances in Food Research, 508
Advances in Free-Radical Chemistry, 502
Advances in Genetics, 510
Advances in Geology, 513
Advances in Geophysics, 491, 513
Advances in Heat Transfer, 517
Advances in Heterocyclic Chemistry, 502
Advances in High Pressure Research,
   533
Advances in Human Genetics, 510
Advances in Hydroscience, 515
Advances in Image Pickup and Display,
   516
Advances in Immunology, 511
Advances in Inorganic Chemistry and
   Radiochemistry, 501
Advances in Insect Physiology, 513
Advances in Lipid Research, 505
Advances in Liquid Crystals, 491
Advances in Machine Tool Design and
   Research, 518
Advances in Macromolecular Chemistry,
   502
Advances in Marine Biology, 505
Advances in Mass Spectrometry, 495
Advances in Materials Research, 517
Advances in Mathematics, 489, 577
Advances in Microbial Ecology, 511
Advances in Microbial Physiology, 511
Advances in Microbiology of the Sea, 513
Advances in Microwaves, 491
Advances in Molten Salt Chemistry, 501
Advances in Nuclear Physics, 495
Advances in Nuclear Science and
   Technology, 518
Advances in Optical and Electron
   Microscopy, 491
Advances in Organometallic Chemistry,
   502
Advances in Parasitology, 513
Advances in Particle Physics, 495
Advances in Photochemistry, 496
Advances in Physical Organic Chemistry,
   502
Advances in Physics, 578
Advances in Plasma Physics, 491
Advances in Polymer Science, 504
Advances in Probability and Related
   Topics, 490
Advances in Protein Chemistry, 502
Advances in Quantum Chemistry, 496
Advances in Quantum Electronics, 491
Advances in Quantum Physics, 491
Advances in Radiation Biology, 505
Advances in Radiation Chemistry, 496
Advances in Reproductive Physiology,
   512
Advances in Solar Energy, 519
Advances in Space Science and
   Technology, 514
Advances in Structural Research by
   Diffraction Methods, 492
Advances in Substance Abuse, 505
Advances in Teratology, 505
Advances in Transport Processes, 519
Advances in Virus Research, 511
Advances in X-ray Analysis, 496
Advances in the Astronautical Sciences,
   514
Advances in the Biosciences, 505
Advances in the Study of Behavior, 505
Aero College Aviation Directory, 399
Aeronautical Chart Catalogue, 319
Aeronautical Engineering, 47
Aeronautical Journal, 584
Aeronautical Quarterly, 584
Aeronautics and Astronautics, 483
Aerospace Bibliography, 47
Aerospace Facts and Figures, 267
Aerospace Research Index, 625
Aerospace Thesis Topics List, 618
Aerospace Yearbook, 443
Africa-Middle East Petroleum Directory,
   430
Agreements Registered with the IAEA,
   336
Agricultural Chemicals and Pesticides,
   227
Agricultural Credit, 38
Agricultural Development Indicators,
   173
Agricultural Economics and Rural Socio.
   Multiling. Thesaurus, 97
Agricultural Engineering, 580
Agricultural Finance Statistics, 265
Agricultural Meteorology, 580

Agricultural Records in Britain, AD 220–1977, 480
Agricultural Research Index, 624
Agricultural Residues, 408
Agricultural Science Review, 580
Agricultural Statistics, 265
Agricultural Terms, 129
Agricultural Terms: English-Chinese, 100
Agriculture in America 1622–1860, 480
Agrindex, 547
Agronomy Journal, 580
Air Almanac, 246
Air Conditioning, Heating and Refrigeration Dictionary, 145
Air Facts and Feats, 263
Air Forces of the World, 411
Air Pollution, 534
Air Pollution Abstracts, 568
Air Pollution Control Association. Journal, 589
Air Pollution Control and Design Handbook, 232
Air Pollution Publications, 60
Air Pollution Sampling and Analysis Deskbook, 262, 284
Air Pollution Technical Publications of US EPA, 594
Air Power, 93
Aircraft Directory, 411
Airman's Information Manual, 276
Airplanes of the World, 1490–1976, 483
Airport Services Manual, 277
Albert Einstein, 455
Albert Einstein, The Human Side, 455
Album of Fluid Motion, 378
Album of Science: Antiquity and the Middle Ages, 465
Album of Science: From Leonardo to Lavoisier, 465
Album of Science: The Nineteenth Century, 465
Aldrich Library of Infrared Spectra, 250
Alexis Lichine's New Encyclopedia of Wines and Spirits, 83
Alkaloids: Chemistry and Physiology, 497, 502
All the Plants of the Bible, 344
Alloys Index, 563
Almanac for Computers, 258

Almanac of the Canning, Freezing, Preserving Industries, 320
Alphabetical List of Compound Names Formulae, and References, 250
Alternative Energy Sources, 90
Alternative Foods, 343
Aluminum, 322
Amateur Astronomer's Handbook, 153
Amateur Naturalist's Handbook, 152
Amateur Radio Advanced-Class License Study Guide, 305
Amateur Radio General-Class License Study Guide, 305
American Scientist, 576
America's Journeys into Space, 461
American Association of Petroleum Geologists. Bulletin, 582
American Astronomical Society. Bulletin, 577
American Breweries, 414
American Bureau of Metal Statistics Yearbook, 267
American Chemical Society Abstracts of Meeting Papers, 608
American Chemical Society-Preprints, 628
American Chemical Society. ACS Symposium Series, 608
American Chemical Society. Journal, 579
American Chemists and Chemical Engineers, 457
American Doctoral Dissertations, 616
American Electrician's Handbook, 196, 302
American Engineers of the Nineteenth Century, 459
American Environmental History, 488
American Ephemeris and Nautical Almanac, 246
American Fisheries, 407
American Geological Literature, 1699 to 1850, 42
American Heritage History of the Automobile in America, 487
American Institute of Biological Sciences Directory of, 399
American Institute of Chemical Engineers. AIChE Symposium, 610
American Institute of Physics Catalog of Journals, Books., 673

American Institute of Physics Handbook, 157
American Journal of Anatomy, 582
American Journal of Botany, 581
American Journal of Mathematical and Management Sciences, 577
American Journal of Physics, 578
American Journal of Physiology, 582
American Mathematical Monthly, 540, 577
American Mathematical Society. Bulletin, 577
American Mathematical Society. Notices, 577
American Mathematical Society. Proceedings, 577, 607
American Mathematical Society. Transactions, 577
American Mathematical Society. Translations, 614
American Medical Ethnobotany, 125
American Medicinal Plants, 344
American Men and Women of Science, 448
American Men and Women of Science, consultants, 448
American Meteorological Society. Bulletin, 583
American Metric Construction Handbook, 190
American Mineralogist, 582
American Museum Guides: Sciences, 397
American Museum of Natural History Guide to Shells, 363
American National Dictionary for Information Processing Sys., 138
American National Standard Vocabulary for Information Proc., 643
American National Standards Institute, 637
American Naturalist, 581
American Nuclear Society Transaction, 588
American Physical Society. Bulletin, 578, 608
American Public Works Association Directory, 415
American Public Works Association Yearbook, 443
American Science and Technology, 29
American Scientists, 575
American Society for Testing and Materials Publications, 673
American Society for Testing and Materials Standards, 637
American Society of Civil Engineers. Division Journals, 585
American Society of Civil Engineers. Proceedings, 585, 611
American Society of Mechanical Engineers. 77th Year Index, 673
American Society of Mechanical Engineers. Transactions, 587
American Statistics Index, 264
American Water Works Association Journal, 589
American Zoologist, 582
Amino Acids, Peptides, and Proteins, 509
Amphibians of North America, 355
Analysis of Polymers, Resins and Additives, 379
Analyst, 579
Analytical Abstracts, 543
Analytical Chemistry, 579
Analytical Methods for Pesticides and Plant Growth Regulator, 508
Analytical and Industrial Molecular Spectroscopy, 579
Anglo-American and German Abbreviations in Data Processing, 95
Anglo-American and German Abbreviations in Environ. Protect., 96
Anglo-American and German Abbreviations in Science and Tech., 96
Animal Identification, 355
Animal Kingdom, 382
Annals of Botany, 581
Annals of Mathematics, 577
Annals of Nuclear Energy, 588
Annals of Physics, 578
Annals of the History of Computing, 586
Annals of the Israel Physical Society, 608
Annotated Bibliographies of Mineral Deposits in Africa, Asia, 43
Annotated Bibliographies of Mineral Deposits in W. Hemisphere, 43
Annotated Bibliography of Canadian Air Pollution Literature, 60
Annotated Bibliography of Exposit. Writing in Math. Sciences, 31

Annotated Bibliography of Technical and Special Dictionaries, 29
Annotated Bibliography of Ecology and Reclamation of., 57
Annotated Bibliography on the History of Data Processing, 486
Annual Bulletin of Electric Energy Statistics for Europe, 268
Annual Bulletin of General Energy Statistics for Europe, 268
Annual Bulletin of Transport Statistics for Europe 1978, 268
Annual Guides to Graduate and Undergraduate Study, 399
Annual Register of Grant Support, 1969–, 619
Annual Report of the Council on Environmental Quality, 265
Annual Report of the Federal Power Commission, 265
Annual Reports in Inorganic and General Syntheses, 501
Annual Reports in Medicinal Chemistry, 497
Annual Reports in Organic Synthesis, 503
Annual Reports on Analytical Atomic Spectroscopyt, 497
Annual Reports on Fermentation Processes, 511
Annual Reports on NMR Spectroscopy, 497
Annual Reports on the Progress of Chemistry, 497
Annual Review in Automatic Programming, 516
Annual Review of Astronomy and Astrophysics, 489
Annual Review of Biochemistry, 499
Annual Review of Biophysics and Bioengineering, 495, 515
Annual Review of Cell Biology, 509
Annual Review of Computer Science, 516
Annual Review of Earth and Planetary Sciences, 513
Annual Review of Ecology and Systematics, 519
Annual Review of Energy, 519
Annual Review of Entomology, 513
Annual Review of Fluid Mechanics, 518
Annual Review of Genetics, 510
Annual Review of Immunology, 511
Annual Review of Material Science, 517
Annual Review of Microbiology, 511
Annual Review of Nuclear Science, 518
Annual Review of Nuclear and Particle Science, 495, 518
Annual Review of Nutrition, 508
Annual Review of Pharmacology and Toxicology, 506
Annual Review of Photochemistry, 503
Annual Review of Physical Chemistry, 497
Annual Review of Physiology, 512
Annual Review of Phytopathology, 508
Annual Review of Plant Physiology, 508
Annual Reviews of Industrial and Engineering Chemistry, 515
Annual Summary of Information on Natural Disasters, 268, 316
Annual Summary of Research in Electronics, 625
Annuals of the International Geophysical Year, 441
Answers Online, 675
Antarctic Bibliography, 43
Antarctic Map Folio Series, 384
Anti-Nuclear Handbook, 223
Antigens, 506
Aphids on the World's Crops, 364
Apparent Places of Fundamental Stars, 377
Apple Computer Directory, 420
Apple II User's Guide, 302
Apple Software Directory, 420
Applications of Energy, 471
Applied Biochemistry and Bioengineering, 527
Applied Biology, 506
Applied Health Physics Abstracts and Notes, 542
Applied Mathematical Sciences, 490
Applied Mechanics Review, 562
Applied Optics, 307, 578
Applied Optics ad Engineering, 523
Applied Polymer Symposia, 610
Applied Science and Technology Index, 555
Applied Solar Energy, 337

Applied Water Resource Systems Planning, 316
Applied and Environmental Microbiology, 581
Appropriate Energy Technology Library Bibliography, 54
Aquarium Encyclopedia, 69
Aquarium Plants, 312
Aquatic Sciences and Fisheries Abstracts, 551
Archaeological Chemistry, 286
Architect's Guide to Energy Conservation, 337
Architect's Data, 190
Architectual Graphic Standards, 643
Architectural Handbook, 190
Architectural and Building Trades Dictionary, 136
Archival Microfilm Packages, 664
Arctic Bibliography, 43
Arctic Institute of North America. Montreal Library Catalog, 9
Army Materials Technology Conference, 612
Art of Natural History, 477
Artificial Intelligence Report, 586
Asbestos Particle Atlas, 378
Asbestos: An Information Resource, 12
Asher's Guide to Botanical Periodicals, 569
Asia-Pacific Petroleum Directory, 430
Asimov's Biographical Encyclopedia of Science and Technology, 449
Asimov's Guide to Science, 307, 465
Aslib Book List, 1
Association for Computing Machinery, 586
Astronautics and Aeronautics, 483, 584
Astronomical Almanac for the Year 1981, 246
Astronomical Catalogues 1951–75, 6
Astronomical Directory, 117
Astronomical Ephemeris, 247
Astronomical Journal, 577
Astronomical Photographic Atlas, 375
Astronomical Society of Pacific. Publications, 577
Astronomy, 117, 522
Astronomy Handbook, 153
Astronomy Data Book, 247

Astronomy and Astrophysics, 577
Astronomy and Astrophysics Abstracts, 540
Astronomy and Telescopes, 153
Astronomy of the Ancients, 469
Astrophysical Abstracts, 540
Astrophysical Formulae, 248
Astrophysical Journal, 577
Astrophysics Today, 629
Astrophysics and Space Science, 577
Atari User's Encyclopedia, 88
Atlantic Oceanographic and Meteorological Laboratories, 629
Atlas and Manual of Plant Pathology, 273
Atlas of Animal Migration, 383
Atlas of Continental Displacement, 384
Atlas of Deep-Sky Spendors, 375
Atlas of Descriptive Histology, 380
Atlas of Distribution of Freshwater Fish Families of World, 383
Atlas of Economic Mineral Deposits, 385
Atlas of Igneous Rocks and Their Textures, 385
Atlas of Infrared Spectroscopy of Clay Minerals and Their., 385
Atlas of Insect Diseases, 380
Atlas of Insect and Plant Viruses, 380
Atlas of Insects Harmful to Forest Trees, 380
Atlas of Landforms, 385
Atlas of Mammalian Chromosomes, 383
Atlas of Marine Use in the North Pacific Region, 389
Atlas of Maritime History, 385
Atlas of Medicinal Plants of Middle America, 381
Atlas of Mercury, 375
Atlas of Metal, 379
Atlas of Metal-Ligand Equilibria in Aqueous Solution, 379
Atlas of Molecular Structures in Biology, 381
Atlas of Orders and Families of Birds, 383
Atlas of Past and Present Pollen maps of Europe, 381
Atlas of Plant Viruses, 381
Atlas of Polymer Damage, 379

Atlas of Primate Gross Anatomy, Baboon, Chimpanzee, and Man, 384
Atlas of Quartz Sand Surface Textures, 386
Atlas of Renewable Energy Resources, 392
Atlas of Representative Stellar Spectra, 375
Atlas of Rock-Forming Minerals in Thin Section, 386
Atlas of Selected Oil and Gas Reservoir Rocks from North America, 392
Atlas of Spectral Data and Physical Constants for Organic., 379
Atlas of Spectral Interferences in ICP Spectroscopy, 378
Atlas of Stereochemistry, 380
Atlas of Textural Patterns of Basalts and Genetic Significance, 386
Atlas of US Trees, 381
Atlas of an Insect Brain, 383
Atlas of the Air, 388
Atlas of the Birds of the Western Palaearctic, 383
Atlas of the Living Resources of the Seas, 383
Atlas of the Netherlands Flora, 381
Atlas of the Planets, 375
Atlas of the World's Railways, 390
Atmospheric Dispersion in Nuclear Power Plant Sitting, 341
Atmospheric Environment, 589
Atom-Molecule Collision Theory, 308
Atomic Energy Levels, 248
Atomic Energy Review, 588
Atomic Energy-Level and Grotian Diagrams, 248
Atomic Handbook, 223
Atomic Physics, 608
Audio Video Market Place 1985, 657
Audio-Visual Equipment Directory, 662
Audiovisual Equipment and Materials, 662
Audiovisual Market Place, 662
Audiovisual Resources in Food and Nutrition, 664
Audobon Society Field Guide to North American Fossils, 370
Audubon, 457

Audubon Illustrated Handbook of American Birds, 177
Audubon Society Book of Insects, 364
Audubon Society Book of Wild Animals, 316
Audubon Society Encyclopedia of Animal Life, 72
Audubon Society Encyclopedia of North American Birds, 72
Audubon Society Field Guide to North America Birds, 358
Audubon Society Field Guide to North American Butterflies, 364
Audubon Society Field Guide to North American Fossils, 370
Audubon Society Field Guide to North American Insects and., 365
Audubon Society Field Guide to North American Mammals, 366
Audubon Society Field Guide to North American Reptiles and., 369
Audubon Society Field Guide to North American Rocks and Minerals, 370
Audubon Society Field Guide to North American Seashore., 355
Audubon Society Field Guide to North American Trees: Eastern, 344
Audubon Society Field Guide to North American Trees: Western, 344
Audubon Society Field Guide to North American Wildflowers, 352
Audubon Society Guide to Attracting Birds, 358
Audubon Society Handbook for Birders, 177
Audubon Society Master Guide to Birding, 358
Audubon's Birds of America, 357
Auerbach Software Reports, 420
Auger Electron Spectroscopy, 33
Auger Electron Spectroscopy Reference Manual, 272
Australian Scientific Societies and Professional Associations, 667
Author Affiliation Index to the Engineering Index Annual, 459
Author Biographies Master Index, 449
Author Index of Mathematical Reviews 1965–1972, 540

Author and Permuted Title Index to Selected Statistical Journals, 540
Auto Electronics Simplified, 306
Automobile Facts and Figures, 268
Automotive Dictionary, 146
Automotive Electrical Reference Manual, 280
Automotive Literature Index, 1947–1976, 562
Automotive Technician's Handbook, 208
Availability of Nuclear Science Conference Literature, 605
Aviation Annual, 443
Aviation Weather and Weather Services, 277
Aviation Week and Space Technology, 584
Aviation, Space, and Environmental Medicine, 584
Aviator's Catalog, 288
Awakening Interest in Science During First Century of Printing, 465
Awards, Honors, and Prizes, 393

BASIC Book, 329
BASIC Handbook, 203
BIOSIS PREVIEWS, 683
BIOSIS Previews Search Guide, 678
BIOTECHNOLOGY, 683
BLL Conference Index, 1964–1973, 603
BLS PRODUCES PRICE INDEX, 684
BRS Brief System Guide Sign-on Procedures, 679
BRS Directory and Database Catalog, 679
BRS System Reference Guide, 679
BS Handbook, 646
Bacteria, 528
Barker Engineering Library Bulletin, 46
Barnes & Noble Thesaurus of Biology, 124
Basic Anatomy, 285
Basic Apple IIc, 303
Basic Auto Repair Manual, 306
Basic Book of Metalworking, 334
Basic Electrical and Electronic Tests and Measurements, 235
Basic Electricity, 297
Basic Guide to Power Electronics, 324

Basic Library List for Four-Year Colleges, 6
Basic Life Sciences, 506
Basic Mathematics, 236
Basic Plant Pathology Methods, 312
Basic Programmers Guide to Pascal, 329
Basic Tables in Chemistry, 241
Basic Television, 305
Basic US Government Micrographic Standards and Specification, 657
Bats of America, 366
Beginner's Computer Dictionary, 139
Beginner's Handbook of Electronics, 196
Behavioural Biology Abstracts, 545
Beilstein Guide, 163
Beilstein's Handbuch der Organischen Chemie, 163
Bell Laboratories Innovation in Telecommunications, 485
Bell Laboratories Record, 655
Bell System Technical Journal, 655
Benchmark Papers in Acoustics, 492
Benchmark Papers in Behavior, 513
Benchmark Papers in Biological Concepts, 506
Benchmark Papers in Ecology, 520
Benchmark Papers in Electrical Engineering and Computer Science, 516
Benchmark Papers in Genetics, 510
Benchmark Papers in Human Physiology, 512
Benchmark Papers in Inorganic Chemistry, 501
Benchmark Papers in Microbiology, 511
Benchmark Papers in Optics, 492
Benchmark Papers in Organic Chemistry, 503
Benchmark Papers in Physical Chemistry and Chemical Physics, 492
Benchmark Papers on Energy, 519
Benchmark Papers in Geology, 513
Bergey's Manual of Determinative Bacteriology, 273
Best Science Books for Children, 4
Best of CP/M Software, 327
Best of IBM PC Software, 327
Beuscher's Law and the Farmer, 311
Bibliographic Atlas of Protein Spectra in Ultraviolet., 35

Bibliographic Encyclopedia of Science and Technology, 449
Bibliographic Guide to Conference Publications, 604
Bibliographic Guide to Governmental Publications—US, 596
Bibliographic Guide to Technology, 29
Bibliographic Guide to Technology, 1978, 11
Bibliographical History of Electricity and Magnetism, 471
Bibliographie Zeitschriftenliteratur Stand der Technikik, 572
Bibliography and Index of Experimental Range and Stopping., 47
Bibliography and Index of Geology, 552
Bibliography and Index, 1845–1977, Great Basin Province., 43
Bibliography and Research Manual of History of Mathematics, 470
Bibliography of "Ab Initio" Molecular Wave Functions, 36
Bibliography of African Ecology, 57
Bibliography of Agricultural Bibliographies 1977, 38
Bibliography of Agricultural Residues, 38
Bibliography of Agricultural 39, 547, 600
Bibliography of American Published Geology, 44
Bibliography of Astronomy, 1881–1898, 469
Bibliography of Astronomy, 1970–1979, 31
Bibliograpy of Bioethics, 36
Bibliography of Birds, 41
Bibliograpy of Books and Pamphlets on History Agriculture, 480
Bibliography of Doctoral Research on Ecology and Environment, 618
Bibliography of Early Modern Algebra, 1500–1800, 31
Bibliography of Energy Conservation in Architecture, 47
Bibliography of Fishes, 41
Bibliography of Fossil Vertebrates, 44
Bibliography of Interlingual Scientific and Technical Dictionary, 100
Bibliography of Literature of North American Climates, 44

Bibliography of Microwave Optical Technology, 33
Bibliography of Natural Radio Emissions from Astronomical., 31
Bibliography of New Mexico Paleontology, 44
Bibliography of Non-Euclidian Geometry, 470
Bibliography of Plant Viruses and Index to Research, 39
Bibliography of Quantitative Ecology, 57
Bibliography of Quantitative Studies Science and Its History, 465
Bibliography of Recreational Mathematics, 31
Bibliography of Reproduction, 42
Bibliography of Space Books and Articles from Non-Aerospace, 47
Bibliography of Statistical Literature, 32, 264
Bibliography of Theses in Geology, 1967–1970, 618
Bibliography of the History of Electronics, 485
Bibliography of the History of Technology, 482
Bibliography of the Philosophy of Science, 1945–1981, 29
Bibliography on Animal Rights and Related Matters, 42
Bibliography on Atomic Energy Levels and Spectra, 33
Bibliography on Atomic Line Shapes and Shifts, 33
Bibliography on Atomic Transition Probabilities, 33
Bibliography on Cold Regions Science and Technology, 44
Bibliography on Corrosion and Protection of Steel in Concrt., 51
Bibliography on Engine Lubricating Oil, 52
Bibliography on Fatigue of Materials, Components and Structures, 51
Bibliography of Marine Geology and Geophysics, 46
Bibliography on Zinc in Biology Systems, 36
Biennial Directory and Information Book, 425

Bio-Base, 449
Bio-Bibliography for the History of Biochemical Sciences, 477
Biochemica et Biophysica Acta, 581
Biochemical Engineering and Biotechnology Handbook, 170
Biochemical Genetics, 581
Biochemical Journal: Part 1, Cellular Aspects, 581
Biochemical Journal: Part 2, Molecular Aspects, 581
Biochemical Title Index, 569
Biochemistry, 581
Biochemistry Abstracts: Amino Acids, Peptides, and Proteins, 545
Biochemistry Abstracts: Nucleic Acids, 543
Biochemistry of Plants, 528
Bioengineering Abstracts, 555
Biographical Dictionaries Master Index, 449
Biographical Dictionary of American Science, 449
Biographicl Dictionary of Botanists, 457
Biographicl Dictionary of Railway Engineers, 462
Biographical Dictionary of Scientists, 450
Biographical Encyclopedia of Scientists, 450
Biographical Index of British Engineers in the 19th Century, 459
Biographical Memoirs of Fellows of the Royal Society, 464
Biographical Notes upon Botanists, 457
Biographisch-Literarisches Handworterbuch zur Geschichte, 450
Biological Abstracts, 545
Bilolgical Abstracts/RRM, 545, 594
Biological Agriculture and Horticulture, 580
Biological Handbooks, 170
Biological Indicators of Environmental Quality, 536
Biological Laboratory Data, 253
Biological Nomenclature, 97
Biological Oceanography Journal, 582
Biological Reviews, 521
Biological Substances, 645
Biological Systems, 295
Biological and Agricultural Index, 547

Biologicke Listg/Biological Review, 580
Biology, 546
Biology Bulletin, 580
Biology Data Book, 253
Biology of the Reptilia, 528
Biomass, 589
Biomedical, Scientific, and Technical Book Reviewing, 1
Biometrika Tables for Statisticians, 238
Bioorganic Chemistry, 527
Biophysical Journal, 581
Biotechnology, 35, 528
Biotechnology Advances, 580
Biotechnology International, 515
Biotechnology News, 580
Bird Families of the World, 358
Birds of the World, 358
Birth of Particle Physics, 472
Black Engineers in the United States, 459
Black's Agricultural Dictionary, 130
Black's Veterinary Dictionary, 127
Blacker-Wood Library of Zoology and Orinthology Dictionary., 9
Blacksmith's Source Book, 51
Blue Book for the Atari Computer, 288
Bluesheets, 679
Boiler Operators Guide, 334
Book Bytes: The User's Guide to 1200 Microcomputer Books, 12
Book of ASTM Standards, 639
Book of American Trade Marks, 321
Books Out of Print 1980–1984, 1
Books in Series, 1
Botanical Abstracts, 548
Botanical Review, 581
Botanical Society of American Yearbook, 441
Bowes and Church's Food Values of Portions Commonly Used, 253
Bowker's Complete Sourcebook of Personal Computing, 288
Bowker's Software in Print, 327
Bowker/Bantam 1984 Complete Sourcebook of Personal Computing, 326
Breeding Birds of Europe, 358
Brief Guide to Sources of Scientific and Technical Info, 14

Brief Subject and Author Index of Papers Published by IME, 673
Brief Subject and Author Index of Papers Published in., 605
Britannical Yearbook of Science and Future, 440
British Astronomical Association, 577
British Books in Print 1985, 1
British Commercial Computer Digest, 560
British Geological Literature, 552
British Journal of Nutrition, 582
British Natural History Books, 1945–1900, 29
British Official Publications, 596
British Ornithologists, 359
British Reports, Translations and Theses, 613, 616
British Standards Yearbook, 639
Buchsbaum's Complete Handbook of Practical Electronic Ref., 258
Builder's Complete Guide to Construction Business Success, 322
Building Construction, 322
Building Construction Cost Data–1978, 268
Building Design and Construction Handbook, 190
Building Science Abstracts, 558
Building Science Laboratory Manual, 297
Building Systems Integration Handbook, 190
Building Technology Publications 1965–1975, 594
Building Technology Publications: Supplement 3, 1978, 11
Builk Carriers in the World Fleet, 332
Bulletin Signaletique, 537
Bulletin des traductions-CNRS, 613
Bulletin of Science, Technology and Society, 576
Bulletin of the Atomic Scientists, 588
Burnham's Celestial Handbook, 153
Business Mini/Micro Software Directory, 421
Business Organizations and Agencies Directory, 394
Busy Person's Guide to Selecting the Right Word Processor, 327
Butterflies, 365

Butterflies of North America, 365
Butterfly and Angelfishes of the World, 73, 316
Buyer's Guide to Microfilm Equipment, Products, and Services, 662
Buyer's Guide to Micrographic Equipment, Products, and Services, 662

CA SEARCH, 684
CAB ABSTRACTS, 684
CAS82/CAS77/CAS72/CAS67, 684
CASSI, 685
CBASIC USER GUIDE, 303
CDI, 685
CEH80/CEH132/CEHINDEX, 685
CHEMCON (CA CONDENSATES), 685
CHEMDEX/CHEMDEX2/CHEMDEX3, 685
CHEMICAL ABSTRACTS: Health and Safety in Chemistry, 712
CHEMICAL EXPOSURE, 686
CHEMICAL INDUSTRY NOTES (CIN), 686
CHEMICAL REGULATIONS AND GUIDELINES SYSTEM (CRGS), 686
CHEMLINE, 712
CHEMNAME, 687
CHEMSEARCH, 687
CHEMSIS, 687
CHEMZERO, 687
CIBA Foundation Symposia, 609
CINDA 81: Index to Literature on Microscopic Neutron Data, 248
CIS INDEX, 688
CLAIMS COMPOUND REGISTRY, 688
CLAIMS/CHEM, 688
CLAIMS/CITATION, 688
CLAIMS/CLASS, 688
CLAIMS/GEM, 689
CLAIMS/UNITERM, 689
CLAIMS/US PATENT ABSTRACTS, 689
CLAIMS/US PATENT ABSTRACTS WEEKLY, 689
CODEN for Periodical Titles, 574
COLD, 689
COMPENDEX, 689
COMPENDEX: Aerospace Engineering, 712

COMPENDEX: Electrical and Computing Engineering, 712
COMPENDIX: Chemical Engineering, 712
COMPUTER DATABASE, 690
CORROSION, 690
CP/M Software Directory, 421
CRC Atlas of Spectral Data and Physical Constants for., 380
CRC Coastal Processes and Erosion, 227
CRC Composite Index for CRC Handbooks, 152
CRC Critical Reviews in Analytical Chemistry, 579
CRC Critical Reviews in Environmental Control, 589
CRC Critical Reviews in Microbiology, 581
CRC Handbook Series in Zoonoses, 171
CRC Handbook of Clinical Chemistry, 163
CRC Handbook of Digital System Design for Scientists and., 204
CRC Handbook of Electrical Resistivities of Binary Metallic, 211
CRC Handbook of Environmental Control, 227
CRC Handbook of Environmental Radiation, 227
CRC Handbook of Flowering, 174
CRC Handbook of Food Additives, 185
CRC Handbook of Fruit Set and Development, 174
CRC Handbook of Laboratory Animal Sciences, 177
CRC Handbook of Laser Science and Technology, 157
CRC Handbook of Lubrication, 215
CRC Handbook of Management of Radiation Protection Programs, 436
CRC Handbook of Mass Spectra of Environmental Contaminants, 227
CRC Handbook of Material Science, 211
CRC Handbook of Microbiology, 170
CRC Handbook of Natural Pesticides, 173
CRC Handbook of Nuclear Reactors Calculations, 223
CRC Handbook of Nutritive Value of Processed Food, 186
CRC Handbook of Physical Properties of Rocks, 212
CRC Handbook of Space Technology, 182
CRC Handbook of Tables for Applied Engineering Science, 244
CRC Handbook of Tables for Mathematics, 236
CRC Handbook of Tables for Organic Compound Identification, 242
CRC Handbook of Tables for Probability and Statistics, 238
CRC Handbook of Tapeworm Identification, 178
CRC Handbook of Tropical Food Crops, 186
CRC Standard Mathematical Tables, 236
CRDS, 690
CRECORD, 690
CRIS/USDA, 690
Cage Bird Identifier, 359
Calculator Handbook, 300
California Energy Directory, 430
California Environment Yearbook and Directory, 445
California Environmental Directory, 436
California Water Atlas, 390
California Water Resources Directory, 436
Cambridge Encyclopedia of Astronomy, 63
Cambridge Encyclopedia of Earth Sciences, 76
Cambridge Photographic Atlas of the Planets, 376
Canadian Books in Print, 2
Canadian Inventors and Innovators, 457
Canadian Mining Journal's Reference Manual and Buyer's Guide, 283
Canadian Oil Industry Directory, 430
Canadian Standards Association. Catalogue of Standards, 640
Cane Sugar Handbook, 186
Carbohydrate Chemistry, 503
Card-a-Lert, 569
Catalog of Books in the American Philosophical Society, 5
Catalog of Cometary Orbits, 342
Catalog of Museum Publications and Media, 664

Catalog of Public Documents of the Congress and All Depts., 596
Catalog of the Conservation Library, Denver Public Library, 12
Catalog of the Farlow Reference Library of Cryptogamic Bot., 8
Catalog of the Imperial College of Tropical Agriculture., 7
Catalog of the Manuscript and Archival Collections and Index, 8
Catalog of the Naval Observatory Library, Washington, DC, 6
Catalog of the Royal Botanic Gardens, Kew, England, 8
Catalogs of the Glaciology Collection, 9
Catalogue and Index of Contributions to N. American Geology, 44
Catalogue of ICAO Publications, 599
Catalogue of Meteorological Data for Research, 254
Catalogue of Publications of World Meteorological Organizations, 599
Catalogue of Scientific Papers, 465
Catalogue of Type Invertebrate, Plant, and Trace Fossils, 243
Catalogue of the Active Volcanoes of the World, 317
Catalogue of the Library of the Royal Entomological Society, 9
Catalogue of the Universe, 376
Catalogue, English Translations of German Standards, 640
Catalogues of the Library of the Marine Biological Association, 9
Catalysis Reviews, 497
Cattle of the World, 366
Cell Biology, 243, 528
Cell Nucleus, 528
Cement Manufacturer's Handbook, 208
Cement and Concrete Reference Catalog, 322
Centennial History of American Society of Mechanical Engineering, 487
Century of Chemistry, 476
Century of DNA, 477
Century of Weather Service, 480
Ceramic Abstracts, 562
Chambers Dictionary of Science and Technology, 114

Changes in the National Electrical Code, 647
Changing Scenes in the National Sciences, 477
Chartered Mechanical Engineer, 587
Check-list of Birds of the World, 359
Checklist of Major US Government Series, 596
Checklist of Names for 3,000 Vascular Plants of Economic., 312
Checklist of North American Plants for Wildlife Biologists, 312
Checklist of US Trees, 345
Chem BUY Direct, 412
Chem Product Index, 405
Chem Sources, 404
Chem Sources–Europe, 412
Chembooks, 35
Chemical Abstracts, 543
Chemical Abstracts Patent Index and Concordance, 634
Chemical Abstracts Service Source Index, 576
Chemical Abstracts Service Source Index 1907–1979, 544
Chemical Analysis of Organo-Metallic Compounds, 503
Chemical Buyer's Directory, 404
Chemical Engineering, 585
Chemical Engineering Drawing Symbols, 257, 321
Chemical Engineering Faculties, 399
Chemical Engineering Monographs, 532
Chemical Engineering News, 579
Chemical Engineering Practice, 532
Chemical Engineering Progress, 585, 603
Chemical Engineering Progress Symposium Series, 610
Chemical Engineers' Handbook, 183
Chemical Equilibria in Carbon-Hydrogen-Oxygen Systems, 244
Chemical Guide to Europe, 404
Chemical Guide to US, 405
Chemical Industries Information Sources, 22
Chemical Industry 1900–1930, 484
Chemical Industry Directory and Who's Who, 412
Chemical Materials Catalog, 405
Chemical Materials for Construction, 259

Chemical Processing and Engineering, 515
Chemical Reviews, 521, 579
Chemical Synonyms and Trade Names, 123
Chemical Tables, 242
Chemical Tables for Laboratory and Industry, 242
Chemical Technician's Ready Reference Handbook, 163
Chemical Technology, 81
Chemical Thermodynamics of Actinide Elements and Compounds, 524
Chemical Titles, 570
Chemical Vapor Deposition, 1960–1980, 48
Chemical Week Buyer's Guide Index, 405
Chemical Zoology, 529
Chemical and Biochemical Applications of Lasers, 499, 527
Chemical and Botanical Guide to Licen Products, 312
Chemical and Engineering News, 585
Chemical and Process Plant, 332
Chemical, Medical and Pharmaceutical Books Printed Before 1800, 476
Chemical-Biological Activities, 569
Chemical/Biological Warfare, 53
Chemicals Technicians' Ready Reference Handbook, 183
Chemist's Companion, 251
Chemistry and Industry, 603
Chemistry of Carbon Compounds, 524
Chemistry of Nitro and Nitroso Groups, 524
Chemo-Reception Abstracts, 544
Chemtech, 585
Chilton's Encyclopedia of Gardening, 69
China's Electronics and Electrical Products, 416
Chinese Herbs, 345
Chinese Petroleum, 54
Chinese-English, English-Chinese Dictionary of Engineering, 101
Choose Your Own Computer, 326
Chromatography, 163
Chromosomal Variation in Man, 37
Chronolog, 679

Chronological Annotated Bibliography of Order Statistics, 29
Ciba-Geigy Journal, 655
Citizens' Energy Directory, 430
Civil Aircraft of the World, 319
Civil Engineer's Reference Book, 189, 257
Civil Engineering, 585
Civil Engineering Hydraulics Abstracts, 558
Civil Engineering Systems, 585
Classed Subject Catalog of the Engineering Societies Library, 11
Classic Mineral Localities of the World, 317
Classical Scientific Papers, 476
Classroom Computer News Directory of Educational Computing., 417
Clay's Handbook of Environmental Health, 227
Clean Air Yearbook, 446
Climates of the States, 243, 255
Climatic Atlas of Tropical Atlantic and Eastern Pacific Ocean, 389
Climatic Atlas of the Indian Ocean, 388
Climatological Data, 255, 265
Clouds of the World, 76
Coal Bibliography and Index, 54
Coal Data, 261
Coal Data Book, 261
Coal Information Sources and Data Bases, 431
Coal Mine Road Technology, 52
Coal in American, 90
Coastal Almanac, 255
Coastal Land Forms and Surface Features, 386
Coating Equipment and Processes, 48
Code of Safety for Fisherman and Fishing Vessels, 644
Coded Workbook of Birds of the World, 295
Cold Spring Harbor Symposia on Quantitative Biology, 609
Cold and Freezer Storage Manual, 277
Coleopterous Insects of Mexico, Central American, the West., 42
Collected Algorithms for CACM, 560, 418
Collected Reprints, 629

Collector's Book of Flourescent Minerals, 370
Collector's Guide to Rocks and Minerals, 370
Collector's Guide to Seashells of the World, 356
College Chemistry Faculties, 400
Color Dictionary of Flowers and Plants for Home and Garden, 125
Color Encyclopedia of gemstones, 77
Color Handbook of House Plants, 174
Color Illustrated Guide Carbonate Rock Constituents., 318
Color Science, 248
Colorful Mineral Identifier, 371
Colour Index, 308
Combat Aircraft of the World, 320
Combined Membership List, 464
Coming to Terms: From Alpha to X-Ray, 115
Comments on Modern Physics, 492
Commercial Food Patents, US, 636
Commissioning Procedures for Nuclear Power Plants, 336
Commodore 64 LOGO, 303
Commodore Microcomputers, 655
Common Insects of North American, 365
Common Plants, 97
Common Plants of the Mid-Atlantic Coast, 345
Communications Standard Dictionary, 144
Communications of the ACM, 586
Compact Dictionary of Exact Science and Technology, 115
Companies Holding Certificates of Authorization Use of CSS, 643
Comparative Anatomy of Domestic Animals, 316
Comparative Properties of Plastics, 257
Compendium of Analytical Nomenclature, 97
Compendium of Meteorology, 531
Compendium of Organic Methods, 524
Compendium of Organic Synthetic Methods, 524
Compendium of Seashells, 363
Compilation, Critical Evaluation and Dist. of Stellar Data, 247

Complete Auto Electric Handbook, 306
Complete Birds of the World, 359
Complete Book of Herbs and Herb Growing, 345
Complete Book of Herbs and Spices, 314
Complete Book of Home Computers, 326
Complete Book of Mushrooms, 345
Complete Book of Terrarium Gardening, 351
Complete Broadcast Antenna Handbook, 201
Complete Checklist of Birds of World, 253
Complete Concrete, Masonry, and Brick Handbook, 191
Complete Dictionary of Wood, 125
Complete Dog Book, 366
Complete Encyclopedia of Horses, 73
Complete Energy Saving Handbook for Homeowners, 223
Complete Family Nature Guide, 309
Complete Food Handbook, 186
Complete Garden, 406
Complete Guide of Herbs and Spices, 345
Complete Guide to All Cats, 366
Complete Guide to Buying Gems, 371
Complete Guide to Compact Disc Player (CD) Troubleshooting, 324
Complete Guide to Computer Camps and Workshops, 329, 418
Complete Guide to Monkeys, Apes and Other Primates, 366
Complete Guide to Water Plants, 346
Complete Handbook of Cacti and Succulents, 346
Complete Handbook of Magnetic Recording, 201
Complete Handbook of Radio Receivers, 202
Complete Handbook of Radio Transmitters, 201
Complete Handbook of Videocassette Recorders, 662
Complete Illustrated Encyclopedia of the World's Aircraft, 80
Complete Index for the Six Volumes of Wild Flowers of US, 548

Complete Multilingual Dictionary of Aviation and Aero. Term., 107
Complete Multilingual Dictionary of Computer Terminology, 107
Complete Trees of North America, 346
Component and Modular Techniques, 191
Composite Index for CRC Handbooks, 537
Composite Materials, 533
Composite Materials Handbook, 212
Comprehensive Biochemistry, 499, 528
Comprehensive Dissertation Index, 1861–1972, 616
Comprehensive Guide to Solar Water Heaters, 338
Comprehensive Inorganic Chemistry, 524
Comprehensive Organic Chemistry, 524
Comprehensive Organometallic Chemistry, 525
Comprehensive Treatise Inorganic and Theoretical Chemistry, 525
Comprehensive Treatise of Electrochemistry, 525
CompuMath Citation Index, 541
Computer Abstracts, 560
Computer Acronyms, Abbreviations, Etc., 96
Computer Aided Design of Digital Systems, 50
Computer Alphabet Book, 139
Computer Books and Serials in Print 1985, 12
Computer Contents, 570
Computer Data and Database Source Book, 675
Computer Dealers, 418
Computer Dictionary, 139
Computer Dictionary and Handbook, 139
Computer Dictionary for Everyone, 140
Computer Directory and Buyers' Guide, 418
Computer Glossary: It's Not Just a Glossary!, 140
Computer Graphics Directory '84, 418
Computer Graphics Glossary, 140
Computer Graphics Marketplace, 418
Computer Industry Abstracts, 561

Computer Industry Annual, 444
Computer Journal, 586
Computer Literature Index, 561
Computer Performance Modeling Handbook, 204
Computer Phone Book, 675
Computer Publishers and Publications 1985–86, 12
Computer Publishers and Publications 1985–86, 572
Computer Resource Guide, 330
Computer Resource Guide for Nonprofits, 419
Computer Science Resources, 24
Computer Simulation 1951–1976, 50
Computer Simulation 1951–1976, 561
Computer Technology, 50
Computer Users's Yearbook, 444
Computer Yearbook and Directory, 444
Computer and Control Abstracts, 560
Computer and Information Systems Abstract Journal, 560
Computer and Structures, 586
Computer-Aided Data Analysis, 330
Computer-Readable Data Bases, 675
Computerist's Handy Databook/Dictionary, 258
Computers and Computing Information Resources Directory, 419
Computers and Humanities, 586
Computers and Information Processing World Index, 561
Computers and People, 586
Computers for Everybody, 330
Computers in Chemical and Biochemical Research, 500
Computext Book Guides, 51, 604
Computing Marketplace, 418
Computing Reporting in Science and Engineering, 655
Computing Reviews, 561
Comtemporary Topics in Immunobiology, 511
Concise Dictionary of Physics, 119
Concise Dictionary of Physics and Related Subjects, 119
Concise Dictionary of Scientific Biography, 450
Concise Encyclopedia of Astronomy, 63

Concise Encyclopedia of Biochemistry, 67
Concise Encyclopedia of Information Technology, 88
Concise Encyclopedia of Solid State Physics, 65
Concise Etymological Dictionary of Chemistry, 121
Concise Herbal Encyclopedia, 274
Concise Illustrated Russian-English Dictionary of Mechanical., 104
Concise Russian-English Chemical Glossary, 104
Concise Soil Mechanics, 191
Concise World Atlas of Geology and Mineral Depostis, 386
Concordance to Darwin's "Origin of Species", 98
Concrete Abstracts, 558
Concrete Bridge Designer's Manual, 278
Concrete Industries Yearbook, 443
Concrete Manual, 278
Concrete Yearbook, 443
Condensed Chemical Dictionary, 122
Conference Proceedings in the IAEA Library, 606
Conference Proceedings of the Smyposium on Computer Architecture, 611
Conference on Analytical Chemistry in Energy Technology, 609
Conifers, 346
Conservation Directory, 436
Consolidated Index to Flora Europaea, 548
Construction Contract Dictionary, 136
Construction Dewatering, 322
Construction Glossary, 136
Construction Information Source and Reference Guide, 23
Construction Inspection, 322
Construction Materials, 212
Construction Materials for Architecture, 322
Construction Regulations Glossary, 136
Constructor Directory, 415
Consulting Engineer, 208
Consulting Engineer Product Bulletin Directory, 415

Consulting Engineering Practice Manual, 276
Consulting Engineers Who's Who and Yearbook, 460
Consumer Handbook of Solar Energy for US and Canada, 338
Consumer's Dictionary of Cosmetic Ingredients, 133
Consumer's Dictionary of Food Additives, 135
Consumer's Energy Handbook, 224
Consumers', Researchers', and Students' Guide to Government Publications, 596
Contact: A Guide to Energy Specialists, 463
Contemporary Topics in Immunochemistry, 500
Contemporary Topics in Molecular Immunology, 512
Control and Dynamic Systems, 518
Control of Noise in Ventilation Systems, 334
Controlled Wildlife, 356
Coping with the Biomedical Literature, 19
Corporate Diagrams and Administrative Personnel of the., 412
Correlation Tables for Structural Determination of Organic., 242
Corrosion Abstracts, 563
Corrosion Control Abstracts, 563
Corrosion Resistant Materials Handbook, 212
Council of Biology Editors Style Manual, 274
Cousteau Almanac of the Environment, 262
Cranes of the World, 359
Critic's Guide to Software for Apple and Apple-Compatible., 327
Critic's Guide to Software for CP/M Computers, 327
Critic's Guide to Software for IBM-Pc and Pc-Compatible., 328
Critical Reports on Applied Chemistry, 525
Crop Production Reports, 265
Crows of the World, 359
Cryogenics Handbook, 412

Crystal Growth Bibliography, 34
Crystal Growth Bibliography Supplement, 34
Crystal Structures, 525
Culpeper's Color Herbal, 346
Cumulative Index to ASCE Publications, 1975–1979, 674
Cumulative Index to Contemporary Acoustical Literature, 542
Cumulative Index to English Translations, 613
Cumulative Index to Science Education, 537
Cumulative Index to the Mossbauer Effect Data Indexes, 576
Cumulative Index: SAE Papers 1965-1973, 673
Cumulative Subject Index to Monthly Catalog US Government Publications, 597
Current Abstracts of Chemistry and Index Chemicus, 544
Current Bibliographic Directory of Arts and Sciences, 450
Current Bibliography of Agriculture in China, 39
Current Chemical Engineering Papers, 570
Current Chemical Papers, 570
Current Contents, 569
Current Genetics, 580
Current Index to Conference Papers, 604
Current Index to Conference Papers in Engineering, 604
Current Index to Statistics, 546
Current Mathematical Publications, 570
Current Papers in Computers and Control, 570
Current Papers in Electrical and Electronics Engineering, 570
Current Papers in Physics, 570
Current Physics Advanced Abstracts, 570
Current Physics Bibliographies, 34
Current Physics Index, 542
Current Physics Microform, 664
Current Physics Reprints, 629
Current Physics Titles, 570
Current Programs, 604
Current Research on Scientific and Technical Information Transfer, 619
Current Technology Index, 555
Current Topics in Bioenergetics, 495
Current Topics in Cellular Regulation, 509
Current Topics in Developmental Biology, 506
Current Topics in Environmental and Toxicological Chemistry, 520
Current Topics in Experimental Endocrinology, 506
Current Topics in Membranes and Transport, 506
Current Topics in Microbiology and Immunology, 512
Cutting for Construction, 191
Cybernetics, 616

D.C. Power Supplies, 324
DATA PROCESSING AND INFORMATION SCIENCE CONTENTS, 691
DATRIX, 616
DEC Microcomputer Directory, 420
DECHEMA Chemistry Data Series, 251
DIN Catalogue of Technical Rules, 640
DIN Standardization Handbooks, 650
DISSERTATION ABSTRACTS ONLINE, 691
DOE ENERGY, 691
Dangerous Plants, 253, 346
Dangerous Properties of Industrial Materials, 257, 332
Dangerous Sea Creatures, 363
Data Book for Pipe Fitters and Pipe Welders, 259
Data Communication for Microcomputers, 303
Data Distionary/Directory Systems, 675
Data Handbook for Clay Materials and Other Nonmetallic Minerals, 212, 259
Data Processing Documentation and Procedures Manual, 280
Data Sources/Hardware, 420
Data Systems Dictionary, 107
Data for Radioactive Waste Management and Nuclear Applications, 262
Database, 676

Database Directory, 675
Database Update, 586
Datapro Who's Who in Microcomputing, 1983, 420
Datapro/McGraw-Hill Guide to Apple Software, 328
Datapro/McGraw-Hill Guide to CP/M Software, 328
Datapro/McGraw-Hill Guide to IBM Personal Computer Software, 328
Datensammlungen in der Physik: Data Compilations in Physics, 249
Davison's Textile Blue Book, 412
Davison's Textile Catalog and Buyer's Guide, 321
Deep Sky Objects, 376
Deep-Sea Research with Oceanographic Literature Review, 554, 583
Defense Mapping Agency Hydrographic Center Catalog of Publications, 10
Derwent Publications, Ltd., 634
Desalination Abstracts, 554
Design Manual for High Temperature Hot Water and Steam Systems, 282
Design Manual for Solar Heating of Buildings and Domestic., 301
Design Manual for Structural Tubing, 305
Design Mix Manual for Concrete Construction, 278
Design Tables for Beams on Elastic Foundations and Related., 244
Design of Operating Systems for Small Computer Systems, 330
Designer's Guide to OSHA, 323, 644
Designer's Handbook of Integrated Circuits, 204
Desk-Top Data Bank, 259
Deuterium and Heavy Water, 53
Developing Collections of US Government Publications, 597
Development and Use of Patent Classification Systems, 631
Developmental Biology, 580
Developments in Toxicology and Environmental Sciences, 520
Diagnostic Manual for the Identification of Insect Pathogens, 295
Dialog Database, 679

Dictionaries of English and Foreign Languages, 100
Dictionary Catalog of Princeton University Plasma Physics Laboratory Library, 6
Dictionary Catalog of the Library of the Massachusetts Horticulture Society, 8
Dictionary Catalog of the Map Division, 373
Dictionary Catalog of the National Agricultural Library, 7
Dictionary Catalog of the Water Resources Center Archives, 10
Dictionary for Automotive Engineering, 146
Dictionary of Agricultural and Food Engineering, 130
Dictionary of Agriculture, 108
Dictionary of American Naval Fighting Ships, 146
Dictionary of Applied Energy Conservation, 147
Dictionary of Astronomy, 117
Dictionary of Astronomy, Space, and Atmospheric Phenomena, 117
Dictionary of Audio-Visual Terms, 657
Dictionary of Automotive Engineering, 108
Dictionary of Basic Military Terms, 147
Dictionary of Biology, 108
Dictionary of Biomedial Acronyms and Abbreviations, 96
Dictionary of Birds in Color, 127
Dictionary of Botany, 126
Dictionary of British and Irish Botanists and Horticulturists, 458
Dictionary of Building, 136
Dictionary of Butterflies and Moths in Color, 128
Dictionary of Ceramic Science and Engineering, 144
Dictionary of Chemical Terminology, 108
Dictionary of Chemistry and Chemical Engineering, 122
Dictionary of Chromatography, 122
Dictionary of Civil Engineering, 137
Dictionary of Civil Engineering and Construction., 101
Dictionary of Computer Graphics, 140

Dictionary of Computers, Data Processing and Telecommunications, 140
Dictionary of Computing, 140
Dictionary of Dairy Terminology, 108
Dictionary of Dangerous Pollutants, Ecology, and Environment, 148
Dictionary of Data Communications, 144
Dictionary of Data Processing, 141
Dictionary of Drying, 145
Dictionary of Earth Sciences, 130
Dictionary of Ecology, Evolution and Systematics, 148
Dictionary of Electrochemistry, 134
Dictionary of Electronics, 102, 137
Dictionary of Energy, 147
Dictionary of Energy Technology, 148
Dictionary of Engineering and Technology, 102
Dictionary of Entomology, 128
Dictionary of Environmental Engineering and Related Sciences, 149
Dictionary of Environmental Terms, 149
Dictionary of Food Ingredients, 135
Dictionary of Gaming, Modelling and Simulation, 118
Dictionary of Gastronomy, 135
Dictioinary of Gemology, 145
Dictionary of Geography, 130
Dictionary of Geological Terms, 130
Dictionary of Geology, 130
Dictionary of Geosciences, 103
Dictionary of Geotechnics, 130
Dictionary of Immunology, 128
Dictionary of Information Technology, 141
Dictionary of Instrument Science, 145
Dictionary of Inventions and Discoveries, 115
Dictionary of Life Sciences, 124
Dictionary of Logistical Terms and Symbols, 141
Dictionary of Materials Testing, 106, 145
Dictionary of Mathematics, 118
Dictionary of Mathematics in Four Languages, 106
Dictionary of Mechanical Engineering, 145
Dictionary of Microbial Taxonomy, 129
Dictionary of Microbiology, 129
Dictionary of Microprocessor Systems, 108
Dictionary of Military Terms, 101
Dictionary of Minicomputing and Microcomputing, 141
Dictionary of Modern Engineering, 103
Dictionary of Mosses, 126
Dictionary of Named Effects and Laws in Chemistry, Physics., 114
Dictionary of New Information Technology Acronyms, 96
Dictionary of Nutrition and Food Technology, 135
Dictionary of Organic Compounds, 122
Dictionary of Organometallic Compounds, 122
Dictionary of Particle Technology, 103
Dictionary of Petroleum Terms, 131
Dictionary of Petrology, 131
Dictionary of Physical Metallurgy, 109
Dictionary of Physical Sciences, 120
Dictionary of Physics and Allied Sciences, 120
Dictionary of Plastics Technology in Four Languages, 109
Dictionary of Public Transport, 151
Dictionary of Report Series Codes, 591
Dictionary of Science and Technology, 101
Dictionary of Science and Technology: English-German, 103
Dictionary of Scientific Biography, 451
Dictionary of Scientific Units, 114
Dictionary of Scientific and Technical Terms, 115
Dictionary of Soil Mechanics and Foundation Engineering, 137
Dictionary of Space Technology, 133
Dictionary of Spectroscopy, 120
Dictionary of Statistical Terms, 118
Dictionary of Statistical, Scientific and Technical Terms, 107
Dictionary of Surface Active Agents, Cosmetics and Toiletries, 109
Dictionary of Symbols of Mathematical Logic, 118
Dictionary of Telecommunications, 144
Dictionary of Terms Used in Safety Profession, 145

Dictionary of Theoretical Concepts in Biology, 124
Dictionary of Thermodynamics, 120
Dictionary of Tools Used in Woodworking and Allied Trades, 305
Dictionary of Tropical American Crops and Their Diseases, 253
Dictionary of Unit Conversion, 114
Dictionary of Useful Plants, 126
Dictionary of Waste and Sewage Engineering, 149
Dictionary of Waste and Waste Treatment, 149
Dictionary of Words About Alcohol, 134
Dictionary of the Environment, 149
Dictionary of the History of Science, 466
Dictionary of the Natural Environment, 149
Die Casting Handbook, 215
Diesel Engine Reference Book, 334
Diesel Spotter's Guide Update, 334
Digest of Literature on Dielectrics, 49
Digital Electronics, 297
Digital Integrated Citcuit DATA Book, 258
Digital Logic and Computer Design, 204
Digital Review, 655
Directory for the Environment, 437
Directory of American Horticulture, 406
Directory of British Associations and Associations in., 667
Directory of Canadian Scientific and Technical Periodicals, 572
Directory of Communication Organizations, 416
Directory of Computer Education and Research, 400, 625
Directory of Computer Software Applications: Energy, 421
Directory of Computerized Data Files and Releated Software, 676
Directory of Computer Software Applications: Mathematics, 421
Directory of Construction Associations, 415
Directory of Corporate Affiliations, 394
Directory of Current Research, 624
Directory of Directories, 393
Directory of Engineering Document Sources, 594

Directory of Engineering Education Institutions, 400
Directory of Engineering Societies and Related Organizations, 667
Directory of Engineering, Scientific and Management., 628
Directory of Engineers in Private Practice, 460
Directory of Environmental Groups in New England, 437
Directory of Environmental Life Scientists, 458
Directory of European Associations, 668
Directory of Fabric Resources for 1975, 412
Directory of Federal Statistics for Local Areas, 265
Directory of Federal Statistics for States, 265
Directory of Federal Technology, 397
Directory of Federal Technology Resources, 397
Directory of Federally Supported Information Analysis Centr., 394
Directory of Food and Nutrition Information Services and., 414
Directory of Foreign Manufacturers in US, 394, 427
Directory of Geoscience Departments, 409
Directory of Government Agencies Safeguarding Consumer., 437
Directory of Governmental Air Pollution Agencies, 439
Directory of Graduate Research, 623
Directory of Industrial Heat Processing and Combustion Equipment, 228, 427
Directory of Industry Data Sources, 410
Directory of Information Management Software, 421
Directory of Information Resources Agriculture and Biology, 408
Directory of Information Resources in US: Biological Scien., 405
Directory of Institutions and Individuals Active in Environ., 437
Directory of Japanese Scientific Periodicals, 573
Directory of Manufacturers in Greater Boston, 394

Directory of Marine Sciences Libraries and Information Centers, 425
Directory of Meteorite Collections and Meteorite Research, 409
Directory of National and International Pollution Monitoring, 627
Directory of Natural History and Related Societies in Brit., 398
Directory of Non-Federal Statistics States and Local Areas, 265
Directory of Non-Governmental Agricultural Organizations, 408
Directory of North American Entomologists and Acarologists, 458
Directory of Nuclear Reactors, 428
Directory of Online Databases, 676
Directory of Paleontologists of the World, 459
Directory of Physics and Astronomy Faculties in North America, 400
Directory of Physics and Astronomy Staff Members, 404
Directory of Physics and Astronomy Staff Members, 464
Directory of Public Domain (and User-Supported) Software, 421
Directory of Published Proceedings, 604
Directory of Publishing Sources, 573
Directory of Research Grants, 619
Directory of Researchers by Discipline, 619
Directory of Scientific Research Institutes in PRC, 619
Directory of Shipowners, Shipbuilders, and Marine Engineers, 426
Directory of Software Publishers, 421
Directory of Soviet Research Organizations, 620
Directory of Special Libraries and Information Centers, 394
Directory of Testing Laboratories, 410
Directory of Textile Plant Processes, 413
Directory of UK Polymer Research, 623
Directory of United Nations Information Systems, 394
Directory of United States Standardization Activities, 638
Directory of Women Mathematicians, 454
Directory of Word Processing Systems, 421
Directory of the Canning, Freezing, Preserving Industries, 413
Directory of the Forest Products Industry, 408
Directory: Who's Who in Nuclear Energy, 463
Discharge of Selected Rivers of the World, 243
Discoverers: Encyclopedia of Explorers and Exploration, 77
Diseases and Pests of Ornamental Plants, 346
Dissertation Abstracts International, 616
Dissertations in Physics, 618
Doctoral Dissertation as an Information Source, 617
Doctoral Scientists and Engineers in US, 451
Documentation Cards, 558
Documentation of Scientific Literature, 14
Dolphins, Whales, and Porpoises, 73
Domestic Water Treatment, 193
Dwarf Rhododendrons, 347
Dyeing of Synthetic Fibers, 636

EBIB, 691
ECOL: Book Catalog of the Environmental Conservation Library, 13
EI ENGINEERING MEETING, 691
EIA Publications Directory, 431
EIA Standards, 647
EIMET, 692
ELECTRIC POWER DATABASE, 692
ELECTRONIC YELLOW PAGES-CONSTRUCTION DIRECTORY, 692
EMBASE (formerly EXCERPTA MEDICA), 692
ENCYCLOPEDIA OF ASSOCIATIONS, 692
ENERGYLINE, 693
ENERGYNET, 693
ENVIRONLINE, 693
ENVIRONMENTAL BIBLIOGRAPHY, 693
EPA, cumulative Bibliography, 59
EPA Index, 566

EPA Publications Bibliography, 59
EPIA, 693
ERA Abstracts, 559
ERIC, 693
ESSA Library Holdings in Oceanography and Marine Meterology, 10
Early American Men of Science, 466
Early American Science, 466
Early Concepts of Energy in Atomic Physics, 472
Early Wildlife Photographers, 458
Earth Science Reviews, 583
Earth Sciences, 44
Earthquake Engineering Reference Index, 552
Earthquake Engineering Research Center Library Catalog, 9
Earthquake Handbook, 180
Earthquakes, Tides, Unidentified Sounds and Related Phenonmena, 318
Earthscape, 284
East African Mammals, 384
Eastern Birds of Prey, 178
Eastern Manufacturers and Industrial Directory, 410
Eastern North America's Wildflowers, 353
Easy Identification Guide to North American Snakes, 369
Easy to Understand Computer Dictionary, 141
Ecological Atlas of Soils of the World, 386
Ecological Studies, 520
Ecology, 580
Ecology Field Glossary, 150
Ecology of Fossils, 318
Ecotechnics, 437
Edge of an Unfamiliar World: A History of Oceanography, 480
Edible Wild Plants, 40
Edmund Sky Guide, 342
Educational Film Locator, 657
Educational Media Catalogs on Microfiche, 658
Educational Media Yearbook, 658
Educational Programs and Faculties Nuclear Science and Engineering, 400
Educators Guide to Free Science Materials, 663

Educators' Guide to Free Films, 658
Efficient Electricity Use, 224
Efficient Energy Use, 338
Eighth Day of Creation, 477
Einstein, 455
Electric Cable Handbook, 196
Electric Motors, 334
Electric Units and Standards, 641
Electrical Distribution Engineering, 324
Electrical Engineer's Reference Book, 280
Electrical Engineering, 585
Electrical Engineering License Review, 302
Electrical Inventions 1745–1976, 472
Electrical Review, 585
Electrical Who's Who, 462
Electrical World, 589
Electrical World Directory of Electrical Utilities, 416
Electrical and Electronic Technologies, 485
Electrical and Electronics Abstracts, 559
Electricity from Glass, 472
Electricity, Magnetism, and Animal Magnetism, 472
Electro-Optical Communications Dictionary, 144
Electroanalytical Abstracts, 557
Electrochemical Society, 579
Electrochemistry, 497
Electron Diffraction 1927–1977, 472
Electron Spectroscopy, 492
Electronic Circuits Notebook, 258
Electronic Communications Applications Handbook, 204
Electronic Components Handbook, 196
Electronic Data Reference Manual, 280
Electronic Design With Off-the-Shelf Integrated Circuits, 281
Electronic Designer's Handbook, 196
Electronic Displays, 324
Electronic Engineering, 585
Electronic Engineers Master, 416
Electronic Engineers' Reference Book, 195
Electronic Filter Design Handbook, 196
Electronic Imaging Techniques, 195
Electronic Inventions 1745-1976, 485

Electronic News Financial Factbook and Directory, 268
Electronic Properties Research Literature Retrieval Guide, 258
Electronic Troubleshooting, 304
Electronic's Designer's Handbook, 196
Electronics, 585
Electronics Buyer's Guide, 417
Electronics Designers' Handbook, 195
Electronics Dictionary, 137
Electronics Engineer's Handbook, 194
Electronics Engineering Master Index, 559
Electronics Engineers' Handbook, 197
Electronics Letters, 586
Electronics Quizbook, 297
Electronics and Communicatioins Abstract Journal, 559
Electronics and Power, 586
Electroplating Engineering Handbook, 208
Elsevier's Dictionary of Automotive Engineering, 109
Elsevier's Dictionary of Botany, I, 109
Elsevier's Dictionary of Botany, II, 110
Elsevier's Dictionary of Brewing, 110
Elsevier's Dictionary of Food Science and Technology, 110
Elsevier's Dictionary of Horticulture in Nine Languages, 110
Elsevier's Dictionary of Measurement and Control Six Languages, 110
Elsevier's Dictionary of Metallurgy and Metal Working, 110
Elsevier's Dictionary of Personal and Office Computing, 111
Elsevier's Dictionary of Tools and Ironware in Six Languages, 111
Elsevier's Dictionary of Weeds of Western Europe, 126
Elsevier's Nautical Dictionary in Six Languages, 111
Elsevier's Oil and Gas Field Dictionary, 111
Elsevier's Telecommunication Dictionary in Six Languages, 111
Employer's Guide to Labor Relations, 330
Encyclopedic Dictionary of Physics, 120
Encyclopedia Botanica, 313

Encyclopedia Dictionary of Mathematics, 119
Encyclopedia Reprints Series, 90, 628
Encyclopedia of Air Warfare, 93
Encyclopedia of Aircraft, 81
Encyclopedia of American Forest and Conservation History, 69
Encyclopedia of Animal Care, 73
Encyclopedia of Animals, 73
Encyclopedia of Aquarium Fishes in Color, 73
Encyclopedia of Architectural Technology, 86
Encyclopedia of Associations, 394, 668
Encyclopedia of Astronomy, 64
Encyclopedia of Aviation, 81
Encyclopedia of Aviculture, 73
Encyclopedia of Beaches and Coastal Environments, 79
Encyclopedia of Biochemistry, 67
Encyclopedia of Biological Sciences, 68
Encyclopedia of Chemical Electrode Potentials, 82
Encyclopedia of Chemical Processing and Design, 82
Encyclopedia of Chemistry, 66
Encyclopedia of Common Natural Ingredients Used in Food., 83
Encyclopedia of Composite Materials and Components, 90
Encyclopedia of Computer Science, 88
Encyclopedia of Computer Science and Engineering, 88
Encyclopedia of Computer Science and Technology, 88
Encyclopedia of Computer Terms, 89
Encyclopedia of Computers and Data Processing, 88
Encyclopedia of Computers and Electronics, 88
Encyclopedia of Earth Sciences, 77
Encyclopedia of Electrochemistry of the Elements, 67
Encyclopedia of Emulsion Technology, 66, 82
Encyclopedia of Energy, 90
Encyclopedia of Energy-Efficient Building Design, 86
Encyclopedia of Engineering, 80

Encyclopedia of Environmental Science and Engineering, 91
Encyclopedia of Fluid Mechanics, 90
Encyclopedia of Food Science, 84
Encyclopedia of Food Technology, 84
Encyclopedia of Food and Nutrition, 83
Encyclopedia of Geochemistry and Environmental Sciences, 77
Encyclopedia of Governmental Advisory Organizations, 395
Encyclopedia of Herbs and Herbalism, 69
Encyclopedia of Home Construction, 301
Encyclopedia of How It's Built, 301
Encyclopedia of Information Systems and Services, 676
Encyclopedia of Integrated Circuits, 87
Encyclopedia of Mammals, 74
Encyclopedia of Marine Invertebrates, 74
Encyclopedia of Materials Science and Engineering, 90
Encyclopedia of Mathematics and Its Applications, 65
Encyclopedia of Microscopy and Microtechnique, 68
Encyclopedia of Mineralogy, 77
Encyclopedia of Minerals and Gemstones, 77
Encyclopedia of Mushrooms, 69
Encyclopedia of Natural Insect and Disease Control, 365
Encyclopedia of North American Railroading, 89
Encyclopedia of North American Wildlife, 74
Encyclopedia of Organic Gardening, 69
Encyclopedia of PVC, 85
Encyclopedia of Paleontology, 78
Encyclopedia of Patent Practice and Inventioin Management, 631
Encyclopedia of Physics, 66
Encyclopedia of Plant Physiology, 70
Encyclopedia of Plastics, Polymers, and Resins, 85
Encyclopedia of Polymer Science and Engineering, 85
Encyclopedia of Sedimentology, 78
Encyclopedia of Semiconductor Technology, 87
Encyclopedia of Shampoo Ingredients, 86
Encyclopedia of Ships and Seafaring, 92
Encyclopedia of Soil Science, 78
Encyclopedia of Statistical Sciences, 65
Encyclopedia of Surfactants, 82
Encyclopedia of Textiles, 86
Encyclopedia of Textiles, Fibers and Non-Woven Fabrics, 86
Encyclopedia of UK and European Patent Law, 631
Encyclopedia of US Air Force Aircraft and Missile Systems, 94
Encyclopedia of Wood, 86
Encyclopedia of World Air Power, 94
Encyclopedia of World Regional Geology, 78
Encyclopedia of the Alkaloids, 67
Encyclopedia of the Animal World, 74
Encyclopedia of the Horse, 74
Encyclopedia of the Terpenoids, 86
Encyclopedia/Handbook of Materials, Parts and Finishes, 89, 212
Encyclopedic Dictionary of Electronic Terms, 137
Encyclopedic Dictionary of Mathematics, 119
Encyclopedic Dictionary of Mathematics for Engineers., 119
Endangered Plant Species of World and Endangered Habitats, 40
Endangered Species, 36
Endangered and Threatened Plants of US, 347
Energy, 54, 534, 589, 618
Energy Abstracts, 565
Energy Abstracts for Policy Analysis, 565
Energy Atlas, 431
Energy Atlas of Asia and Far East, 392
Energy Bibliography and Index, 55, 565
Energy Conservation Idea Handbook, 224
Energy Conservation Standards Building Design, Construction, 647
Energy Conservation in Buildings 1973-1983, 338
Energy Conversion and Management, 589
Energy Costs and Costing, 55
Energy Crisis, 261

Energy Deskbook, 224
Energy Dictionary, 148
Energy Directory Updates, 431
Energy Factbook, 261
Energy Handbook, 224
Energy Index, 565
Energy Information Abstracts, 566
Energy Information Guide, 25
Energy Information Handbook, 244
Energy Information Locator, 432
Energy Management, 291
Energy Management Handbook, 224
Energy Managers' Handbook, 225
Energy Policy, 589
Energy Policy and Third World Development, 338
Energy Policy-Making, 55
Energy Products Specification Guide, 338
Energy Reference Handbook, 225
Energy Research Abstracts, 566
Energy Research Programs, 626
Energy Resources, 338
Energy Resources of the US Geological Survey, 265
Energy Review, 566
Energy Saver's Handbook for Town and City People, 225
Energy Saving Handbook for Homes, Businesses and Institutes, 225
Energy Sourcebook, 291
Energy Sources, 589
Energy Statistics, 25
Energy Statistics Yearbook, 269
Energy Systems and Policy, 589
Energy Technology Handbook, 225
Energy Terminology, 111
Energy Today, 589
Energy Use in the United States by State and Region, 269
Energy and Congress, 55
Energy and Environment Bibliography, 58
Energy and Environment Checklist, 58
Energy and Environment Information Resource Guide, 26
Energy and the Environment, 589
Energy and the Social Sciences, 55
Energy for Industry and Commerce, 431
Energy in the World Economy, 268

Energy-Efficient Products and Systems, 415
Energy-Scientific, Technical and Socioeconomic Bibliography, 55
Energy: A Multimedia Guide for Children and Young Adults, 665
Energy: Guide to Organizations and Information Resources US, 431
Energy: Historical Development of the Concept, 472
Energy: Sources of Print and Non-Print Materials, 56, 665
Engineer, 584
Engineer Buyers Guide, 410
Engineer's Guide to Solar Energy, 339
Engineer-in-Training Review Manual, 276
Engineering, 584
Engineering College Research and Graduate Study, 625
Engineering Compendium on Radiation Shielding, 223
Engineering Data Book, 256
Engineering Eponyms, 47, 52, 181, 256
Engineering Foundation, 443
Engineering Geological Maps, 373, 386
Engineering Geology, 585
Engineering Index Monthly and Author, 555
Engineering Issues, 585
Engineering Journal, 584
Engineering Manual, 276
Engineering Mathematics Handbook, 155
Engineering News Record, 585
Engineering Properties of Soils and Rocks, 255
Engineering Properties of Steel, 260
Engineering Sciences Data, 259
Engineering Specifications and Statistical Issue, 650
Engineering Tables and Data, 244
Engineering and Industrial Graphics Handbook, 209
Engineering and Mining Journal Operating Handbook, 220
Engineering and Technology Degrees, 400
Engineering and Technology Graduates, 400

Engineers Salaries, 400
Engineers' Joint Council Annual Report, 442
English-Chinese Dictionary of Engineering and Technology, 101
English-French Petroleum Dictionary, 101
English-Russian Dictionary of Applied Geophysics, 104
English-Russian Dictionary of Refrigeration and Low-temp., 105
English-Russian Physics Dictionary, 105
English-Russian Polytechnical Dictionary, 105
English-Russian Reliability and Quality Control Dictionary, 105
English-Spanish, Spanish-English Encyclopedic Dictionary of Technical Terms, 107
English/French Paints and Coatings Vocabulary, 102
Entomology, 20
Environment Abstracts, 567
Environment Film Review, 665
Environment International, 590
Environment Regulation Handbook, 228
Environment and Behavior, 58
Environment and Man, 520
Environmental Assessment and Impact Statement Handbook, 228
Environmental Biology for Engineers, 341
Environmental Economics, 26
Environmental Effects of Materials and Equipment, 567
Environmental Engineering and Sanitation, 291
Environmental Glossary, 150
Environmental Impact Analysis Handbook, 228
Environmental Impact Assessment, 58
Environmental Impact Data Book, 262
Environmental Impact Statement Directory, 437
Environmental Impact Statement Glossary, 150
Environmental Law, 58
Environmental Law Handbook, 228
Environmental Management Handbook for the Hydrcarbon Proc., 228

Environmental Planning, 26, 59
Environmental Pollution, 589, 627
Environmental Quality, 446
Environmental Quality Abstracts, 567
Environmental Quality and Safety, 534
Environmental Radiation Measurement, 651
Environmental Regulation Handbook, 232
Environmental Science, 535
Environmental Science Handbook for Architects and Builders, 191
Environmental Science and Technology, 590
Environmental Standard Review Plans, 651
Environmental Toxicology, 27
Environmental Values, 59
Environmental and Civil Programs of Federal Government, 627
Enzyme Nomenclature, 98
Enzymes, 529
Ephemeris of Sun, Polaris and Other Selected Stars, 247
Equalant I & Equalant II: Oceanographic Atlas, 390
Equilibrium Properties of Fluid Mixtures-2, 35
Equipment Design Handbook for Refineries and Chemical Plants, 183
Ergonomics, 587
Essays in Biochemistry, 500
Essays in Chemistry, 497
Essays in History of Geology, 481
Essays in Physics, 523
Essays in the History of Mechanics, 473
Essentials of Mathematics, 307
Estimated Federal Expenditures on Domestic Transportation., 266
Estuarine Hydrography and Sedimentation, 180
European Chemical Industry Handbook, 183
European Journal of Physics, 578
European Offshore Oil and Gas Yearbook, 446
European Patents Handbook, 631
European Petroleum Directory, 432
European Research Centres, 398
European Research Index, 395

European Sources of Scientific and Technical Information, 14
Europlastics Yearbook, 413
Eusidic Database Guide 1983, 676
Evaluation of the Microbiology Standards for Drinking Water, 644
Everyone's Guide to Better Food and Nutrition, 315
Evolution and Genetics of Populations, 529
Evolutionary Biology, 510
Evolutionary History of the Primates, 477
Examination of Water for Pollution Control, 232
Excavation Handbook, 191
Excerpta Botanica, 548
Excerpta Medica, 609
Exercise Manual in Immunology, 295
Exercises in Computer Systems Analysis, 297
Exercises in Physical Geology, 296
Exobiology: A Research Guide, 37
Exotic Plant Manual, 274
Experiment in Fluids, 587
Experimental Botany, 508
Experimental Chemical Thermodynamics, 525
Experimental Foods Laboratory Manual, 297
Experiments in College Physics, 293
Experiments in Physics, 293
Experiments in Physiology, 295
Exploring Laser Light, 293
Exploring the History of Nuclear Physics, 473
External Man-Induced Events in Relation to Nuclear Power., 336
Exxon Aviation News Digest, 655

FAA Statistical Handbook of Civil Aviation, 601
FEDERAL RESEARCH IN PROGRESS, 694
FEDREG, 694
FLUIDEX, 694
FOOD SCIENCE AND TECHNOLOGY ABSTRACTS, 694
FOODS ADLIBRA, 694
FOREST, 695

FOUNDATION DIRECTORY, 695
FOUNDATION GRANTS INDEX, 695
FSTA, 695
Fabric Almanac, 257
Facts on File Dictionary of Astronomy, 117
Facts on File Dictionary of Biology, 124
Facts on File Dictionary of Botany, 126
Facts on File Dictionary of Chemistry, 123
Facts on File Dictionary of Mathematics, 119
Facts on File Dictionary of Microcomputers, 142
Facts on File Dictionary of Physics, 120
Facts on File Dictionary of Telecommunications, 144
Facts on File Yearbook, 440
Factual Data Banks in Agriculture, 253
Fairchild's Dictionary of Textiles, 134
Families of Flowering Plants Arranged to a New System, 351
Family Computers under $200, 326
Famous Names in Engineering, 460
Farm Builder's Handbook, 173, 191
Farm Journal, 580
Farm and Garden Index, 547
Farwell's Rules of the Nautical Road, 255
Fatal and Injury Accident Rates Federal-Aid and Other Hgwy., 266
Feature Films on 8mm, 16mm, and Videotape, 658
Federal Aviation Regulations for Aviation Mechanics, 320
Federal Data Base Finder, 592, 677
Federal Energy Information Sources and Data Bases, 25, 432
Federal Funds for Research Development and Other Scientific Activities, 620
Federal Information Processing Standards Index, 640
Federal Scientific and Technical Communication Activities, 398
Federal Statistical Directory, 264
Federal Technology Catalog 1983, 592
Federation Proceedings, 609
Federation of European Biochemical Societies, 609

Fenaroli's Handbook of Flavor Ingredients, 186
Fern Growers Manual, 274
Ferns and Allied Plants, 174
Ferromagnetic-Core Design and Application Handbook, 204
Fiber Optics and Lightwave Communications Standard Dictionary, 144
Fiberglass-Reinforced Plastics Deskbook, 212
Field Book of the Skies, 342
Field Crop Diseases Handbook, 173
Field Engineer's Manual, 278
Field Guide in Color to Insects, 365
Field Guide to Berries and Berrylike Fruits, 347
Field Guide to Birds of USSR, 360
Field Guide to Dangerous Animals of North America, 367
Field Guide to Dinosaurs, 356
Field Guide to Edible Wild Plants of Eastern and Central., 347
Field Guide to Medicinal Wild Plants, 347
Field Guide to North American Edible Wild Plants, 347
Field Guide to Orchids of North America, 351
Field Guide to Pacific Coast Fishes of North America, 363
Field Guide to Poisonous Plants and Mushrooms of North America, 348
Field Guide to Stars and Planets, 342
Field Guide to Tropical and Subtropical Plants, 348
Field Guide to the Atmosphere, 370
Field Guide to the Birds, 359
Field Guide to the Birds of Australia, 359
Field Guide to the Birds of North America, 360
Field Guide to the Birds of South-East Asia, 360
Field Guide to the Whales, Porpoises and Seals, 367
Field and Laboratory Guide to Tree Pathology, 348
Fieldbook of Natural History, 356

Fieldbook of Pacific Northwest Sea Creatures, 356
Fiesers' Reagents for Organic Synthesis, 251
Fifty Years of Electronic Components— 1921–1971, 485
Films in the Mathematics Classroom, 665
Films in the Sciences—Reviews and Recommendations, 664
Finding Answers in Science and Technology, 14
Finding Birds Around the World, 360
Fine Structure of Parasitic Protozoa, 384
Finishing Handbook and Directory, 209
Fire Protection Guide on Hazardous Materials, 332
Fire Protection in Nuclear Power Plants, 336
First Aid Manual for Chemical Accidents, 272
Fish Physiology, 529
Fishes of the World, 128, 363
Fishing Industry Index International, 551
Five Kingdoms: Illustrated Guide to Phyla of Life on Earth, 309
Flammarion Book of Astronomy, 272
Flight Directory of British Aviation, 411
Flight International, 584
Flow Measurement Engineering Handbook, 215
Flower and Fruit Prints of the 18th and Early 19th Centuries, 8
Flowering Plants of the World, 254
Flowers of the Wild: Ontario and the Great Lakes Region, 353
Flowmeters, 331
Fluid Flow Measurements Abstracts, 563
Fluid Flow Pocket Handbook, 184
Fluid Power Abstracts, 558
Fluidics Feedbacks, 563
Flying, 584
Food, 98
Food Chemicals Codex, 644
Food Industries Manual, 277
Food Ingredients Directory, 413
Food Science and Technology, 37
Food Science and Technology Abstracts, 557

Food Service Market for Frozen Foods, 315
Food and Nutrition, 582
Food and Nutrition Bibliography, 41
Food and Nutrition Encyclopedia, 84
Food and Nutrition News, 582
Food: An Authoritative and Visual History and Dictionary, 136
Foods and Food Production Encyclopedia, 84
For Good Measure, 249
Foreign Patents, 632
Forest History Museums of the World, 409
Forest Land Use, 39
Forest Service Organizational Directory, 409
Forestry Abstracts, 549
Forestry Handbook, 174
Forestry Theses Accepted by Colleges and Universities in US, 618
Formulary of Paints and Other Coatings, 257
Forthcoming International Scientific and Technical Conference, 602
Fortran 77, 297
Fossils for Amateurs, 371
Foundation Directory, 395
Foundation Grants to Individuals, 395
Foundations of Mathematical Biology, 529
Foundry Yearbook, 444
Fowler's Mechanical Engineer's Pocket Book, 282
Franklin Institute Journal, 576
Fraser Williams FINE Chemical Directory, 713
Free Software Catalog and Directory, 422
Free and Inexpensive Materials, 653
Freeze-Drying Biological Specimens, 295
French-English Chemical Terminology, 102
French-English Science and Technology Dictionary, 102
French-English and English-French Dictionary of Technical., 102
Fresh Meat Processing, 637
Freshwater Wetlands, 313

Freshwater and Terrestrial Radioecology, 59
From Dits to Bits, 486
From Spark to Satellite, 486
From X-Rays to Quarks, 476
Frozen Foods, 484
Fruit Processing, 637
Fuel and Energy Abstracts, 566
Fuel from Farms, 339
Fundamental Experiments in Microbiology, 295
Fundamentals Handbook of Electrical and Computer Engineering, 195, 204
Fundamentals of Mathematics, 522
Fungi, 529
Fusion Power Report, 578

GEOARCHIVE, 695
GEOREF, 695
GEOREF Thesaurus and Guide to Indexing, 98
GMT World Register of Oceanographic Products and Services, 409
GPO MONTHLY CATALOG, 695
GRANT INFORMATION SYSTEM, 696
GSA Yearbook, 441
Galileo at Work, 455
Garden Book, 323
Gary Null's Nutrition Sourcebook for the '80's, 286
Gas Abstracts, 566
Gas Tables, 245
Gebbie House Magazine Directory, 653
Gems, 371
Gemstone and Mineral Data Book, 180
General Information Concerning Patents, 632
General Patents Index, 634
General Physics Advance Abstracts, 571
General Relativity and Gravitation, 473
General Relativity: An Einstein Centenary Survey, 473
General Science Index, 537
Genetic Engineering and Biotechnology Firms, 406
Genetic Engineering, DNA and Cloning, 37
Genetic Engineering/Biotechnology Patents, 636

Genetic Engineering/Biotechnology Sourcebook, 286
Genetical Research, 581
Genetics, 171, 581
Genetics Abstracts, 546
Geo Abstracts, 553
Geochemical Tables, 242
Geodex System/Structural Information Service, 553
Geographical Atlas of World Weeds, 382
Geography and Earth Sciences Publications, 9
Geologic Names of the United States Through 1975, 131
Geologic Reference Sources, 21, 45
Geological Abstracts, 553
Geological Magazine, 583
Geological Society of America, 1888–1930, 481
Geological Society of America. Bulletin, 583
Geological Society of America. Proceedings, 610
Geological Society. Journal, 583
Geological World Atlas, 386
Geologists and the History of Geology, 45
Geomorphology Laboratory Manual with Report Forms, 297
Geophysical Abstracts, 553
Geophysical Directory, 404, 409
Geophysics Abstracts, 553
Geosciences and Oceanography, 409
Geotechnical Abstracts, 553
Geotechnical Engineering Investigation Manual, 278
Geothermal Resources and Technology in US, 432
Geothermal World Directory, 432
Geotitles Weekly, 571
German-English Science Dictionary, 103
Gerrish's Technical Dictionary, 115
Getting Food From Water, 310
Giant Handbook of Electronic Circuits, 197
Glaciers, 601
Glass, 533
Glossary for Horticultural Crops, 126
Glossary for Radiologic Technologists, 121
Glossary of Agricultural Terms, 107
Glossary of Air Pollution, 150
Glossary of Astronomy and Astrophysics, 118
Glossary of Automotive Terminology, 102
Glossary of Chemical Terms, 123
Glossary of Environment with French and German Equivalents, 150
Glossary of Genetics and Cytogenetics, 124
Glossary of Geographical Terms, 131
Glossary of Inland Fishery Terms, 128
Glossary of Marine Technology Terms, 131
Glossary of Mineral Species 1983, 131
Glossary of Packaging Terms, 134
Glossary of Soil Micromorphology, 112
Glossary of Transport, 112
Glossary of Wood, 134
Gmelins Handbuch der Anorganischen Chemie, 164
Government Production Prime Contractors Directory, 415
Government Publications, 597
Government Publications Reviews, 597
Government Reference Books, 597
Government Reports Announcements and Index, 537, 592
Government Research Centers Directory, 620
Government-Wide Index Federal Research and Development Report, 592
Governmental Organization for Regulation of Nuclear Power., 336
Graduate Assistantship Directory in the Computer Science, 400
Graduate Programs in Physics, Astronomy and Related Fields, 400
Graduate Texts in Mathematics, 490
Grafter's Handbook, 174
Grant Data Quarterly, 620
Grants and Awards, 620
Grass Systematics, 348
Grasses, 348
Gray Herbarium Index, 549
Great Botanical Gardens of the World, 406
Great Chemists, 457

Great Engineers and Pioneers in Technology, 460
Great International Disaster Book, 255
Green's Functions and Transfer Functions Handbook, 155
Ground Water Manual, 275
Ground Water Pollution, 60
Growing Up with Science, 62
Growth of Crystals, 525
Grzimek's Animal Life Encyclopedia, 74
Grzimek's Encyclopedia of Ecology, 92
Grzimek's Encyclopedia of Ethology, 74
Grzimek's Encyclopedia of Evolution, 75
Guide Specification for Highway Constructions, 647
Guide for Safety in the Chemical Laboratory, 309
Guide to "Referativnyi Zhurnal", 538
Guide to American Scientific and Technical Directories, 393
Guide to Atlases, 373
Guide to Basic Information Sources in Chemistry, 18
Guide to Basic Information Sources in Engineering, 21
Guide to Beilstein's Handbook, 164
Guide to Bird Behavior, 316
Guide to Chemical Abstracts, 536
Guide to Classification in Geology, 318
Guide to Dialog Searching, 679
Guide to Ecology Information and Organizations, 27
Guide to European Sources of Technical Information, 398
Guide to Field Identification Wildflowers of North America, 353
Guide to Gas Chromatography Literature, 18, 48
Guide to Good Programming Practice, 329
Guide to Identifying and Classifying Yeasts, 310
Guide to Information Services in Marine Technology, 24
Guide to Literature and Patents Concerning PVC Technology, 636
Guide to Literature on Civil Engineering, 23
Guide to Literature on Nuclear Engineering, 24
Guide to Mammals of Plains States, 367
Guide to Manuscripts in the National Agricultural Library, 7
Guide to Microforms in Print, 658
Guide to Micrographic Equipment, 663
Guide to Non-Ferrous Metals and Their Markets, 333
Guide to North American Waterfowl, 360
Guide to Nuclear Power Technology, 336
Guide to Observing Insect Lives, 316
Guide to Popular Government Publications, 597
Guide to Port Entry 1984–1985, 332
Guide to Publications and Subsequent Investigations of Deep, 46
Guide to Quality Control, 331
Guide to Reference Material, 15
Guide to Reprints, 628
Guide to Safe Handling of Radioactive Wastes Nuclear Power., 336
Guide to Science and Technology in UK, 398
Guide to Scientific Instruments, 307
Guide to Scientific and Technical Journals in Translation, 615
Guide to Searching the Biological Literature, 19
Guide to Site and Environmental Planning, 341
Guide to Sources for Agricultural and Biological Research, 310
Guide to Special Issues and Indexes of Periodicals, 536, 653
Guide to Specifications and Standards of Federal Government, 638
Guide to Theses and Dissertations, 617
Guide to US Government Directories, 393
Guide to US Government Maps, 373
Guide to US Government Publications, 597
Guide to US Government Scientific and Technical Resources, 592
Guide to US Government Statistics, 264, 597
Guide to US Scientific and Technical Resources, 15
Guide to World Science Series, 668

Guide to Writing Better Technical Papers, 307
Guide to the Chemicals Used in Crop Production, 311
Guide to the Energy Industries, 339
Guide to the HPLC Literature, 18
Guide to the Identification of the Genera of Bacteria, 310
Guide to the Literature of Astronomy, 17
Guide to the National Electrical Code, 324, 648
Guide to the Safe Use of Food Additives, 315
Guide to the Study of Animal Populations, 316
Guide to the Wines of the United States, 315
Guide to the World's Abstracting and Indexing Services in., 536
Guidebook for Technology Assessment and Impact Analysis, 319
Guidebook to Biochemistry, 309
Guidebook to Departments in Mathematical Sciences in US., 401
Guidebook to Mechanism in Organic Chemistry, 309
Guidebook to Nuclear Reactors, 288
Guidebook to Organic Synthesis, 309
Guidelines for Format and Production of Scientific and Technical Reports, 592
Guides to International Organizations, 395
Guinness Aircraft Facts and Feats, 263
Guinness Book of Animal Facts and Feats, 178
Guinness Book of Astronomy, 247
Guinness Book of Ships and Shipping, 263
Gulls: A Guide to Identification, 360

HEILBRON, 696
HRIS Abstracts, 558
Hacker's Dictionary, 142
Hadronic Journal, 578
Hamlyn Nature Guide to Fossils, 318
Handbook for Electronics Engineering Technicians, 197
Handbook for Radio Engineering Managers, 202

Handbook in Applied Meterology, 180
Handbook of Acoustical Enclosures and Barriers, 158
Handbook of Active Filters, 197
Handbook of Adhesives, 213
Handbook of Advanced Robotics, 216
Handbook of Advanced Wastewater Treatment, 228
Handbook of Aerosol Technology, 184
Handbook of Agricultural Occupations, 311
Handbook of Air Conditioning, Heating, and Ventilating, 216
Handbook of Air Pollution Analysis, 232
Handbook of Air Pollution Technology, 232
Handbook of Airline Statistics, 266
Handbook of Analysis of Synthetic Polymers and Plastics, 187
Handbook of Analytical Control of Iron and Steel Production, 220
Handbook of Analytical Derivatization Reactions, 164
Handbook of Anion Determination, 164
Handbook of Applicable Mathematics, 156
Handbook of Applied Hydraulics, 189
Handbook of Applied Mathematics, 156
Handbook of Applied Mathematics. Selected Results & Methods, 156
Handbook of Architectural Details for Commercial Buildings, 192
Handbook of Architectural and Civil Drafting, 192
Handbook of Artificial Intelligence, 205
Handbook of Astronomy, Astrophysics and Geophysics, 154
Handbook of Atomic Data, 249
Handbook of Automated Analysis, Continuous Flow Techniques, 164
Handbook of Basic Electronic Troubleshooting, 197
Handbook of Batteries and Fuel Cells, 197
Handbook of Bimolecular and Termolecular Gas Reactions, 165
Handbook of Biochemistry and Molecular Biology, 171
Handbook of Birds of Europe, Middle East and North Africa, 179

Handbook of Canadian Mammals, 178
Handbook of Chemical Engineering Calculations, 184
Handbook of Chemical Equilibria in Analytical Chemistry, 165
Handbook of Chemical Lasers, 158
Handbook of Chemical Microscopy, 165
Handbook of Chemical Property Estimation Methods, 165
Handbook of Chemical Synonyms and Trade Names, 98
Handbook of Chemistry Specialities, 188
Handbook of Chemistry and Physics, 152
Handbook of Comparative World Steel Standards, 650
Handbook of Components for Electronics, 197
Handbook of Composite Construction Engineering, 192
Handbook of Composites, 213
Handbook of Computational Chemistry, 165
Handbook of Computers and Computing, 205
Handbook of Construction Equipment Maintenance, 192
Handbook of Construction Management and Organization, 192
Handbook of Construction Resources and Support Services, 415
Handbook of Dam Engineering, 193
Handbook of Data on Organic Compounds, 165
Handbook of Decomposition Methods in Analytical Chemistry, 165
Handbook of Digital IC Applications, 205
Handbook of Dimensional Measurements, 209
Handbook of Electrical Circuit Designs, 198
Handbook of Electronic Communications, 198
Handbook of Electronic Formulas, Symbols and Definitions, 198
Handbook of Electronic Systems Design, 205
Handbook of Electronic Tables and Formulas, 245

Handbook of Electronics Calculations for Engineers and Technicians, 198
Handbook of Electronics Industry Cost Estimating Data, 258
Handbook of Electronics Packing Design and Engineering, 198
Handbook of Electropainting Technology, 205
Handbook of Elementary Physics, 158
Handbook of Energy Conservation for Mechanical Systems., 225
Handbook of Energy Technology and Economics, 225
Handbook of Engineering Fundamentals, 181
Handbook of Engineering Geomorphology, 181
Handbook of Engineering Management, 182
Handbook of Environmental Control, 229
Handbook of Environmental Data and Ecological Parameters, 229
Handbook of Environmental Data on Organic Chemicals, 166, 229
Handbook of Environmental Engineering, 229
Handbook of Environmental Health and Safety, 229
Handbook of Enzyme Inhibitors, 166
Handbook of Essential Formulae and Data Heat Transfer for Engineers, 184
Handbook of Ethological Methods, 171
Handbook of Experiences in Design and Installation of Solar., 226
Handbook of Experimental Immunology, 171
Handbook of Fiber Optics, 158
Handbook of Fillers and Reinforcements for Plastics, 187
Handbook of Flame Spectroscopy, 158
Handbook of Fluids in Motion, 158
Handbook of Food and Nutrition, 187
Handbook of Freshwater Fishery Biology, 178
Handbook of Genetics, 171
Handbook of Geochemistry, 166
Handbook of Geology in Civil Engineering, 189
Handbook of Hawaiian Weeds, 353

Handbook of Highway Engineering, 193
Handbook of Hydraulics for the Solution of Hydraulic., 193
Handbook of Hypergeometric Integrals, 156
Handbook of Industrial Energy Analysis, 226
Handbook of Industrial Engineering, 209
Handbook of Industrial Robotics, 216
Handbook of Industrial Waste Disposal, 229
Handbook of Industrial Wastes Pretreatment, 230
Handbook of Industrial Water Conditioning, 184
Handbook of Intermediary Metabolism of Aromatic Compounds, 166
Handbook of International Alloy Compositions and Designation, 220
Handbook of Iron Meteorites, 181
Handbook of Laboratory Distillation, 166
Handbook of Laboratory Waste Disposal, 209
Handbook of Lipid Research, 166
Handbook of Machinery Adhesives, 216
Handbook of Marine Science, 181
Handbook of Materials Handling, 213
Handbook of Mathematical Functions and Formulas, 236
Handbook of Mathematical Tables and Formulas, 156, 236
Handbook of Measurement Science, 156, 214
Handbook of Mechanical and Electrical Systems for Buildings, 198
Handbook of Mechanics, Materials and Structures, 214, 216
Handbook of Metal Ligand Heats and Related Thermodynamic., 167
Handbook of Metal Treatments and Testing, 221
Handbook of Microbiology, 172
Handbook of Microcircuit Design and Application, 205
Handbook of Microcomputer-Based Instrumentation and Controls, 205
Handbook of Microprocessors, Microcomputers and Minicomputers, 206
Handbook of Mining and Tunnelling Machinery, 216
Handbook of Modern Analytical Instruments, 167
Handbook of Modern Electrical Wiring, 199
Handbook of Modern Electrical and Electronic Engineering, 195
Handbook of Modern Electrical and Electronic Engineering, 199
Handbook of Modern Solid-State Amplifiers, 202
Handbook of Multiphase Systems, 158
Handbook of Municipal Administration and Engineering, 230
Handbook of Neurochemistry, 167
Handbook of Noise Assessment, 230
Handbook of Noise Control, 230
Handbook of Noise and Vibration Control, 216
Handbook of North American Birds, 178
Handbook of Nuclear Data for Neutron Activation Analysis, 249
Handbook of Numerical and Statistical Techniques, 157
Handbook of Oceangraphic Tables, 243
Handbook of Oceanographic Engineering Materials, 213
Handbook of Oil Industry Terms and Phrases, 184
Handbook of Operations Research, 157
Handbook of Optical Holography, 159
Handbook of Optics, 159
Handbook of Organic Reagents in Inorganic Analysis, 167
Handbook of Organic Waste Conversion, 184
Handbook of Package Engineering, 209
Handbook of Phycological Methods, 175
Handbook of Physical Calculations, 159
Handbook of Physiology, 172
Handbook of Plant Virus Infections, 175
Handbook of Plastic Product Design Engineering, 187
Handbook of Plastics Testing Technology, 188
Handbook of Pollution Control Management, 232
Handbook of Powder Science and Technology, 213

Handbook of Power Generation, 199
Handbook of Practical Electrical Design, 199
Handbook of Practical Gear Design, 216
Handbook of Practical Microcomputer Troubleshooting, 206
Handbook of Practical Organic Micro-Analysis, 167
Handbook of Precision Engineering, 215
Handbook of Precision Sheet, Strip and Foil, 217
Handbook of Pressure-Sensitive Adhesive Technology, 217
Handbook of Protein Sequence Analysis, 172
Handbook of Proton Ionization Heats and Related Thermodyn., 251
Handbook of Public Speaking for Scientists and Engineers, 152
Handbook of Radio Sources, 154
Handbook of Radioimmunoassay, 172
Handbook of Reactive Chemical Hazards, 167
Handbook of Refractory Compounds, 213
Handbook of Remote Control and Automation Techniques, 217
Handbook of Scientific and Technical Awards, 152
Handbook of Seagrass Biology, 175
Handbook of Semiconductor and Bubble Memories, 206
Handbook of Sensory Physiology, 172
Handbook of Separation Techniques for Chemical Engineers, 185
Handbook of Silicone Rubber Fabrication, 185
Handbook of Simplified Electrical Wiring Design, 199
Handbook of Simplified Solid-State Circuit Design, 199
Handbook of Software Maintenance, 206
Handbook of Soil Mechanics, 194
Handbook of Solar Flare Monitoring and Propagation Forecast., 154
Handbook of Solid Waste Management, 230
Handbook of Solid-State Devices, 206
Handbook of Space Astronomy and Astrophysics, 154

Handbook of Spectroscopy, 159
Handbook of Stack Sampling and Analysis, 230
Handbook of Stainless Steels, 221
Handbook of Statistical Methods, 157
Handbook of Structural Concrete, 192
Handbook of Sugars, 187
Handbook of Superalloys, 221
Handbook of Surfaces and Interfaces, 159
Handbook of Synfuels Technology, 226
Handbook of Technical Writing Practices, 276
Handbook of Terms Used in Algebra and Analysis, 119
Handbook of Textile Fibers, Dyes, and Finishes, 189
Handbook of Textile Fibres, 189
Handbook of Thermochemical Data Compounds and Aqueous., 168
Handbook of Thermodynamic Tables and Charts, 168
Handbook of Toxic Fungal Metabolites, 172
Handbook of Toxic and Hazardous Chemicals, 168
Handbook of Transformer Design and Applications, 199
Handbook of Tropical Foods, 187
Handbook of Tropical Forage Grasses, 175
Handbook of US Colorants for Food, Drugs and Cosmetics, 185
Handbook of Utilization of Aquatic Plants, 175
Handbook of Valves, Piping and Pipelines, 217
Handbook of Wastewater Collection and Treatment, 230
Handbook of Water Resources and Pollution Control, 233
Handbook of Water-Soluble Gums and Resins, 188
Handbook of X-ray and Ultraviolet Photoelectron Spectroscopy, 160
Handbook of the Birds of India and Pakistan, 179
Handbook of the British Astronomical Association, 154

Handbook of the Engineering Sciences, 182
Handbook on Industrial Robotics, 217
Handbook on Mechanical Properties of Rocks, 213
Handbook on Physics and Chemistry of Rare Earths, 168
Handbook on Plasma Instabilities, 160
Handbook on Semiconductors, 159
Handbooks and Tables in Science and Technology, 30
Handbooks of Patent Technology, 632
Handbuch der Astrophysik, 66
Handbuch der Physik, 66
Handbuch der Physik: Index, 66
Harper & Row's Complete Field Guide to North American Wild., 356
Hart's Rocky Mountain Mining Directory, 1983, 409
Harvard University Museum of Comparative Zoology. Library., 9
Hawkweed Passive Solar House Book, 323
Hazardous Chemicals Data Book, 251
Hazardous Materials Spills Handbook, 231
Hazardous Waste Management Directory, 437
Hazardous Waste in America, 92
Hazards in Chemical Laboratory, 168
Healing Plants, 175
Health Physics Research Abstracts, 622
Health, Food and Nutrition in Third-World Development, 315
Healthy Garden Book, 352
Heat Bibliography, 1948/52–, 34
Heat Exchanger Design Handbook, 217
Heat Exchanger Sourcebook, 288
Heat Pipe Technology, 52
Heat Transfer Engineering, 587
Heat Transfer Pocket Handbook, 217
Heat and Mass Transfer Data Book, 249
Heat and Power From the Sun, 56
Heilbron's Dictionary of Organic Compounds, 123
Heliocopters of the World, 483
Henderson's Dictionary of Biological Terms, 124
Herbage Abstracts, 549
Herbs, 40, 348

Here Come the Clones!, 326
Heredity, 581
Hi-Tech Data Base Buyer's Guide, 677
High School Mathematics Library, 32
Highway Research Abstracts, 558
Highway Research in Progress, 515
Highway Statistics, 266
Hillier Colour Dictionary of Trees and Shrubs, 126
Histological Methods and Terminology in Dictionary Form, 124
Historic American Engineering Record Catalog, 1976, 599
Historical Catalogue of Scientists and Scientific Books, 466
Historical Development of Quantum Theory, 473
Historical Geology, 297
Historical Studies in the Physical Sciences, 473
History and Philosophy of Technology, 482
History of American Archaeology, 481
History of American Ecology, 478
History of Analytical Chemistry, 477
History of Ancient Mathematical Astronomy, 469
History of Astronomy, 469
History of Biochemistry, 477
History of Botany, 1860–1900, 478
History of British Geology, 481
History of Building Types, 484
History of Calculus of Variations 17th Through 19th Century, 470
History of Cave Science, 481
History of Chemical Engineering, 484
History of Chemical Technology, 48
History of Classical Physics, 34
History of Computing in the Twentieth Century, 486
History of Control Engineering 1800–1930, 482
History of Earth Sciences during Scientific and Industrial., 481
History of Engineering and Science in Bell System, 486
History of Entomology, 478
History of Fishes, 478
History of Genetics, 478
History of Man-Powered Flight, 484

History of Mathematics, 471
History of Microtechnique, 486
History of Modern Astronomy and Astrophysics, 470
History of Modern Physics, 34
History of Modern Physics, 1800–1950, 473
History of Physics, 473
History of Programming Languages, 486
History of Public Works in US, 1776–1976, 485
History of Science, 466
History of Science and Technology in US, 466
History of Science and Technology in US. Vol. 2, 467
History of Scientific and Technical Periodicals, 573
History of Study of Landforms or Development Geomorphology, 481
History of Technology, 482
History of Technology and Invention, 483
History of Theory of Structures in the 19th Century, 487
History of Tribology, 487
History of Twentieth Century Physics, 474
History of the British Petroleum Company, 484
History of the Earth, 481
History of the Life Sciences, 478
History of the Quantum Theory, 474
Home Wind Power, 339
Horizons in Biochemistry and Biophysics, 500
Hormonal Proteins and Peptides, 529
Horticultural Abstracts, 549
Horticultural Handbook, 176
Horticultural Reviews, 508
Hortus Third, 127
House Plants Indoor/Outdoors, 352
How Birds Work, 361
How It Works, 62
How to Build a Solar Heater, 302
How to Debug Your Personal Computer, 303
How to Design, Build and Use Electronic Control Systems, 304
How to Find Chemical Information, 18

How to Find Out About Patents, 632
How to Know the Aquatic Plants, 300
How to Know the Ferns and Fern Allies, 300
How to Know the Insects, 300
How to Know the Lichens, 300
How to Know the Seed Plants, 300
How to Know the Trees, 301
How to Know the Weeds, 301
How to Name an Inorganic Substance, 98
How to Patent Without a Lawyer, 632
How to Remove Pollutants and Toxic Materials from Air and W., 341
How to Save Energy and Cut Costs in Existing Industrial., 302
How to Solve General Chemistry Problems, 294
How to Write and Publish a Scientific Paper, 299
Human Adaptability, 623
Human Biology, 580
Human Ecology, 27
Human Food Uses, 41
Human and Mammalian Cytogenetics, 479
Huntia, 40
Hydata, 571
Hydraulic Pumps and Motors, 334
Hydraulic Research in US and Canada, 416
Hydro-Abstracts, 567
Hydrogen Energy, 56
Hydrogeologic Atlas of the People's Republic of China, 391
Hyman Series in Invertebrate Biology, 530

IAEA Bibliographical Series, 53
IAEA's Safety Standards and Measures, 644
IBM Journal of Research and Development, 655
IBM PC PASCAL, 303
IBM PC-Compatible Computer Directory, 420
IBM Software Directory, 422
IBM Systems Journal, 586
IC Schematic Sourcemaster, 289
ICP Software Directory, 422

ICP Software Directory: Business Applications for Microcomputers, 422
IEA: The Directory of Instruments, Electronics, Automation, 417
IEC Multilingual Dictionary of Electricity, 112
IEEE International Conference on Engineering in Ocean Environments, 610
IEEE Proceedings, 586, 611
IEEE Spectrum, 586
IEEE Standard Atlas Test Language, 648
IEEE Standard Dictionary of Electrical and Electronics Terms, 138, 643
IEEE Standard Logical Link Control—Local Area Networks Stan., 649
IEEE Standard Techniques for High-Voltage Testing, 648
IEEE Standards for Local Area Networks, 648
IEEE Transactions, 586
IEEE Transactions on Aerospace and Electronics Systems, 584
IEEE Transactions of Power Apparatus and Systems, 589
IINPADOC, 634
IMM Abstracts, 563
INFORMATION SCIENCE ABSTRACTS, 696
INIS, 99
INIS ATOMINDEX, 564
INSPEC, 697
INSPEC List of Journals, 575
INSPEC-ELEC/COMP, 697
INSPEC-PHYSICS, 697
INTERNATIONAL PHARMEUTICAL ABSTRACTS, 697
IRIS, 698
IRIS Cumulative Bibliography, 30
ISI Atlas of Science: Biochemistry and Molecular Biology, 68
ISI's Who is Publishing in Science, 451
ISIS, 576
ISIS Cumulative Bibliography, 467
ISIS Cumulative Bibliography: Personalities and Institutions, 467
ISMEC, 697
ISMEC Bulletin, 571
ITIS: Environmental Health and Safety, 712

Identification of Flowering Plant Families, 351
Identification of Molecular Spectra, 251, 294
Identification of Textile Materials, 321
Illustrated Computer Dictionary, 142
Illustrated Dictionary of Building, 137
Illustrated Dictionary of Electronics, 138
Illustrated Dictionary of Microcomputer Technology, 142
Illustrated Dinosaur Dictionary, 128
Illustrated Encyclopedia of Astronomy and Space, 64
Illustrated Encyclopedia of Aviation, 81
Illustrated Encyclopedia of Birds, 75
Illustrated Encyclopedia of General Aviation, 81
Illustrated Encyclopedia of Indoor Plants, 70
Illustrated Encyclopedia of North American Locomotives, 439
Illustrated Encyclopedia of Ships, Boats, Vessels, and., 93
Illustrated Encyclopedia of Solid-State Circuits and Applications, 87
Illustrated Encyclopedia of Space Technology, 64
Illustrated Encyclopedia of Succulents, 70
Illustrated Encyclopedia of Trees, Timbers, and Forests., 70
Illustrated Encyclopedia of the Mineral Kingdom, 79
Illustrated Encyclopedia of the Universe, 64
Illustrated Encyclopedia of the World's Automobiles, 93
Illustrated Encyclopedia of the World's Modern Locomotives, 439
Illustrated Encyclopedia of the World's Rockets and Missiles, 94
Illustrated Encyclopedic Dictionary of Building and Construction, 137
Illustrated Encyclopedic Dictionary of Electronic Circuits, 87
Illustrated Fact Book of Science, 307
Illustrated Glossary for Solar and Solar-Terrestrial Physics, 121
Illustrated Glossary of Process Equipment, 112

Illustrated Guide to Fossil Collecting, 371
Illustrated Guide to Home Retrofit for Energy Savings, 339
Illustrated Guide to Pollen Analysis, 313
Illustrated Guide to Textiles, 322
Illustrated Guide to Wine, 315
Illustrated Guidebook to Electronics Devices and Circuits, 324
Illustrated Handbook of Electronic Tables, Symbols, Measurements, 258
Illustrated History of the Herbals, 479
Illustrated Petroleum Reference Dictionary, 131
Illustrated Reference on Cacti and Other Succulents, 70
Illustrated Science and Invention Encyclopedia, 301
Images of the Earth, 488
Implementation of Long-Term Environmental Radiation Standards, 651
Index Aeronautics, 556
Index Bergeyana, 310
Index Herbarium, 550
Index Kewenis, 550
Index Kewenis. Supplementum, 550
Index Medicus, 546
Index Muscorum, 550
Index Translationum, 613
Index and Directory of US Industry Standards, 638
Index of Administrative Publications, 592
Index of Conference Proceedings Received, 604
Index of Current Literature on Coal Mining and Applied Subjects, 563
Index of Fungi, 549
Index of Generic Names of Fossil Plants, 318
Index of Human Ecology, 546
Index of International Standards, 640
Index of Mathematical Papers, 541
Index of Mathematical Tables, 236
Index of Reviews in Organic Chemistry, 544
Index of SNAME Publications, 1961–1969, 674
Index of Specifications and Standards, 640
Index of State Geological Survey Publications., 553
Index of Trademarks Issued from US Patent Office, 633
Index of US Nuclear Standards, 651
Index of Vibrational Spectra of Inorganic and Organic Compositions, 251
Index to 16mm Educational Films, 659
Index to 35mm Educational Filmstrips, 659
Index to 8mm Motion Cartridges, 658
Index to ASTM Technical Papers and Reports, 594
Index to American Botanical Literature, 1886–1966, 549
Index to Book Reviews in the Sciences, 538
Index to Classification, 633
Index to Conferences Assigned CONF-Numbers, 606
Index to Conferences Relating to Nuclear Science, 606
Index to Ecology, 567
Index to Educational Overhead Transparencies, 659
Index to Educational Records, 658
Index to Educational Slides, 658
Index to Educational Videotapes, 658
Index to Grass Species, 549
Index to IEEE Publications, 559
Index to Illustrations of Living Things Outside North America, 550
Index to Patents Issued from US Patent and Trademark Office, 633
Index to Plant Distribution Maps in North American Period., 550
Index to Producers and Distributors, 663
Index to Reviews, Symposia Volumes, and Monographs in., 544
Index to Scientific Reviews, 538, 620
Index to Scientific and Technical Proceedings, 538, 605
Index to Statistics and Probability, 541
Index to Theses Accepted for Higher Degrees in the Universities, 617
Index to Translations Selected by the American Mathematical Society, 614
Index to US Government Periodicals, 538, 593

Index to US Voluntary Engineering Standards, 640
Index to the Gray Herbarium of Harvard University, 348
Index to the Hazardous Materials Regulations, 644
Index to the Literature of Magnetism, 542
Index to the US Patent Classification, 633
Index-Abstracts to SAE Transactions and Literature, 674
Indexed Bibliography of Office of Research and Development., 595
Indoor Trees, 352
Indoor Water Gardener's How-to-Handbook, 301
Induction Heating Handbook, 221
Industrial Aerodynamics Abstracts, 556
Industrial Air Pollution Handbook, 233
Industrial Energy Conservation, 226
Industrial Engineering, 587
Industrial Engineering Terminology, 134
Industrial Enzymology, 286
Industrial Hazard and Safety Handbook, 209
Industrial Heat Exchangers, 331
Industrial Lubrication, 217
Industrial Management, 587
Industrial Microbiology, 310
Industrial New Product Development, 282
Industrial Noise Control Handbook, 231
Industrial Research Laboratories of US, 620
Industrial Research in Britain, 410, 625
Industrial Robotics Handbook, 218
Industrial Safety Handbook, 210
Industrial Solvents Handbook, 168
Industrial Source Sampling, 231
Industrial Wastewater Management Handbook, 231
Industrial and Commercial Power System Technical Conf. Rec., 611
Industrial and Engineering Chemistry. Quarterlies, 585
InfoWorld, 586
Information Resource/Data Dictionary Systems, 142
Information Resources for Engineers and Scientists, 15
Information Resources in Toxicology, 22
Information Science Abstracts, 538
Information Sources in Agriculture and Food Science, 19
Information Sources in Architecture, 23
Information Sources in Biotechnology, 22
Information Sources in Power Engineering, 26
Information Sources in Transportation, Material Management, 61
Information Sources in the History of Science and Medicine, 467
Information Sources on Bioconversion of Agricultural Wastes, 19
Information Sources on Industrial Maintenance and Repair, 24
Information Sources on Non-Conventional Sources of Energy, 26
Information Sources on the Natural and Synthetic Rubber Industries, 22
Information Sources: Physical Sciences and Engineering, 15
Information on International Research and Development Activities, 626
Infrared Handbook, 160
Infrared Spectra Handbook of Priority Poll. and Toxic Chem., 160
Infrared and Millimeter Waves, 492
Innovation in Public Transportation, 439
Inorganic Chemistry, 579
Inorganic Chemistry Concepts, 501, 526
Inorganic Chemistry of the Main Group Elements, 501
Inorganic Syntheses, 501
Insect Pests of Farm, Garden and Orchard, 179, 365
Insecticide and Fungicide Handbook for Crop Production, 173
Inside the Animal World, 75
Institution of Civil Engineers Proceedings, 611
Institution of Civil Engineers Yearbook, 444
Institution of Mechanical Engineers Proceedings, 612
Instrument Pilot Handbook, 182

Instrumental Liquid Chromatography, 294
Instruments and Control Systems, 584
Insulation/Circuits, 417
Integer Programming and Related Areas, 32
Integrated Circuits Application Handbook, 206
Integrated Energy Vocabulary, 148
Integrated System of Classification of Flowering Plants, 176
Interavia ABC: World Directory of Aviation and Astronautics, 411
Intergovernmental Oceanographic Commission Technical Series, 595
International Abstracts of Biological Sciences, 546
International Aerospace Abstracts, 556
International Astronomical Union, 607
International Atomic Energy Agency Publications, 599
International Bibliography of Alternative Energy Sources, 56
International Bibliography of Automatic Control, 49
International Bibliography of Directories, 393
International Bio-Energy Directory, 432
International Biotechnology Directory 1984, 406
International Books in Print 1985, 2
International Butterfly Book, 366
International Catalogue of Scientific Literature, 6, 467
International Chemistry Directory, 405
International Citrus Crops Bibliography, 39
International Civil Engineering Abstracts, 558
International Classification of Patents, 632
International Cloud Atlas, 389
International Compendium on Numerical Data Projects, 248
International Conference on Atomic Physics, 608
International Conference on Peaceful Uses of Atomic Energy, 612
International Conference on Recent Advances in Biomedical E., 610

International Congress Calendar, 602
International Congress of Pure and Applied Chemistry, 609
International Copper Information, 563
International Corn Bibliography, 39
International Council of the Aeronautical Sciences, 611
International Countermeasures Handbook, 200
International Critical Tables of Numerical Data, Physics., 234
International Data Series B, 251
International Dictionary of Heating, Ventilating, and Air., 146
International Directory of Agricultural Engineering Institute, 409
International Directory of Botanical Gardens, 407
International Directory of Certified Radioactive Materials, 429
International Directory of Genetic Services, 406
International Directory of Marine Scientists, 459
International Directory of Research and Development Scientists, 451
International Directory of the Nonwoven Fabrics Industry, 413
International Encyclopedia of Aviation, 81
International File of Micrographics Equipment and Accessories, 663
International Foundation Directory, 396
International Frequency List, 324
International Geological Congress Proceedings, 610
International Geophysics Series, 514
International Glossary of Technical Terms for Pulp and., 134
International Guide to Screw Threads, 650
International Handbook of Aerospace Awards and Trophies, 183
International Horseman's Dictionary, 128
International Hydrographic Organization Yearbook, 444
International Index of Laboratory Animals, 551
International Index of Patents, 634

International Index to Film Periodicals, 659
International Index to Multimedia Information, 659
International Journal for Numerical and Analytical Methods., 585
International Journal of Computer Mathematics, 577
International Journal of Engineering Science, 584
International Journal of Environmental Studies, 590
International Journal of Mechanical Sciences, 587
International Journal of Mini and Microcomputers, 586
International Journal of Production Research, 587
International Journal of Robotics Research, 586
International Journal of Solar Energy, 589
International Journal of Thermophysics, 578
International List of Selected, Supplementary and Auxiliary Ships, 269
International Maps and Atlases in Print, 374
International Maritime Dangerous Goods Code, 644
International Maritime Dictionary, 112
International Meeting Reports, 605
International Microcomputer Software Directory, 422
International Micrographics Source Book, 659
International Operations Handbook for Measurements of., 233
International Organizations, 396
International Patents Digest of Foamed Plastics, 635
International Petroleum Abstracts, 566
International Petroleum Encyclopedia, 91
International Petroleum Yearbook, 446
International Physics and Astronomy Directory, 404
International Rayon and Synthetic Fibres Statistical Yearbook, 269
International Research Centers Directory, 396
International Review of Biochemistry, 500
International Review of Connective Tissue Research, 509
International Review of Cytology, 509
International Review of Experimental Pathology, 506
International Review of General and Experimental Zoology, 513
International Review of Neurobiology, 506
International Robotics Yearbook, 444
International Safety Standard Guideline for the Operation., 649
International Serials Catalogue, 4
International Series in Pure and Applied Mathematics, 490
International Shipping and Shipbuilding Directory, 426
International Soybean Bibliography, 39
International Stratigraphic Guide, 643
International Union on Pure and Applied Chemistry Conference, 611
International Who's Who, 451
International Who's Who in Energy and Nuclear Sciences, 432, 463
International Zoo Yearbook, 442
Internationale Bibliographie der Fachadressbuecher, 393
Into the Deep: The History of Man's Underwater Exploration, 481
Introduction and Guide to Marine Bluegreen Algae, 310
Introduction to Canadian Amphibians and Reptiles, 369
Introduction to International System Units with Conversion., 235
Introduction to Isaac Newton's "Principia", 455
Introduction to Physiology, 530
Introduction to US Public Documents, 597
Introduction to the History of Science, 467
Introduction to the History of Virology, 479
Introductory Guide to Information Sources in Physics, 18

Introductory Soil Science, 297
Intuitive IC Electronics, 298
Inventor's Patent Handbook, 632
Inventory of Power Plants in US, 433
Inventory of Published Letters to and from Physicists, 35
Ion Exchange in Water Treatment, 331
Ion Implantations in Microelectronics, 49
Iractical Applications of Data Communications, 325
Iron Castings Handbook, 218
Irregular Serials and Annuals, 489
Isaac Newton's Papers and Letters on Natural Philosophy., 455
Island Waterfowl, 361
Isocyanate Polymers, 632
Isophotometric Atlas of Comets, 377
Isotopes of Water, 45

JANAF Thermochemical Tables, 242
JASA. Journal of the American Statistical Association, 577
JPT. Journal of Petroleum Technology, 589
Jane's Aerospace Dictionary, 133
Jane's All the World's Aircraft, 320, 443
Jane's Armour and Artillery, 444
Jane's Aviation Annual, 443
Jane's Avionics, 443
Jane's Dictionary of Military Terms, 147
Jane's Dictionary of Naval Terms, 147
Jane's Fighting Ships, 426, 447
Jane's Freight Containers, 446
Jane's Military Communications, 447
Jane's Ocean Technology, 426
Jane's Surface Skimmers, 447
Jane's Surface Skimmers: Hovercraft and Hydrofoils, 426
Jane's Urban Transport Systems, 447
Jane's World Railways, 439, 447
Japan Chemical Directory, 1985, 413
Japan EBG Japan Electronics Buyers' Guide, 325
Japan's 100 Leaders in Electric and Electronic Industries, 417
Japanese Mathematics, 32
Japanese Physics, 576
Japanese Scientific and Technical Literature, 15
Jet Pumps And Ejectors, 52

Jewels of the Plains, 353
John Crear Library (Chicago) Catalog, 6
John Crear Library Classified Subject Catalog, 6
Journal of Agricultural and Food Chemistry, 580
Journal of Aircraft, 584
Journal of Applied Nutrition, 582
Journal of Applied Mathematics and Mechanics, 587
Journal of Applied Mechanics, 587
Journal of Applied Physics, 578
Journal of Applied Physics and Applied Physics Letters, 576
Journal of Applied Physiology, 582
Journal of Astronautical Sciences, 584
Journal of Atmospheric Sciences, 583
Journal of Bacteriology, 582
Journal of Biological Chemistry, 581
Journal of Cereal Science, 580
Journal of Chemical Education, 579
Journal of Chemical Information and Computer Sciences, 579
Journal of Chemical Physics, 578
Journal of Chemical Technology and Biotechnology, 585
Journal of Chemical and Engineering Data, 579, 585
Journal of Chromatography Library, 498
Journal of Climate and Applied Meteorology, 583
Journal of Computational Chemistry, 579
Journal of Dairy Science, 580
Journal of Electronic Materials, 586
Journal of Engineering and Applied Science, 584
Journal of Engineering for Industry, 588
Journal of Engineering for Power, 589
Journal of Environment Science and Health, 590
Journal of Environmental Sciences, 590
Journal of Experimental Biology, 580
Journal of Experimental Botany, 581
Journal of Experimental Zoology, 582
Journal of Fluid Mechanics, 588
Journal of General Microbiology, 582
Journal of General Physiology, 582
Journal of Geology, 583
Journal of Geophysical Research, 583

Journal of Heat Transfer, 588
Journal of Heredity, 581
Journal of Lubrication Technology, 588
Journal of Marine Research, 583
Journal of Mathematical Physics, 578
Journal of Microbiological Methods, 580
Journal of Molecular Biology, 581
Journal of Morphology, 582
Journal of Nuclear Materials, 588
Journal of Nuclear Science and Technology, 588
Journal of Nutrition, 582
Journal of Nutrition Education, 582
Journal of Organic Chemistry, 579
Journal of Petrology, 583
Journal of Physical Chemistry, 579
Journal of Physical Oceanography, 583
Journal of Physics, 578
Journal of Physiology, 582
Journal of Research, 601
Journal of Robotic Systems, 586
Journal of Spacecraft and Rockets, 584
Journal of Structural Geology, 583
Journal of Structural Mechanics, 588
Journal of Systems and Software, 586
Journal of Telecommunications Networks, 586
Journal of Theoretical Biology, 580
Journal of VLSI and Computer Systems, 586
Jungles, 313

KIRK-OTHMER ENCYCLOPEDIA OF CHEMICAL TECHNOLOGY, 703
Kaiman's Encyclopedia of Energy Topics, 91
Kempe's Engineer's Yearbook, 442
Kepler, 470
Key Abstracts, 559
Key Definitions in Astronomy, 118
Key Works to Fauna and Flora of British Isles and NW Europe, 38
Kingzett's Chemical Encyclopedia, 67
Kirk-Othmer Concise Encyclopedia of Chemical Technology, 82
Kirk-Othmer Encyclopedia of Chemical Technology, 82
Knitting Times Yearbook, 442
Knott's Handbook for Vegetable Growers, 176

Know Your Woods, 323

LC Science Tracer Bullet, 15
LIBCON/E, 698
LIBCON/F, 698
LIBCON/S, 698
LIFE SCIENCES COLLECTION, 698
LISA (LIBRARY AND INFORMATION SCIENCE ABSTRACTS), 699
Laboratory Animal Science, 551
Laboratory Exercises in Microbiology, 295
Laboratory Experiments for Microprocessor Systems, 298
Laboratory Handbook of Chromatographic and Allied Methods, 168
Laboratory Handbook of Paper and Thin-Layer Chromatography, 169
Laboratory Manual and Study Guide for Anatomy and Physiology, 296
Laboratory Manual for Fundamentals of Chemistry, 294
Laboratory Manual for General Botany, 296
Laboratory Manual for Photographic Science, 272
Laboratory Manual for Physics in the Modern World, 293
Laboratory Manual for Schools and Colleges, 293
Laboratory Manual in Food Chemistry, 298
Laboratory Manual of Cell Biology, 296
Laboratory Manual of General Ecology, 296
Laboratory Manual of Physical Chemistry, 294
Laboratory Manual on Use of Radiotracer Techniques, 298
Laboratory Techniques in Biochemistry and Molecular Biology, 509
Laboratory Text for Organic Chemistry, 286
Laboratory Training Manual on Use of Nuclear Techniques., 296, 298
Laboratory Waste Disposal Manual, 298
Laboratory and Field Manual of Ecology, 284
Lagoon Information Source Book, 291

Land Drilling and Oilwell Servicing Contractors Directory, 433
Land Use Planning Abstracts, 567
Landmark Experiments in Twentieth-Century Physics, 293, 474
Landmarks in Science, 451
Landolf-Bornstein Numerical Data and Functional Relationships, 247
Lange's Handbook of Chemistry, 169
Larousse Dictionary of Wines of the World, 136
Larousse Encyclopedia of the Animal World, 75
Larousse Guide to Astronomy, 342
Larousse Guide to Birds of Britain and Europe, 361
Larousse Guide to Horses and Ponies of World, 367
Larousse Guide to Minerals, Rocks and Fossils, 371
Laser Abstracts, 542
Laser Crystals, 35
Laser Experimenter's Handbook, 160
Laser Handbook, 160
Laser Safety Handbook, 160
Laser and Particle Beams, 578
Lasers, 632
Latin American Petroleum Directory, 433
Law and Science, 30
Leaders of American Conservation, 464
Leather Technical Dictionary, 135
Lecture Notes in Biomathematics, 490
Lecture Notes in Computer Science, 517
Lecture Notes in Mathematics, 490
Lecture Notes in Physics, 492
Lecture Notes in Pure and Applied Mathematics, 490
Legal Care for Your Software, 328
Lexicon of Terms Relating to the Assessment and., 150
Libraries, Information Centers and Databases in Science., 396
Library Index to Military Periodicals, 564
Library of Congress Catalog, 659
Light, Life and Chemical Change, 286
Lighting Handbook, 200
Limnology and Oceanography, 583
Linear Integrated Circuit Data Book, 258

Lineman's and Cableman's Handbook, 200
Liquid Fertilizers, 635
Liquified Natural Gas, 56
List of Available Publications of US Dept. of Agriculture, 7
List of Bureau of Mines Publications and Articles, 600
List of Journals Indexed by the National Agricultural Library, 575
List of Radio and Radar Astronomy Observatories, 403
List of Serial Publications in British Museum of Natural History, 573
List of Serials with Coden, Title Abbreviations, New., 575
Lists of International Biological Standards, 645
Literature Guide to the GLC of Body Fluids, 36
Literature Survey of Comm. Satellite Systems and Technology, 50
Literature in Digital Signal Processing, 49
Literature of Matrix Chemistry, 19
Literature of the Life Sciences, 19
Literature on the History of Physics in the 20th Century, 474
Lloyd's Register of American Yachts, 426
Lloyd's Register of Shipping Statistical Tables, 269
Lloyd's Register of Shipping-list of Shipowners, 426
Locating US Government Information, 598
Location Key to Foreign Language Dictionaries, 100
London Mathematical Society Lecture Note Series, 490
London Mathematical Society Monographs, 522
Longman Dictionary of Scientific Usage, 115
Longman Illustrated Science Dictionary, 115
Lotus, 655
Lyons' Encyclopedia of Valves, 89

MATHFILE, 699
MDF/I, 699

MEDLARS, 712
MEDLINE, 699, 712
MENTAL HEALTH ABSTRACTS, 700
MENU-THE INTERNATIONAL SOFTWARE DATABASE, 700
METADEX, 700
METEOROLOGICAL AND GEOASTROPHYSICAL, 700
MICROCOMPUTER INDEX, 701
MICROSEARCH, 701
MIT Abstracts of Theses Accepted in Partial Fulfillment, 617
MIT Wavelength Tables, 239
MOSFET Technologies, 50
MacIntosh: A Concise Guide to Applications Software, 422
Machine Tools, 534
Machine and Tool Directory, 427
Machinery Noise, 335
Machinery's Handbook, 218
Macmillan Book of Natural Herbs, 274
Macmillan Encyclopedia of Architects, 87
Macmillan Encyclopedia of Architecture and Technical Change, 87
Macmillan Illustrated Animal Encyclopedia, 75
Macromolecular Chemistry, 498
Magazine Industry Market Place, 654
Magazine of Concrete Research, 585
Magnetism Diagrams for Transition Metal Ions, 260
Mainly on Patents, 630
Maintenance Engineering Handbook, 210
Major Activities in the Atomic Energy Program, 429
Makers of Mathematics, 454
Mammal Species of the World, 179
Mammals of North America, 367
Mammals of the Northern Great Plains, 367
Mammals of the World, 368
Mammals—Their Latin Names Explained, 368
Man and the Environment Information Guide Series, 27
Manpower Development for Nuclear Power, 336
Manual for Environmental Impact Evaluation, 284

Manual for Maintenance Inspection of Bridges, 279
Manual for Operational Amplifier Users, 305
Manual for the Organization of Scientific Congresses, 272
Manual of Analytical Quality Control for Pesticides and., 284
Manual of Applied Geology for Engineers, 275
Manual of Classification of Patents, 630
Manual of Clinical Biology, 274
Manual of Clinical Laboratory Methods, 274
Manual of Clinical Microbiology, 274
Manual of Concrete Practice, 279
Manual of Economic Analysis of Chemical Processes, 277
Manual of Energy Saving in Existing Buildings and Plants, 279
Manual of Food Quality Control, 277
Manual of Geology for Civil Engineers, 279
Manual of Grey Water Treatment Practice, 284
Manual of Highway Road Materials and Design, 279
Manual of Mammalogy, 275
Manual of Methods for Fish Stock Assessment, 275
Manual of Methods for Fisheries Resource Survey and Appraisal, 275
Manual of Mineralogy, 276, 283
Manual of Neotropical Birds, 361
Manual of New Mineral Names, 283
Manual of Patent Examining Procedures, 630
Manual of Precast Concrete Construction, 279
Manual of Soil Laboratory Testing, 279
Manual of Steel Construction, 283
Manual of Symbols and Terminology for Physiochemical Quan., 272
Manual of Underwater Photography, 273
Manual of Water Utility Operations, 279
Manual on Decontamination of Surfaces, 284
Manual on Oil Production, 284
Manual on Urban Air Quality Management, 285

Manual on the Global Data-Processing System, 276
Map Collections in the US and Canada, 374
Maps and Geological Publications of US, 374
Maps on File, 374
Marconi Instruments Contact, 655
Marine Bio Lab and Woods Hole Ocean Institute Library Catalog, 10
Marine Biology, 583
Marine Chemistry, 583
Marine Directory, 426
Marine Ecology, 530
Marine Engineering/Log, 447
Marine Engineering/Shipbuilding Abstracts, 562
Marine Fisheries Abstracts, 551
Marine Geology, 583
Marine Life, 76
Marine Research, 624
Marine Science Content Tables, 571
Marine Science Instrumentation, 535
Marine Technology Society, 583
Marine Technology Society Annual Meeting, 610
Mariner's Annual Ordering Guide, 332
Marks' Standard Handbook for Mechanical Engineers, 215
Marshall Cavendish Illustrated Encyclopedia of Gardening, 70
Masers and Lasers, 474
Mass Spectrometry of Priority Pollutants, 262
Master Abstracts, 617
Master Handbook of Acoustics, 161
Master Handbook of Electronic Tables and Formulas, 239
Master Handbook of IC Circuits, 206
Master Handbook of Microprocessor Chips, 207
Master Index to Materials and Properties, 562
Master List of Nonstellar Optical Astronomical Objects, 235
Master Tables for Electromagnetic Depth Sounding Interpretations, 239
Masters Theses in Pure and Applied Sciences Accepted by., 617
Materials Engineering, 584

Materials Handbook, 214
Materials Handling Handbook, 214
Materials Research Symposium, 612
Materials Science Research, 517
Materials and Technology, 90
Mathematical Methods in the Applied Sciences, 577
Mathematical Papers of Isaac Newton, 471
Mathematical Reviews, 541
Mathematical Reviews Annual Index, 541
Mathematical Sciences Professional Directory, 1984, 403
Mathematical Tables, 237
Mathematical Tables Project, 237
Mathematics Computation, 237
Mathematics Dictionary, 119
Mathematics Encyclopedia, 65
Mathematics Into Type, 273
Mathematics Magazine, 578
Mathematics for Technical Occupations, 293
Mathematics in Science and Engineering, 522
Mathematics of Computation, 578
McClane's Field Guide to Saltwater Fishes of North America, 363
McGraw-Hill Computer Handbook, 207
McGraw-Hill Concise Encyclopedia of Science and Technology, 62
McGraw-Hill Dictionary of Chemistry, 123
McGraw-Hill Dictionary of Earth Sciences, 132
McGraw-Hill Dictionary of Electronics and Computer Technology, 142
McGraw-Hill Dictionary of Engineering, 133
McGraw-Hill Dictionary of Physics and Mathematics, 121
McGraw-Hill Dictionary of Science and Engineering, 116
McGraw-Hill Dictionary of Scientific and Technical Terms, 116
McGraw-Hill Dictionary of the Life Sciences, 125
McGraw-Hill Dictionary of Astronomy, 64

McGraw-Hill Encyclopedia of Chemistry, 67
McGraw-Hill Encyclopedia of Electronics and Computers, 89
McGraw-Hill Encyclopedia of Energy, 91
McGraw-Hill Encyclopedia of Engineering, 80
McGraw-Hill Encyclopedia of Environmental Science, 92
McGraw-Hill Encyclopedia of Food, Agriculture and Nutrition, 84
McGraw-Hill Encyclopedia of Ocean and Atmospheric Sciences, 79
McGraw-Hill Encyclopedia of Physics, 66
McGraw-Hill Encyclopedia of Science and Technology, 62
McGraw-Hill Encyclopedia of Scientific and Technical Terms, 62
McGraw-Hill Encyclopedia of World Biography, 451
McGraw-Hill Encyclopedia of the Geological Sciences, 79
McGraw-Hill Modern Scientists and Engineers, 452
McGraw-Hill Yearbook of Science and Technology, 440
McGraw-Hill's Compilation of Data Communications Standards, 649
McGraw-Hill's Leaders in Electronics, 462
McGraw-Hill's National Electrical Code Handbook, 200
Mechanic's Guide to Electronic Emission Control and Tune-Up, 335
Mechanical Engineer's Catalog and Product Directory, 427
Mechanical Engineer's Handbook, 215
Mechanical Engineer's Reference Book, 283
Mechanical Engineer's Reference Tables, 245
Mechanical Engineering, 487, 588, 603
Mechanical Engineering for Professional Engineers' Examination, 335
Mechanical Engineers in America Born Prior to 1861, 463
Mechanical Fastening of Plastics, 333
Mechanical Standards, 650
Mechanical Technician's Handbook, 218
Mechanical World Yearbook, 445
Mechanics of Materials, 588
Mechanisms and Dynamics of Machinery, 259
Media Review Digest, 659
Medical Physics Handbook, 161
Medicinal Plants of North Africa, 176
Medicines from the Earth, 349
Meetings on Atomic Energy, 603
Membrane Proteins, 296
Men of Mathematics, 454
Men of Space, 461
Merchant Vessels of US, 426
Merck Index, 68
Messier Catalogue, 342
Metal Abstracts, 564
Metal Bulletin Handbook, 221
Metal Finishing, 428
Metal Finishing Guidebook-Directory, 445
Metal Progress Databook, 260
Metalcutting, 289
Metals Handbook, 221
Metals Joining Manual, 283
Metals Reference Book, 260
Metalworking Directory, 428
Meteorites, Classification and Properties, 318
Meteorological and Geoastrophysical Abstracts, 554
Methods in Cell Biology, 509
Methods in Computational Physics, 492
Methods in Enzymology, 510
Methods in Membrane Biology, 506
Methods in Microbiology, 512
Methods of Biochemistry Analysis, 500
Methods of Enzymatic Analysis, 296
Methods of Experimental Physics, 492, 523
Methods of Modern Mathematical Physics, 523
Methods of Surface Analysis, 493
Metric Architectural Drawing, 280
Metric Guide to Mechanical Design and Drafting, 335
Metric Manual, 273
Metric Standards for Engineering, 641
Metric System Guide, 273
Metric System and Metric Conversion, 235
Metrication Handbook, 642

Micro Software Today, 587
MicroUse Directory: Software, 327, 422
Microbial Toxins, 530
Microbiology, 512
Microbiology Abstracts, 546
Microcomputer Applications, 587
Microcomputer Buyer's Guide, 326
Microcomputer Dictionary, 142
Microcomputer Directory: Applications Educational Settings, 422
Microcomputer Marketplace, 419
Microcomputer Review, 587
Microcomputer Software Buyer's Guide, 328
Microcomputer Users' Handbook, 207, 303
Microcomputers in Education, 207
Microcomputing Periodicals, 573
Microelectronic Engineering, 586
Microelectronic Packaging, 50
Microelectronics Dictionary, 104
Microfilm Review, 659
Microfilm Source Book, 659
Microforms, 659
Microprocessor Applications Handbook, 207
Microprocessor Applications in Science and Medicine, 51
Microprocessor Data Book, 259
Microprocessor Handbook, 207
Microprocessor-Microcomputer Technology, 281
Microprogrammer's Market 1985, 422
Microtomist's Formulary and Guide, 310
Microwave Journal, 655
Military Aircraft of the World, 320
Military Science Index, 564
Military Small Arms of the 20th Century, 94
Mineral Atlas of the Pacific Northwest, 387
Mineral Exploration, Mining and Processing Patents, 636
Mineral Facts and Problems, 600
Mineral Names: What Do They Mean? 132
Mineral Processing Technology, 283
Mineral Tables, 244
Mineralogical Abstracts, 564
Mineralogical Record, 583

Minerals Handbook, 181
Minerals Yearbook, 445, 600
Minerals and Gemstones, 371
Minerals and Rocks, 372
Minerals, Lands and Geology for Common Defence and., 481
Minerals, Rocks, and Fossils, 372
Mini- and Microcomputer Control in Industrial Processes, 207
Minicomputers, 422
Mining Chemicals Handbook, 222
Model State Packaging and Labelling Regulations, 642
Modern American Herbal, 349
Modern Chemical Engineering, 521
Modern Dictionary of Electronics, 138
Modern Electronics Circuits Reference Manual, 281
Modern Encyclopedia of Wine, 84
Modern Military Dictionary, 147
Modern Oscilloscope Handbook, 161
Modern Plastics Encyclopedia, 413
Modified Nucleosides of Transfer RNA, 37
Molecular Structures and Dimensions, 38
Molecular and Crystal Structure Models, 308
Molecules, Measurements, Meanings, 296
Monographic Series, 4
Monthly Catalog, 613
Monthly Catalog of US Government Publications, 598
Monthly Checklist of State Publications, 598
Monthly Climatic Data for World, 266
Monthly Review of Technical Literature, 571
Monthly Weather Review, 521
Mossbauer Effect Data Index, 249
Motor Trade Handbook, 218
Motor Truck Facts, 269
Motor Vehicle Emissions, 61
Mount St. Helens, 45
Moving the Earth, 298
Multicomponent Alloy Construction Bibliography 1955–1973, 52
Multilingual Compendium of Plant Diseases, 113
Multilingual Computer Dictionary, 113
Multilingual Dictionary of Concrete, 113

Multilingual Energy Dictionary, 113
Municipal Government Reference Sources, 598
Mushroom Trailguide, 349
Mushrooms and Toadstools, 349
Mycology Guidebook, 349
Mycoplasmas, 530
Mycotoxic Fungi, Mycotoxins, Mycotoxicoses, 176
Mysterious Universe, 154

NALCO Water Treatment Handbook, 231
NASA Continuing Bibliography Series, 47
NASA Patents Abstracts Bibliography, 636
NASA Specifications and Standards, 646
NASA Technical Translations, 615
NASA Thesaurus, 99
NATIONAL FOUNDATIONS, 701
NATO Advanced Study Institutes Series, 605
NATO Conference Series, 605
NFPA Handbook of the National Electrical Code, 200
NICSEM/MIMIS, 701
NLL Announcement Bulletin, 2
NMAC Catalog, 665
NONFERROUS METAL ABSTRACTS, 701
NSB Standard Reference Materials Catalog, 640
NSF Factbook, 246
NTIS, 701
NTIS: Aeronautics, Aerospace, and Astronomy, 712
NTIS: Computers, Communication, and Electronics, 712
NTIS: Medicine, Health Care, and Biology, 712
NUC/CODES, 702
NURSING AND ALLIED HEALTH (CINAHL), 702
Nachweise von Vebersetzungen, 613
National Air Monitoring Program, 266
National Atlas of US, 601
National Biomedial Research Directory, 623
National Bureau of Standards, 638
National Construction Estimator, 269
National Directory of Manufacturers' Representatives, 427
National Electrical Code, 648
National Electrical Code Handbook, 648
National Electrical Code Reference Book, 648
National Electrical Code and Blueprint Reading, 648
National Electrical Safety Code, 417, 648
National Faculty Directory, 401
National Fire Codes, 645
National Fire Protection Association, Catalog, Publications, 673
National Geographic Book of Mammals, 384
National Geographic Picture Atlas of Our Universe, 376
National Highway Safety Bureau Corporate Author Authority List, 601
National Highway Saftety Bureau Subject Category List, 601
National Historic Mechanical Engineering Landmarks, 487
National Patterns of R & D Resources, 620
National Physical Laboratory-A History, 474
National Science Policy and Organization of Research in., 398
National Service Data-Advance, 259
National Solar Energy Education Directory, 401
National Trade and Professional Associations of US and Canada, 654
National Trade and Professional Associations of the US, 668
National Transportation Statistics, 266
National and International Standardization Radiation Dosimetry, 652
Nationwide Survey of Resource Recovery Activities, 438
Natural Environment, 59
Natural Resource Commodities, 269
Natural Science Books in English, 479
Natural Vegetation of North America, 349
Natural Wonders of the World, 387
Naturalist's Color Guide, 311

Naturalists, 458
Naturalists' Directory International, 407, 458
Nature, 576, 603
Nature of Light, 474
Nature/Science Annual, 440
Nautical Almanac, 247
Naval Terms Dictionary, 147
Navigation Afloat, 285
Neblette's Handbook of Photography and Reprography, 161
Neils Bohr, Collected Works, 474
Neutron Activation Tables, 239
Neutron Nuclear Data Evaluation, 260
Neutron Standard Reference Data, 260
Never at Rest: A Biography of Isaac Newton, 455
New Car Dealers, 440
New Civil Engineer, 585
New Concise British Flora, 176
New Enclyclopedia of Science, 63
New Field Book of Reptiles and Amphibians, 369
New Generation Computing, 587
New Guide to Popular Government Publications for Libraries., 598
New Guide to the Planets, 342
New Guide to the Stars, 343
New International Dictionary of Refrigeration, 113
New Larousse Encyclopedia of Animal Life, 76
New Larousse Gastronomique, 85
New Literature on Automation, 561
New Penguin Dictionary of Electronics, 138
New Research Centers, 396
New Scientist, 576
New Table of Indefinite Integrals, Computer Processed, 245
New Technical Books, 2
New York Botanical Garden Illustrated Encyclopedia of Horticulture, 71
New York Times Atlas of the World, 387
New York Times Book of Annuals and Perennials, 351
New York Times Book of Vegetable Gardening, 349
Newnes Radio and Electronics Engineer's Pocket Book, 202

Newton and Newtoniana, 456
No-Nonsense Guide to Food and Nutrition, 315
Nobel Lectures in Molecular Biology, 1933-1975, 459
Nobel Lectures, Including Presentation Speeches and, 452
Nobel Prize Winners in Chemistry, 457
Nobel Symposia, 605
Noise Control, 231
Noise Control for Engineers, 341
Noise Evaluation Tables in Digitalized Form, 239
Noise Pollution, 27
Nomenclature of Organic Chemistry, 99
Nomina Anatomica Avium: Annotated Dictionary of Birds, 129
Non-Book Materials in Libraries, 660
Non-Ferrous Metal Data, 269, 445
Nonwoven Fabric Technology, 636
North America Online Directory, 677
North American Forest History, 19
North American Horticulture, 668
North American Online Directory, 419
North American Range Plants, 313
North American Trees (Exclusive of Mexico and Tropical US), 177
North-Holland Mathematics Studies, 522
North-Holland Series in Applied Mathematics and Mechanics, 522
Northeast and Great Lakes Wind Atlas, 389
Northwestern University Transportation Center Catalog, 11
Notes on the History of Nutrition Research, 479
Nuclear Data in Science and Technology, 260
Nuclear Energy, 588
Nuclear Engineering International, 588
Nuclear Engineering and Design, 588
Nuclear Fusion, 589
Nuclear News, 588, 603
Nuclear Physics, 578
Nuclear Power Debate, 25
Nuclear Power Issue, 429
Nuclear Power Stations, 429
Nuclear Proliferation Factbook, 261
Nuclear Reactors Built, Being Built, or Planned in US, 429

Nuclear Research Index, 429
Nuclear Research Report, 625
Nuclear Resources, 270
Nuclear Safeguards Technology 1978, 652
Nuclear Safety, 588
Nuclear Science Abstracts, 565
Nuclear Science and Engineering, 588
Nuclear Science and Technology, 534
Nuclear Standards News, 652
Nuclear Tables, 240
Nuclear Technology, 588
Nuclear Technology/Fusion, 588
Nuclear Waste Management Abstracts, 567
Nuclear/IEEE Standards, 652
Numeric Databases, 677
Nutrition Abstracts and Reviews, 546
Nutrition Almanac, 315
Nutrition News, 582
Nutrition Reviews, 582
Nutritional Quality Index of Foods, 254
Nutritive Value of American Foods, 254

OCCUPATIONAL SAFETY AND HEALTH (NIOSH), 702
OCEANIC ABSTRACTS, 702
ONLINE CHRONICLE, 703
ONTAP COMPENDEX, 702
OP AMP Network Design Manual, 303
OPC Preprint Index, 628
OPD Chemical Buyers Directory, 405
Observatories of the World, 403
Observatory, 577
Observer's Book of Aircraft, 183
Observer's Handbook, 154
Ocean Almanac, 255
Ocean Data Resources, 255
Ocean Engineering, 584
Ocean Engineering Information Series, 46
Ocean Engineering and Oceanography Technical Literature Collection, 10
Ocean Industry, 584
Ocean Research Index, 554, 624
Ocean Research Institute Collected Reprints, 629
Ocean World Encyclopedia, 80
Ocean World of Jaques Cousteau, 76
Ocean Yearbook, 447

Ocean and Marine Dictionary, 132
Oceanic Abstracts, 554
Oceanic Patents, 635
Oceanographic Atlas of the Bering Sea Basin, 390
Oceanographic Index, 555
Oceanography Information Resources 70, 426
Oceanography and Marine Biology, 514
Oceans, 390
Oceans of the World, 46
Ocran's Acronyms, 132
Office Automation and Word Processing Buyer's Guide, 423
Official Directory of Data Processing, 419
Official Directory of Industrial and Commercial Traffic Exec., 462
Official Gazette of US Patent and Trademark Office, 633
Official Methods of Analysis, 169
Offshore Abstracts, 555
Offshore Contractors and Equipment Directory, 433
Offshore Oil and Gas, 21
Offshore Oil and Gas Yearbook, 446
Offshore Petroleum Engineering, 22, 56
Oil Terms, 132
Oil and Gas Resources, 261
Oil: A Plain Man's Guide to the World Energy Crisis, 339
Oilfields of the World, 339, 433
Old Farm Tools and Machinery, 480
Omni Online Database Directory, 677
On the History of Statistics and Probability, 471
On-Line Process Analyzers, 169
Online Bibliographic Databases, 678
Online Database Search Services Directory, 678
Online International Command Chart, 678
Online Micro-Software Guide and Directory, 423
Online Searching, 678
Operating Handbook of Mineral Processing, 222
Operational Limits and Conditions for Nuclear Power Plants, 337
Operations Research, 587
Operations Research Handbook, 157

Optical Industry and Systems Directory, 404
Optical Society of America, 578
Opto-Electronics/Fiber Optics Applications Manual, 281
Orbit, 679
Orbit User Manual, 679
Orbital Remote Sensing of Coastal and Offshore Environments, 285
Orders and Families of Recent Mammals of the World, 254
Organic Chemistry, 503
Organic Electronic Spectral Data, 252
Organic Nomenclature: A Programmed Study Guide, 99
Organic Photochemistry, 503
Organic Reaction Mechanisms, 503
Organic Reactions, 503
Organic Synthesis, 503
Organometallic Chemistry, 503
Organophosphorus Chemistry, 504
Origins in Acoustics, 475
Our Magnificent Earth, 387
Outliers in Statistical Data, 238
Outline of Plant Classification, 350
Owls of Britain and Europe, 361
Oxford Book of Insects, 366
Oxford Companion to Ships and the Sea, 93
Oxford Encyclopedia of Trees of the World, 71
Oxide Handbook, 169
Ozone Chemistry and Technology, 49

P/E NEWS, 704
PACKAGING SCIENCE AND TECHNOLOGY ABSTRACTS, 703
PAPERCHEM, 703
PATLAW, 703
PC Clearinghouse Software Directory, 423
PC News Watch, 571
PC Telemart/VANLOVES IBM Software Directory, 423
PCI Design Handbook, 192
PCjr Data File Programming, 303
PESTDOC/PESTDOCII/PESTDOC UDB, 704
PHARMACEUTICAL NEWS INDEX, 704
PIE, 704
PIE: Publications Indexed for Engineering, 555, 575
PLASDOC-Plastics and Polymers Patents Documentation, 655
POLLUTION, 704
POWER, 704
PSYCALERT, 705
PSYCINFO, 705
PTS DOMESTIC STATISTICS, 705
PTS EIS PLANTS, 705
PTS F&S INDEXES (FUNK & SCOTT), 705
PTS INTERNATIONAL STATISTICS, 705
PTS MARKET ABSTRACTS (CMA & EMA), 705
PTS MARKET DEFENSE AND TECHNOLOGY, 705
PTS US FORECASTS, 706
PTS US TIME SERIES, 706
Pac-Finder System 34/36, 424
Pacific Aerospace Index, 557
Package Engineering, 655
Paint Handbook, 210
Paint/Coatings Dictionary, 135
Paints and Coatings Handbook, 210
Paleobiology of the Invertebrates, 319
Palomar Observatory Sky Atlas, 377
Pandex Current Index to Scientific and Technical Literature, 539
Paperbound Books in Print, 2
Papermaking Fibers, 391
Parent Compound Handbook, 169
Parent/Teacher's Microcomputing Sourcebook for Children, 289
Particle Atlas, 378
Passive Solar Energy Book, 292, 339
Patent Abstract Series No. 5, 636
Patent Abstracts, 636
Patent Yourself!, 632
Patenting in the Biological Sciences, 635
Patents for Chemical Inventions, 635
Patty's Industrial Hygiene and Toxicology, 210
Penguin Book of Tables, 237
Penguin Dictionary of Biology, 125
Penguin Dictionary of Geology, 132
Penguin Dictionary of Microprocessors, 142

Penguin Dictionary of Physics, 121
Penguin Dictionary of Science, 116
Periodical Title Abbreviations, 575
Permanent Magnet Design and Application Handbook, 214
Permanent Magnets, 308
Personal Computer Book, 326
Personal Computer Buyers Guide, 327
Personal Computer Glossary, 143
Personal Computers A-Z, 143
Personal Computing, 587
Perspectives in Mathematical Logic, 522
Perspectives in Structural Chemistry, 526
Pesticide Book, 83, 252
Pesticide Handbook-Entoma, 185
Pesticide Manual, 278
Pesticides Guide, 321
Pesticides Process Encyclopedia, 83
Petroleum Abstracts, 566
Petroleum Dictionary, 133
Petroleum Training Directory, 401
Phase Diagrams, 249, 533
Philips Technical Review, 655
Photochemistry, 498
Photographic Techniques in Scientific Research, 523
Photosynthesis Bibliography, 40
Physical Acoustics, 493
Physical Chemistry, 526
Physical Properties, 252
Physical Review, 578
Physical Review Abstracts, 542
Physical Review Letters, 579
Physical Review and Physical Review Letters, 576
Physical Reviews, 521
Physical Sciences Data, 493
Physical Sciences Dictionary, 116
Physical Techniques in Biological Research, 530
Physical and Chemical Properties of Water, 36
Physical and Mathematical Tables, 237
Physicists, 475
Physics Abstracts, 543
Physics Demonstration Experiments, 293
Physics Laboratory Manual, 294
Physics Nobel Lectures Including Presentation Speeches and., 456
Physics Today, 579, 603

Physics Today Buyers Guide, 404
Physics and Chemistry in Space, 493
Physics for Engineers and Scientists, 308
Physics of Quantum Electronics, 493
Physiological Ecology, 535
Physiological Reviews, 582
Physiological Zoology, 582
Phytoplankton Manual, 275
Pictorial Guide to Fossils, 319
Pictorial Guide to the Moon, 343
Pictorial Guide to the Planets, 343
Pictured-key Nature Series, 506
Pilot Study on the Use of Scientific Literature by Scientist, 15
Pioneers of Science, 456
Pioneers of Science: Nobel Prize Winners in Physics, 456
Planet Guidebook, 343
Planet Jupiter, 155
Planet We Live On, 79
Planets and Planetariums, 470
Plant Disease, 530
Plant Engineer's Handbook of Formulas, Charts, and Tables, 245
Plant Engineering, 587
Plant Engineers and Managers' Guide to Energy Conservation, 339
Plant Management and Engineering, 587
Plant Pathology, 531
Plant Physiology, 531, 581
Plant Propagation for the Amateur Gardener, 313
Plant Science Catalog, 8
Plasticizers, 321
Plastics & Rubber: World Sources of Information, 22
Plastics Book List, 49
Plastics Engineering Handbook, 188
Plastics Materials Guide, 333
Plastics World: Directory of Plastics Industry, 413
Plumbing Dictionary, 146
Pocket Encyclopedia of Cacti and Succulents in Color, 71
Pocket Encyclopedia of Modern Roses, 71
Pocket Guide to Chemical Hazards, 341
Pocket Guide to Indoor Plants, 352
Pocket Guide to Spirits and Liquors, 315
Pocket Guidebook to Astronomy, 343

Poisonous and Venomous Marine Animals of World, 363
Pollution Abstracts, 568
Pollution Control in the Petroleum Industry, 637
Pollution Research Index, 627
Pollution Technology Review, 520
Polymer Additives, 321
Polymer Handbook, 188
Polymer Melt Rheology, 321
Polymer Preprints, 628
Polymer Symposia, 609
Popular Circuits Ready-Reference, 325
Popular Encyclopedia of Plants, 71
Popular Marine Fish for Your Aquarium, 363
Popular Tropical Fish for Your Aquarium, 364
Port Development, 194
Ports of the World, 426
Post's Pulp & Paper Directory, 413
Poultry Processing, 637
Powder Diffraction File: Inorganic, 252
Powder Diffraction File: Organic, 252
Power, 589
Power Electronics, 305
Power Engineering, 589
Power Reactors in Member States, 430
Practical Electrical Wiring, 200, 302
Practical Guide to Computer Communications and Networking, 330
Practical Guide to Computer Methods for Engineers, 330
Practical Guide to Molecular Cloning, 311
Practical Instrumentation Handbook, 161
Practical Introduction to Computer Graphics, 330
Practical Inventor's Handbook, 182
Practical Invertebrate Zoology, 296
Practical Oscilloscope Handbook, 161
Practical PASCAL for Microcomputers, 303
Practical Protein Biochemistry, 170
Practical Video, 660
Practicing Scientist's Handbook, 152
Preliminary Directory, 407
Preparation of Nuclear Targets, 53
Pressure Vessel Handbook, 218
Previews, 660

Primates of the World, 42
Princeton Guide to Microforms, 660
Princeton Telephone Guide to Microforms, 660
Principles of Home Inspection, 323
Printed Books, 1481–1900, in the Horticultural Society of NY, 8
Printed Circuits Handbook, 201
Printing Presses, 467
Probabilistic Analysis and Related Topics, 490
Proceedings (Trudy) of the P. N. Lebedey Physics Institute, 615
Proceedings in Print, 605
Proceedings of Annual Conference on Application X-ray Analysis, 608
Proceedings of IEEE Annual Conference on Nuclear and Space., 612
Proceedings of National Conference on Technology for Energy., 612
Proceedings of the International Conference on Data Bases, 611
Proceedings of the International School of Physics, 608
Proceedings of the Offshore Technology Conference, 611
Product Safety and Liability, 210
Product Standards Index, 646
Professional Engineer, 584
Professional Engineers Income and Salary Survey, 401
Professional Handbook of Building Construction, 192
Professional Income of Engineers, 401
Professionals in Chemistry, 457
Program Design and Construction, 329
Programmer's ANSI COBOL Reference Manual, 281
Programmer's Markey, 1984, 424
Progress in Aerospace Sciences, 514
Progress in Analytical Chemistry, 498
Progress in Biomass Conversion, 519
Progress in Bioorganic Chemistry, 500
Progress in Biophysics and Molecular Biology, 495
Progress in Construction Science and Technology, 516
Progress in Control Engineering, 517
Progress in Cybernetics and Systems Research, 517

Progress in Dielectrics, 517
Progress in Electrochemistry of Organic Compounds, 504
Progress in Heat and Mass Transfer, 493
Progress in High Temperature Physics and Chemistry, 523
Progress in Industrial Microbiology, 512
Progress in Inorganic Chemistry, 501
Progress in Low Temperature Physics, 493
Progress in Materials Science, 517
Progress in Mathematics, 490
Progress in Molecular and Subcellular Biology, 510
Progress in Neurobiology, 506
Progress in Nuclear Energy, 518
Progress in Nuclear Magnetics Resource Spectroscopy, 493
Progress in Nuclear Science, 495
Progress in Nucleic Acid Research and Molecular Biology, 507
Progress in Oceanography, 514
Progress in Optics, 493
Progress in Organic Chemistry, 504
Progress in Pesticide Biochemistry, 500
Progress in Physical Organic Chemistry, 504
Progress in Polymer Science, 504
Progress in Quantum Electronics, 493
Progress in Solid State Chemistry, 498
Progress in Surface Science, 493
Progress in Surface and Membrane Science, 507
Progress in Theoretical Biology, 507
Progress in Thin Layer Chromatography and Related Methods, 498
Progress in Total Synthesis, 504
Progress in Toxicology, 507
Progress in the Chemistry of Fats and Other Lipids, 500
Projected Pulp and Paper Mills in the World 1979–1989, 270
Projects in the Industrial Pollution Control Program, 627
Prominent Scientists, 452
Protection System and Related Features in Nuclear Power Pt., 337
Proteins, 507
Public Access to Government Information, 598

Public Regulation of Site Selection for Nuclear Power Plants, 54
Publication of the NBS, 641
Publications Catalog of the Institution of Electrical Engineers, 673
Publications of the Geological Survey, 601
Publications of the National Bureau of Standards, 598
Publications of the World Meteorological Organization, 600
Publications on Toxic Substances, 600
Pugh's Dictionary of Acronyms and Abbreviations, 96
Pump Handbook, 218
Pure and Applied Mathematics, 522
Pure and Applied Physics, 523
Pure and Applied Science Books: 1876–1982, 30, 2

Quality Assurance in Manufacture of Items for Nuclear Power., 337
Quality Technology Handbook, 211
Quantitative Chemical Analysis, 294
Quantities and Units of Measurement, 114
Quantities, Units, and Symbols, 642
Quarterly Journal of Mathematics, 578
Quarterly Oil Statistics, 270
Quarterly Review of Biology, 521
Quarterly Reviews of Biophysics, 581
Quick Frozen Foods Directory of Frozen Food Processors, 415
Quick Reference Guides, 304
Quick Reference Manual for Silicon Integrated Circuit Tech., 281

RAE Table of Earth Satellites, 240
RAE Table of Earth Satellites 1957–1982, 245
RAPRA Abstracts, 557
RC Active Filter Design Handbook, 201
RCA Cos-mos Integrated Circuits Manual, 654
RCA Electro-Optics Handbook, 654
RCA Linear Integrated Circuits, 654
RCA Power Transistor Manual, 654
RCA Power Transistors, 654
RCA RF Power Transistor Manual, 655
RCA Receiving Tube Manual, 655

RCA Review, 586
RCA Review, 655
RCA Silicon Controlled Rectifier, 655
RCA Solid State Hobby Circuits Manual, 655
RCA Solid State Servicing, 655
RCA Transistor, Thyristor, and Diode Manual, 655
Regulations for Safe Transport of Radioactive Materials, 644
RF Radiometer Handbook, 162
RIBA Directory of Manufacturers, 416
RINGDOG (includes UDB, Ring 6475), 706
ROBOTICS, 706
RSC: Current Biotechnology Abstracts, 713
Radio Amateur's Handbook, 202
Radio Control Handbook, 202
Radio Control Manual—Systems, Circuits, Construction, 281
Radio Handbook, 202
Radio Propagation Handbook, 203
Radio and Electronic Engineer, 586
Radio, TV and Audio Technical Reference Book, 203
Radio-Electronic Master, 325
Radioactivity, 475
Radioactivity in America, 475
Railroad Names, 440
Railway Directory and Yearbook, 447
Railways Atlas of the World, 391
Rainbow Prehistoric Life Encyclopedia, 79
Raintree Illustrated Science Encyclopedia, 63
Rand McNally Atlas of the Oceans, 390
Rand McNally Concise Atlas of the Earth, 387
Rand McNally Encyclopedia of Transportation, 93
Rand McNally Encyclopedia of World Rivers, 80
Rand McNally New Concise Atlas of the Universe, 376
Range Science, 16
Rare Earth Research Conference, 610
Reactivity and Structure, 498
Reactor Handbook, 223
Reactor Safety, 54

Reader's Digest Encyclopedia of Garden Plants and Flowers, 71
Reader's Digest Illustrated Guide to Gardening, 352
Reader's Guide to Microcomputer Books, 24
Reagent Chemicals, 646
Reagents for Organic Synthesis, 252
Recent Advances in Biochemistry, 500
Recent Advances in Phytochemistry, 498
Recent Awards in Engineering, 584
Recent Developments in Separation Science, 526
Recent Progress in Hormone Research, 507
Record of Conference Papers of the Petroleum and Chemical., 611
Referativnyi Zhurnal, 539
Reference Data for Acoustic Noise Control, 249
Reference Guide to Practical Electronics, 325
Reference Manual for Telecommunications, 282
Reference Sources for Small and Medium-sized Libraries, 2
Refrigeration Processes, 162
Refrigeration and Air Conditioning Yearbook, 445
Regional Chemical Atlas: Orkney, 387
Regional Conference Series in Mathematics, 607
Register of Environmental Engineering Graduate Programs, 401
Regulations for Safe Transport of Radioactive Materials, 645
Reinforced Concrete Designer's Handbook, 193
Report Availability Notice, 595
Report to Congress, Disposal of Hazardous Wastes, 601
Reports on Progress in Physics, 521, 579
Reports on the Progress of Applied Chemistry, 498
Reproduction of Marine Invertebrates, 531
Reptiles of North America, 370
Research Catalog of the Library, 4
Research Centers Directory, 396, 621

Research Contracts in the Life Sciences, 624
Research Experience in Plant Physiology, 275
Research Fields in Physics at UK Universities and Polytech., 623
Research Grant Index, 621
Research Highlights, 624
Research Reports, 267
Research and Development in Federal Budget, 621
Research in Biological and Medical Sciences, 624
Research in British Universities, Polytechnics and Colleges, 621
Research in Protozoology, 513
Research on Transport Economics, 611
Researchers' Guide to Iron Ore, 53
Residue Reviews, 507
Resources for the History of Physics, 475
Reston Encyclopedia of Biomedical Engineering Terms, 68, 125
Results and Problems in Cell Differentiation, 510
Retrospective Index to Theses of Great Britain and Ireland, 543
Reverse Acronyms, Initialisms, and Abbreviations Dictionary, 95
Review of Agricultural Policies in OECD Member Countries, 312
Review of Scientific Instruments, 579
Review of Textile Progress, 515
Review of the US EPA Environmental Research Outlook, 627
Review of the World Wheat Situation, 508
Reviews in Biochemical Toxicology, 500
Reviews in Molecular Chemistry, 504
Reviews in Polymer Technology, 515
Reviews of Geophysics and Space Physics, 521
Reviews of Geophysics and Space Physics, 583
Reviews of Modern Physics, 521, 579
Reviews of Physiology, Biochemistry, and Pharmacology, 512
Reviews of Plasma Physics, 493
Revolution in Miniature, 475
Rheology Abstracts, 557
Rice Abstracts, 580

Road and Bridge Construction Handbook, 193
Road and Track Illustrated Dictionary, 137
Robotics Sourcebook and Dictionary, 143
Robotics and Automation Today, 24
Robotics and CAD/CAM Marketplace 1985, 24
Robotics, 1960–1983, 51
Rock Gardens and Water Plants in Color, 351
Rock Hunter's Guide, 372
Rocket, 484
Rockets, Missiles and Spacecraft of National Air & Space Museum, 320
Rocky Mountain Energy Directory, 433
Rocky Mountain Wildflowers, 354
Rodale's Color Handbook of Garden Insects, 179
Rodale's Encyclopedia of Indoor Gardening, 71
Rodd's Chemistry of Carbon Compounds, 526
Roses, 40
Roster of US Government Research and Development Contracts., 625
Roster of Women and Minority Engineering Students, 401
Royal Astronomical Society, 577
Royal Meteorological Society, 583
Royal Society Mathematical Tables, 237
Royal Society Proceedings, 576
Royal Society of Chemistry. Journal, 579
Royal Society of Chemistry. Reviews, 579
Rubber Red Book, 413
Rules and Regulations for the Construction and Classific., 649
Rules for Building and Classing Steel Barges for Offshore., 649
Rules for Building and Classing Steel Vessels, 649
Rules for Construction and Classification of Mobil Offshore., 649
Russian English Index to Scientific Apparatus Nomenclature, 105
Russian-English Chemical and Polytechnical Dictionary, 105
Russian-English Dictionary of the Mathematical Sciences, 105

Russian-English Glossary of Fishing and Related Marine Terms, 105
Russian-English Oil-Field Dictionary, 106
Russian-English Polytechnical Dictionary, 106
Russian-English Translators Dictionary, 106
Rutherford and Physics at Turn of Century, 456, 475

SAE, 706
SAE Aerospace Index of Aerospace Standards, 646
SAE Handbook, 219
SAE Motor Vehicle, Safety and Environmental Terminology, 151
SAE Transactions and Literature Developed, 52
SAO Russian Translation Series, 615
SCISEARCH, 706
SDC Search Service Quick Reference Guide, 679
SDF, 706
SHE: Subject Headings for Engineering, 556
SI Metric Handbook, 642
SI Units in Engineering and Technology, 642
SI: International System of Units, 235
SIAM Journal on Scientific and Statistical Computing, 587
SIAM Review, 578
SOCIAL SCISEARCH, 707
SOCIOLOGICAL ABSTRACTS, 708
SOVIET SCIENCE AND TECHNOLOGY, 707
SPIN, 707
SPINES, 99
SPINES Thesaurus, 99
SSIE, 707
SSIE CURRENT RESEARCH, 707
SSIE Science Newsletter, 621
STANDARDS, 708
STANDARDS & SPECIFICATIONS, 708
STAR, 556
SUPERINDEX, 708
Sadtler Handbook of Infrared Spectra, 161
Safety and Operational Guidelines for Undersea Vehicles, 332
Safety in Nuclear Power Plant Siting, 337
Salaries and Income of Certified Engineering Technicians, 402
Salaries of Engineering Technicians and Technologists, 402
Salaries of Engineers in Education—1980, 402
Salaries of Engineers in Government, 402
Salaries of Scientists, Engineers, and Technicians, 402
Sanyo's Trilingual Glossary of Chemical Terms, 113
Sausage Processing, 637
Scented Flora of the World, 350
Science, 577, 603
Science Abstracts, 539
Science Abstracts Part A: Physics Abstracts, 635
Science Books, 3
Science Books and Films, 2, 664
Science Citation Index, 539
Science Fair Project Index 1973–1980, 539
Science Fiction Book Review Index, 5
Science Information Available from the AEC, 25
Science Information Resources, 16
Science News Yearbook, 441
Science Policy, 116
Science Research Abstracts Journal, 539
Science Research in Progress, 622
Science Since 1500, 468
Science Software Quarterly, 12
Science Year, 441
Science and Civilization in China, 468
Science and Engineering Literature, 16
Science and Engineering Software, 424
Science and Government Report International Almanac, 595
Science and Technology, 16
Science and Technology Research in Progress 1972–1973, 621
Science and Technology for Development, 113
Science and Technology in Latin America, 398

Science and Technology in Middle East, 399
Science for Society, 30
Science in America, 467
Science in America Since 1820, 467
Science in the Middle Ages, 468
Science, Engineering, and Humanities Doctorates in US, 402
Scientific American, 576
Scientific American Cumulative Index, 1948–1978, 540
Scientific Books, Libraries, and Collectors, 5, 468
Scientific Elite, 452
Scientific Illustrations, 311
Scientific Journal, 573
Scientific Journals in the United States, 573
Scientific Maps and Atlases, 374
Scientific Meetings, 602
Scientific Periodicals, 573
Scientific Quotations, 246
Scientific Research in British Universities and Colleges, 622
Scientific Thought, 1900–1960, 468
Scientific Words, 116
Scientific and Technical Books and Serials in Print, 3, 5
Scientific and Technical Information Resources, 5
Scientific and Technical Information Sources, 16
Scientific and Technical Journals, 573
Scientific and Technical Research Centres in Australia, 399, 622
Scientific, Engineering and Medical Societies Publications in Print, 3, 672
Scientists and Engineers in Federal Government, 452
Scientists and Inventors, 453
Scientists and the Sea, 1650–1900, 481
Scripps Institution of Oceanography, 482
Scripps Institution of Oceanography Library. University Library Catalog, 11
Sea, 532
Sea Grant Newsletter Index, 568
Sea Technology Handbook/Directory, 427
Sea: A Select Bibliography on the Legal, Political., 46

Sea: Economic and Technological Aspects, 46
Sea: Legal and Political Aspects, 46
Seabirds, 361
Seabirds of the World, 362
Search for Environmental Ethics, 59
Searching for Foreign Patents, 633
Seawater Corrosion Handbook, 222
Seaweed Handbook, 354
Seaweeds, 314
Secondary Plant Products, 72
Security Dictionary, 143
Seed Abstracts, 580
Selected Data Resources, 399
Selected Information Resources on Electronics, 417
Selected Information Resources on Plastics, Polymers, and., 413
Selected Information Resources on Pulp and Paper, 414
Selected Information Resources on Science and Mathematics., 402
Selected Information Resources on Solar Energy, 434
Selected Information Resource on Solid Wastes, 438
Selected Information Resources on Textiles, 414
Selected Information Resources on Wood Products, 416
Selected Instrumental Films, 665
Selected Papers 1945–1980, 456
Selected Publications on the Environment, 601
Selected References on Environmental Quality as It Relates., 60
Selected Sources of Information on Corrosion and Corrosion., 414
Selected Sources of Information on Inventions and Product., 410
Selected Tables in Mathematical Statistics, 238
Selected Titles in Chemistry, 6
Selected Values of Chemical Thermodynamic Properties, 242
Selected Water Resources Abstracts, 568
Selecting Instructional Media, 660
Selection and Use of Thermoplastics, 321
Semiconductor Laser Diodes, 162

Semiconductors and Semimetals, 517, 523
Sensor and Analyzer Handbook, 211
Separation Procedures in Inorganic Analysis, 170
Serial Atlas of the Marine Environment, 392
Serials Directory, 489, 574
Shallow Ground Disposal of Radioactive Wastes, 341
Sharks of North American Waters, 364
Shell Aviation News, 655
Ship Abstracts, 562
Shock and Vibration Handbook, 162
Short History of Twentieth-Century Technology, 468
Shrubs in the Landscape, 350
Sierra Club Handbook of Whales and Dolphins, 179
Signet Book of American Wine, 316
Simon and Schuster's Complete Guide to Freshwater & Marine., 364
Simon and Schuster's Guide to Garden Flowers, 351
Simon and Schuster's Guide to Rocks and Minerals, 372
Simon and Schuster's Guide to Shells, 357
Simon and Schuster's Guide to Cats, 368
Simulation, 587
Sisson and Grossman's—The Anatomy of the Domestic Animals, 531
Site Selection and Evaluation for Nuclear Power Plants, 341
Skinner's Mining International Yearbook, 445
Skinner's Oil and Gas International Yearbook, 446
Sky Atlas 2000.0, 377
Sky Catalogue 2000.0, 247
Sky and Telescope, 577
Skyguide, 343
Small Business Computers, 327
Smith's Guide to the Literature of the Life Sciences, 20
Smithells Metals Reference Book, 333
Smithsonian Annual of Flight, 443
Smithsonian Book of Invention, 468
Smithsonian Meteorological Tables, 244
Smithsonian Physical Tables, 240

Societal Directions and Alternatives, 27
Society for Experimental Biology, 609
Society for General Microbiology, 609
Society of Petroleum Engineers Journal, 589
Sociology of Sciences, 30
Software, 587
Software Catalog, 424
Software Encyclopedia, 424
Software Engineering Standards, 649
Software Finder, 328
Software Maintenance Guidebook, 328
Software Marketplace, 424
Software Reliability Guidebook, 328
Software Reviews on File, 12
Software Tools Directory, 424
Software Writer's Marketplace, 425
Software: Engineering, 1984, 424
Soil Science, 581
Soil Survey Reports, 600
Soils and Fertilizer, 548
Solar Age Catalog, 246
Solar Age Resource Book, 340
Solar Census, 463
Solar Decision Book, 340
Solar Energy, 56, 589, 595
Solar Energy Almanac, 261
Solar Energy Application in Buildings, 340
Solar Energy Books, 56
Solar Energy Dictionary, 148
Solar Energy Directory, 434
Solar Energy Experiments, 299
Solar Energy Handbook, 226
Solar Energy Index, 566
Solar Energy Research and Development in European Community, 534
Solar Energy Source Book, 292
Solar Energy Technology Handbook, 227
Solar Energy and Research Directory, 626
Solar Energy in Buildings for Engineering, Architecture., 299
Solar Energy in Developing Countries, 340
Solar Greenhouse Book, 340
Solar Products Specifications Guide, 340
Solar System, 522
Solar Thermal Energy Utilization, 57

Solid State Physics, 494
Solid State Physics Literature Guides, 35
Solid Waste Handbook, 232
Solid Waste Information Retrieval System Accession Bulletin, 571
Solid-State Abstracts Journal, 562
Solid-State Circuit Design Users' Manual, 282
Source Book I: Small Systems Software and Services Source., 290
Source Book for Farm Energy Alternatives, 292
Source Book for Food Scientists, 287
Source Book in Astronomy and Astrophysics, 287
Source Book in Astronomy, 1900–1950, 470
Source Book in Mathematics, 471
Source Book in Flavors, 289
Source Book of Food Enzymology, 289
Source Book of Brazing and Brazing Technology, 289
Source Book on Food and Nutrition, 289
Source Book on Industrial Alloy and Engineering Data, 290
Source Book on Materials for Elevated-Temperature Applications, 290
Source Book on Powder Metallurgy, 290
Source Book: Small Systems Software and Services Source Book, 290
Source Index Quarterly, 544, 575
Sourcebook for Programmable Calculators, 286
Sourcebook of Experiments for the Teaching of Microbiology, 287
Sourcebook of Titanium Alloy Superconductivity, 288
Sourcebook on Asbestos Diseases, 292
Sourcebook on the Environment, 27
Sourcebook on the Production of Electricity from Geothermal Energy, 292
Sources for the History of Science 1660–1914, 469
Sources in the History of Mathematics and Physical Sciences, 469
Sources of Construction Information, 23
Sources of Information in Transportation, 28
Sources of Information in Water Resources, 21
Sources of Invention, 287
Sources of Serials, 5
South American Land Birds, 382
Soviet Abstracts, 616
Soviet Astronomy, 577
Soviet Atomic Energy, 588
Soviet Physics,—JETP, 579
Space Research, 610
Space Research in the UK, 625
Space Science Reviews, 584
Specifications for Identity Food Colours, Flavouring Agents, 645
Spectral and Chemical Characterization of Organic Compounds, 170
Spectroscopic Data, 240
Spectroscopic References to Polyatomic Molecules, 170
Spices, Condiments, Teas, Coffees, and Other Delicacies, 415
Spray Drying Handbook, 188
Spring Designer's Handbook, 219
Springer Series in Chemical Physics, 494
Springer Series in Computational Physics, 494
Springer Series in Electrophysics, 494
Springer Series in Solid-State Sciences, 494
Springer Tracts in Modern Physics, 494
Standard Boiler Room Questions and Answers, 290
Standard Dictionary of Electrical and Electronics Terms, 138
Standard Forms of Building Contract, 323
Standard Graphical Symbols, 643
Standard Guide to Cat Breeds, 368
Standard Handbook for Civil Engineering, 189
Standard Handbook for Electrical Engineers, 195
Standard Handbook for Mechanical Engineers, 215
Standard Handbook of Engineering Calculations, 182
Standard Handbook of Fastening and Joining, 219
Standard Handbook of Machine Design, 219

Standard Handbook of Plant
  Engineering, 211
Standard Handbook of Textiles, 189
Standard Methods for Examination of
  Water and Wastewater, 285
Standard Methods of Chemical Analysis,
  526
Standard Plant Operator's Questions and
  Answers, 290
Standard Plant Operators' Manual, 282
Standard Specifications for Highway
  Bridges, 647
Standard Specifications for Movable
  Highway Bridges, 647
Standard Specifications for
  Transportation Materials, 647
Standard Specifications for Welding of
  Structural Steel., 647
Standard Terms of the Energy Economy,
  114
Standard for Classification and
  Application of Welded Joints, 650
Standardized Development of Computer
  Software, 649
Standards and Practices of
  Instrumentation, 651
Standards and Specifications Information
  Sources, 638
Standards for Engineering Qualifications,
  638
Standards in Building Codes, 647
Standards in Nuclear Science and
  Technology, 639
Star Atlas, 377
Star Maps for Beginners, 377
Stars and Stellar Systems, 378
State Government Reference
  Publications, 598
State of Food and Agriculture, 1976, 508
Statistical Abstract of the United States,
  267
Statistical Analysis of Geological Data,
  256
Statistical Data Analysis Handbook, 157
Statistical Sources, 264
Statistical Tables, 239
Statistical Tables for Biological,
  Agricultural and Medical., 239
Statistical Tables for Social, Biological
  and Physical Sciences, 235

Statistical Theory and Methods Abstracts,
  541
Statistics and Econometrics, 17
Steam and Air Tables in SI Units, 235
Steel Castings Handbook, 222
Steel Pipelines 1, 651
Steel Selection, 333
Steel Transformation Diagrams, 391
Storm Data for the United States, 262
Story of the Royal Horticultural Society,
  479
Strange Universe, 287
Stratigraphic Atlas of North and Central
  America, 388
Strength of Materials, 533
Structural Design Guide to the ACI
  Building Code, 647
Structural Engineer, 585
Structural Engineering Handbook, 194
Structural Engineering for Professional
  Engineers' Examin., 323
Structural Welding Code, 651
Structural and Construction Design
  Manual, 280
Studies in Environmental Science, 520
Studies in History of Biology, 479
Studies in Logic and the Foundations of
  Mathematics, 490
Studies in Mathematical Education, 490
Studies in Organic Chemistry, 504
Studies in the History of Modern
  Science, 469
Studies in the History of Statistics and
  Probability, 471
Subject Collections, 17
Subject Directory of Special Libraries
  and Information Centers, 397
Subtidal Marine Invertebrates of North
  America, 666
Succulent Flora of Southern Africa, 314
Summmary of Research in Science
  Education—1983, 622
Sun Power, 57
Sunrise and Sunset Tables for Key Cities
  and Weather., 240
Surface Vehicle Sound Measurement
  Procedures, 219
Surface Water Data, 256
Surfactant Science Series, 498
Survey of Biological Progress, 507

Survey of Progress in Chemistry, 498
Surveys in High-Energy Physics, 579
Swans of the World, 362
Swift's Directory of Educational Software for the IBM PC, 425
Swift's Educational Software Directory, 425
Swift's Educational Software Directory for Corvus Networks, 425
Switchgear and Control Handbook, 201
Sybex Personal Computer Dictionary, 143
Symposia Mathematics, 490
Symposia of the Society for Experimental Biology, 609
Symposium in Pure Mathematics, 607
Symposium on Naval Hydrodynamics, 612
Synergy, 434
Synfuels Project Directory, 434
Synonymized Checklist of Vascular Flora of US, Canada and., 350
Synopsis and Classification of Living Organisms, 127
Synthetic Fuels Data Handbook, 227
Synthetic Fuels Research, 57
Synthetic Methods of Organic Chemistry, 442, 504
Systematic Identification of Organic Compounds, 294
Systematic Zoology, 582
Systems Research, 587
Systems Troubleshooting Handbook, 208

TELEGEN, 708
TOXLINE, 712
TRINET COMPANY DATABASE, 709
TRINET ESTABLISHMENT DATABASE, 709
TRIS, 709
TROPAG, 709
TSCA INITIAL INVENTORY, 709
TULSA, 709
Table of Isotopes, 240
Table of Laser Lines in Gases and Vapors, 241
Table of Molecular Weights, 241
Table of Series and Products, 237
Tables of Integrals, Series, and Products, 237

Tables of Physical and Chemical Constants and Some Math., 235
Tables of Standard Electrode Potentials, 241
Tables of the F-E Related Distribution Algorithms, 238
Tables of Wavenumbers for Calibration of Infrared Spectrum., 241
Tables on the Thermophysical Properties of Liquids and Gases, 241
Tabulation of Infrared Spectral Data, 241
Tanaka's Cyclopedia of Edible Plants of the World, 72
Taxonomic Literature, 20
Teaching Manual on Food and Nutrition for Non-Science Majors, 300
Technical Books in Print, 3
Technical Abstract Bulletin, 593
Technical Abstracts, 556
Technical Americana, 483
Technical Book Review Index, 3
Technical Data on Fuel, 235
Technical Dictionary for Automotive Engineering, 146
Technical Editor's and Secretary's Desk Guide, 299
Technical Evaluation of Bids for Nuclear Power Plants, 337
Technical Guide to Computer-Communications Interface Standards, 649
Technical Information Sources, 631, 639
Technical Literature Abstracts, 556
Technical Memo, 679
Technical News Bulletin, 560
Technical Report Standards, 593
Technical Writer's Handbook, 153
Technician Education Yearbook, 402, 441
Technician's Handbook of Plastics, 188
Techniques in Mineral Exploration, 283
Techniques of Chemistry, 499, 526
Techniques of Metals Research, 533
Techniques of Organic Chemistry, 527
Techniques of Physics, 494
Technology Book Guide, 47
Technology Policy and Development, 290
Technology and Culture, 584

Telecommunications Systems and Services Directory, 678
Telecommunication Transmission Handbook, 203
Telecommunications Systems and Services Directory, 417
Television Interference Manual, 305
Television and Management, 660
Terry Robard's New Book of Wine, 316
Textile Industries Buyers Guide, 322
Textile Manual, 278
Textile Technology Digest, 557, 708
Textile World's Leaders in the Textile Industry, 461
Theoretical Chemical Engineering Abstracts, 557
Theoretical Chemistry, 499
Thermal Engineering, 589
Thermal Insulation Handbook, 219
Thermal Theory of Cyclones, 482
Thermochemical Properties of Inorganic Substances, 252
Thermodynamic and Thermophysical Properties of Combustion., 260
Thermophysical Properties Research Literature Retrieval., 250
Thermophysical Properties of Matter, 250
Thesaurus of Agricultural Terms as Used in Bibliography., 99
Thesaurus of Information Science Terminology, 100
Thesaurus of Metallurgical Terms, 100
Theses and Dissertations as Information Sources, 617
Thin Layer Chromatography Abstracts, 1971–1973, 544
Thirty Years of Fusion, 476
This Wild Abyss, 454, 470
Thorburn's Birds, 362
Three Mile Island Sourcebook, 292
Timber Construction Manual, 280
Times Atlas of the Moon, 376
Times Atlas of the Oceans, 390
Timescale: An Atlas of the Fourth Dimension, 379
Timex-Sinclair 1983 Directory, 425
Tissues and Organs, 382
Title List of Documents Made Publicly Available, 595
Titles in Series, 5
Tom Brown's Guide to Wild Edible and Medicinal Plants, 314
Tool and Manufacturing Engineers Handbook, 219
Tools and How to Use Them, 305
Topics and Terms in Environmental Problems, 92
Topics in Applied Physics, 494
Topics in Bioelectrochemistry and Bioenergetics, 507
Topics in Carbon-13 NMR Spectroscopy, 499
Topics in Inorganic and General Chemistry, 499
Topics in Lipid Chemistry, 500
Topics in Stereochemistry, 499
Topics on Nucleic Acid Structure, 527
Total Book of House Plants, 352
Tower's International Microprocessor Selector, 208
Toxic Substances Control Sourcebook, 292
Toxic Substances Sourcebook, 28
Toxic and Hazardous Industrial Chemical Safety Manual for., 285
Toxicity of Chemicals and Pulping Wastes to Fish, 61
Traffic Noise, 61
Train-Watcher's Guide to North American Railroads, 335
Traite de Zoologie, 531
Transactions of Society of Naval Architects and Marine Eng., 612
Transactions of the Annual Marine Technology Society Confer., 611
Transdex, 613
Transguide, 28
Transistor Substitution Handbook, 201
Transition Metal Chemistry, 502
Translation Bulletin, 615
Translation and Translators, 614
Translations Journals, 615
Translations Register-Index, 613
Translations of Mathematical Monographs, 615
Transliterated Dictionary of the Russian Language, 106
Transmission Line Reference Book, 291
Transportation Research Abstracts, 559

Transportation System Management, 61
Transportation-Logistics Dictionary, 151
Treatise on Adhesion and Adhesives, 532
Treatise on Analytical Chemistry, 527
Treatise on Invertebrate Paleontology, 532
Treatise on Materials Science and Technology, 533
Treatise on Solid State Chemistry, 527
Tree Key, 314
Trees and Shrubs Hardy in British Isles, 314
Trees and Shrubs of the United States, 41
Trees of North American and Europe, 350
Trees of the World, 350
Tropical Cyclones of the North Atlantic Ocean, 270
Tropical Fish Identifier, 364
Troubleshooting Microprocessors and Digital Logic, 304
True Visual Magnitude Photographic Star Atlas, 378
Tube Caddy, 325
Tube Substitution Handbook, 201
Tunnel Engineering Handbook, 194
Turbomachines, 335
Twentieth Century Physics, 476
Two Hundred Years of Flight in America, 484
Two Hundred Years of Geology in America, 482

UFO Encyclopedia, 64
UFO's and Related Subjects, 48
UK Offshore Oil and Gas Yearbook, 446
UNDOC, 598
US Civil Aircraft Register, 601
US Department of the Interior Library. Dictionary Catalog, 10
US Directory of Marine Scientists, 427, 459
US ERDA. Material Sciences Program, 626
US Energy Information Administration. Quarterly Coal Report, 595
US Federal Supply Service, 641
US Food Laws, Regulations and Standards, 645
US Geological Survey Library Catalog, 10
US Government Manual, 599
US Government Research and Development Reports, 593
US Government Scientific and Technical Periodicals, 574, 599
US Observatories, 403
US Patent Previews, 633
US Plastics in Building and Construction, 323
US and the Global Environment, 438
USA Oil Industry Directory, 434
USA Standard Glossary of Terms in Nuclear Science and Tech., 644
USCLASS, 710
USPA/USP77/USP70, 710
USSR Agriculture Atlas, 267
Ullmanns Encyklopadie der Technischen Chemie, 83
Ulrich's International Periodicals, 574
Unclassified Publications of Lincoln Laboratory, 593
Underground Coal Gasification, 291
Understanding Chemical Patents, 635
Underwater Construction and Mining, 53
Unesco Annual Summary of Information on Natural Disasters, 446
Unesco List of Documents and Publications, 1972–1976, 599
Unesco Statistical Yearbook, 270
Uniform Building Code and 1975 Accumulative Supplement, 647
Union Catalog of Maps, 374
United Kingdom Mineral Statistics, 270
United States Energy Atlas, 392
Uniterm Index to US Chemical and Chemically Related Patents, 636
Units of Weight and Measure, 235, 642
Universal Encyclopedia of Mathematics, 65
University Curricula in the Marine Sciences and Related Fld., 402
University of London Theses and Dissertations Accepted, 618
Unix System Guidebook, 329

Urban Mass Transportation Abstracts, 559
Use Guide to COBOL 85, 304
Use of Biological Literature, 20
Use of Chemical Literature, 19
Use of Earth Sciences Literature, 21
Use of Engineering Literature, 22
Use of Mathematical Literature, 17
Use of Physics Literature, 18
Use of Reports in Literature, 593
User's Guide to Computer Peripherals, 330
User's Guide to Selecting Electronic Components, 325
User's Guide to Standard Microfiche Formats, 660
User's Guide to Vacuum Technology, 308
User's Guidebook to Digital CMOS Integrated Circuits, 330
User's Handbook of Semiconductor Memories, 208
Users' Guide to Microcomputer Buzzwords, 143
Using Commercial Fertilizers, 312
Using DEC Personal Computers, 304
Using Personal Computers in Public Agencies, 304
Using the Biological Literature, 20
Using the Mathematical Literature, 17
Utilization of Microorganisms in Meat Processing, 173

VETDOC UDB/VETDOC, 710
VHF/UHF Manual, 282
VLSI System Design, 299
VNR Color Dictionary of Minerals and Gemstones, 133
VNR Color Dictionary of Mushrooms, 127
VNR Concise Encyclopedia of Mathematics, 65
VNR Dictionary of Ships and Sea, 151
VNR Metric Handbook of Architectural Standards, 642
Vacuum Special Issue, 428
Van Nostrand Reinhold Encyclopedia of Chemistry, 67
Van Nostrand's Scientific Encyclopedia, 63

Vapor-Liquid Equilibrium Data Bibliography, 51
Vector Analysis, 307
Vegetable Growing Handbook, 174
Vegetable Processing, 637
Venomous Arthropod Handbook, 180
Vertical File Index, 654
Veterinary Multilingual Thesaurus, 100
Victorian Science and Religion, 31
Video Guide, 661
Video Programs Index, 661
Video Register, 661
Video Source Book, 661
Video User's Handbook, 661
Video and Digital Electronic Displays, 325
Video in Libraries, 660
Video-based Information Systems, 661
Videodisc Book, 419, 662
Viewpoints in Biology, 521
Virology Abstracts, 547
Virus, 480
Vistas in Astronomy, 489
Visual Display Terminals, 282
Vitamins and Hormones, 501
Vocabulary of Organic Chemistry, 123

WATERNET, 710
WELDASEARCH, 711
WILSONLINE Guide and Documentation, 679
WILSONLINE Quick Reference Guide, 679
WMO Bulletin, 583
WORLD ALUMINUM ABSTRACTS, 711
WORLD PATENTS INDEX, 711
WORLD TEXTILES, 711
WPI/WPIL, 711
Wading Birds of the World, 362
Walker's Mammals of the World, 368
Washington Information Workbook, 593
Waste Recycling and Pollution Control Handbook, 233
Wastewater Management, 60
Water—A Comprehensive Treatise, 532
Water Engineering and Management, 590
Water Nautralist, 319
Water Pollution, 28, 535

Water Pollution Control, 439, 590
Water Pollution Control Federation, 590
Water Quality Abstracts, 568
Water Reactor Safety Research Program, 625
Water Research, 590
Water Resources, 46
Water Resources Bulletin, 590
Water Resources Research, 590
Water Resources Research Catalog, 625
Water Resources of the World, 270
Water Treatment Handbook, 194
Water Treatment Plant Design for the Practicing Engineer, 331
Water and Waste Treatment Data Book, 263
Water, Air, and Soil Pollution, 590
Water-in-Plants Bibliography, 41
Waterfowl, 362
Way Things Work, 306
Wear, 588
Weather, 583
Weather Almanac, 256
Weather Atlas of US, 388
Weather Data Handbook for HVAC and Cooling Equipment Design, 220
Weather Handbook, 181
Weather of the United States Cities, 256
Weatherwise, 583
Webb Society Deep-Sky Observer's Handbook, 155
Weed Abstracts, 550
Weed Control Handbook, 177
Weeds, 354
Weekly Abstract Newsletter, 569, 594
Welding 1, 651
Welding Design and Fabrication Data Book, 260
Welding Handbook, 220
Welding Manual for Engineering Steel Forgings, 283
Welding Research Council Yearbook, 445
Westcott's Plant Disease Handbook, 177
Western Electric Engineer, 655
Whale Manual '78, 180
What Every Engineer Should Know About Patents, 633
Where You Live May Be Hazardous to Your Health, 263

Where to Find Free Programs for TRS-80, Apple, or IBM Microcomputers, 425
Whisper and the Vision, 454
Whittington's Dictionary of Plastics, 135
Who Knows? Selected Information Resources on Science Education, 402
Who Was Who in American History, 453
Who is Publishing in Science, 453
Who is Who: Directory of Agricultural Engineers Available., 460
Who's Who and Guide to the Electrical Industry, 463
Who's Who in Architecture, from 1400 to the Present, 462
Who's Who in Atoms, 463
Who's Who in Aviation, 461
Who's Who in Computer Education and Research, 462
Who's Who in Consulting, 460
Who's Who in Electronics, 462
Who's Who in Engineering, 461
Who's Who in Frontier Science and Technology, 453
Who's Who in Ocean and Freshwater Science, 459
Who's Who in Science in Europe, 453
Who's Who in Technology, 460
Who's Who in Technology Today, 461
Who's Who of British Engineers, 460
Who's Who of British Scientists, 454
Whole World Oil Directory, 434
Wild Flowers of the United States, 354
Wild Garden, 355
Wild Geese of the World, 362
Wild Orchids of Britain and Europe, 354
Wildflower Folklore, 354
Wildflowers and Weeds, 354
Wildflowers of North America, 354
Wildfowl of the World, 362
Wildlife Abstracts, 551
Wildlife Observer's Guidebook, 357
Wildlife in Danger, 76
Wildlife of the Deserts, 357
Wildlife of the Rivers, 357
Wiley Engineer's Desk Reference, 319
Wiley Metric Guide, 235
Windpower, 340
Winemaker's Encyclopedia, 85
Wines of America, 316

Wire and Wire Products, 325
Woldman's Engineering Alloys, 428
Women and Mathematics, Science and Engineering, 454
Women in Mathematics, 454
Women in Physics, 457
Wood Handbook, 600
Wood Structures, 324
Workbench Guide to Practical Solid State Electronics, 325
Works of Science and the History Behind Them, 469
Workshop on Wind Energy Conversion Systems, 612
World Agriculture Economics and Rural Sociology Abstracts, 548
World Aluminum Abstracts, 564
World Atlas of Agriculture, 382
World Atlas of Geology and Mineral Deposits, 388
World Atlas of Geomorphic Features, 388
World Atlas of Horses and Ponies, 384
World Aviation Directory, 411
World Bibliography of Bibliographies and Bibliographic Cat., 537
World Calendar of Forthcoming Meetings, 603
World Catalogue of Very Large Floods, 256
World Cement Directory, 1980, 416
World Coal Industry Report and Directory, 434
World Crops, 581
World Destroyed, 487
World Dictionaries in Print, 95
World Directory Hydrobiological and Fisheries Institutions, 407
World Directory of Collections of Cultures of Microorganisms, 406
World Directory of Energy Information, 435
World Directory of Engineering Schools, 403
World Directory of Environmental Education Programs, 403
World Directory of Environmental Organizations, 438
World Directory of Environmental Research Centers, 438

World Directory of Map Collections, 374
World Directory of Mathematicians, 454
World Directory of National Parks and Other Protected Areas, 438
World Directory of Research Projects, Studies, and Courses., 622
World Encyclopedia of Food, 85
World Energy Book, 91, 392
World Energy Directory, 626
World Energy Supplies, 262, 271
World Environment, 488
World Environmental Directory, 438
World Fisheries Abstracts, 551
World Food Book, 271
World Guide to Abbreviations of Organizations, 95, 397
World Guide to Batter-Powered Road Transportation, 335
World Guide to Scientific Associations and Learned Societies, 668
World Guide to Trade Associations, 669
World Index of Plastics Standards, 646
World Index of Strategic Minerals, 554
World List Scientific Periodicals Published in 1900–1960, 574
World List of Aquatic Sciences and Fisheries Serial Titles, 8, 574
World List of Forestry Schools, 403
World List of Mammalian Species, 368
World Map of Desertification, 388
World Meetings Outside USA and Canada, 603
World Meetings: United States and Canada, 603
World Metric Standards for Engineering, 642
World Mines Register, 428
World Motor Vehicle Data, 448
World Motor Vehicle Data Book, 263
World Nuclear Directory, 625
World Ocean Atlas, 390
World Palaeontological Collections, 45
World Patents Index, 634
World Pipelines and International Directory of Pipeline., 435
World Power Data, 267, 589
World Radio TV Handbook, 203
World Register of Dams, 416
World Road Statistics, 271
World Space Directory, 411

World Survey of Climatology, 532
World Survey of Major Facilities in Controlled Fusion Res., 430
World Textile Abstracts, 557
World Transindex, 614
World Transindex: Announcing Translations in All Fields., 614
World Vegetables, 275
World Wheat Statistics, 254
World Who's Who in Science, 453
World of Zoos, 407
World's Whales, 369
World's Worst Weeds, 311
Worldwide Chemical Directory, 414
Worldwide Directory: Offshore Contractors & Equipment, 435
Worldwide Guide to Equivalent Nonferrous Metals and Alloys, 333
Worldwide Oilfield Service, Supply and Manufacturers Directory, 435
Worldwide Petrochemical Directory, 435
Worldwide Petrochemical Industry, 435
Worldwide Pipeline and Contractors Directory, 435
Worldwide Refining and Gas Processing Directory, 435
Worldwide Synthetic Fuels and Alternate Energy Directory, 436
Worterbuch Technischer Begriffe mit 4,300 Definitionen., 104
Worterbuch der Elektronik, English-Deutsch, 104
Wyman's Gardening Encyclopedia, 72

Yearbook of Agriculture, 442
Yearbook of Astronomy, 442
Yearbook of Consumer Electronics, 444
Yearbook of Fishery Statistics, 271
Yearbook of Forest Products, 271, 442
Yearbook of International Congress Proceedings, 605
Yearbook of International Organizations, 397, 441
Yearbook of Science and Future, 441
Yeasts, 531

Z8000 CPU User's Reference Manual, 304
ZOOLOGICAL RECORD, 711

Zentralblatt fur Mathematik und Ihre Grenzgebiete, 542
Zinc and Its Alloys and Compounds, 222
Zoobooks, 42
Zoological Record, 552
Zoophysiology, 520
Zoos and Aquariums in Americas, 408